D1919729

FOURTH EDITION

Introduction to Data Compression

The Morgan Kaufmann Series in Multimedia Information and Systems
Series Editor, Edward A. Fox, Virginia Polytechnic University

Introduction to Data Compression, Third Edition
Khalid Sayood

Understanding Digital Libraries, Second Edition
Michael Lesk

Bioinformatics: Managing Scientific Data
Zoe Lacroix and Terence Critchlow

How to Build a Digital Library
Ian H. Witten and David Bainbridge

Digital Watermarking
Ingemar J. Cox, Matthew L. Miller, and Jeffrey A. Bloom

Readings in Multimedia Computing and Networking
Edited by Kevin Jeffay and HongJiang Zhang

Introduction to Data Compression, Second Edition
Khalid Sayood

Multimedia Servers: Applications, Environments, and Design
Dinkar Sitaram and Asit Dan

Managing Gigabytes: Compressing and Indexing Documents and Images, Second Edition
Ian H. Witten, Alistair Moffat, and Timothy C. Bell

Digital Compression for Multimedia: Principles and Standards
Jerry D. Gibson, Toby Berger, Tom Lookabaugh, Dave Lindbergh, and Richard L. Baker

Readings in Information Retrieval
Edited by Karen Sparck Jones and Peter Willett

FOURTH EDITION

Introduction to Data Compression

Khalid Sayood
University of Nebraska

ELSEVIER

AMSTERDAM • BOSTON • HEIDELBERG • LONDON
NEW YORK • OXFORD • PARIS • SAN DIEGO
SAN FRANCISCO • SINGAPORE • SYDNEY • TOKYO

Morgan Kaufmann is an imprint of Elsevier

MK MORGAN KAUFMANN

Acquiring Editor: Andrea Dierna
Development Editor: Meagan White
Project Manager: Danielle S. Miller
Designer: Eric DeCicco

Morgan Kaufmann is an imprint of Elsevier
225 Wyman Street, Waltham, MA 02451, USA

Library of Congress Cataloging-in-Publication Data
Sayood, Khalid.
Introduction to data compression / Khalid Sayood. – 4th ed.
p. cm.
ISBN 978-0-12-415796-5
1. Data compression (Telecommunication) 2. Coding theory. I. Title.
TK5102.92.S39 2012
005.74'6–dc23
2012023803

British Library Cataloguing-in-Publication Data
A catalogue record for this book is available from the British Library.

ISBN: 978-0-12-415796-5

Printed and bound by CPI Group (UK) Ltd, Croydon, CR0 4YY

To Füsun

Contents

Preface xvii

1 Introduction **1**
1.1 Compression Techniques 3
1.1.1 Lossless Compression 4
1.1.2 Lossy Compression 5
1.1.3 Measures of Performance 5
1.2 Modeling and Coding 6
1.3 Summary 10
1.4 Projects and Problems 10

2 Mathematical Preliminaries for Lossless Compression **13**
2.1 Overview 13
2.2 A Brief Introduction to Information Theory 13
2.2.1 Derivation of Average Information ★ 20
2.3 Models 25
2.3.1 Physical Models 25
2.3.2 Probability Models 25
2.3.3 Markov Models 26
2.3.4 Composite Source Model 29
2.4 Coding 29
2.4.1 Uniquely Decodable Codes 30
2.4.2 Prefix Codes 33
2.4.3 The Kraft-McMillan Inequality ★ 34
2.5 Algorithmic Information Theory 37
2.6 Minimum Description Length Principle 38
2.7 Summary 39
2.8 Projects and Problems 40

3 Huffman Coding **43**
3.1 Overview 43
3.2 The Huffman Coding Algorithm 43
3.2.1 Minimum Variance Huffman Codes 47
3.2.2 Canonical Huffman Codes 50
3.2.3 Length-Limited Huffman Codes 52
3.2.4 Optimality of Huffman Codes ★ 55

3.2.5 Length of Huffman Codes ★ 56
3.2.6 Extended Huffman Codes ★ 58
3.2.7 Implementation of Huffman Codes 61
3.3 Nonbinary Huffman Codes ★ 65
3.4 Adaptive Huffman Coding 67
3.4.1 Update Procedure 68
3.4.2 Encoding Procedure 71
3.4.3 Decoding Procedure 73
3.5 Golomb Codes 75
3.6 Rice Codes 76
3.6.1 CCSDS Recommendation for Lossless Compression 77
3.7 Tunstall Codes 79
3.8 Applications of Huffman Coding 81
3.8.1 Lossless Image Compression 81
3.8.2 Text Compression 83
3.8.3 Audio Compression 85
3.9 Summary 86
3.10 Projects and Problems 87

4 Arithmetic Coding 91
4.1 Overview 91
4.2 Introduction 91
4.3 Coding a Sequence 93
4.3.1 Generating a Tag 94
4.3.2 Deciphering the Tag 101
4.4 Generating a Binary Code 102
4.4.1 Uniqueness and Efficiency of the Arithmetic Code 103
4.4.2 Algorithm Implementation 106
4.4.3 Integer Implementation 111
4.5 Adaptive Arithmetic Coding 119
4.6 Binary Arithmetic Coding 120
4.6.1 The QM Coder 125
4.6.2 The MQ Coder 125
4.6.3 The M Coder 126
4.7 Comparison of Huffman and Arithmetic Coding 127
4.8 Applications 130
4.9 Summary 131
4.10 Projects and Problems 131

5 Dictionary Techniques 135
5.1 Overview 135
5.2 Introduction 135
5.3 Static Dictionary 136
5.3.1 Digram Coding 137
5.4 Adaptive Dictionary 139

5.4.1 The LZ77 Approach 139
5.4.2 The LZ78 Approach 143
5.5 Applications 150
5.5.1 File Compression–UNIX compress 151
5.5.2 Image Compression–The Graphics Interchange Format (GIF) 151
5.5.3 Image Compression–Portable Network Graphics (PNG) 152
5.5.4 Compression over Modems–V.42 bis 153
5.6 Beyond Compression—Lempel-Ziv Complexity ★ 156
5.7 Summary 158
5.8 Projects and Problems 159

6 Context-Based Compression 163
6.1 Overview 163
6.2 Introduction 163
6.3 Prediction with Partial Match (*ppm*) 165
6.3.1 The Basic Algorithm 165
6.3.2 The Escape Symbol 170
6.3.3 Length of Context 172
6.3.4 The Exclusion Principle 173
6.4 The Burrows-Wheeler Transform 174
6.4.1 Move-to-Front Coding 177
6.5 Associative Coder of Buyanovsky (ACB) 178
6.6 Dynamic Markov Compression 179
6.7 Summary 182
6.8 Projects and Problems 182

7 Lossless Image Compression 183
7.1 Overview 183
7.2 Introduction 183
7.2.1 The Old JPEG Standard 184
7.3 CALIC 186
7.4 JPEG-LS 190
7.5 Prediction Using Conditional Averages 192
7.6 Multiresolution Approaches 193
7.6.1 Progressive Image Transmission 193
7.7 Facsimile Encoding 198
7.7.1 Run-Length Coding 199
7.7.2 CCITT Group 3 and 4–Recommendations T.4 and T.6 200
7.7.3 JBIG 203
7.7.4 JBIG2-T.88 209
7.8 MRC-T.44 211
7.9 Summary 213
7.10 Projects and Problems 214

8 Mathematical Preliminaries for Lossy Coding **217**
8.1 Overview 217
8.2 Introduction 217
8.3 Distortion Criteria 220
8.3.1 The Human Visual System 223
8.3.2 Auditory Perception 224
8.4 Information Theory Revisited ★ 225
8.4.1 Conditional Entropy 225
8.4.2 Average Mutual Information 228
8.4.3 Differential Entropy 229
8.5 Rate Distortion Theory ★ 232
8.6 Models 240
8.6.1 Probability Models 240
8.6.2 Linear System Models 243
8.6.3 Physical Models 248
8.7 Summary 248
8.8 Projects and Problems 249

9 Scalar Quantization **251**
9.1 Overview 251
9.2 Introduction 251
9.3 The Quantization Problem 252
9.4 Uniform Quantizer 257
9.5 Adaptive Quantization 268
9.5.1 Forward Adaptive Quantization 269
9.5.2 Backward Adaptive Quantization 271
9.6 Nonuniform Quantization 277
9.6.1 *pdf*-Optimized Quantization 278
9.6.2 Companded Quantization 282
9.7 Entropy-Coded Quantization 287
9.7.1 Entropy Coding of Lloyd-Max Quantizer Outputs 288
9.7.2 Entropy-Constrained Quantization ★ 289
9.7.3 High-Rate Optimum Quantization ★ 289
9.8 Summary 292
9.9 Projects and Problems 293

10 Vector Quantization **295**
10.1 Overview 295
10.2 Introduction 295
10.3 Advantages of Vector Quantization over Scalar Quantization 298
10.4 The Linde-Buzo-Gray Algorithm 304
10.4.1 Initializing the LBG Algorithm 309
10.4.2 The Empty Cell Problem 315
10.4.3 Use of LBG for Image Compression 315
10.5 Tree-Structured Vector Quantizers 320

10.5.1 Design of Tree-Structured Vector Quantizers 323
10.5.2 Pruned Tree-Structured Vector Quantizers 324
10.6 Structured Vector Quantizers 324
10.6.1 Pyramid Vector Quantization 326
10.6.2 Polar and Spherical Vector Quantizers 327
10.6.3 Lattice Vector Quantizers 328
10.7 Variations on the Theme 332
10.7.1 Gain-Shape Vector Quantization 332
10.7.2 Mean-Removed Vector Quantization 332
10.7.3 Classified Vector Quantization 333
10.7.4 Multistage Vector Quantization 334
10.7.5 Adaptive Vector Quantization 335
10.8 Trellis-Coded Quantization 337
10.9 Summary 340
10.10 Projects and Problems 342

11 Differential Encoding 345
11.1 Overview 345
11.2 Introduction 345
11.3 The Basic Algorithm 348
11.4 Prediction in DPCM 352
11.5 Adaptive DPCM 357
11.5.1 Adaptive Quantization in DPCM 358
11.5.2 Adaptive Prediction in DPCM 358
11.6 Delta Modulation 361
11.6.1 Constant Factor Adaptive Delta Modulation (CFDM) 363
11.6.2 Continuously Variable Slope Delta Modulation 364
11.7 Speech Coding 365
11.7.1 G.726 366
11.8 Image Coding 369
11.9 Summary 371
11.10 Projects and Problems 371

12 Mathematical Preliminaries for Transforms, Subbands, and Wavelets 373
12.1 Overview 373
12.2 Introduction 373
12.3 Vector Spaces 374
12.3.1 Dot or Inner Product 375
12.3.2 Vector Space 375
12.3.3 Subspace 377
12.3.4 Basis 377
12.3.5 Inner Product–Formal Definition 379
12.3.6 Orthogonal and Orthonormal Sets 379
12.4 Fourier Series 380
12.5 Fourier Transform 382

12.5.1 Parseval's Theorem 384
12.5.2 Modulation Property 384
12.5.3 Convolution Theorem 385
12.6 Linear Systems 385
12.6.1 Time Invariance 386
12.6.2 Transfer Function 386
12.6.3 Impulse Response 387
12.6.4 Filter 388
12.7 Sampling 390
12.7.1 Ideal Sampling–Frequency Domain View 390
12.7.2 Ideal Sampling–Time Domain View 392
12.8 Discrete Fourier Transform 394
12.9 Z-Transform 396
12.9.1 Tabular Method 399
12.9.2 Partial Fraction Expansion 399
12.9.3 Long Division 403
12.9.4 Z-Transform Properties 404
12.9.5 Discrete Convolution 404
12.10 Summary 406
12.11 Projects and Problems 407

13 Transform Coding 409
13.1 Overview 409
13.2 Introduction 409
13.3 The Transform 414
13.4 Transforms of Interest 418
13.4.1 Karhunen-Loéve Transform 418
13.4.2 Discrete Cosine Transform 420
13.4.3 Discrete Sine Transform 423
13.4.4 Discrete Walsh-Hadamard Transform 423
13.5 Quantization and Coding of Transform Coefficients 424
13.5.1 Operational Rate-Distortion Bit Allocation 428
13.6 Application to Image Compression—JPEG 432
13.6.1 The Transform 432
13.6.2 Quantization 432
13.6.3 Coding 434
13.6.4 Format—JFIF 438
13.7 Application to Audio Compression—The MDCT 440
13.8 Summary 443
13.9 Projects and Problems 444

14 Subband Coding 447
14.1 Overview 447
14.2 Introduction 447
14.3 Filters 452

14.3.1 Some Filters Used in Subband Coding 456
14.4 The Basic Subband Coding Algorithm 459
14.4.1 Analysis 459
14.4.2 Quantization and Coding 461
14.4.3 Synthesis 461
14.5 Design of Filter Banks ★ 462
14.5.1 Downsampling ★ 463
14.5.2 Upsampling ★ 466
14.6 Perfect Reconstruction Using Two-Channel Filter Banks ★ 467
14.6.1 Two-Channel PR Quadrature Mirror Filters ★ 470
14.6.2 Power Symmetric FIR Filters ★ 472
14.7 M-Band Quadrature Mirror Filter Banks ★ 474
14.8 The Polyphase Decomposition ★ 477
14.9 Bit Allocation 482
14.10 Application to Speech Coding—G.722 484
14.11 Application to Audio Coding—MPEG Audio 485
14.12 Application to Image Compression 486
14.12.1 Decomposing an Image 488
14.12.2 Coding the Subbands 490
14.13 Summary 492
14.14 Projects and Problems 493

15 Wavelets 497
15.1 Overview 497
15.2 Introduction 497
15.3 Wavelets 500
15.4 Multiresolution Analysis and the Scaling Function 504
15.5 Implementation Using Filters 510
15.5.1 Scaling and Wavelet Coefficients 513
15.5.2 Families of Wavelets 516
15.6 Biorthogonal Wavelets 516
15.7 Lifting 523
15.8 Summary 527
15.9 Projects and Problems 528

16 Wavelet-Based Image Compression 529
16.1 Overview 529
16.2 Introduction 529
16.3 Embedded Zerotree Coder 532
16.4 Set Partitioning in Hierarchical Trees 540
16.5 JPEG 2000 547
16.5.1 Color Component Transform 548
16.5.2 Tiling 549
16.5.3 Wavelet Transform 549
16.5.4 Quantization 551

16.5.5 Tier I Coding 552
16.5.6 Tier II Coding 559
16.5.7 JPEG 2000 bitstream 561
16.6 Summary 568
16.7 Projects and Problems 568

17 Audio Coding 569
17.1 Overview 569
17.2 Introduction 569
17.2.1 Spectral Masking 570
17.2.2 Temporal Masking 571
17.2.3 Psychoacoustic Model 572
17.3 MPEG Audio Coding 573
17.3.1 Layer I Coding 573
17.3.2 Layer II Coding 575
17.3.3 Layer III Coding—mp3 577
17.4 MPEG Advanced Audio Coding 581
17.4.1 MPEG-2 AAC 582
17.4.2 MPEG-4 AAC 586
17.5 Dolby AC-3 (Dolby Digital) 587
17.5.1 Bit Allocation 588
17.6 Other Standards 589
17.7 Summary 589

18 Analysis/Synthesis and Analysis by Synthesis Schemes 591
18.1 Overview 591
18.2 Introduction 591
18.3 Speech Compression 593
18.3.1 The Channel Vocoder 594
18.3.2 The Linear Predictive Coder (Government Standard LPC-10) 596
18.3.3 Code-Excited Linear Predicton (CELP) 603
18.3.4 Sinusoidal Coders 606
18.3.5 Mixed Excitation Linear Prediction (MELP) 608
18.4 Wideband Speech Compression—ITU-T G.722.2 611
18.5 Coding of Speech for Internet Applications 613
18.5.1 iLBC 613
18.5.2 G.729 618
18.5.3 SILK 621
18.6 Image Compression 623
18.7 Summary 631
18.8 Projects and Problems 632

19 Viedo Compression 633
19.1 Overview 633

19.2 Introduction 633
19.3 Motion Compensation 635
19.4 Video Signal Representation 638
19.5 ITU-T Recommendation H.261 644
19.5.1 Motion Compensation 644
19.5.2 The Loop Filter 646
19.5.3 The Transform 647
19.5.4 Quantization and Coding 647
19.5.5 Rate Control 649
19.6 Model-Based Coding 649
19.7 Asymmetric Applications 650
19.8 The MPEG-1 Video Standard 652
19.9 The MPEG-2 Video Standard—H.262 655
19.10 ITU-T Recommendation H.263 658
19.10.1 Unrestricted Motion Vector Mode 660
19.10.2 Syntax-Based Arithmetic Coding Mode 660
19.10.3 Advanced Prediction Mode 661
19.10.4 PB-Frames and Improved PB-Frames Mode 661
19.10.5 Advanced Intra Coding Mode 661
19.10.6 Deblocking Filter Mode 661
19.10.7 Reference Picture Selection Mode 662
19.10.8 Temporal, SNR, and Spatial Scalability Mode 662
19.10.9 Reference Picture Resampling 662
19.10.10 Reduced-Resolution Update Mode 662
19.10.11 Alternative Inter VLC Mode 662
19.10.12 Modified Quantization Mode 663
19.10.13 Enhanced Reference Picture Selection Mode 663
19.11 ITU-T Recommendation H.264, MPEG-4 Part 10, Advanced Video Coding 664
19.11.1 Motion-Compensated Prediction 664
19.11.2 The Transform 665
19.11.3 Intra Prediction 666
19.11.4 Quantization 666
19.11.5 Coding 668
19.12 MPEG-4 Part 2 669
19.13 Packet Video 670
19.13.1 ATM Networks 671
19.13.2 Compression Issues in ATM Networks 671
19.13.3 Compression Algorithms for Packet Video 672
19.14 Summary 673
19.15 Projects and Problems 674

A Probability and Random Processes **675**
A.1 Probability 675

A.1.1 Frequency of Occurrence 675
A.1.2 A Measure of Belief 676
A.1.3 The Axiomatic Approach 678
A.2 Random Variables 680
A.3 Distribution Functions 681
A.4 Expectation 683
A.4.1 Mean 685
A.4.2 Second Moment 685
A.4.3 Variance 685
A.5 Types of Distribution 685
A.5.1 Uniform Distribution 685
A.5.2 Gaussian Distribution 686
A.5.3 Laplacian Distribution 686
A.5.4 Gamma Distribution 686
A.6 Stochastic Process 687
A.7 Projects and Problems 689

B A Brief Review of Matrix Concepts 691
B.1 A Matrix 691
B.2 Matrix Operations 692

C The Root Lattices 697

Bibliography 699

Index 717

Preface

Data compression has been an enabling technology for the information revolution, and as this revolution has changed our lives, data compression has become a more and more ubiquitous, if often invisible, presence. From mp3 players, to smartphones, to digital television and movies, data compression is an integral part of almost all information technology. This incorporation of compression into more and more of our lives also points to a certain degree of maturation and stability of the technology. This maturity is reflected in the fact that there are fewer differences between each edition of this book. In the second edition we added new techniques that had been developed since the first edition of this book came out. In the third edition we added a chapter on audio compression, a topic that had not been adequately covered in the second edition. In this edition we have tried to do the same with wavelet-based compression, in particular with the increasingly popular JPEG 2000 standard. There are now two chapters dealing with wavelet-based compression, one devoted exclusively to wavelet-based image compression algorithms. We have also filled in details that were left out from previous editions, such as a description of canonical Huffman codes and more information on binary arithmetic coding. We have also added descriptions of techniques that have been motivated by the Internet, such as the speech coding algorithms used for Internet applications.

All this has yet again enlarged the book. However, the intent remains the same: to provide an introduction to the art or science of data compression. There is a tutorial description of most of the popular compression techniques followed by a description of how these techniques are used for image, speech, text, audio, and video compression. One hopes the size of the book will not be intimidating. Once you open the book and begin reading a particular section we hope you will find the content easily accessible. If some material is not clear write to me at sayood@datacompression.unl.edu with specific questions and I will try and help (homework problems and projects are completely your responsibility).

Audience

If you are designing hardware or software implementations of compression algorithms, or need to interact with individuals engaged in such design, or are involved in development of multimedia applications and have some background in either electrical or computer engineering, or computer science, this book should be useful to you. We have included a large number of examples to aid in self-study. We have also included discussion of various multimedia standards. The intent here is not to provide all the details that may be required to implement a standard but to provide information that will help you follow and understand the standards documents. The final authority is always the standards document.

Course Use

The impetus for writing this book came from the need for a self-contained book that could be used at the senior/graduate level for a course in data compression in either electrical engineering, computer engineering, or computer science departments. There are problems and project ideas after most of the chapters. A solutions manual is available from the publisher. Also at *datacompression.unl.edu* we provide links to various course homepages, which can be a valuable source of project ideas and support material.

The material in this book is too much for a one-semester course. However, with judicious use of the starred sections, this book can be tailored to fit a number of compression courses that emphasize various aspects of compression. If the course emphasis is on lossless compression, the instructor could cover most of the sections in the first seven chapters. Then, to give a taste of lossy compression, the instructor could cover Sections 1–5 of Chapter 9, followed by Chapter 13 and its description of JPEG, and Chapter 19, which describes video compression approaches used in multimedia communications. If the class interest is more attuned to audio compression, then instead of Chapters 13 and 19, the instructor could cover Chapters 14 and 17. If the latter option is taken, depending on the background of the students in the class, Chapter 12 may be assigned as background reading. If the emphasis is to be on lossy compression, the instructor could cover Chapter 2, the first two sections of Chapter 3, Sections 4 and 6 of Chapter 4 (with a cursory overview of Sections 2 and 3), Chapter 8, selected parts of Chapter 9, and Chapters 10 through 16. At this point depending on the time available and the interests of the instructor and the students, portions of the remaining three chapters can be covered. I have always found it useful to assign a term project in which the students can follow their own interests as a means of covering material that is not covered in class but is of interest to the student.

Approach

In this book, we cover both lossless and lossy compression techniques with applications to image, speech, text, audio, and video compression. The various lossless and lossy coding techniques are introduced with just enough theory to tie things together. The necessary theory is introduced just before we need it. Therefore, there are three *mathematical preliminaries* chapters. In each of these chapters, we present the mathematical material needed to understand and appreciate the techniques that follow.

Although this book is an introductory text, the word *introduction* may have a different meaning for different audiences. We have tried to accommodate the needs of different audiences by taking a dual-track approach. Wherever we felt there was material that could enhance the understanding of the subject being discussed but could still be skipped without seriously hindering your understanding of the technique, we marked those sections with a star ($\star$). If you are primarily interested in understanding how the various techniques function, especially if you are using this book for self-study, we recommend you skip the starred sections, at least in a first reading. Readers who require a slightly more theoretical approach should use the starred sections. Except for the starred sections, we have tried to keep the mathematics to a minimum.

Learning from This Book

I have found that it is easier for me to understand things if I can see examples. Therefore, I have relied heavily on examples to explain concepts. You may find it useful to spend more time with the examples if you have difficulty with some of the concepts.

Compression is still largely an art and to gain proficiency in an art we need to get a "feel" for the process. We have included software implementations for most of the techniques discussed in this book, along with a large number of data sets. The software and data sets can be obtained from *datacompression.unl.edu*. The programs are written in C and have been tested on a number of platforms. The programs should run under most flavors of UNIX machines and, with some slight modifications, under other operating systems as well.

You are strongly encouraged to use and modify these programs to work with your favorite data in order to understand some of the issues involved in compression. A useful and achievable goal should be the development of your own compression package by the time you have worked through this book. This would also be a good way to learn the trade-offs involved in different approaches. We have tried to give comparisons of techniques wherever possible; however, different types of data have their own idiosyncrasies. The best way to know which scheme to use in any given situation is to try them.

Content and Organization

The organization of the chapters is as follows: We introduce the mathematical preliminaries necessary for understanding lossless compression in Chapter 2; Chapters 3 and 4 are devoted to coding algorithms, including Huffman coding, arithmetic coding, Golomb-Rice codes, and Tunstall codes. Chapters 5 and 6 describe many of the popular lossless compression schemes along with their applications. The schemes include LZW, *ppm*, BWT, and DMC, among others. In Chapter 7 we describe a number of lossless image compression algorithms and their applications in a number of international standards. The standards include the JBIG standards and various facsimile standards.

Chapter 8 is devoted to providing the mathematical preliminaries for lossy compression. Quantization is at the heart of most lossy compression schemes. Chapters 9 and 10 are devoted to the study of quantization. Chapter 9 deals with scalar quantization, and Chapter 10 deals with vector quantization. Chapter 11 deals with differential encoding techniques, in particular differential pulse code modulation (DPCM) and delta modulation. Included in this chapter is a discussion of the CCITT G.726 standard.

Chapter 12 is our third mathematical preliminaries chapter. The goal of this chapter is to provide the mathematical foundation necessary to understand some aspects of the transform, subband, and wavelet-based techniques that are described in the next four chapters. As in the case of the previous mathematical preliminaries chapters, not all material covered is necessary for everyone. We describe the JPEG standard in Chapter 13, the CCITT G.722 international standard in Chapter 14, and EZW, SPIHT, and JPEG 2000 in Chapter 16.

Chapter 17 is devoted to audio compression. We describe the various MPEG audio compression schemes in this chapter including the scheme popularly known as mp3.

Chapter 18 covers techniques in which the data to be compressed are analyzed, and a model for the generation of the data is transmitted to the receiver. The receiver uses this model to synthesize the data. These analysis/synthesis and analysis by synthesis schemes include linear predictive schemes used for low-rate speech coding and the fractal compression technique. We describe the federal government LPC-10 standard. Code-excited linear prediction (CELP) is a popular example of an analysis by synthesis scheme. We also discuss three CELP-based standards (Federal Standard 1016, the international standard G.728, and the wideband speech compression standard G.722.2) as well as the 2.4 kbps mixed excitation linear prediction (MELP) technique. We have also included an introduction to three speech compression standards currently in use for speech compression for Internet applications: the Internet Low Bitrate Codec, the ITU-T G.729 standard, and SILK.

Chapter 19 deals with video coding. We describe popular video coding techniques via description of various international standards, including H.261, H.264, and the various MPEG standards.

A Personal View

For me, data compression is more than a manipulation of numbers; it is the process of discovering structures that exist in the data. In the 9th century, the poet Omar Khayyam wrote

> The moving finger writes, and having writ,
> moves on; not all thy piety nor wit,
> shall lure it back to cancel half a line,
> nor all thy tears wash out a word of it.
>
> *(The Rubaiyat of Omar Khayyam)*

To explain these few lines would take volumes. They tap into a common human experience so that in our mind's eye, we can reconstruct what the poet was trying to convey centuries ago. To understand the words we not only need to know the language, we also need to have a model of reality that is close to that of the poet. The genius of the poet lies in identifying a model of reality that is so much a part of our humanity that centuries later and in widely diverse cultures, these few words can evoke volumes.

Data compression is much more limited in its aspirations, and it may be presumptuous to mention it in the same breath as poetry. But there is much that is similar to both endeavors. Data compression involves identifying models for the many different types of structures that exist in different types of data and then using these models, perhaps along with the perceptual framework in which these data will be used, to obtain a compact representation of the data. These structures can be in the form of patterns that we can recognize simply by plotting the data, or they might be structures that require a more abstract approach to comprehend. Often, it is not the data but the structure within the data that contains the information, and the development of data compression involves the discovery of these structures.

In *The Long Dark Teatime of the Soul* by Douglas Adams, the protagonist finds that he can enter Valhalla (a rather shoddy one) if he tilts his head in a certain way. Appreciating the structures that exist in data sometimes require us to tilt our heads in a certain way. There are an infinite number of ways we can tilt our head and, in order not to get a pain in the neck (carrying our analogy to absurd limits), it would be nice to know some of the ways that will generally lead to a profitable result. One of the objectives of this book is to provide you with a frame of reference that can be used for further exploration. I hope this exploration will provide as much enjoyment for you as it has given to me.

Acknowledgments

It has been a lot of fun writing this book. My task has been made considerably easier and the end product considerably better because of the help I have received. Acknowledging that help is itself a pleasure.

The first edition benefitted from the careful and detailed criticism of Roy Hoffman from IBM, Glen Langdon from the University of California at Santa Cruz, Debra Lelewer from California Polytechnic State University, Eve Riskin from the University of Washington, Ibrahim Sezan from Kodak, and Peter Swaszek from the University of Rhode Island. They provided detailed comments on all or most of the first edition. Nasir Memon from Polytechnic University, Victor Ramamoorthy then at S3, Grant Davidson at Dolby Corporation, Hakan Caglar, who was then at TÜBITAK in Gebze, and Allen Gersho from the University of California at Santa Barbara reviewed parts of the manuscript.

For the second edition Steve Tate at the University of North Texas, Sheila Horan at New Mexico State University, Edouard Lamboray at Oerlikon Contraves Group, Steven Pigeon at the University of Montreal, and Jesse Olvera at Raytheon Systems reviewed the entire manuscript. Emin Anarım of Boğaziçi University and Hakan Çağlar helped me with the development of the chapter on wavelets. Mark Fowler provided extensive comments on Chapters 12–15, correcting mistakes of both commission and omission. Tim James, Devajani Khataniar, and Lance Pérez also read and critiqued parts of the new material in the second edition. Chloeann Nelson, along with trying to stop me from splitting infinitives, also tried to make the first two editions of the book more user-friendly. The third edition benefitted from the critique of Rob Maher, now at Montana State, who generously gave of his time to help with the chapter on audio compression.

Since the appearance of the first edition, various readers have sent me their comments and critiques. I am grateful to all who sent me comments and suggestions. I am especially grateful to Roberto Lopez-Hernandez, Dirk vom Stein, Christopher A. Larrieu, Ren Yih Wu, Humberto D'Ochoa, Roderick Mills, Mark Elston, and Jeerasuda Keesorth for pointing out errors and suggesting improvements to the book. I am also grateful to the various instructors who have sent me their critiques. In particular I would like to thank Bruce Bomar from the University of Tennessee, K.R. Rao from the University of Texas at Arlington, Ralph Wilkerson from the University of Missouri–Rolla, Adam Drozdek from Duquesne University, Ed Hong and Richard Ladner from the University of Washington, Lars Nyland from the Colorado School of Mines, Mario Kovac from the University of Zagreb, Jim Diamond of Acadia University, and Haim Perlmutter from Ben-Gurion University. Paul Amer, from the University of Delaware, has been one of my earliest, most consistent, and most welcome critics. His courtesy is greatly appreciated.

Frazer Williams and Mike Hoffman, from my department at the University of Nebraska, provided reviews for the first edition of the book. Mike has continued to provide me with guidance and has read and critiqued the new chapters in every edition of the book including this one. I rely heavily on his insights and his critique and would be lost without him. It is nice to have friends of his intellectual caliber and generosity.

The improvement and changes in this edition owe a lot to Mark Fowler from SUNY Binghamton and Pierre Jouvelet from the Ecole Superieure des Mines de Paris. Much of the new material was added because Mark thought that it should be there. He provided detailed guidance both during the planning of the changes and during their implementation. Pierre provided me with the most thorough critique I have ever received for this book. His insight into all aspects of compression and his willingness to share them has significantly improved this book. The chapter on wavelet image compression benefitted from the review of Mike Marcellin of the University of Arizona. Mike agreed to look at the chapter while in the midst of end-of-semester crunch, which is an act of friendship those in the teaching profession will appreciate. Mike is a gem. Pat Worster edited many of the chapters and tried to teach me the proper use of the semi-colon, and to be a bit more generous with commas. The book reads a lot better because of her attention. With all this help one would expect a perfect book. The fact that it is not is a reflection of my imperfection.

Rick Adams formerly at Morgan Kaufmann convinced me that I had to revise this book. Andrea Dierna inherited the book and its recalcitrant author and somehow, in a very short time, got reviews, got revisions—got things working. Meagan White had the unenviable task of getting the book ready for production, and still allowed me to mess up her schedule. Danielle Miller was the unfailingly courteous project manager who kept the project on schedule despite having to deal with an author who was bent on not keeping on schedule. Charles Roumeliotis was the copy editor. He caught many of my mistakes that I would never have caught; both I and the readers owe him a lot.

Most of the examples in this book were generated in a lab set up by Andy Hadenfeldt. James Nau helped me extricate myself out of numerous software puddles giving freely of his time. In my times of panic, he has always been just an email or voice mail away. The current denizens of my lab, the appropriately named Occult Information Lab, helped me in many ways small and big. Sam Way tried (and failed) to teach me Python and helped me out with examples. Dave Russell, who had to teach out of this book, provided me with very helpful criticism, always gently, with due respect to my phantom grey hair. Discussions with Ufuk Nalbantoglu about the more abstract aspects of data compression helped clarify things for me.

I would like to thank the various "models" for the data sets that accompany this book and were used as examples. The individuals in the images are Sinan Sayood, Sena Sayood, and Elif Sevuktekin. The female voice belongs to Pat Masek.

This book reflects what I have learned over the years. I have been very fortunate in the teachers I have had. David Farden, now at North Dakota State University, introduced me to the area of digital communication. Norm Griswold, formerly at Texas A&M University, introduced me to the area of data compression. Jerry Gibson, now at the University of California at Santa Barbara, was my Ph.D. advisor and helped me get started on my professional career. The world may not thank him for that, but I certainly do.

I have also learned a lot from my students at the University of Nebraska and Boğaziçi University. Their interest and curiosity forced me to learn and kept me in touch with the broad field that is data compression today. I learned at least as much from them as they learned from me.

Much of this learning would not have been possible but for the support I received from NASA. The late Warner Miller and Pen-Shu Yeh at the Goddard Space Flight Center and Wayne Whyte at the Lewis Research Center were a source of support and ideas. I am truly grateful for their helpful guidance, trust, and friendship.

Our two boys, Sena and Sinan, graciously forgave my evenings and weekends at work. They were tiny (witness the images) when I first started writing this book. They are young men now, as gorgeous to my eyes now as they have always been, and "the book" has been their (sometimes unwanted) companion through all these years. For their graciousness and for the great joy they have given me, I thank them.

Above all the person most responsible for the existence of this book is my partner and closest friend Füsun. Her support and her friendship gives me the freedom to do things I would not otherwise even consider. She centers my universe, is the color of my existence, and, as with every significant endeavor that I have undertaken since I met her, this book is at least as much hers as it is mine.

1

Introduction

In the last decade, we have been witnessing a transformation—some call it a revolution—in the way we communicate, and the process is still under way. This transformation includes the ever-present, ever-growing Internet; the explosive development of mobile communications; and the ever-increasing importance of video communication. Data compression is one of the enabling technologies for each of these aspects of the multimedia revolution. It would not be practical to put images, let alone audio and video, on websites if it were not for data compression algorithms. Cellular phones would not be able to provide communication with increasing clarity were it not for compression. The advent of digital TV would not be possible without compression. Data compression, which for a long time was the domain of a relatively small group of engineers and scientists, is now ubiquitous. Make a call on your cell phone, and you are using compression. Surf on the Internet, and you are using (or wasting) your time with assistance from compression. Listen to music on your MP3 player or watch a DVD, and you are being entertained courtesy of compression.

So what is data compression, and why do we need it? Most of you have heard of JPEG and MPEG, which are standards for representing images, video, and audio. Data compression algorithms are used in these standards to reduce the number of bits required to represent an image or a video sequence or music. In brief, data compression is the art or science of representing information in a compact form. We create these compact representations by identifying and using structures that exist in the data. Data can be characters in a text file, numbers that are samples of speech or image waveforms, or sequences of numbers that are generated by other processes. The reason we need data compression is that more and more of the information that we generate and use is in digital form—consisting of numbers represented by bytes of data. And the number of bytes required to represent multimedia data can be huge. For example, in order to digitally represent 1 second of video without compression (using the CCIR 601 format described in Chapter 18), we need more than 20 megabytes, or 160 megabits. If we consider the number of seconds in a movie, we can easily see why we would need compression. To represent 2 minutes of uncompressed CD-quality music (44,100 samples per second, 16 bits per sample) requires more than 84 million bits. Downloading music from a website at these rates would take a long time.

As human activity has a greater and greater impact on our environment, there is an ever-increasing need for more information about our environment, how it functions, and what we are doing to it. Various space agencies from around the world, including the European Space Agency (ESA), the National Aeronautics and Space Administration (NASA), the Canadian Space Agency (CSA), and the Japan Aerospace Exploration Agency (JAXA), are collaborating on a program to monitor global change that will generate half a terabyte of data per *day* when it is fully operational. New sequencing technology is resulting in ever-increasing database sizes containing genomic information while new medical scanning technologies could result in the generation of petabytes[1] of data.

Given the explosive growth of data that needs to be transmitted and stored, why not focus on developing better transmission and storage technologies? This is happening, but it is not enough. There have been significant advances that permit larger and larger volumes of information to be stored and transmitted without using compression, including CD-ROMs, optical fibers, Asymmetric Digital Subscriber Lines (ADSL), and cable modems. However, while it is true that both storage and transmission capacities are steadily increasing with new technological innovations, as a corollary to Parkinson's First Law,[2] it seems that the need for mass storage and transmission increases at least twice as fast as storage and transmission capacities improve. Then there are situations in which capacity has not increased significantly. For example, the amount of information we can transmit over the airwaves will always be limited by the characteristics of the atmosphere.

An early example of data compression is Morse code, developed by Samuel Morse in the mid-19th century. Letters sent by telegraph are encoded with dots and dashes. Morse noticed that certain letters occurred more often than others. In order to reduce the average time required to send a message, he assigned shorter sequences to letters that occur more frequently, such as e ($\cdot$) and a ($\cdot -$), and longer sequences to letters that occur less frequently, such as q ($- - \cdot -$) and j ($\cdot - - -$). This idea of using shorter codes for more frequently occurring characters is used in Huffman coding, which we will describe in Chapter 3.

Where Morse code uses the frequency of occurrence of single characters, a widely used form of Braille code, which was also developed in the mid-19th century, uses the frequency of occurrence of words to provide compression [1]. In Braille coding, 2×3 arrays of dots are used to represent text. Different letters can be represented depending on whether the dots are raised or flat. In Grade 1 Braille, each array of six dots represents a single character. However, given six dots with two positions for each dot, we can obtain 2^6, or 64, different combinations. If we use 26 of these for the different letters, we have 38 combinations left. In Grade 2 Braille, some of these leftover combinations are used to represent words that occur frequently, such as "and" and "for." One of the combinations is used as a special symbol indicating that the symbol that follows is a word and not a character, thus allowing a large number of words to be

[1] mega: 10^6, giga: 10^9, tera: 10^{12}, peta: 10^{15}, exa: 10^{18}, zetta: 10^{21}, yotta: 10^{24}

[2] Parkinson's First Law: "Work expands so as to fill the time available," in *Parkinson's Law and Other Studies in Administration*, by Cyril Northcote Parkinson, Ballantine Books, New York, 1957.

represented by two arrays of dots. These modifications, along with contractions of some of the words, result in an average reduction in space, or compression, of about 20% [1].

Statistical structure is being used to provide compression in these examples, but that is not the only kind of structure that exists in the data. There are many other kinds of structures existing in data of different types that can be exploited for compression. Consider speech. When we speak, the physical construction of our voice box dictates the kinds of sounds that we can produce. That is, the mechanics of speech production impose a structure on speech. Therefore, instead of transmitting the speech itself, we could send information about the conformation of the voice box, which could be used by the receiver to synthesize the speech. An adequate amount of information about the conformation of the voice box can be represented much more compactly than the numbers that are the sampled values of speech. Therefore, we get compression. This compression approach is currently being used in a number of applications, including transmission of speech over cell phones and the synthetic voice in toys that speak. An early version of this compression approach, called the *vocoder* (*vo*ice *coder*), was developed by Homer Dudley at Bell Laboratories in 1936. The vocoder was demonstrated at the New York World's Fair in 1939, where it was a major attraction. We will revisit the vocoder and this approach to compression of speech in Chapter 18.

These are only a few of the many different types of structures that can be used to obtain compression. The structure in the data is not the only thing that can be exploited to obtain compression. We can also make use of the characteristics of the user of the data. Many times, for example, when transmitting or storing speech and images, the data are intended to be perceived by a human, and humans have limited perceptual abilities. For example, we cannot hear the very high frequency sounds that dogs can hear. If something is represented in the data that cannot be perceived by the user, is there any point in preserving that information? The answer is often "no." Therefore, we can make use of the perceptual limitations of humans to obtain compression by discarding irrelevant information. This approach is used in a number of compression schemes that we will visit in Chapters 13, 14, and 17.

Before we embark on our study of data compression techniques, let's take a general look at the area and define some of the key terms and concepts we will be using in the rest of the book.

1.1 Compression Techniques

When we speak of a compression technique or compression algorithm,[3] we are actually referring to two algorithms. There is the compression algorithm that takes an input $\mathcal{X}$ and generates a representation $\mathcal{X}_c$ that requires fewer bits, and there is a reconstruction algorithm that operates on the compressed representation $\mathcal{X}_c$ to generate the reconstruction $\mathcal{Y}$. These operations are shown schematically in Figure 1.1. We will follow convention and refer to both the compression and reconstruction algorithms together to mean the compression algorithm.

[3] The word *algorithm* comes from the name of an early 9th-century Arab mathematician, Al-Khwarizmi, who wrote a treatise entitled *The Compendious Book on Calculation by* al-jabr *and* al-muqabala, in which he explored (among other things) the solution of various linear and quadratic equations via rules or an "algorithm." This approach became known as the method of Al-Khwarizmi. The name was changed to *algoritni* in Latin, from which we get the word *algorithm*.The name of the treatise also gave us the word *algebra* [2].

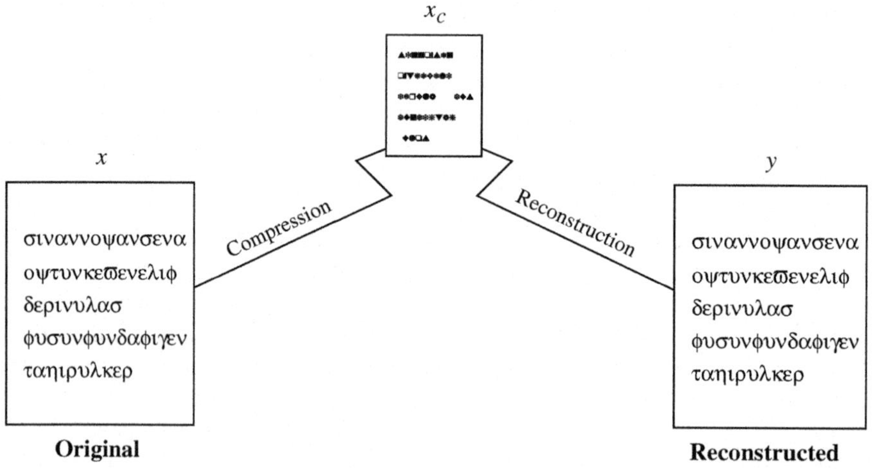

FIGURE 1.1 Compression and reconstruction.

Based on the requirements of reconstruction, data compression schemes can be divided into two broad classes: *lossless* compression schemes, in which $\mathcal{Y}$ is identical to $\mathcal{X}$, and *lossy* compression schemes, which generally provide much higher compression than lossless compression but allow $\mathcal{Y}$ to be different from $\mathcal{X}$.

1.1.1 Lossless Compression

Lossless compression techniques, as their name implies, involve no loss of information. If data have been losslessly compressed, the original data can be recovered exactly from the compressed data. Lossless compression is generally used for applications that cannot tolerate any difference between the original and reconstructed data.

Text compression is an important area for lossless compression. It is very important that the reconstruction is identical to the original text, as very small differences can result in statements with very different meanings. Consider the sentences "Do *not* send money" and "Do *now* send money." A similar argument holds for computer files and for certain types of data such as bank records.

If data of any kind are to be processed or "enhanced" later to yield more information, it is important that the integrity be preserved. For example, suppose we compressed a radiological image in a lossy fashion, and the difference between the reconstruction $\mathcal{Y}$ and the original $\mathcal{X}$ was visually undetectable. If this image was later enhanced, the previously undetectable differences may cause the appearance of artifacts that could seriously mislead the radiologist. Because the price for this kind of mishap may be a human life, it makes sense to be very careful about using a compression scheme that generates a reconstruction that is different from the original.

Data obtained from satellites often are processed later to obtain different numerical indicators of vegetation, deforestation, and so on. If the reconstructed data are not identical to the original data, processing may result in "enhancement" of the differences. It may not be possible to go back and obtain the same data over again. Therefore, it is not advisable to allow for any differences to appear in the compression process.

There are many situations that require compression where we want the reconstruction to be identical to the original. There are also a number of situations in which it is possible to relax this requirement in order to get more compression. In these situations, we look to lossy compression techniques.

1.1.2 Lossy Compression

Lossy compression techniques involve some loss of information, and data that have been compressed using lossy techniques generally cannot be recovered or reconstructed exactly. In return for accepting this distortion in the reconstruction, we can generally obtain much higher compression ratios than is possible with lossless compression.

In many applications, this lack of exact reconstruction is not a problem. For example, when storing or transmitting speech, the exact value of each sample of speech is not necessary. Depending on the quality required of the reconstructed speech, varying amounts of loss of information about the value of each sample can be tolerated. If the quality of the reconstructed speech is to be similar to that heard on the telephone, a significant loss of information can be tolerated. However, if the reconstructed speech needs to be of the quality heard on a compact disc, the amount of information loss that can be tolerated is much lower.

Similarly, when viewing a reconstruction of a video sequence, the fact that the reconstruction is different from the original is generally not important as long as the differences do not result in annoying artifacts. Thus, video is generally compressed using lossy compression.

Once we have developed a data compression scheme, we need to be able to measure its performance. Because of the number of different areas of application, different terms have been developed to describe and measure the performance.

1.1.3 Measures of Performance

A compression algorithm can be evaluated in a number of different ways. We could measure the relative complexity of the algorithm, the memory required to implement the algorithm, how fast the algorithm performs on a given machine, the amount of compression, and how closely the reconstruction resembles the original. In this book we will mainly be concerned with the last two criteria. Let us take each one in turn.

A very logical way of measuring how well a compression algorithm compresses a given set of data is to look at the ratio of the number of bits required to represent the data before compression to the number of bits required to represent the data after compression. This ratio is called the *compression ratio*. Suppose storing an image made up of a square array of 256×256 pixels requires 65,536 bytes. The image is compressed and the compressed version requires 16,384 bytes. We would say that the compression ratio is 4:1. We can also represent the compression ratio by expressing the reduction in the amount of data required as a percentage of the size of the original data. In this particular example, the compression ratio calculated in this manner would be 75%.

Another way of reporting compression performance is to provide the average number of bits required to represent a single sample. This is generally referred to as the *rate*. For example, in the case of the compressed image described above, if we assume 8 bits per byte (or pixel), the average number of bits per pixel in the compressed representation is 2. Thus, we would say that the rate is 2 bits per pixel.

In lossy compression, the reconstruction differs from the original data. Therefore, in order to determine the efficiency of a compression algorithm, we have to have some way of quantifying the difference. The difference between the original and the reconstruction is often called the *distortion*. (We will describe several measures of distortion in Chapter 8.) Lossy techniques are generally used for the compression of data that originate as analog signals, such as speech and video. In compression of speech and video, the final arbiter of quality is human. Because human responses are difficult to model mathematically, many approximate measures of distortion are used to determine the quality of the reconstructed waveforms. We will discuss this topic in more detail in Chapter 8.

Other terms that are also used when talking about differences between the reconstruction and the original are *fidelity* and *quality*. When we say that the fidelity or quality of a reconstruction is high, we mean that the difference between the reconstruction and the original is small. Whether this difference is a mathematical difference or a perceptual difference should be evident from the context.

1.2 Modeling and Coding

While reconstruction requirements may force the decision of whether a compression scheme is to be lossy or lossless, the exact compression scheme we use will depend on a number of different factors. Some of the most important factors are the characteristics of the data that need to be compressed. A compression technique that will work well for the compression of text may not work well for compressing images. Each application presents a different set of challenges.

There is a saying attributed to Bob Knight, the former basketball coach at Indiana University and Texas Tech University: "If the only tool you have is a hammer, you approach every problem as if it were a nail." Our intention in this book is to provide you with a large number of tools that you can use to solve a particular data compression problem. It should be remembered that data compression, if it is a science at all, is an experimental science. The approach that works best for a particular application will depend to a large extent on the redundancies inherent in the data.

The development of data compression algorithms for a variety of data can be divided into two phases. The first phase is usually referred to as *modeling*. In this phase, we try to extract information about any redundancy that exists in the data and describe the redundancy in the form of a model. The second phase is called *coding*. A description of the model and a "description" of how the data differ from the model are encoded, generally using a binary alphabet. The difference between the data and the model is often referred to as the *residual*.

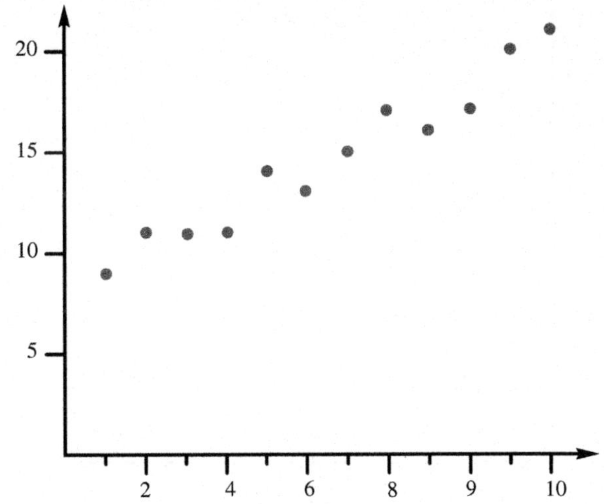

FIGURE 1.2 **A sequence of data values.**

In the following three examples, we will look at three different ways that data can be modeled. We will then use the model to obtain compression.

Example 1.2.1:

Consider the following sequence of numbers $\{x_1, x_2, x_3, \ldots\}$:

9	11	11	11	14	13	15	17	16	17	20	21

If we were to transmit or store the binary representations of these numbers, we would need to use 5 bits per sample. However, by exploiting the structure in the data, we can represent the sequence using fewer bits. If we plot these data as shown in Figure 1.2, we see that the data seem to fall on a straight line. A model for the data could, therefore, be a straight line given by the equation

$$\hat{x}_n = n + 8 \qquad \text{n} = 1, 2, \ldots$$

The structure in this particular sequence of numbers can be characterized by an equation. Thus, $\hat{x}_1 = 9$, while $x_1 = 9$, $\hat{x}_2 = 10$, while $x_2 = 11$, and so on. To make use of this structure, let's examine the difference between the data and the model. The difference (or residual) is given by the sequence

$$e_n = x_n - \hat{x}_n : 010 - 11 - 101 - 1 - 111$$

The residual sequence consists of only three numbers $\{-1, 0, 1\}$. If we assign a code of 00 to -1, a code of 01 to 0, and a code of 10 to 1, we need to use 2 bits to represent each element of the residual sequence. Therefore, we can obtain compression by transmitting or storing the parameters of the model and the residual sequence. The encoding can be exact if the required compression is to be lossless, or approximate if the compression can be lossy. ♦

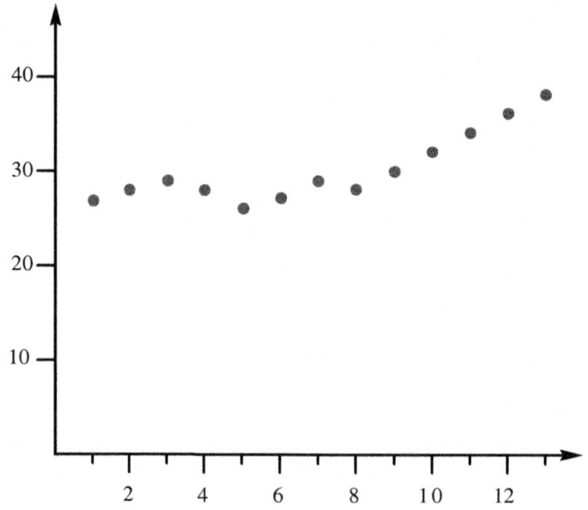

FIGURE 1.3 **A sequence of data values.**

The type of structure or redundancy that existed in these data follows a simple law. Once we recognize this law, we can make use of the structure to *predict* the value of each element in the sequence and then encode the residual. Structure of this type is only one of many types of structure.

Example 1.2.2:

Consider the following sequence of numbers:

27	28	29	28	26	27	29	28	30	32	34	36	38

The sequence is plotted in Figure 1.3.

The sequence does not seem to follow a simple law as in the previous case. However, each value in this sequence is close to the previous value. Suppose we send the first value, then in place of subsequent values we send the difference between it and the previous value. The sequence of transmitted values would be

27	1	1	-1	-2	1	2	-1	2	2	2	2	2

Like the previous example, the number of distinct values has been reduced. Fewer bits are required to represent each number, and compression is achieved. The decoder adds each received value to the previous decoded value to obtain the reconstruction corresponding to the received value. Techniques that use the past values of a sequence to *predict* the current value and then encode the error in prediction, or residual, are called *predictive coding* schemes. We will discuss lossless predictive compression schemes in Chapter 7 and lossy predictive coding schemes in Chapter 11.

Assuming both encoder and decoder know the model being used, we would still have to send the value of the first element of the sequence. ♦

A very different type of redundancy is statistical in nature. Often we will encounter sources that generate some symbols more often than others. In these situations, it will be advantageous to assign binary codes of different lengths to different symbols.

Example 1.2.3:

Suppose we have the following sequence:

$$a\not{b}barrayaran\not{b}barray\not{b}ran\not{b}far\not{b}faar\not{b}faaar\not{b}away$$

which is typical of all sequences generated by a source ($\not{b}$ denotes a blank space). Notice that the sequence is made up of eight different symbols. In order to represent eight symbols, we need to use 3 bits per symbol. Suppose instead we used the code shown in Table 1.1. Notice that we have assigned a codeword with only a single bit to the symbol that occurs most often (a) and correspondingly longer codewords to symbols that occur less often. If we substitute the codes for each symbol, we will use 106 bits to encode the entire sequence. As there are 41 symbols in the sequence, this works out to approximately 2.58 bits per symbol. This means we have obtained a compression ratio of 1.16:1. We will study how to use statistical redundancy of this sort in Chapters 3 and 4.

TABLE 1.1 A code with codewords of varying length.

a	1
$\not{b}$	001
b	01100
f	0100
n	0111
r	000
w	01101
y	0101

♦

When dealing with text, along with statistical redundancy, we also see redundancy in the form of words that repeat often. We can take advantage of this form of redundancy by constructing a list of these words and then representing them by their position in the list. This type of compression scheme is called a *dictionary* compression scheme. We will study these schemes in Chapter 5.

Often the structure or redundancy in the data becomes more evident when we look at groups of symbols. We will look at compression schemes that take advantage of this in Chapters 4 and 10.

Finally, there will be situations in which it is easier to take advantage of the structure if we decompose the data into a number of components. We can then study each component separately and use a model appropriate to that component. We will look at such schemes in Chapters 13, 14, 15, and 16.

There are a number of different ways to characterize data. Different characterizations will lead to different compression schemes. We will study these compression schemes in the upcoming chapters and use a number of examples that should help us understand the relationship between the characterization and the compression scheme.

With the increasing use of compression, there has also been an increasing need for standards. Standards allow products developed by different vendors to communicate. Thus, we can compress something with products from one vendor and reconstruct it using the products of a different vendor. The different international standards organizations have responded to this need, and a number of standards for various compression applications have been approved. We will discuss these standards as applications of the various compression techniques.

Finally, compression is still largely an art, and to gain proficiency in an art, you need to get a feel for the process. To help, we have developed software implementations of most of the techniques discussed in this book and have also provided the data sets used for developing the examples in this book. Details on how to obtain these programs and data sets are provided in the Preface. You should use these programs on your favorite data or on the data sets provided in order to understand some of the issues involved in compression. We would also encourage you to write your own software implementations of some of these techniques, as very often the best way to understand how an algorithm works is to implement the algorithm.

1.3 Summary

In this chapter, we have introduced the subject of data compression. We have provided some motivation for why we need data compression and defined some of the terminology used in this book. Additional terminology will be defined as needed. We have briefly introduced the two major types of compression algorithms: lossless compression and lossy compression. Lossless compression is used for applications that require an exact reconstruction of the original data, while lossy compression is used when the user can tolerate some differences between the original and reconstructed representations of the data. An important element in the design of data compression algorithms is the modeling of the data. We have briefly looked at how modeling can help us in obtaining more compact representations of the data. We have described some of the different ways we can view the data in order to model it. The more ways we have of looking at the data, the more successful we will be in developing compression schemes that take full advantage of the structures in the data.

1.4 Projects and Problems

1. Use the compression utility on your computer to compress different files. Study the effect of the original file size and type on the ratio of the compressed file size to the original file size.

2. Take a few paragraphs of text from a popular magazine and compress them by removing all words that are not essential for comprehension. For example, in the sentence, "This is the dog that belongs to my friend," we can remove the words *is*, *the*, *that*, and *to* and still convey the same meaning. Let the ratio of the words removed to the total number of words in the original text be the measure of redundancy in the text. Repeat the experiment using paragraphs from a technical journal. Can you make any quantitative statements about the redundancy in the text obtained from different sources?

2

Mathematical Preliminaries for Lossless Compression

2.1 Overview

The treatment of data compression in this book is not very mathematical. (For a more mathematical treatment of some of the topics covered in this book, see [3–6].) However, we do need some mathematical preliminaries to appreciate the compression techniques we will discuss. Compression schemes can be divided into two classes, lossy and lossless. Lossy compression schemes involve the loss of some information, and data that have been compressed using a lossy scheme generally cannot be recovered exactly. Lossless schemes compress the data without loss of information, and the original data can be recovered exactly from the compressed data. In this chapter, some of the ideas in information theory that provide the framework for the development of lossless data compression schemes are briefly reviewed. We will also look at some ways to model the data that lead to efficient coding schemes. We have assumed some knowledge of probability concepts (see Appendix A for a brief review of probability and random processes).

2.2 A Brief Introduction to Information Theory

Although the idea of a quantitative measure of information has been around for a while, the person who pulled everything together into what is now called information theory was Claude Elwood Shannon [3], an electrical engineer at Bell Labs. Shannon defined a quantity called *self-information*. Suppose we have an event A, which is a set of outcomes of some random experiment. If $P(A)$ is the probability that the event A will occur, then the self-information associated with A is given by

$$i(A) = \log_b \frac{1}{P(A)} = -\log_b P(A) \tag{1}$$

Note that we have not specified the base b of the log function. We will discuss the choice of the base later in this section. The use of the logarithm to obtain a measure of information was not an arbitrary choice as we shall see in Section 2.2.1. But first let's see if the use of a logarithm in this context makes sense from an intuitive point of view. Recall that $\log(1)=0$, and $-\log(x)$ increases as x decreases from one to zero. Therefore, if the probability of an event is low, the amount of self-information associated with it is high; if the probability of an event is high, the information associated with it is low. Even if we ignore the mathematical definition of information and simply use the definition we use in everyday language, this makes some intuitive sense. The barking of a dog during a burglary is a high-probability event and, therefore, does not contain too much information. However, if the dog did not bark during a burglary, this is a low-probability event and contains a lot of information. (Obviously, Sherlock Holmes understood information theory!)[1] Although this equivalence of the mathematical and semantic definitions of information holds true most of the time, it does not hold all of the time. For example, a totally random string of letters will contain more information (in the mathematical sense) than a well-thought-out treatise on information theory.

Another property of this mathematical definition of information that makes intuitive sense is that the information obtained from the occurrence of two independent events is the sum of the information obtained from the occurrence of the individual events. Suppose A and B are two independent events. The self-information associated with the occurrence of both event A *and* event B is, by Equation (1),

$$i(AB) = \log_b \frac{1}{P(AB)}$$

as A and B are independent,

$$P(AB) = P(A)P(B)$$

and

$$\begin{aligned} i(AB) &= \log_b \frac{1}{P(A)P(B)} \\ &= \log_b \frac{1}{P(A)} + \log_b \frac{1}{P(B)} \\ &= i(A) + i(B) \end{aligned}$$

The unit of information depends on the base of the log. If we use log base 2, the unit is *bits*; if we use log base e, the unit is *nats*; and if we use log base 10, the unit is *hartleys*. In general, if we do not explicitly specify the base of the log we will be assuming a base of 2.

---

[1] *Silver Blaze* by Arthur Conan Doyle.

Because the logarithm base 2 probably does not appear on your calculator, let's briefly review logarithms. Recall that

$$\log_b x = a$$

means that

$$b^a = x$$

Therefore, if we want to take the log base 2 of x

$$\log_2 x = a \Rightarrow 2^a = x$$

we want to find the value of a. We can take the natural log (log base e), which we will write as *ln*, or log base 10 of both sides (which do appear on your calculator). Then

$$\ln(2^a) = \ln x \Rightarrow a \ln 2 = \ln x$$

and

$$a = \frac{\ln x}{\ln 2}$$

Example 2.2.1:

Let H and T be the outcomes of flipping a coin. If the coin is fair, then

$$P(H) = P(T) = \frac{1}{2}$$

and

$$i(H) = i(T) = 1 \text{ bit}$$

If the coin is not fair, then we would expect the information associated with each event to be different. Suppose

$$P(H) = \frac{1}{8}, \quad P(T) = \frac{7}{8}$$

Then

$$i(H) = 3 \text{ bits}, \quad i(T) = 0.193 \text{ bits}$$

At least mathematically, the occurrence of a head conveys much more information than the occurrence of a tail. As we shall see later, this has certain consequences for how the information conveyed by these outcomes should be encoded. ♦

If we have a set of independent events A_i, which are sets of outcomes of some experiment $\mathcal{S}$, such that

$$\bigcup A_i = S$$

where S is the sample space, then the average self-information associated with the random experiment is given by

$$H = \sum P(A_i)i(A_i) = -\sum P(A_i) \log_b P(A_i)$$

This quantity is called the *entropy* associated with the experiment. One of the many contributions of Shannon was that he showed that if the experiment is a source that puts out symbols A_i from a set $\mathcal{A}$, then the entropy is a measure of the average number of binary symbols needed to code the output of the source. Shannon showed that the best that a lossless compression scheme can do is to encode the output of a source with an average number of bits equal to the entropy of the source.

The set of symbols $\mathcal{A}$ is often called the *alphabet* for the source, and the symbols are referred to as *letters*. In our definition of entropy we have assumed that a general source $\mathcal{S}$ with alphabet $\mathcal{A} = \{1, 2, \ldots, m\}$ generates a sequence $\{X_1, X_2, \ldots\}$, and the elements in the sequence are generated independently. Thus each letter appears as a surprise. In practice this is not necessarily the case and there may be considerable dependence between letters. These dependencies will affect the entropy of the source. In later sections we will look at specific ways to model these dependencies for various sources of interest. However, in order to make a general statement about the effect of these dependencies on the entropy of stationary sources we need a general approach that will capture all dependencies. One way to capture dependencies is to look at the joint distributions of longer and longer sequences generated by the source. Consider the n-length most likely sequences from three very different texts shown in Table 2.1 for $n = 1, 2, 3, 4$. We can see that for n small, all we get is the inherent structure of the English language. However, as we increase n to 10 we can identify the particular text simply by looking at the five most probable sequences. That is, as we increase n we capture more and more of the structure of the sequence. Define G_n as

$$G_n = -\sum_{i_1=1}^{i_1=m}\sum_{i_2=1}^{i_2=m}\cdots\sum_{i_n=1}^{i_n=m} P(X_1 = i_1, X_2 = i_2, \ldots, X_n = i_n)$$
$$\times \log P(X_1 = i_1, X_2 = i_2, \ldots, X_n = i_n)$$

This quantity will denote the amount of information contained in n-tuples from the source. The per-letter information can be obtained by normalizing G_n as

$$H_n = \frac{1}{n}G_n$$

If we plot this quantity for n from 1 to 12 for the book *Wealth of Nations* we obtain the values shown in Figure 2.1. We can see that H_n is converging to a particular value. Shannon showed [3] that for a stationary source, in the limit this value will converge to the entropy:

$$H(\mathcal{S}) = \lim_{n\to\infty} \frac{1}{n} H_n \tag{2}$$

If each element in the sequence is independent and identically distributed (*iid*), then we can show that

$$G_n = -n \sum_{i_1=1}^{i_1=m} P(X_1 = i_1) \log P(X_1 = i_1) \tag{3}$$

and the equation for the entropy becomes

$$H(S) = -\sum P(X_1) \log P(X_1) \tag{4}$$

TABLE 2.1 **The most probable five sequences of lengths 1, 2, 3, and 10 from *Peter Pan* by J.M. Barrie, *The Communist Manifesto* by K. Marx and F. Engle, and *The Wealth of Nations* by A. Smith. (All text files obtained from the Gutenberg Project.)**

$n = 1$			$n = 2$			$n = 3$			$n = 10$		
ƀ	ƀ	ƀ	eƀ	eƀ	eƀ	ƀth	ƀth	ƀth	theyƀwere	ƀbourgeois	therefore,
e	e	e	he	th	ƀt	the	the	the	.ƀdarlingƀ	bourgeoisi	ƀwhichƀthe
t	t	t	ƀt	ƀt	th	heƀ	heƀ	heƀ	,"ƀheƀsaid	ourgeoisie	ƀtheƀgreat
a	o	o	th	he	he	and	ƀof	ƀof	ƀtheyƀwere	proletaria	ƀtheƀsame
h	i	a	dƀ	s/ƀ	ƀa	ndƀ	ofƀ	ofƀ	mrs.ƀdarli	ebourgeoi	herefore,ƀ

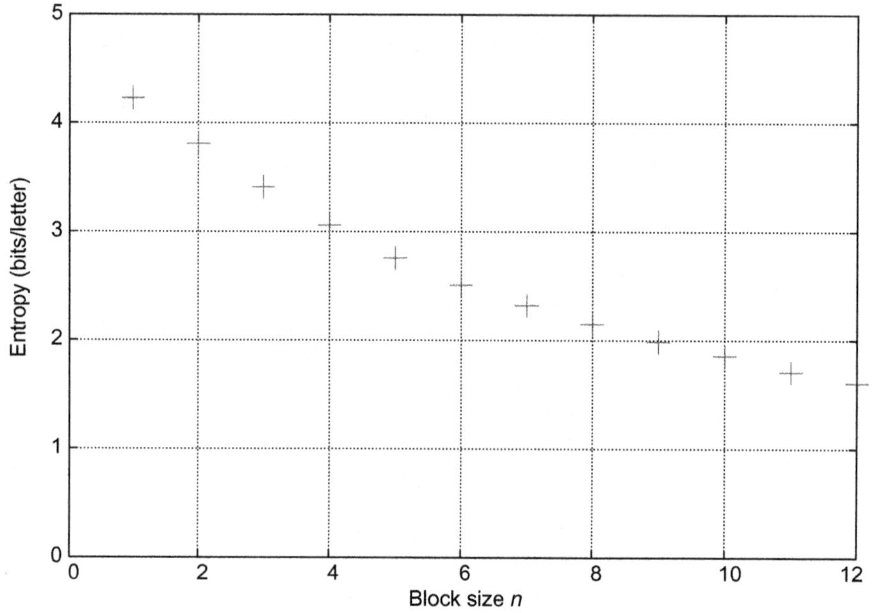

FIGURE 2.1 H_n **in bits per letter for** $n = 1, \ldots, 12$ **for *Wealth of Nations*.**

For most sources, Equations (2) and (4) are not identical. If we need to distinguish between the two, we will call the quantity computed in (4) the *first-order entropy* of the source, while the quantity in (2) will be referred to as the *entropy* of the source.

In general, it is not possible to know the entropy for a physical source, so we have to estimate the entropy. The estimate of the entropy depends on our assumptions about the structure of the source sequence.

Consider the following sequence:

1 2 3 2 3 4 5 4 5 6 7 8 9 8 9 10

Assuming the frequency of occurrence of each number is reflected accurately in the number of times it appears in the sequence, we can estimate the probability of occurrence of each symbol as follows:

$$P(1) = P(6) = P(7) = P(10) = \frac{1}{16}$$

$$P(2) = P(3) = P(4) = P(5) = P(8) = P(9) = \frac{2}{16}$$

Assuming the sequence is *iid*, the entropy for this sequence is the same as the first-order entropy defined in (4). The entropy can then be calculated as

$$H = -\sum_{i=1}^{10} P(i) \log_2 P(i).$$

With our stated assumptions, the entropy for this source is 3.25 bits. This means that the best scheme we could find for coding this sequence could only code it at 3.25 bits/sample.

However, if we assume that there was sample-to-sample correlation between the samples and we remove the correlation by taking differences of neighboring sample values, we arrive at the *residual* sequence

$$1\ 1\ 1\ {-1}\ 1\ 1\ 1\ {-1}\ 1\ 1\ 1\ 1\ 1\ {-1}\ 1\ 1\ 1$$

This sequence is constructed using only two values with probabilities: $P(1)=\frac{13}{16}$ and $P(-1)=\frac{3}{16}$. The entropy in this case is 0.70 bits per symbol. Of course, knowing only this sequence would not be enough for the receiver to reconstruct the original sequence. The receiver must also know the process by which this sequence was generated from the original sequence. The process depends on our assumptions about the structure of the sequence. These assumptions are called the *model* for the sequence. In this case, the model for the sequence is

$$x_n = x_{n-1} + r_n$$

where x_n is the nth element of the original sequence and r_n is the nth element of the residual sequence. This model is called a *static* model because its parameters do not change with n. A model whose parameters change or adapt with n to the changing characteristics of the data is called an *adaptive* model.

We see that knowing something about the structure of the data can help to "reduce the entropy." We have put "reduce the entropy" in quotes because the entropy of the source is a measure of the amount of information generated by the source. As long as the information generated by the source is preserved (in whatever representation), the entropy remains the same. What we are reducing is our estimate of the entropy. The "actual" structure of the data in practice is generally unknowable, but anything we can learn about the data can help us to estimate the actual source entropy. Theoretically, as seen in Equation (2), we accomplish this in our definition of the entropy by picking larger and larger blocks of data to calculate the probability over, letting the size of the block go to infinity.

Consider the following contrived sequence:

$$1\ 2\ 1\ 2\ 3\ 3\ 3\ 3\ 1\ 2\ 3\ 3\ 3\ 3\ 1\ 2\ 3\ 3\ 1\ 2$$

Obviously, there is some structure to this data. However, if we look at it one symbol at a time, the structure is difficult to extract. Consider the probabilities: $P(1) = P(2) = \frac{1}{4}$, and $P(3) = \frac{1}{2}$. The entropy is 1.5 bits/symbol. This particular sequence consists of 20 symbols; therefore, the total number of bits required to represent this sequence is 30. Now let's take the same sequence and look at it in blocks of two. Obviously, there are only two symbols, 1 2, and 3 3. The probabilities are $P(1\ 2) = \frac{1}{2}$, $P(3\ 3) = \frac{1}{2}$, and the entropy is 1 bit/symbol. As there are 10 such symbols in the sequence, we need a total of 10 bits to represent the entire sequence—a reduction of a factor of three. The theory says we can always extract the structure of the data by taking larger and larger block sizes; in practice, there are limitations to this approach. To avoid these limitations, we try to obtain an accurate model for the data and code the source with respect to the model. In Section 2.3, we describe some of the models commonly used in lossless compression algorithms. But before we do that, let's make a slight detour and see a more rigorous development of the expression for average information. While the explanation is interesting, it is not really necessary for understanding much of what we will study in this book and can be skipped.

2.2.1 Derivation of Average Information ★

We start with the properties we want in our measure of average information. We will then show that requiring these properties in the information measure leads inexorably to the particular definition of average information, or entropy, that we have provided earlier.

Given a set of independent events $A_1, A_2, \ldots, A_n$ with probability $p_i = P(A_i)$, we desire the following properties in the measure of average information H:

1. We want H to be a continuous function of the probabilities p_i. That is, a small change in p_i should only cause a small change in the average information.
2. If all events are equally likely, that is, $p_i = 1/n$ for all i, then H should be a monotonically increasing function of n. The more possible outcomes there are, the more information should be contained in the occurrence of any particular outcome.
3. Suppose we divide the possible outcomes into a number of groups. We indicate the occurrence of a particular event by first indicating the group it belongs to, then indicating which particular member of the group it is. Thus, we get some information first by knowing which group the event belongs to; and then we get additional information by learning which particular event (from the events in the group) has occurred. The information associated with indicating the outcome in multiple stages should not be any different than the information associated with indicating the outcome in a single stage. For example, suppose we have an experiment with three outcomes, A_1, A_2, and A_3, with corresponding probabilities, p_1, p_2, and p_3. The average information associated with this experiment is simply a function of the probabilities:

$$H = H(p_1, p_2, p_3)$$

Let's group the three outcomes into two groups:

$$B_1 = \{A_1\}, \quad B_2 = \{A_2, A_3\}$$

The probabilities of the events B_i are given by

$$q_1 = P(B_1) = p_1, \quad q_2 = P(B_2) = p_2 + p_3$$

If we indicate the occurrence of an event A_i by first declaring which group the event belongs to and then declaring which event occurred, the total amount of average information would be given by

$$H = H(q_1, q_2) + q_1 H\left(\frac{p_1}{q_1}\right) + q_2 H\left(\frac{p_2}{q_2}, \frac{p_3}{q_2}\right)$$

We require that the average information computed either way be the same.

In his classic paper, Shannon showed that the only way all of these conditions could be satisfied was if

$$H = -K \sum p_i \log p_i$$

where K is an arbitrary positive constant, and the base of the logarithm is irrelevant as it only impacts the value of K. Let's review his proof as it appears in the appendix of his paper [3].

Suppose we have an experiment with $n = k^m$ equally likely outcomes. The average information $H(\frac{1}{n}, \frac{1}{n}, \ldots, \frac{1}{n})$ associated with this experiment is a function of n. In other words,

$$H\left(\frac{1}{n}, \frac{1}{n}, \ldots, \frac{1}{n}\right) = A(n)$$

We can indicate the occurrence of an event from k^m events by a series of m choices from k equally likely possibilities. For example, consider the case of $k = 2$ and $m = 3$. There are eight equally likely events; therefore, $H(\frac{1}{8}, \frac{1}{8}, \ldots, \frac{1}{8}) = A(8)$.

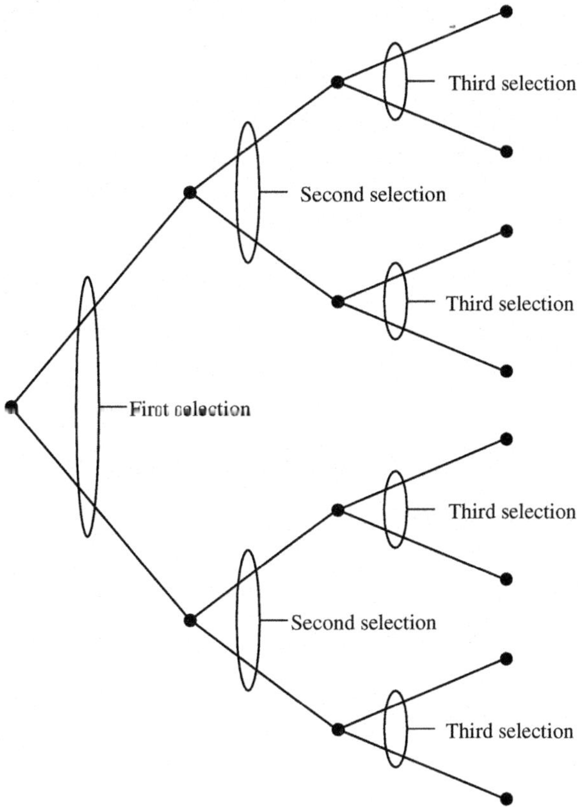

FIGURE 2.2 **A possible way of identifying the occurrence of an event.**

We can indicate the occurrence of any particular event, as shown in Figure 2.2. In this case, we have a sequence of three selections. Each selection is between two equally likely possibilities. Therefore,

$$\begin{aligned} H\left(\frac{1}{8},\frac{1}{8},\ldots,\frac{1}{8}\right) &= A(8) \\ &= H\left(\frac{1}{2},\frac{1}{2}\right)+\frac{1}{2}\left[H\left(\frac{1}{2},\frac{1}{2}\right)+\frac{1}{2}H\left(\frac{1}{2},\frac{1}{2}\right)\right. \\ &\qquad\left.+\frac{1}{2}H\left(\frac{1}{2},\frac{1}{2}\right)\right] \\ &\qquad+\frac{1}{2}\left[H\left(\frac{1}{2},\frac{1}{2}\right)+\frac{1}{2}H\left(\frac{1}{2},\frac{1}{2}\right)\right. \\ &\qquad\left.+\frac{1}{2}H\left(\frac{1}{2},\frac{1}{2}\right)\right] \\ &= 3H\left(\frac{1}{2},\frac{1}{2}\right) \\ &= 3A(2). \end{aligned} \tag{5}$$

In other words,

$$A(8) = 3A(2)$$

(The rather odd way of writing the right-hand side of Equation (5) is to show how the terms correspond to the branches of the tree shown in Figure 2.2.) We can generalize this for the case of $n = k^m$ as

$$A(n) = A(k^m) = mA(k)$$

Similarly, for j^l choices,

$$A(j^l) = lA(j)$$

We can pick l arbitrarily large (more on this later) and then choose m so that

$$k^m \leqslant j^l \leqslant k^{(m+1)}$$

Taking logarithms of all terms, we get

$$m \log k \leqslant l \log j \leqslant (m+1)\log k$$

Now divide through by $l \log k$ to get

$$\frac{m}{l} \leqslant \frac{\log j}{\log k} \leqslant \frac{m}{l} + \frac{1}{l}$$

Recall that we picked l arbitrarily large. If l is arbitrarily large, then $\frac{1}{l}$ is arbitrarily small. This means that the upper and lower bounds of $\frac{\log j}{\log k}$ can be made arbitrarily close to $\frac{m}{l}$ by picking l arbitrarily large. Another way of saying this is

$$\left|\frac{m}{l} - \frac{\log j}{\log k}\right| < \epsilon$$

where ϵ can be made arbitrarily small. We will use this fact to find an expression for $A(n)$ and hence for $H(\frac{1}{n},\ldots,\frac{1}{n})$.

To do this we use our second requirement that $H(\frac{1}{n}, \ldots, \frac{1}{n})$ be a monotonically increasing function of n. As

$$H\left(\frac{1}{n}, \ldots, \frac{1}{n}\right) = A(n)$$

this means that $A(n)$ is a monotonically increasing function of n. If

$$k^m \leqslant j^l \leqslant k^{m+1}$$

then in order to satisfy our second requirement

$$A(k^m) \leqslant A(j^l) \leqslant A(k^{m+1})$$

or

$$mA(k) \leqslant lA(j) \leqslant (m+1)A(k)$$

Dividing through by $lA(k)$, we get

$$\frac{m}{l} \leqslant \frac{A(j)}{A(k)} \leqslant \frac{m}{l} + \frac{1}{l}$$

Using the same arguments as before, we get

$$\left|\frac{m}{l} - \frac{A(j)}{A(k)}\right| < \epsilon$$

where ϵ can be made arbitrarily small.

Now $\frac{A(j)}{A(k)}$ is at most a distance of ϵ away from $\frac{m}{l}$, and $\frac{\log j}{\log k}$ is at most a distance of ϵ away from $\frac{m}{l}$. Therefore, $\frac{A(j)}{A(k)}$ is at most a distance of 2ϵ away from $\frac{\log j}{\log k}$:

$$\left|\frac{A(j)}{A(k)} - \frac{\log j}{\log k}\right| < 2\epsilon$$

We can pick ϵ to be arbitrarily small, and j and k are arbitrary. The only way this inequality can be satisfied for arbitrarily small ϵ and arbitrary j and k is for $A(j) = K\log(j)$, where K is an arbitrary constant. In other words,

$$H = K\log(n)$$

Up to this point we have only looked at equally likely events. We now make the transition to the more general case of an experiment with outcomes that are not equally likely. We do that by considering an experiment with $\sum n_i$ equally likely outcomes that are grouped in n unequal groups of size n_i with rational probabilities (if the probabilities are not rational, we approximate them with rational probabilities and use the continuity requirement):

$$p_i = \frac{n_i}{\sum_{j=1}^{n} n_j}$$

Given that we have $\sum n_i$ equally likely events, from the development above we have

$$H = K\log\left(\sum n_j\right) \tag{6}$$

If we indicate an outcome by first indicating which of the n groups it belongs to, and second indicating which member of the group it is, then by our earlier development the average information H is given by

$$H = H(p_1, p_2, \ldots, p_n) + p_1 H\left(\frac{1}{n_1}, \ldots, \frac{1}{n_1}\right) + \cdots + p_n H\left(\frac{1}{n_n}, \ldots, \frac{1}{n_n}\right) \quad (7)$$

$$= H(p_1, p_2, \ldots, p_n) + p_1 K \log n_1 + p_2 K \log n_2 + \cdots + p_n K \log n_n \quad (8)$$

$$= H(p_1, p_2, \ldots, p_n) + K \sum_{i=1}^{n} p_i \log n_i \quad (9)$$

Equating the expressions in Equations (6) and (9), we obtain

$$K \log\left(\sum n_j\right) = H(p_1, p_2, \ldots, p_n) + K \sum_{i=1}^{n} p_i \log n_i$$

or

$$H(p_1, p_2, \ldots, p_n) = K \log\left(\sum n_j\right) - K \sum_{i=1}^{n} p_i \log n_i$$

$$= -K\left[\sum_{i=1}^{n} p_i \log n_i - \log\left(\sum_{j=1}^{n} n_j\right)\right]$$

$$= -K\left[\sum_{i=1}^{n} p_i \log n_i - \log\left(\sum_{j=1}^{n} n_j\right)\sum_{i=1}^{n} p_i\right] \quad (10)$$

$$= -K\left[\sum_{i=1}^{n} p_i \log n_i - \sum_{i=1}^{n} p_i \log\left(\sum_{j=1}^{n} n_j\right)\right]$$

$$= -K\sum_{i=1}^{n} p_i\left[\log n_i - \log\left(\sum_{j=1}^{n} n_j\right)\right]$$

$$= -K\sum_{i=1}^{n} p_i \log \frac{n_i}{\sum_{j=1}^{n} n_j} \quad (11)$$

$$= -K\sum p_i \log p_i \quad (12)$$

where, in Equation (10) we have used the fact that $\sum_{i=1}^{n} p_i = 1$. By convention, we pick K to be 1, and we have the formula

$$H = -\sum p_i \log p_i$$

Note that this formula is a natural outcome of the requirements we imposed in the beginning. It was not artificially forced in any way. Therein lies the beauty of information theory. Like the laws of physics, its laws are intrinsic in the nature of things. Mathematics is simply a tool to express these relationships.

2.3 Models

As we saw in Section 2.2, having a good model for the data can be useful in estimating the entropy of the source. As we will see in later chapters, good models for sources lead to more efficient compression algorithms. In general, in order to develop techniques that manipulate data using mathematical operations, we need to have a mathematical model for the data. Obviously, the better the model (i.e., the closer the model matches the aspects of reality that are of interest to us), the more likely it is that we will come up with a satisfactory technique. There are several approaches to building mathematical models.

2.3.1 Physical Models

If we know something about the physics of the data generation process, we can use that information to construct a model. For example, in speech-related applications, knowledge about the physics of speech production can be used to construct a mathematical model for the sampled speech process. Sampled speech can then be encoded using this model. We will discuss speech production models in more detail in Chapter 8 and Chapter 18.

Models for certain telemetry data can also be obtained through knowledge of the underlying process. For example, if residential electrical meter readings at hourly intervals were to be coded, knowledge about the living habits of the populace could be used to determine when electricity usage would be high and when the usage would be low. Then instead of the actual readings, the difference (residual) between the actual readings and those predicted by the model could be coded.

In general, however, the physics of data generation is simply too complicated to understand, let alone use to develop a model. Where the physics of the problem is too complicated, we can obtain a model based on empirical observation of the statistics of the data.

2.3.2 Probability Models

The simplest statistical model for the source is to assume that each letter that is generated by the source is independent of every other letter, and each occurs with the same probability. We could call this the *ignorance model*, as it would generally be useful only when we know nothing about the source. (Of course, that *really* might be true, in which case we have a rather unfortunate name for the model!) The next step up in complexity is to keep the independence assumption, but remove the equal probability assumption and assign a probability of occurrence to each letter in the alphabet. For a source that generates letters from an alphabet $\mathcal{A} = \{a_1, a_2, \ldots, a_M\}$, we can have a *probability model* $\mathcal{P} = \{P(a_1), P(a_2), \ldots, P(a_M)\}$.

Given a probability model (and the independence assumption), we can compute the entropy of the source using Equation (4). As we will see in the following chapters using the probability model, we can also construct some very efficient codes to represent the letters in $\mathcal{A}$. Of course, these codes are only efficient if our mathematical assumptions are in accord with reality.

If the assumption of independence does not fit with our observation of the data, we can generally find better compression schemes if we discard this assumption. When we discard the independence assumption, we have to come up with a way to describe the dependence of elements of the data sequence on each other.

2.3.3 Markov Models

One of the most popular ways of representing dependence in the data is through the use of Markov models, named after the Russian mathematician Andrei Andrevich Markov (1856–1922). For models used in lossless compression, we use a specific type of Markov process called a *discrete time Markov chain.* Let $\{x_n\}$ be a sequence of observations. This sequence is said to follow a kth-order Markov model if

$$P(x_n|x_{n-1},\ldots,x_{n-k}) = P(x_n|x_{n-1},\ldots,x_{n-k},\ldots) \tag{13}$$

In other words, knowledge of the past k symbols is equivalent to the knowledge of the entire past history of the process. The values taken on by the set $\{x_{n-1},\ldots,x_{n-k}\}$ are called the *states* of the process. If the size of the source alphabet is l then the number of states is l^k. The most commonly used Markov model is the first-order Markov model, for which

$$P(x_n|x_{n-1}) = P(x_n|x_{n-1},x_{n-2},x_{n-3},\ldots) \tag{14}$$

Equations (13) and (14) indicate the existence of dependence between samples. However, they do not describe the form of the dependence. We can develop different first-order Markov models depending on our assumption about the form of the dependence between samples.

If we assumed that the dependence was introduced in a linear manner, we could view the data sequence as the output of a linear filter driven by white noise. The output of such a filter can be given by the difference equation

$$x_n = \rho x_{n-1} + \epsilon_n \tag{15}$$

where ϵ_n is a white noise process. This model is often used when developing coding algorithms for speech and images.

The use of the Markov model does not require the assumption of linearity. For example, consider a binary image. The image has only two types of pixels, white pixels and black pixels. We know that the appearance of a white pixel as the next observation depends, to some extent, on whether the current pixel is white or black. Therefore, we can model the pixel process as a discrete time Markov chain. Define two states S_w and S_b (S_w would correspond to the case where the current pixel is a white pixel, and S_b corresponds to the case where the current pixel is a black pixel). We define the transition probabilities $P(w/b)$ and $P(b/w)$ and the probability of being in each state $P(S_w)$ and $P(S_b)$. The Markov model can then be represented by the state diagram shown in Figure 2.3.

The entropy of a finite state process with states S_i is simply the average value of the entropy at each state:

$$H = \sum_{i=1}^{M} P(S_i)H(S_i) \tag{16}$$

For our particular example of a binary image

$$H(S_w) = -P(b|w)\log P(b|w) - P(w|w)\log P(w|w)$$

where $P(w|w) = 1 - P(b|w)$. $H(S_b)$ can be calculated in a similar manner.

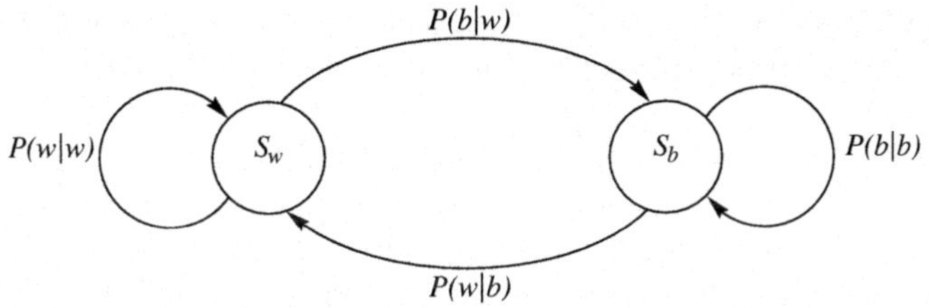

FIGURE 2.3 **A two-state Markov model for binary images.**

Example 2.3.1: Markov Model

To see the effect of modeling on the estimate of entropy, let us calculate the entropy for a binary image, first using a simple probability model and then using the finite state model described above. Let us assume the following values for the various probabilities:

$$P(S_w) = 30/31 \quad P(S_b) = 1/31$$
$$P(w|w) = 0.99 \quad P(b|w) = 0.01 \quad P(b|b) = 0.7 \quad P(w|b) = 0.3$$

Then the entropy using a probability model and the *iid* assumption is

$$H = -0.8 \log 0.8 - 0.2 \log 0.2 = 0.206 \text{ bits}$$

Now using the Markov model

$$H(S_b) = -0.3 \log 0.3 - 0.7 \log 0.7 = 0.881 \text{ bits}$$

and

$$H(S_w) = -0.01 \log 0.01 - 0.99 \log 0.99 = 0.081 \text{ bits}$$

which, using Equation (16), results in an entropy for the Markov model of 0.107 bits, about half of the entropy obtained using the *iid* assumption. ♦

Markov Models in Text Compression

As expected, Markov models are particularly useful in text compression, where the probability of the next letter is heavily influenced by the preceding letters. In fact, the use of Markov models for written English appears in the original work of Shannon [3]. In current text compression literature, the kth-order Markov models are more widely known as *finite context models*, with the word *context* being used for what we have earlier defined as state.

Consider the word *preceding*. Suppose we have already processed *precedin* and are going to encode the next letter. If we take no account of the context and treat each letter as a surprise, the probability of the letter g occurring is relatively low. If we use a first-order Markov model or single-letter context (that is, we look at the probability model given n), we can see that the probability of g would increase substantially. As we increase the context size (go from n to in to din and so on), the probability of the alphabet becomes more and more skewed, which results in lower entropy.

Shannon used a second-order model for English text consisting of the 26 letters and one space to obtain an entropy of 3.1 bits/letter [4]. Using a model where the output symbols were words rather than letters brought down the entropy to 2.4 bits/letter. Shannon then used predictions generated by people (rather than statistical models) to estimate the upper and lower bounds on the entropy of the second-order model. For the case where the subjects knew the 100 previous letters, he estimated these bounds to be 1.3 and 0.6 bits/letter, respectively.

The longer the context, the better its predictive value. However, if we were to store the probability model with respect to all contexts of a given length, the number of contexts would grow exponentially with the length of context. Furthermore, given that the source imposes some structure on its output, many of these contexts may correspond to strings that would never occur in practice. Consider a context model of order four (the context is determined by the last four symbols). If we take an alphabet size of 95, the possible number of contexts is 95^4—more than 81 million!

This problem is further exacerbated by the fact that different realizations of the source output may vary considerably in terms of repeating patterns. Therefore, context modeling in text compression schemes tends to be an adaptive strategy in which the probabilities for different symbols in the different contexts are updated as they are encountered. However, this means that we will often encounter symbols that have not been encountered before for any of the given contexts (this is known as the *zero frequency problem*). The larger the context, the more often this will happen. This problem could be resolved by sending a code to indicate that the following symbol was being encountered for the first time, followed by a prearranged code for that symbol. This would significantly increase the length of the code for the symbol on its first occurrence (in the given context). However, if this situation did not occur too often, the overhead associated with such occurrences would be small compared to the total number of bits used to encode the output of the source. Unfortunately, in context-based encoding, the zero frequency problem is encountered often enough for overhead to be a problem, especially for longer contexts. Solutions to this problem are presented by the *ppm* (prediction with partial match) algorithm and its variants (described in detail in Chapter 6).

Briefly, the *ppm* algorithm first attempts to find if the symbol to be encoded has a nonzero probability with respect to the maximum context length. If this is so, the symbol is encoded and transmitted. If not, an escape symbol is transmitted, the context size is reduced by one, and the process is repeated. This procedure is repeated until a context is found with respect to which the symbol has a nonzero probability. To guarantee that this process converges, a null context is always included with respect to which all symbols have equal probability. Initially, only the shorter contexts are likely to be used. However, as more and more of the source output is processed, the longer contexts, which offer better prediction, will be used more often. The probability of the escape symbol can be computed in a number of different ways leading to different implementations [1].

The use of Markov models in text compression is a rich and active area of research. We describe some of these approaches in Chapter 6 (for more details, see [1]).

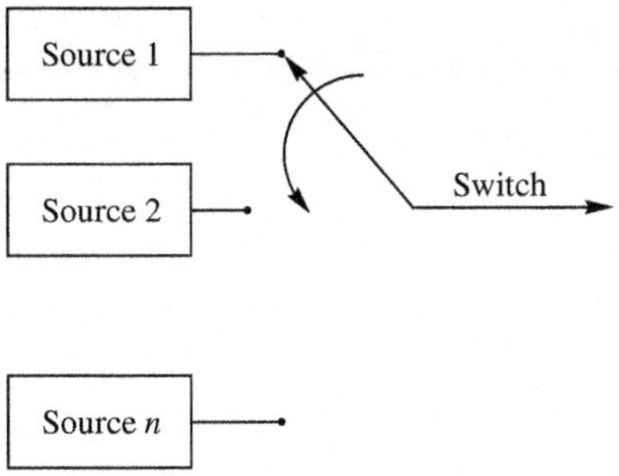

FIGURE 2.4 A composite source.

2.3.4 Composite Source Model

In many applications, it is not easy to use a single model to describe the source. In such cases, we can define a *composite source*, which can be viewed as a combination or composition of several sources, with only one source being *active* at any given time. A composite source can be represented as a number of individual sources $\mathcal{S}_i$, each with its own model $\mathcal{M}_i$ and a switch that selects a source $\mathcal{S}_i$ with probability P_i (as shown in Figure 2.4). This is an exceptionally rich model and can be used to describe some very complicated processes. We will describe this model in more detail when we need it.

2.4 Coding

When we talk about *coding* in this chapter (and through most of this book), we mean the assignment of binary sequences to elements of an alphabet. The set of binary sequences is called a *code*, and the individual members of the set are called *codewords*. An *alphabet* is a collection of symbols called *letters*. For example, the alphabet used in writing most books consists of the 26 lowercase letters, 26 uppercase letters, and a variety of punctuation marks. In the terminology used in this book, a comma is a letter. The ASCII code for the letter a is 1000011, the letter A is coded as 1000001, and the letter "," is coded as 0011010. Notice that the ASCII code uses the same number of bits to represent each symbol. Such a code is called a *fixed-length code*. If we want to reduce the number of bits required to represent different messages, we need to use a different number of bits to represent different symbols. If we use fewer bits to represent symbols that occur more often, on the average we would use fewer bits per symbol. The average number of bits per symbol is often called the *rate* of the code. The idea of using fewer bits to represent symbols that occur more often is the same idea that is used in Morse code: the codewords for letters that occur more frequently are shorter than for letters that occur less frequently. For example, the codeword for E is $\cdot$, while the codeword for Z is —·· [5].

TABLE 2.2 Four different codes for a four-letter alphabet.

Letters	Probability	Code 1	Code 2	Code 3	Code 4
a_1	0.5	0	0	0	0
a_2	0.25	0	1	10	01
a_3	0.125	1	00	110	011
a_4	0.125	10	11	111	0111
	Average length	1.125	1.25	1.75	1.875

2.4.1 Uniquely Decodable Codes

The average length of the code is not the only important point in designing a "good" code. Consider the following example adapted from [6]. Suppose our source alphabet consists of four letters a_1, a_2, a_3, and a_4, with probabilities $P(a_1) = \frac{1}{2}$, $P(a_2) = \frac{1}{4}$, and $P(a_3) = P(a_4) = \frac{1}{8}$. The entropy for this source is 1.75 bits/symbol. Consider the codes for this source in Table 2.2.

The average length l for each code is given by

$$l = \sum_{i=1}^{4} P(a_i)n(a_i)$$

where $n(a_i)$ is the number of bits in the codeword for letter a_i and the average length is given in bits/symbol. Based on the average length, Code 1 appears to be the best code. However, to be useful, a code should have the ability to transfer information in an unambiguous manner. This is obviously not the case with Code 1. Both a_1 and a_2 have been assigned the codeword 0. When a 0 is received, there is no way to know whether an a_1 was transmitted or an a_2. We would like each symbol to be assigned a *unique* codeword.

At first glance, Code 2 does not seem to have the problem of ambiguity; each symbol is assigned a distinct codeword. However, suppose we want to encode the sequence $a_2\, a_1\, a_1$. Using Code 2, we would encode this with the binary string 100. However, when the string 100 is received at the decoder, there are several ways in which the decoder can decode this string. The string 100 can be decoded as $a_2 a_1 a_1$, or as $a_2 a_3$. This means that once a sequence is encoded with Code 2, the original sequence cannot be recovered with certainty. In general, this is not a desirable property for a code. We would like *unique decodability* from the code; that is, any given sequence of codewords can be decoded in one, and only one, way.

We have already seen that Code 1 and Code 2 are not uniquely decodable. How about Code 3? Notice that the first three codewords all end in a 0. In fact, a 0 always denotes the termination of a codeword. The final codeword contains no 0s and is 3 bits long. Because all other codewords have fewer than three 1s and terminate in a 0, the only way we can get three 1s in a row is as a code for a_4. The decoding rule is simple. Accumulate bits until you get a 0 or until you have three 1s. There is no ambiguity in this rule, and it is reasonably easy to see that this code is uniquely decodable. With Code 4 we have an even simpler condition. Each codeword starts with a 0, and the only time we see a 0 is in the beginning of a codeword.

TABLE 2.3 Code 5. A code that is uniquely decodable but not instantaneous.

Letter	Codeword
a_1	0
a_2	01
a_3	11

Therefore, the decoding rule is to accumulate bits until you see a 0. The bit before the 0 is the last bit of the previous codeword.

There is a slight difference between Code 3 and Code 4. In the case of Code 3, the decoder knows the moment a code is complete. In Code 4, we have to wait till the beginning of the next codeword before we know that the current codeword is complete. Because of this property, Code 3 is called an *instantaneous* code. Although Code 4 is not an instantaneous code, it is almost that.

While this property of instantaneous or near-instantaneous decoding is a nice property to have, it is not a requirement for unique decodability. Consider the code shown in Table 2.3. Let's decode the string 011111111111111111. In this string, the first codeword is either 0 corresponding to a_1 or 01 corresponding to a_2. We cannot tell which one until we have decoded the whole string. Starting with the assumption that the first codeword corresponds to a_1, the next eight pairs of bits are decoded as a_3. However, after decoding eight a_3s, we are left with a single (dangling) 1 that does not correspond to any codeword. On the other hand, if we assume the first codeword corresponds to a_2, we can decode the next 16 bits as a sequence of eight a_3s, and we do not have any bits left over. The string can be uniquely decoded. In fact, Code 5, while it is certainly not instantaneous, is uniquely decodable.

We have been looking at small codes with four letters or less. Even with these, it is not immediately evident whether the code is uniquely decodable or not. In deciding whether larger codes are uniquely decodable, a systematic procedure would be useful. Actually, we should include a caveat with that last statement. Later in this chapter we will include a class of variable-length codes that are always uniquely decodable, so a test for unique decodability may not be that necessary. You might wish to skip the following discussion for now, and come back to it when you find it necessary.

Before we describe the procedure for deciding whether a code is uniquely decodable, let's take another look at our last example. We found that we had an incorrect decoding because we were left with a binary string (1) that was not a codeword. If this had not happened, we would have had two valid decodings. For example, consider the code shown in Table 2.4. Let's encode the sequence a_1 followed by eight a_3s using this code. The coded sequence is 01010101010101010. The first bit is the codeword for a_1. However, we can also decode it as the first bit of the codeword for a_2. If we use this (incorrect) decoding, we decode the next seven pairs of bits as the codewords for a_2. After decoding seven a_2s, we are left with a single 0 that we decode as a_1. Thus, the incorrect decoding is also a valid decoding, and this code is not uniquely decodable.

TABLE 2.4 **Code 6. A code that is not uniquely decodable.**

Letter	Codeword
a_1	0
a_2	01
a_3	10

A Test for Unique Decodability ★

In the previous examples, in the case of the uniquely decodable code, the binary string left over after we had gone through an incorrect decoding was not a codeword. In the case of the code that was not uniquely decodable, in the incorrect decoding what was left was a valid codeword. Based on whether the dangling suffix is a codeword or not, we get the following test [7,8].

We start with some definitions. Suppose we have two binary codewords a and b, where a is k bits long, b is n bits long, and $k < n$. If the first k bits of b are identical to a, then a is called a *prefix* of b. The last $n - k$ bits of b are called the *dangling suffix* [7]. For example, if $a = 010$ and $b = 01011$, then a is a prefix of b and the dangling suffix is 11.

Construct a list of all the codewords. Examine all pairs of codewords to see if any codeword is a prefix of another codeword. Whenever you find such a pair, add the dangling suffix to the list unless you have added the same dangling suffix to the list in a previous iteration. Now repeat the procedure using this larger list. Continue in this fashion until one of the following two things happens:

1. You get a dangling suffix that is a codeword.
2. There are no more unique dangling suffixes.

If you get the first outcome, the code is not uniquely decodable. However, if you get the second outcome, the code is uniquely decodable.

Let's see how this procedure works with a couple of examples.

Example 2.4.1:

Consider Code 5. First list the codewords:

$$\{0, 01, 11\}$$

The codeword 0 is a prefix for the codeword 01. The dangling suffix is 1. There are no other pairs for which one element of the pair is the prefix of the other. Let us augment the codeword list with the dangling suffix:

$$\{0, 01, 11, 1\}$$

Comparing the elements of this list, we find 0 is a prefix of 01 with a dangling suffix of 1. But we have already included 1 in our list. Also, 1 is a prefix of 11. This gives us a dangling suffix of 1, which is already in the list. There are no other pairs that would generate a dangling suffix, so we cannot augment the list any further. Therefore, Code 5 is uniquely decodable. ♦

Example 2.4.2:

Consider Code 6. First list the codewords:

$$\{0, 01, 10\}$$

The codeword 0 is a prefix for the codeword 01. The dangling suffix is 1. There are no other pairs for which one element of the pair is the prefix of the other. Augmenting the codeword list with 1, we obtain the list

$$\{0, 01, 10, 1\}$$

In this list, 1 is a prefix for 10. The dangling suffix for this pair is 0, which is the codeword for a_1. Therefore, Code 6 is not uniquely decodable. ♦

2.4.2 Prefix Codes

The test for unique decodability requires examining the dangling suffixes initially generated by codeword pairs in which one codeword is the prefix of the other. If the dangling suffix is itself a codeword, then the code is not uniquely decodable. One type of code in which we will never face the possibility of a dangling suffix being a codeword is a code in which no codeword is a prefix of the other. In this case, the set of dangling suffixes is the null set, and we do not have to worry about finding a dangling suffix that is identical to a codeword. A code in which no codeword is a prefix to another codeword is called a *prefix code*. A simple way to check if a code is a prefix code is to draw the rooted binary tree corresponding to the code. Draw a tree that starts from a single node (the *root node*) and has a maximum of two possible branches at each node. One of these branches corresponds to a 1 and the other branch corresponds to a 0. In this book, we will adopt the convention that when we draw a tree with the root node at the top, the left branch corresponds to a 0 and the right branch corresponds to a 1. Using this convention, we can draw the binary tree for Code 2, Code 3, and Code 4 as shown in Figure 2.5.

Note that apart from the root node, the trees have two kinds of nodes—nodes that give rise to other nodes and nodes that do not. The first kind of nodes are called *internal nodes*,

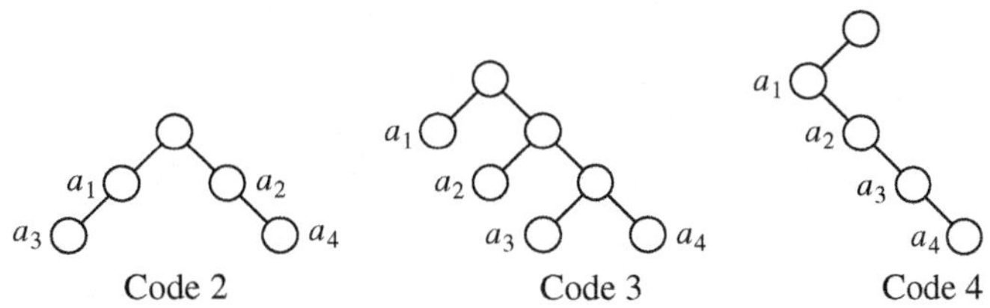

FIGURE 2.5 Binary trees for three different codes.

and the second kind are called *external nodes* or *leaves*. In a prefix code, the codewords are only associated with the external nodes. A code that is not a prefix code, such as Code 4, will have codewords associated with internal nodes. The code for any symbol can be obtained by traversing the tree from the root to the external node corresponding to that symbol. Each branch on the way contributes a bit to the codeword: a 0 for each left branch and a 1 for each right branch.

It is nice to have a class of codes whose members are so clearly uniquely decodable. However, are we losing something if we restrict ourselves to prefix codes? Could it be that if we do not restrict ourselves to prefix codes, we can find shorter codes? Fortunately for us the answer is no. For any nonprefix uniquely decodable code, we can always find a prefix code with the same codeword lengths. We prove this in the next section.

2.4.3 The Kraft-McMillan Inequality ★

The particular result we look at in this section consists of two parts. The first part provides a necessary condition on the codeword lengths of uniquely decodable codes. The second part shows that we can always find a prefix code that satisfies this necessary condition. Therefore, if we have a uniquely decodable code that is not a prefix code, we can always find a prefix code with the same codeword lengths.

Theorem *Let $\mathcal{C}$ be a code with N codewords with lengths $l_1, l_2, \ldots, l_N$. If $\mathcal{C}$ is uniquely decodable, then*

$$K(\mathcal{C}) = \sum_{i=1}^{N} 2^{-l_i} \leqslant 1$$

This inequality is known as the Kraft-McMillan inequality.

Proof The proof works by looking at the nth power of $K(\mathcal{C})$. If $K(\mathcal{C})$ is greater than one, then $K(\mathcal{C})^n$ should grow exponentially with n. If it does not grow exponentially with n, then this is proof that $\sum_{i=1}^{N} 2^{-l_i} \leqslant 1$.

Let n be an arbitrary integer. Then

$$\left[\sum_{i=1}^{N} 2^{-l_i}\right]^n = \left(\sum_{i_1=1}^{N} 2^{-l_{i_1}}\right)\left(\sum_{i_2=1}^{N} 2^{-l_{i_2}}\right)\cdots\left(\sum_{i_n=1}^{N} 2^{-l_{i_n}}\right) \quad (17)$$

$$= \sum_{i_1=1}^{N}\sum_{i_2=1}^{N}\cdots\sum_{i_n=1}^{N} 2^{-(l_{i_1}+l_{i_2}+\cdots+l_{i_n})} \quad (18)$$

The exponent $l_{i_1} + l_{i_2} + \cdots + l_{i_n}$ is simply the length of n codewords from the code $\mathcal{C}$. The smallest value that this exponent can take is greater than or equal to n, which would be the case if all codewords were 1 bit long. If

$$l = \max\{l_1, l_2, \ldots, l_N\}$$

then the largest value that the exponent can take is less than or equal to nl. Therefore, we can write this summation as

$$K(\mathcal{C})^n = \sum_{k=n}^{nl} A_k 2^{-k}$$

where A_k is the number of combinations of n codewords that have a combined length of k. Let's take a look at the size of this coefficient. The number of possible distinct binary sequences of length k is 2^k. If this code is uniquely decodable, then each sequence can represent one and only one sequence of codewords. Therefore, the number of possible combinations of codewords whose combined length is k cannot be greater than 2^k. In other words,

$$A_k \leqslant 2^k$$

This means that

$$K(\mathcal{C})^n = \sum_{k=n}^{nl} A_k 2^{-k} \leqslant \sum_{k=n}^{nl} 2^k 2^{-k} = nl - n + 1 \tag{19}$$

But if $K(\mathcal{C})$ is greater than one, it will grow exponentially with n, while $n(l-1)+1$ can only grow linearly. So if $K(\mathcal{C})$ is greater than one, we can always find an n large enough that the inequality (19) is violated. Therefore, for a uniquely decodable code $\mathcal{C}$, $K(\mathcal{C})$ is less than or equal to one. □

This part of the Kraft-McMillan inequality provides a necessary condition for uniquely decodable codes. That is, if a code is uniquely decodable, the codeword lengths have to satisfy the inequality. The second part of this result is that if we have a set of codeword lengths that satisfy the inequality, we can always find a prefix code with those codeword lengths. The proof of the assertion presented here is adapted from [9].

Theorem *Given a set of integers $l_1, l_2, \ldots, l_N$ that satisfy the inequality*

$$\sum_{i=1}^{N} 2^{-l_i} \leqslant 1$$

we can always find a prefix code with codeword lengths $l_1, l_2, \ldots, l_N$.

Proof We will prove this assertion by developing a procedure for constructing a prefix code with codeword lengths $l_1, l_2, \ldots, l_N$ that satisfy the given inequality. We will assume, without loss of generality, that $l_1 \leqslant l_2 \leqslant \ldots \leqslant l_N$.

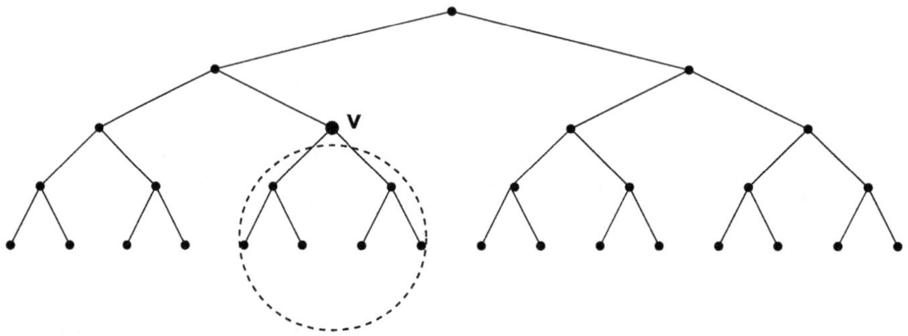

FIGURE 2.6 A full binary tree of depth four.

Before we build the code let us briefly look at binary trees. Consider the full binary tree of depth four shown in Figure 2.6. The number of leaf nodes on this tree is $2^4 = 8$. In fact the number of leaf nodes in a full binary tree of depth m is 2^m. We will construct our code by assigning vertices at depth l_i as codewords. In order for this code to be a prefix code when we assign a code to a vertex within the tree we cannot assign a codeword to any leaves belonging to the subtree rooted at that vertex. In effect, we have to prune the subtree rooted at that vertex. For example if we assign a codeword to the vertex **v** indicated in the figure we have to remove the subtree shown in the dashed circle. In the figure the vertex **v** is at depth two. The removal of the corresponding subtree results in the removal of four leaf nodes. In general we can see that in a full binary tree of depth m, a vertex at depth k is the root of a subtree with 2^{m-k} leaves.

Given the set of lengths $l_1, l_2, \ldots, l_N$, define

$$l = \max\{l_1, l_2, \ldots, l_N\}$$

Construct a full binary tree of depth l. This tree has 2^l leaves, and hence the possibility of having 2^l codewords of length l. Let's assign a codeword to a vertex v_1 at depth l_1. The path from the root node of the tree to this vertex will be a binary code of length l_1. As we mentioned earlier, in order for this codeword to be part of a prefix code we need to prune the subtree rooted at node v_1. This will result in a loss of 2^{l-l_1} leaf nodes from the full binary tree of depth l. Assign the next codeword to a vertex v_2 at a depth of l_2 and prune the subtree rooted at v_2. Continuing in this fashion we will obtain a prefix code with lengths $l_1, l_2, \ldots, l_N$ as long as we don't use up more than 2^l leaf nodes. As a codeword of length l_i results in the loss of 2^{l-l_i} leaf nodes from the full tree, the number of leaf nodes needed to build a code with codeword lengths $l_1, l_2, \ldots, l_N$ is given by $\sum_{i=1}^{N} 2^{l-l_i}$. But

$$\sum_{i=1}^{N} 2^{l-l_i} = 2^l \sum_{i=1}^{N} 2^{-l_i} \leqslant 2^l$$

where the last inequality is because of our assumption that the lengths satisfy the Kraft-McMillan inequality. Thus, given a set of codeword lengths satisfying the Kraft McMillan inequality we can always construct a prefix code with codewords having those lengths. □

Therefore, if we have a uniquely decodable code, the codeword lengths have to satisfy the Kraft-McMillan inequality. And, given codeword lengths that satisfy the Kraft-McMillan inequality, we can always find a prefix code with those codeword lengths. Thus, by restricting ourselves to prefix codes, we are not in danger of overlooking nonprefix uniquely decodable codes that have a shorter average length.

2.5 Algorithmic Information Theory

The theory of information described in the previous sections is intuitively satisfying and has useful applications. However, when dealing with real world data, it does have some theoretical difficulties. Suppose you were given the task of developing a compression scheme for use with a specific set of documents. We can view the entire set as a single long string. You could develop models for the data. Based on these models you could calculate probabilities using the relative frequency approach. These probabilities could then be used to obtain an estimate of the entropy and, thus, an estimate of the amount of compression available. All is well except for a "fly in the ointment." The string you have been given is fixed. There is nothing probabilistic about it. There is no abstract source that will generate different sets of these specific documents at different times. So how can we talk about the entropies without pretending that reality is somehow different from what it actually is? Unfortunately, it is not clear that we can. Our definition of entropy requires the existence of an abstract source. Our estimate of the entropy is still useful. It will give us a very good idea of how much compression we can get. So, practically speaking, information theory comes through. However, theoretically it seems there is some pretending involved. Algorithmic information theory is a different way of looking at information that has not been as useful in practice (and, therefore, we will not be looking at it a whole lot) but it gets around this theoretical problem. At the heart of algorithmic information theory is a measure called *Kolmogorov complexity*. This measure, while it bears the name of one person, was actually discovered independently by three people: R. Solomonoff, who was exploring machine learning; the Russian mathematician A.N. Kolmogorov; and G. Chaitin, who was in high school when he came up with this idea.

The Kolmogorov complexity $K(x)$ of a sequence x is the size of the program needed to generate x. In this size we include all inputs that might be needed by the program. We do not specify the programming language because it is always possible to translate a program in one language to a program in another language at fixed cost. If x was a sequence of all ones, a highly compressible sequence, the program would simply be a print statement in a loop. On the other extreme, if x were a random sequence with no structure, then the only program that could generate it would contain the sequence itself. The size of the program would be slightly larger than the sequence itself. Thus, there is a clear correspondence between the size of the smallest program that can generate a sequence and the amount of compression that can be obtained. Kolmogorov complexity seems to be the ideal measure to use in data compression. The problem is we do not know of any systematic way of computing or closely approximating Kolmogorov complexity. Clearly, any program that can generate a particular sequence is an upper bound for the Kolmogorov complexity of the sequence. However, we have no way of determining a lower bound. Thus, while the notion of Kolmogorov complexity is more

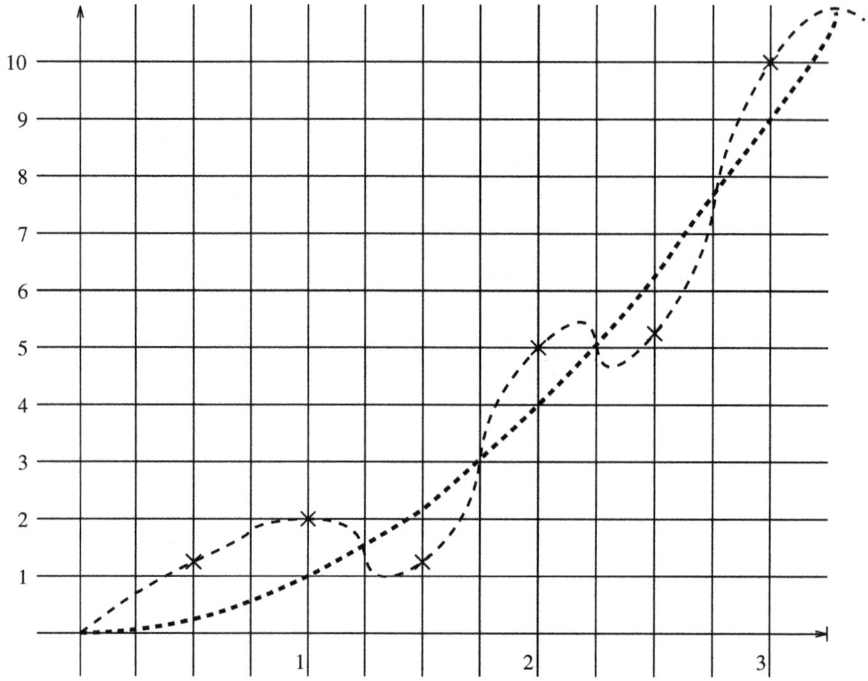

FIGURE 2.7 An illustration of the minimum description length (MDL) principle.

satisfying theoretically than the notion of entropy when compressing sequences, in practice it is not yet as helpful. However, given the active interest in these ideas it is quite possible that they will result in more practical applications.

2.6 Minimum Description Length Principle

One of the more practical offshoots of Kolmogorov complexity is the minimum description length (MDL) principle. The first discoverer of Kolmogorov complexity, Ray Solomonoff, viewed the concept of a program that would generate a sequence as a way of modeling the data. Independent from Solomonoff but inspired nonetheless by the ideas of Kolmogorov complexity, Jorma Risannen in 1978 [10] developed the modeling approach commonly known as MDL.

Let M_j be a model from a set of models $\mathcal{M}$ that attempt to characterize the structure in a sequence x. Let D_{M_j} be the number of bits required to describe the model M_j. For example, if the set of models $\mathcal{M}$ can be represented by a (possibly variable) number of coefficients, then the description of M_j would include the number of coefficients and the value of each coefficient. Let $R_{M_j}(x)$ be the number of bits required to represent x with respect to the model M_j. The minimum description length would be given by

$$\min_j (D_{M_j} + R_{M_j}(x))$$

which is the minimum number of bits required to represent both the model and the sequence with respect to the model. Consider the example illustrated in Figure 2.7, where the Xs represent data values. Suppose the set of models $\mathcal{M}$ is the set of kth order polynomials. We have also sketched two polynomials that could be used to model the data. Clearly, the higher-order polynomial does a much "better" job of modeling the data in the sense that the model exactly describes the data. To describe the higher order polynomial, we need to specify the value of each coefficient. The coefficients have to be exact if the polynomial is to exactly model the data requiring a large number of bits. The quadratic model, on the other hand, does not fit any of the data values. However, its description is very simple and the data values are either $+1$ or -1 away from the quadratic. So we could exactly represent the data by sending the coefficients of the quadratic $(1, 0)$ and 1 bit per data value to indicate whether each data value is $+1$ or -1 away from the quadratic. In this case, from a compression point of view, using the worse model actually gives better compression.

2.7 Summary

In this chapter, we learned some of the basic definitions of information theory. This was a rather brief visit, and we will revisit the subject in Chapter 8. However, the coverage in this chapter will be sufficient to take us through the next five chapters. The concepts introduced in this chapter allow us to estimate the number of bits we need to represent the output of a source given the probability model for the source. The process of assigning a binary representation to the output of a source is called coding. We have introduced the concepts of unique decodability and prefix codes, which we will use in the next two chapters when we describe various coding algorithms. We also looked, rather briefly, at different approaches to modeling. If we need to understand a model in more depth later in the book, we will devote more attention to it at that time. However, for the most part, the coverage of modeling in this chapter will be sufficient to understand methods described in the next four chapters.

Further Reading

1. A very readable book on information theory and its applications in a number of fields is *Symbols, Signals, and Noise—The Nature and Process of Communications*, by J.R. Pierce [11].
2. Another good introductory source for the material in this chapter is Chapter 6 of *Coding and Information Theory*, by R.W. Hamming [5].
3. Various models for text compression are described very nicely and in more detail in *Text Compression*, by T.C. Bell, J.G. Cleary, and I.H. Witten [1].
4. For a more thorough and detailed account of information theory, the following books are especially recommended (the first two are my personal favorites): *Information Theory*, by R.B. Ash [117]; *Transmission of Information*, by R.M. Fano [20]; *Information Theory and Reliable Communication*, by R.G. Gallagher [21]; *Entropy and Information*

Theory, by R.M. Gray [118]; and *Elements of Information Theory*, by T.M. Cover and J.A. Thomas [38].
5. Kolmogorov complexity is addressed in detail in *An Introduction to Kolmogorov Complexity and Its Applications*, by M. Li and P. Vitanyi [12].
6. A very readable overview of Kolmogorov complexity in the context of lossless compression can be found in the chapter "Complexity Measures," by S.R. Tate, in the *Lossless Compression Handbook* [13].
7. Various aspects of the minimum description length principle are discussed in *Advances in Minimum Description Length* edited by P. Grunwald, I.J. Myung, and M.A. Pitt [14]. Included in this book is a very nice introduction to the minimum description length principle by Peter Grunwald [15].

2.8 Projects and Problems

1. Suppose X is a random variable that takes on values from an M-letter alphabet. Show that $0 \leqslant H(X) \leqslant \log_2 M$.
2. Show that for the case where the elements of an observed sequence are *iid*, the entropy is equal to the first-order entropy.
3. Given an alphabet $\mathcal{A} = \{a_1, a_2, a_3, a_4\}$, find the first-order entropy in the following cases:
   (a) $P(a_1) = P(a_2) = P(a_3) = P(a_4) = \frac{1}{4}$
   (b) $P(a_1) = \frac{1}{2}$, $P(a_2) = \frac{1}{4}$, $P(a_3) = P(a_4) = \frac{1}{8}$
   (c) $P(a_1) = 0.505$, $P(a_2) = \frac{1}{4}$, $P(a_3) = \frac{1}{8}$, and $P(a_4) = 0.12$
4. Suppose we have a source with a probability model $P = \{p_0, p_1, \ldots, p_m\}$ and entropy H_P. Suppose we have another source with probability model $Q = \{q_0, q_1, \ldots, q_m\}$ and entropy H_Q, where

$$q_i = p_i \qquad i = 0, 1, \ldots, j-2, j+1, \ldots, m$$

   and

$$q_j = q_{j-1} = \frac{p_j + p_{j-1}}{2}$$

   How is H_Q related to H_P (greater, equal, or less)? Prove your answer.
5. Consider the following sequence:

$$ATGCTTAAGCTGCTTAACCTGAAGCTTCCGCTGAAGAACCTG$$
$$CTGAACCCGCTTAAGCTGAACCTTCTGAAGCTTAACCTGCTT$$

   (a) Estimating the probabilities from the sequence, compute the first-, second-, third-, and fourth-order entropy for this sequence.
   (b) Based on these entropies, can you infer how this sequence is structured?
6. There are several image and speech files among the accompanying data sets.

(a) Write a program to compute the first-order entropy of some of the image and speech files.

(b) Pick one of the image files and compute its second-order entropy. Comment on the difference between the first- and second-order entropies.

(c) Compute the entropy of the differences between neighboring pixels for the image you used in part (b). Comment on what you discover.

7. Conduct an experiment to see how well a model can describe a source.

(a) Write a program that randomly selects letters from the 26-letter alphabet $\{a, b, \ldots, z\}$ and forms four-letter words. Form 100 such words and see how many of these words make sense.

(b) Among the accompanying data sets is a file called `4letter.words`, which contains a list of four-letter words. Using this file, obtain a probability model for the alphabet. Now repeat part (a) generating the words using the probability model. To pick letters according to a probability model, construct the cumulative density function (*cdf*) $F_X(x)$ (see Appendix A for the definition of *cdf*). Using a uniform pseudorandom number generator to generate a value r, where $0 \leqslant r < 1$, pick the letter x_k if $F_X(x_k - 1) \leqslant r < F_X(x_k)$. Compare your results with those of part (a).

(c) Repeat (b) using a single-letter context.

(d) Repeat (b) using a two-letter context.

8. Determine whether the following codes are uniquely decodable:

(a) $\{0, 01, 11, 111\}$

(b) $\{0, 01, 110, 111\}$

(c) $\{0, 10, 110, 111\}$

(a) $\{1, 10, 110, 111\}$

9. Using a text file compute the probabilities of each letter p_i.

(a) Assume that we need a codeword of length $\lceil \log_2 \frac{1}{p_i} \rceil$ to encode the letter i. Determine the number of bits needed to encode the file.

(b) Compute the conditional probabilities $P(i/j)$ of a letter i given that the previous letter is j. Assume that we need $\lceil \log_2 \frac{1}{P(i/j)} \rceil$ to represent a letter i that follows a letter j. Determine the number of bits needed to encode the file.

10. Consider a set of experiments in which you flip a biased coin with the probability of a head being 0.4. In the first experiment you flip the coin once, and you have two possible outcomes. In the second experiment you flip the coin three times. In this experiment you have eight possible outcomes (HHH, HHT, etc.).

(a) What is the information associated with each outcome in the first experiment? What is the average information associated with the first experiment?

(b) What is the probability of each of the eight outcomes in the second experiment? What is the average amount of information associated with the second experiment?

(c) What is the relationship between the average information associated with the two experiments?

3

Huffman Coding

3.1 Overview

In this chapter we describe a very popular coding algorithm called the Huffman coding algorithm. We first present a procedure for building Huffman codes when the probability model for the source is known, and then we introduce a procedure for building codes when the source statistics are unknown. We also describe a few techniques for code design that are in some sense similar to the Huffman coding approach. Finally, we give some examples of using the Huffman code for image compression, audio compression, and text compression.

3.2 The Huffman Coding Algorithm

This technique was developed by David Huffman as part of a class assignment; the class was the first ever in the area of information theory and was taught by Robert Fano at MIT [16]. The codes generated using this technique or procedure are called *Huffman codes*. These codes are prefix codes and are optimum for a given model (set of probabilities).

The Huffman procedure is based on two observations regarding optimum prefix codes.

1. In an optimum code, symbols that occur more frequently (have a higher probability of occurrence) will have shorter codewords than symbols that occur less frequently.
2. In an optimum code, the two symbols that occur least frequently will have the same length.

It is easy to see that the first observation is correct. If symbols that occur more often had codewords that were longer than the codewords for symbols that occurred less often, the average number of bits per symbol would be larger than if the conditions were reversed. Therefore, a code that assigns longer codewords to symbols that occur more frequently cannot be optimum.

To see why the second observation holds true, consider the following situation. Suppose an optimum code $\mathcal{C}$ exists in which the two codewords corresponding to the two least probable symbols do not have the same length. Suppose the longer codeword is k bits longer than the shorter codeword. Because this is a prefix code, the shorter codeword cannot be a prefix of the longer codeword. This means that even if we drop the last k bits of the longer codeword, the two codewords would still be distinct. As these codewords correspond to the least probable symbols in the alphabet, no other codeword can be longer than these codewords; therefore, there is no danger that the shortened codeword would become the prefix of some other codeword. Furthermore, by dropping these k bits we obtain a new code that has a shorter average length than $\mathcal{C}$. But this violates our initial contention that $\mathcal{C}$ is an optimal code. Therefore, for an optimal code the second observation also holds true.

The Huffman procedure adds a simple requirement to these two observations. This requirement is that the codewords corresponding to the two lowest probability symbols differ only in the last bit. Assume that γ and δ are the two least probable symbols in an alphabet. If the codeword for γ is $\mathbf{m} * 0$, the codeword for δ would be $\mathbf{m} * 1$. Here, $\mathbf{m}$ is a string of 1s and 0s, and $*$ denotes concatenation.

This requirement does not violate our two observations and leads to a very simple encoding procedure. We describe this procedure with the help of the following example.

Example 3.2.1: Design of a Huffman Code

Let us design a Huffman code for a source that puts out letters from an alphabet $\mathcal{A} = \{a_1, a_2, a_3, a_4, a_5\}$ with $P(a_1) = P(a_3) = 0.2$, $P(a_2) = 0.4$, and $P(a_4) = P(a_5) = 0.1$. The entropy for this source is 2.122 bits/symbol. To design the Huffman code, we first sort the letters in a descending probability order as shown in Table 3.1. Here $c(a_i)$ denotes the codeword for a_i.

The two symbols with the lowest probability are a_4 and a_5. Therefore, we can assign their codewords as

TABLE 3.1 The initial five-letter alphabet.

Letter	Probability	Codeword
a_2	0.4	$c(a_2)$
a_1	0.2	$c(a_1)$
a_3	0.2	$c(a_3)$
a_4	0.1	$c(a_4)$
a_5	0.1	$c(a_5)$

TABLE 3.2 **The reduced four-letter alphabet.**

Letter	Probability	Codeword
a_2	0.4	$c(a_2)$
a_1	0.2	$c(a_1)$
a_3	0.2	$c(a_3)$
a'_4	0.2	α_1

TABLE 3.3 **The reduced three-letter alphabet.**

Letter	Probability	Codeword
a_2	0.4	$c(a_2)$
a'_3	0.4	α_2
a_1	0.2	$c(a_1)$

$$c(a_4) = \alpha_1 * 0$$
$$c(a_5) = \alpha_1 * 1$$

where α_1 is a binary string, and $*$ denotes concatenation.

We now define a new alphabet A' with a four-letter alphabet a_1, a_2, a_3, a'_4, where a'_4 is composed of a_4 and a_5 and has a probability $P(a'_4) = P(a_4) + P(a_5) = 0.2$. We sort this new alphabet in descending order to obtain Table 3.2.

In this alphabet, a_3 and a'_4 are the two letters at the bottom of the sorted list. We assign their codewords as

$$c(a_3) = \alpha_2 * 0$$
$$c(a'_4) = \alpha_2 * 1$$

but $c(a'_4) = \alpha_1$. Therefore,

$$\alpha_1 = \alpha_2 * 1$$

which means that

$$c(a_4) = \alpha_2 * 10$$
$$c(a_5) = \alpha_2 * 11$$

At this stage, we again define a new alphabet A'' that consists of three letters a_1, a_2, a'_3, where a'_3 is composed of a_3 and a'_4 and has a probability $P(a'_3) = P(a_3) + P(a'_4) = 0.4$. We sort this new alphabet in descending order to obtain Table 3.3.

In this case, the two least probable symbols are a_1 and a'_3. Therefore,

$$c(a'_3) = \alpha_3 * 0$$
$$c(a_1) = \alpha_3 * 1$$

TABLE 3.4 The reduced two-letter alphabet.

Letter	Probability	Codeword
a_3''	0.6	α_3
a_2	0.4	$c(a_2)$

TABLE 3.5 Huffman code for the original five-letter alphabet.

Letter	Probability	Codeword
a_2	0.4	1
a_1	0.2	01
a_3	0.2	000
a_4	0.1	0010
a_5	0.1	0011

But $c(a_3') = \alpha_2$. Therefore,

$$\alpha_2 = \alpha_3 * 0$$

which means that

$$c(a_3) = \alpha_3 * 00$$
$$c(a_4) = \alpha_3 * 010$$
$$c(a_5) = \alpha_3 * 011$$

Again we define a new alphabet, this time with only two letters a_3'' and a_2. Here a_3'' is composed of the letters a_3' and a_1 and has probability $P(a_3'') = P(a_3') + P(a_1) = 0.6$. We now have Table 3.4.

As we have only two letters, the codeword assignment is straightforward:

$$c(a_3'') = 0$$
$$c(a_2) = 1$$

which means that $\alpha_3 = 0$, which in turn means that

$$c(a_1) = 01$$
$$c(a_3) = 000$$
$$c(a_4) = 0010$$
$$c(a_5) = 0011$$

The Huffman code is given by Table 3.5. The procedure can be summarized as shown in Figure 3.1. ♦

The average length for this code is

$$l = .4 \times 1 + .2 \times 2 + .2 \times 3 + .1 \times 4 + .1 \times 4 = 2.2 \text{ bits/symbol}$$

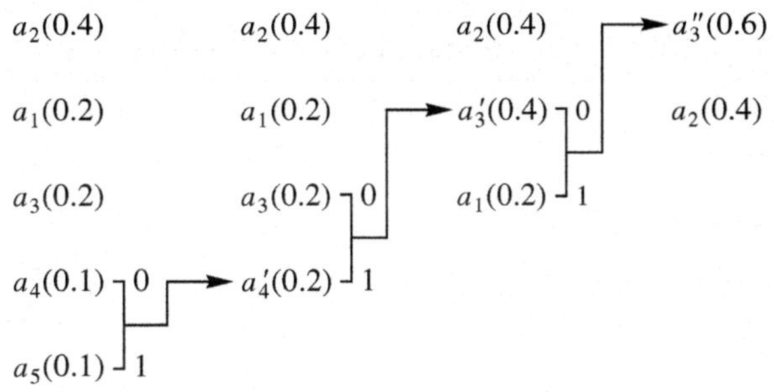

FIGURE 3.1 **The Huffman encoding procedure. The symbol probabilities are listed in parentheses.**

A measure of the efficiency of this code is its *redundancy*—the difference between the entropy and the average length. In this case, the redundancy is 0.078 bits/symbol. The redundancy is zero when the probabilities are negative powers of two.

An alternative way of building a Huffman code is to use the fact that the Huffman code, by virtue of being a prefix code, can be represented as a binary tree in which the external nodes or leaves correspond to the symbols. The Huffman code for any symbol can be obtained by traversing the tree from the root node to the leaf corresponding to the symbol, adding a 0 to the codeword every time the traversal takes us over an upper branch and a 1 every time the traversal takes us over a lower branch.

We build the binary tree starting at the leaf nodes. We know that the codewords for the two symbols with smallest probabilities are identical except for the last bit. This means that the traversal from the root to the leaves corresponding to these two symbols must be the same except for the last step. This in turn means that the leaves corresponding to the two symbols with the lowest probabilities are offspring of the same node. Once we have connected the leaves corresponding to the symbols with the lowest probabilities to a single node, we treat this node as a symbol of a reduced alphabet. The probability of this symbol is the sum of the probabilities of its offspring. We can now sort the nodes corresponding to the reduced alphabet and apply the same rule to generate a parent node for the nodes corresponding to the two symbols in the reduced alphabet with lowest probabilities. Continuing in this manner, we end up with a single node, which is the root node. To obtain the code for each symbol, we traverse the tree from the root to each leaf node, assigning a 0 to the upper branch and a 1 to the lower branch. This procedure, as applied to the alphabet of Example 3.2.1, is shown in Figure 3.2. Notice the similarity between Figures 3.1 and 3.2. This is not surprising, as they are a result of viewing the same procedure in two different ways.

3.2.1 Minimum Variance Huffman Codes

By performing the sorting procedure in a slightly different manner, we could have found a different Huffman code. In the first re-sort, we could place a_4' higher in the list, as shown in Table 3.6.

TABLE 3.6 Reduced four-letter alphabet.

Letter	Probability	Codeword
a_2	0.4	$c(a_2)$
a'_4	0.2	α_1
a_1	0.2	$c(a_1)$
a_3	0.2	$c(a_3)$

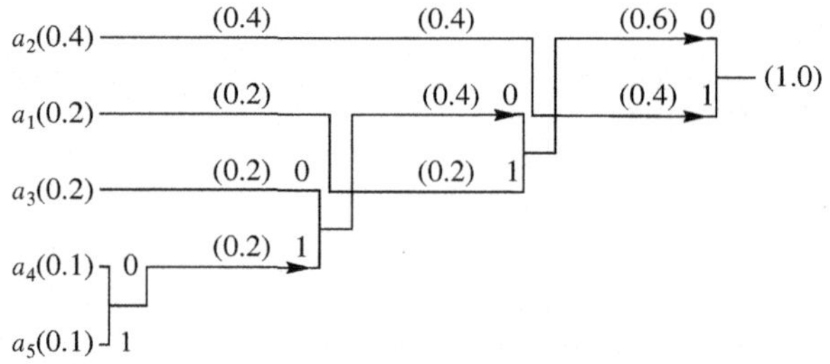

FIGURE 3.2 Building the binary Huffman tree.

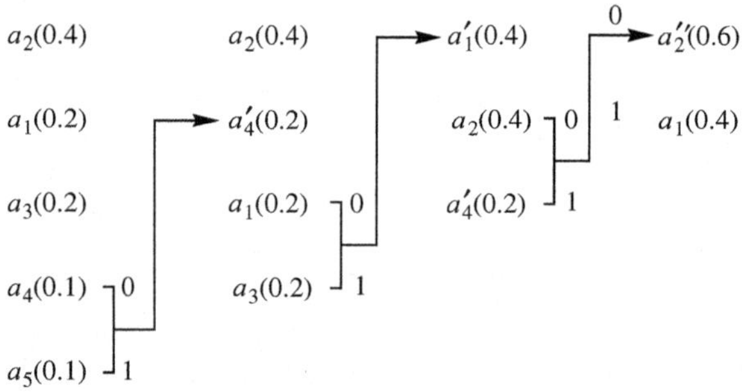

FIGURE 3.3 The minimum variance Huffman encoding procedure.

Now combine a_1 and a_3 into a'_1, which has a probability of 0.4. Sorting the alphabet a_2, a'_4, and a'_1 and putting a'_1 as far up the list as possible, we get Table 3.7. Finally, by combining a_2 and a'_4 and re-sorting, we get Table 3.8. If we go through the unbundling procedure, we get the codewords in Table 3.9. The procedure is summarized in Figure 3.3. The average length of the code is

$$l = .4 \times 2 + .2 \times 2 + .2 \times 2 + .1 \times 3 + .1 \times 3 = 2.2 \text{ bits/symbol.}$$

The two codes are identical in terms of their redundancy. However, the variance of the length of the codewords is significantly different. This can be clearly seen from Figure 3.4.

Remember that in many applications, although you might be using a variable-length code, the available transmission rate is generally fixed. For example, if we were going to transmit symbols from the alphabet we have been using at 10,000 symbols per second, we might ask for a transmission capacity of 22,000 bits per second. This means that during each second, the

TABLE 3.7 Reduced three-letter alphabet.

Letter	Probability	Codeword
a'_1	0.4	α_2
a_2	0.4	$c(a_2)$
a'_4	0.2	α_1

TABLE 3.8 Reduced two-letter alphabet.

Letter	Probability	Codeword
a'_2	0.6	α_3
a'_1	0.4	α_2

TABLE 3.9 Minimum variance Huffman code.

Letter	Probability	Codeword
a_1	0.2	10
a_2	0.4	00
a_3	0.2	11
a_4	0.1	010
a_5	0.1	011

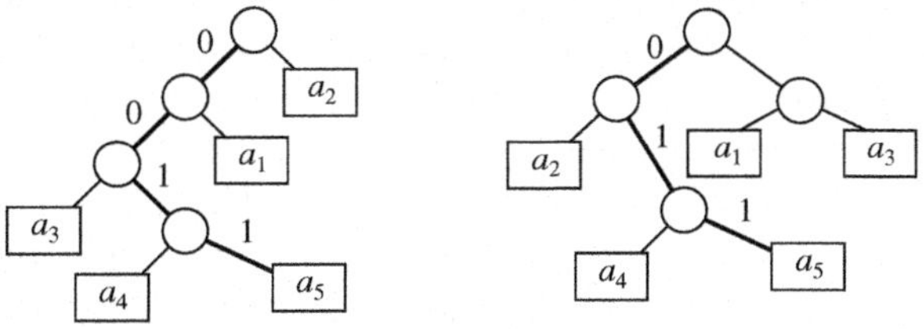

FIGURE 3.4 Two Huffman trees corresponding to the same probabilities.

channel expects to receive 22,000 bits, no more and no less. As the bit generation rate will vary around 22,000 bits per second, the output of the source coder is generally fed into a buffer. The purpose of the buffer is to smooth out the variations in the bit generation rate. However, the buffer has to be of finite size, and the greater the variance in the codewords, the more difficult the buffer design problem becomes. Suppose that the source we are discussing generates a string of a_4s and a_5s for several seconds. If we are using the first code, this means that we will be generating bits at a rate of 40,000 bits per second. For each second, the buffer has to store 18,000 bits. On the other hand, if we use the second code, we would be generating 30,000 bits per second, and the buffer would have to store 8000 bits for every second this condition persisted. If we have a string of a_2s instead of a string of a_4s and a_5s, the first code would result in the generation of 10,000 bits per second. Remember that the channel will still be expecting 22,000 bits every second, so somehow we will have to make up a deficit of 12,000

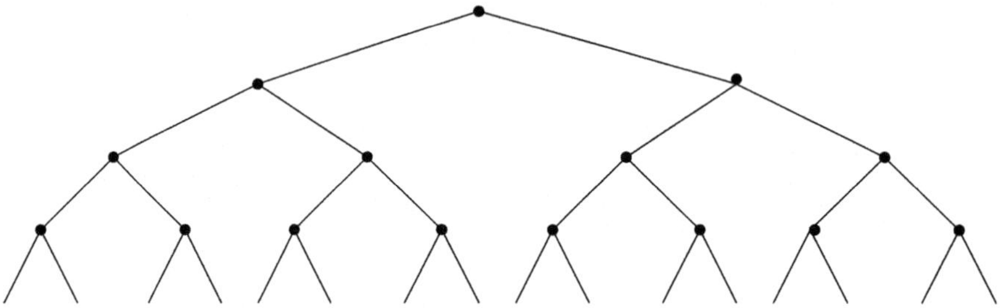

FIGURE 3.5 **A binary tree of depth four.**

bits per second. The same situation using the second code would lead to a deficit of 2000 bits per second. Thus, it seems reasonable to elect to use the second code instead of the first. To obtain the Huffman code with minimum variance, we always put the combined letter as high in the list as possible.

3.2.2 Canonical Huffman Codes

To be effective, a Huffman code needs to reflect the statistics of the source being encoded. When encoding sources with widely varying statistics, it is sometimes necessary to include the Huffman code along with the compressed representation of the source. Depending on the alphabet size, this can be a substantial burden, perhaps even cancelling out any advantages of compression. Therefore, it is useful to develop Huffman codes that can be stored in an efficient manner. Consider the Huffman code in Table 3.5. One way to store this code is to write the code in lexicographic order of the symbols. Thus, we first write the code for a_1, then the code for a_2, and so on. Each codeword could be represented by the length of the codeword followed by the codeword. Thus the code would be stored as [2, 01, 1, 1, 3, 000, 4, 0010, 5, 0011]. This is certainly manageable for the Huffman code of a small alphabet, but it is clearly going to have an impact on compression performance when we have Huffman codes for large alphabets.

We can substantially reduce the storage requirement for the code by using a version of the Huffman code known as the *canonical Huffman code*. Before we describe how to construct a canonical Huffman code, we examine a property of Huffman trees. Consider a Huffman tree such as one of the trees in Figure 3.4. For convenience, let us use the tree on the left. Notice that like all Huffman trees, this is a fully populated tree. That is, there is a codeword assigned to each leaf of the tree. What we will show is that if we are only given the lengths of the codewords moving from left to right on the tree we can reconstruct the code. For example, in this tree the lengths are 3, 4, 4, 2, and 1. In order to regenerate the code from the lengths, we begin with a tree of depth four (the longest codeword), as shown in Figure 3.5.

We then prune away the branches starting on the left, to match the lengths of the codewords. We first prune away the leftmost two branches, as shown in Figure 3.6, to leave a leaf at a depth of three. This corresponds to the first codeword. The next two codewords have a length of four, so we leave the next two leaves at a depth of four. The next codeword has a length of two so we prune all the lower branches as indicated in Figure 3.6, leaving a leaf at depth two. Finally, we prune most of the right half of the tree to get a leaf corresponding to the codeword of length one. The pruned tree can be populated by zeros on the left branches and

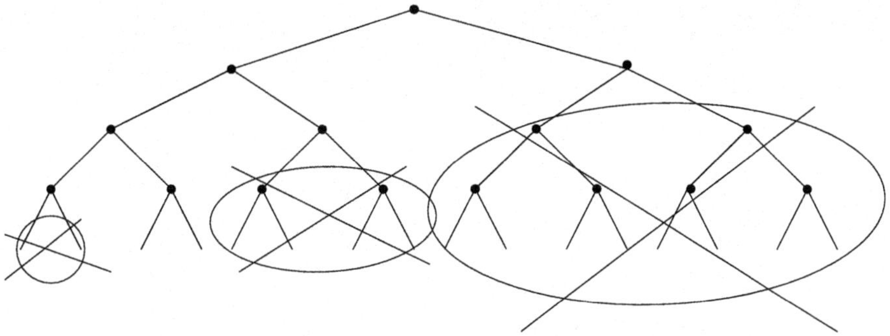

FIGURE 3.6 **Pruning the binary tree to get codewords of a specified length.**

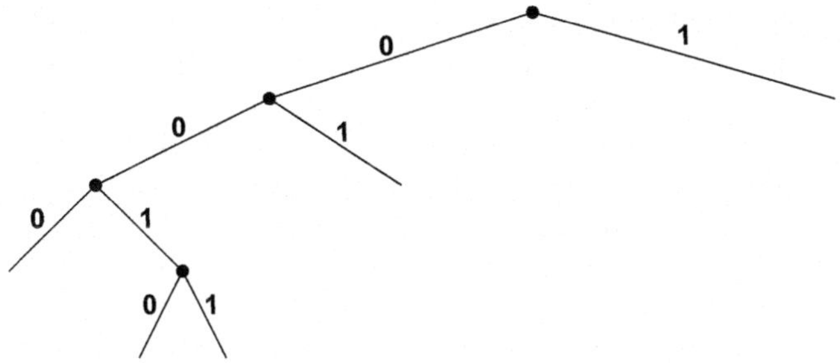

FIGURE 3.7 **Generating the code.**

ones on the right branches to generate the code as shown in Figure 3.7. From the figure, we can see that the codewords are 000, 0010, 0011, 01, and 1. So, just knowing the codeword lengths in a particular order permits us to regenerate the Huffman code. The problem is that we do not know which codeword belongs to which letter. The canonical procedure provides us with a way to generate a code that implicitly contains that information. In order to embed this additional information, we need some additional constraints to our Huffman coding procedure. For example, the *deflate* algorithm in zlib imposes the following conditions on a Huffman code [17]:

1. All codes of a given length have lexicographically consecutive values in the same order as the symbols they represent.
2. Shorter codes lexicographically precede longer codes.

Recall that when we generated Huffman codes, we could choose to assign zeros and ones to the branches as we wished. These constraints can be viewed as taking that freedom away.

We can incorporate these constraints into the Huffman coding procedure, or we can generate a Huffman code using the procedure described previously and transform the code thus produced into a canonical Huffman code. It is simpler to use the latter approach. In order to do this, we will begin by designing a Huffman code. From this design, we will extract the lengths of the codewords. We will use these lengths with the canonical constraints to design the code.

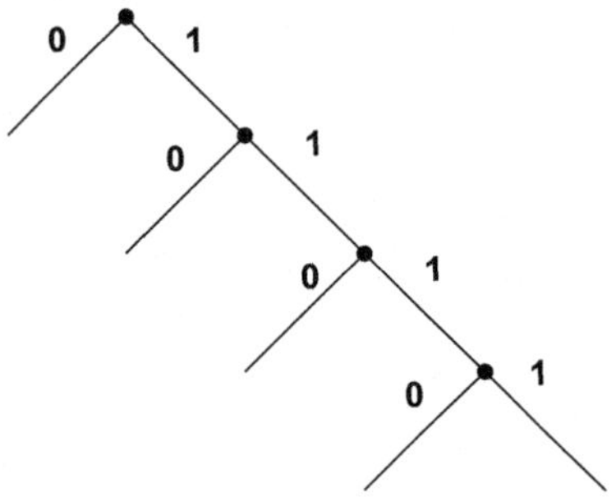

FIGURE 3.8 Tree for the canonical Huffman code.

We illustrate this procedure using the example we have been considering in this section. Our alphabet consists of five letters $\{a_1, a_2, a_3, a_4, a_5\}$. The lengths of the corresponding Huffman codewords are $\{2, 1, 3, 4, 4\}$. We generate the codewords from the shortest to the longest keeping in mind the constraint that the shorter codewords lexicographically precede longer codewords. The shortest codeword is that assigned to a_2. The lexicographically smallest codeword of length one is 0, so the codeword for a_2 is 0. The codewords of length two have to be of the form 1x. There is only one codeword of length two required, that being the codeword for a_1, so we assign the codeword 10 to a_1. The codewords of length three now have to be of the form 11x. We only need a single codeword of length three, the codeword for a_3; therefore, the codeword for a_3 is 110. There are two codewords, a_4 and a_5, of length four. Therefore, the codeword for a_4 is 1110, and the codeword for a_5 is 1111. This code can now be encoded by just sending the lengths. Unlike the previous code we know exactly which codeword belongs to which letter in the alphabet. This is because of the lexicographic constraints. As can be seen from the tree for this code in Figure 3.8, the tree increases in depth as we move from left to right. We know from the sequence of lengths that the shortest codeword is for the letter a_2, the next longest codeword is for the letter a_1, and so on. Thus simply encoding the list of lengths is sufficient for storing the Huffman code. For an alphabet of size five, this is not a significant saving; however, when the alphabet size is on the order of 256, the savings can be very significant.

3.2.3 Length-Limited Huffman Codes

The Huffman coding algorithm tries to minimize the *average* length of codewords. However, there are no limits on the *maximum* length of an individual codeword. This is not necessarily a problem when dealing with limited alphabet sizes. However, if the alphabet size is relatively large there is a possibility of getting very long Huffman codes. This could result in codewords that do not fit in a single machine word. This in turn can lead to inefficiencies in the coding process. These problems can be avoided by adding an additional constraint to the variable length code design problem: a requirement that all codewords be less than or equal to some maximum codeword length l_{max}. If m is the size of the source alphabet then clearly l_{max} has to be greater than or equal to $\lceil \log_2 m \rceil$ for the code to be a valid code.

The most widely used algorithm for constructing length-limited Huffman codes is the *package-merge* algorithm due to Larmore and Hirchberg [18]. The package-merge algorithm is more general than what we describe here, our only interest being its application to designing length-limited Huffman codes. Our description is based on the work of Turpin and Moffat [19]. We will use a couple of facts from our prior discussions to design length-limited Huffman codes. First, as we have seen in the previous section, all we need to obtain a Huffman code is the length of the codewords. Second, as we showed in Chapter 2, for an alphabet of size m the lengths of the codewords $\{l_1, l_2, \cdots, l_m\}$ in any variable length code have to satisfy the Kraft-McMillan inequality

$$\sum_{i=1}^{m} 2^{-l_i} \leqslant 1$$

and, given a set of l_i that satisfy the Kraft-McMillan inequality, we can always find a uniquely decodable code with codewords of length l_i.

Suppose our alphabet consists of the letters $\{a_1, a_2, \ldots, a_m\}$ with probabilities $\{p_1, p_2, \ldots, p_m\}$. We will assume that the letters have been sorted based on their probabilities such that $p_1 \leqslant p_2 \leqslant \cdots \leqslant p_m$. Our design process will involve incrementing the lengths l_i until the Kraft-McMillan inequality is satisfied while at the same time keeping the average length $\bar{l}$ as small as we can. We start with assigning zero bits for all lengths. Clearly, this will result in the lowest possible length of zero. However, it certainly does not give us a uniquely decodable code. Not that we need to justify the latter point mathematically, but if we did, we could compute the Kraft-McMillan sum as

$$\sum_{i=1}^{m} 2^{-l_i} = \sum_{i=1}^{m} 2^{0} = m$$

which is clearly greater than one. We can decrement this sum by 0.5 by incrementing the codeword for one of the letters to one. We want to pick the codeword that will have the minimal impact on the average codeword length. As the letters are sorted in nondecreasing probability order and the average codeword length is given by

$$\bar{l} = \sum_{i=1}^{m} p_i l_i$$

we will increase l_1 to one. This will give us an average codeword length of p_1 and a Kraft-McMillan sum of $m - 0.5$. Actually, we could have picked any of the letters as we will need to increment each length to at least one if we are to have any hope of reducing the Kraft-McMillan sum to one. However, once we have incremented each codeword length to one how do we decide which codeword lengths to increment further? The package-merge algorithm is an iterative algorithm that solves this problem by generating a list of choices that can be sorted in order of increasing cost. Each iteration consists of two steps operating on the list, a packaging step and a merge step. In the packaging step the list from the previous step is partitioned, or packaged, into groups of two neighboring items. The cost of a package is the sum of the costs of the items in the package where the cost of a letter is the probability of that letter. If the number of items is odd the item with the highest cost is discarded. In the merge step the packages from the previous step are merged with the list of items that consists of the individual letters. This list is then sorted in order of increasing cost. After $l_{max} - 1$ iterations the lowest cost $2m - 2$ items are selected. The length of the codeword for each letter is equal to the number of items containing that letter. As there are $2m - 2$ items, each of which decrements the Kraft-McMillan sum by 0.5, we end up with a set of lengths $\{l_i\}$ such that

$$\sum_{i=1}^{m} l_i = 1$$

Example 3.2.2:

Let us design a length-limited Huffman code for a source that puts out letters from an alphabet $\mathcal{A} = \{a_1, a_2, a_3, a_4, a_5, a_6\}$ with $P(a_1) = 0.05$, $P(a_2) = 0.1$, $P(a_3) = 0.15$, $P(a_4) = P(a_5) = 0.2$, and $P(a_6) = 0.3$. The entropy for this source is 2.409 bits/symbol. Using the standard Huffman procedure we can generate the code shown in Table 3.10 with an average length $\bar{l} = 2.45$. The longest codewords are four bits long. Let us design a code with the added constraint that $l_{max} = 3$.

We begin with the list of letters. The costs are listed next to the letters:

$$L_0 = [a_1(0.05), a_2(0.1), a_3(0.15), a_4(0.2), a_5(0.2), a_6(0.3)]$$

We then proceed with the first packaging step. The packages are denoted by $a_{jkl}(p)$ where the subscripts indicate the letters in the package and the quantity in the argument is the total cost of the item. In the first packaging step we form packages by grouping the letters two-by-two to form

$$\text{Package}_1 : [a_{12}(.15), a_{34}(.35), a_{56}(.5)]$$

We then merge these packages with our original list to get the merged list:

$$\text{Merge}_1\text{: } [a_1(0.05), a_2(0.1), a_3(0.15), a_{12}(0.15), a_4(0.2), a_5(0.2), a_6(0.3), a_{34}(0.35), a_{56}(0.5)]$$

To get a code limited to three bits we need to go through one more iteration. In the next packaging step we take the items in the previous merged list and group them two-by-two:

$$\text{Package}_2 : [a_{12}(0.15), a_{312}(0.3), a_{45}(0.4), a_{634}(0.65)]$$

TABLE 3.10 Huffman code.

Letter	Probability	Codeword
a_1	0.05	0100
a_2	0.1	0101
a_3	0.15	011
a_4	0.2	10
a_5	0.2	11
a_6	0.3	00

TABLE 3.11 Length-limited Huffman code.

Letter	Probability	Codeword
a_1	0.05	100
a_2	0.1	101
a_3	0.15	110
a_4	0.2	111
a_5	0.2	00
a_6	0.3	01

Finally, we go through one more merge operation in which we merge the packages with the original list of letters:

$$\text{Merge}_2 : [a_1(0.05), a_2(0.1), a_3(0.15), a_{12}(0.15), a_4(0.2), a_5(0.2), \\ a_6(0.3), a_{312}(0.3), a_{45}(0.4), a_{634}(0.65)]$$

We now pick the ten items ($2 \times 6 - 2 = 10$) with the lowest cost and count the number of times each letter appears in the ten items. The letter a_1 appears in a_1, a_{12}, and a_{312}. Therefore, the number of bits in the codeword for a_1 is three. Continuing in this fashion we obtain the lengths of the codewords as {3, 3, 3, 3, 2, 2}. A code with these lengths is shown in Table 3.11. The average codeword length is 2.5 bits. Comparing this code with the Huffman code in Table 3.10, the cost of limiting the length of the longest codeword to three bits is $2.5 - 2.45 = 0.05$. ♦

3.2.4 Optimality of Huffman Codes ★

The optimality of Huffman codes can be proven rather simply by first writing down the necessary conditions that an optimal code has to satisfy and then showing that satisfying these conditions necessarily leads to designing a Huffman code. The proof we present here is based on the proof shown in [20] and is obtained for the binary case (for a more general proof, see [20]).

The necessary conditions for an optimal variable-length binary code are as follows:

- **Condition 1:** Given any two letters, a_j and a_k, if $P[a_j] \geqslant P[a_k]$, then $l_j \leqslant l_k$, where l_j is the number of bits in the codeword for a_j.
- **Condition 2:** The two least probable letters have codewords with the same maximum length l_m.

We have provided the justification for these two conditions in the opening sections of this chapter.

- **Condition 3:** In the tree corresponding to the optimum code, there must be two branches stemming from each intermediate node.

If there were any intermediate node with only one branch coming from that node, we could remove it without affecting the decipherability of the code while reducing its average length.

- **Condition 4:** Suppose we change an intermediate node into a leaf node by combining all of the leaves descending from it into a composite word of a reduced alphabet. Then, if the original tree was optimal for the original alphabet, the reduced tree would be optimal for the reduced alphabet.

If this condition were not satisfied, we could find a code with a smaller average code length for the reduced alphabet and then simply expand the composite word again to get a new code tree that would have a shorter average length than our original "optimum" tree. This would contradict our statement about the optimality of the original tree.

In order to satisfy conditions 1, 2, and 3, the two least probable letters would have to be assigned codewords of maximum length l_m. Furthermore, the leaves corresponding to these letters arise from the same intermediate node. This is the same as saying that the codewords for these letters are identical except for the last bit. Consider the common prefix as the codeword for the composite letter of a reduced alphabet. Since the code for the reduced alphabet needs to be optimum for the code of the original alphabet to be optimum, we follow the same procedure again. To satisfy the necessary conditions, the procedure needs to be iterated until we have a reduced alphabet of size one. But this is exactly the Huffman procedure. Therefore, the necessary conditions above, which are all satisfied by the Huffman procedure, are also sufficient conditions.

3.2.5 Length of Huffman Codes ★

We have said that the Huffman coding procedure generates an optimum code, but we have not said what the average length of an optimum code is. The length of any code will depend on a number of things, including the size of the alphabet and the probabilities of individual letters. In this section, we will show that the optimal code for a source $\mathcal{S}$, and hence the Huffman code for the source $\mathcal{S}$, has an average code length $\bar{l}$ bounded below by the entropy and bounded above by the entropy plus 1 bit. In other words,

$$H(\mathcal{S}) \leqslant \bar{l} < H(\mathcal{S}) + 1 \tag{1}$$

In order for us to do this, we will need to use the Kraft-McMillan inequality introduced in Chapter 2. Recall that the first part of this result, due to McMillan, states that if we have a uniquely decodable code $\mathcal{C}$ with K codewords of length $\{l_i\}_{i=1}^{K}$, then the following inequality holds:

$$\sum_{i=1}^{K} 2^{-l_i} \leqslant 1 \tag{2}$$

Example 3.2.3:

Examining the code generated in Example 3.2.1 (Table 3.5), the lengths of the codewords are $\{1, 2, 3, 4, 4\}$. Substituting these values into the left-hand side of Equation (2), we get

$$2^{-1} + 2^{-2} + 2^{-3} + 2^{-4} + 2^{-4} = 1$$

which satisfies the Kraft-McMillan inequality.

If we use the minimum variance code (Table 3.9), the lengths of the codewords are $\{2, 2, 2, 3, 3\}$. Substituting these values into the left-hand side of Equation (2), we get

$$2^{-2} + 2^{-2} + 2^{-2} + 2^{-3} + 2^{-3} = 1$$

which again satisfies the inequality. ♦

The second part of this result, due to Kraft, states that if we have a sequence of positive integers $\{l_i\}_{i=1}^K$ that satisfies (2), then there exists a uniquely decodable code whose codeword lengths are given by the sequence $\{l_i\}_{i=1}^K$.

Using this result, we will now show the following:

1. The average codeword length $\bar{l}$ of an optimal code for a source $\mathcal{S}$ is greater than or equal to $H(\mathcal{S})$.
2. The average codeword length $\bar{l}$ of an optimal code for a source $\mathcal{S}$ is strictly less than $H(\mathcal{S}) + 1$.

For a source $\mathcal{S}$ with alphabet $\mathcal{A} = \{a_1, a_2, \ldots a_K\}$, and probability model $\{P(a_1), P(a_2), \ldots, P(a_K)\}$, the average codeword length is given by

$$\bar{l} = \sum_{i=1}^{K} P(a_i) l_i$$

Therefore, we can write the difference between the entropy of the source $H(\mathcal{S})$ and the average length as

$$\begin{aligned}
H(\mathcal{S}) - \bar{l} &= -\sum_{i=1}^{K} P(a_i) \log_2 P(a_i) - \sum_{i=1}^{K} P(a_i) l_i \\
&= \sum_{i=1}^{K} P(a_i) \left(\log_2 \left[\frac{1}{P(a_i)} \right] - l_i \right) \\
&= \sum_{i=1}^{K} P(a_i) \left(\log_2 \left[\frac{1}{P(a_i)} \right] - \log_2 [2^{l_i}] \right) \\
&= \sum_{i=1}^{K} P(a_i) \log_2 \left[\frac{2^{-l_i}}{P(a_i)} \right] \\
&\leq \log_2 \left[\sum_{i=1}^{K} 2^{-l_i} \right]
\end{aligned}$$

The last inequality is obtained using Jensen's inequality, which states that if $f(x)$ is a concave (convex cap, convex $\cap$) function, then $E[f(X)] \leqslant f(E[X])$. The log function is a concave function.

As the code is an optimal code $\sum_{i=1}^{K} 2^{-l_i} \leqslant 1$, therefore

$$H(\mathcal{S}) - \bar{l} \leqslant 0 \tag{3}$$

We will prove the upper bound by showing that there exists a uniquely decodable code with average codeword length less than $H(\mathcal{S}) + 1$. Therefore, if we have an optimal code, this code must have an average length that is less than $H(\mathcal{S}) + 1$.

Given a source, alphabet, and probability model as before, define

$$l_i = \left\lceil \log_2 \frac{1}{P(a_i)} \right\rceil$$

where $\lceil x \rceil$ is the smallest integer greater than or equal to x. For example, $\lceil 3.3 \rceil = 4$ and $\lceil 5 \rceil = 5$. Thus,

$$\lceil x \rceil = x + \epsilon \qquad \text{where } 0 \leqslant \epsilon < 1$$

Therefore,

$$\log_2 \frac{1}{P(a_i)} \leqslant l_i < \log_2 \frac{1}{P(a_i)} + 1 \tag{4}$$

From the left inequality of (4), we can see that

$$2^{-l_i} \leqslant P(a_i)$$

Accordingly,

$$\sum_{i=1}^{K} 2^{-l_i} \leqslant \sum_{i=1}^{K} P(a_i) = 1$$

and by the Kraft-McMillan inequality, there exists a uniquely decodable code with codeword lengths $\{l_i\}$. The average length of this code can be upper-bounded by using the right inequality of (4):

$$\bar{l} = \sum_{i=1}^{K} P(a_i) l_i < \sum_{i=1}^{K} P(a_i) \left[\log_2 \frac{1}{P(a_i)} + 1 \right]$$

or

$$\bar{l} < H(\mathcal{S}) + 1$$

We can see from the way the upper bound was derived that this is a rather loose upper bound. In fact, it can be shown that if $p_{\max}$ is the largest probability in the probability model, then for $p_{\max} \geqslant 0.5$, the upper bound for the Huffman code is $H(\mathcal{S}) + p_{\max}$, while for $p_{\max} < 0.5$, the upper bound is $H(\mathcal{S}) + p_{\max} + 0.086$. Obviously, this is a much tighter bound than the one we derived above. The derivation of this bound takes some time (see [21] for details).

3.2.6 Extended Huffman Codes ★

In applications where the alphabet size is large, $p_{\max}$ is generally quite small, and the amount of deviation from the entropy, especially in terms of a percentage of the rate, is quite small. However, in cases where the alphabet is small and the probability of occurrence of the different letters is skewed, the value of $p_{\max}$ can be quite large; and the Huffman code can become rather inefficient when compared to the entropy.

TABLE 3.12 Huffman code for the alphabet $\mathcal{A}$.

Letter	Codeword
a_1	00
a_2	11
a_3	10

Example 3.2.4:

Consider a source that puts out *iid* letters from the alphabet $\mathcal{A} = \{a_1, a_2, a_3\}$ with the probability model $P(a_1) = 0.8$, $P(a_2) = 0.02$, and $P(a_3) = 0.18$. The entropy for this source is 0.816 bits/symbol. A Huffman code for this source is shown in Table 3.12.

The average length for this code is 1.2 bits/symbol. The difference between the average code length and the entropy, or the redundancy, for this code is 0.384 bits/symbol, which is 47% of the entropy. This means that to code this sequence, we would need 47% more bits than the minimum required. ♦

We can sometimes reduce the coding rate by blocking more than one symbol together. To see how this can happen, consider a source S that emits a sequence of letters from an alphabet $\mathcal{A} = \{a_1, a_2, \ldots, a_m\}$. Each element of the sequence is generated independently of the other elements in the sequence. The entropy for this source is given by

$$H(S) = -\sum_{i=1}^{m} P(a_i) \log_2 P(a_i)$$

We know that we can generate a Huffman code for this source with rate R such that

$$H(S) \leqslant R < H(S) + 1 \tag{5}$$

We have used the looser bound here; the same argument can be made with the tighter bound. Notice that we have used "rate R" to denote the number of bits per symbol. This is a standard convention in the data compression literature. However, in the communication literature, the word "rate" often refers to the number of bits per second.

Suppose we now encode the sequence by generating one codeword for every n symbols. As there are m^n combinations of n symbols, we will need m^n codewords in our Huffman code. We could generate this code by viewing the m^n symbols as letters of an *extended alphabet*

$$\mathcal{A}^{(n)} = \{\overbrace{a_1 a_1 \ldots a_1}^{\text{n times}}, a_1 a_1 \ldots a_2, \ldots, a_1 a_1 \ldots a_m, a_1 a_1 \ldots a_2 a_1, \ldots, a_m a_m \ldots a_m\}$$

from a source $S^{(n)}$. Let us denote the rate for the new source as $R^{(n)}$. Then we know that

$$H(S^{(n)}) \leqslant R^{(n)} < H(S^{(n)}) + 1 \tag{6}$$

$R^{(n)}$ is the number of bits required to code n symbols. Therefore, the number of bits required per symbol, R, is given by

$$R = \frac{1}{n} R^{(n)}$$

The number of bits per symbol can be bounded as

$$\frac{H(S^{(n)})}{n} \leqslant R < \frac{H(S^{(n)})}{n} + \frac{1}{n}$$

In order to compare this to (5), and see the advantage we get from encoding symbols in blocks instead of one at a time, we need to express $H(S^{(n)})$ in terms of $H(S)$. This turns out to be a relatively easy (although somewhat messy) thing to do.

$$\begin{aligned}
H(S^{(n)}) &= -\sum_{i_1=1}^{m}\sum_{i_2=1}^{m}\cdots\sum_{i_n=1}^{m} P(a_{i_1}, a_{i_2}, \ldots a_{i_n}) \log[P(a_{i_1}, a_{i_2}, \ldots a_{i_n})] \\
&= -\sum_{i_1=1}^{m}\sum_{i_2=1}^{m}\cdots\sum_{i_n=1}^{m} P(a_{i_1})P(a_{i_2})\ldots P(a_{i_n}) \log[P(a_{i_1})P(a_{i_2})\ldots P(a_{i_n})] \\
&= -\sum_{i_1=1}^{m}\sum_{i_2=1}^{m}\cdots\sum_{i_n=1}^{m} P(a_{i_1})P(a_{i_2})\ldots P(a_{i_n}) \sum_{j=1}^{n} \log[P(a_{i_j})] \\
&= -\sum_{i_1=1}^{m} P(a_{i_1}) \log[P(a_{i_1})] \left\{\sum_{i_2=1}^{m}\cdots\sum_{i_n=1}^{m} P(a_{i_2})\ldots P(a_{i_n})\right\} \\
&\quad -\sum_{i_2=1}^{m} P(a_{i_2}) \log[P(a_{i_2})] \left\{\sum_{i_1=1}^{m}\sum_{i_3=1}^{m}\cdots\sum_{i_n=1}^{m} P(a_{i_1})P(a_{i_3})\ldots P(a_{i_n})\right\} \\
&\quad \vdots \\
&\quad -\sum_{i_n=1}^{m} P(a_{i_n}) \log[P(a_{i_n})] \left\{\sum_{i_1=1}^{m}\sum_{i_2=1}^{m}\cdots\sum_{i_{n-1}=1}^{m} P(a_{i_1})P(a_{i_2})\ldots P(a_{i_{n-1}})\right\}
\end{aligned}$$

The $n - 1$ summations in braces in each term sum to one. Therefore,

$$\begin{aligned}
H(S^{(n)}) &= -\sum_{i_1=1}^{m} P(a_{i_1}) \log[P(a_{i_1})] - \sum_{i_2=1}^{m} P(a_{i_2}) \log[P(a_{i_2})] - \cdots - \sum_{i_n=1}^{m} P(a_{i_n}) \log[P(a_{i_n})] \\
&= nH(S)
\end{aligned}$$

and we can write (6) as

$$H(S) \leqslant R \leqslant H(S) + \frac{1}{n} \tag{7}$$

By comparing this to (5), we can see that encoding the output of the source in longer blocks of symbols *guarantees* a rate closer to the entropy. Note that all we are talking about here is a bound or guarantee about the rate. As we have seen in the previous chapter, there are a number of situations in which we can achieve a rate *equal* to the entropy with a block length of one!

Example 3.2.5:

For the source described in the previous example, we will generate a codeword for every *two* symbols, instead of generating a codeword for every symbol. If we look at the source sequence two at a time, the number of possible symbol pairs, or size of the extended alphabet, is $3^2 = 9$. The extended alphabet, probability model, and Huffman code for this example are shown in Table 3.13.

TABLE 3.13 The extended alphabet and corresponding Huffman code

Letter	Probability	Code
a_1a_1	0.64	0
a_1a_2	0.016	10101
a_1a_3	0.144	11
a_2a_1	0.016	101000
a_2a_2	0.0004	10100101
a_2a_3	0.0036	1010011
a_3a_1	0.1440	100
a_3a_2	0.0036	10100100
a_3a_3	0.0324	1011

The average codeword length for this extended code is 1.7228 bits/symbol. However, each symbol in the extended alphabet corresponds to two symbols from the original alphabet. Therefore, in terms of the original alphabet, the average codeword length is $1.7228/2 = 0.8614$ bits/symbol. This redundancy is about 0.045 bits/symbol, which is only about 5.5% of the entropy. ♦

We see that by coding blocks of symbols together, we can reduce the redundancy of Huffman codes. In the previous example, two symbols were blocked together to obtain a rate reasonably close to the entropy. Blocking two symbols together means the alphabet size goes from m to m^2, where m was the size of the initial alphabet. In this case, m was three, so the size of the extended alphabet was nine. This size is not an excessive burden for most applications. However, if the probabilities of the symbols were more unbalanced, then it would require blocking many more symbols together before the redundancy was lowered to acceptable levels. As we block more and more symbols together, the size of the alphabet grows exponentially, and the Huffman coding scheme becomes impractical. Under these conditions, we need to look at techniques other than Huffman coding. One approach that is very useful in these conditions is *arithmetic coding*. We will discuss this technique in some detail in the next chapter.

3.2.7 Implementation of Huffman Codes

Huffman encoding is relatively simple to implement using a table lookup. Decoding is another matter. If speed is the only factor, we can implement the decoder using table lookup as well. However, for the decoder, the table has to be of size 2^N where N is the length of the longest

TABLE 3.14 **A nine-letter alphabet and corresponding Huffman code.**

Letter	Code
A	0
B	10101
C	11
D	101000
E	10100101
F	1010011
G	100
H	10100100
I	1011

code. Consider the code in Table 3.13. The longest codeword is eight bits long. This means we need a table of size 256, whereas the code consists of only nine codewords. This is a significant amount of wasted storage. And the waste grows exponentially with the length of the longest codeword. Another option is to store the code as a binary tree. This is certainly memory efficient, however, decoding each symbol requires traversing the tree, which may be computationally inefficient.

Most implementations of Huffman decoding find a middle ground by combining aspects of a table-based decoder and a tree-based decoder. The decoder used in the *inflate* algorithm, which is part of both gzip and zlib, uses such a hybrid method. One way of thinking about the decoder is to view it as a state machine. The state machine takes as its input a specified number of bits, where the number of input bits may be different in each state. Having received these bits, the state machine moves to a new state, and in the process may output a decoded letter. Let's describe the algorithm using an example. Suppose we have a source with the alphabet and code shown in Table 3.14. A tabular implementation of the decoder would require a table of size 256. Instead, we will implement a hierarchy of three tables with the first table being of size 16 and the others being of size 4. Each of the 16 values points to either a decoded letter and the number of bits used to decode the letter (number of bits "to be gobbled" in the inimitable words of Gailly and Adler [22]), or a subsidiary table. The decoding operation can be visualized in terms of the state machine shown in Figure 3.9. In this implementation, the decoding process can be represented using a machine with three states. In state I, the decoder uses a four-bit input as an index into a lookup table; while in state II and state III, the decoder uses a two-bit index into the lookup table. The decoder moves from state I to state II when it receives an input of 1010 and from state II to state III when it receives an input of 01. In the figure, the outcome of other patterns of input are shown as $nnnn/(L, m)$ where $nnnn$ represents the binary index, L denotes the letter decoded, and m indicates the number of bits used in the decoding process. In the figure, x denotes a "don't care," that is, a bit that can be either a 0 or a 1.

The lookup tables corresponding to these states are shown in Tables 3.15–3.17. Notice that in most cases, the letter A is decoded and one bit is used up. This is to be expected as A is the most probable letter with the single bit codeword 0. As eight of the entries in the table begin with 0, it makes sense that eight of the entries will decode to A. After each of these

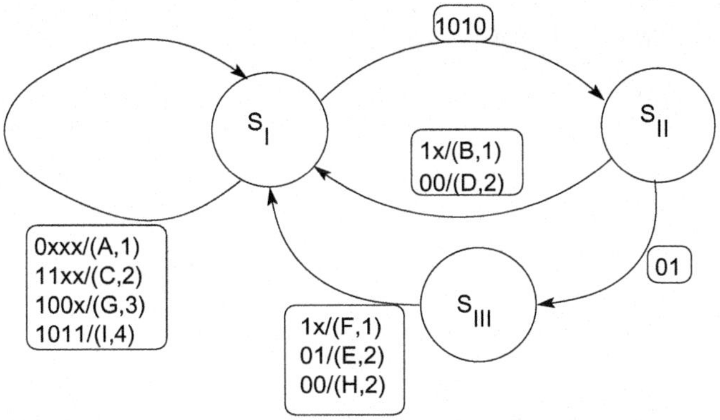

FIGURE 3.9 **A state machine description of Huffman decoding.**

TABLE 3.15 **Table I.**

Letter	Code
0000	$A, 1$
0001	$A, 1$
0010	$A, 1$
0011	$A, 1$
0100	$A, 1$
0101	$A, 1$
0110	$A, 1$
0111	$A, 1$
1000	$G, 3$
1001	$G, 3$
1010	Table II
1011	$I, 4$
1100	$C, 2$
1101	$C, 2$
1110	$C, 2$
1111	$C, 2$

entries is decoded, the decoder shifts out (gobbles up) the leftmost bit and shifts in another bit from the received bitstream before decoding the next letter using the same table. One of the entries, 1010, points to a second table referred to as Table II. In this case, the decoder moves to state II, shifts out all four bits, reads in two new bits, and uses them as a lookup index for Table II shown in Table 3.16. If these two bits are 01 the decoder moves to state III and uses the following two bits from the received bitstream as an index into the lookup table shown in Table 3.17. If these bits are not 01, the decoder uses them as an index into Table 3.16, decodes the corresponding letter, shifts out the number of bits indicated by the table and returns to state I. In state III the decoder decodes the letter indicated by the lookup table, shifts out the corresponding number of bits, and returns to state I.

TABLE 3.16 Table II.

Letter	Code
00	$D, 2$
01	Table III
10	$B, 1$
11	$B, 1$

TABLE 3.17 Table III.

Letter	Code
00	$H, 2$
01	$E, 2$
10	$F, 1$
11	$F, 1$

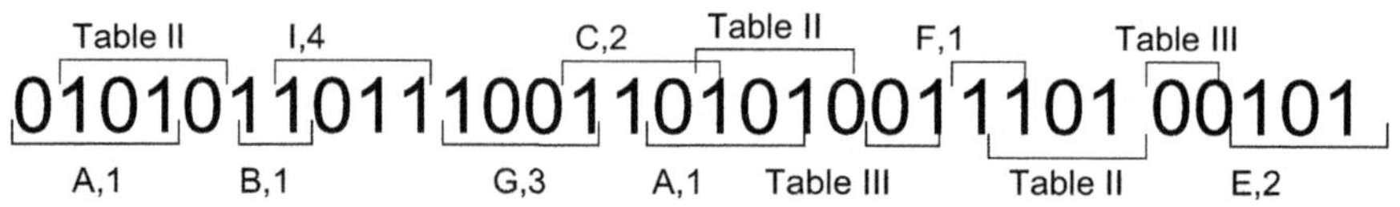

FIGURE 3.10 Decoding of the example binary string.

To see how the decoding process works let's decode the following bitstream:

0101011011100110101001110100101

We begin in state I with the first four bits 0101. Using this as an index to Table 3.15, we decode the letter A, shift out one bit, and shift in another bit to get a new index of 1010. According to Table 3.15, this is a pointer to Table 3.16, and the decoder moves to state II. Therefore, we read the next two bits from the bitstream and use them as an index to Table 3.16. The next two bits are 11 so we decode a B, shift out a bit, and move back to state I. We then read three more bits to make up the index into Table 3.15. The next three bits are 011 giving us an index of 1011 into Table 3.15. We decode this as an I and read the next four bits, 1001. From Table 3.15, this is decoded as a G for an expenditure of three bits. We read three more bits to get an index of 1101. This decodes as a C with an expenditure of two bits. The process continues as shown in Figure 3.10 resulting in the decoded sequence ABIGCAFE.

We can see in this example that we saved some memory by going from a 256 element lookup table to three lookup tables with 24 entries. In order to obtain these savings, we paid in terms of computation. In the case of the single lookup table, one lookup is sufficient to decode a letter. In this case, when we encounter the letters A, C, G, or I, we use one table lookup. For the letters D and B, we use two table lookups, and for the letters H, E, and F, we need three lookups. Thus, on the average, we need $1 \times (0.64 + 0.144 + 0.144 + 0.0324) + 2 \times (0.016 + 0.016) + 3 \times (0.0036 + 0.0036 + 0.0004) = 1.0472$ lookups per letter. This is only slightly more lookups than we would have needed had we used a single lookup table.

3.3 Nonbinary Huffman Codes ★

The binary Huffman coding procedure can be easily extended to the nonbinary case where the code elements come from an m-ary alphabet, and m is not equal to two. Recall that we obtained the Huffman algorithm based on the following observations about an optimum binary prefix code:

1. Symbols that occur more frequently (have a higher probability of occurrence) will have shorter codewords than symbols that occur less frequently.
2. The two symbols that occur least frequently will have the same length.

Also recall that an additional requirement is that the two symbols with the lowest probability differ only in the last position.

We can obtain a nonbinary Huffman code in almost exactly the same way. The obvious thing to do would be to modify the second observation to read "The m symbols that occur least frequently will have the same length," and also modify the additional requirement to read "The m symbols with the lowest probability differ only in the last position."

However, we run into a small problem with this approach. Consider the design of a ternary Huffman code for a source with a six-letter alphabet. Using the rules described above, we would first combine the three letters with the lowest probability into a composite letter. This would give us a reduced alphabet with four letters. However, combining the three letters with lowest probability from this alphabet would result in a further reduced alphabet consisting of only two letters. We have three values to assign and only two letters. Instead of combining three letters at the beginning, we could have combined two letters. This would result in a reduced alphabet of size five. If we combined three letters from this alphabet, we would end up with a final reduced alphabet size of three. Finally, we could combine two letters in the second step, which would again result in a final reduced alphabet of size three. Which alternative should we choose?

Recall that the symbols with the lowest probability will have the longest codeword. Furthermore, all of the symbols that we combine together into a composite symbol will have codewords of the same length. This means that all letters we combine together at the very first stage will have codewords that have the same length, and these codewords will be the longest of all the codewords. If at some stage we are allowed to combine less than m symbols, the logical place to do this would be in the very first stage.

In the general case of an m-ary code and an M-letter alphabet, how many letters should we combine in the first phase? Let m' be the number of letters that are combined in the first phase. Then m' is the number between two and m, such that m' modulo $(m-1)$ is equal to M modulo $(m-1)$.

Example 3.3.1:

Generate a ternary Huffman code for a source with a six-letter alphabet and a probability model $P(a_1) = P(a_3) = P(a_4) = 0.2$, $P(a_5) = 0.25$, $P(a_6) = 0.1$, and $P(a_2) = 0.05$. In this case $m = 3$, therefore m' is either 2 or 3.

$$6 \pmod 2 = 0, \qquad 2 \pmod 2 = 0, \qquad 3 \pmod 2 = 1$$

TABLE 3.18 Sorted six-letter alphabet.

Letter	Probability	Codeword
a_5	0.25	$c(a_5)$
a_1	0.20	$c(a_1)$
a_3	0.20	$c(a_3)$
a_4	0.20	$c(a_4)$
a_6	0.10	$c(a_6)$
a_2	0.05	$c(a_2)$

TABLE 3.19 Reduced five-letter alphabet.

Letter	Probability	Codeword
a_5	0.25	$c(a_5)$
a_1	0.20	$c(a_1)$
a_3	0.20	$c(a_3)$
a_4	0.20	$c(a_4)$
a_6'	0.15	α_1

Since $6 \pmod 2) = 2 \pmod 2)$, $m' = 2$. Sorting the symbols in probability order results in Table 3.18.

As m' is 2, we can assign the codewords of the two symbols with the lowest probability as

$$c(a_6) = \alpha_1 * 0$$

$$c(a_2) = \alpha_1 * 1$$

where α_1 is a ternary string and $*$ denotes concatenation. The reduced alphabet is shown in Table 3.19.

Now we combine the three letters with the lowest probability into a composite letter a_3' and assign their codewords as

$$c(a_3) = \alpha_2 * 0$$

$$c(a_4) = \alpha_2 * 1$$

$$c(a_6') = \alpha_2 * 2$$

But $c(a_6') = \alpha_1$. Therefore,

$$\alpha_1 = \alpha_2 * 2$$

which means that

$$c(a_6) = \alpha_2 * 20$$

$$c(a_2) = \alpha_2 * 21$$

Sorting the reduced alphabet, we have Table 3.20. Thus, $\alpha_2 = 0$, $c(a_5) = 1$, and $c(a_1) = 2$. Substituting for α_2, we get the codeword assignments in Table 3.21.

The tree corresponding to this code is shown in Figure 3.11. Notice that at the lowest level of the tree, we have only two codewords. If we had combined three letters at the first step and combined two letters at a later step, the lowest level would have contained three codewords, and a longer average code length would result (see Problem 7 at the end of this chapter). ♦

TABLE 3.20 Reduced three-letter alphabet.

Letter	Probability	Codeword
a_3'	0.55	α_2
a_5	0.25	$c(a_5)$
a_1	0.20	$c(a_1)$

TABLE 3.21 Ternary code for six-letter alphabet.

Letter	Probability	Codeword
a_1	0.20	2
a_2	0.05	021
a_3	0.20	00
a_4	0.20	01
a_5	0.25	1
a_6	0.10	020

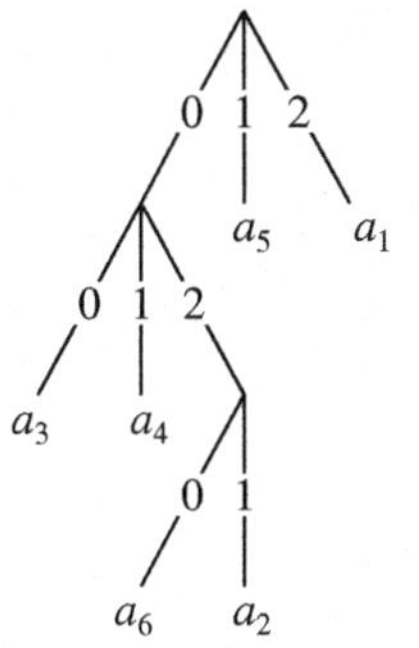

FIGURE 3.11 Code tree for the nonbinary Huffman code.

3.4 Adaptive Huffman Coding

Huffman coding requires knowledge of the probabilities of the source sequence. If this knowledge is not available, Huffman coding becomes a two-pass procedure: the statistics are collected in the first pass, and the source is encoded in the second pass. In order to convert this algorithm into a one-pass procedure, Faller [23] and Gallagher [21] independently developed adaptive algorithms to construct the Huffman code based on the statistics of the symbols already encountered. These were later improved by Knuth [24] and Vitter [25].

Theoretically, if we wanted to encode the $(k+1)$th symbol using the statistics of the first k symbols, we could recompute the code using the Huffman coding procedure each time a symbol is transmitted. However, this would not be a very practical approach due to the large amount of computation involved—hence, the adaptive Huffman coding procedures.

The Huffman code can be described in terms of a binary tree similar to the ones shown in Figure 3.4. The squares denote the external nodes or leaves and correspond to the symbols in the source alphabet. The codeword for a symbol can be obtained by traversing the tree from the root to the leaf corresponding to the symbol, where 0 corresponds to a left branch and 1 corresponds to a right branch. In order to describe how the adaptive Huffman code works, we add two other parameters to the binary tree: the *weight* of each leaf, which is written as a number inside the node, and a *node number*. The weight of each external node is simply the number of times the symbol corresponding to the leaf has been encountered. The weight of each internal node is the sum of the weights of its offspring. The node number y_i is a unique number assigned to each internal and external node. If we have an alphabet of size n, then the $2n - 1$ internal and external nodes can be numbered as $y_1, \ldots, y_{2n-1}$ such that if x_j is the weight of node y_j, we have $x_1 \leqslant x_2 \leqslant \cdots \leqslant x_{2n-1}$. Furthermore, the nodes y_{2j-1} and y_{2j} are offspring of the same parent node, or siblings, for $1 \leqslant j < n$, and the node number for the parent node is greater than y_{2j-1} and y_{2j}. These last two characteristics are called the *sibling property*, and any tree that possesses this property is a Huffman tree [21].

In the adaptive Huffman coding procedure, neither transmitter nor receiver knows anything about the statistics of the source sequence at the start of transmission. The tree at both the transmitter and the receiver consists of a single node that corresponds to all symbols not yet transmitted (NYT) and has a weight of zero. As transmission progresses, nodes corresponding to symbols transmitted are added to the tree, and the tree is reconfigured using an update procedure. Before the beginning of transmission, a fixed code for each symbol is agreed upon between transmitter and receiver. A simple (short) code is as follows:

If the source has an alphabet $(a_1, a_2, \ldots, a_m)$ of size m, then pick e and r such that $m = 2^e + r$ and $0 \leqslant r < 2^e$. The letter a_k is encoded as the $(e + 1)$-bit binary representation of $k - 1$, if $1 \leqslant k \leqslant 2r$; else, a_k is encoded as the e-bit binary representation of $k - r - 1$. For example, suppose $m = 26$, then $e = 4$, and $r = 10$. The symbol a_1 is encoded as 00000, the symbol a_2 is encoded as 00001, and the symbol a_{22} is encoded as 1011.

When a symbol is encountered for the first time, the code for the NYT node is transmitted, followed by the fixed code for the symbol. A node for the symbol is then created, and the symbol is taken out of the NYT list.

Both transmitter and receiver start with the same tree structure. The updating procedure used by both transmitter and receiver is identical. Therefore, the encoding and decoding processes remain synchronized.

3.4.1 Update Procedure

The update procedure requires that the nodes be in a fixed order. This ordering is preserved by numbering the nodes. The largest node number is given to the root of the tree, and the smallest number is assigned to the NYT node. The numbers from the NYT node to the root of the tree are assigned in increasing order from left to right and from lower level to upper level. The set of nodes with the same weight makes up a *block*. Figure 3.12 is a flowchart of the updating procedure.

The function of the update procedure is to preserve the sibling property. In order that the update procedures at the transmitter and receiver both operate with the same information, the

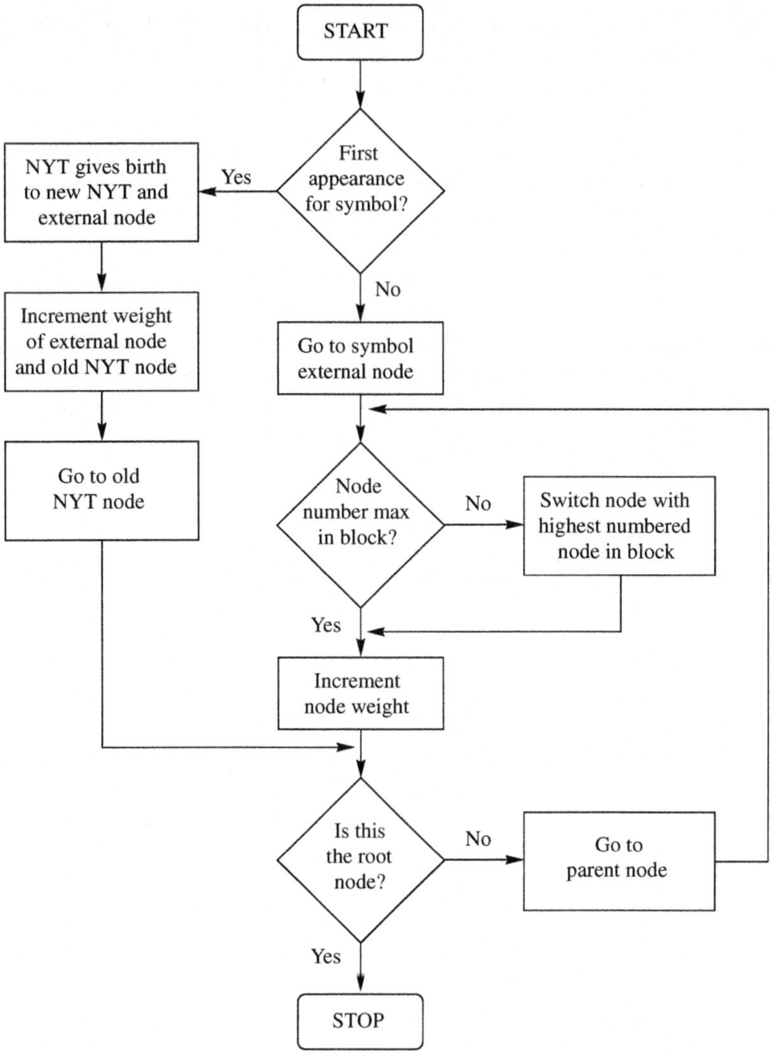

FIGURE 3.12 Update procedure for the adaptive Huffman coding algorithm.

tree at the transmitter is updated after each symbol is encoded, and the tree at the receiver is updated after each symbol is decoded. The procedure operates as follows.

After a symbol has been encoded or decoded, the external node corresponding to the symbol is examined to see if it has the largest node number in its block. If the external node does not have the largest node number, it is exchanged with the node that has the largest node number in the block, as long as the node with the higher number is not the parent of the node being updated. The weight of the external node is then incremented. If we did not exchange the nodes before the weight of the node was incremented, it is very likely that the ordering required by the sibling property would be destroyed. Once we have incremented the weight of the node, we have adapted the Huffman tree at that level. We then turn our attention to the next level by examining the parent node of the node whose weight was incremented to see if it has the largest number in its block. If it does not, it is exchanged with the node with the largest number in the block. Again, an exception to this is when the node with the higher node number is the parent of the node under consideration. Once an exchange has taken place (or it has been determined that there is no need for an exchange), the weight of the parent node is incremented. We then proceed to a new parent node and the process is repeated. This process continues until the root of the tree is reached.

If the symbol to be encoded or decoded has occurred for the first time, a new external node with a weight of zero is assigned to the symbol and a new NYT node is appended to the tree. Both the new external node and the new NYT node are offsprings of the old NYT node. We increment the weight of the new external node by one. As the old NYT node is the parent of the new external node, we increment its weight by one and then go on to update all the other nodes until we reach the root of the tree.

Example 3.4.1: Update Procedure

Assume we are encoding the message [a a r d v a r k], where our alphabet consists of the 26 lowercase letters of the English alphabet.

The updating process is shown in Figure 3.13. We begin with only the NYT node. The total number of nodes in this tree will be $2 \times 26 - 1 = 51$, so we start numbering backwards from 51 with the number of the root node being 51. The first letter to be transmitted is a. As a does not yet exist in the tree, we send a binary code 00000 for a and then add a to the tree. The NYT node gives birth to a new NYT node and a terminal node corresponding to a. The weight of the terminal node will be higher than the NYT node, so we assign the number 49 to the NYT node and 50 to the terminal node corresponding to the letter a. The second letter to be transmitted is also a. This time the transmitted code is 1. The node corresponding to a has the highest number (if we do not consider its parent), so we do not need to swap nodes. The next letter to be transmitted is r. This letter does not have a corresponding node on the tree, so we send the codeword for the NYT node, which is 0 followed by the index of r, which is 10001. The NYT node gives birth to a new NYT node and an external node corresponding to r. Again, no update is required. The next letter to be transmitted is d, which is also being sent for the first time. We again send the code for the NYT node, which is now 00 followed by the index for d, which is 00011. The NYT node again gives birth to two new nodes. However, an update is still not required. This changes with the transmission of the next letter, v, which has also not yet been encountered. Nodes 43 and 44 are added to the tree, with 44 as the terminal node corresponding to v. We examine the grandparent node of v (node 47) to see if it has the largest number in its block. As it does not, we swap it with node 48, which has the largest number in its block. We then increment node 48 and move to its parent, which is node 49. In the block containing node 49, the largest number belongs to node 50. Therefore, we swap nodes 49 and 50 and then increment node 50. We then move to the parent node of node 50, which is node 51. As this is the root node, all we do is increment node 51. ♦

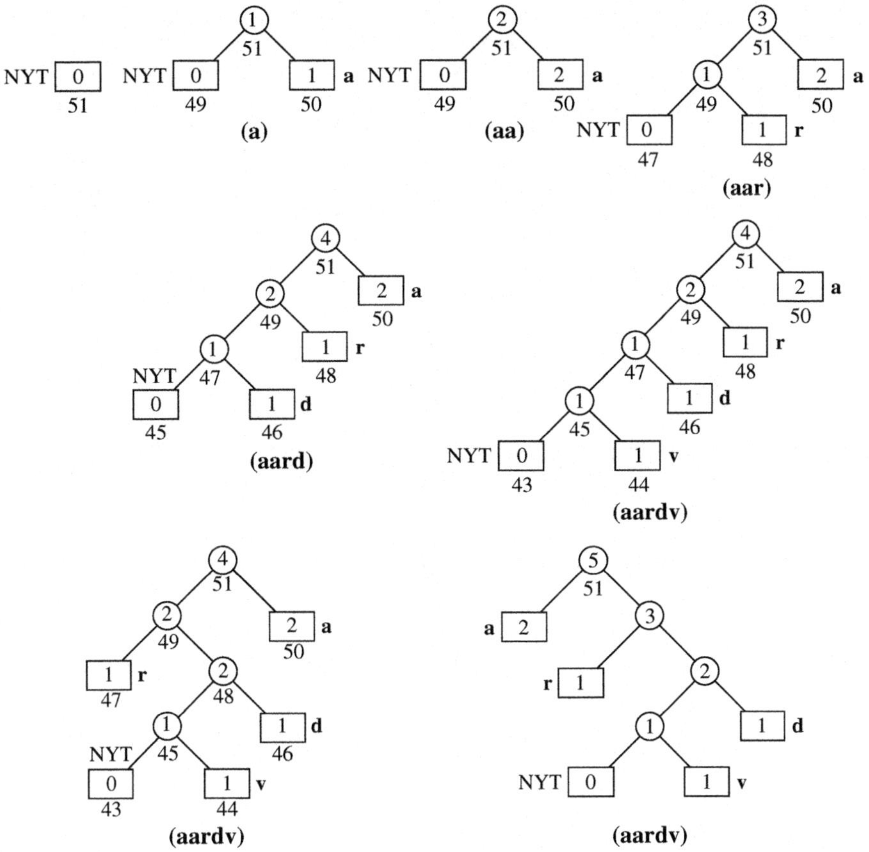

FIGURE 3.13 **Adaptive Huffman tree after [a a r d v] is processed.**

3.4.2 Encoding Procedure

The flowchart for the encoding procedure is shown in Figure 3.14. Initially, the tree at both the encoder and decoder consists of a single node, the NYT node. Therefore, the codeword for the very first symbol that appears is a previously agreed-upon fixed code. After the very first symbol, whenever we have to encode a symbol that is being encountered for the first time, we send the code for the NYT node, followed by the previously agreed-upon fixed code for the symbol. The code for the NYT node is obtained by traversing the Huffman tree from the root to the NYT node. This alerts the receiver to the fact that the symbol whose code follows does not as yet have a node in the Huffman tree. If a symbol to be encoded has a corresponding node in the tree, then the code for the symbol is generated by traversing the tree from the root to the external node corresponding to the symbol.

To see how the coding operation functions, we use the same example that was used to demonstrate the update procedure.

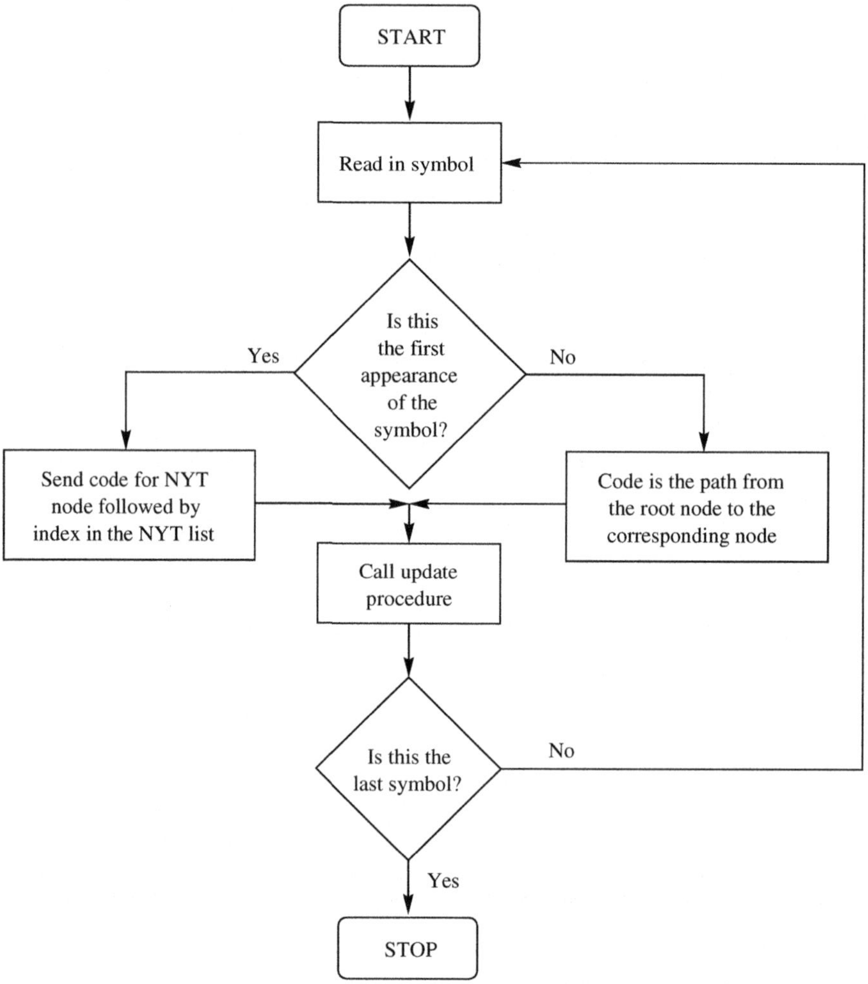

FIGURE 3.14 **Flowchart of the encoding procedure.**

Example 3.4.2: Encoding Procedure

In Example 3.4.1 we used an alphabet consisting of 26 letters. In order to obtain our prearranged code, we have to find m and e such that $2^e + r = 26$, where $0 \leqslant r < 2^e$. It is easy to see that the values of $e = 4$ and $r = 10$ satisfy this requirement.

The first symbol encoded is the letter a. As a is the first letter of the alphabet, $k = 1$. As 1 is less than 20, a is encoded as the 5-bit binary representation of $k - 1$, or 0, which is 00000. The Huffman tree is then updated as shown in the figure. The NYT node gives birth to an external node corresponding to the element a and a new NYT node. As a has occurred once, the external node corresponding to a has a weight of one. The weight of the NYT node is zero. The internal node also has a weight of one, as its weight is the sum of the weights of its offspring. The next symbol is again a. As we have an external node corresponding to symbol a, we simply traverse the tree from the root node to the external node corresponding to a in order to find the codeword. This traversal consists of a single right branch. Therefore, the Huffman code for the symbol a is 1.

After the code for a has been transmitted, the weight of the external node corresponding to a is incremented, as is the weight of its parent. The third symbol to be transmitted is r. As this is the first appearance of this symbol, we send the code for the NYT node followed by the previously arranged binary representation for r. If we traverse the tree from the root to the NYT node, we get a code of 0 for the NYT node. The letter r is the 18th letter of the alphabet; therefore, the binary representation of r is 10001. The code for the symbol r becomes 010001. The tree is again updated as shown in the figure, and the coding process continues with symbol d. Using the same procedure for d, the code for the NYT node, which is now 00, is sent, followed by the index for d, resulting in the codeword 0000011. The next symbol v is the 22nd symbol in the alphabet. As this is greater than 20, we send the code for the NYT node followed by the 4-bit binary representation of $22 - 10 - 1 = 11$. The code for the NYT node at this stage is 000, and the 4-bit binary representation of 11 is 1011; therefore, v is encoded as 0001011. The next symbol is a, for which the code is 0, and the encoding proceeds. ♦

3.4.3 Decoding Procedure

The flowchart for the decoding procedure is shown in Figure 3.15. As we read in the received binary string, we traverse the tree in a manner identical to that used in the encoding procedure. Once a leaf is encountered, the symbol corresponding to that leaf is decoded. If the leaf is the NYT node, then we check the next e bits to see if the resulting number is less than r. If it is less than r, we read in another bit to complete the code for the symbol. The index for the symbol is obtained by adding one to the decimal number corresponding to the e- or $e+1$-bit binary string. Once the symbol has been decoded, the tree is updated and the next received bit is used to start another traversal down the tree. To see how this procedure works, let us decode the binary string generated in the previous example.

Example 3.4.3: Decoding Procedure

The binary string generated by the encoding procedure is

$$000001010001000001100010110$$

Initially, the decoder tree consists only of the NYT node. Therefore, the first symbol to be decoded must be obtained from the NYT list. We read in the first 4 bits, 0000, as the value of e is four. The 4 bits 0000 correspond to the decimal value of 0. As this is less than the value of r, which is 10, we read in one more bit for the entire code of 00000. Adding one to the decimal value corresponding to this binary string, we get the index of the received symbol as 1. This is the index for a; therefore, the first letter is decoded as a. The tree is now updated as shown in Figure 3.13. The next bit in the string is 1. This traces a path from the root node to the external node corresponding to a. We decode the symbol a and update the tree. In this case, the update consists only of incrementing the weight of the external node corresponding

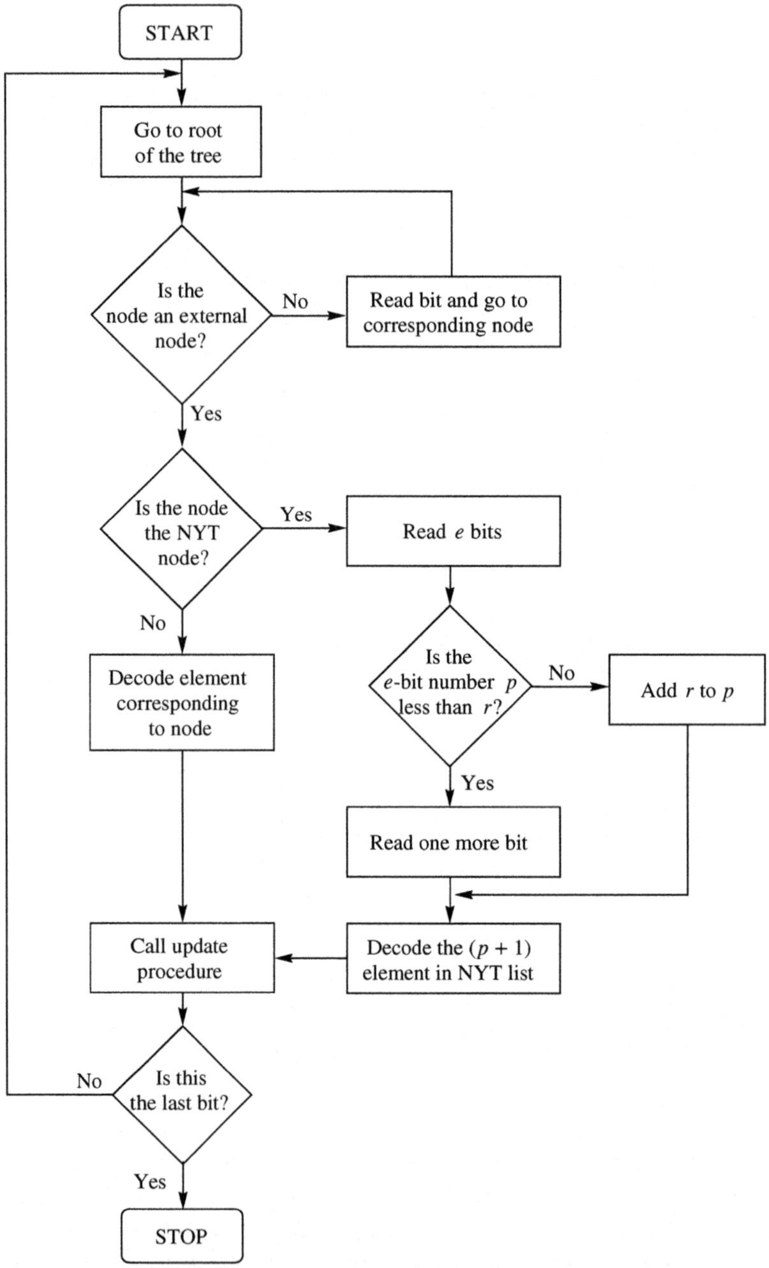

FIGURE 3.15 Flowchart of the decoding procedure.

to a. The next bit is a 0, which traces a path from the root to the NYT node. The next 4 bits, 1000, correspond to the decimal number 8, which is less than 10, so we read in one more bit to get the 5-bit word 10001. The decimal equivalent of this 5-bit word plus one is 18, which is the index for r. We decode the symbol r and then update the tree. The next 2 bits, 00, again trace a path to the NYT node. We read the next 4 bits, 0001. Since this corresponds to the decimal number 1, which is less than 10, we read another bit to get the 5-bit word 00011. To get the index of the received symbol in the NYT list, we add one to the decimal value of this 5-bit word. The value of the index is 4, which corresponds to the symbol d. Continuing in this fashion, we decode the sequence $aardva$. ♦

Although the Huffman coding algorithm is one of the best-known variable-length coding algorithms, there are some other lesser-known algorithms that can be very useful in certain situations. In particular, the Golomb-Rice codes and the Tunstall codes are becoming increasingly popular. We describe these codes in the following sections.

3.5 Golomb Codes

The Golomb-Rice codes belong to a family of codes designed to encode integers with the assumption that the larger an integer, the lower its probability of occurrence. The simplest code for this situation is the *unary* code. The unary code for a positive integer n is simply n 1s followed by a 0. Thus, the code for 4 is 11110, and the code for 7 is 11111110. The unary code is the same as the Huffman code for the semi-infinite alphabet $\{1, 2, 3, \ldots\}$ with probability model

$$P[k] = \frac{1}{2^k}$$

Because the Huffman code is optimal, the unary code is also optimal for this probability model.

Although the unary code is optimal in very restricted conditions, we can see that it is certainly very simple to implement. One step higher in complexity are a number of coding schemes that split the integer into two parts, representing one part with a unary code and the other part with a different code. An example of such a code is the Golomb code. Other examples can be found in [26].

The Golomb code is described in a succinct paper [27] by Solomon Golomb, which begins "Secret Agent 00111 is back at the Casino again, playing a game of chance, while the fate of mankind hangs in the balance." Agent 00111 requires a code to represent runs of success in a roulette game, and Golomb provides it! The Golomb code is actually a family of codes parameterized by an integer $m > 0$. In the Golomb code with parameter m, we represent an integer $n > 0$ using two numbers q and r, where

$$q = \left\lfloor \frac{n}{m} \right\rfloor$$

and

$$r = n - qm$$

TABLE 3.22 **Golomb code for $m = 5$.**

n	q	r	Codeword	n	q	r	Codeword
0	0	0	0000	08	1	3	101100
1	0	1	0010	09	1	4	101110
2	0	2	0100	10	2	0	110000
3	0	3	0110	11	2	1	110010
4	0	4	0111	12	2	2	110100
5	1	0	1000	13	2	3	110110
6	1	1	1001	14	2	4	110111
7	1	2	1010	15	3	0	111000

$\lfloor x \rfloor$ is the integer part of x. In other words, q is the quotient and r is the remainder when n is divided by m. The quotient q can take on values $0, 1, 2, \ldots$ and is represented by the unary code of q. The remainder r can take on the values $0, 1, 2, \ldots, m-1$. If m is a power of two, we use the $\log_2 m$-bit binary representation of r. If m is not a power of two, we could still use $\lceil \log_2 m \rceil$ bits, where $\lceil x \rceil$ is the smallest integer greater than or equal to x. We can reduce the number of bits required if we use the $\lfloor \log_2 m \rfloor$-bit binary representation of r for the first $2^{\lceil \log_2 m \rceil} - m$ values, and the $\lceil \log_2 m \rceil$-bit binary representation of $r + 2^{\lceil \log_2 m \rceil} - m$ for the rest of the values.

Example 3.5.1: Golomb Code

Let's design a Golomb code for $m = 5$. As

$$\lceil \log_2 5 \rceil = 3 \quad \text{and} \quad \lfloor \log_2 5 \rfloor = 2$$

the first $8 - 5 = 3$ values of r (that is, $r = 0, 1, 2$) will be represented by the 2-bit binary representation of r, and the next two values (that is, $r = 3, 4$) will be represented by the 3-bit representation of $r + 3$. The quotient q is always represented by the unary code for q. Thus, the codeword for 3 is 0110, and the codeword for 21 is 1111001. The codewords for $n = 0, \ldots, 15$ are shown in Table 3.22. ♦

It can be shown that the Golomb code is optimal for the probability model

$$P(n) = p^{n-1}q, \qquad q = 1 - p$$

when

$$m = \left\lceil -\frac{1}{\log_2 p} \right\rceil.$$

3.6 Rice Codes

The Rice code was originally developed by Robert F. Rice (he called it the Rice machine) [28,29] and later extended by Pen-Shu Yeh and Warner Miller [30]. The Rice code can be viewed as an adaptive Golomb code. In the Rice code, a sequence of nonnegative integers (which might have been obtained from the preprocessing of other data) is divided into blocks of J integers apiece. Each block is then coded using one of several options, most of which are a form of Golomb codes. Each block is encoded with each of these options, and the option resulting in the least number of coded bits is selected. The particular option used is indicated by an identifier attached to the code for each block.

The easiest way to understand the Rice code is to examine one of its implementations. We will study the implementation of the Rice code in the recommendation for lossless compression from the Consultative Committee on Space Data Standards (CCSDS).

3.6.1 CCSDS Recommendation for Lossless Compression

As an application of the Rice algorithm, let's briefly look at the algorithm for lossless data compression recommended by CCSDS. The algorithm consists of a preprocessor (modeling step) and a binary coder (coding step). The preprocessor removes correlation from the input and generates a sequence of nonnegative integers. This sequence has the property that smaller values are more probable than larger values. The binary coder generates a bitstream to represent the integer sequence. The binary coder is our main focus at this point.

The preprocessor functions as follows: given a sequence $\{y_i\}$, for each y_i we generate a prediction $\hat{y}_i$. A simple way to generate a prediction would be to take the previous value of the sequence to be a prediction of the current value of the sequence:

$$\hat{y}_i = y_{i-1}$$

We will look at more sophisticated ways of generating a prediction in Chapter 7. We then generate a sequence whose elements are the difference between y_i and its predicted value $\hat{y}_i$:

$$d_i = y_i - \hat{y}_i$$

The d_i value will have a small magnitude when our prediction is good and a large value when it is not. Assuming an accurate modeling of the data, the former situation is more likely than the latter. Let $y_{\max}$ and $y_{\min}$ be the largest and smallest values that the sequence $\{y_i\}$ takes on. It is reasonable to assume that the value of $\hat{y}_i$ will be confined to the range $[y_{\min}, y_{\max}]$. Define

$$T_i = \min\{y_{\max} - \hat{y}_i, \hat{y}_i - y_{\min}\} \tag{8}$$

The sequence $\{d_i\}$ can be converted into a sequence of nonnegative integers $\{x_i\}$ using the following mapping:

$$x_i = \begin{cases} 2d_i & 0 \leqslant d_i \leqslant T_i \\ 2\,|d_i| - 1 & -T_i \leqslant d_i < 0 \\ T_i + |d_i| & otherwise \end{cases} \tag{9}$$

The value of x_i will be small whenever the magnitude of d_i is small. Therefore, the value of x_i will be small with higher probability. The sequence $\{x_i\}$ is divided into segments with each segment being further divided into blocks of size J. It is recommended by CCSDS that J have a value of 16. Each block is then coded using one of the following options. The coded block is transmitted along with an identifier that indicates which particular option was used.

TABLE 3.23 Code used for zero block option.

Number of All-Zero Blocks	Codeword
1	1
2	01
3	001
4	0001
5	000001
6	0000001
⋮	⋮
63	$\overbrace{000\cdots0}^{63\ 0\text{s}}1$
ROS	00001

- **Fundamental sequence:** This is a unary code. A number n is represented by a sequence of n 0s followed by a 1 (or a sequence of n 1s followed by a 0).
- **Split sample options:** These options consist of a set of codes indexed by a parameter m. The code for a k-bit number n using the mth split sample option consists of the m least significant bits of k followed by a unary code representing the $k - m$ most significant bits. For example, suppose we wanted to encode the 8-bit number 23 using the third split sample option. The 8-bit representation of 23 is 00010111. The three least significant bits are 111. The remaining bits (00010) correspond to the number 2, which has a unary code 001. Therefore, the code for 23 using the third split sample option is 111011. Notice that different values of m will be preferable for different values of x_i, with higher values of m used for higher-entropy sequences.
- **Second extension option:** The second extension option is useful for sequences with low entropy—when, in general, many of the values of x_i will be zero. In the second extension option, the sequence is divided into consecutive pairs of samples. Each pair is used to obtain an index γ using the following transformation:

$$\gamma = \frac{1}{2}(x_i + x_{i+1})(x_i + x_{i+1} + 1) + x_{i+1} \tag{10}$$

  and the value of γ is encoded using a unary code. The value of γ is an index to a lookup table with each value of γ corresponding to a pair of values x_i, x_{i+1}.
- **Zero block option:** The zero block option is used when one or more of the blocks of x_i are zero—generally when we have long sequences of y_i that have the same value. In this case the number of zero blocks are transmitted using the code shown in Table 3.23. The ROS code is used when the last five or more blocks in a segment are all zero.

The Rice code has been used in several space applications, and variations of the Rice code have been proposed for a number of different applications.

TABLE 3.24 A 2-bit Tunstall code.

Sequence	Codeword
AAA	00
AAB	01
AB	10
B	11

3.7 Tunstall Codes

Most of the variable-length codes that we look at in this book encode letters from the source alphabet using codewords with varying numbers of bits: codewords with fewer bits for letters that occur more frequently and codewords with more bits for letters that occur less frequently. The Tunstall code is an important exception. In the Tunstall code, all codewords are of equal length. However, each codeword represents a different number of letters. An example of a 2-bit Tunstall code for an alphabet $\mathcal{A} = \{A, B\}$ is shown in Table 3.24. The main advantage of a Tunstall code is that errors in codewords do not propagate, unlike other variable-length codes, such as Huffman codes, in which an error in one codeword will cause a series of errors to occur.

Example 3.7.1:

Let's encode the sequence $AAABAABAABAABAAA$ using the code in Table 3.24. Starting at the left, we can see that the string AAA occurs in our codebook and has a code of 00. We then code B as 11, AAB as 01, and so on. We finally end up with the code 001101010100 for the sequence. ♦

The design of a code that has a fixed codeword length but a variable number of symbols per codeword should satisfy the following conditions:

1. We should be able to parse a source output sequence into sequences of symbols that appear in the codebook.
2. We should maximize the average number of source symbols represented by each codeword.

In order to understand what we mean by the first condition, consider the code shown in Table 3.25. Let's encode the same sequence $AAABAABAABAABAAA$ as in the previous example using the code in Table 3.25. We first encode AAA with the code 00. We then encode B with 11. The next three symbols are AAB. However, there are no codewords corresponding to this sequence of symbols. Thus, this sequence is unencodable using this particular code—not a desirable situation.

Tunstall [31] gives a simple algorithm that fulfills these conditions. The algorithm is as follows:

Suppose we want an n-bit Tunstall code for a source that generates *iid* letters from an alphabet of size N. The number of codewords is 2^n. We start with the N letters of the source

TABLE 3.25 A 2-bit (non-Tunstall) code.

Sequence	Codeword
AAA	00
ABA	01
AB	10
B	11

alphabet in our codebook. Remove the entry in the codebook that has the highest probability and add the N strings obtained by concatenating this letter with every letter in the alphabet (including itself). This will increase the size of the codebook from N to $N(N-1)$. The probabilities of the new entries will be the product of the probabilities of the letters concatenated to form the new entry. Now look through the $N+(N-1)$ entries in the codebook and find the entry that has the highest probability, keeping in mind that the entry with the highest probability may be a concatenation of symbols. Each time we perform this operation we increase the size of the codebook by $N-1$. Therefore, this operation can be performed K times, where

$$N + K(N-1) \leqslant 2^n$$

Example 3.7.2: Tunstall Codes

Let us design a 3-bit Tunstall code for a memoryless source with the following alphabet:

$$\mathcal{A} = \{A, B, C\}$$
$$P(A) = 0.6, \quad P(B) = 0.3, \quad P(C) = 0.1$$

We start out with the codebook and associated probabilities shown in Table 3.26. Since the letter A has the highest probability, we remove it from the list and add all two-letter strings beginning with A as shown in Table 3.27. After one iteration, we have five entries in our codebook. Going through one more iteration will increase the size of the codebook by two, and we will have seven entries, which is still less than the final codebook size. Going through another iteration after that would bring the codebook size to nine, which is greater than the maximum size of eight. Therefore, we will go through just one more iteration. Looking through the entries in Table 3.27, the entry with the highest probability is AA. Therefore, at the next step we remove AA and add all extensions of AA as shown in Table 3.28. The final 3-bit Tunstall code is shown in Table 3.28. ♦

TABLE 3.26 Source alphabet and associated probabilities.

Letter	Probability
A	0.60
B	0.30
C	0.10

TABLE 3.27 **The codebook after one iteration.**

Sequence	Probability
B	0.30
C	0.10
AA	0.36
AB	0.18
AC	0.06

TABLE 3.28 **A 3-bit Tunstall code.**

Sequence	Code
B	000
C	001
AB	010
AC	011
AAA	100
AAB	101
AAC	110

3.8 Applications of Huffman Coding

In this section, we describe some applications of Huffman coding. As we progress through the book, we will describe more applications, since Huffman coding is often used in conjunction with other coding techniques.

3.8.1 Lossless Image Compression

A simple application of Huffman coding to image compression would be to generate a Huffman code for the set of values that any pixel may take. For monochrome images, this set usually consists of integers from 0 to 255. Examples of such images are contained in the accompanying data sets. The four that we will use in the examples in this book are shown in Figure 3.16.

We will make use of one of the programs from the accompanying software (see Preface) to generate a Huffman code for each image and then encode the image using the Huffman code. The results for the four images in Figure 3.16 are shown in Table 3.29. The Huffman code is stored along with the compressed image as the code will be required by the decoder to reconstruct the image.

The original (uncompressed) image representation uses 8 bits/pixel. The image consists of 256 rows of 256 pixels, so the uncompressed representation uses 65,536 bytes. The compression ratio is simply the ratio of the number of bytes in the uncompressed representation to the number of bytes in the compressed representation. The number of bytes in the compressed representation includes the number of bytes needed to store the Huffman code. Notice that the compression ratio is different for different images. This can cause some problems in

FIGURE 3.16 Test images.

TABLE 3.29 Compression using Huffman codes on pixel values.

Image Name	Bits/Pixel	Total Size (bytes)	Compression Ratio
Sena	7.01	57,504	1.14
Sensin	7.49	61,430	1.07
Earth	4.94	40,534	1.62
Omaha	7.12	58,374	1.12

certain applications where it is necessary to know in advance how many bytes will be needed to represent a particular data set.

The results in Table 3.29 are somewhat disappointing because we get a reduction of only about $\frac{1}{2}$ to 1 bit/pixel after compression. For some applications, this reduction is acceptable. For example, if we were storing hundreds of thousands of images in an archive, a reduction of 1 bit/pixel saves many gigabytes in disk space. However, we can do better. Recall that when we first talked about compression, we said that the first step for any compression algorithm was to model the data so as to make use of the structure in the data. In this case, we have made absolutely no use of the structure in the data.

TABLE 3.30 Compression using Huffman codes on pixel difference values.

Image Name	Bits/Pixel	Total Size (bytes)	Compression Ratio
Sena	4.02	32,968	1.99
Sensin	4.70	38,541	1.70
Earth	4.13	33,880	1.93
Omaha	6.42	52,643	1.24

TABLE 3.31 Compression using adaptive Huffman codes on pixel difference values.

Image Name	Bits/Pixel	Total Size (bytes)	Compression Ratio
Sena	3.93	32,261	2.03
Sensin	4.63	37,896	1.73
Earth	4.82	39,504	1.66
Omaha	6.39	52,321	1.25

From a visual inspection of the test images, we can clearly see that the pixels in an image are heavily correlated with their neighbors. We could represent this structure with the crude model $\hat{x}_n = x_{n-1}$. The residual would be the difference between neighboring pixels. If we carry out this differencing operation and use the Huffman coder on the residuals, the results are as shown in Table 3.30. As we can see, using the structure in the data resulted in substantial improvement.

The results in Tables 3.29 and 3.30 were obtained using a two-pass system, in which the statistics were collected in the first pass and a Huffman table was generated using these statistics. The image was then encoded in the second pass. Instead of using a two-pass system, we could have used a one-pass adaptive Huffman coder. The results for this are given in Table 3.31.

Notice that there is little difference between the performance of the adaptive Huffman code and the two-pass Huffman coder. In addition, the fact that the adaptive Huffman coder can be used as an online or real-time coder makes the adaptive Huffman coder a more attractive option in many applications. However, the adaptive Huffman coder is more vulnerable to errors and may also be more difficult to implement. In the end, the particular application will determine which approach is more suitable.

3.8.2 Text Compression

Text compression seems natural for Huffman coding. In text, we have a discrete alphabet that, in a given class, has relatively stationary probabilities. For example, the probability model for a particular novel will not differ significantly from the probability model for another novel. Similarly, the probability model for a set of C programs is not going to be much different than

TABLE 3.32 Probabilities of occurrence of the letters in the English alphabet in the U.S. Constitution.

Letter	Probability	Letter	Probability
A	0.057305	N	0.056035
B	0.014876	O	0.058215
C	0.025775	P	0.021034
D	0.026811	Q	0.000973
E	0.112578	R	0.048819
F	0.022875	S	0.060289
G	0.009523	T	0.078085
H	0.042915	U	0.018474
I	0.053475	V	0.009882
J	0.002031	W	0.007576
K	0.001016	X	0.002264
L	0.031403	Y	0.011702
M	0.015892	Z	0.001502

TABLE 3.33 Probabilities of occurrence of the letters in the English alphabet in this chapter.

Letter	Probability	Letter	Probability
A	0.049855	N	0.048039
B	0.016100	O	0.050642
C	0.025835	P	0.015007
D	0.030232	Q	0.001509
E	0.097434	R	0.040492
F	0.019754	S	0.042657
G	0.012053	T	0.061142
H	0.035723	U	0.015794
I	0.048783	V	0.004988
J	0.000394	W	0.012207
K	0.002450	X	0.003413
L	0.025835	Y	0.008466
M	0.016494	Z	0.001050

the probability model for a different set of C programs. The probabilities in Table 3.32 are the probabilities of the 26 letters (upper- and lowercase) obtained for the U.S. Constitution and are representative of English text. The probabilities in Table 3.33 were obtained by counting the frequency of occurrences of letters in an earlier version of this chapter. While the two documents are substantially different, the two sets of probabilities are very much alike.

We encoded the earlier version of this chapter using Huffman codes that were created using the probabilities of occurrence obtained from the chapter. The file size dropped from about 70,000 bytes to about 43,000 bytes with Huffman coding.

While this reduction in file size is useful, we could have obtained better compression if we had first removed the structure existing in the form of correlation between the symbols in the file. Obviously, there is a substantial amount of correlation in this text. For example, *Huf* is always followed by *fman*! Unfortunately, this correlation is not amenable to simple numerical models, as was the case for the image files. However, there are other somewhat more complex techniques that can be used to remove the correlation in text files. We will look more closely at these in Chapters 5 and 6.

3.8.3 Audio Compression

Another class of data that is very suitable for compression is CD-quality audio data. The audio signal for each stereo channel is sampled at 44.1 kHz, and each sample is represented by 16 bits. This means that the amount of data stored on one CD is enormous. If we want to transmit this data, the amount of channel capacity required would be significant. Compression is definitely useful in this case. In Table 3.34 we show, for a variety of audio material, the file size, the entropy, the estimated compressed file size if a Huffman coder is used, and the resulting compression ratio.

The three segments used in this example represent a wide variety of audio material, from a symphonic piece by Mozart to a folk rock piece by Cohn. Even though the material is varied, Huffman coding can lead to some reduction in the capacity required to transmit this material.

Note that we have only provided the *estimated* compressed file sizes. The estimated file size in bits was obtained by multiplying the entropy by the number of samples in the file. We used this approach because the samples of 16-bit audio can take on 65,536 distinct values, and, therefore, the Huffman coder would require 65,536 distinct (variable-length) codewords. In most applications, a codebook of this size would not be practical. There is a way of handling large alphabets, called recursive indexing, that we will describe in Chapter 9. There is also some recent work [11] on using a Huffman tree in which leaves represent sets of symbols with the same probability. The codeword consists of a prefix that specifies the set followed by a suffix that specifies the symbol within the set. This approach can accommodate relatively large alphabets.

As with the other applications, we can obtain an increase in compression if we first remove the structure from the data. Audio data can be modeled numerically. In later chapters we will examine more sophisticated modeling approaches. For now, let us use the very simple model that was used in the image-coding example; that is, each sample has the same value as the previous sample. Using this model, we obtain the difference sequence. The entropy of the difference sequence is shown in Table 3.35.

TABLE 3.34 Huffman coding of 16-bit CD-quality audio.

File Name	Original File Size (bytes)	Entropy (bits)	Estimated Compressed File Size (bytes)	Compression Ratio
Mozart	939,862	12.8	725,420	1.30
Cohn	402,442	13.8	349,300	1.15
Mir	884,020	13.7	759,540	1.16

TABLE 3.35 Huffman coding of differences of 16-bit CD-quality audio.

File Name	Original File Size (bytes)	Entropy of Differences (bits)	Estimated Compressed File Size (bytes)	Compression Ratio
Mozart	939,862	09.7	569,792	1.65
Cohn	402,442	10.4	261,590	1.54
Mir	884,020	10.9	602,240	1.47

Note that there is a further reduction in the file size: the compressed file sizes are about 60% of the original files. Further reductions can be obtained by using more sophisticated models.

Many of the lossless audio compression schemes, including FLAC (Free Lossless Audio Codec), Apple's ALAC or ALE, *Shorten* [32], *Monkey's Audio*, and the MPEG-4 ALS [33] algorithms, use a linear predictive model to remove some of the structure from the audio sequence and then use Rice coding to encode the residuals. Most others, such as *AudioPak* [34] and *OggSquish*, use Huffman coding to encode the residuals.

3.9 Summary

In this chapter we began our exploration of data compression techniques with a description of the Huffman coding technique and several other related techniques. The Huffman coding technique and its variants are some of the most commonly used coding approaches. We will encounter modified versions of Huffman codes when we look at compression techniques for text, image, and video. In this chapter we described how to design Huffman codes and discussed some of the issues related to Huffman codes. We also described how adaptive Huffman codes work and looked briefly at some of the places where Huffman codes are used. We will see more of these in future chapters.

To explore further applications of Huffman coding, you can use the programs `huff_enc`, `huff_dec`, and `adap_huff` to generate your own Huffman codes for your favorite applications.

Further Reading

1. A detailed and very accessible overview of Huffman codes is provided in "Huffman Coding," by S. Pigeon [35], in *Lossless Compression Handbook.*
2. Details about nonbinary Huffman codes and a much more theoretical and rigorous description of variable-length codes can be found in *The Theory of Information and Coding*, Volume 3 of *Encyclopedia of Mathematics and Its Application*, by R.J. McEliece [9].
3. The tutorial article "Data Compression" in the September 1987 issue of *ACM Computing Surveys*, by D.A. Lelewer and D.S. Hirschberg [36], along with other material, provides a very nice brief coverage of the material in this chapter.

4. A somewhat different approach to describing Huffman codes can be found in *Data Compression—Methods and Theory*, by J.A. Storer [37].

5. A more theoretical but very readable account of variable-length coding can be found in *Elements of Information Theory*, by T.M. Cover and J.A. Thomas [38].

6. Although the book *Coding and Information Theory*, by R.W. Hamming [5], is mostly about channel coding, Huffman codes are described in some detail in Chapter 4.

3.10 Projects and Problems

1. The probabilities in Tables 3.32 and 3.33 were obtained using the program `countalpha` from the accompanying software. Use this program to compare probabilities for different types of text, C programs, messages on Internet forums, and so on. Comment on any differences you might see and describe how you would tailor your compression strategy for each type of text.

2. Use the programs `huff_enc` and `huff_dec` to do the following (in each case use the codebook generated by the image being compressed):

   (a) Code the Sena, Sensin, and Omaha images.
   (b) Write a program to take the difference between adjoining pixels, and then use `huffman` to code the difference images.
   (c) Repeat (a) and (b) using `adap_huff`.

   Report the resulting file sizes for each of these experiments and comment on the differences.

3. Using the programs `huff_enc` and `huff_dec`, code the Bookshelf1 and Sena images using the codebook generated by the Sensin image. Compare the results with the case where the codebook was generated by the image being compressed.

4. A source emits letters from an alphabet $\mathcal{A} = \{a_1, a_2, a_3, a_4, a_5\}$ with probabilities $P(a_1) = 0.15$, $P(a_2) = 0.04$, $P(a_3) = 0.26$, $P(a_4) = 0.05$, and $P(a_5) = 0.50$.

   (a) Calculate the entropy of this source.
   (b) Find a Huffman code for this source. Item(c)Find the average length of the code in (b) and its redundancy.

5. For an alphabet $\mathcal{A} = \{a_1, a_2, a_3, a_4\}$ with probabilities $P(a_1) = 0.1$, $P(a_2) = 0.3$, $P(a_3) = 0.25$, and $P(a_4) = 0.35$, find a Huffman code using the following:

   (a) The first procedure outlined in this chapter
   (b) The minimum variance procedure

   Comment on the difference in the Huffman codes.

6. In many communication applications, it is desirable that the number of 1s and 0s transmitted over the channel be about the same. However, if we look at Huffman codes,

many of them seem to have many more 1s than 0s or vice versa. Does this mean that Huffman coding will lead to inefficient channel usage? For the Huffman code obtained in Problem 3, find the probability that a 0 will be transmitted over the channel. What does this probability say about the question posed above?

7. For the source in Example 3.3.1, generate a ternary code by combining three letters in the first and second steps and two letters in the third step. Compare with the ternary code obtained in the example.

8. In Example 3.4.1, we showed how the tree develops when the sequence $aardv$ is transmitted. Continue this example with the next letters in the sequence, ark.

9. The Monte Carlo approach is often used for studying problems that are difficult to solve analytically. Let's use this approach to study the problem of buffering when using variable-length codes. We will simulate the situation in Example 3.2.1 and study the time to overflow and underflow as a function of the buffer size. In our program, we will need a random number generator, a set of seeds to initialize the random number generator, a counter B to simulate the buffer occupancy, a counter T to keep track of the time, and a value N, which is the size of the buffer. Input to the buffer is simulated by using the random number generator to select a letter from our alphabet. The counter B is then incremented by the length of the codeword for the letter. The output to the buffer is simulated by decrementing B by 2 except when T is divisible by 5. For values of T divisible by 5, decrement B by 3 instead of 2 (why?). Keep incrementing T, each time simulating an input and an output, until either $B \geqslant N$, corresponding to a buffer overflow, or $B < 0$, corresponding to a buffer underflow. When either of these events happens, record what happened and when, and restart the simulation with a new seed. Do this with at least 100 seeds.
Perform this simulation for a number of buffer sizes ($N = 100, 1000, 10{,}000$), and the two Huffman codes obtained for the source in Example 3.2.1. Describe your results in a report.

10. While the variance of lengths is an important consideration when choosing between two Huffman codes that have the same average lengths, it is not the only consideration. Another consideration is the ability to recover from errors in the channel. In this problem we will explore the effect of error on two equivalent Huffman codes.

(a) For the source and Huffman code of Example 3.2.1 (Table 3.5), encode the sequence

$$a_2 a_1 a_3 a_2 a_1 a_2$$

Suppose there was an error in the channel and the first bit was received as a 0 instead of a 1. Decode the received sequence of bits. How many characters are received in error before the first correctly decoded character?

(b) Repeat using the code in Table 3.9.

(c) Repeat parts (a) and (b) with the error in the third bit.

11. (This problem was suggested by P.F. Swaszek.)

(a) For a binary source with probabilities $P(0) = 0.9$, $P(1) = 0.1$, design a Huffman code for the source obtained by blocking m bits together, $m = 1, 2, \ldots, 8$. Plot the average lengths versus m. Comment on your result.

(b) Repeat for $P(0) = 0.99$, $P(1) = 0.01$.

You can use the program `huff_enc` to generate the Huffman codes.

12. Encode the following sequence of 16 values using the Rice code with $J = 8$ and one split sample option:

$$32, 33, 35, 39, 37, 38, 39, 40, 40, 40, 40, 39, 40, 40, 41, 40$$

For prediction use the previous value in the sequence

$$\hat{y}_i = y_{i-1}$$

and assume a prediction of zero for the first element of the sequence.

13. For an alphabet $\mathcal{A} = \{a_1, a_2, a_3\}$ with probabilities $P(a_1) = 0.7$, $P(a_2) = 0.2$, $P(a_3) = 0.1$, design a 3-bit Tunstall code.

14. Write a program for encoding images using the Rice algorithm. Use eight options, including the fundamental sequence, five split sample options, and the two low-entropy options. Use $J = 16$. For prediction use either the pixel to the left or the pixel above. Encode the Sena image using your program. Compare your results with the results obtained by Huffman coding the differences between pixels.

4

Arithmetic Coding

4.1 Overview

In the previous chapter, we saw one approach to generating variable-length codes. In this chapter, we see another, increasingly popular, method of generating variable-length codes called *arithmetic coding*. Arithmetic coding is especially useful when dealing with sources with small alphabets, such as binary sources, and alphabets with highly skewed probabilities. It is also a very useful approach when, for various reasons, the modeling and coding aspects of lossless compression are to be kept separate. In this chapter, we look at the basic ideas behind arithmetic coding, study some of the properties of arithmetic codes, and describe an implementation.

4.2 Introduction

In the last chapter, we studied the Huffman coding method, which guarantees a coding rate R within 1 bit of the entropy H. Recall that the coding rate is the average number of bits used to represent a symbol from a source, and, for a given probability model, the entropy is the lowest rate at which the source can be coded. We can tighten this bound somewhat. It has been shown [21] that the Huffman algorithm will generate a code whose rate is within $p_{max} + 0.086$ of the entropy, where p_{max} is the probability of the most frequently occurring symbol. We noted in the last chapter that, in applications where the alphabet size is large, p_{max} is generally quite small, and the amount of deviation from the entropy, especially in terms of a percentage of the rate, is quite small. However, in cases where the alphabet is small and the probability of occurrence of the different letters is skewed, the value of p_{max} can be quite large, and the Huffman code can become rather inefficient when compared to the entropy. One way to avoid this problem is to block more than one symbol together and generate an extended Huffman code. Unfortunately, this approach does not always work.

Example 4.2.1:

Consider a source that puts out independent, identically distributed (*iid*) letters from the alphabet $\mathcal{A} = \{a_1, a_2, a_3\}$ with the probability model $P(a_1) = 0.95$, $P(a_2) = 0.02$, and $P(a_3) = 0.03$. The entropy for this source is 0.335 bits/symbol. A Huffman code for this source is given in Table 4.1.

TABLE 4.1 Huffman code for three-letter alphabet.

Letter	Codeword
a_1	0
a_2	11
a_3	10

The average length for this code is 1.05 bits/symbol. The difference between the average code length and the entropy, or the redundancy, for this code is 0.715 bits/symbol, which is 213% of the entropy. This means that to code this sequence, we would need more than twice the number of bits promised by the entropy.

Recall Example 3.2.4. Here also we can group the symbols in blocks of two. The extended alphabet, probability model, and code can be obtained as shown in Table 4.2. The average rate for the extended alphabet is 1.222 bits/symbol, which in terms of the original alphabet is 0.611 bits/symbol. As the entropy of the source is 0.335 bits/symbol, the additional rate over the entropy is still about 82% of the entropy! By continuing to block symbols together, we find that the redundancy drops to acceptable values when we block eight symbols together. The corresponding alphabet size for this level of blocking is 6561! A code of this size is impractical for a number of reasons. Storage of a code like this requires memory that may not be available for many applications. While it may be possible to design reasonably efficient encoders, decoding a Huffman code of this size would be a highly inefficient and time-consuming procedure. Finally, if there were some perturbation in the statistics, and some of the assumed probabilities changed slightly, this would have a major impact on the efficiency of the code. ♦

We can see that it is more efficient to generate codewords for groups or sequences of symbols rather than to generate a separate codeword for each symbol in a sequence. However, this approach becomes impractical when we try to obtain Huffman codes for long sequences of symbols. In order to find the Huffman codeword for a particular sequence of length m, we need codewords for all possible sequences of length m. This fact causes an exponential growth in the size of the codebook. We need a way of assigning codewords to *particular*

TABLE 4.2 Huffman code for extended alphabet.

Letter	Probability	Code
a_1a_1	0.9025	0
a_1a_2	0.0190	111
a_1a_3	0.0285	100
a_2a_1	0.0190	1101
a_2a_2	0.0004	110011
a_2a_3	0.0006	110001
a_3a_1	0.0285	101
a_3a_2	0.0006	110010
a_3a_3	0.0009	110000

sequences without having to generate codes for all sequences of that length. The arithmetic coding technique fulfills this requirement.

In arithmetic coding, a unique identifier or tag is generated for the sequence to be encoded. This tag corresponds to a binary fraction, which becomes the binary code for the sequence. In practice, the generation of the tag and the binary code are the same process. However, the arithmetic coding approach is easier to understand if we conceptually divide the approach into two phases. In the first phase, a unique identifier or tag is generated for a given sequence of symbols. This tag is then given a unique binary code. A unique arithmetic code can be generated for a sequence of length m without the need for generating codewords for all sequences of length m. This is unlike the situation for Huffman codes. In order to generate a Huffman code for a sequence of length m, where the code is not a concatenation of the codewords for the individual symbols, we need to obtain the Huffman codes for all sequences of length m.

4.3 Coding a Sequence

In order to distinguish a sequence of symbols from another sequence of symbols we need to tag it with a unique identifier. One possible set of tags for representing sequences of symbols are the numbers in the unit interval [0,1). Because the number of numbers in the unit interval is infinite, it should be possible to assign a unique tag to each distinct sequence of symbols. In order to do this, we need a function that will map sequences of symbols into the unit interval. A function that maps random variables, and sequences of random variables, into the unit interval is the cumulative distribution function (*cdf*) of the random variable associated with the source. This is the function we will use in developing the arithmetic code. (If you are not familiar with random variables and cumulative distribution functions or need to refresh your memory, you may wish to look at Appendix A.)

The use of the cumulative distribution function to generate a binary code for a sequence has a rather interesting history. Shannon, in his original 1948 paper [3], mentioned an approach using the cumulative distribution function when describing what is now known as the Shannon-Fano code. Peter Elias, another member of Fano's first information theory class at MIT (this class also included Huffman), came up with a recursive implementation for this idea. However, he never published it, and we only know about it through a mention in a 1963 book on information theory by Abramson [39]. Abramson described this coding approach in a note to a chapter. In another book on information theory by Jelinek [40] in 1968, the idea of arithmetic coding is further developed, this time in an appendix, as an example of variable-length coding. Modern arithmetic coding owes its birth to the independent discoveries in 1976 of Pasco [41] and Rissanen [42] who solved the problem of finite precision. Finally, several papers appeared that provided practical arithmetic coding algorithms, the most well known of which is the paper by Rissanen and Langdon [43].

Before we begin our development of the arithmetic code, we need to establish some notation. Recall that a random variable maps the outcomes, or sets of outcomes, of an experiment to values on the real number line. For example, in a coin-tossing experiment, the random variable could map a head to zero and a tail to one (or it could map a head to 2367.5 and a tail to -192). To use this technique, we need to map the source symbols or letters to numbers. For convenience, in the discussion in this chapter we will use the mapping

$$X(a_i) = i \qquad a_i \in \mathcal{A} \tag{1}$$

where $\mathcal{A} = \{a_1, a_2, \ldots, a_m\}$ is the alphabet for a discrete source and X is a random variable. This mapping means that given a probability model $\mathcal{P}$ for the source, we also have a probability density function for the random variable

$$P(X = i) = P(a_i)$$

and the cumulative density function can be defined as

$$F_X(i) = \sum_{k=1}^{i} P(X = k)$$

Notice that for each symbol a_i with a nonzero probability, we have a distinct value of $F_X(i)$. We will use this fact in what follows to develop the arithmetic code. Our development may be more detailed than what you are looking for, at least on the first reading. If so, skip or skim Sections 4.3.1–4.4.1 and go directly to Section 4.4.2.

4.3.1 Generating a Tag

The procedure for generating the tag works by reducing the size of the interval in which the tag resides as more and more elements of the sequence are received.

We start out by first dividing the unit interval into subintervals of the form $[F_X(i-1), F_X(i))$, $i = 1, \ldots, m$. Because the minimum value of the *cdf* is zero and the maximum value is one, this exactly partitions the unit interval. We associate the subinterval $[F_X(i-1), F_X(i))$

with the symbol a_i. The appearance of the first symbol in the sequence restricts the interval containing the tag to one of these subintervals. Suppose the first symbol was a_k. Then the interval containing the tag value will be the subinterval $[F_X(k-1), F_X(k))$. This subinterval is now partitioned in exactly the same proportions as the original interval. That is, the jth interval corresponding to the symbol a_j is given by $[F_X(k-1) + F_X(j-1) \times (F_X(k) - F_X(k-1), F_X(k-1) + F_X(j) \times (F_X(k) - F_X(k-1))]$. So if the second symbol in the sequence is a_j, then the interval containing the tag value becomes $[F_X(k-1) + F_X(j-1) \times (F_X(k) - F_X(k-1), F_X(k-1) + F_X(j) \times (F_X(k) - F_X(k-1)))]$. Each succeeding symbol causes the tag to be restricted to a subinterval that is further partitioned in the same proportions. This process can be more clearly understood through an example.

Example 4.3.1:

Consider a three-letter alphabet $\mathcal{A} = \{a_1, a_2, a_3\}$ with $P(a_1) = 0.7$, $P(a_2) = 0.1$, and $P(a_3) = 0.2$. Using the mapping of Equation (1), $F_X(1) = 0.7$, $F_X(2) = 0.8$, and $F_X(3) = 1$. This partitions the unit interval as shown in Figure 4.1.

The partition in which the tag resides depends on the first symbol of the sequence being encoded. For example, if the first symbol is a_1, the tag lies in the interval [0.0, 0.7); if the first symbol is a_2, the tag lies in the interval [0.7, 0.8); and if the first symbol is a_3, the tag lies in the interval [0.8, 1.0). Once the interval containing the tag has been determined, the rest of the unit interval is discarded, and this restricted interval is again divided in the same proportions as the original interval. Suppose the first symbol was a_1. The tag would be contained in the subinterval [0.0, 0.7). This subinterval would then be subdivided into exactly the same proportions as the original interval, yielding the subintervals [0.0, 0.49), [0.49, 0.56), and [0.56, 0.7). The first partition, as before, would correspond to the symbol a_1, the second partition would correspond to the symbol a_2, and the third partition [0.56, 0.7) would correspond to

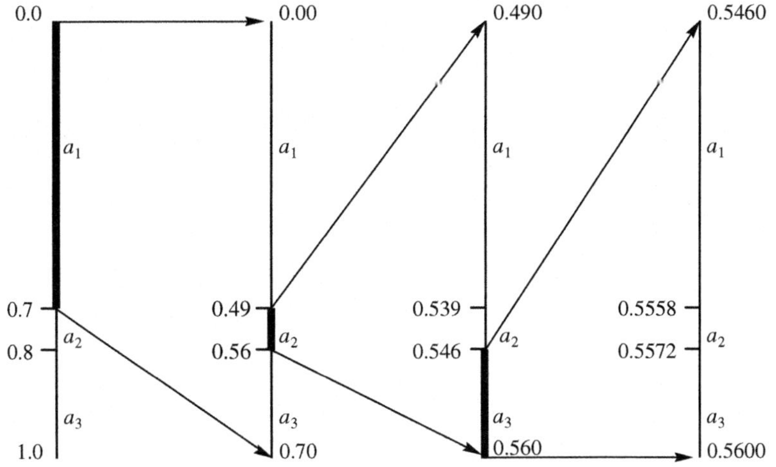

FIGURE 4.1 **Restricting the interval containing the tag for the input sequence $\{a_1, a_2, a_3, \ldots\}$.**

the symbol a_3. Suppose the second symbol in the sequence was a_2. The tag value would then be restricted to lie in the interval [0.49, 0.56). We would then partition this interval in the same proportion as the original interval to obtain the subintervals [0.49, 0.539) corresponding to the symbol a_1, [0.539, 0.546) corresponding to the symbol a_2, and [0.546, 0.56) corresponding to the symbol a_3. If the third symbol was a_3, the tag would be restricted to the interval [0.546, 0.56), which could then be subdivided further. This process is described graphically in Figure 4.1.

Notice that the appearance of each new symbol restricts the tag to a subinterval that is disjoint from any other subinterval that may have been generated using this process. For the sequence beginning with $a_1, a_2, a_3, \ldots$, by the time the third symbol a_3 is received, the tag has been restricted to the subinterval [0.546, 0.56). If the third symbol had been a_1 instead of a_3, the tag would have resided in the subinterval [0.49, 0.539), which is disjoint from the subinterval [0.546, 0.56). Even if the two sequences are identical from this point on (one starting with a_1, a_2, a_3 and the other beginning with a_1, a_2, a_1), the tag interval for the two sequences will always be disjoint. ♦

As we can see, the interval in which the tag for a particular sequence resides is disjoint from all intervals in which the tag for any other sequence may reside. As such, any member of this interval can be used as a tag. One popular choice is the lower limit of the interval; another possibility is the midpoint of the interval. For the moment, let's use the midpoint of the interval as the tag.

In order to see how the tag generation procedure works mathematically, we start with sequences of length one. Suppose we have a source that puts out symbols from some alphabet $\mathcal{A} = \{a_1, a_2, \ldots, a_m\}$. We can map the symbols $\{a_i\}$ to real numbers $\{i\}$. Define $\overline{T}_X(a_i)$ as

$$\overline{T}_X(a_i) = \sum_{k=1}^{i-1} P(X = k) + \frac{1}{2}P(X = i) \tag{2}$$

$$= F_X(i-1) + \frac{1}{2}P(X = i) \tag{3}$$

For each a_i, $\overline{T}_X(a_i)$ will have a unique value. This value can be used as a unique tag for a_i.

Example 4.3.2:

Consider a simple dice-throwing experiment with a fair die. The outcomes of a roll of the die can be mapped into the numbers $\{1, 2, \ldots, 6\}$. For a fair die

$$P(X = k) = \frac{1}{6} \quad \text{for } k = 1, 2, \ldots, 6$$

Therefore, using (3) we can find the tag for $X = 2$ as

$$\overline{T}_X(2) = P(X = 1) + \frac{1}{2}P(X = 2) = \frac{1}{6} + \frac{1}{12} = 0.25$$

and the tag for $X = 5$ as

$$\bar{T}_X(5) = \sum_{k=1}^{4} P(X = k) + \frac{1}{2}P(X = 5) = 0.75$$

The tags for all other outcomes are shown in Table 4.3 where the overbar denotes repeating decimal values.

TABLE 4.3 **Toss for outcomes in a dice-throwing experment.**

Outcome	Tag
1	$0.08\overline{33}$
3	$0.41\overline{66}$
4	$0.58\overline{33}$
6	$0.91\overline{66}$

♦

As we can see from the example above, giving a unique tag to a sequence of length one is an easy task. This approach can be extended to longer sequences by imposing an order on the sequences. We need to impose an order on the sequences because we will assign a tag to a particular sequence $\mathbf{x}_i$ as

$$\bar{T}_{\mathbf{X}}^{(m)}(\mathbf{x}_i) = \sum_{\mathbf{y} < \mathbf{x}_i} P(\mathbf{y}) + \frac{1}{2}P(\mathbf{x}_i) \quad (4)$$

where $\mathbf{y} < \mathbf{x}$ means that $\mathbf{y}$ precedes $\mathbf{x}$ in the ordering, and the superscript denotes the length of the sequence.

An easy ordering to use is *lexicographic ordering*. In lexicographic ordering, the ordering of letters in an alphabet induces an ordering on the words constructed from this alphabet. The ordering of words in a dictionary is a good (maybe the original) example of lexicographic ordering. *Dictionary order* is sometimes used as a synonym for lexicographic order.

Example 4.3.3:

We can extend Example 4.3.2 so that the sequence consists of two rolls of a die. Using the ordering scheme described above, the outcomes (in order) would be 11, 12, 13, ..., 66. The tags can then be generated using Equation (4). For example, the tag for the sequence 1 3 would be

$$\bar{T}_X(13) = P(\mathbf{x} = 11) + P(\mathbf{x} = 12) + 1/2P(\mathbf{x} = 13) \quad (5)$$

$$= 1/36 + 1/36 + 1/2(1/36) \quad (6)$$

$$= 5/72 \quad (7)$$

♦

Notice that to generate the tag for the sequence 1 3 we did not have to generate a tag for every other possible message. However, based on Equation (4) and Example 4.3.3, we need to know the probability of every sequence that is "less than" the sequence for which the tag is being generated. The requirement that the probability of all sequences of a given length be explicitly calculated can be as prohibitive as the requirement that we have codewords for all sequences of a given length. Fortunately, we shall see that to compute a tag for a given sequence of symbols, all we need is the probability of individual symbols, or the probability model.

Recall that, given our construction, the interval containing the tag value for a given sequence is disjoint from the intervals containing the tag values of all other sequences. This means that any value in this interval would be a unique identifier for x_i. Therefore, to fulfill our initial objective of uniquely identifying each sequence, it would be sufficient to compute the upper and lower limits of the interval containing the tag and select any value in that interval. The upper and lower limits can be computed recursively as shown in the following example.

Example 4.3.4:

We will use the alphabet of Example 4.3.2 and find the upper and lower limits of the interval containing the tag for the sequence 3 2 2. Assume that we are observing 3 2 2 in a sequential manner; that is, first we see 3, then 2, and then 2 again. After each observation, we will compute the upper and lower limits of the interval containing the tag of the sequence observed to that point. We will denote the upper limit by $u^{(n)}$ and the lower limit by $l^{(n)}$, where n denotes the length of the sequence.

We first observe 3. Therefore,

$$u^{(1)} = F_X(3), \qquad l^{(1)} = F_X(2)$$

We then observe 2, and the sequence is $\mathbf{x} = 32$. Therefore,

$$u^{(2)} = F_X^{(2)}(32), \qquad l^{(2)} = F_X^{(2)}(31)$$

We can compute these values as follows:

$$\begin{aligned} F_X^{(2)}(32) = {} & P(\mathbf{x} = 11) + P(\mathbf{x} = 12) + \cdots + P(\mathbf{x} = 16) \\ & + P(\mathbf{x} = 21) + P(\mathbf{x} = 22) + \cdots + P(\mathbf{x} = 26) \\ & + P(\mathbf{x} = 31) + P(\mathbf{x} = 32) \end{aligned}$$

But,

$$\sum_{i=1}^{i=6} P(\mathbf{x} = ki) = \sum_{i=1}^{i=6} P(x_1 = k, x_2 = i) = P(x_1 = k)$$

where $\mathbf{x} = x_1 x_2$. Therefore,

$$\begin{aligned} F_X^{(2)}(32) &= P(x_1 = 1) + P(x_1 = 2) + P(\mathbf{x} = 31) + P(\mathbf{x} = 32) \\ &= F_X(2) + P(\mathbf{x} = 31) + P(\mathbf{x} = 32) \end{aligned}$$

However, assuming each roll of the dice is independent of the others,

$$P(\mathbf{x}=31) = P(x_1 = 3)P(x_2 = 1)$$

and

$$P(\mathbf{x}=32) = P(x_1 = 3)P(x_2 = 2)$$

Therefore,

$$\begin{aligned} P(\mathbf{x}=31) + P(\mathbf{x}=32) &= P(x_1 = 3)(P(x_2 = 1) + P(x_2 = 2)) \\ &= P(x_1 = 3)F_X(2) \end{aligned}$$

Noting that

$$P(x_1 = 3) = F_X(3) - F_X(2)$$

we can write

$$P(\mathbf{x}=31) + P(\mathbf{x}=32) = (F_X(3) - F_X(2))F_X(2)$$

and

$$F_X^{(2)}(32) = F_X(2) + (F_X(3) - F_X(2))F_X(2)$$

We can also write this as

$$u^{(2)} = l^{(1)} + (u^{(1)} - l^{(1)})F_X(2)$$

We can similarly show that

$$F_X^{(2)}(31) = F_X(2) + (F_X(3) - F_X(2))F_X(1)$$

or

$$l^{(2)} = l^{(1)} + (u^{(1)} - l^{(1)})F_X(1)$$

The third element of the observed sequence is 2, and the sequence is $\mathbf{x} = 322$. The upper and lower limits of the interval containing the tag for this sequence are

$$u^{(3)} = F_X^{(3)}(322), \qquad l^{(3)} = F_X^{(3)}(321)$$

Using the same approach as above, we find that

$$\begin{aligned} F_X^{(3)}(322) &= F_X^{(2)}(31) + (F_X^{(2)}(32) - F_X^{(2)}(31))F_X(2) \\ F_X^{(3)}(321) &= F_X^{(2)}(31) + (F_X^{(2)}(32) - F_X^{(2)}(31))F_X(1) \end{aligned} \qquad (8)$$

or

$$\begin{aligned} u^{(3)} &= l^{(2)} + (u^{(2)} - l^{(2)})F_X(2) \\ l^{(3)} &= l^{(2)} + (u^{(2)} - l^{(2)})F_X(1) \end{aligned}$$

♦

In general, we can show that for any sequence $\mathbf{x} = (x_1 x_2 \ldots x_n)$

$$l^{(n)} = l^{(n-1)} + (u^{(n-1)} - l^{(n-1)})F_X(x_n - 1) \tag{9}$$

$$u^{(n)} = l^{(n-1)} + (u^{(n-1)} - l^{(n-1)})F_X(x_n) \tag{10}$$

Notice that throughout this process we did not explicitly need to compute any joint probabilities.

If we are using the midpoint of the interval for the tag, then

$$\overline{T}_X(\mathbf{x}) = \frac{u^{(n)} + l^{(n)}}{2}$$

Therefore, the tag for any sequence can be computed in a sequential fashion. The only information required by the tag generation procedure is the *cdf* of the source, which can be obtained directly from the probability model.

Example 4.3.5: Generating a Tag

Consider the source in Example 3.2.4. Define the random variable $X(a_i) = i$. Suppose we wish to encode the sequence **1 3 2 1**. From the probability model, we know that

$$F_X(k) = 0,\ k \leqslant 0, \quad F_X(1) = 0.8, \quad F_X(2) = 0.82, \quad F_X(3) = 1, \quad F_X(k) = 1,\ k > 3$$

We can use Equations (9) and (10) sequentially to determine the lower and upper limits of the interval containing the tag. Initializing $u^{(0)}$ to 1, and $l^{(0)}$ to 0, the first element of the sequence, **1**, results in the following update:

$$l^{(1)} = 0 + (1 - 0)0 = 0$$

$$u^{(1)} = 0 + (1 - 0)(0.8) = 0.8$$

That is, the tag is contained in the interval [0, 0.8). The second element of the sequence is **3**. Using the update equations we get

$$l^{(2)} = 0 + (0.8 - 0)F_X(2) = 0.8 \times 0.82 = 0.656$$

$$u^{(2)} = 0 + (0.8 - 0)F_X(3) = 0.8 \times 1.0 = 0.8$$

Therefore, the interval containing the tag for the sequence **1 3** is [0.656,0.8). The third element, **2**, results in the update equations

$$l^{(3)} = 0.656 + (0.8 - 0.656)F_X(1) = 0.656 + 0.144 \times 0.8 = 0.7712$$

$$u^{(3)} = 0.656 + (0.8 - 0.656)F_X(2) = 0.656 + 0.144 \times 0.82 = 0.77408$$

and the interval for the tag is [0.7712, 0.77408). Continuing with the last element, the upper and lower limits of the interval containing the tag are

$$l^{(4)} = 0.7712 + (0.77408 - 0.7712)F_X(0) = 0.7712 + 0.00288 \times 0.0 = 0.7712$$

$$u^{(4)} = 0.7712 + (0.77408 - 0.1152)F_X(1) = 0.7712 + 0.00288 \times 0.8 = 0.773504$$

and the tag for the sequence **1 3 2 1** can be generated as

$$\overline{T}_X(1321) = \frac{0.7712 + 0.773504}{2} = 0.772352$$

♦

Notice that each succeeding interval is contained in the preceding interval. If we examine the equations used to generate the intervals, we see that this will always be the case. This property will be used to decipher the tag. An undesirable consequence of this process is that the intervals get smaller and smaller and require higher precision as the sequence gets longer. To combat this problem, a rescaling strategy needs to be adopted. In Section 4.4.2, we will describe a simple rescaling approach that takes care of this problem.

4.3.2 Deciphering the Tag

We have spent a considerable amount of time showing how a sequence can be assigned a unique tag, given a minimal amount of information. However, the tag is useless unless we can also decipher it with minimal computational cost. Fortunately, deciphering the tag is as simple as generating it. We can see this most easily through an example.

Example 4.3.6: Deciphering a Tag

Given the tag obtained in Example 4.3.5, let's try to obtain the sequence represented by the tag. We will try to mimic the encoder in order to do the decoding. The tag value is 0.772352. The interval containing this tag value is a subset of every interval obtained in the encoding process. Our decoding strategy will be to decode the elements in the sequence in such a way that the upper and lower limits $u^{(k)}$ and $l^{(k)}$ will always contain the tag value for each k. We start with $l^{(0)} = 0$ and $u^{(0)} = 1$. After decoding the first element of the sequence x_1, the upper and lower limits become

$$\begin{aligned} l^{(1)} &= 0 + (1-0)F_X(x_1 - 1) = F_X(x_1 - 1) \\ u^{(1)} &= 0 + (1-0)F_X(x_1) = F_X(x_1) \end{aligned}$$

In other words, the interval containing the tag is $[F_X(x_1 - 1), F_X(x_1))$. We need to find the value of x_1 for which 0.772352 lies in the interval $[F_X(x_1 - 1), F_X(x_1))$. If we pick $x_1 = \mathbf{1}$, the interval is [0, 0.8). If we pick $x_1 = \mathbf{2}$, the interval is [0.8, 0.82), and if we pick $x_1 = \mathbf{3}$, the interval is [0.82, 1.0). As 0.772352 lies in the interval [0.0, 0.8], we choose $x_1 = \mathbf{1}$. We now repeat this procedure for the second element x_2, using the updated values of $l^{(1)}$ and $u^{(1)}$:

$$\begin{aligned} l^{(2)} &= 0 + (0.8 - 0)F_X(x_2 - 1) = 0.8F_X(x_2 - 1) \\ u^{(2)} &= 0 + (0.8 - 0)F_X(x_2) = 0.8F_X(x_2) \end{aligned}$$

If we pick $x_2 = \mathbf{1}$, the updated interval is [0, 0.64), which does not contain the tag. Therefore, x_2 cannot be **1**. If we pick $x_2 = \mathbf{2}$, the updated interval is [0.64, 0.656), which also does not contain the tag. If we pick $x_2 = \mathbf{3}$, the updated interval is [0.656, 0.8), which does contain the tag value of 0.772352. Therefore, the second element in the sequence is **3**. Knowing the second element of the sequence, we can update the values of $l^{(2)}$ and $u^{(2)}$ and find the element x_3, which will give us an interval containing the tag:

$$\begin{aligned} l^{(3)} &= 0.656 + (0.8 - 0.656)F_X(x_3 - 1) = 0.656 + 0.144 \times F_X(x_3 - 1) \\ u^{(3)} &= 0.656 + (0.8 - 0.656)F_X(x_3) = 0.656 + 0.144 \times F_X(x_3) \end{aligned}$$

However, the expressions for $l^{(3)}$ and $u^{(3)}$ are cumbersome in this form. To make the comparisons more easily, we could subtract the value of $l^{(2)}$ from both the limits and the tag. That is, we find the value of x_3 for which the interval $[0.144 \times F_X(x_3 - 1), 0.144 \times F_X(x_3))$ contains $0.772352 - 0.656 = 0.116352$. Or, we could make this even simpler and divide the residual tag value of 0.116352 by 0.144 to get 0.808 and find the value of x_3 for which 0.808 falls in the interval $[F_X(x_3 - 1), F_X(x_3))$. We can see that the only value of x_3 for which this is possible is **2**. Substituting **2** for x_3 in the update equations, we can update the values of $l^{(3)}$ and $u^{(3)}$. We can now find the element x_4 by computing the upper and lower limits as

$$l^{(4)} = 0.7712 + (0.77408 - 0.7712)F_X(x_4 - 1) = 0.7712 + 0.00288 \times F_X(x_4 - 1)$$
$$u^{(4)} = 0.7712 + (0.77408 - 0.1152)F_X(x_4) = 0.7712 + 0.00288 \times F_X(x_4)$$

Again, we can subtract $l^{(3)}$ from the tag to get $0.772352 - 0.7712 = 0.001152$ and find the value of x_4 for which the interval $[0.00288 \times F_X(x_4 - 1), 0.00288 \times F_X(x_4))$ contains 0.001152. To make the comparisons simpler, we can divide the residual value of the tag by 0.00288 to get 0.4 and find the value of x_4 for which 0.4 is contained in $[F_X(x_4 - 1), F_X(x_4))$. We can see that the value is $x_4 = \mathbf{1}$, and we have decoded the entire sequence. Note that we knew the length of the sequence beforehand, and, therefore, we knew when to stop. ♦

From the example above, we can deduce an algorithm that can decipher the tag.

1. Initialize $l^{(0)} = 0$ and $u^{(0)} = 1$.
2. For each k find $t^* = (tag - l^{(k-1)})/(u^{(k-1)} - l^{(k-1)})$.
3. Find the value of x_k for which $F_X(x_k - 1) \leqslant t^* < F_X(x_k)$.
4. Update $u^{(k)}$ and $l^{(k)}$.
5. Continue until the entire sequence has been decoded.

There are two ways to know when the entire sequence has been decoded. The decoder may know the length of the sequence, in which case the deciphering process is stopped when that many symbols have been obtained. The second way to know if the entire sequence has been decoded is that a particular symbol is denoted as an end-of-transmission symbol. The decoding of this symbol would bring the decoding process to a close.

4.4 Generating a Binary Code

Using the algorithm described in the previous section, we can obtain a tag for a given sequence **x**. However, the *binary code* for the sequence is what we really want to know. We want to find a binary code that will represent the sequence **x** in a unique and efficient manner.

We have said that the tag forms a unique representation for the sequence. This means that the binary representation of the tag forms a unique binary code for the sequence. However, we have placed no restrictions on what values in the unit interval the tag can take. The binary representation of some of these values would be infinitely long, in which case, although the code is unique, it may not be efficient. To make the code efficient, the binary representation has to be truncated. But if we truncate the representation, is the resulting code still unique? Finally, is the resulting code efficient? How far or how close is the average number of bits per symbol from the entropy? We will examine all these questions in the next section.

Even if we show the code to be unique and efficient, the method described to this point is highly impractical. In Section 4.4.2, we will describe a more practical algorithm for generating the arithmetic code for a sequence. We will give an integer implementation of this algorithm in Section 4.4.3.

4.4.1 Uniqueness and Efficiency of the Arithmetic Code

$\overline{T}_X(x)$ is a number in the interval [0, 1). A binary code for $\overline{T}_X(x)$ can be obtained by taking the binary representation of this number and truncating it to $l(x) = \lceil \log \frac{1}{P(x)} \rceil + 1$ bits. Recall that the binary representations of decimal numbers in the interval [0, 1) are obtained as the negative powers of two. The decimal equivalent of the binary number $.b_1b_2b_3 \ldots b_k$ is $b_1 2^{-1} + b_2 2^{-2} + b_3 2^{-3} + \cdots + b_k 2^{-k}$. Thus $.101 = 2^{-1} + 2^{-3}$.

Example 4.4.1:

Consider a source $\mathcal{A}$ that generates letters from an alphabet of size four,

$$\mathcal{A} = \{a_1, a_2, a_3, a_4\}$$

with probabilities

$$P(a_1) = \frac{1}{2}, \quad P(a_2) = \frac{1}{4}, \quad P(a_3) = \frac{1}{8}, \quad P(a_4) = \frac{1}{8}$$

A binary code for this source can be generated as shown in Table 4.4. The quantity $\overline{T}_x$ is obtained using Equation (3). The binary representation of $\overline{T}_x$ is truncated to $\lceil \log \frac{1}{P(x)} \rceil + 1$ bits to obtain the binary code.

TABLE 4.4 A binary code for a four-letter alphabet.

Symbol	F_X	$\overline{T}_X$	In Binary	$\lceil \log \frac{1}{P(x)} \rceil + 1$	Code
1	.500	.2500	.0100	2	01
2	.750	.6250	.1010	3	101
3	.875	.8125	.1101	4	1101
4	1.000	.9375	.1111	4	1111

♦

We will show that a code obtained in this fashion is a uniquely decodable code. We first show that this code is unique, and then we will show that it is uniquely decodable. We will use $\lfloor \overline{T}_X(\mathbf{x}) \rfloor_{l(\mathbf{x})}$ to denote the truncation of $\overline{T}_X(x)$ to l(**x**) bits.

Recall that while we have been using $\bar{T}_X(x)$ as the tag for a sequence $\mathbf{x}$, any number in the interval $[F_X(\mathbf{x}-1), F_X(\mathbf{x}))$ would be a unique identifier. Therefore, to show that the code $\lfloor \bar{T}_X(\mathbf{x}) \rfloor_{l(\mathbf{x})}$ is unique, all we need to do is show that it is contained in the interval $[F_X(\mathbf{x}-1), F_X(\mathbf{x}))$. Because we are truncating the binary representation of $\bar{T}_X(\mathbf{x})$ to obtain $\lfloor \bar{T}_X(\mathbf{x}) \rfloor_{l(\mathbf{x})}$, $\lfloor \bar{T}_X(\mathbf{x}) \rfloor_{l(\mathbf{x})}$ is less than or equal to $\bar{T}_X(\mathbf{x})$. More specifically,

$$0 \leqslant \bar{T}_X(\mathbf{x}) - \lfloor \bar{T}_X(\mathbf{x}) \rfloor_{l(\mathbf{x})} < \frac{1}{2^{l(\mathbf{x})}} \tag{11}$$

As $\bar{T}_X(\mathbf{x})$ is strictly less than $F_X(\mathbf{x})$,

$$\lfloor \bar{T}_X(\mathbf{x}) \rfloor_{l(\mathbf{x})} < F_X(\mathbf{x})$$

To show that $\lfloor \bar{T}_X(\mathbf{x}) \rfloor_{l(\mathbf{x})} \geqslant F_X(\mathbf{x}-1)$, note that

$$\begin{aligned}
\frac{1}{2^{l(\mathbf{x})}} &= \frac{1}{2^{\lceil \log \frac{1}{P(x)} \rceil + 1}} \\
&< \frac{1}{2^{\log \frac{1}{P(x)} + 1}} \\
&= \frac{1}{2 \frac{1}{P(x)}} \\
&= \frac{P(\mathbf{x})}{2}
\end{aligned}$$

From (3) we have

$$\frac{P(\mathbf{x})}{2} = \bar{T}_X(\mathbf{x}) - F_X(\mathbf{x}-1)$$

Therefore,

$$\bar{T}_X(\mathbf{x}) - F_X(\mathbf{x}-1) > \frac{1}{2^{l(\mathbf{x})}} \tag{12}$$

Combining (11) and (12), we have

$$\lfloor \bar{T}_X(\mathbf{x}) \rfloor_{l(\mathbf{x})} > F_X(\mathbf{x}-1) \tag{13}$$

Therefore, the code $\lfloor \bar{T}_X(\mathbf{x}) \rfloor_{l(\mathbf{x})}$ is a unique representation of $\bar{T}_X(\mathbf{x})$.

To show that this code is uniquely decodable, we will show that the code is a prefix code; that is, no codeword is a prefix of another codeword. Because a prefix code is always uniquely decodable, this is enough to show that an arithmetic code is uniquely decodable. Given a number a in the interval $[0, 1)$ with an n-bit binary representation $[b_1 b_2 \ldots b_n]$, for any other number b to have a binary representation with $[b_1 b_2 \ldots b_n]$ as the prefix, b has to lie in the interval $[a, a + \frac{1}{2^n})$. (See Problem 1 at the end of this chapter.)

If $\mathbf{x}$ and $\mathbf{y}$ are two distinct sequences, we know that $\lfloor \bar{T}_X(\mathbf{x}) \rfloor_{l(\mathbf{x})}$ and $\lfloor \bar{T}_X(\mathbf{y}) \rfloor_{l(\mathbf{y})}$ lie in two *disjoint* intervals, $[F_X(\mathbf{x}-1), F_X(\mathbf{x}))$ and $[F_X(\mathbf{y}-1), F_X(\mathbf{y}))$. Therefore, if we can show that for any sequence $\mathbf{x}$ the interval $[\lfloor \bar{T}_X(\mathbf{x}) \rfloor_{l(\mathbf{x})}, \lfloor \bar{T}_X(\mathbf{x}) \rfloor_{l(\mathbf{x})} + \frac{1}{2^{l(\mathbf{x})}})$ lies entirely within the interval $[F_X(\mathbf{x}-1), F_X(\mathbf{x}))$, the code for one sequence cannot be the prefix for the code for another sequence.

We have already shown that $\lfloor \bar{T}_X(\mathbf{x}) \rfloor_{l(\mathbf{x})} > F_X(\mathbf{x}-1)$. Therefore, all we need to do is show that

$$F_X(\mathbf{x}) - \lfloor \bar{T}_X(\mathbf{x}) \rfloor_{l(\mathbf{x})} > \frac{1}{2^{l(\mathbf{x})}}$$

This is true because

$$\begin{aligned} F_X(\mathbf{x}) - \lfloor \bar{T}_X(\mathbf{x}) \rfloor_{l(\mathbf{x})} &> F_X(\mathbf{x}) - \bar{T}_X(\mathbf{x}) \\ &= \frac{P(\mathbf{x})}{2} \\ &> \frac{1}{2^{l(\mathbf{x})}} \end{aligned}$$

This code is prefix free; and by taking the binary representation of $\bar{T}_X(\mathbf{x})$ and truncating it to $l(x) = \lceil \log \frac{1}{P(x)} \rceil + 1$ bits, we obtain a uniquely decodable code.

Although the code is uniquely decodable, how efficient is it? We have shown that the number of bits $l(\mathbf{x})$ required to represent $F_X(\mathbf{x})$ with enough accuracy such that the code for different values of $\mathbf{x}$ is distinct is

$$l(\mathbf{x}) = \left\lceil \log \frac{1}{P(\mathbf{x})} \right\rceil + 1$$

Remember that $l(\mathbf{x})$ is the number of bits required to encode the *entire* sequence $\mathbf{x}$. So, the average length of an arithmetic code for a sequence of length m is given by

$$l_{A^{(m)}} = \sum P(\mathbf{x}) l(\mathbf{x}) \tag{14}$$

$$= \sum P(\mathbf{x}) \left[\left\lceil \log \frac{1}{P(\mathbf{x})} \right\rceil + 1 \right] \tag{15}$$

$$< \sum P(\mathbf{x}) \left[\log \frac{1}{P(\mathbf{x})} + 1 + 1 \right] \tag{16}$$

$$= -\sum P(\mathbf{x}) \log P(\mathbf{x}) + 2 \sum P(\mathbf{x}) \tag{17}$$

$$= H(X^m) + 2 \tag{18}$$

Given that the average length is always greater than the entropy, the bounds on $l_{A^{(m)}}$ are

$$H(X^{(m)}) \leqslant l_{A^{(m)}} < H(X^{(m)}) + 2.$$

The length per symbol, l_A, or rate of the arithmetic code is $\frac{l_{A^{(m)}}}{m}$. Therefore, the bounds on l_A are

$$\frac{H(X^{(m)})}{m} \leqslant l_A < \frac{H(X^{(m)})}{m} + \frac{2}{m} \tag{19}$$

We have shown in Chapter 3 that for *iid* sources

$$H(X^{(m)}) = mH(X) \tag{20}$$

Therefore,

$$H(X) \leqslant l_A < H(X) + \frac{2}{m} \tag{21}$$

By increasing the length of the sequence, we can guarantee a rate as close to the entropy as we desire.

4.4.2 Algorithm Implementation

In Section 4.3.1, we developed a recursive algorithm for the boundaries of the interval containing the tag for the sequence being encoded as

$$l^{(n)} = l^{(n-1)} + (u^{(n-1)} - l^{(n-1)})F_X(x_n - 1) \tag{22}$$

$$u^{(n)} = l^{(n-1)} + (u^{(n-1)} - l^{(n-1)})F_X(x_n) \tag{23}$$

where x_n is the value of the random variable corresponding to the nth observed symbol, $l^{(n)}$ is the lower limit of the tag interval at the nth iteration, and $u^{(n)}$ is the upper limit of the tag interval at the nth iteration.

Before we can implement this algorithm, there is one major problem we have to resolve. Recall that the rationale for using numbers in the interval [0, 1) as a tag was that there are an infinite number of numbers in this interval. However, in practice, the number of numbers that can be uniquely represented on a machine is limited by the maximum number of digits (or bits) we can use for representing the number. Consider the values of $l^{(n)}$ and $u^{(n)}$ in Example 4.3.5. As n gets larger, these values come closer and closer together. This means that in order to represent all of the subintervals uniquely, we need increasing precision as the length of the sequence increases. In a system with finite precision, the two values are bound to converge, and we will lose all information about the sequence from the point at which the two values converged. To avoid this situation, we need to rescale the interval. However, we have to do it in a way that will preserve the information that is being transmitted. We would also like to perform the encoding *incrementally*—that is, to transmit portions of the code as the sequence is being observed, rather than wait until the entire sequence has been observed before transmitting the first bit. The algorithm we describe in this section takes care of the problems of synchronized rescaling and incremental encoding.

As the interval becomes narrower, we have three possibilities:

1. The interval is entirely confined to the lower half of the unit interval [0, 0.5).
2. The interval is entirely confined to the upper half of the unit interval [0.5, 1.0).
3. The interval straddles the midpoint of the unit interval.

We will look at the third case a little later in this section. First, let us examine the first two cases. Once the interval is confined to either the upper or lower half of the unit interval, it is forever confined to that half of the unit interval. The most significant bit of the binary representation of all numbers in the interval [0, 0.5) is 0, and the most significant bit of the binary representation of all numbers in the interval [0.5, 1] is 1. Therefore, once the interval gets restricted to either the upper or lower half of the unit interval, the most significant bit of the tag is fully determined. Thus, without waiting to see what the rest of the sequence looks like, we can indicate to the decoder whether the tag is confined to the upper or lower half of the unit interval by sending a 1 for the upper half and a 0 for the lower half. The bit that we send is also the first bit of the tag.

Once the encoder and decoder know which half contains the tag, we can ignore the half of the unit interval not containing the tag and concentrate on the half containing the tag. As our arithmetic is of finite precision, we can do this best by mapping the half interval containing the tag to the full [0,1) interval. The mappings required are

$$E_1 : [0, 0.5) \to [0, 1); \qquad E_1(x) = 2x \tag{24}$$
$$E_2 : [0.5, 1) \to [0, 1); \qquad E_2(x) = 2(x - 0.5) \tag{25}$$

As soon as we perform either of these mappings, we lose all information about the most significant bit. However, this should not matter because we have already sent that bit to the decoder. We can now continue with this process, generating another bit of the tag every time the tag interval is restricted to either half of the unit interval. This process of generating the bits of the tag without waiting to see the entire sequence is called incremental encoding.

Example 4.4.2: Tag Generation with Scaling

Let's revisit Example 4.3.5. Recall that we wish to encode the sequence **1321**. The probability model for the source is $P(a_1) = 0.8$, $P(a_2) = 0.02$, $P(a_3) = 0.18$. Initializing $u^{(0)}$ to 1, and $l^{(0)}$ to 0, the first element of the sequence, **1**, results in the following update:

$$l^{(1)} = 0 + (1 - 0)0 = 0$$
$$u^{(1)} = 0 + (1 - 0)(0.8) = 0.8$$

The interval [0, 0.8) is not confined to either the upper or lower half of the unit interval, so we proceed.

The second element of the sequence is **3**. This results in the update

$$l^{(2)} = 0 + (0.8 - 0)F_X(2) = 0.8 \times 0.82 = 0.656$$
$$u^{(2)} = 0 + (0.8 - 0)F_X(3) = 0.8 \times 1.0 = 0.8$$

The interval [0.656, 0.8) is contained entirely in the upper half of the unit interval, so we send the binary code 1 and rescale:

$$l^{(2)} = 2 \times (0.656 - 0.5) = 0.312$$
$$u^{(2)} = 2 \times (0.8 - 0.5) = 0.6$$

The third element, **2**, results in the following update equations:

$$l^{(3)} = 0.312 + (0.6 - 0.312)F_X(1) = 0.312 + 0.288 \times 0.8 = 0.5424$$
$$u^{(3)} = 0.312 + (0.8 - 0.312)F_X(2) = 0.312 + 0.288 \times 0.82 = 0.54816$$

The interval for the tag is [0.5424,0.54816), which is contained entirely in the upper half of the unit interval. We transmit a 1 and go through another rescaling:

$$l^{(3)} = 2 \times (0.5424 - 0.5) = 0.0848$$
$$u^{(3)} = 2 \times (0.54816 - 0.5) = 0.09632$$

This interval is contained entirely in the lower half of the unit interval, so we send a 0 and use the E_1 mapping to rescale:

$$l^{(3)} = 2 \times (0.0848) = 0.1696$$
$$u^{(3)} = 2 \times (0.09632) = 0.19264$$

The interval is still contained entirely in the lower half of the unit interval, so we send another 0 and go through another rescaling:

$$l^{(3)} = 2 \times (0.1696) = 0.3392$$
$$u^{(3)} = 2 \times (0.19264) = 0.38528$$

Because the interval containing the tag remains in the lower half of the unit interval, we send another 0 and rescale one more time:

$$l^{(3)} = 2 \times 0.3392 = 0.6784$$
$$u^{(3)} = 2 \times 0.38528 = 0.77056$$

Now the interval containing the tag is contained entirely in the upper half of the unit interval. Therefore, we transmit a 1 and rescale using the E_2 mapping:

$$l^{(3)} = 2 \times (0.6784 - 0.5) = 0.3568$$
$$u^{(3)} = 2 \times (0.77056 - 0.5) = 0.54112$$

At each stage, we are transmitting the most significant bit that is the same in both the upper and lower limit of the tag interval. If the most significant bits in the upper and lower limit are the same, then the value of this bit will be identical to the most significant bit of the tag. Therefore, by sending the most significant bits of the upper and lower endpoint of the tag whenever they are identical, we are actually sending the binary representation of the tag. The rescaling operations can be viewed as left shifts, which make the second most significant bit the most significant bit.

Continuing with the last element, the upper and lower limits of the interval containing the tag are

$$l^{(4)} = 0.3568 + (0.54112 - 0.3568)F_X(0) = 0.3568 + 0.18422 \times 0.0 = 0.3568$$
$$u^{(4)} = 0.3568 + (0.54112 - 0.3568)F_X(1) = 0.3568 + 0.18422 \times 0.8 = 0.504256$$

At this point, if we wished to stop encoding, all we need to do is inform the receiver of the final status of the tag value. We can do so by sending the binary representation of any value in the final tag interval. Generally, this value is taken to be $l^{(n)}$. In this particular example, it is convenient to use the value of 0.5. The binary representation of 0.5 is .10 . . . . Thus, we would transmit a 1 followed by as many 0s as required by the word length of the implementation being used. ♦

Notice that the tag interval size at this stage is approximately 64 times the size it was when we were using the unmodified algorithm. Therefore, this technique solves the finite precision problem. As we shall soon see, the bits that we have been sending with each mapping constitute the tag itself, which satisfies our desire for incremental encoding. The binary sequence generated during the encoding process in the previous example is 1100011. We could simply treat this as the binary expansion of the tag. A binary number .1100011 corresponds to the decimal number 0.7734375. Looking back to Example 4.3.5, notice that this number lies within the final tag interval. Therefore, we could use this to decode the sequence.

However, we would like to do incremental decoding as well as incremental encoding. This raises three questions:

1. How do we start decoding?
2. How do we continue decoding?
3. How do we stop decoding?

The second question is the easiest to answer. Once we have started decoding, all we have to do is mimic the encoder algorithm. That is, once we have started decoding, we know how to continue decoding. To begin the decoding process, we need to have enough information to decode the first symbol unambiguously. In order to guarantee unambiguous decoding, the number of bits received should point to an interval smaller than the smallest tag interval. Based on the smallest tag interval, we can determine how many bits we need before we start the decoding procedure. We will demonstrate this procedure in Example 4.4.4. First let's look at other aspects of decoding using the message from Example 4.4.2.

Example 4.4.3:

We will use a word length of 6 for this example. Note that because we are dealing with real numbers, this word length may not be sufficient for a different sequence. As in the encoder, we start with initializing $u^{(0)}$ to 1 and $l^{(0)}$ to 0. The sequence of received bits is $110001100\ldots0$. The first 6 bits correspond to a tag value of 0.765625, which means that the first element of the sequence is **1**, resulting in the following update:

$$\begin{aligned} l^{(1)} &= 0 + (1-0)0 = 0 \\ u^{(1)} &= 0 + (1-0)(0.8) = 0.8 \end{aligned}$$

The interval [0, 0.8) is not confined to either the upper or lower half of the unit interval, so we proceed. The tag 0.765625 lies in the top 18% of the interval [0, 0.8); therefore, the second element of the sequence is **3**. Updating the tag interval we get

$$\begin{aligned} l^{(2)} &= 0 + (0.8-0)F_X(2) = 0.8 \times 0.82 = 0.656 \\ u^{(2)} &= 0 + (0.8-0)F_X(3) = 0.8 \times 1.0 = 0.8 \end{aligned}$$

The interval [0.656, 0.8) is contained entirely in the upper half of the unit interval. At the encoder, we sent the bit 1 and rescaled. At the decoder, we will shift 1 out of the receive buffer and move the next bit in to make up the 6 bits in the tag. We will also update the tag interval, resulting in

$$\begin{aligned} l^{(2)} &= 2 \times (0.656 - 0.5) = 0.312 \\ u^{(2)} &= 2 \times (0.8 - 0.5) = 0.6 \end{aligned}$$

while shifting a bit to give us a tag of 0.546875. When we compare this value with the tag interval, we can see that this value lies in the 80–82% range of the tag interval, so we decode the next element of the sequence as **2**. We can then update the equations for the tag interval as

$$l^{(3)} = 0.312 + (0.6 - 0.312)F_X(1) = 0.312 + 0.288 \times 0.8 = 0.5424$$
$$u^{(3)} = 0.312 + (0.8 - 0.312)F_X(2) = 0.312 + 0.288 \times 0.82 = 0.54816$$

As the tag interval is now contained entirely in the upper half of the unit interval, we rescale using E_2 to obtain

$$l^{(3)} = 2 \times (0.5424 - 0.5) = 0.0848$$
$$u^{(3)} = 2 \times (0.54816 - 0.5) = 0.09632$$

We also shift out a bit from the tag and shift in the next bit. The tag is now 000110. The interval is contained entirely in the lower half of the unit interval. Therefore, we apply E_1 and shift another bit. The lower and upper limits of the tag interval become

$$l^{(3)} = 2 \times (0.0848) = 0.1696$$
$$u^{(3)} = 2 \times (0.09632) = 0.19264$$

and the tag becomes 001100. The interval is still contained entirely in the lower half of the unit interval, so we shift out another 0 to get a tag of 011000 and go through another rescaling:

$$l^{(3)} = 2 \times (0.1696) = 0.3392$$
$$u^{(3)} = 2 \times (0.19264) = 0.38528$$

Because the interval containing the tag remains in the lower half of the unit interval, we shift out another 0 from the tag to get 110000 and rescale one more time:

$$l^{(3)} = 2 \times 0.3392 = 0.6784$$
$$u^{(3)} = 2 \times 0.38528 = 0.77056$$

Now the interval containing the tag is contained entirely in the upper half of the unit interval. Therefore, we shift out a 1 from the tag and rescale using the E_2 mapping:

$$l^{(3)} = 2 \times (0.6784 - 0.5) = 0.3568$$
$$u^{(3)} = 2 \times (0.77056 - 0.5) = 0.54112$$

Now we compare the tag value to the tag interval to decode our final element. The tag is 100000, which corresponds to 0.5. This value lies in the first 80% of the interval, so we decode this element as **1**. ♦

If the tag interval is entirely contained in the upper or lower half of the unit interval, the scaling procedure described will prevent the interval from continually shrinking. Now we consider the case where the diminishing tag interval straddles the midpoint of the unit interval. As our trigger for rescaling, we check to see if the tag interval is contained in the interval [0.25, 0.75).

This will happen when $l^{(n)}$ is greater than 0.25 and $u^{(n)}$ is less than 0.75. When this happens, we double the tag interval using the following mapping:

$$E_3 : [0.25, 0.75) \rightarrow [0, 1); \qquad E_3(x) = 2(x - 0.25) \tag{26}$$

We have used a 1 to transmit information about an E_2 mapping, and a 0 to transmit information about an E_1 mapping. How do we transfer information about an E_3 mapping to the decoder? We use a somewhat different strategy in this case. At the time of the E_3 mapping, we do not send any information to the decoder; instead, we simply record the fact that we have used the E_3 mapping at the encoder. Suppose that after this, the tag interval gets confined to the upper half of the unit interval. At this point we would use an E_2 mapping and send a 1 to the receiver. Note that the tag interval at this stage is at least twice what it would have been if we had not used the E_3 mapping. Furthermore, the upper limit of the tag interval would have been less than 0.75. Therefore, if the E_3 mapping had not taken place right before the E_2 mapping, the tag interval would have been contained entirely in the lower half of the unit interval. At this point we would have used an E_1 mapping and transmitted a 0 to the receiver. In fact, the effect of the earlier E_3 mapping can be mimicked at the decoder by following the E_2 mapping with an E_1 mapping. At the encoder, right after we send a 1 to announce the E_2 mapping, we send a 0 to help the decoder track the changes in the tag interval at the decoder. If the first rescaling after the E_3 mapping happens to be an E_1 mapping, we do exactly the opposite. That is, we follow the 0 announcing an E_1 mapping with a 1 to mimic the effect of the E_3 mapping at the encoder.

What happens if we have to go through a series of E_3 mappings at the encoder? We simply keep track of the number of E_3 mappings and then send that many bits of the opposite variety after the first E_1 or E_2 mapping. If we went through three E_3 mappings at the encoder, followed by an E_2 mapping, we would transmit a 1 followed by three 0s. On the other hand, if we went through an E_1 mapping after the E_3 mappings, we would transmit a 0 followed by three 1s. Since the decoder mimics the encoder, the E_3 mappings are also applied at the decoder when the tag interval is contained in the interval [0.25, 0.75).

4.4.3 Integer Implementation

We have described a floating-point implementation of arithmetic coding. Let us now repeat the procedure using integer arithmetic and generate the binary code in the process.

Encoder Implementation

The first thing we have to do is decide on the word length to be used. Given a word length of m, we map the important values in the [0,1) interval to the range of 2^m binary words. The point 0 gets mapped to

$$\overbrace{00\ldots0}^{m\text{times}}$$

and 1 gets mapped to

$$\overbrace{11\ldots1}^{m\text{times}}$$

The value of 0.5 gets mapped to

$$1\overbrace{00\ldots0}^{m-1\text{times}}$$

The update equations remain almost the same as Equations (9) and (10). As we are going to do integer arithmetic, we need to replace $F_X(x)$ in these equations.

Define n_j as the number of times the symbol j occurs in a sequence of length *Total_Count*. Then $F_X(k)$ can be estimated by

$$F_X(k) = \frac{\sum_{i=1}^{k} n_i}{Total\_Count}. \tag{27}$$

If we now define

$$Cum\_Count(k) = \sum_{i=1}^{k} n_i$$

we can write Equations (9) and (10) as

$$l^{(n)} = l^{(n-1)} + \left\lfloor \frac{(u^{(n-1)} - l^{(n-1)} + 1) \times Cum\_Count(x_n - 1)}{Total\_Count} \right\rfloor \tag{28}$$

$$u^{(n)} = l^{(n-1)} + \left\lfloor \frac{(u^{(n-1)} - l^{(n-1)} + 1) \times Cum\_Count(x_n)}{Total\_Count} \right\rfloor - 1 \tag{29}$$

where x_n is the nth symbol to be encoded, $\lfloor x \rfloor$ is the largest integer less than or equal to x, and the addition and subtraction of one is to handle the effects of the integer arithmetic. Note that in order to reduce effects of finite precision arithmetic it is a good idea to perform the multiplication in the numerator prior to the division.

The word length m has to be large enough to accommodate all the upper and lower limits of all the subintervals, which means that there should be enough values to unambiguously represent each entry of the *Cum_Count* array. As the maximum number of distinct values is *Total_Count* this means that the number of values that need to be represented is *Total_Count* and, therefore, we need to pick m such that

$$2^m > Total\_Count$$

or

$$m > \lceil \log_2(Total\_Count) \rceil$$

However, this may not be sufficient as often the active interval, that is the interval $[l^{(n)}, u^{(n)}]$ is only a portion of the total range available. As all the subintervals need to be contained within the active interval at any given time, what we need to do is determine the smallest size the active interval can be and then make sure that m is large enough to contain *Total_Count* different values in this limited range. So what is the smallest the active interval can be? At first sight it might seem that the smallest the active interval can be is about half the maximum range because as soon as the upper limit slips below the halfway mark or the lower limit slips above the halfway mark we double the interval. However, on closer examination, we can see that this is not the case. When the upper limit is barely above the halfway mark, the lower limit is not required to be at 0. In fact, the lower limit can be just below the quarter range without any rescaling being triggered. However, the moment the lower limit goes above the quarter mark an E_3 rescaling is triggered and the interval is redoubled one or more times. Thus the smallest the active range can be is a quarter of the total range. This means that we need to accommodate *Total_Count* values in a quarter of the total range available, that is

$$\frac{1}{4}2^m > Total\_Count$$

or

$$\begin{aligned} m &> \lceil \log_2(4 \times Total\_Count) \rceil \\ &= 2 + \lceil \log_2(Total\_Count) \rceil \end{aligned}$$

Because of the way we mapped the endpoints and the halfway points of the unit interval, when both $l^{(n)}$ and $u^{(n)}$ are in either the upper half or lower half of the interval, the leading bit of $u^{(n)}$ and $l^{(n)}$ will be the same. If the leading or most significant bit (MSB) is 1, then the tag interval is contained entirely in the upper half of the $[00\ldots0, 11\ldots1]$ interval. If the MSB is 0, then the tag interval is contained entirely in the lower half. Applying the E_1 and E_2 mappings is a simple matter. All we do is shift out the MSB and then shift in a 1 into the integer code for $u^{(n)}$ and a 0 into the code for $l^{(n)}$. For example, suppose m was 6, $u^{(n)}$ was 54, and $l^{(n)}$ was 33. The binary representations of $u^{(n)}$ and $l^{(n)}$ are 110110 and 100001, respectively. Notice that the MSB for both endpoints is 1. Following the procedure above, we would shift out (and transmit or store) the 1, and shift in 1 for $u^{(n)}$ and 0 for $l^{(n)}$, obtaining a new value for $u^{(n)}$ as 101101, or 45, and a new value for $l^{(n)}$ as 000010, or 2. This is equivalent to performing the E_2 mapping. We can see how the E_1 mapping would also be performed using the same operation

To see if the E_3 mapping needs to be performed, we monitor the second most significant bit of $u^{(n)}$ and $l^{(n)}$. When the second most significant bit of $u^{(n)}$ is 0 and the second most significant bit of $l^{(n)}$ is 1, this means that the tag interval lies in the middle half of the $[00\ldots0, 11\ldots1]$ interval. To implement the E_3 mapping, we complement the second most significant bit in $u^{(n)}$ and $l^{(n)}$, and shift left, shifting in a 1 in $u^{(n)}$ and a 0 in $l^{(n)}$. We also keep track of the number of E_3 mappings in Scale3.

We can summarize the encoding algorithm using the following pseudocode:

Initialize l and u.
Get symbol.

$$l \leftarrow l + \left\lfloor \frac{(u - l + 1) \times Cum\_Count(x-1)}{Total\_Count} \right\rfloor$$

$$u \leftarrow l + \left\lfloor \frac{(u - l + 1) \times Cum\_Count(x)}{Total\_Count} \right\rfloor - 1$$

```
while (MSB of u and l are both equal to b or E3 condition holds)
if (MSB of u and l are both equal to b)
  {
    send b
    shift l to the left by 1 bit and shift 0 into LSB
    shift u to the left by 1 bit and shift 1 into LSB
    while(Scale3 > 0)
      {
        send complement of b
        decrement Scale3
        }
  }
if (E3 condition holds)
   {
     shift l to the left by 1 bit and shift 0 into LSB
     shift u to the left by 1 bit and shift 1 into LSB
     complement (new) MSB of l and u
     increment Scale3
     }
```

To see how all this functions together, let's look at an example.

Example 4.4.4:

We will encode the sequence **1 3 2 1** with the parameters shown in Table 4.5. First we need to select the word length m. Note that *Cum_Count(1)* and *Cum_Count(2)* differ by only 1. Recall that the values of *Cum_Count* will get translated to the endpoints of the subintervals. We want to make sure that the value we select for the word length will allow enough range for it to be possible to represent the smallest difference between the endpoints of intervals. We always rescale whenever the interval gets small. In order to make sure that the endpoints of the intervals always remain distinct, we need to make sure that all values in the range from 0 to $Total\_Count$, which is the same as *Cum_Count(3)*, are uniquely represented in the smallest range an interval under consideration can be without triggering a rescaling. The interval is smallest without triggering a rescaling when $l^{(n)}$ is just below the midpoint of the interval and $u^{(n)}$ is at three-quarters of the interval, or when $u^{(n)}$ is right at the midpoint of the interval and $l^{(n)}$ is just below a quarter of the interval. That is, the smallest the interval $[l^{(n)}, u^{(n)}]$ can be is one-quarter of the total available range of 2^m values. Thus, m should be large enough to uniquely accommodate the set of values between 0 and *Total_Count*.

For this example, this means that the total interval range has to be greater than 200. A value of $m = 8$ satisfies this requirement.With this value of m we have

$$l^{(0)} = 0 = (00000000)_2 \tag{30}$$

$$u^{(0)} = 255 = (11111111)_2 \tag{31}$$

where $(\cdots)_2$ is the binary representation of a number.

TABLE 4.5 **Values of some of the parameters for arithmetic coding example.**

$Count(1) = 40$	$Cum\_Count(0) = 0$	Scale3 = 0
$Count(2) = 1$	$Cum\_Count(1) = 40$	
$Count(3) = 9$	$Cum\_Count(2) = 41$	
$Total\_Count = 50$	$Cum\_Count(3) = 50$	

The first element of the sequence to be encoded is **1**. Using Equations (9) and (10)

$$l^{(1)} = 0 + \left\lfloor \frac{256 \times Cum\_Count(0)}{50} \right\rfloor = 0 = (00000000)_2 \tag{32}$$

$$u^{(1)} = 0 + \left\lfloor \frac{256 \times Cum\_Count(1)}{50} \right\rfloor - 1 = 203 = (11001011)_2 \tag{33}$$

The next element of the sequence is **3**:

$$l^{(2)} = 0 + \left\lfloor \frac{204 \times Cum\_Count(2)}{50} \right\rfloor = 167 = (10100111)_2 \tag{34}$$

$$u^{(2)} = 0 + \left\lfloor \frac{204 \times Cum\_Count(3)}{50} \right\rfloor - 1 = 203 = (11001011)_2 \tag{35}$$

The MSBs of $l^{(2)}$ and $u^{(2)}$ are both 1. Therefore, we shift this value out and send it to the decoder. All other bits are shifted left by 1 bit, giving

$$l^{(2)} = (01001110)_2 = 78 \tag{36}$$

$$u^{(2)} = (10010111)_2 = 151 \tag{37}$$

Notice that while the MSBs of the limits are different, the second MSB of the upper limit is 0, while the second MSB of the lower limit is 1. This is the condition for the E_3 mapping. We complement the second MSB of both limits and shift 1 bit to the left, shifting in a 0 as the least significant bit (LSB) of $l^{(2)}$ and a 1 as the LSB of $u^{(2)}$. This gives us

$$l^{(2)} = (00011100)_2 = 28 \tag{38}$$

$$u^{(2)} = (10101111)_2 = 175 \tag{39}$$

We also increment Scale3 to a value of 1.

The next element in the sequence is **2**. Updating the limits, we have

$$l^{(3)} = 28 + \left\lfloor \frac{148 \times Cum\_Count(1)}{50} \right\rfloor = 146 = (10010010)_2 \tag{40}$$

$$u^{(3)} = 28 + \left\lfloor \frac{148 \times Cum\_Count(2)}{50} \right\rfloor - 1 = 148 = (10010100)_2 \tag{41}$$

The two MSBs are identical, so we shift out a 1 and shift left by 1 bit:

$$l^{(3)} = (00100100)_2 = 36 \tag{42}$$

$$u^{(3)} = (00101001)_2 = 41 \tag{43}$$

As Scale3 is 1, we transmit a 0 and decrement Scale3 to 0. The MSBs of the upper and lower limits are both 0, so we shift out and transmit 0:

$$l^{(3)} = (01001000)_2 = 72 \tag{44}$$
$$u^{(3)} = (01010011)_2 = 83 \tag{45}$$

Both MSBs are again 0, so we shift out and transmit 0:

$$l^{(3)} = (10010000)_2 = 144 \tag{46}$$
$$u^{(3)} = (10100111)_2 = 167 \tag{47}$$

Now both MSBs are 1, so we shift out and transmit a 1. The limits become

$$l^{(3)} = (00100000)_2 = 32 \tag{48}$$
$$u^{(3)} = (01001111)_2 = 79 \tag{49}$$

Once again the MSBs are the same. This time we shift out and transmit a 0.

$$l^{(3)} = (01000000)_2 = 64 \tag{50}$$
$$u^{(3)} = (10011111)_2 = 159 \tag{51}$$

Now the MSBs are different. However, the second MSB for the lower limit is 1 while the second MSB for the upper limit is 0. This is the condition for the E_3 mapping. Applying the E_3 mapping by complementing the second MSB and shifting 1 bit to the left, we get

$$l^{(3)} = (00000000)_2 = 0 \tag{52}$$
$$u^{(3)} = (10111111)_2 = 191 \tag{53}$$

We also increment Scale3 to 1.

The next element in the sequence to be encoded is **1**. Therefore,

$$l^{(4)} = 0 + \left\lfloor \frac{192 \times Cum\_Count(0)}{50} \right\rfloor = 0 = (00000000)_2 \tag{54}$$
$$u^{(4)} = 0 + \left\lfloor \frac{192 \times Cum\_Count(1)}{50} \right\rfloor - 1 = 152 = (10011000)_2 \tag{55}$$

The encoding continues in this fashion. To this point we have generated the binary sequence 1100010. If we wish to terminate the encoding at this point, we have to send the current status of the tag. This can be done by sending the value of the lower limit $l^{(4)}$. As $l^{(4)}$ is 0, we will end up sending eight 0s. However, Scale3, at this point, is 1. Therefore, after we send the first 0 from the value of $l^{(4)}$, we need to send a 1 before sending the remaining seven 0s. The final transmitted sequence is 1100010010000000. ♦

Decoder Implementation

Once we have the encoder implementation, the decoder implementation is easy to describe. As mentioned earlier, once we have started decoding all we have to do is mimic the encoder algorithm. Let us first describe the decoder algorithm using pseudocode and then study its implementation using Example 4.4.5.

Decoder Algorithm

Initialize l and u.
Read the first m bits of the received bitstream into tag t.
$k = 0$
while $\left(\left\lfloor \frac{(t - l + 1) \times Total\_Count - 1}{u - l + 1} \right\rfloor \geqslant Cum\_Count(k) \right)$
$k \leftarrow k + 1$
Decode symbol x.
$l \leftarrow l + \left\lfloor \frac{(u - l + 1) \times Cum\_Count(x - 1)}{Total\_Count} \right\rfloor$
$u \leftarrow l + \left\lfloor \frac{(u - l + 1) \times Cum\_Count(x)}{Total\_Count} \right\rfloor - 1$
while (MSB of u and l are both equal to b or E_3 condition holds)
if (MSB of u and l are both equal to b)
  {
  shift l to the left by 1 bit and shift 0 into LSB
  shift u to the left by 1 bit and shift 1 into LSB
  shift t to the left by 1 bit and read next bit from received bitstream into LSB
  }
if (E_3 condition holds)
  {
  shift l to the left by 1 bit and shift 0 into LSB
  shift u to the left by 1 bit and shift 1 into LSB
  shift t to the left by 1 bit and read next bit from received bitstream into LSB
  complement (new) MSB of l, u, and t
  }

Example 4.4.5:

After encoding the sequence in Example 4.4.4, we ended up with the following binary sequence: 1100010010000000. Treating this as the received sequence and using the parameters from Table 4.5, let us decode this sequence. Using the same word length, 8, we read in the first 8 bits of the received sequence to form the tag t:

$$t = (11000100)_2 = 196$$

We initialize the lower and upper limits as

$$l = (00000000)_2 = 0$$
$$u = (11111111)_2 = 255$$

To begin decoding, we compute

$$\left\lfloor \frac{(t-l+1)\times Total\_Count - 1}{u-l+1} \right\rfloor = \left\lfloor \frac{197\times 50 - 1}{255-0+1} \right\rfloor = 38$$

and compare this value to

$$Cum\_Count = \begin{bmatrix} 0 \\ 40 \\ 41 \\ 50 \end{bmatrix}$$

Since

$$0 \leqslant 38 < 40$$

we decode the first symbol as **1**. Once we have decoded a symbol, we update the lower and upper limits:

$$l = 0 + \left\lfloor \frac{256\times Cum\_Count[0]}{Total\_Count} \right\rfloor = 0 + \left\lfloor 256\times \frac{0}{50} \right\rfloor = 0$$
$$u = 0 + \left\lfloor \frac{256\times Cum\_Count[1]}{Total\_Count} \right\rfloor - 1 = 0 + \left\lfloor 256\times \frac{40}{50} \right\rfloor - 1 = 203$$

or

$$l = (00000000)_2$$
$$u = (11001011)_2$$

The MSBs of the limits are different, and the E_3 condition does not hold. Therefore, we continue decoding without modifying the tag value. To obtain the next symbol, we compare

$$\left\lfloor \frac{(t-l+1)\times Total\_Count - 1}{u-l+1} \right\rfloor$$

which is 48, against the *Cum_Count* array:

$$Cum\_Count[2] \leqslant 48 < Cum\_Count[3]$$

Therefore, we decode **3** and update the limits:

$$l = 0 + \left\lfloor \frac{204\times Cum\_Count[2]}{Total\_Count} \right\rfloor = 0 + \left\lfloor 204\times \frac{41}{50} \right\rfloor = 167 = (1010011)_2$$
$$u = 0 + \left\lfloor \frac{204\times Cum\_Count[3]}{Total\_Count} \right\rfloor - 1 = 0 + \left\lfloor 204\times \frac{50}{50} \right\rfloor - 1 = 203 = (11001011)_2$$

As the MSB of u and l are the same, we shift the MSB out and read in a 0 for the LSB of l and a 1 for the LSB of u. We mimic this action for the tag as well, shifting the MSB out and reading in the next bit from the received bitstream as the LSB:

$$l = (01001110)_2$$
$$u = (10010111)_2$$
$$t = (10001001)_2$$

Examining l and u we can see we have an E_3 condition. Therefore, for l, u, and t, we shift the MSB out, complement the new MSB, and read in a 0 as the LSB of l, a 1 as the LSB of u, and the next bit in the received bitstream as the LSB of t. We now have

$$\begin{aligned} l &= (00011100)_2 = 28 \\ u &= (10101111)_2 = 175 \\ t &= (10010010)_2 = 146 \end{aligned}$$

To decode the next symbol, we compute

$$\left\lfloor \frac{(t - l + 1) \times Total\_Count - 1}{u - l + 1} \right\rfloor = 40$$

Since $40 \leqslant 40 < 41$, we decode **2**.

Updating the limits using this decoded symbol, we get

$$\begin{aligned} l &= 28 + \left\lfloor \frac{(175 - 28 + 1) \times 40}{50} \right\rfloor = 146 = (10010010)_2 \\ u &= 28 + \left\lfloor \frac{(175 - 28 + 1) \times 41}{50} \right\rfloor - 1 = 148 = (10010100)_2 \end{aligned}$$

We can see that we have quite a few bits to shift out. However, notice that the lower limit l has the same value as the tag t. Furthermore, the remaining received sequence consists entirely of 0s. Therefore, we will be performing identical operations on numbers that are the same, resulting in identical numbers. This will result in the final decoded symbol being **1**. We knew this was the final symbol to be decoded because only four symbols had been encoded. In practice this information has to be conveyed to the decoder. ♦

4.5 Adaptive Arithmetic Coding

We have seen how to construct arithmetic coders when the distribution of the source, in the form of cumulative counts, is available. In many applications such counts are not available a priori. It is a relatively simple task to modify the algorithms discussed so that the coder learns the distribution as the coding progresses. A straightforward implementation is to start out with a count of 1 for each letter in the alphabet. We need to have a count of at least 1 for each symbol, because if we do not, we will have no way of encoding the symbol when it is first encountered. This assumes that we know nothing about the distribution of the source. If we do know something about the distribution of the source, we can let the initial counts reflect our knowledge.

After coding is initiated, the count for each letter encountered is incremented *after* that letter has been encoded. The cumulative count table is updated accordingly. It is very important that the updating take place after the encoding; otherwise the decoder will not be using the same cumulative count table as the encoder to perform the decoding. At the decoder, the count and cumulative count tables are updated after each letter is decoded.

In the case of the static arithmetic code, we picked the size of the word based on Total_Count, the total number of symbols to be encoded. In the adaptive case, we may not know ahead of time what the total number of symbols is going to be. In this case we have to pick the word length independent of the total count. However, given a word length m we know that we can only accommodate a total count of 2^{m-2} or less. Therefore, during the encoding and decoding processes when the total count approaches 2^{m-2}, we have to go through a rescaling, or renormalization, operation. A simple rescaling operation is to divide all counts by 2 and round up the result so that no count gets rescaled to zero. This periodic rescaling can have an added benefit in that the count table better reflects the local statistics of the source.

4.6 Binary Arithmetic Coding

In many applications the alphabet itself is binary. While at first sight this may seem odd (Why would one need an encoder to encode a binary sequence?), a little bit of thought makes the reasoning apparent. Consider a binary source that puts out one symbol with probability 0.125 and the other with probability 0.875. If we directly encode one of the letters as a 0 and the other as a 1, we would get a rate of 1 bit/symbol. However, the entropy of the source is 0.543; so by directly encoding the output of the source as a 0 or a 1, we are encoding at almost twice the rate we need to. In order to achieve the entropy, we need to encode the more probable symbol at a fractional bit rate, which is exactly what arithmetic coding allows us to do; hence the popularity of the arithmetic code for encoding binary sources with highly skewed probabilities. The source can be binary by its nature, such as bilevel documents; or the binary input may be the binary representation of nonbinary data, as is the case for the Context Adaptive Binary Arithmetic Coder (CABAC) in the H.264 video coding standard.

As for arithmetic coding in general, the binary arithmetic coder also requires a probability model. However, because there are only two letters in the alphabet, the probability model consists of a single number, namely the probability of one of the symbols; the probability of the other symbol is simply one minus the specified probability. Because we only need a single number to represent the probability model, it is easy to use multiple arithmetic codes to encode a source where different models represent different contexts. This results in a much more accurate modeling of the source than would have been possible with a single model, which in turn leads to better compression.

Example 4.6.1:

Consider a binary sequence generated by scanning a document where the white pixels are represented by a 0 and the black pixels by a 1. Clearly the probability of a pixel being white or black depends heavily on whether the neighboring pixel is white or black. For this example we will use a first order context, so we have two sets of *Count* and *Cum_Count* tables:

$Count^0(0) = 8$	$Cum\_Count^0(-1) = 0$	$Count^1(0) = 2$	$Cum\_Count^1(-1) = 0$
$Count^0(1) = 2$	$Cum\_Count^0(0) = 8$	$Count^1(1) = 8$	$Cum\_Count^1(0) = 2$
$Total\_Count^0 = 10$	$Cum\_Count^0(1) = 10$	$Total\_Count^1(0) = 10$	$Cum\_Count^1(1) = 10$

where the superscript indicates whether the previous pixel was a 0 or a 1. We will always assume that for each row there is an imaginary white pixel to the left of the leftmost pixel. Therefore, we will always begin our encoding using the Cum_Count^0 array. After encoding the first pixel all other pixels will be encoded using the Cum_Count array corresponding to the previous pixel. Assume we wish to encode the sequence

$$000000111111$$

Notice that the marginal probabilities of 1 and 0 are 1/2 so if we used an arithmetic coder that did not take the context into account we would end up with a 1 bit/pixel encoding. Using the conditional probability tables we hope to do better. We will use a word length m of six:

$$l^{(0)} = 0 = (000000)_2 \tag{56}$$

$$u^{(0)} = 63 = (111111)_2 \tag{57}$$

The first bit to be encoded is 0:

$$\begin{aligned} l^{(1)} &= 0 + \left\lfloor \frac{64 \times Cum\_Count^0(-1)}{10} \right\rfloor = \left\lfloor \frac{64 \times 0}{10} \right\rfloor = 0 \\ &= (000000)_2 \\ u^{(1)} &= 0 + \left\lfloor \frac{64 \times Cum\_Count^0(0)}{10} \right\rfloor - 1 = \left\lfloor \frac{64 \times 8}{10} \right\rfloor - 1 = 50 \\ &= (110010)_2 \end{aligned}$$

The second bit to be encoded is also 0:

$$\begin{aligned} l^{(2)} &= 0 + \left\lfloor \frac{51 \times Cum\_Count^0(-1)}{10} \right\rfloor = \left\lfloor \frac{51 \times 0}{10} \right\rfloor = 0 \\ &= (000000)_2 \\ u^{(2)} &= 0 + \left\lfloor \frac{51 \times Cum\_Count^0(0)}{10} \right\rfloor - 1 = \left\lfloor \frac{51 \times 8}{10} \right\rfloor - 1 = 39 \\ &= (100111)_2 \end{aligned}$$

The third bit to be encoded is also 0:

$$\begin{aligned} l^{(3)} &= 0 + \left\lfloor \frac{40 \times Cum\_Count^0(-1)}{10} \right\rfloor = \left\lfloor \frac{40 \times 0}{10} \right\rfloor = 0 \\ &= (\mathbf{0}00000)_2 \\ u^{(3)} &= 0 + \left\lfloor \frac{40 \times Cum\_Count^0(0)}{10} \right\rfloor - 1 = \left\lfloor \frac{40 \times 8}{10} \right\rfloor - 1 = 31 \\ &= (\mathbf{0}11111)_2 \end{aligned}$$

The MSBs of $l^{(3)}$ and $u^{(3)}$ are both 0. Therefore, we shift this value out and send it to the decoder. All other bits are shifted to the left and we get

$$\begin{aligned} l^{(3)} &= (000000)_2 = 0 \\ u^{(3)} &= (111111)_2 = 63 \end{aligned}$$

Encoding another 0, we obtain

$$\begin{aligned}
l^{(4)} &= 0 + \left\lfloor \frac{64 \times Cum_Count^0(-1)}{10} \right\rfloor = \left\lfloor \frac{64 \times 0}{10} \right\rfloor = 0 \\
&= (000000)_2 \\
u^{(4)} &= 0 + \left\lfloor \frac{64 \times Cum_Count^0(0)}{10} \right\rfloor - 1 = \left\lfloor \frac{64 \times 8}{10} \right\rfloor - 1 = 50 \\
&= (110010)_2
\end{aligned}$$

Continuing with the next two 0s we get

$$\begin{aligned}
l^{(5)} &= 0 + \left\lfloor \frac{64 \times Cum_Count^0(-1)}{10} \right\rfloor = \left\lfloor \frac{51 \times 0}{10} \right\rfloor = 0 \\
&= (000000)_2 \\
u^{(5)} &= 0 + \left\lfloor \frac{64 \times Cum_Count^0(0)}{10} \right\rfloor - 1 = \left\lfloor \frac{51 \times 8}{10} \right\rfloor - 1 = 39 \\
&= (100111)_2 \\
l^{(6)} &= 0 + \left\lfloor \frac{40 \times Cum_Count^0(-1)}{10} \right\rfloor = \left\lfloor \frac{40 \times 0}{10} \right\rfloor = 0 \\
&= (\mathbf{0}00000)_2 \\
u^{(6)} &= 0 + \left\lfloor \frac{40 \times Cum_Count^0(0)}{10} \right\rfloor - 1 = \left\lfloor \frac{40 \times 8}{10} \right\rfloor - 1 = 31 \\
&= (\mathbf{0}11111)_2
\end{aligned}$$

The MSBs of $l^{(6)}$ and $u^{(6)}$ are both 0, therefore, we shift out a 0 from both $l^{(6)}$ and $u^{(6)}$:

$$\begin{aligned}
l^{(6)} &= (000000)_2 = 0 \\
u^{(6)} &= (111111)_2 = 63
\end{aligned}$$

The next bit to be encoded is a 1. However the bit prior to it was a 0, therefore we use the Cum_Count^0 tables:

$$\begin{aligned}
l^{(7)} &= 0 + \left\lfloor \frac{64 \times Cum_Count^0(0)}{10} \right\rfloor = \left\lfloor \frac{64 \times 8}{10} \right\rfloor = 51 \\
&= (110011)_2 \\
u^{(7)} &= 0 + \left\lfloor \frac{64 \times Cum_Count^0(1)}{10} \right\rfloor - 1 = \left\lfloor \frac{64 \times 10}{10} \right\rfloor - 1 = 63 \\
&= (111111)_2
\end{aligned}$$

The first two MSBs of the upper and lower limit are the same, so we shift out two bits, 11, from both the upper and the lower limit:

$$\begin{aligned}
l^{(7)} &= (001100)_2 = 12 \\
u^{(7)} &= (111111)_2 = 63
\end{aligned}$$

The next bit to be encoded is also a 1. However, as the bit prior to that was a 1 we use the Cum_Count^1 tables:

$$\begin{aligned} l^{(8)} &= 12 + \left\lfloor \frac{52 \times Cum\_Count^1(0)}{10} \right\rfloor = \left\lfloor \frac{52 \times 2}{10} \right\rfloor = 22 \\ &= (010110)_2 \\ u^{(8)} &= 12 + \left\lfloor \frac{52 \times Cum\_Count^1(1)}{10} \right\rfloor - 1 = \left\lfloor \frac{52 \times 10}{10} \right\rfloor - 1 = 63 \\ &= (111111)_2 \end{aligned}$$

Encoding the next two 1s we obtain

$$\begin{aligned} l^{(9)} &= 22 + \left\lfloor \frac{42 \times Cum\_Count^1(0)}{10} \right\rfloor = \left\lfloor \frac{42 \times 2}{10} \right\rfloor = 30 \\ &= (011110)_2 \\ u^{(9)} &= 22 + \left\lfloor \frac{42 \times Cum\_Count^1(1)}{10} \right\rfloor - 1 = \left\lfloor \frac{42 \times 10}{10} \right\rfloor - 1 = 63 \\ &= (111111)_2 \\ l^{(10)} &= 30 + \left\lfloor \frac{34 \times Cum\_Count^1(0)}{10} \right\rfloor = \left\lfloor \frac{34 \times 2}{10} \right\rfloor = 36 \\ &= (100100)_2 \\ u^{(10)} &= 30 + \left\lfloor \frac{34 \times Cum\_Count^1(1)}{10} \right\rfloor - 1 = \left\lfloor \frac{34 \times 10}{10} \right\rfloor - 1 = 63 \\ &= (111111)_2 \end{aligned}$$

The MSBs of the upper and lower limits are both equal to 1 so we shift out 1:

$$\begin{aligned} l^{(10)} &= (001000)_2 = 8 \\ u^{(10)} &= (111111)_2 = 63 \end{aligned}$$

Encoding the next two 1s we get

$$\begin{aligned} l^{(11)} &= 8 + \left\lfloor \frac{56 \times Cum\_Count^1(0)}{10} \right\rfloor = \left\lfloor \frac{56 \times 2}{10} \right\rfloor = 19 \\ &= (001011)_2 \\ u^{(11)} &= 8 + \left\lfloor \frac{56 \times Cum\_Count^1(1)}{10} \right\rfloor - 1 = \left\lfloor \frac{56 \times 10}{10} \right\rfloor - 1 = 63 \\ &= (111111)_2 \\ l^{(12)} &= 19 + \left\lfloor \frac{45 \times Cum\_Count^1(0)}{10} \right\rfloor = \left\lfloor \frac{45 \times 2}{10} \right\rfloor = 28 \\ &= (011100)_2 \\ u^{(11)} &= 19 + \left\lfloor \frac{45 \times Cum\_Count^1(1)}{10} \right\rfloor - 1 = \left\lfloor \frac{45 \times 10}{10} \right\rfloor - 1 = 63 \\ &= (111111)_2 \end{aligned}$$

We have encoded twelve bits, and the sequence of bits generated by the encoder until this point is 00111. In other words we have a coding rate of 5/12 bits/pixel, which is less than half the coding rate we would have achieved had we used a single *Cum_Count* table. At this point if we wish to terminate the encoding we would have to incur the overhead of transmitting the bits in the lower limit. The six bits of the lower limit would be a significant overhead for this toy sequence. However, in practice, when the input sequence is much longer than twelve, the six-bit overhead would be negligible. We leave the decoding of this sequence as an exercise for the reader. ♦

Furthermore, the simple nature of the coder allows for approximations that result in simple and fast implementations. We will look at three applications of the binary coder including the QM coder used in the JBIG standard for encoding bilevel images and the M (or modulo) coder, which is a part of the coder CABAC used in the H.264 video coding standard.

Before we describe the particular implementations, let us take a general view of binary arithmetic coding. In our description of arithmetic coding, we updated the tag interval by updating the endpoints of the interval, $u^{(n)}$ and $l^{(n)}$. We could just as well have kept track of one endpoint and the size of the interval. This is the approach adopted in many of the binary coders, which track the lower end of the tag interval $l^{(n)}$ and the size of the interval $A^{(n)}$, where

$$A^{(n)} = u^{(n)} - l^{(n)} \tag{58}$$

The tag for a sequence is the binary representation of $l^{(n)}$.

We can obtain the update equation for $A^{(n)}$ by subtracting Equation (9) from Equation (10) and substituting $A^{(n)}$ for $u^{(n)} - l^{(n)}$.

$$A^{(n)} = A^{(n-1)}(F_X(x_n) - F_X(x_n - 1)) \tag{59}$$

$$= A^{(n-1)}P(x_n) \tag{60}$$

Substituting $A^{(n)}$ for $u^{(n)} - l^{(n)}$ in Equation (9), we get the update equation for $l^{(n)}$:

$$l^{(n)} = l^{(n-1)} + A^{(n-1)}F_X(x_n - 1) \tag{61}$$

Instead of dealing directly with the 0s and 1s put out by the source, many of the binary coders map them into a More Probable Symbol (MPS) and Less Probable Symbol (LPS). If 0 represents black pixels and 1 represents white pixels, then in a mostly black image, 0 will be the MPS, whereas in an image with mostly white regions 1 will be the MPS. Denoting the probability of occurrence of the LPS for the context C by q_c and mapping the MPS to the lower subinterval, the occurrence of an MPS symbol results in the update equations

$$l^{(n)} = l^{(n-1)} \tag{62}$$

$$A^{(n)} = A^{(n-1)}(1 - q_c) \tag{63}$$

while the occurrence of an LPS symbol results in the update equations

$$l^{(n)} = l^{(n-1)} + A^{(n-1)}(1 - q_c) \tag{64}$$

$$A^{(n)} = A^{(n-1)}q_c \tag{65}$$

4.6.1 The QM Coder

Until this point, the binary coder looks very much like the arithmetic coder described earlier in this chapter. To make the implementation simpler and computationally more efficient, the Joint Bi-level Image Experts Group (JBIG) recommended several deviations from the standard arithmetic coding algorithm for the version of the arithmetic coder used in the JBIG algorithm for compression of bi-level images. The update equations involve multiplications, which are expensive in both hardware and software. In the QM coder, the multiplications are avoided by assuming that $A^{(n)}$ has a value close to 1, and multiplication with $A^{(n)}$ can be approximated by multiplication with 1. Therefore, the update equations become

$$\text{For MPS :}$$
$$l^{(n)} = l^{(n-1)} \tag{66}$$
$$A^{(n)} = 1 - q_c \tag{67}$$
$$\text{For LPS :}$$
$$l^{(n)} = l^{(n-1)} + (1 - q_c) \tag{68}$$
$$A^{(n)} = q_c \tag{69}$$

In order not to violate the assumption on $A^{(n)}$ whenever the value of $A^{(n)}$ drops below 0.75, the QM coder goes through a series of rescalings until the value of $A^{(n)}$ is greater than or equal to 0.75. The rescalings take the form of repeated doubling, which corresponds to a left shift in the binary representation of $A^{(n)}$. To keep all parameters in sync, the same scaling is also applied to $l^{(n)}$. The bits shifted out of the buffer containing the value of $l^{(n)}$ make up the encoder output. Looking at the update equations for the QM coder, we can see that a rescaling will occur every time an LPS occurs. Occurrence of an MPS may or may not result in a rescale, depending on the value of $A^{(n)}$.

The probability q_c of the LPS for context C is updated each time a rescaling takes place and the context C is active. An ordered list of values for q_c is listed in a table. Every time a rescaling occurs, the value of q_c is changed to the next lower or next higher value in the table, depending on whether the rescaling was caused by the occurrence of an LPS or an MPS.

In a nonstationary situation, the symbol assigned to LPS may actually occur more often than the symbol assigned to MPS. This condition is detected when $q_c > (A^{(n)} - q_c)$. In this situation, the assignments are reversed; the symbol assigned the LPS label is assigned the MPS label and vice versa. The test is conducted every time a rescaling takes place.

The decoder for the QM coder operates in much the same way as the decoder described in this chapter, mimicking the encoder operation.

4.6.2 The MQ Coder

The MQ coder is a variant of the QM coder. Unlike the QM coder the MQ coder assigns the lower interval to the LPS and the upper interval to the MPS. The update equations in this case without any approximations would become

$$\text{For MPS :}$$
$$l^{(n)} = l^{(n-1)} + A^{(n-1)} q_c \tag{70}$$
$$A^{(n)} = A^{(n-1)}(1 - q_c) \tag{71}$$
$$\text{For LPS :}$$
$$l^{(n)} = l^{(n-1)} \tag{72}$$
$$A^{(n)} = A^{(n-1)} q_c \tag{73}$$

However, as in the case of the QM coder, we wish to avoid multiplication; therefore, with the same assumption that $A^{(n)}$ has a value close to one we modify the update equations to

$$\text{For MPS :}$$
$$l^{(n)} = l^{(n-1)} + q_c \tag{74}$$
$$A^{(n)} = A^{(n-1)} - q_c \tag{75}$$
$$\text{For LPS :}$$
$$l^{(n)} = l^{(n-1)} \tag{76}$$
$$A^{(n)} = q_c \tag{77}$$

The adaptation in the MQ coder is modeled by a state machine. In practice, the A and l registers are assigned 16 bits of precision. When the value of A falls below 0x8000, it is left shifted until the value reaches or exceeds 0x8000. The same operation is performed on the register where l is stored, and the bits shifted out of the l register become the output codewords of the arithmetic coder.

4.6.3 The M Coder

The M coder is another variant of the QM coder in which the multiply operation is replaced with a table lookup. To better understand the approximations used by the M coder, let us rewrite the update equations without approximation. The occurrence of an MPS symbol results in the update equations

$$l^{(n)} = l^{(n-1)} \tag{78}$$
$$A^{(n)} = A^{(n-1)} - A^{(n-1)} q_c \tag{79}$$

while the occurrence of an LPS symbol results in the update equations

$$l^{(n)} = l^{(n-1)} + A^{(n-1)} - A^{(n-1)} q_c \tag{80}$$
$$A^{(n)} = A^{(n-1)} q_c \tag{81}$$

Notice that the only multiplications are between the estimate of the range $A^{(n-1)}$ and the probability of the LPS q_c. The M coder gets rid of the costly multiplication operation by allowing the range and the probability to take on only a specified number of values and then replacing the multiplication by a table lookup indexed by the quantized values of the range and the LPS probability [44]. Given a minimum value A_{min} and a maximum value A_{max} for the range A, the range is restricted to four values:

$$A = A_{min} + (k+1)\Delta A, \ k = 0, 1, 2, 3$$

where

$$\Delta A = \frac{A_{max} - A_{min}}{4}$$

The LPS probability q_c can take on one of 64 possible values q_m where

$$q_m = \alpha q_{m-1}, \ m = 1, 2, \cdots, 63 \quad (82)$$

where

$$\alpha = \left(\frac{0.01875}{0.5}\right)^{\frac{1}{63}} \text{ and } q_0 = 0.5$$

In the update equation, instead of multiplying the range and the lower limit, the range is mapped to the closest of the four quantized ranges, the corresponding index k along with the index m of the probability of the LPS are used as pointers into a lookup table, and the product is read from the table.

In order to make the coder adaptive, all we need to do is update the value of the LPS probability as we see more data. If we see more occurrences of MPS, we should decrease the LPS probability. If we encounter more occurrences of LPS, we should increase the LPS probability. The M coder needs to do this while keeping the LPS probability restricted to the 64 allowed probabilities generated by Equation (82). This can be done by simply incrementing or decrementing the index m of q_m. When an MPS is encountered, we increment the index and when an LPS is encountered, we decrement the index. The index is incremented by one each time an MPS is encountered until it reaches the maximum allowed value where it remains. When an LPS is encountered, the index is decremented by a variable amount until the index reaches 0. At this point the LPS probability is one half and further occurrence of LPS indicates that this symbol is no longer the least probable and therefore, the value of MPS is changed. In other words, the MPS and LPS symbols can be swapped. The M coder forms the core of the CABAC coder, which is part of the H.264 standard for video compression.

4.7 Comparison of Huffman and Arithmetic Coding

We have described a new coding scheme that, although more complicated than Huffman coding, allows us to code *sequences* of symbols. How well this coding scheme works depends on how it is used. Let's first try to use this code for encoding sources for which we know the Huffman code.

Looking at Example 4.4.1, the average length for this code is

$$l = 2 \times 0.5 + 3 \times 0.25 + 4 \times 0.125 + 4 \times 0.125 \quad (83)$$

$$= 2.75 \text{ bits/symbol} \quad (84)$$

TABLE 4.6 Arithmetic code for two-symbol sequences.

Message	$P(x)$	$\overline{T}_X(x)$	$\overline{T}_X(x)$ in Binary	$\lceil \log \frac{1}{P(x)} \rceil + 1$	Code
11	.25	.125	.001	3	001
12	.125	.3125	.0101	4	0101
13	.0625	.40625	.01101	5	01101
14	.0625	.46875	.01111	5	01111
21	.125	.5625	.1001	4	1001
22	.0625	.65625	.10101	5	10101
23	.03125	.703125	.101101	6	101101
24	.03125	.734375	.101111	6	101111
31	.0625	.78125	.11001	5	11001
32	.03125	.828125	.110101	6	110101
33	.015625	.8515625	.1101101	7	1101101
34	.015625	.8671875	.1101111	7	1101111
41	.0625	.90625	.11101	5	11101
42	.03125	.953125	.111101	6	111101
43	.015625	.9765625	.1111101	7	1111101
44	.015625	.984375	.1111111	7	1111111

Recall from Section 2.4 that the entropy of this source was 1.75 bits/symbol and the Huffman code achieved this entropy. Obviously, arithmetic coding is not a good idea if you are going to encode your message one symbol at a time. Let's repeat the example with messages consisting of two symbols. (Note that we are only doing this to demonstrate a point. In practice, we would not code sequences this short using an arithmetic code.)

Example 4.7.1:

If we encode two symbols at a time, the resulting code is shown in Table 4.6.

The average length per message is 4.5 bits. Therefore, using two symbols at a time we get a rate of 2.25 bits/symbol (certainly better than 2.75 bits/symbol, but still not as good as the best rate of 1.75 bits/symbol). However, we see that as we increase the number of symbols per message, our results get better and better. ♦

How many samples do we have to group together to make the arithmetic coding scheme perform better than the Huffman coding scheme? We can get some idea by looking at the bounds on the coding rate.

Recall that the bounds on the average length l_A of the arithmetic code are

$$H(X) \leqslant l_A \leqslant H(X) + \frac{2}{m}$$

It does not take many symbols in a sequence before the coding rate for the arithmetic code becomes quite close to the entropy. However, recall that for Huffman codes, if we block m symbols together, the coding rate is

$$H(X) \leqslant l_H \leqslant H(X) + \frac{1}{m}$$

The advantage seems to lie with the Huffman code, although the advantage decreases with increasing m. However, remember that to generate a codeword for a sequence of length m using the Huffman procedure requires building the entire code for all possible sequences of length m. If the original alphabet size was k, then the size of the codebook would be k^m. Taking relatively reasonable values of $k = 16$ and $m = 20$ gives a codebook size of 16^{20}! This is obviously not a viable option. For the arithmetic coding procedure, we do not need to build the entire codebook. Instead, we simply obtain the code for the tag corresponding to a given sequence. Therefore, it is entirely feasible to code sequences of length 20 or much more. In practice, we can make m large for the arithmetic coder and not for the Huffman coder. This means that for most sources we can get rates closer to the entropy by using arithmetic coding than by using Huffman coding. The exceptions are sources whose probabilities are powers of two. In these cases, the single-letter Huffman code achieves the entropy, and we cannot do any better with arithmetic coding, no matter how long a sequence we pick.

The amount of gain also depends on the source. Recall that for Huffman codes, we are guaranteed to obtain rates within $0.086 + p_{\max}$ of the entropy, where $p_{\max}$ is the probability of the most probable letter in the alphabet. If the alphabet size is relatively large and the probabilities are not too skewed, the maximum probability $p_{\max}$ is generally small. In these cases, the advantage of arithmetic coding over Huffman coding is small, and it might not be worth the extra complexity to use arithmetic coding rather than Huffman coding. However, there are many sources, such as facsimile, in which the alphabet size is small, and the probabilities are highly unbalanced. In these cases, the use of arithmetic coding is generally worth the added complexity.

Another major advantage of arithmetic coding is that it is easy to implement a system with multiple arithmetic codes. This may seem contradictory, as we have claimed that arithmetic coding is more complex than Huffman coding. However, it is the computational machinery that causes the increase in complexity. Once we have the computational machinery to implement one arithmetic code, all we need to implement more than a single arithmetic code is the availability of more probability tables. If the alphabet size of the source is small, as in the case of a binary source, there is very little added complexity indeed. In fact, as we shall see in the next section, it is possible to develop multiplication-free arithmetic coders that are quite simple to implement (nonbinary multiplication-free arithmetic coders are described in [45]).

Finally, it is much easier to adapt arithmetic codes to changing input statistics. All we need to do is estimate the probabilities of the input alphabet. This can be done by keeping a count of the letters as they are coded. There is no need to preserve a tree, as with adaptive Huffman codes. Furthermore, there is no need to generate a code a priori, as in the case of Huffman coding. This property allows us to separate the modeling and coding procedures in a manner that is not very feasible with Huffman coding. This separation permits greater flexibility in the design of compression systems, which can be used to great advantage.

4.8 Applications

Arithmetic coding is used in a variety of lossless and lossy compression applications. It is a part of many international standards. In the area of multimedia there are a few principal organizations that develop standards. The International Standards Organization (ISO) and the International Electrotechnical Commission (IEC) are industry groups that work on multimedia standards, while the International Telecommunications Union (ITU), which is part of the United Nations, works on multimedia standards on behalf of the member states of the United Nations. Quite often these institutions work together to create international standards. In later chapters, we will be looking at a number of these standards, and we will see how arithmetic coding is used in image compression, audio compression, and video compression standards. For now, let us look at the lossless compression example from the previous chapter.

In Tables 4.7 and 4.8, we show the results of using adaptive arithmetic coding to encode the same test images that were previously encoded using Huffman coding. We have included the compression ratios obtained using Huffman code from the previous chapter for comparison. Comparing these values to those obtained in the previous chapter, we can see very little change. The reason is that because the alphabet size for the images is quite large, the value of $p_{\max}$ is quite small, and the Huffman coder performs very close to the entropy.

As we mentioned before, a major advantage of arithmetic coding over Huffman coding is the ability to separate the modeling and coding aspects of the compression approach. In terms of image coding, this allows us to use a number of different models that take advantage of local properties. For example, we could use different decorrelation strategies in regions of the image that are quasi-constant and will, therefore, have differences that are small, and in regions where there is a lot of activity, causing the presence of larger difference values.

TABLE 4.7 Compression using adaptive arithmetic coding of pixel values.

Image Name	Bits/ Pixel	Total Size (bytes)	Compression Ratio (arithmetic)	Compression Ratio (Huffman)
Sena	6.52	53,431	1.23	1.16
Sensin	7.12	58,306	1.12	1.27
Earth	4.67	38,248	1.71	1.67
Omaha	6.84	56,061	1.17	1.14

TABLE 4.8 Compression using adaptive arithmetic coding of pixel differences.

Image Name	Bits/ Pixel	Total Size (bytes)	Compression Ratio (arithmetic)	Compression Ratio (Huffman)
Sena	3.89	31,847	2.06	2.08
Sensin	4.56	37,387	1.75	1.73
Earth	3.92	32,137	2.04	2.04
Omaha	6.27	51,393	1.28	1.26

4.9 Summary

In this chapter, we introduced the basic ideas behind arithmetic coding. We have shown that the arithmetic code is a uniquely decodable code that provides a rate close to the entropy for long stationary sequences. This ability to encode sequences directly instead of as a concatenation of the codes for the elements of the sequence makes this approach more efficient than Huffman coding for alphabets with highly skewed probabilities. We have looked in some detail at the implementation of the arithmetic coding approach.

The arithmetic coding results in this chapter were obtained by using the program provided by Witten, Neal, and Cleary [46]. This code can be used (with some modifications) for exploring different aspects of arithmetic coding (see the following problems).

Further Reading

1. The book *Text Compression*, by T.C. Bell, J.G. Cleary, and I.H. Witten [1], contains a very readable section on arithmetic coding, complete with pseudocode and C code.
2. A thorough treatment of various aspects of arithmetic coding can be found in the excellent chapter "Arithmetic Coding," by Amir Said [47] in the *Lossless Compression Handbook*.
3. There is an excellent tutorial article by G.G. Langdon, Jr. [48] in the March 1984 issue of the *IBM Journal of Research and Development*.
4. The separate model and code paradigm is explored in a precise manner in the context of arithmetic coding in a paper by J.J. Rissanen and G.G. Langdon [49].
5. The separation of modeling and coding is exploited in a very nice manner in an early paper by G.G. Langdon and J.J. Rissanen [50].
6. Various models for text compression that can be used effectively with arithmetic coding are described by T.G. Bell, I.H. Witten, and J.G. Cleary [51] in an article in the *ACM Computing Surveys*.
7. The coder used in the JBIG algorithm is a descendant of the Q coder, described in some detail in several papers in the November 1988 issue of the *IBM Journal of Research and Development*.

4.10 Projects and Problems

1. Given a number a in the interval [0, 1) with an n-bit binary representation $[b_1 b_2 \ldots b_n]$, show that for any other number b to have a binary representation with $[b_1 b_2 \ldots b_n]$ as the prefix, b has to lie in the interval $[a, a + \frac{1}{2^n})$.
2. The binary arithmetic coding approach specified in the JBIG standard can be used for coding grayscale images via *bit plane encoding*. In bit plane encoding, we combine the most significant bits for each pixel into one bit plane, the next most significant bits into another bit plane, and so on. Use the function `extrctbp` to obtain eight bit planes for the `sena.img` and `omaha.img` test images, and encode them using arithmetic coding. Use the low-resolution contexts shown in Figure 7.13.

TABLE 4.9 **Probability model for Problems 5 and 6.**

Letter	Probability
a_1	0.2
a_2	0.3
a_3	0.5

3. Bit plane encoding is more effective when the pixels are encoded using a *Gray code*. The Gray code assigns numerically adjacent values binary codes that differ by only 1 bit. To convert from the standard binary code $b_0 b_1 b_2 \ldots b_7$ to the Gray code $g_0 g_1 g_2 \ldots g_7$, we can use the equations

$$g_0 = b_0$$
$$g_k = b_k \oplus b_{k-1}$$

Convert the test images `sena.img` and `omaha.img` to a Gray code representation and bit plane encode. Compare with the results for the non-Gray-coded representation.

4. In Example 4.4.4, repeat the encoding using $m = 6$. Comment on your results.

5. Given the probability model in Table 4.9, find the real valued tag for the sequence $a_1 a_1 a_3 a_2 a_3 a_1$.

6. For the probability model in Table 4.9, decode a sequence of length 10 with the tag 0.63215699.

7. Consider the frequency counts shown in Table 4.10:

(a) What is the word length required for unambiguous encoding?

(b) Find the binary code for the sequence *abacabb*.

(c) Decode the code you obtained to verify that your encoding was correct.

8. Generate a binary sequence of length L with $P(0) = 0.8$, and use the arithmetic coding algorithm to encode it. Plot the difference of the rate in bits/symbol and the entropy as a function of L. Comment on the effect of L on the rate.

9. Decode the bitstream generated in Example 4.6.1. Do not forget to append the bits in the lower limit to terminate the bitstream.

10. Generate a random binary sequence of length 100,000 with $P(0) = 0.75$.

TABLE 4.10 **Frequency counts for Problem 7.**

Letter	Count
a	37
b	38
c	25

(a) Encode this sequence using a binary arithmetic coder as described by the update Equations (60) and (61). Note the size of the compressed sequence.

(b) Implement the QM coder described in Section 4.6.1 and encode the binary sequence using this coder. Compare the length of the compressed sequence. Comment on any differences.

11. Generate a binary sequence of length 100,000 using the model $P[0|0] = P[1|1] = 0.9$.

(a) Encode this sequence using a binary arithmetic coder as described by the update Equations (60) and (61). Note the size of the compressed sequence.

(b) Encode the sequence using a context-based arithmetic coder where the context is determined by the previous two bits. Compare the difference in the size of the compressed sequences.

5

Dictionary Techniques

5.1 Overview

In the previous two chapters we looked at coding techniques that assume a source that generates a sequence of independent symbols. As most sources are correlated to start with, the coding step is generally preceded by a decorrelation step. In this chapter we will look at techniques that incorporate the structure in the data in order to increase the amount of compression. These techniques—both static and adaptive (or dynamic)—build a list of commonly occurring patterns and encode these patterns by transmitting their index in the list. They are most useful with sources that generate a relatively small number of patterns quite frequently, such as text sources and computer commands. We discuss applications to text compression, modem communications, and image compression.

5.2 Introduction

In many applications, the output of the source consists of recurring patterns. A classic example is a text source in which certain patterns or words recur constantly. Also, there are certain patterns that simply do not occur, or if they do occur, it is with great rarity. For example, *sgiomlaireached* might be an annoying habit but the word probably occurs in a very small fraction of the text sources in existence.

A very reasonable approach to encoding such sources is to keep a list, or *dictionary*, of frequently occurring patterns. When these patterns appear in the source output, they are encoded with a reference to the dictionary. If the pattern does not appear in the dictionary, then it can be encoded using some other less efficient method. In effect, we are splitting the input into two classes: frequently occurring patterns and infrequently occurring patterns. For this technique to be effective, the class of frequently occurring patterns, and hence the size of the dictionary, must be much smaller than the number of all possible patterns.

Suppose we have a particular text that consists of four-character words, three characters from the 26 lowercase letters of the English alphabet followed by a punctuation mark. Suppose our source alphabet consists of the 26 lowercase letters of the English alphabet and the punctuation marks comma, period, exclamation mark, question mark, semicolon, and colon. In other words, the size of the input alphabet is 32. If we were to encode the text source one character at a time, treating each character as an equally likely event, we would need 5 bits per character. Treating all $32^4 (= 2^{20} = 1,048,576)$ four-character patterns as equally likely, we have a code that assigns 20 bits to each four-character pattern. Let us now put the 256 most likely four-character patterns into a dictionary. The transmission scheme works as follows. Whenever we want to send a pattern that exists in the dictionary, we will send a 1-bit flag, say, a 0, followed by an 8-bit index corresponding to the entry in the dictionary. If the pattern is not in the dictionary, we will send a 1 followed by the 20-bit encoding of the pattern. If the pattern we encounter is not in the dictionary, we will actually use more bits than in the original scheme, 21 instead of 20. But if it is in the dictionary, we will send only 9 bits. The utility of our scheme will depend on the percentage of the words we encounter that are in the dictionary. We can get an idea about the utility of our scheme by calculating the average number of bits per pattern. If the probability of encountering a pattern from the dictionary is p, then the average number of bits per pattern R is given by

$$R = 9p + 21(1 - p) = 21 - 12p \tag{5.1}$$

For our scheme to be useful, R should have a value less than 20. This happens when $p \geqslant 0.084$. This does not seem like a very large number. However, note that if all patterns were occurring in an equally likely manner, the probability of encountering a pattern from the dictionary would be less than 0.00025!

We do not simply want a coding scheme that performs slightly better than the simple-minded approach of coding each pattern as equally likely; we would like to improve the performance as much as possible. In order for this to happen, p should be as large as possible. This means that we should carefully select patterns that are most likely to occur as entries in the dictionary. To do this, we have to have a pretty good idea about the structure of the source output. If we do not have information of this sort available to us prior to the encoding of a particular source output, we need to acquire this information somehow when we are encoding. If we feel we have sufficient prior knowledge, we can use a *static* approach; if not, we can take an *adaptive* approach. We will look at both these approaches in this chapter.

5.3 Static Dictionary

Choosing a static dictionary technique is most appropriate when considerable prior knowledge about the source is available. This technique is especially suitable for use in specific applications. For example, if the task were to compress the student records at a university, a static dictionary approach may be the best. This is because we know ahead of time that certain words and phrases such as "Name" and "Student ID" are going to appear in almost all of the records. Other words such as "Sophomore," "credits," and so on will occur quite often. Depending on the location of the university, certain digits in Social Security numbers are more likely to occur. For example, in Nebraska most student ID numbers used to begin with the digits 505. In fact, most entries will be of a recurring nature. In this situation, it is highly efficient to design a compression scheme based on a static dictionary containing the recurring patterns. Similarly, there could be a number of other situations in which an application-specific or data-specific static-dictionary-based coding scheme would be the most efficient. It should be noted that these schemes would work well only for the applications and data they were designed for. If these schemes were to be used with different applications, they may cause an expansion of the data instead of compression.

A static dictionary technique that is less specific to a single application is *digram coding*. We describe this in the next section.

5.3.1 Digram Coding

One of the more common forms of static dictionary coding is digram coding. In this form of coding, the dictionary consists of all letters of the source alphabet followed by as many pairs of letters, called *digrams*, as can be accommodated by the dictionary. For example, suppose we were to construct a dictionary of size 256 for digram coding of all printable ASCII characters. The first 95 entries of the dictionary would be the 95 printable ASCII characters. The remaining 161 entries would be the most frequently used pairs of characters.

The digram encoder reads a two-character input and searches the dictionary to see if this input exists in the dictionary. If it does, the corresponding index is encoded and transmitted. If it does not, the first character of the pair is encoded. The second character in the pair then becomes the first character of the next digram. The encoder reads another character to complete the digram, and the search procedure is repeated.

Example 5.3.1:

Suppose we have a source with a five-letter alphabet $\mathcal{A} = \{a, b, c, d, r\}$. Based on knowledge about the source, we build the dictionary shown in Table 5.1.

TABLE 5.1 A sample dictionary.

Code	Entry	Code	Entry
000	a	100	r
001	b	101	ab
010	c	110	ac
011	d	111	ad

TABLE 5.2 Thirty most frequently occurring pairs of characters in a 41,364-character-long LaTeX document.

Pair	Count	Pair	Count
*e*ƀ	1128	*ar*	314
ƀ*t*	838	*at*	313
ƀƀ	823	ƀ*w*	309
th	817	*te*	296
he	712	ƀ*s*	295
in	512	*d*ƀ	272
*s*ƀ	494	ƀ*o*	266
er	433	*io*	257
ƀ*a*	425	*co*	256
*t*ƀ	401	*re*	247
en	392	ƀ$	246
on	385	*r*ƀ	239
*n*ƀ	353	*di*	230
ti	322	*ic*	229
ƀ*i*	317	*ct*	226

Suppose we wish to encode the sequence

$$abracadabra$$

The encoder reads the first two characters *ab* and checks to see if this pair of letters exists in the dictionary. It does and is encoded using the codeword 101. The encoder then reads the next two characters *ra* and checks to see if this pair occurs in the dictionary. It does not, so the encoder sends out the code for *r*, which is 100, then reads in one more character, *c*, to make the two-character pattern *ac*. This does exist in the dictionary and is encoded as 110. Continuing in this fashion, the remainder of the sequence is coded. The output string for the given input sequence is 101100110111101100000. ♦

A list of the 30 most frequently occurring pairs of characters in an earlier version of this chapter is shown in Table 5.2. For comparison, the 30 most frequently occurring pairs of characters in a set of C programs is shown in Table 5.3.

In these tables, ƀ corresponds to a space and *nl* corresponds to a new line. Notice how different the two tables are. It is easy to see that a dictionary designed for compressing LaTeX documents would not work very well when compressing C programs. However, we generally want techniques that will be able to compress a variety of source outputs. If we want to compress computer files, we do not want to change techniques based on the content of the file. Rather, we would like the technique to *adapt* to the characteristics of the source output. We discuss adaptive-dictionary-based techniques in the next section.

TABLE 5.3 Thirty most frequently occurring pairs of characters in a collection of C programs containing 64,983 characters.

Pair	Count	Pair	Count
ƀƀ	5728	st	442
nlƀ	1471	le	440
; nl	1133	ut	440
in	985	f(	416
nt	739	ar	381
= ƀ	687	or	374
ƀi	662	rƀ	373
tƀ	615	en	371
ƀ =	612	er	358
);	558	ri	357
, ƀ	554	at	352
nlnl	506	pr	351
ƀf	505	te	349
eƀ	500	an	348
ƀ*	444	lo	347

5.4 Adaptive Dictionary

Most adaptive-dictionary-based techniques have their roots in two landmark papers by Jacob Ziv and Abraham Lempel in 1977 [52] and 1978 [53]. These papers provide two different approaches to adaptively building dictionaries, and each approach has given rise to a number of variations. The approaches based on the 1977 paper are said to belong to the LZ77 family (also known as LZ1), while the approaches based on the 1978 paper are said to belong to the LZ78, or LZ2, family. The transposition of the initials is a historical accident and is a convention we will observe in this book. In the following sections, we first describe an implementation of each of the adaptive-dictionary-based compression approaches followed by some of the more well-known variations.

5.4.1 The LZ77 Approach

In the LZ77 approach, the dictionary is simply a portion of the previously encoded sequence. The encoder examines the input sequence through a sliding window, as shown in Figure 5.1. The window consists of two parts, a *search buffer* that contains a portion of the recently encoded sequence and a *lookahead buffer* that contains the next portion of the sequence to be encoded. In Figure 5.1, the search buffer contains eight symbols, while the lookahead buffer contains seven symbols. In practice, the size of the buffers is significantly larger; however, for the purpose of explanation, we will keep the buffer sizes small.

To encode the sequence in the lookahead buffer, the encoder moves a search pointer back through the search buffer until it encounters a match to the first symbol in the lookahead

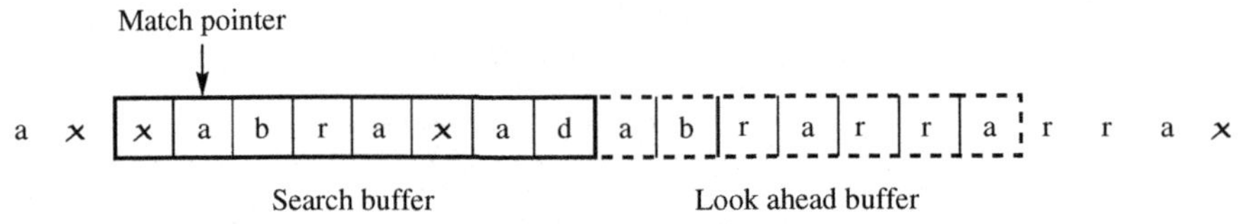

FIGURE 5.1 **Encoding using the LZ77 approach.**

buffer. The distance of the pointer from the lookahead buffer is called the *offset*. The encoder then examines the symbols following the symbol at the pointer location to see if they match consecutive symbols in the lookahead buffer. The number of consecutive symbols in the search buffer that match consecutive symbols in the lookahead buffer, starting with the first symbol, is called the length of the match. The encoder searches the search buffer for the longest match. Once the longest match has been found, the encoder encodes it with a triple $\langle o, l, c \rangle$, where o is the offset, l is the length of the match, and c is the codeword corresponding to the symbol in the lookahead buffer that follows the match. For example, in Figure 5.1 the pointer is pointing to the beginning of the longest match. The offset o in this case is 7, the length of the match l is 4, and the symbol in the lookahead buffer following the match is r.

The reason for sending the third element in the triple is to take care of the situation where no match for the symbol in the lookahead buffer can be found in the search buffer. In this case, the offset and match-length values are set to 0, and the third element of the triple is the code for the symbol itself.

If the size of the search buffer is S, the size of the window (search and lookahead buffers) is W, and the size of the source alphabet is A, then the number of bits needed to code the triple using fixed-length codes is $\lceil \log_2 S \rceil + \lceil \log_2 W \rceil + \lceil \log_2 A \rceil$. Notice that the second term is $\lceil \log_2 W \rceil$, not $\lceil \log_2 S \rceil$. The reason for this is that the length of the match can actually exceed the length of the search buffer. We will see how this happens in Example 5.4.1.

In the following example, we will look at three different possibilities that may be encountered during the coding process.

1. There is no match for the next character to be encoded in the window.
2. There is a match.
3. The matched string extends inside the lookahead buffer.

Example 5.4.1: The LZ77 Approach

Suppose the sequence to be encoded is

$$\ldots cabracadabrarrarrad \ldots$$

Suppose the length of the window is 13, the size of the lookahead buffer is 6, and the current condition is

cabraca	*dabrar*

with *dabrar* in the lookahead buffer. We look back in the already encoded portion of the window to find a match for d. As we can see, there is no match, so we transmit the triple

$\langle 0, 0, C(d)\rangle$. The first two elements of the triple show that there is no match to d in the search buffer, while $C(d)$ is the code for the character d. This seems like a wasteful way to encode a single character, and we will have more to say about this later.

For now, let's continue with the encoding process. As we have encoded a single character, we move the window by one character. Now the contents of the buffer are

abracad	*abrarr*

with *abrarr* in the lookahead buffer. Looking back from the current location, we find a match to a at an offset of two. The length of this match is one. Looking further back, we have another match for a at an offset of four; again the length of the match is one. Looking back even further in the window, we have a third match for a at an offset of seven. However, this time the length of the match is four (see Figure 5.2). So we encode the string *abra* with the triple $\langle 7, 4, C(r)\rangle$, and move the window forward by five characters. The window now contains the following characters:

adabrar	*rarrad*

Now the lookahead buffer contains the string *rarrad*. Looking back in the window, we find a match for r at an offset of one and a match length of one, and a second match at an offset of three with a match length of what at first appears to be three. It turns out we can use a match length of five instead of three.

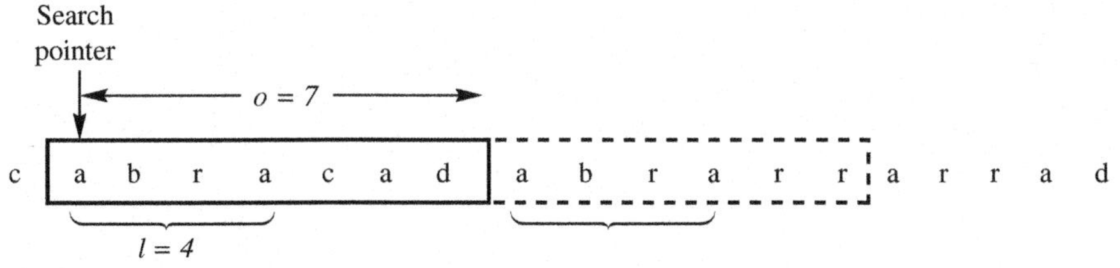

FIGURE 5.2 The encoding process.

Why this is so will become clearer when we decode the sequence. To see how the decoding works, let us assume that we have decoded the sequence *cabraca* and we receive the triples $\langle 0, 0, C(d)\rangle$, $\langle 7, 4, C(r)\rangle$, and $\langle 3, 5, C(d)\rangle$. The first triple is easy to decode; there was no match within the previously decoded string, and the next symbol is d. The decoded string is now *cabracad*. The first element of the next triple tells the decoder to move the copy pointer back seven characters, and copy four characters from that point. The decoding process works as shown in Figure 5.3.

Finally, let's see how the triple $\langle 3, 5, C(d)\rangle$ gets decoded. We move back three characters and start copying. The first three characters we copy are *rar*. The copy pointer moves once again, as shown in Figure 5.4, to copy the recently copied character r. Similarly, we copy the next character a. Even though we started copying only three characters back, we end up decoding five characters. Notice that the match only has to *start* in the search buffer; it can extend into the lookahead buffer. In fact, if the last character in the lookahead buffer had been r instead of d, followed by several more repetitions of *rar*, the entire sequence of repeated *rar*s could have been encoded with a single triple. ♦

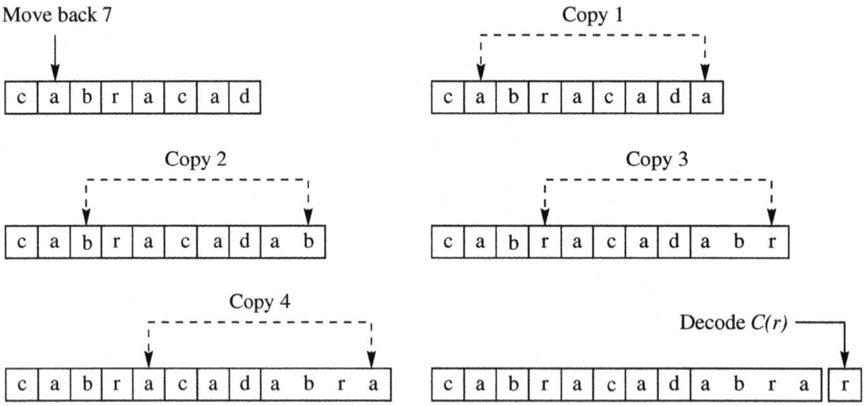

FIGURE 5.3 **Decoding the triple ⟨7, 4, *C*(*r*)⟩.**

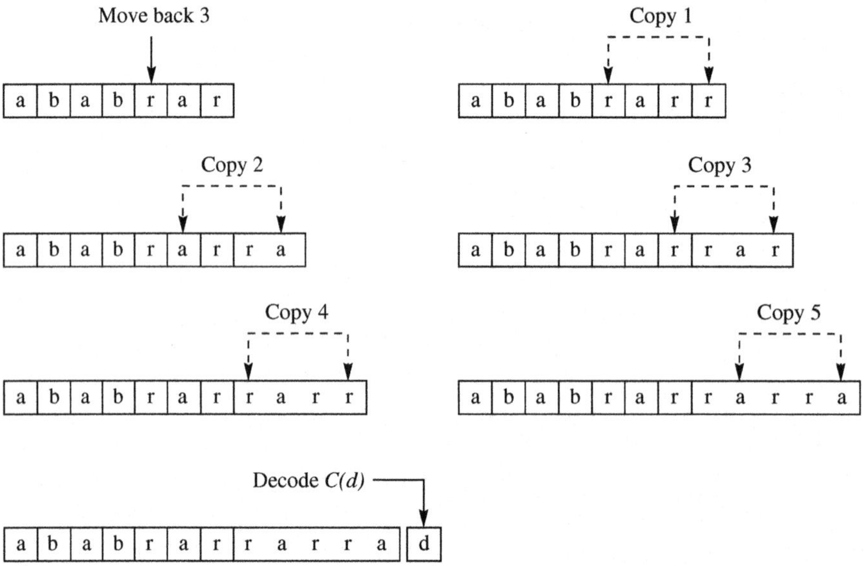

FIGURE 5.4 **Decoding the triple ⟨3, 5, *C*(*d*)⟩.**

As we can see, the LZ77 scheme is a very simple adaptive scheme that requires no prior knowledge of the source and seems to require no assumptions about the characteristics of the source. The authors of this algorithm showed that the performance of this algorithm asymptotically approached the best that could be obtained by using a scheme that had full knowledge about the statistics of the source. While this may be true asymptotically, in practice there are a number of ways of improving the performance of the LZ77 algorithm as described here. Furthermore, by using the recent portions of the sequence, there is an assumption of sorts being used here—that is, that patterns recur “close” together. As we shall see, the authors removed this "assumption" and came up with an entirely different adaptive-dictionary-based scheme in LZ78. Before we get to that, let us look at the different variations of the LZ77 algorithm.

Variations on the LZ77 Theme

There are a number of ways that the LZ77 scheme can be made more efficient, and most of these have appeared in the literature. Many of the improvements deal with the efficient encoding of the triples. In the description of the LZ77 algorithm, we assumed that the triples were encoded using a fixed-length code. However, if we were willing to accept more complexity, we could encode the triples using variable-length codes. As we saw in earlier chapters, these codes can be adaptive or, if we were willing to use a two-pass algorithm, they can be semiadaptive. Popular compression packages, such as PKZip, Zip, LHarc, PNG, gzip, and ARJ, all use an LZ77-based algorithm followed by a variable-length coder.

Other variations on the LZ77 algorithm include varying the size of the search and lookahead buffers. To make the search buffer large requires the development of more effective search strategies. Such strategies can be implemented more effectively if the contents of the search buffer are stored in a manner conducive to fast searches.

The simplest modification to the LZ77 algorithm, and one that is used by most variations of the LZ77 algorithm, is to eliminate the situation where we use a triple to encode a single character. Use of a triple is highly inefficient, especially if a large number of characters occur infrequently. The modification to get rid of this inefficiency is simply the addition of a flag bit, to indicate whether what follows is the codeword for a single symbol. By using this flag bit we also get rid of the necessity for the third element of the triple. Now all we need to do is to send a pair of values corresponding to the offset and length of match. This modification to the LZ77 algorithm, due to J.A. Storer and T.G. Syzmanski, is referred to as LZSS [54,55].

5.4.2 The LZ78 Approach

The LZ77 approach implicitly assumes that like patterns will occur close together. It makes use of this structure by using the recent past of the sequence as the dictionary for encoding. However, this means that any pattern that recurs over a period longer than that covered by the coder window will not be captured. The worst-case situation would be where the sequence to be encoded was periodic with a period longer than the search buffer. Consider Figure 5.5.

This is a periodic sequence with a period of nine. If the search buffer had been just one symbol longer, this sequence could have been significantly compressed. As it stands, none of the new symbols will have a match in the search buffer and will have to be represented by separate codewords. As this involves sending along overhead (a 1-bit flag for LZSS and a triple for the original LZ77 algorithm), the net result will be an expansion rather than a compression.

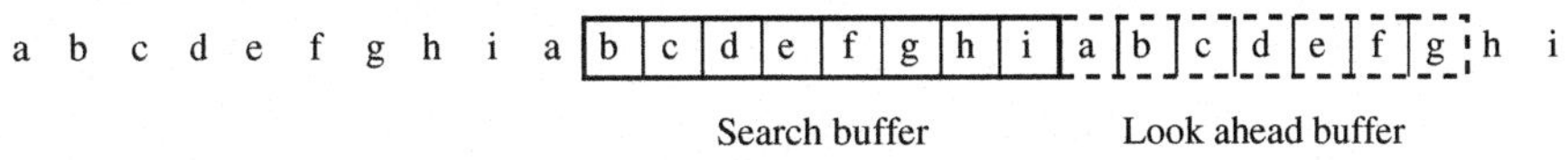

FIGURE 5.5 The Achilles' heel of LZ77.

TABLE 5.4 The initial dictionary.

Index	Entry
1	w
2	a
3	b

Although this is an extreme situation, there are less drastic circumstances in which the finite view of the past would be a drawback. The LZ78 algorithm solves this problem by dropping the reliance on the search buffer and keeping an explicit dictionary. This dictionary has to be built at both the encoder and decoder, and care must be taken that the dictionaries are built in an identical manner. The inputs are coded as a double $\langle i, c\rangle$, with i being an index corresponding to the dictionary entry that was the longest match to the input and c being the code for the character in the input following the matched portion of the input. As in the case of LZ77, the index value of 0 is used in the case of no match. This double then becomes the newest entry in the dictionary. Thus, each new entry into the dictionary is one new symbol concatenated with an existing dictionary entry. To see how the LZ78 algorithm works, consider the following example.

Example 5.4.2: The LZ78 Approach

Let us encode the following sequence using the LZ78 approach:[1]

$$wabba\not{b}wabba\not{b}wabba\not{b}wabba\not{b}woo\not{b}woo\not{b}woo.$$

where $\not{b}$ stands for space. Initially, the dictionary is empty, so the first few symbols encountered are encoded with the index value set to 0. The first three encoder outputs are $\langle 0, C(w)\rangle$, $\langle 0, C(a)\rangle$, and $\langle 0, C(b)\rangle$, and the dictionary looks like Table 5.4.

The fourth symbol is a b, which is the third entry in the dictionary. If we append the next symbol, we would get the pattern ba, which is not in the dictionary, so we encode these two symbols as $\langle 3, C(a)\rangle$ and add the pattern ba as the fourth entry in the dictionary. Continuing in this fashion, the encoder output and the dictionary develop as in Table 5.5. Notice that the entries in the dictionary generally keep getting longer, and if this particular sentence was repeated often, as it is in the song, after a while the entire sentence would be an entry in the dictionary. ♦

While the LZ78 algorithm has the ability to capture patterns and hold them indefinitely, it also has a rather serious drawback. As seen from the example, the dictionary keeps growing without bound. In a practical situation, we would have to stop the growth of the dictionary at some stage and then either prune it back or treat the encoding as a fixed dictionary scheme. We will discuss some possible approaches when we study applications of dictionary coding.

[1] "The Monster Song" from *Sesame Street*.

TABLE 5.5 Development of the dictionary.

	Dictionary	
Encoder Output	Index	Entry
$\langle 0, C(w)\rangle$	01	w
$\langle 0, C(a)\rangle$	02	a
$\langle 0, C(b)\rangle$	03	b
$\langle 3, C(a)\rangle$	04	ba
$\langle 0, C(\not b)\rangle$	05	$\not b$
$\langle 1, C(a)\rangle$	06	wa
$\langle 3, C(b)\rangle$	07	bb
$\langle 2, C(\not b)\rangle$	08	$a\not b$
$\langle 6, C(b)\rangle$	09	wab
$\langle 4, C(\not b)\rangle$	10	$ba\not b$
$\langle 9, C(b)\rangle$	11	$wabb$
$\langle 8, C(w)\rangle$	12	$a\not b w$
$\langle 0, C(o)\rangle$	13	o
$\langle 13, C(\not b)\rangle$	14	$o\not b$
$\langle 1, C(o)\rangle$	15	wo
$\langle 14, C(w)\rangle$	16	$o\not b w$
$\langle 13, C(o)\rangle$	17	oo

Variations on the LZ78 Theme—The LZW Algorithm

There are a number of ways the LZ78 algorithm can be modified, and as is the case with the LZ77 algorithm, anything that can be modified probably has been. The most well-known modification, one that initially sparked much of the interest in the LZ algorithms, is a modification by Terry Welch known as LZW [56]. Welch proposed a technique for removing the necessity of encoding the second element of the pair $\langle i, c\rangle$. That is, the encoder would only send the index to the dictionary. In order to do this, the dictionary has to be primed with all the letters of the source alphabet. The input to the encoder is accumulated in a pattern p as long as p is contained in the dictionary. If the addition of another letter a results in a pattern $p * a$ ($*$ denotes concatenation) that is not in the dictionary, then the index of p is transmitted to the receiver, the pattern $p * a$ is added to the dictionary, and we start another pattern with the letter a. The LZW algorithm is best understood with an example. In the following two examples, we will look at the encoder and decoder operations for the same sequence used to explain the LZ78 algorithm.

Example 5.4.3: The LZW Algorithm—Encoding

We will use the sequence previously used to demonstrate the LZ78 algorithm as our input sequence:

$$wabba\not b wabba\not b wabba\not b wabba\not b woo\not b woo\not b woo$$

Assuming that the alphabet for the source is $\{\not b, a, b, o, w\}$, the LZW dictionary initially looks like Table 5.6.

TABLE 5.6 Initial LZW dictionary.

Index	Entry
1	$\not b$
2	a
3	b
4	o
5	w

TABLE 5.7 Constructing the 12th entry of the LZW dictionary.

Index	Entry
01	$\not b$
02	a
03	b
04	o
05	w
06	wa
07	ab
08	bb
09	ba
10	$a\not b$
11	$\not b$ w
12	$w\ldots$

The encoder first encounters the letter w. This "pattern" is in the dictionary so we concatenate the next letter to it, forming the pattern wa. This pattern is not in the dictionary, so we encode w with its dictionary index 5, add the pattern wa to the dictionary as the sixth element of the dictionary, and begin a new pattern starting with the letter a. As a is in the dictionary, we concatenate the next element b to form the pattern ab. This pattern is not in the dictionary, so we encode a with its dictionary index value 2, add the pattern ab to the dictionary as the seventh element of the dictionary, and start constructing a new pattern with the letter b. We continue in this manner, constructing two-letter patterns, until we reach the letter w in the second $wabba$. At this point, the output of the encoder consists entirely of indices from the initial dictionary: 5 2 3 3 2 1. The dictionary looks like Table 5.7. (The 12th entry in the dictionary is still under construction.) The next symbol in the sequence is a. Concatenating this to w, we get the pattern wa. This pattern already exists in the dictionary (item 6), so we read the next symbol, which is b. Concatenating this to wa, we get the pattern wab. This pattern does not exist in the dictionary, so we include it as the 12th entry in the dictionary and start a new pattern with the symbol b. We also encode wa with its index value of 6. Notice that after a series of two-letter entries, we now have a three-letter entry. As the encoding progresses, the length of the entries keeps increasing. The longer entries in the dictionary indicate that the dictionary is capturing more of the structure in the sequence. The dictionary at the end of the

TABLE 5.8 **The LZW dictionary for encoding wabba$\not b$wabba$\not b$wabba$\not b$wabba$\not b$woo$\not b$woo$\not b$woo.**

Index	Entry	Index	Entry
01	$\not b$	14	$a\not b w$
02	a	15	$wabb$
03	b	16	$ba\not b$
04	o	17	$\not b wa$
05	w	18	abb
06	wa	19	$ba\not b w$
07	ab	20	wo
08	bb	21	oo
09	ba	22	$o\not b$
10	$a\not b$	23	$\not b wo$
11	$\not b w$	24	$oo\not b$
12	wab	25	$\not b woo$
13	bba		

encoding process is shown in Table 5.8. Notice that the 12th through the 19th entries are all either three or four letters in length. Then we encounter the pattern woo for the first time and we drop back to two-letter patterns for three more entries, after which we go back to entries of increasing length.

The encoder output sequence is 5 2 3 3 2 1 6 8 10 12 9 11 7 16 5 4 4 11 21 23 4. ♦

Example 5.4.4: The LZW Algorithm—Decoding

In this example we will take the encoder output from the previous example and decode it using the LZW algorithm. The encoder output sequence in the previous example was

5 2 3 3 2 1 6 8 10 12 9 11 7 16 5 4 4 11 21 23 4

This becomes the decoder input sequence. The decoder starts with the same initial dictionary as the encoder (Table 5.6).

The index value 5 corresponds to the letter w, so we decode w as the first element of our sequence. At the same time, we begin construction of the next element of the dictionary in order to mimic the dictionary construction procedure of the encoder. We start with the letter w. This pattern exists in the dictionary, so we do not add it to the dictionary and continue with the decoding process. The next decoder input is 2, which is the index corresponding to the letter a. We decode an a and concatenate it with our current pattern to form the pattern wa. As this does not exist in the dictionary, we add it as the sixth element of the dictionary and start a new pattern beginning with the letter a. The next four inputs 3 3 2 1 correspond to the letters $bba\not b$ and generate the dictionary entries ab, bb, ba, and $a\not b$. The dictionary now looks like Table 5.9, where the 11th entry is under construction.

The next input is 6, which is the index of the pattern wa. Therefore, we decode a w and an a. We first concatenate w to the existing pattern, which is $\not b$ and form the pattern $\not b w$. As $\not b w$

TABLE 5.9 Constructing the 11th entry of the LZW dictionary while decoding.

Index	Entry
01	$\not b$
02	a
03	b
04	o
05	w
06	wa
07	ab
08	bb
09	ba
10	$a\not b$
11	$\not b \ldots$

does not exist in the dictionary, it becomes the 11th entry. The new pattern now starts with the letter w. We had previously decoded the letter a, which we now concatenate to w to obtain the pattern wa. This pattern is contained in the dictionary, so we decode the next input, which is 8. This corresponds to the entry bb in the dictionary. We decode the first b and concatenate it to the pattern wa to get the pattern wab. This pattern does not exist in the dictionary, so we add it as the 12th entry in the dictionary and start a new pattern with the letter b. Decoding the second b and concatenating it to the new pattern, we get the pattern bb. This pattern exists in the dictionary, so we decode the next element in the sequence of encoder outputs. Continuing in this fashion, we can decode the entire sequence. Notice that the dictionary being constructed by the decoder is identical to that constructed by the encoder. ♦

There is one particular situation in which the method of decoding the LZW algorithm described above breaks down. Suppose we had a source with an alphabet $\mathcal{A} = \{a, b\}$, and we were to encode the sequence beginning with $abababab\ldots$. The encoding process is still the same. We begin with the initial dictionary shown in Table 5.10 and end up with the final dictionary shown in Table 5.11.

The transmitted sequence is 1235 .... This looks like a relatively straightforward sequence to decode. However, when we try to do so, we run into a snag. Let us go through the decoding process and see what happens.

TABLE 5.10 Initial dictionary for *abababab*.

Index	Entry
1	a
2	b

TABLE 5.11 Final dictionary for *abababab*.

Index	Entry
1	*a*
2	*b*
3	*ab*
4	*ba*
5	*aba*
6	*abab*
7	*b*...

TABLE 5.12 Constructing the fourth entry of the dictionary while decoding.

Index	Entry
1	*a*
2	*b*
3	*ab*
4	*b*...

TABLE 5.13 Constructing the fifth entry (stage one).

Index	Entry
1	*a*
2	*b*
3	*ab*
4	*ba*
5	*a*...

We begin with the same initial dictionary as the encoder (Table 5.10). The first two elements in the received sequence 1235 ... are decoded as a and b, giving rise to the third dictionary entry ab, and the beginning of the next pattern to be entered in the dictionary, b. The dictionary at this point is shown in Table 5.12.

The next input to the decoder is 3. This corresponds to the dictionary entry ab. Decoding each in turn, we first concatenate a to the pattern under construction to get ba. This pattern is not contained in the dictionary, so we add this to the dictionary (keep in mind, we have not used the b from ab yet), which now looks like Table 5.13.

The new entry starts with the letter a. We have only used the first letter from the pair ab. Therefore, we now concatenate b to a to obtain the pattern ab. This pattern is contained in the dictionary, so we continue with the decoding process. The dictionary at this stage looks like Table 5.14.

TABLE 5.14 Constructing the fifth entry (stage two).

Index	Entry
1	a
2	b
3	ab
4	ba
5	$ab\ldots$

TABLE 5.15 Completion of the fifth entry.

Index	Entry
1	a
2	b
3	ab
4	ba
5	aba
6	$a\ldots$

The first four entries in the dictionary are complete, while the fifth entry is still under construction. However, the very next input to the decoder is 5, which corresponds to the incomplete entry! How do we decode an index for which we do not as yet have a complete dictionary entry?

The situation is actually not as bad as it looks. (Of course, if it were, we would not now be studying LZW.) While we may not have a fifth entry for the dictionary, we do have the beginnings of the fifth entry, which is $ab\ldots$. Let us, for the moment, pretend that we do indeed have the fifth entry and continue with the decoding process. If we had a fifth entry, the first two letters of the entry would be a and b. Concatenating a to the partial new entry we get the pattern aba. This pattern is not contained in the dictionary, so we add this to our dictionary, which now looks like Table 5.15. Notice that we now have the fifth entry in the dictionary, which is aba. We have already decoded the ab portion of aba. We can now decode the last letter a and continue on our merry way.

This means that the LZW decoder has to contain an exception handler to handle the special case of decoding an index that does not have a corresponding complete entry in the decoder dictionary.

5.5 Applications

Since the publication of Terry Welch's article [56], there has been a steadily increasing number of applications that use some variant of the LZ78 algorithm. Among the LZ78 variants, by far the most popular is the LZW algorithm. In this section, we describe two applications of LZW:

GIF and V.42 bis. While the LZW algorithm was initially the algorithm of choice, patent concerns led to increasing use of the LZ77 algorithm. The most popular implementation of the LZ77 algorithm is the *deflate* algorithm initially designed by Phil Katz. It is part of the popular zlib library developed by Jean-Loup Gailly and Mark Adler. Jean-Loup Gailly also used *deflate* in the widely used *gzip* algorithm. The *deflate* algorithm is also used in PNG, which we describe below.

5.5.1 File Compression—UNIX `compress`

The UNIX `compress` command is one of the earlier applications of LZW. The size of the dictionary is adaptive. We start with a dictionary of size 512. This means that the transmitted codewords are 9 bits long. Once the dictionary has filled up, the size of the dictionary is doubled to 1024 entries. The codewords transmitted at this point have 10 bits. The size of the dictionary is progressively doubled as it fills up. In this way, during the earlier part of the coding process when the strings in the dictionary are not very long, the codewords used to encode them also have fewer bits. The maximum size of the codeword, $b_{\max}$, can be set by the user to between 9 and 16, with 16 bits being the default. Once the dictionary contains $2^{b_{\max}}$ entries, `compress` becomes a static dictionary coding technique. At this point the algorithm monitors the compression ratio. If the compression ratio falls below a threshold, the dictionary is flushed, and the dictionary building process is restarted. This way, the dictionary always reflects the local characteristics of the source.

5.5.2 Image Compression—The Graphics Interchange Format (GIF)

The Graphics Interchange Format (GIF) was developed by CompuServe Information Service to encode graphical images. It is another implementation of the LZW algorithm and is very similar to the `compress` command. The compressed image is stored with the first byte being the minimum number of bits b per pixel in the original image. For the images we have been using as examples, this would be eight. The binary number 2^b is defined to be the *clear code*. This code is used to reset all compression and decompression parameters to a start-up state. The initial size of the dictionary is 2^{b+1}. When this fills up, the dictionary size is doubled, as was done in the `compress` algorithm, until the maximum dictionary size of 4096 is reached. At this point, the compression algorithm behaves like a static dictionary algorithm. The codewords from the LZW algorithm are stored in blocks of characters. The characters are 8 bits long, and the maximum block size is 255. Each block is preceded by a header that contains the block size. The block is terminated by a block terminator consisting of eight 0s. The end of the compressed image is denoted by an end-of-information code with a value of $2^b + 1$. This codeword should appear before the block terminator.

GIF has become quite popular for encoding all kinds of images, both computer-generated and "natural" images. While GIF works well with computer-generated graphical images and pseudocolor or color-mapped images, it is generally not the most efficient way to losslessly compress images of natural scenes, photographs, satellite images, and so on. In Table 5.16, we give the file sizes for the GIF-encoded test images. For comparison, we also include the file sizes for arithmetic coding the original images and arithmetic coding the differences.

TABLE 5.16 Comparison of GIF with arithmetic coding.

Image	GIF	Arithmetic Coding of Pixel Values	Arithmetic Coding of Pixel Differences
Sena	51,085	53,431	31,847
Sensin	60,649	58,306	37,126
Earth	34,276	38,248	32,137
Omaha	61,580	56,061	51,393

Notice that even if we account for the extra overhead in the GIF files, for these images GIF barely holds its own even with simple arithmetic coding of the original pixels. While this might seem odd at first, if we examine the image on a pixel level, we see that there are very few repetitive patterns compared to a text source. Some images, like the Earth image, contain large regions of constant values. In the dictionary coding approach, these regions become single entries in the dictionary. Therefore, for images like these, the straightforward dictionary coding approach does hold its own. However, for most other images, it would probably be preferable to perform some preprocessing to obtain a sequence more amenable to dictionary coding. The Portable Network Graphics (PNG) standard described next develops an appropriate preprocessor to take advantage of the fact that in natural images the pixel-to-pixel variation is generally small. We will also revisit this subject in Chapter 7.

5.5.3 Image Compression—Portable Network Graphics (PNG)

The PNG standard is one of the first standards to be collaboratively developed over the Internet. The impetus for it was an announcement in December 1994 by Unisys (which had acquired the patent for LZW from Sperry) and CompuServe that they would start charging royalties to authors of software that included support for GIF. The announcement resulted in an uproar in the segment of the compression community that formed the core of the Usenet group comp.compression. The community decided that a patent-free replacement for GIF should be developed, and within three months, PNG was born. (For a more detailed history of PNG as well as software and much more, go to the PNG website maintained by Greg Roelof, http://www.libpng.org/pub/png/.)

Unlike GIF, the compression algorithm used in PNG is based on LZ77. In particular, it is based on the *deflate* [17] implementation of LZ77. This implementation allows for match lengths of between 3 and 258. At each step, the encoder examines three bytes. If it cannot find a match of at least three bytes, it puts out the first byte and examines the next three bytes. So, at each step it either puts out the value of a single byte, or *literal*, or the pair $< match\ length, offset >$. The alphabets of the *literal* and *match length* are combined to form an alphabet of size 286 (indexed by 0–285). The indices 0–255 represent literal bytes, and the index 256 is an end-of-block symbol. The remaining 29 indices represent codes for ranges of lengths between 3 and 258, as shown in Table 5.17. The table shows the index, the number of selector bits to follow the index, and the lengths represented by the index and selector bits. For example, the index 277 represents the range of lengths from 67 to 82. To specify which of the 16 values has actually occurred, the code is followed by 4 selector bits.

The index values are represented using a Huffman code. The Huffman code is specified in Table 5.18.

The *offset* can take on values between 1 and 32,768. These values are divided into 30 ranges. The 30 range values are encoded using a Huffman code (different from the Huffman code for the *literal* and *length* values) and the code is followed by a number of selector bits to specify the particular distance within the range.

We have mentioned earlier that in natural images there is not a great deal of repetition of sequences of pixel values. However, pixel values that are spatially close also tend to have values that are similar. The PNG standard makes use of this structure by estimating the value of a pixel based on its causal neighbors and subtracting this estimate from the pixel. The difference modulo 256 is then encoded in place of the original pixel. There are four different ways of getting the estimate (five if you include no estimation), and PNG allows the use of a different method of estimation for each row. The first way is to use the pixel from the row above as the estimate. The second method is to use the pixel to the left as the estimate. The third method uses the average of the pixel above and the pixel to the left. The final method is a bit more complex. An initial estimate of the pixel is first made by adding the pixel to the left and the pixel above and subtracting the pixel to the upper left. Then the pixel that is closest to the initial estimate (upper, left, or upper left) is taken as the estimate. A comparison of the performance of PNG and GIF on our standard image set is shown in Table 5.19. The PNG method clearly outperforms GIF.

5.5.4 Compression over Modems—V.42 bis

The ITU-T Recommendation V.42 bis is a compression standard devised for use over a telephone network along with error-correcting procedures described in CCITT Recommendation V.42. This algorithm is used in modems connecting computers to remote users. The algorithm described in this recommendation operates in two modes, a transparent mode and a compressed mode. In the transparent mode, the data are transmitted in uncompressed form, while in the compressed mode an LZW algorithm is used to provide compression.

The reason for the existence of two modes is that at times the data being transmitted do not have repetitive structure and, therefore, cannot be compressed using the LZW algorithm. In this case, the use of a compression algorithm may even result in expansion. In these situations, it is better to send the data in an uncompressed form. A random data stream would cause the dictionary to grow without any long patterns as elements of the dictionary. This means that most of the time the transmitted codeword would represent a single letter from the source alphabet. As the dictionary size is much larger than the source alphabet size, the number of bits required to represent an element in the dictionary is much more than the number of bits required to represent a source letter. Therefore, if we tried to compress a sequence that does not contain repeating patterns, we would end up with more bits to transmit than if we had not performed any compression. Data without repetitive structure are often encountered when a previously compressed file is transferred over the telephone lines.

TABLE 5.17 Codes for representations of *match length* [17].

Index	# of Selector Bits	Length	Index	# of Selector Bits	Length	Index	# of Selector Bits	Length
257	0	3	267	1	15,16	277	4	67–82
258	0	4	268	1	17,18	278	4	83–98
259	0	5	269	2	19–22	279	4	099–114
260	0	6	270	2	23–26	280	4	115–130
261	0	7	271	2	27–30	281	5	131–162
262	0	8	272	2	31–34	282	5	163–194
263	0	9	273	3	35–42	283	5	195–226
264	0	10	274	3	43–50	284	5	227–257
265	1	11, 12	275	3	51–58	285	0	258
266	1	13, 14	276	3	59–66			

TABLE 5.18 Huffman codes for the *match length* alphabet [17].

Index Ranges	# of Bits	Binary Codes
0–143	8	00110000 through 10111111
144–255	9	110010000 through 111111111
256–279	7	0000000 through 0010111
280–287	8	11000000 through 11000111

TABLE 5.19 Comparison of PNG with GIF and arithmetic coding.

Image	PNG	GIF	Arithmetic Coding of Pixel Values	Arithmetic Coding of Pixel Differences
Sena	31,577	51,085	53,431	31,847
Sensin	34,488	60,649	58,306	37,126
Earth	26,995	34,276	38,248	32,137
Omaha	50,185	61,580	56,061	51,393

TABLE 5.20 Control codewords in compressed mode.

Codeword	Name	Description
0	ETM	Enter transparent mode
1	FLUSH	Flush data
2	STEPUP	Increment codeword size

The V.42 bis recommendation suggests periodic testing of the output of the compression algorithm to see if data expansion is taking place. The exact nature of the test is not specified in the recommendation.

In the compressed mode, the system uses LZW compression with a variable-size dictionary. The initial dictionary size is negotiated at the time a link is established between the transmitter and receiver. The V.42 bis recommendation suggests a value of 2048 for the dictionary size. It specifies that the minimum size of the dictionary is to be 512. Suppose the initial negotiations result in a dictionary size of 512. This means that our codewords that are indices into the dictionary will be 9 bits long. Actually, the entire 512 indices do not correspond to input strings; three entries in the dictionary are reserved for control codewords. These codewords in the compressed mode are shown in Table 5.20.

When the numbers of entries in the dictionary exceed a prearranged threshold C_3, the encoder sends the STEPUP control code, and the codeword size is incremented by 1 bit. At the same time, the threshold C_3 is also doubled. When all available dictionary entries are filled, the algorithm initiates a reuse procedure. The location of the first string entry in the dictionary is maintained in a variable N_5. Starting from N_5, a counter C_1 is incremented until it finds a dictionary entry that is not a prefix to any other dictionary entry. The fact that this entry is not a prefix to another dictionary entry means that this pattern has not been encountered since it was created. Furthermore, because of the way it was located, this pattern has been around the longest among patterns of this kind. This reuse procedure enables the algorithm to prune the dictionary of strings that may have been encountered in the past but have not been encountered recently, on a continual basis. In this way, the dictionary is always matched to the current source statistics.

To reduce the effect of errors, the CCITT recommends setting a maximum string length. This maximum length is negotiated at link setup. The CCITT recommends a range of 6–250, with a default value of 6.

The V.42 bis recommendation avoids the need for an exception handler for the case where the decoder receives a codeword corresponding to an incomplete entry by forbidding the use of the last entry in the dictionary. Instead of transmitting the codeword corresponding to the last entry, the recommendation requires the sending of the codewords corresponding to the constituents of the last entry. In the example used to demonstrate this quirk of the LZW algorithm, the V.42 bis recommendation would have forced us to send the codewords 3 and 1 instead of transmitting the codeword 5.

5.6 Beyond Compression—Lempel-Ziv Complexity ★

In Chapter 2, we described the notion of Kolmogorov complexity, noting that the Kolmogorov complexity of a sequence was not something that we could necessarily compute. Lempel and Ziv provided a definition of complexity in 1976 [57] that is computable. This notion views complexity as the level of diversity in a sequence. Thus, a sequence that consists of repetitions of the same letter would not be a complex sequence, while a sequence that appears random would be complex. The development of this definition of complexity formed the basis for the two compression approaches that were to follow in the next two years. Since then, this definition of complexity has had an impact in areas other than compression. We will take a brief detour into these applications as they confirm our assertion that data compression algorithms are ways of modeling and understanding how information is organized in data.

The computation of Lempel-Ziv (LZ) complexity is performed in a manner very similar to the LZ77 algorithm. The basic idea is to parse the input sequence into phrases that do not exist in the past of the sequence. Thus, a phrase is built up by copying from the past of the sequence until we add a letter that makes the phrase distinct from anything in the past of the sequence. Let's see how this works by using a familiar phrase, *cabracadabrarrarrad*. We begin with an empty history so the first letter c is a unique phrase, as are the letters a, b, and r. The next letter in the sequence is a, which has occurred in the past, so we add the next letter

c to form the new phrase ac. Continuing in this manner we get the following parsing of the sequence:

$$c \cdot a \cdot b \cdot r \cdot ac \cdot ad \cdot abrar \cdot rarrad$$

The number of distinct phrases in this parsing, which in this case is eight, is the LZ complexity of this sequence. We can see how the LZ77 and LZ78 adaptive dictionary compression followed from this method of computing complexity. However, beyond compression this approach also provides a method for computing similarity between sequences.

Consider three sequences:

$$R : abracadabra$$
$$Q : rrarracadab$$
$$S : arbacbacarc$$

Parsing these sequences we get

$$R : a \cdot b \cdot r \cdot ac \cdot ad \cdot abra$$
$$Q : r \cdot ra \cdot rrac \cdot ad \cdot ab$$
$$S : a \cdot r \cdot b \cdot ac \cdot baca \cdot rc$$

If we denote the LZ complexity of a sequence by $c(\cdot)$, then we have $c(R) = 6$, $c(Q) = 5$, and $c(S) = 6$. One way to compare the relatedness of two sequences is to concatenate them and see how that affects the complexity. Let's compare the relatedness of sequence R with sequence Q and sequence S. Sequence RQ is

$$abracadabrarrarracadab$$

If we parse this sequence, we get

$$a \cdot b \cdot r \cdot ac \cdot ad \cdot abrar \cdot rarrac \cdot adab$$

or $c(RQ) = 8$. In other words, the concatenation of Q to R raised the complexity from six to eight. If we now concatenate sequence R with sequence S we get sequence RS:

$$abracadabraarbacbacarc$$

Parsing this sequence we obtain

$$a \cdot b \cdot r \cdot ac \cdot ad \cdot abraa \cdot rb \cdot acb \cdot acar \cdot c$$

which gives us an LZ complexity $c(RS) = 10$. Thus in this case the complexity increased from six to ten, or twice as much as the increase when we concatenated sequences R and Q. Based on this, we can claim that sequence R is closer to sequence Q than it is to sequence S.

One way of looking at the change in complexity is to note that the parsing of the appended sequence (S or Q) was conducted using the dictionary of sequence R. Thus the increase in complexity is a measure of the number of additional words required to represent the appended sequence. This idea has been used for automatic language recognition and authorship attribution [58]. By using a known piece of text as the original sequence (sequence R in the example here), we capture patterns specific to the text. When we append different fragments of text to the original by monitoring the increase in the LZ complexity (perhaps normalized by the size of the text), we can get an idea of how close the texts we are examining are to the original. This can give us an indication of which texts have been authored by the same individual, and it can provide an easy method for automatically detecting the language of a piece of text. The Lempel-Ziv approach thus provides a model for the original. The amount of increase in complexity is an indication of how closely the text being examined fits the model of the original text.

One of the richest sources of information is the DNA sequence of living organisms. By comparing the same gene from different organisms, we can estimate the evolutionary relationship between organisms. Model testing of the type described above has been shown to be highly effective in estimating evolutionary distances between organisms [59,60]. Keeping within the realm of bioinformatics, one of the ways in which biologists attempt to understand the function of various biological molecules, such as proteins, is to compare the composition of proteins with unknown functions with proteins with known functions. LZ complexity has been shown to be a valuable tool [61–63] for comparing DNA and protein sequences.

LZ complexity has also been used in modeling and analyzing various biomedical signals [64]. Changes in LZ complexity have been proposed as a means of detecting epileptic seizures [65] and monitoring of brain activity during anesthesia [66] and in patients with Alzheimer's disease [67]. LZ complexity has also been used to analyze cardiac signals [68]. In short, the concept of LZ complexity has ramifications well outside the area of compression and is a useful tool for understanding signals of all sorts.

5.7 Summary

In this chapter, we have introduced techniques that keep a dictionary of recurring patterns and transmit the index of those patterns instead of the patterns themselves in order to achieve compression. There are a number of ways the dictionary can be constructed.

- In applications where certain patterns consistently recur, we can build application-specific static dictionaries. Care should be taken not to use these dictionaries outside their area of intended application. Otherwise, we may end up with data expansion instead of data compression.
- The dictionary can be the source output itself. This is the approach used by the LZ77 algorithm. When using this algorithm, there is an implicit assumption that recurrence of a pattern is a local phenomenon.
- This assumption is removed in the LZ78 approach, which dynamically constructs a dictionary from patterns observed in the source output.

Dictionary-based algorithms are being used to compress all kinds of data; however, care should be taken with their use. This approach is most useful when structural constraints restrict the frequently occurring patterns to a small subset of all possible patterns. This is the case with text, as well as computer-to-computer communication.

Further Reading

1. *Text Compression*, by T.C. Bell, J.G. Cleary, and I.H. Witten [1], provides an excellent exposition of dictionary-based coding techniques.
2. *The Data Compression Book*, by M. Nelson and J.-L. Gailley [69], also does a good job of describing the Ziv-Lempel algorithms. There is also a very nice description of some of the software implementation aspects.
3. *Data Compression*, by G. Held and T.R. Marshall [70], contains a description of digram coding under the name "diatomic coding." The book also includes BASIC programs that help in the design of dictionaries.
4. The PNG algorithm is described in a very accessible manner in "PNG Lossless Compression," by G. Roelofs [71] in the *Lossless Compression Handbook*.
5. A more in-depth look at dictionary compression is provided in "Dictionary-Based Data Compression: An Algorithmic Perspective," by S.C. Şahinalp and N.M. Rajpoot [72] in the *Lossless Compression Handbook*.

5.8 Projects and Problems

1. To study the effect of dictionary size on the efficiency of a static dictionary technique, we can modify Equation (5.1) so that it gives the rate as a function of both p and the dictionary size M. Plot the rate as a function of p for different values of M, and discuss the trade-offs involved in selecting larger or smaller values of M.
2. Design and implement a digram coder for text files of interest to you.

   (a) Study the effect of the dictionary size and the size of the text file being encoded on the amount of compression.

   (b) Use the digram coder on files that are not similar to the ones you used to design the digram coder. How much does this affect your compression?

3. Given an initial dictionary consisting of the letters $a\ b\ r\ y\ \not{b}$, encode the following message using the LZW algorithm: $a\not{b}bar\not{b}array\not{b}by\not{b}barrayar\not{b}bay$.
4. A sequence is encoded using the LZW algorithm and the initial dictionary shown in Table 5.21.

   (a) The output of the LZW encoder is the following sequence:

6	3	4	5	2	3	1	6	2	9	11	16	12	14	4	20	10	8	23	13

   Decode this sequence.

   (b) Encode the decoded sequence using the same initial dictionary. Does your answer match the sequence given above?

TABLE 5.21 **Initial dictionary for Problem 4.**

Index	Entry
1	a
2	$\not b$
3	h
4	i
5	s
6	t

TABLE 5.22 **Initial dictionary for Problem 5.**

Index	Entry
1	a
2	$\not b$
3	r
4	t

5. A sequence is encoded using the LZW algorithm and the initial dictionary shown in Table 5.22.

(a) The output of the LZW encoder is the following sequence:

3	1	4	6	8	4	2	1	2	5	10	6	11	13	6

Decode this sequence.

(b) Encode the decoded sequence using the same initial dictionary. Does your answer match the sequence given above?

6. Encode the following sequence using the LZ77 algorithm:

$$barrayar\not b bar\not b by\not b barrayar\not b bay$$

Assume you have a window size of 30 with a lookahead buffer of size 15. Furthermore, assume that $C(a) = 1$, $C(b) = 2$, $C(\not b) = 3$, $C(r) = 4$, and $C(y) = 5$.

7. A sequence is encoded using the LZ77 algorithm. Given that $C(a) = 1$, $C(\not b) = 2$, $C(r) = 3$, and $C(t) = 4$, decode the following sequence of triples:

$$\langle 0, 0, 3\rangle \langle 0, 0, 1\rangle \langle 0, 0, 4\rangle \langle 2, 8, 2\rangle \langle 3, 1, 2\rangle \langle 0, 0, 3\rangle \langle 6, 4, 4\rangle \langle 9, 5, 4\rangle$$

Assume that the size of the window is 20 and the size of the lookahead buffer is 10. Encode the decoded sequence and make sure you get the same sequence of triples.

8. Given the following initial dictionary and the received sequence below, build an LZW dictionary *and* decode the transmitted sequence.
Received sequence: 4, 5, 3, 1, 2, 8, 2, 7, 9, 7, 4
Initial dictionary:

Index	Entry
1	S
2	ƀ
3	I
4	T
5	H

9. Given the following initial dictionary and the received sequence below, build an LZW dictionary and decode the transmitted sequence.
Received sequence: 3, 2, 1, 5, 3, 4, 1, 6, 5
Initial dictionary:

Index	Entry
1	I
2	N
3	O
4	P

10. Use the Lempel-Ziv algorithm to classify paragraphs from plays by William Shakespeare and George Bernard Shaw. Compress the Hamlet soliloquy (*To be or not to be, that is the question ....*) and use that as the reference sequence to compress different pieces of text from Shakespeare's plays *Hamlet* and *Julius Caeser*, and Shaw's play *Caesar and Cleopatra*. Use the compressed file sizes to make your classification decision. Plot the percentage of correct decisions against the size of the text used for classification.

6

Context-Based Compression

6.1 Overview

In this chapter, we present a number of techniques that use minimal prior assumptions about the statistics of the data. Instead they use the context of the data being encoded and the past history of the data to provide more efficient compression. We will look at a number of schemes that are principally used for the compression of text. These schemes use the context in which the data occurs in different ways.

6.2 Introduction

In Chapters 3 and 4, we learned that we get more compression when the message that is being coded has a more skewed set of probabilities. By "skewed" we mean that certain symbols occur with much higher probability than others in the sequence to be encoded. So it makes sense to look for ways to represent the message that would result in greater skew. One very effective way to do so is to look at the probability of occurrence of a letter in the context in which it occurs. That is, we do not look at each symbol in a sequence as if it had just happened out of the blue. Instead, we examine the history of the sequence before determining the likely probabilities of different values that the symbol can take.

In the case of English text, Shannon [4] showed the role of context in two very interesting experiments. In the first, a portion of text was selected and a subject (possibly his wife, Mary Shannon) was asked to guess each letter. If she guessed correctly, she was told that she was correct and moved on to the next letter. If she guessed incorrectly, she was told the correct answer and again moved on to the next letter. Here is a result from one of these experiments. The dashes represent the letters that were correctly guessed.

Actual Text	T H E R O O M W A S N O T V E R Y L I G H T A S M A L L O B L O N G
Subject Performance	_ _ _ _ R O O _ _ _ _ _ _ N O T _ V _ _ _ _ _ _ I _ _ _ _ _ _ S M _ _ _ _ O B L _ _ _

Notice that there is a good chance that the subject will guess the letter, especially if the letter is at the end of a word or if the word is clear from the context. If we now represent the original sequence by the subject performance, we would get a very different set of probabilities for the values that each element of the sequence takes on. The probabilities are definitely much more skewed in the second row: the "letter" _ occurs with high probability. If a mathematical twin of the subject were available at the other end, we could send the "reduced" sentence in the second row and have the twin go through the same guessing process to come up with the original sequence.

In the second experiment, the subject was allowed to continue guessing until she had guessed the correct letter and the number of guesses required to correctly predict the letter was noted. Again, the subject guessed correctly most of the time, resulting in 1 being the most probable number. The existence of a mathematical twin at the receiving end would allow this skewed sequence to represent the original sequence to the receiver. Shannon used his experiments to come up with upper and lower bounds for the English alphabet (1.3 bits per letter and 0.6 bits per letter, respectively).

The difficulty with using these experiments is that the human subject was much better at predicting the next letter in a sequence than any mathematical predictor we can develop. Grammar is hypothesized to be innate to humans [73], in which case development of a predictor as efficient as a human for language is not possible in the near future. However, the experiments do provide an approach to compression that is useful for compression of all types of sequences, not simply language representations.

If a sequence of symbols being encoded does not consist of independent occurrences of the symbols, then knowledge of which symbols have occurred in the neighborhood of the symbol being encoded will give us a much better idea of the value of the symbol being encoded. If we know the context in which a symbol occurs, we can guess what the value of the symbol is with a much greater likelihood of success. This is just another way of saying that, given the context, some symbols will occur with much higher probability, or much lower, than without the context. That is, the probability distribution given the context is more skewed. If the context is known to both encoder and decoder, we can use this skewed distribution to perform the encoding, thus increasing the level of compression. The decoder can use its knowledge of the context to determine the distribution to be used for decoding. If we can somehow group like contexts together, it is quite likely that the symbols following these contexts will be the same, allowing for the use of some very simple and efficient compression strategies. We can see that context can play an important role in enhancing compression, and in this chapter we will look at several different ways of using context.

Consider the encoding of the word *probability*. Suppose we have already encoded the first four letters, and we want to code the fifth letter, a. If we ignore the first four letters, the probability of the letter a is about 0.06. If we use the information that the previous letter is b, this reduces the probability of several letters, such as q and z, occurring and boosts the probability of an a occurring. In this example, b would be the first-order context for a, ob would be the second-order context for a, and so on. Using more letters to define the context in which a occurs, or higher-order contexts, will generally increase the probability of the occurrence of a in this example and, hence, reduce the number of bits required to encode its occurrence. Therefore, what we would like to do is to encode each letter using the probability of its occurrence with respect to a context of high order.

If we want to have probabilities with respect to all possible high-order contexts, this might be an overwhelming amount of information. Consider an alphabet of size M. The number of first-order contexts is M, the number of second-order contexts is M^2, and so on. Therefore, if we wanted to encode a sequence from an alphabet of size 256 using contexts of order 5, we would need 256^5, or about 1.09951×10^{12} probability distributions! This is not a practical alternative. A set of algorithms that resolve this problem in a very simple and elegant way is based on the *prediction with partial match* approach. We will describe this in the next section.

6.3 Prediction with Partial Match (*ppm*)

The best-known context-based algorithm is the *ppm* algorithm, first proposed by Cleary and Witten [74] in 1984. It has not been as popular as the various Ziv-Lempel-based algorithms mainly because of the faster execution speeds of the latter algorithms. Lately, with the development of more efficient variants, *ppm*-based algorithms are becoming increasingly popular.

The idea of the *ppm* algorithm is elegantly simple. We would like to use large contexts to determine the probability of the symbol being encoded. However, the use of large contexts would require us to estimate and store an extremely large number of conditional probabilities, which might not be feasible. Instead of estimating these probabilities ahead of time, we can reduce the burden by estimating the probabilities as the coding proceeds. This way we only need to store those contexts that have occurred in the sequence being encoded. This is a much smaller number than the number of all possible contexts. While this mitigates the problem of storage, it also means that, especially at the beginning of an encoding, we will need to code letters that have not occurred previously in this context. In order to handle this situation, the source coder alphabet always contains an escape symbol, which is used to signal that the letter to be encoded has not been seen in this context.

6.3.1 The Basic Algorithm

The basic algorithm initially attempts to use the largest context. The size of the largest context is predetermined. If the symbol to be encoded has not previously been encountered in this context, an escape symbol is encoded and the algorithm attempts to use the next smaller context. If the symbol has not occurred in this context either, the size of the context is further reduced. This process continues until either we obtain a context that has previously been encountered with this symbol or we arrive at the conclusion that the symbol has not been encountered previously in *any* context. In this case, we use probability of $1/M$ to encode the symbol, where M is the size of the source alphabet. For example, when coding the a of *probability*,

TABLE 6.1 Count array for −1 order context.

Letter	Count	*Cum_Count*
t	1	1
h	1	2
i	1	3
s	1	4
e	1	5
b̸	1	6
Total_Count		6

we would first attempt to see if the string *proba* has previously occurred—that is, if *a* had previously occurred in the context of *prob*. If not, we would encode an escape and see if *a* had occurred in the context of *rob*. If the string *roba* had not occurred previously, we would again send an escape symbol and try the context *ob*. Continuing in this manner, we would try the context *b*, and failing that, we would see if the letter *a* (with a zero-order context) had occurred previously. If *a* was being encountered for the first time, we would use a model in which all letters occur with equal probability to encode *a*. This equiprobable model is sometimes referred to as the context of order −1.

As the development of the probabilities with respect to each context is an adaptive process, the count corresponding to a symbol is updated each time the symbol is encountered. The number of counts to be assigned to the escape symbol is not obvious, and a number of different approaches have been used. One approach used by Cleary and Witten is to give the escape symbol a count of one, thus inflating the total count by one. Cleary and Witten call this method of assigning counts Method A and the resulting algorithm *ppma*. We will describe some of the other ways of assigning counts to the escape symbol later in this section.

Before we delve into some of the details, let's work through an example to see how all this works together. As we will be using arithmetic coding to encode the symbols, you might wish to refresh your memory of the arithmetic coding algorithms.

Example 6.3.1:

Let's encode the sequence

$$thisb\!\!/isb\!\!/theb\!\!/tithe$$

Assuming we have already encoded the initial seven characters *thisb̸is*, the various counts and *Cum_Count* arrays to be used in the arithmetic coding of the symbols are shown in Tables 6.1–6.4. In this example, we are assuming that the longest context length is two. This is a rather small value and is used here to keep the size of the example reasonably small. A more common value for the longest context length is five.

We will assume that the word length for arithmetic coding is six. Thus, $l = 000000$ and $u = 111111$. As *thisb̸is* has already been encoded, the next letter to be encoded is *b̸*. The second-order context for this letter is *is*. Looking at Table 6.4, we can see that the letter *b̸* is the first letter in this context with a *Cum_Count* value of 1. As the *Total_Count* in this case

TABLE 6.2 Count array for zero-order context.

Letter	Count	*Cum_Count*
t	1	1
h	1	2
i	2	4
s	2	6
$\not b$	1	7
$\langle Esc \rangle$	1	8
Total_Count		8

is 2, the update equations for the lower and upper limits are

$$l = 0 + \left\lfloor (63 - 0 + 1) \times \frac{0}{2} \right\rfloor = 0 = 000000$$

$$u = 0 + \left\lfloor (63 - 0 + 1) \times \frac{1}{2} \right\rfloor - 1 = 31 = 011111$$

As the MSBs of both l and u are the same, we shift that bit out and shift a 0 into the LSB of l and a 1 into the LSB of u. The transmitted sequence, lower limit, and upper limit after the update are

$$\begin{aligned} \text{Transmitted sequence}: &\quad 0 \\ l: &\quad 000000 \\ u: &\quad 111111 \end{aligned}$$

We also update the counts in Tables 6.2–6.4.

The next letter to be encoded in the sequence is t. The second-order context is $s\not b$. Looking at Table 6.4, we can see that t has not appeared before in this context. We therefore encode an escape symbol. Using the counts listed in Table 6.4, we update the lower and upper limits:

$$l = 0 + \left\lfloor (63 - 0 + 1) \times \frac{1}{2} \right\rfloor = 32 = 100000$$

$$u = 0 + \left\lfloor (63 - 0 + 1) \times \frac{2}{2} \right\rfloor - 1 = 63 = 111111$$

Again, the MSBs of l and u are the same, so we shift the bit out and shift 0 into the LSB of l and 1 into u, restoring l to a value of 0 and u to a value of 63. The transmitted sequence is now 01. After transmitting the escape, we look at the first-order context of t, which is $\not b$. Looking at Table 6.3, we can see that t has not previously occurred in this context. To let the decoder know this, we transmit another escape. Updating the limits, we get

$$l = 0 + \left\lfloor (63 - 0 + 1) \times \frac{1}{2} \right\rfloor = 32 = 100000$$

$$u = 0 + \left\lfloor (63 - 0 + 1) \times \frac{2}{2} \right\rfloor - 1 = 63 = 111111$$

TABLE 6.3 Count array for first-order contexts.

Context	Letter	Count	*Cum_Count*
t	h	1	1
	$\langle Esc \rangle$	1	2
	Total_Count		2
h	i	1	1
	$\langle Esc \rangle$	1	2
	Total_Count		2
i	s	2	2
	$\langle Esc \rangle$	1	3
	Total_Count		3
$\not b$	i	1	1
	$\langle Esc \rangle$	1	2
	Total_Count		2
s	$\not b$	1	1
	$\langle Esc \rangle$	1	2
	Total_Count		2

TABLE 6.4 Count array for second-order contexts.

Context	Letter	Count	*Cum_Count*
th	i	1	1
	$\langle Esc \rangle$	1	2
	Total_Count		2
hi	s	1	1
	$\langle Esc \rangle$	1	2
	Total_Count		2
is	$\not b$	1	1
	$\langle Esc \rangle$	1	2
	Total_Count		2
$s\not b$	i	1	1
	$\langle Esc \rangle$	1	2
	Total_Count		2
$\not b i$	s	1	1
	$\langle Esc \rangle$	1	2
	Total_Count		2

TABLE 6.5 Updated count array for zero-order context.

Letter	Count	*Cum_Count*
t	1	1
h	1	2
i	2	4
s	2	6
$\not b$	2	8
$\langle Esc \rangle$	1	9
Total_Count		9

As the MSBs of l and u are the same, we shift the MSB out and shift 0 into the LSB of l and 1 into the LSB of u. The transmitted sequence is now 011. Having escaped out of the first-order contexts, we examine Table 6.5, the updated version of Table 6.2, to see if we can encode t using a zero-order context. Indeed we can, and using the Cum_Count array, we can update l and u:

$$l = 0 + \left\lfloor (63 - 0 + 1) \times \frac{0}{9} \right\rfloor = 0 = 000000$$

$$u = 0 + \left\lfloor (63 - 0 + 1) \times \frac{1}{9} \right\rfloor - 1 = 6 = 000110$$

The three most significant bits of both l and u are the same, so we shift them out. After the update we get

Transmitted sequence:	011000
l :	000000
u :	110111

The next letter to be encoded is h. The second-order context $\not b t$ has not occurred previously, so we move directly to the first-order context t. The letter h has occurred previously in this context, so we update l and u and obtain

Transmitted sequence :	0110000
l :	000000
u :	110101

The method of encoding should now be clear. At this point the various counts are as shown in Tables 6.6–6.8. ♦

Now that we have an idea of how the *ppm* algorithm works, let's examine some of the variations.

TABLE 6.6 Count array for zero-order context.

Letter	Count	*Cum_Count*
t	2	2
h	2	4
i	2	6
s	2	8
ƀ	2	10
⟨*Esc*⟩	1	11
Total_Count		11

TABLE 6.7 Count array for first-order contexts.

Context	Letter	Count	*Cum_Count*
t	*h*	2	2
	⟨*Esc*⟩	1	3
	Total_Count		3
h	*i*	1	1
	⟨*Esc*⟩	1	2
	Total_Count		2
i	*s*	2	2
	⟨*Esc*⟩	1	3
	Total_Count		3
ƀ	*i*	1	1
	t	1	2
	⟨*Esc*⟩	1	3
	Total_Count		3
s	ƀ	2	2
	⟨*Esc*⟩	1	3
	Total_Count		3

6.3.2 The Escape Symbol

In our example we used a count of one for the escape symbol, thus inflating the total count in each context by one. Cleary and Witten call this Method A, and the corresponding algorithm is referred to as *ppma*. There is really no obvious justification for assigning a count of one to the escape symbol. For that matter, there is no obvious method of assigning counts to the escape symbol. There have been various methods reported in the literature.

Another method described by Cleary and Witten is to reduce the counts of each symbol by one and assign these counts to the escape symbol. For example, suppose in a given sequence *a* occurs 10 times in the context of *prob*, *l* occurs 9 times, and *o* occurs 3 times in the same

TABLE 6.8 Count array for second-order contexts.

Context	Letter	Count	*Cum_Count*
th	*i*	1	1
	⟨*Esc*⟩	1	2
	Total_Count		2
hi	*s*	1	1
	⟨*Esc*⟩	1	2
	Total_Count		2
is	*b̸*	2	2
	⟨*Esc*⟩	1	3
	Total_Count		3
sb̸	*i*	1	1
	t	1	2
	⟨*Esc*⟩	1	3
	Total_Count		3
b̸i	*s*	1	1
	⟨*Esc*⟩	1	2
	Total_Count		2
b̸t	*h*	1	1
	⟨*Esc*⟩	1	2
	Total_Count		2

TABLE 6.9 Counts using Method A.

Context	Symbol	Count
prob	*a*	10
	l	09
	o	03
	⟨*Esc*⟩	01
Total_Count		23

context (e.g., *problem, proboscis*, etc.). In Method A, we assign a count of one to the escape symbol, resulting in a total count of 23, which is one more than the number of times *prob* has occurred. This situation is shown in Table 6.9.

In this second method, known as Method B, we reduce the count of each of the symbols a, l, and o by one and give the escape symbol a count of three, resulting in the counts shown in Table 6.10.

The reasoning behind this approach is that if in a particular context more symbols can occur, there is a greater likelihood that there is a symbol in this context that has not occurred

TABLE 6.10 Counts using Method B.

Context	Symbol	Count
prob	a	09
	l	08
	o	02
	$\langle Esc \rangle$	03
Total_Count		22

TABLE 6.11 Counts using Method C.

Context	Symbol	Count
prob	a	10
	1	09
	o	03
	$\langle Esc \rangle$	03
Total_Count		25

before. This increases the likelihood that the escape symbol will be used. Therefore, we should assign a higher probability to the escape symbol.

A variant of Method B, appropriately named Method C, was proposed by Moffat [75]. In Method C, the count assigned to the escape symbol is the number of symbols that have occurred in that context. In this respect, Method C is similar to Method B. The difference comes in the fact that, instead of "robbing" this from the counts of individual symbols, the total count is inflated by this amount. This situation is shown in Table 6.11.

While there is some variation in performance depending on the characteristics of the data being encoded, Method C, on average, seems to provide the best performance of the three methods for assigning counts to the escape symbol.

6.3.3 Length of Context

It would seem that as far as the maximum length of the contexts is concerned, more is better. However, this is not necessarily true. A longer maximum length will usually result in a higher probability if the symbol to be encoded has a non zero count with respect to that context. However, a long maximum length also means a higher probability of long sequences of escapes, which in turn can increase the number of bits used to encode the sequence. If we plot the compression performance versus maximum context length, we see an initial sharp increase in performance until some value of the maximum length, followed by a steady drop as the maximum length is further increased. The value at which we see a downturn in performance changes depending on the characteristics of the source sequence.

An alternative to the policy of a fixed maximum length is used in the algorithm *ppm** [76]. This algorithm uses the fact that long contexts that give only a single prediction are seldom followed by a new symbol. If *mike* has always been followed by y in the past, it will probably not be followed by ƀ the next time it is encountered. Contexts that are always followed by the same symbol are called *deterministic* contexts. The *ppm** algorithm first looks for the longest deterministic context. If the symbol to be encoded does not occur in that context, an escape symbol is encoded and the algorithm defaults to the maximum context length. This approach seems to provide a small but significant amount of improvement over the basic algorithm. Currently, the best variant of the *ppm** algorithm is the *ppmz* algorithm by Charles Bloom. Details of the *ppmz* algorithm as well as implementations of the algorithm can be found at http://www.cbloom.com/src/ppmz.html.

6.3.4 The Exclusion Principle

The basic idea behind arithmetic coding is the division of the unit interval into subintervals, each of which represents a particular letter. The smaller the subinterval, the more bits are required to distinguish it from other subintervals. If we can reduce the number of symbols to be represented, the number of subintervals goes down as well. This in turn means that the sizes of the subintervals increase, leading to a reduction in the number of bits required for encoding. The exclusion principle used in *ppm* provides this kind of reduction in rate. Suppose we have been compressing a text sequence and come upon the sequence *proba*, and suppose we are trying to encode the letter a. Suppose also that the state of the two-letter context *ob* and the one-letter context b are as shown in Table 6.12.

First, we attempt to encode a with the two-letter context. As a does not occur in this context, we issue an escape symbol and reduce the size of the context. Looking at the table for the one-letter context b, we see that a does occur in this context with a count of 4 out of a total possible count of 21. Notice that other letters in this context include l and o. However, by sending the escape symbol in the context of *ob*, we have already signalled to the decoder that the symbol being encoded is not any of the letters that have previously been encountered in the context of *ob*. Therefore, we can increase the size of the subinterval corresponding to a by temporarily removing l and o from the table. Instead of using Table 6.12, we use Table 6.13 to encode a. This exclusion of symbols from contexts on a temporary basis can result in cumulatively significant savings in terms of rate.

TABLE 6.12 Counts for exclusion example.

Context	Symbol	Count
ob	l	10
	o	03
	$\langle Esc \rangle$	02
Total_Count		15
b	l	05
	o	03
	a	04
	r	02
	e	02
	$\langle Esc \rangle$	05
Total_Count		21

TABLE 6.13 Modified table used for exclusion example.

Context	Symbol	Count
b	a	4
	r	2
	e	2
	$\langle Esc \rangle$	3
Total_Count		11

You may have noticed that we keep talking about small but significant savings. In lossless compression schemes, there is usually a basic principle, such as the idea of prediction with partial match, followed by a host of relatively small modifications. The importance of these modifications should not be underestimated, because together they often provide the margin of compression that makes a particular scheme competitive.

6.4 The Burrows-Wheeler Transform

The Burrows-Wheeler transform (BWT) algorithm also uses the context of the symbol being encoded, but in a very different way, for lossless compression. The transform that is a major part of this algorithm was developed by Wheeler in 1983. However, the BWT compression algorithm, which uses this transform, saw the light of day in 1994 [77]. Unlike most of the previous algorithms we have looked at, the BWT algorithm requires that the entire sequence to be coded be available to the encoder before the coding takes place. Also, unlike most of the previous algorithms, the decoding procedure is not immediately evident once we know the encoding procedure. We will first describe the encoding procedure. If it is not clear how this particular encoding can be reversed, bear with us and we will get to it.

The algorithm can be summarized as follows. Given a sequence of length N, we create $N - 1$ other sequences where each of these $N - 1$ sequences is a cyclic shift of the original sequence. These N sequences are arranged in lexicographic order. The encoder then transmits the sequence of length N created by taking the last letter of each sorted, cyclically shifted, sequence. This sequence of last letters L, and the position of the original sequence in the sorted list, are coded and sent to the decoder. As we shall see, this information is sufficient to recover the original sequence.

We start with a sequence of length N and end with a representation that contains $N + 1$ elements. However, this sequence has a structure that makes it highly amenable to compression. In particular we will use a method of coding called move to front (*mtf*), which is particularly effective on the type of structure exhibited by the sequence L.

Before we describe the *mtf* approach, let us work through an example to generate the L sequence.

Example 6.4.1:

Let's encode the sequence

thisb̸isb̸the

We start with all the cyclic permutations of this sequence. As there are a total of 11 characters, there are 11 permutations, shown in Table 6.14.

Now let's sort these sequences in lexicographic (dictionary) order (Table 6.15). The sequence of the last letters L in this case is

L : *sshtthb̸iib̸e*

Notice how like letters have come together. If we had a longer sequence of letters, the *runs* of like letters would have been even longer. The *mtf* algorithm, which we will describe later, takes advantage of these runs.

The original sequence appears as sequence number 10 in the sorted list, so the encoding of the sequence consists of the sequence L and the index value 10. ♦

Now that we have an encoding of the sequence, let's see how we can decode the original sequence by using the sequence L and the index to the original sequence in the sorted list. The important thing to note is that all the elements of the initial sequence are contained in L. We just need to figure out the permutation that will let us recover the original sequence.

The first step in obtaining the permutation is to generate the sequence F consisting of the first element of each row. That is simple to do because we lexicographically ordered the sequences. Therefore, the sequence F is simply the sequence L in lexicographic order. In our example this means that F is given as

F : *b̸b̸ehhiisstt*

We can use L and F to generate the original sequence. Look at Table 6.15 containing the cyclically shifted sequences sorted in lexicographic order. Because each row is a cyclical shift, the letter in the first column of any row is the letter appearing after the last column in the row in the original sequence. If we know that the original sequence is in the kth row, then we can begin unraveling the original sequence starting with the kth element of F.

TABLE 6.14 Permutations of *thisb̸isb̸the*.

00	*t*	*h*	*i*	*s*	*b̸*	*i*	*s*	*b̸*	*t*	*h*	*e*
01	*h*	*i*	*s*	*b̸*	*i*	*s*	*b̸*	*t*	*h*	*e*	*t*
02	*i*	*s*	*b̸*	*i*	*s*	*b̸*	*t*	*h*	*e*	*t*	*h*
03	*s*	*b̸*	*i*	*s*	*b̸*	*t*	*h*	*e*	*t*	*h*	*i*
04	*b̸*	*i*	*s*	*b̸*	*t*	*h*	*e*	*t*	*h*	*i*	*s*
05	*i*	*s*	*b̸*	*t*	*h*	*e*	*t*	*h*	*i*	*s*	*b̸*
06	*s*	*b̸*	*t*	*h*	*e*	*t*	*h*	*i*	*s*	*b̸*	*i*
07	*b̸*	*t*	*h*	*e*	*t*	*h*	*i*	*s*	*b̸*	*i*	*s*
08	*t*	*h*	*e*	*t*	*h*	*i*	*s*	*b̸*	*i*	*s*	*b̸*
09	*h*	*e*	*t*	*h*	*i*	*s*	*b̸*	*i*	*s*	*b̸*	*t*
10	*e*	*t*	*h*	*i*	*s*	*b̸*	*i*	*s*	*b̸*	*t*	*h*

TABLE 6.15 Sequences sorted into lexicographic order.

0	$\not b$	i	s	$\not b$	t	h	e	t	h	i	s
1	$\not b$	t	h	e	t	h	i	s	$\not b$	i	s
2	e	t	h	i	s	$\not b$	i	s	$\not b$	t	h
3	h	e	t	h	i	s	$\not b$	i	s	$\not b$	t
4	h	i	s	$\not b$	i	s	$\not b$	t	h	e	t
5	i	s	$\not b$	i	s	$\not b$	t	h	e	t	h
6	i	s	$\not b$	t	h	e	t	h	i	s	$\not b$
7	s	$\not b$	i	s	$\not b$	t	h	e	t	h	i
8	s	$\not b$	t	h	e	t	h	i	s	$\not b$	i
9	t	h	e	t	h	i	s	$\not b$	i	s	$\not b$
10	t	h	i	s	$\not b$	i	s	$\not b$	t	h	e

Example 6.4.2:

In our example

$$F = \begin{bmatrix} \not b \\ \not b \\ e \\ h \\ h \\ i \\ i \\ s \\ s \\ t \\ t \end{bmatrix} \quad L = \begin{bmatrix} s \\ s \\ h \\ t \\ t \\ h \\ \not b \\ i \\ i \\ \not b \\ e \end{bmatrix}$$

the original sequence is sequence number 10, so the first letter in the original sequence is $F[10] = t$. To find the letter following t, we look for t in the array L. There are two t's in L. Which should we use? The t in F that we are working with is the lower of two t's, so we pick the lower of two t's in L. This is $L[4]$. Therefore, the next letter in our reconstructed sequence is $F[4] = h$. The reconstructed sequence to this point is th. To find the next letter, we look for h in the L array. Again there are two h's. The h at $F[4]$ is the lower of two h's in F, so we pick the lower of the two h's in L. This is the fifth element of L, so the next element in our decoded sequence is $F[5] = i$. The decoded sequence to this point is thi. The process continues as depicted in Figure 6.1 to generate the original sequence. ♦

Why go through all this trouble? After all, we are going from a sequence of length N to another sequence of length N plus an index value. It appears that we are actually causing expansion instead of compression. The answer is that the sequence L can be compressed much more efficiently than the original sequence. Even in our small example, we have runs of like symbols. This will happen a lot more when N is large. Consider a large sample of text that has been cyclically shifted and sorted in a list A. Consider all the rows of A beginning with

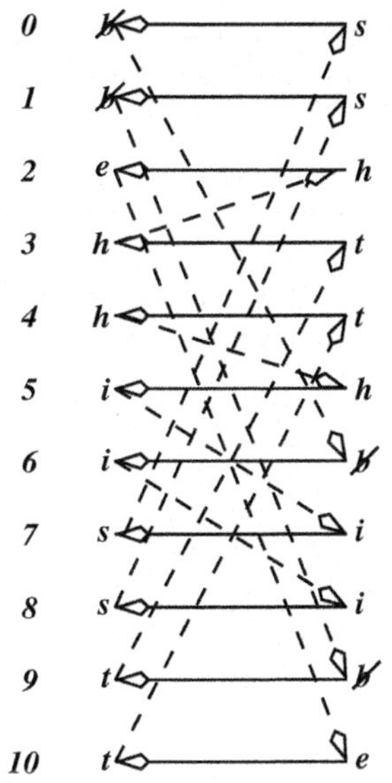

FIGURE 6.1 **Decoding process.**

heb̸. With high probability, *heb̸* would be preceded by t. Therefore, in L we would get a long run of t's.

6.4.1 Move-to-Front Coding

A coding scheme that takes advantage of long runs of identical symbols is move-to-front (mtf) coding. In this coding scheme, we start with some initial listing of the source alphabet. The symbol at the top of the list is assigned the number 0, the next one is assigned the number 1, and so on. The first time a particular symbol occurs, the number corresponding to its place in the list is transmitted. Then it is moved to the top of the list. If we have a run of this symbol, we transmit a sequence of 0s. This way, long runs of different symbols get transformed to a large number of 0s. Applying this technique to our example does not produce very impressive results due to the small size of the sequence, but we can see how the technique functions.

Example 6.4.3:

Let's encode $L = sshtthb\!\!/iib\!\!/e$. Let's assume that the source alphabet is given by

$$\mathcal{A} = \{b\!\!/, e, h, i, s, t\}$$

We start out with the assignment

0	1	2	3	4	5
b̸	e	h	i	s	t

The first element of L is s, which gets encoded as a 4. We then move s to the top of the list, which gives us

0	1	2	3	4	5
s	b̸	e	h	i	t

The next s is encoded as 0. Because s is already at the top of the list, we do not need to make any changes. The next letter is h, which we encode as 3. We then move h to the top of the list:

0	1	2	3	4	5
h	s	b̸	e	i	t

The next letter is t, which gets encoded as 5. Moving t to the top of the list, we get

0	1	2	3	4	5
t	h	s	b̸	e	i

The next letter is also a t, so that gets encoded as a 0.

Continuing in this fashion, we get the sequence

$$4\ 0\ 3\ 5\ 0\ 1\ 3\ 5\ 0\ 1\ 5$$

As we warned, the results are not too impressive with this small sequence, but we can see how we would get large numbers of 0s and small values if the sequence to be encoded was longer. These long sequences of 0s could then be efficiently encoded using schemes specifically designed for encoding long repetitions of the same symbol, such as *run-length coding*, which is described in the next chapter. ♦

6.5 Associative Coder of Buyanovsky (ACB)

A different approach to using context for compression is employed by the eponymous compression utility developed by George Buyanovsky. The details of this very efficient coder are not well known; however, the way the context is used is interesting and we will briefly describe this aspect of ACB. More detailed descriptions are available in [78,79]. The ACB coder develops a sorted dictionary of all encountered contexts. In this it is similar to other context-based encoders. However, it also keeps track of the *content* of these contexts. The content of a context is what appears after the context. In a traditional left-to-right reading of text, the contexts are unbounded to the left and the contents to the right (to the limits of text that has already been encoded). When encoding the coder searches for the longest match to the current context reading right to left. This is, again, not an unusual thing to do. What is interesting is what the coder does after the best match is found. Instead of simply examining the *content* corresponding to the best matched context, the coder also examines the *contents* of the coders in the neighborhood of the best matched contexts. Fenwick [78] describes this process as first finding an anchor point then searching the *contents* of the neighboring contexts for the best match. The location of the anchor point is known to both the encoder and the decoder. The location of the best *content* match is signaled to the decoder by encoding the offset δ of the context of this *content* from the anchor point. We have not specified what we mean by "best" match. The coder takes the utilitarian approach that the best match is the one that ends up providing the most compression. Thus, a longer match farther away from the anchor may not be as advantageous as a shorter match closer to the anchor because of the number of bits required to encode δ. The length of the match λ is also sent to the decoder.

The interesting aspect of this scheme is that it moves away from the idea of exactly matching the past. It provides a much richer environment and flexibility to enhance the compression and will, hopefully, provide a fruitful avenue for further research.

6.6 Dynamic Markov Compression

Quite often the probabilities of the value that the next symbol in a sequence takes on depend not only on the current value but on past values as well. The *ppm* scheme relies on this longer-range correlation. The *ppm* scheme, in some sense, reflects the application, that is, text compression, for which it is most used. Dynamic Markov compression (DMC), introduced by Cormack and Horspool [80], uses a more general framework to take advantage of relationships and correlations, or contexts, that extend beyond a single symbol.

Consider the sequence of pixels in a scanned document. The sequence consists of runs of black and white pixels. If we represent black by 0 and white by 1, we have runs of 0s and 1s. If the current value is 0, the probability that the next value is 0 is higher than if the current value was 1. The fact that we have two different sets of probabilities is reflected in the two-state model shown in Figure 6.2. Consider state A. The probability of the next value being 1 changes depending on whether we reached state A from state B or from state A itself. We can have the model reflect this by *cloning* state A, as shown in Figure 6.3, to create state A'. Now if we see a white pixel after a run of black pixels, we go to state A'. The probability that the next value will be 1 is very high in this state. This way, when we estimate probabilities for the next pixel value, we take into account not only the value of the current pixel but also the value of the previous pixel.

This process can be continued as long as we wish to take into account longer and longer histories. "As long as we wish" is a rather vague statement when it comes to implementing

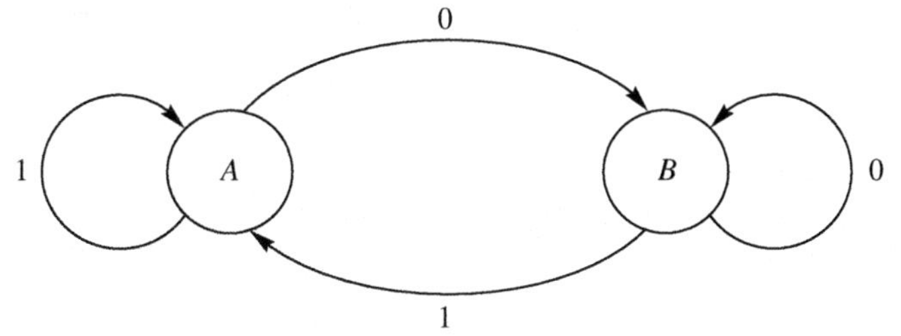

FIGURE 6.2 **A two-state model for binary sequences.**

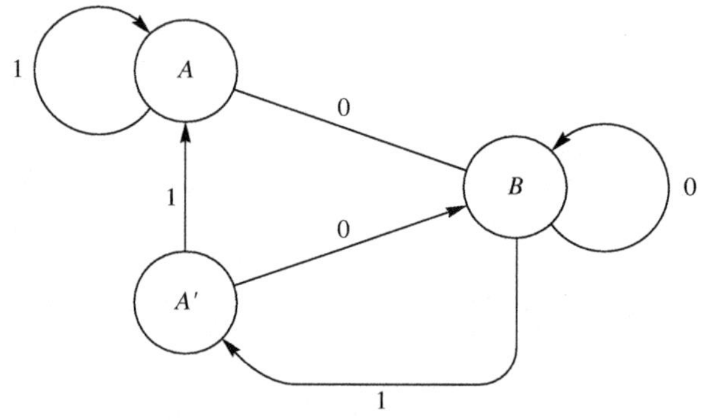

FIGURE 6.3 **A three-state model obtained by cloning.**

the algorithm. In fact, we have been rather vague about a number of implementation issues. We will attempt to rectify the situation.

There are a number of issues that need to be addressed in order to implement this algorithm:

1. What is the initial number of states?
2. How do we estimate probabilities?
3. How do we decide when a state needs to be cloned?
4. What do we do when the number of states becomes too large?

Let's answer each question in turn.

We can start the encoding process with a single state with two self-loops for 0 and 1. This state can be cloned to two and then a higher number of states. In practice, it has been found that, depending on the particular application, it is more efficient to start with a larger number of states than one.

The probabilities from a given state can be estimated by simply counting the number of times a 0 or a 1 occurs in that state divided by the number of times the particular state is occupied. For example, if in state V the number of times a 0 occurs is denoted by n_0^V and the number of times a 1 occurs is denoted by n_1^V, then

$$P(0|V) = \frac{n_0^V}{n_0^V + n_1^V}$$

$$P(1|V) = \frac{n_1^V}{n_0^V + n_1^V}$$

What if a 1 has never previously occurred in this state? This approach would assign a probability of zero to the occurrence of a 1. This means that there will be no subinterval assigned to the possibility of a 1 occurring, and when it does occur, we will not be able to

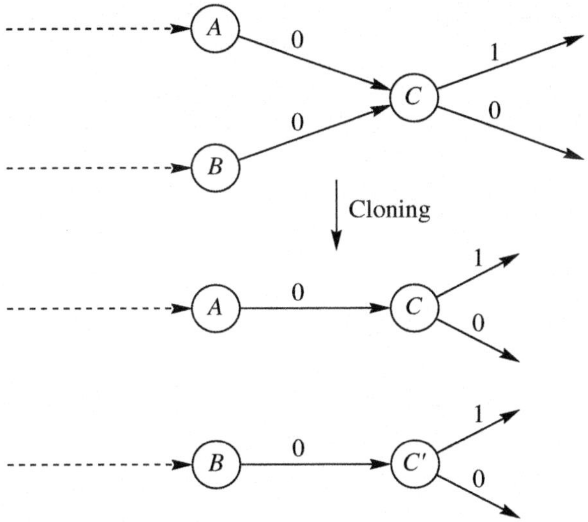

FIGURE 6.4 **The cloning process.**

represent it. In order to avoid this, instead of counting from zero, we start the count of 1s and 0s with a small number c and estimate the probabilities as

$$P(0|V) = \frac{n_0^V + c}{n_0^V + n_1^V + 2c}$$

$$P(1|V) = \frac{n_1^V + c}{n_0^V + n_1^V + 2c}$$

Whenever we have two branches leading to a state, it can be cloned. And, theoretically, cloning is never harmful. By cloning we are providing additional information to the encoder. This might not reduce the rate, but it should never result in an increase in the rate. However, cloning does increase the complexity of the coding, and hence the decoding, process. In order to control the increase in the number of states, we should only perform cloning when there is a reasonable expectation of a reduction in rate. We can do this by making sure that both paths leading to the state being considered for cloning are used often enough. Consider the situation shown in Figure 6.4. Suppose the current state is A and the next state is C. As there are two paths entering C, C is a candidate for cloning. Cormack and Horspool suggest that C be cloned if $n_0^A > T_1$ and $n_0^B > T_2$, where T_1 and T_2 are threshold values set by the user. If there are more than three paths leading to a candidate for cloning, then we check that both the number of transitions from the current state is greater than T_1 and the number of transitions from all other states to the candidate state is greater than T_2.

Finally, what do we do when, for practical reasons, we cannot accommodate any more states? A simple solution is to restart the algorithm. In order to make sure that we do not start from ground zero every time, we can train the initial state configuration using a certain number of past inputs.

6.7 Summary

The context in which a symbol occurs can be very informative about the value that the symbol takes on. If this context is known to the decoder, then this information need not be encoded; it can be inferred by the decoder. In this chapter we have looked at several creative ways in which the knowledge of the context can be used to provide compression.

Further Reading

1. The basic *ppm* algorithm is described in detail in *Text Compression*, by T.C. Bell, J.G. Cleary, and I.H. Witten [1].
2. For an excellent description of Burrows-Wheeler coding, including methods of implementation and improvements to the basic algorithm, see "Burrows-Wheeler Compression," by P. Fenwick [81] in the *Lossless Compression Handbook*.
3. The ACB algorithm is described in "Symbol Ranking and ACB Compression," by P. Fenwick [78] in the *Lossless Compression Handbook*, and in *Data Compression: The Complete Reference*, by D. Salomon [79]. The chapter by Fenwick also explores compression schemes based on Shannon's experiments.

6.8 Projects and Problems

1. Decode the bitstream generated in Example 6.3.1. Assume you have already decoded $this\not{b}is$ and Tables 6.1–6.4 are available to you.
2. Consider the sequence $the\not{b}beta\not{b}cat\not{b}ate\not{b}the\not{b}ceta\not{b}hat$:
   (a) Encode the sequence using the *ppma* algorithm and an adaptive arithmetic coder. Assume a six-letter alphabet $\{h, e, t, a, c, \not{b}\}$.
   (b) Decode the encoded sequence.
3. Consider the sequence $eta\not{b}ceta\not{b}and\not{b}beta\not{b}ceta$:
   (a) Encode the sequence using the Burrows-Wheeler transform and move-to-front coding.
   (b) Decode the encoded sequence.
4. A sequence is encoded using the Burrows-Wheeler transform. Given $L = elbkkee$, and index = 5 (we start counting from 1, not 0), find the original sequence.
5. Consider the sequence $onionopinion$:
   (a) Encode the sequence using the *ppma* algorithm and an adaptive arithmetic coder. Assume a four-letter alphabet $\{i, n, o, p\}$.
   (b) Decode the encoded sequence.

7

Lossless Image Compression

7.1 Overview

In this chapter, we examine a number of schemes used for lossless compression of images. We will look at schemes for compression of grayscale and color images as well as schemes for compression of binary images. Among these schemes are several that are a part of international standards.

7.2 Introduction

To this point in the book, we have focused on compression techniques. Although some of them may apply to some preferred applications, the focus has been on the technique rather than on the application. However, there are certain techniques for which it is impossible to separate the technique from the application. This is because the techniques rely upon the properties or characteristics of the application. Therefore, we have several chapters in this book that focus on particular applications. In this chapter, we will examine techniques specifically geared toward lossless image compression. Later chapters will examine speech, audio, and video compression.

In earlier chapters, we observed that a more skewed set of probabilities for the message being encoded results in better compression. In Chapter 6, we saw how the use of context to obtain a skewed set of probabilities can be especially effective when encoding text. We can also transform the sequence (in an invertible fashion) into another sequence that has the desired property in other ways. For example, consider the following sequence:

1	2	5	7	2	−2	0	−5	−3	−1	1	−2	−7	−4	−2	1	3	4

If we consider this sample to be fairly typical of the sequence, we can see that the probability of any given number being in the range from −7 to 7 is about the same. If we were to encode this sequence using a Huffman or arithmetic code, we would use almost 4 bits per symbol.

Instead of encoding this sequence directly, we could do the following: add two to the previous number in the sequence and send the difference between the current element in the sequence and this *predicted* value. The transmitted sequence would be

1	−1	1	0	−7	−4	0	−7	0	0	0	−5	−7	1	0	1	0	−1

This method uses a rule (add two) and the history (value of the previous symbol) to generate the new sequence. If the rule by which this *residual sequence* was generated is known to the decoder, it can recover the original sequence from the residual sequence. The length of the residual sequence is the same as the original sequence. However, notice that the residual sequence is much more likely to contain 0s, 1s, and −1s than other values. That is, the probability of 0, 1, and −1 will be significantly higher than the probabilities of other numbers. This, in turn, means that the entropy of the residual sequence will be low, and, therefore, the residual sequence can be encoded using fewer bits.

We used a particular method of prediction in this example (add two to the previous element of the sequence) that was specific to this sequence. In order to get the best possible performance, we need to find the prediction approach that is best suited to the particular data we are dealing with. We will look at several prediction schemes used for lossless image compression in the following sections.

7.2.1 The Old JPEG Standard

The Joint Photographic Experts Group (JPEG) is a joint International Standards Organization/International Telecommunication Union (ISO/ITU) committee responsible for developing standards for continuous-tone still-picture coding. The more famous standard produced by this group is the lossy image compression standard. However, at the time of the creation of the famous JPEG standard, the committee also created a lossless standard [82]. It is more or less obsolete, having been overtaken by the much more efficient JPEG-LS standard described later in this chapter. However, the old JPEG standard is still useful as a first step into examining predictive coding in images.

The old JPEG lossless still compression standard [82] provides eight different predictive schemes from which the user can select. The first scheme makes no prediction. The next seven are listed below. Three of the seven are one-dimensional predictors, and four are two-dimensional prediction schemes. Here, $I(i, j)$ is the (i, j)th pixel of the original image, and $\widehat{I}(i, j)$ is the predicted value for the (i, j)th pixel.

$$1 \quad \hat{I}(i,j) = I(i-1,j) \tag{1}$$
$$2 \quad \hat{I}(i,j) = I(i,j-1) \tag{2}$$
$$3 \quad \hat{I}(i,j) = I(i-1,j-1) \tag{3}$$
$$4 \quad \hat{I}(i,j) = I(i,j-1) + I(i-1,j) - I(i-1,j-1) \tag{4}$$
$$5 \quad \hat{I}(i,j) = I(i,j-1) + (I(i-1,j) - I(i-1,j-1))/2 \tag{5}$$
$$6 \quad \hat{I}(i,j) = I(i-1,j) + (I(i,j-1) - I(i-1,j-1))/2 \tag{6}$$
$$7 \quad \hat{I}(i,j) = (I(i,j-1) + I(i-1,j))/2 \tag{7}$$

Different images can have different structures that can best be exploited by one of these eight modes of prediction. If compression is performed in a non-real-time environment—for example, for the purposes of archiving—all eight modes of prediction can be tried, and the one that gives the most compression is used. The mode used to perform the prediction can be stored in a 3-bit header along with the compressed file. We encoded our four test images using the various JPEG modes. The residual images were encoded using adaptive arithmetic coding. The results are shown in Table 7.1.

The best results—that is, the smallest compressed file sizes—are indicated in bold in the table. From these results, we can see that a different JPEG predictor is the best for the different images. In Table 7.2, we compare the best JPEG results with the file sizes obtained using GIF and PNG. Note that PNG also uses predictive coding with four possible predictors, where each row of the image can be encoded using a different predictor. The PNG approach is described in Chapter 5.

From this comparison, even if we take into account the overhead associated with GIF, we can see that the predictive approaches are generally better suited to lossless image compression than the dictionary-based approach when the images are "natural" grayscale images. The situation is different when the images are graphic images or pseudocolor images. A possible exception could be the Earth image. The best compressed file size using the second JPEG mode and adaptive arithmetic coding is 32,137 bytes, compared to 34,276 bytes using GIF. The difference between the file sizes is not significant. We can see the reason by looking at the Earth image. Note that a significant portion of the image is the background, which is of a constant value. In dictionary coding, this would result in some very long entries that would provide significant compression. We can see that if the ratio of background to foreground were just a little different in this image, the dictionary method in GIF might have outperformed the JPEG approach. The PNG approach which allows the use of a different predictor (or no predictor) on each row, prior to dictionary coding, significantly outperforms both GIF and JPEG on this image.

TABLE 7.1 Compressed file size (in bytes) of the residual images obtained using the various JPEG prediction modes.

Image	JPEG 0	JPEG 1	JPEG 2	JPEG 3	JPEG 4	JPEG 5	JPEG 6	JPEG 7
Sena	53,431	37,220	31,559	38,261	31,055	**29,742**	33,063	32,179
Sensin	58,306	41,298	37,126	43,445	**32,429**	33,463	35,965	36,428
Earth	38,248	32,295	**32,137**	34,089	33,570	33,057	33,072	32,672
Omaha	56,061	**48,818**	51,283	53,909	53,771	53,520	52,542	52,189

TABLE 7.2 Comparison of the file sizes obtained using JPEG lossless compression, GIF, and PNG.

Image	Best JPEG	GIF	PNG
Sena	31,055	51,085	31,577
Sensin	32,429	60,649	34,488
Earth	32,137	34,276	26,995
Omaha	48,818	61,341	50,185

7.3 CALIC

The Context Adaptive Lossless Image Compression (CALIC) scheme, which came into being in response to a call for proposals for a new lossless image compression scheme in 1994 [83,84], uses both context and prediction of the pixel values. The CALIC scheme actually functions in two modes, one for grayscale images and another for bi-level images. In this section, we will concentrate on the compression of grayscale images.

In an image, a given pixel generally has a value close to one of its neighbors. Which neighbor has the closest value depends on the local structure of the image. Depending on whether there is a horizontal or vertical edge in the neighborhood of the pixel being encoded, the pixel above, or the pixel to the left, or some weighted average of neighboring pixels may give the best prediction. How close the prediction is to the pixel being encoded depends on the surrounding texture. In a region of the image with a great deal of variability, the prediction is likely to be further from the pixel being encoded than in the regions with less variability.

In order to take into account all of these factors, the algorithm has to make a determination about the environment of the pixel to be encoded. The information used to make this determination has to be available to both the encoder and decoder.

Let's take up the question of the presence of vertical or horizontal edges in the neighborhood of the pixel being encoded. To help our discussion, we will refer to Figure 7.1. In this figure, the pixel to be encoded has been marked with an *X*. The pixel above is called the north pixel, the pixel to the left is the west pixel, and so on. Note that when pixel *X* is being encoded, all other marked pixels (*N, W, NW, NE, WW, NN, NE,* and *NNE*) are available to both the encoder and decoder.

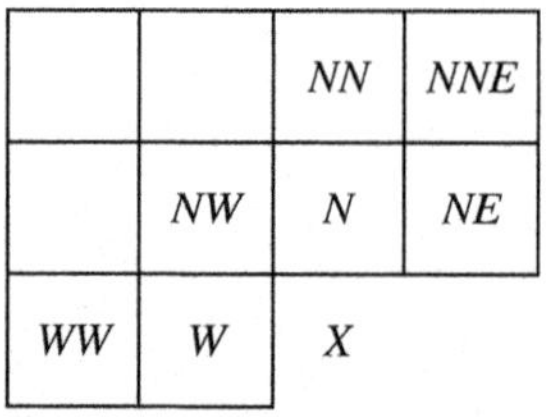

FIGURE 7.1 Labeling the neighbors of pixel X.

We can get an idea of what kinds of boundaries may or may not be in the neighborhood of X by computing

$$d_h = |W - WW| + |N - NW| + |NE - N|$$
$$d_v = |W - NW| + |N - NN| + |NE - NNE|.$$

The relative values of d_h and d_v are used to obtain the initial prediction of the pixel X. This initial prediction is then refined by taking other factors into account. If the value of d_h is much higher than the value of d_v, this will mean there is a large amount of horizontal variation, and it would be better to pick N to be the initial prediction. If, on the other hand, d_v is much larger than d_h, this would mean that there is a large amount of vertical variation, and the initial prediction is taken to be W. If the differences are more moderate or smaller, the predicted value is a weighted average of the neighboring pixels.

The exact algorithm used by CALIC to form the initial prediction is given by the following pseudocode:

```
if d_h - d_v > 80
  X̂ ← N
else if d_v - d_h > 80
  X̂ ← W
else
{
    X̂ ← (N + W)/2 + (NE - NW)/4
    if d_h - d_v > 32
      X̂ ← (X̂ + N)/2
    else if d_v - d_h > 32
      X̂ ← (X̂ + W)/2
    else if d_h - d_v > 8
      X̂ ← (3X̂ + N)/4
    else if d_v - d_h > 8
      X̂ ← (3X̂ + W)/4
}
```

Using the information about whether the pixel values are changing by large or small amounts in the vertical or horizontal direction in the neighborhood of the pixel being encoded results in a good initial prediction. In order to refine this prediction, we need some information about the interrelationships of the pixels in the neighborhood. Using this information, we can generate an offset or refinement to our initial prediction. We quantify the information about the neighborhood by first forming the vector

$$[N, W, NW, NE, NN, WW, 2N - NN, 2W - WW]$$

We then compare each component of this vector with our initial prediction $\hat{X}$. If the value of the component is less than the prediction, we replace the value with a 1; otherwise we replace it with a 0. Thus, we end up with an eight-component binary vector. If each component of the binary vector was independent, we would end up with 256 possible vectors. However, because of the dependence of various components, we actually have 144 possible configurations. We also compute a quantity that incorporates the vertical and horizontal variations and the previous error in prediction by

$$\delta = d_h + d_v + 2\left|N - \widehat{N}\right| \tag{8}$$

where $\widehat{N}$ is the predicted value of N. This range of values of δ is divided into four intervals, each represented by 2 bits. These four possibilities, along with the 144 texture descriptors, create $144 \times 4 = 576$ contexts for X. As the encoding proceeds, we keep track of how much prediction error is generated in each context and offset our initial prediction by that amount. This results in the final predicted value.

Once the prediction is obtained, the difference between the pixel value and the prediction (the prediction error or residual) has to be encoded. While the prediction process outlined above removes a lot of the structure that was in the original sequence, there is still some structure left in the residual sequence. We can take advantage of some of this structure by coding the residual in terms of its context. The context of the residual is taken to be the value of δ defined in Eq. (8). In order to reduce the complexity of the encoding, rather than using the actual value as the context, CALIC uses the range of values in which δ lies as the context. Thus:

$$\begin{aligned}
0 \leqslant \delta < q_1 &\Rightarrow \text{Context 1}\\
q_1 \leqslant \delta < q_2 &\Rightarrow \text{Context 2}\\
q_2 \leqslant \delta < q_3 &\Rightarrow \text{Context 3}\\
q_3 \leqslant \delta < q_4 &\Rightarrow \text{Context 4}\\
q_4 \leqslant \delta < q_5 &\Rightarrow \text{Context 5}\\
q_5 \leqslant \delta < q_6 &\Rightarrow \text{Context 6}\\
q_6 \leqslant \delta < q_7 &\Rightarrow \text{Context 7}\\
q_7 \leqslant \delta < q_8 &\Rightarrow \text{Context 8}
\end{aligned}$$

The values of q_1–q_8 can be prescribed by the user.

If the original pixel values lie between 0 and $M - 1$, the differences or prediction residuals will lie between $-(M - 1)$ and $M - 1$. Even though most of the differences will have a magnitude close to zero, for arithmetic coding we still have to assign a count to all possible symbols. This means a reduction in the size of the intervals assigned to values that do occur, which in turn means using a larger number of bits to represent these values. The CALIC algorithm attempts to resolve this problem in a number of ways. Let's describe these using an example.

Consider the sequence

$$x_n : 0, 7, 4, 3, 5, 2, 1, 7$$

We can see that all of the numbers lie between 0 and 7, a range of values that would require 3 bits to represent. Now suppose we predict a sequence element by the previous element in the sequence. The sequence of differences

$$r_n = x_n - x_{n-1}$$

is given by

$$r_n : 0, 7, -3, -1, 2, -3, -1, 6$$

If we were given this sequence, we could easily recover the original sequence by using

$$x_n = x_{n-1} + r_n.$$

However, the prediction residual values r_n lie in the $[-7, 7]$ range. That is, the alphabet required to represent these values is almost twice the size of the original alphabet. However, if we look closely we can see that the value of r_n actually lies between $-x_{n-1}$ and $7 - x_{n-1}$. The smallest value that r_n can take on occurs when x_n has a value of 0, in which case r_n will have a value of $-x_{n-1}$. The largest value that r_n can take on occurs when x_n is 7, in which case r_n has a value of $7 - x_{n-1}$. In other words, given a particular value for x_{n-1}, the number of different values that r_n can take on is the same as the number of values that x_n can take on. Generalizing from this, we can see that if a pixel takes on values between 0 and $M - 1$, then given a predicted value $\widehat{X}$, the difference $X - \widehat{X}$ will take on values in the range $-\widehat{X}$ to $M - 1 - \widehat{X}$. We can use this fact to map the difference values into the range $[0, M - 1]$, using the following mapping:

$$\begin{array}{rcl}
0 & \rightarrow & 0 \\
1 & \rightarrow & 1 \\
-1 & \rightarrow & 2 \\
2 & \rightarrow & 3 \\
\vdots & & \vdots \\
-\widehat{X} & \rightarrow & 2\widehat{X} \\
\widehat{X}+1 & \rightarrow & 2\widehat{X}+1 \\
\widehat{X}+2 & \rightarrow & 2\widehat{X}+2 \\
\vdots & & \vdots \\
M-1-\widehat{X} & \rightarrow & M-1
\end{array}$$

where we have assumed that $\widehat{X} \leqslant (M - 1)/2$.

Another approach used by CALIC to reduce the size of its alphabet is to use a modification of a technique called *recursive indexing* [85] described in Chapter 9. Recursive indexing is a technique for representing a large range of numbers using only a small set. It is easiest to explain using an example. Suppose we want to represent positive integers using only the integers between 0 and 7—that is, a representation alphabet of size 8. Recursive indexing works as follows. If the number to be represented lies between 0 and 6, we simply represent it by that number. If the number to be represented is greater than or equal to 7, we first send the number 7, subtract 7 from the original number, and repeat the process. We keep repeating the process until the remainder is a number between 0 and 6. Thus, for example, 9 would be represented by 7 followed by a 2, and 17 would be represented by two 7s followed by a 3. The decoder, when it sees a number between 0 and 6, will decode it at its face value, and when it sees 7, it will keep accumulating the values until a value between 0 and 6 is received. This method of representation followed by entropy coding has been shown to be optimal for sequences that follow a geometric distribution [86].

In CALIC, the representation alphabet is different for different coding contexts. For each coding context k, we use an alphabet $A_k = \{0, 1, \ldots, N_k\}$. Furthermore, if the residual occurs in context k, then the first number that is transmitted is coded with respect to context k; if further recursion is needed, we use the $k+1$ context.

We can summarize the CALIC algorithm as follows:

1. Find the initial prediction $\widehat{X}$.
2. Compute the prediction context.
3. Refine the prediction by removing the estimate of the bias in that context.
4. Update the bias estimate.
5. Obtain the residual and remap it so the residual values lie between 0 and $M-1$, where M is the size of the initial alphabet.
6. Find the coding context k.
7. Code the residual using the coding context.

All these components working together have kept CALIC as the state of the art in lossless image compression. However, we can get almost as good a performance if we simplify some of the more involved aspects of CALIC. We study such a scheme in the next section.

7.4 JPEG-LS

The JPEG-LS standard looks more like CALIC than the old JPEG standard. When the initial proposals for the new lossless compression standard were compared, CALIC was rated first in six of the seven categories of images tested. Motivated by some aspects of CALIC, a team from Hewlett-Packard (HP) proposed a much simpler predictive coder, under the name LOCO-I (for low complexity), that still performed close to CALIC [87].

As in CALIC, the standard has both a lossless and a lossy mode. We will not describe the lossy coding procedures.

The initial prediction is obtained using the following algorithm:

$$
\begin{array}{l}
\text{if } NW \geqslant \max(W, N) \\
\widehat{X} = \max(W, N) \\
\text{else} \\
\{ \\
\quad \text{if } NW \leqslant \min(W, N) \\
\quad \widehat{X} = \min(W, N) \\
\quad \text{else} \\
\quad \widehat{X} = W + N - NW \\
\}
\end{array}
$$

This prediction approach is a variation of Median Adaptive Prediction [88], in which the predicted value is the median of the N, W, and NW pixels. The initial prediction is then refined using the average value of the prediction error in that particular context.

The contexts in JPEG-LS also reflect the local variations in pixel values. However, they are computed differently from CALIC. First, measures of differences D_1, D_2, and D_3 are computed as follows:

$$\begin{aligned} D_1 &= NE - N \\ D_2 &= N - NW \\ D_3 &= NW - W. \end{aligned}$$

The values of these differences define a three-component context vector **Q**. The components of **Q** (Q_1, Q_2, and Q_3) are defined by the following mappings:

$$\begin{aligned} D_i \leqslant -T_3 &\Rightarrow Q_i = -4 \\ -T_3 < D_i \leqslant -T_2 &\Rightarrow Q_i = -3 \\ -T_2 < D_i \leqslant -T_1 &\Rightarrow Q_i = -2 \\ -T_1 < D_i \leqslant 0 &\Rightarrow Q_i = -1 \\ D_i = 0 &\Rightarrow Q_i = 0 \\ 0 < D_i \leqslant T_1 &\Rightarrow Q_i = 1 \\ T_1 < D_i \leqslant T_2 &\Rightarrow Q_i = 2 \\ T_2 < D_i \leqslant T_3 &\Rightarrow Q_i = 3 \\ T_3 < D_i &\Rightarrow Q_i = 4 \end{aligned} \tag{9}$$

where T_1, T_2, and T_3 are positive coefficients that can be defined by the user. Given nine possible values for each component of the context vector, this results in $9 \times 9 \times 9 = 729$ possible contexts. In order to simplify the coding process, the number of contexts is reduced by replacing any context vector **Q** whose first nonzero element is negative by $-$**Q**. Whenever this happens, a variable *SIGN* is also set to -1; otherwise, it is set to +1. This reduces the number of contexts to 365. The vector **Q** is then mapped into a number between 0 and 364. (The standard does not specify the particular mapping to use.)

The variable *SIGN* is used in the prediction refinement step. The correction is first multiplied by *SIGN* and then added to the initial prediction.

The prediction error r_n is mapped into an interval that is the same size as the range occupied by the original pixel values. The mapping used in JPEG-LS is as follows:

$$\begin{aligned} r_n < -\frac{M}{2} &\Rightarrow r_n \leftarrow r_n + M \\ r_n > \frac{M}{2} &\Rightarrow r_n \leftarrow r_n - M \end{aligned}$$

Finally, the prediction errors are encoded using adaptively selected codes based on Golomb codes, which have also been shown to be optimal for sequences with a geometric distribution. In Table 7.3, we compare the performance of the old and new JPEG standards and CALIC. The results for the new JPEG scheme were obtained using a software implementation courtesy of HP.

We can see that for most of the images, the new JPEG standard performs very close to CALIC and outperforms the old standard by 6% to 18%. The only case where the performance is not as good is for the Omaha image. While the performance improvement in these examples

TABLE 7.3 **Comparison of the file sizes obtained using the new and old JPEG lossless compression standards and CALIC.**

Image	Old JPEG	New JPEG	CALIC
Sena	31,055	27,339	26,433
Sensin	32,429	30,344	29,213
Earth	32,137	26,088	25,280
Omaha	48,818	50,765	48,249

may not be very impressive, we should keep in mind that we are picking the best result out of eight for the old JPEG. In practice, this would mean trying all eight JPEG predictors and picking the best. On the other hand, both CALIC and the new JPEG standard are single-pass algorithms. Furthermore, because of the ability of both CALIC and the new standard to function in multiple modes, both perform very well on compound documents, which may contain images along with text.

7.5 Prediction Using Conditional Averages

Both of the predictive schemes we have looked at for lossless image coding obtain their predictive power from the fact that neighboring pixels tend to be alike. However, this assumption is not necessary for developing a predictive scheme. A somewhat different idea is used in schemes like *ppm*, which assume that a similar context will result in a similar value. To obtain a prediction using this approach, one would try to find a pixel neighborhood similar to the neighborhood of the pixel being encoded and use that pixel as a predictor for the pixel being encoded. Using the labeling of Figure 7.1, we look for a pixel in the history of the image whose neighbors are the same as the neighbors of X in order to find the best predictor for pixel X. The value of that pixel can then be used as a predictor for the pixel X. We briefly describe a scheme based on this idea [89,90], which we will call *conditional average prediction (cap)*. This can be used in place of the initial predictor in JPEG-LS.

The *ppm* approach implicitly relies on the fact that textual information contains many exact repeats. As this situation is not duplicated in natural images, the algorithm used in *ppm* cannot be applied directly to the problem of generating predictions. Fortunately, while we do not have exact repeats as in textual data, our objectives are also not the same. We are not looking for an exact match, rather we are looking for a close value that can be used as a prediction.

In the *cap* approach, we define sets of pixels in the neighborhood to be the contexts for which we will look for a match. For example, a context of size 4 consists of the W, NW, N, and NE pixels, and a context of size 3 consists of the W, NW, and N pixels. In order to generate a prediction, the encoder looks for matches to the context in the already encoded portion of the image. A match can be defined rigidly, that is, each pixel in the context has to be exactly matched, or it can be defined more loosely, that is, two pixels are said to be matched if the absolute difference between pixel values is less than a threshold. To guard against the possibility of a bad prediction, the algorithm requires that at least five matches be observed

TABLE 7.4 **Prediction error entropies and corresponding file sizes obtained using the *cap* prediction.**

Image	Residual Entropy	*cap*
Sena	3.41	27,934
Sensin	3.81	31,252
Earth	3.20	26,238
Omaha	6.03	49,937

before a prediction can be generated. If more than five matches are observed, the algorithm takes the average of the pixels that have the matching contexts as the predicted value. If the larger context cannot garner five or more matches, the algorithm shifts to the next smaller context in a manner similar to *ppm*. If there are not sufficient matches for the smallest context allowed, the algorithm uses the version of the median adaptive predictor used in JPEG-LS as a default. Note that the decoder does not have to be informed about which context is being used as the context selection is based on information available to both the encoder and decoder. Using this approach, we obtain the results shown in Table 7.4.

7.6 Multiresolution Approaches

Our final predictive image compression scheme is perhaps not as competitive as the other schemes. However, it is an interesting algorithm because it approaches the problem from a slightly different point of view.

Multiresolution models generate representations of an image with varying spatial resolution. This usually results in a pyramid-like representation of the image, with each layer of the pyramid serving as a prediction model for the layer immediately below.

One of the more popular of these techniques is known as HINT (Hierarchical INTerpolation) [91]. The specific steps involved in HINT are as follows. First, residuals corresponding to the pixels labeled Δ in Figure 7.2 are obtained using linear prediction, and they are transmitted. Then, the intermediate pixels ($\circ$) are estimated by linear interpolation, and the error in estimation is then transmitted. Then, the pixels X are estimated from Δ and $\circ$, and the estimation error is transmitted. Finally, the pixels labeled $*$ and then $\bullet$ are estimated from known neighbors, and the errors are transmitted. The reconstruction process proceeds in a similar manner.

One use of a multiresolution approach is in progressive image transmission. We describe this application in the next section.

7.6.1 Progressive Image Transmission

The last few years have seen a very rapid increase in the amount of information stored as images, especially remotely sensed images (such as images from weather and other satellites) and medical images (such as computerized axial tomography (CAT) scans, magnetic resonance

Δ	$\bullet$	X	$\bullet$	Δ	$\bullet$	X	$\bullet$	Δ
$\bullet$	$*$	$\bullet$	$*$	$\bullet$	$*$	$\bullet$	$*$	$\bullet$
X	$\bullet$	$\circ$	$\bullet$	X	$\bullet$	$\circ$	$\bullet$	X
$\bullet$	$*$	$\bullet$	$*$	$\bullet$	$*$	$\bullet$	$*$	$\bullet$
Δ	$\bullet$	X	$\bullet$	Δ	$\bullet$	X	$\bullet$	Δ
$\bullet$	$*$	$\bullet$	$*$	$\bullet$	$*$	$\bullet$	$*$	$\bullet$
X	$\bullet$	$\circ$	$\bullet$	X	$\bullet$	$\circ$	$\bullet$	X
$\bullet$	$*$	$\bullet$	$*$	$\bullet$	$*$	$\bullet$	$*$	$\bullet$
Δ	$\bullet$	X	$\bullet$	Δ	$\bullet$	X	$\bullet$	Δ

FIGURE 7.2 The HINT scheme for hierarchical prediction.

images (MRI), and mammograms). It is not enough to have information. We also need to make these images accessible to individuals who can make use of them. There are many issues involved with making large amounts of information accessible to a large number of people. In this section, we will look at one particular issue—transmitting these images to remote users. (For a more general look at the problem of managing large amounts of information, see [92].)

Suppose a user wants to browse through a number of images in a remote database. The user is connected to the database via a 1 Mbps link. Suppose the size of the images is 2048×1536, and on average, users have to look through 30 images before finding the image for which they are looking. If these images were monochrome with 8 bits per pixel, this process would take close to 15 minutes, which is not very practical. Even if we compressed these images before transmission, lossless compression, on average, gives us about a 2:1 compression. This would only cut the transmission in half, which still makes the approach cumbersome. A better alternative is to send an approximation of each image first, which does not require too many bits but is still sufficiently accurate to give users an idea of what the image looks like. If users find the image to be of interest, they can request a further refinement of the approximation or the complete image. This approach is called *progressive image transmission*.

Example 7.6.1:

A simple progressive transmission scheme is to divide the image into blocks and then send a representative pixel for the block. The receiver replaces each pixel in the block with the representative value. In this example, the representative value is the value of the pixel in the top-left corner. Depending on the size of the block, the amount of data that would need to be transmitted could be substantially reduced. For example, to transmit a 2048×1536 image at 8 bits per pixel over a 1 Mbps link takes about 25 seconds. Using a block size of 8×8 and using the top-left pixel in each block as the representative value, we can approximate the 1024×1024 image with a 128×128 subsampled image. Using 8 bits per pixel and a 1 Mbps link, the time required to transmit this approximation to the image takes less than a third of a second.

Assuming that this approximation was sufficient to let the user decide whether a particular image was the desired image, the time required now to look through 30 images becomes on the order of seconds instead of the 15 minutes mentioned earlier. If the approximation using a block size of 8×8 does not provide enough resolution to make a decision, the user can ask for a refinement. The transmitter can then divide the 8×8 block into four 4×4 blocks. The pixel at the upper-left corner of the upper-left block was already transmitted as the representative pixel for the 8×8 block, so we need to send three more pixels for the other three 4×4 blocks. This takes a little more than a second, so even if the user had to request a finer approximation every third image, this would only increase the total search time by a very small amount. To see what these approximations look like, we have taken the Sena image and encoded it using different block sizes. The results are shown in Figure 7.3. The lowest-resolution image, shown in the top left, is a 32×32 image. The top-right image is a 64×64 image. The bottom-left image is a 128×128 image, and the bottom-right image is the 256×256 original.

FIGURE 7.3 **Sena image coded using different block sizes for progressive transmission. Top row: block size 8×8 and block size 4×4. Bottom row: block size 2×2 and original image.**

Notice that even with a block size of 8, the image is clearly recognizable as a person. Therefore, if the user was looking for a house, they would probably skip over this image after seeing the first approximation. If the user was looking for a picture of a person, they could still make decisions based on the second approximation.

Finally, when an image is built line by line, the eye tends to follow the scan line. With the progressive transmission approach, the user gets a more global view of the image very early in the image formation process. Consider the images in Figure 7.4. The images on the left are the 8×8, 4×4, and 2×2 approximations of the Sena image. On the right, we show how much of the image we would see in the same amount of time if we used the standard line-by-line raster scan order. ♦

We would like the first approximations that we transmit to use as few bits as possible yet be accurate enough to allow the user to make a decision to accept or reject the image with a certain degree of confidence. As these approximations are lossy, many progressive transmission schemes use well-known lossy compression schemes in the first pass.

The more popular lossy compression schemes, such as transform coding, tend to require a significant amount of computation. As the decoders for most progressive transmission schemes have to function on a wide variety of platforms, they are generally implemented in software and need to be simple and fast. This requirement has led to the development of a number of progressive transmission schemes that do not use lossy compression for their initial approximations. Most of these schemes have a form similar to the one described in Example 7.6.1, and they are generally referred to as *pyramid schemes* because of the manner in which the approximations are generated and the image is reconstructed.

When we use the pyramid form, we still have a number of ways to generate the approximations. One of the problems with the simple approach described in Example 7.6.1 is that if the pixel values vary a lot within a block, the "representative" value may not be very representative. To prevent this from happening, we could represent the block by some sort of an average or composite value. For example, suppose we start out with a 512×512 image. We first divide the image into 2×2 blocks and compute the integer value of the average of each block [93,94]. The integer values of the averages would constitute the penultimate approximation. The approximation to be transmitted prior to that can be obtained by taking the average of 2×2 averages and so on, as shown in Figure 7.5.

Using the simple technique in Example 7.6.1, we ended up transmitting the same number of values as the original number of pixels. However, when we use the mean of the pixels as our approximation, after we have transmitted the mean values at each level, we still have to transmit the actual pixel values. The reason is that when we take the integer part of the average, we end up throwing away information that cannot be retrieved. To avoid this problem of data expansion, we can transmit the sum of the values in the 2×2 block. Then we only need to transmit three more values to recover the original four values. With this approach, although we would be transmitting the same number of values as the number of pixels in the image, we might still end up sending more bits because representing all possible values of the sum would require transmitting 2 more bits than was required for the original value. For example, if the pixels in the image can take on values between 0 and 255, which can be represented by 8 bits, their sum will take on values between 0 and 1024, which would require 10 bits. If we

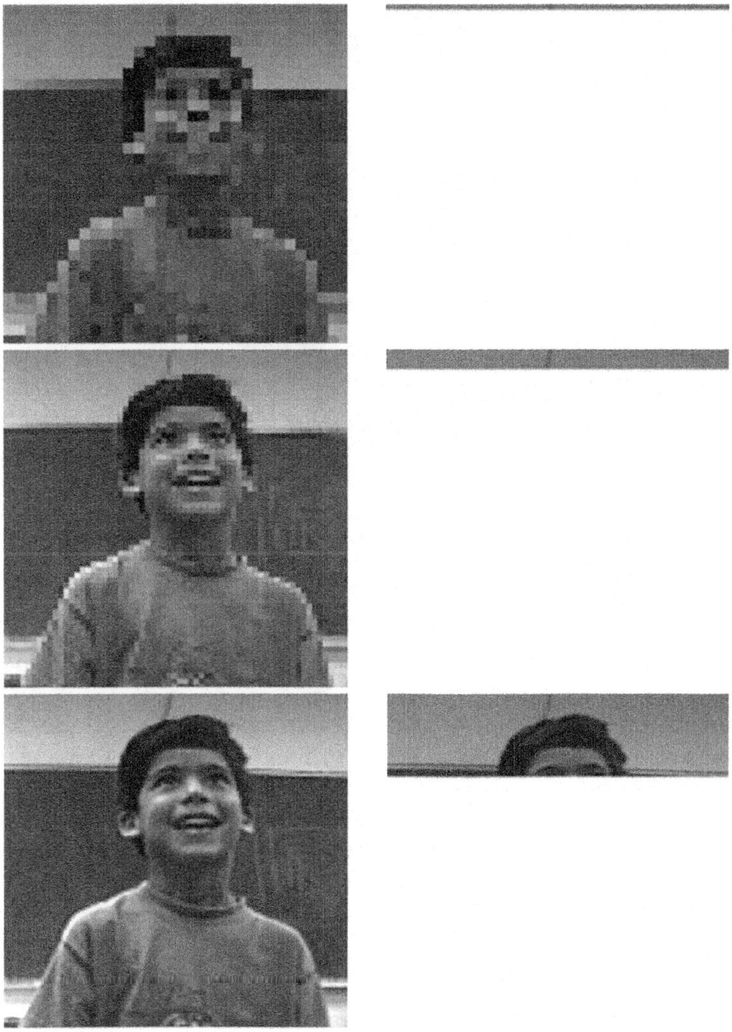

FIGURE 7.4 **Comparison between the received image using progressive transmission and using the standard raster scan order.**

are allowed to use entropy coding, we can remove the problem of data expansion by using the fact that the neighboring values in each approximation are heavily correlated, as are values in different levels of the pyramid. This means that differences between these values can be efficiently encoded using entropy coding. By doing so, we end up getting compression instead of expansion.

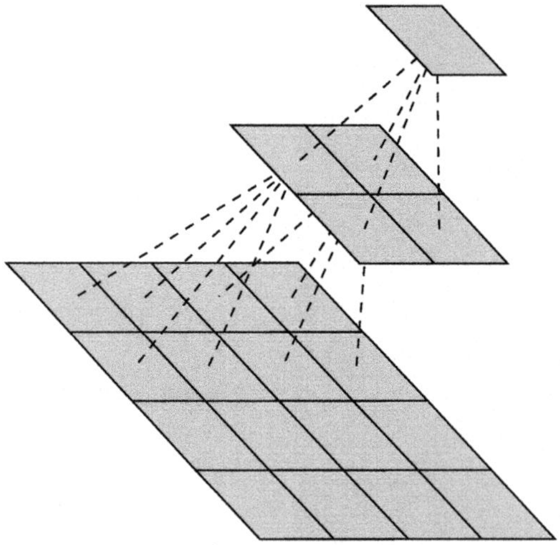

FIGURE 7.5 The pyramid structure for progressive transmission.

Instead of taking the arithmetic average, we could also form some sort of weighted average. The general procedure would be similar to that described above. (For one of the more well-known weighted average techniques, see [95].)

The representative value does not have to be an average. We could use the pixel values in the approximation at the lower levels of the pyramid as indices into a lookup table. The lookup table could be designed to preserve important information such as edges. The problem with this approach would be the size of the lookup table. If we were using 2×2 blocks of 8-bit values, the lookup table would have 2^{32} values, which is too large for most applications. The size of the table could be reduced if the number of bits per pixel was lower or if, instead of taking 2×2 blocks, we used rectangular blocks of size 2×1 and 1×2 [96].

Finally, we do not have to build the pyramid one layer at a time. After sending the lowest-resolution approximations, we can use some measure of information contained in a block to decide whether it should be transmitted [97]. One possible measure could be the difference between the largest and smallest intensity values in the block. Another might be to look at the maximum number of similar pixels in a block. Using an information measure to guide the progressive transmission of images allows the user to see portions of the image first that are visually more significant.

7.7 Facsimile Encoding

One of the earliest applications of lossless compression in the modern era has been the compression of facsimile, or fax. In facsimile transmission, a page is scanned and converted into a sequence of black or white pixels. The requirements of how fast the facsimile of an A4 document (210×297 mm) must be transmitted have changed over the last two decades. The

Comité Consultatif International Télephoniqué et Télégraphique (CCITT) (now ITU-T) has issued a number of recommendations based on the speed requirements at a given time. The CCITT classifies the apparatus for facsimile transmission into four groups. Although several considerations are used in this classification, if we only consider the time to transmit an A4-size document over phone lines, the four groups can be described as follows:

- **Group 1:** This apparatus is capable of transmitting an A4-size document in about 6 minutes over phone lines using an analog scheme. The apparatus is standardized in recommendation T.2.
- **Group 2:** This apparatus is capable of transmitting an A4-size document over phone lines in about 3 minutes. A Group 2 apparatus also uses an analog scheme and, therefore, does not use data compression. The apparatus is standardized in recommendation T.3.
- **Group 3:** This apparatus uses a digitized binary representation of the facsimile. Because it is a digital scheme, it can and does use data compression and is capable of transmitting an A4-size document in about a minute. The apparatus is standardized in recommendation T.4.
- **Group 4:** This apparatus has the same speed requirement as Group 3. The apparatus is standardized in recommendations T.6, T.503, T.521, and T.563.

With the arrival of the Internet, facsimile transmission has changed as well. Given the wide range of rates and "apparatus" used for digital communication, it makes sense to focus more on protocols than on apparatus. The newer recommendations from the ITU provide standards for compression that are more or less independent of apparatus.

Later in this chapter, we will look at the compression schemes described in ITU-T recommendations T.4, T.6, T.82 (JBIG), T.88 (JBIG2), and T.44 (MRC). We begin with a look at an earlier technique for facsimile called *run-length coding*, which still survives as part of the T.4 recommendation.

7.7.1 Run-Length Coding

The model that gives rise to run-length coding is the Capon model [98], a two-state Markov model with states S_w and S_b (S_w corresponds to the case where the pixel that has just been encoded is a white pixel, and S_b corresponds to the case where the pixel that has just been encoded is a black pixel). The transition probabilities $P(w|b)$ and $P(b|w)$ and the probability of being in each state $P(S_w)$ and $P(S_b)$ completely specify this model. For facsimile images, $P(w|w)$ and $P(w|b)$ are generally significantly higher than $P(b|w)$ and $P(b|b)$. The Markov model is represented by the state diagram shown in Figure 7.6.

The entropy of a finite state process with states S_i is given by Equation (2.16). Recall that in Example 2.3.1, the entropy using a probability model and the *iid* assumption was significantly more than the entropy using the Markov model.

Let us try to interpret what the model says about the structure of the data. The highly skewed nature of the probabilities $P(b|w)$ and $P(w|w)$, and to a lesser extent $P(w|b)$ and $P(b|b)$, says that once a pixel takes on a particular color (black or white), it is highly likely that the following pixels will also be of the same color. So, rather than code the color of each

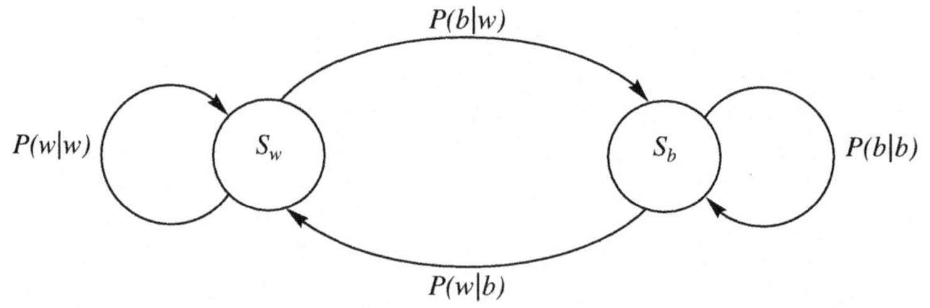

FIGURE 7.6 The Capon model for binary images.

pixel separately, we can simply code the length of the runs of each color. For example, if we had 190 white pixels followed by 30 black pixels, followed by another 210 white pixels, we would code the sequence 190, 30, 210, along with an indication of the color of the first string of pixels, instead of coding the 430 pixels individually. Coding the lengths of runs instead of coding individual values is called run-length coding.

7.7.2 CCITT Group 3 and 4—Recommendations T.4 and T.6

The recommendations for Group 3 facsimile include two coding schemes. One is a one-dimensional scheme in which the coding on each line is performed independently of any other line. The other is two-dimensional; the coding of one line is performed using the line-to-line correlations.

The one-dimensional coding scheme is a run-length coding scheme in which each line is represented as a series of alternating white runs and black runs. The first run is always a white run. If the first pixel is a black pixel, then we assume that we have a white run of length zero.

Runs of different lengths occur with different probabilities; therefore, they are coded using a variable-length code. The approach taken in CCITT standards T.4 and T.6 is to use a Huffman code to encode the run lengths. However, the number of possible runlengths is extremely large, and it is simply not feasible to build a codebook that large. Therefore, instead of generating a Huffman code for each run length r_l, the run length is expressed in the form

$$r_l = 64 \times m + t \qquad \text{for } \; t = 0, 1, \ldots, 63, \; \text{and } m = 1, 2, \ldots, 27. \tag{10}$$

When we have to represent a run length r_l, instead of finding a code for r_l, we use the corresponding codes for m and t. The codes for t are called the *terminating codes*, and the codes for m are called the *make-up codes*. If $r_l < 63$, we only need to use a terminating code. Otherwise, both a make-up code and a terminating code are used. For the range of m and t given here, we can represent lengths of 1728, which is the number of pixels per line in an A4-size document. However, if the document is wider, the recommendations provide for this with an optional set of 13 codes. Except for the optional codes, there are separate codes for black and white run lengths. This coding scheme is generally referred to as a *modified Huffman (MH)* scheme. Note that the Huffman codes are static codes based on probability models obtained using a set of documents that the standards committee considered to be typical.

In the two-dimensional scheme, instead of reporting the run lengths, which in terms of our Markov model are the lengths of time we remain in one state, we report the transition times when we move from one state to another state. Look at Figure 7.7. We can encode this in two ways. We can say that the first row consists of a sequence of runs 0, 2, 3, 3, 8, and the second row consists of runs of lengths 0, 1, 8, 3, 4 (notice the first runs of length zero). Or, we can encode the location of the pixel values that occur at a transition from white to black or black to white. The first pixel is an imaginary white pixel assumed to be to the left of the first actual pixel. Therefore, if we were to code transition locations, we would encode the first row as 1, 3, 6, 9 and the second row as 1, 2, 10, 13.

Generally, rows of a facsimile image are heavily correlated. Therefore, it would be easier to code the transition points with reference to the previous line than to code each one in terms of its absolute location, or even its distance from the previous transition point. This is the basic idea behind the recommended two-dimensional coding scheme. This scheme is a modification of a two-dimensional coding scheme called the *Relative Element Address Designate* (READ) code [99,100] and is often referred to as *Modified READ* (MR). The READ code was the Japanese proposal to the CCITT for the Group 3 standard.

To understand the two-dimensional coding scheme, we need some definitions.

a_0: This is the last pixel whose value is known to both the encoder and decoder. At the beginning of encoding each line, a_0 refers to an imaginary white pixel to the left of the first actual pixel. While it is often a transition pixel, it does not have to be.

a_1: This is the first transition pixel to the right of a_0. By definition, its color should be the opposite of a_0. The location of this pixel is known only to the encoder.

a_2: This is the second transition pixel to the right of a_0. Its color should be the opposite of a_1, which means it has the same color as a_0. The location of this pixel is also known only to the encoder.

b_1: This is the first transition pixel on the line above the line currently being encoded to the right of a_0 whose color is the opposite of a_0. As the line above is known to both encoder and decoder, as is the value of a_0, the location of b_1 is also known to both the encoder and decoder.

b_2: This is the first transition pixel to the right of b_1 on the line above the line currently being encoded.

For the pixels in Figure 7.7, if the second row is the one being currently encoded, and if we have encoded the pixels up to the second pixel, the assignment of the different pixels is shown in Figure 7.8. The pixel assignments for a slightly different arrangement of black and white pixels are shown in Figure 7.9.

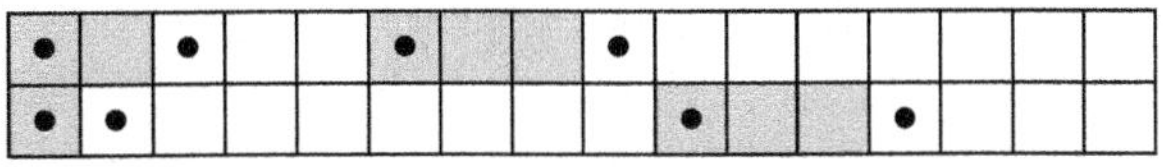

FIGURE 7.7 **Two rows of an image. The transition pixels are marked with a dot.**

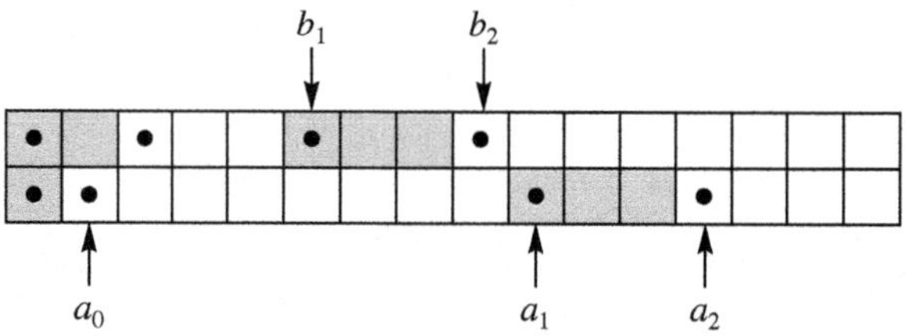

FIGURE 7.8 **Two rows of an image. The transition pixels are marked with a dot.**

If b_1 and b_2 lie between a_0 and a_1, we call the coding mode used the *pass mode*. The transmitter informs the receiver about the situation by sending the code 0001. Upon receipt of this code, the receiver knows that from the location of a_0 to the pixel right below b_2, all pixels are of the same color. If this had not been true, we would have encountered a transition pixel. As the first transition pixel to the right of a_0 is a_1 and as b_2 occurs before a_1, no transitions have occurred, and all pixels from a_0 to right below b_2 are the same color. At this time, the last pixel known to both the transmitter and receiver is the pixel below b_2. Therefore, this now becomes the new a_0, and we find the new positions of b_1 and b_2 by examining the row above the one being encoded and continue with the encoding process.

If a_1 is detected before b_2 by the encoder, we do one of two things. If the distance between a_1 and b_1 (the number of pixels from a_1 to right under b_1) is less than or equal to three, then we send the location of a_1 with respect to b_1, move a_0 to a_1, and continue with the coding process. This coding mode is called the *vertical mode*. If the distance between a_1 and b_1 is large, the coding mode is called the *horizontal mode*. In his mode, we essentially revert to the one-dimensional technique and send the distances between a_0 and a_1 and a_1 and a_2, using the modified Huffman code. Let us look at exactly how this is accomplished.

In the vertical mode, if the distance between a_1 and b_1 is zero (that is, a_1 is exactly under b_1), we send the code 1. If a_1 is to the right of b_1 by one pixel (as in Figure 7.9), we send the code 011. If a_1 is to the right of b_1 by two or three pixels, we send the code 000011 or 0000011, respectively. If a_1 is to the left of b_1 by one, two, or three pixels, we send the code 010, 000010, or 0000010, respectively.

In the horizontal mode, we first send the code 001 to inform the receiver about the mode, and we then send the modified Huffman codewords corresponding to the run length from a_0 to a_1, and a_1 to a_2.

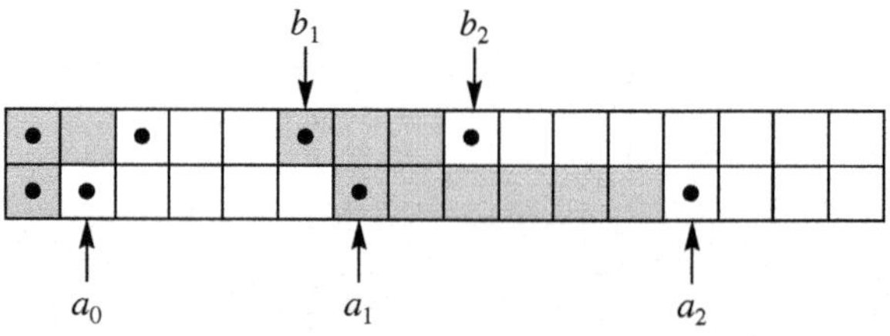

FIGURE 7.9 **Two rows of an image. The transition pixels are marked with a dot.**

As the encoding of a line in the two-dimensional algorithm is based on the previous line, an error in one line could conceivably propagate to all other lines in the transmission. To prevent this from happening, the T.4 recommendations contain the requirement that after each line is coded with the one-dimensional algorithm, at most $K - 1$ lines will be coded using the two-dimensional algorithm. For standard vertical resolution, $K = 2$, and for high resolution, $K = 4$.

The Group 4 encoding algorithm, as standardized in CCITT recommendation T.6, is identical to the two-dimensional encoding algorithm in recommendation T.4. The main difference between T.6 and T.4 from the compression point of view is that T.6 does not have a one-dimensional coding algorithm, which means that the restriction described in the previous paragraph is also not present. This slight modification of the modified READ algorithm has earned the name *modified modified READ* (MMR)!

7.7.3 JBIG

The JBIG image compression standard provides the user with the option of either sequential encoding, progressive encoding, or something in between. In the sequential mode the pixels in the image are encoded and decoded in raster scan order, that is left to right and top to bottom. In the progressive mode the image to be encoded can be first decomposed into images of varying resolution. In some applications we may not always need to view an image at full resolution. For example, if we are looking at the layout of a page, we may not need to know what each word or letter on the page is. The JBIG standard allows for the generation of progressively lower-resolution images. If the user is interested in some gross patterns in the image (for example, if they are interested in seeing if there are any figures on a particular page) they can request a lower-resolution image, which can be transmitted using fewer bits. Once the lower-resolution image is available, the user can decide whether a higher-resolution image is necessary.

The JBIG standard also permits an in-between mode called the progressive compatible sequential mode. In this mode, the image is divided into stripes that consist of rows of the image. The image is encoded stripe by stripe in a sequential manner. However, each stripe is encoded in a progressive manner.

Once the mode has been selected, the JBIG encoding process proceeds with resolution reduction (in the progressive modes), redundancy removal, and arithmetic coding of the residuals. We briefly describe each of these steps below.

Resolution Reduction

The JBIG specification recommends generating one lower-resolution pixel for each 2×2 block in the higher-resolution image. The number of lower-resolution images (called layers) is not specified by JBIG. A straightforward method for generating lower-resolution images is to replace every 2×2 block of pixels with the average value of the four pixels, thus reducing the resolution by two in both the horizontal and vertical directions. This approach works well as long as three of the four pixels are either black or white. However, when we have two pixels of each kind, we run into trouble; consistently replacing the four pixels with either a white or

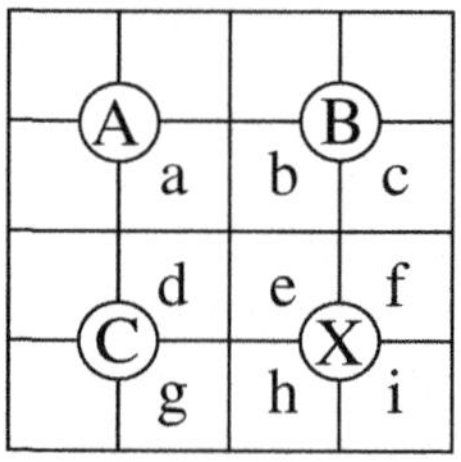

FIGURE 7.10 **Pixels used to determine the value of a lower-level pixel.**

black pixel causes a severe loss of detail, and randomly replacing with a black or white pixel introduces a considerable amount of noise into the image [92].

Instead of simply taking the average of every 2×2 block, the JBIG specification provides a table-based method for resolution reduction. The table is indexed by the neighboring pixels shown in Figure 7.10, in which the circles represent the lower-resolution layer pixels and the squares represent the higher-resolution layer pixels.

Each pixel contributes a bit to the index. The table is formed by computing the expression

$$4e + 2(b + d + f + h) + (a + c + g + i) - 3(B + C) - A$$

If the value of this expression is greater than 4.5, pixel X is tentatively declared to be 1. The table has certain exceptions to this rule to reduce the amount of edge smearing, generally encountered in a filtering operation. There are also exceptions that preserve periodic patterns and dither patterns.

Redundancy Removal

There are two kinds of redundancy removal available in the JBIG standard. These are known as *typical prediction* and *deterministic prediction*. Typical prediction exploits the fact that many bi-level images have large regions that are the same color. This being the case, knowledge of the low-resolution pixel value is sufficient to determine the value of the high-resolution pixel for the progressive modes. Deterministic prediction operates on the concept that given the fact that the low-resolution images were constructed from the high-resolution images, knowledge of the values of the low-resolution pixels is sufficient to determine the high-resolution pixels exactly under certain conditions.

Typical Prediction Figure 7.11 indicates the relationship between various high- and low-resolution pixels. In the figure the squares correspond to the high-resolution pixels and the circles correspond to low-resolution pixels. Notice that there is one low-resolution pixel corresponding to each of the four (2×2) high-resolution pixels.

Many of the pixels in a bi-level document occur in regions of constant color, and many of the pixels in a document image occur in regions of white pixels. For these regions, pixels in both high-resolution and low-resolution images have the same color. Therefore, if the low-resolution pixels are available, the high-resolution pixels need not be transmitted. The typical prediction is a labeling of the high-resolution pixels that do not need to be transmitted. The

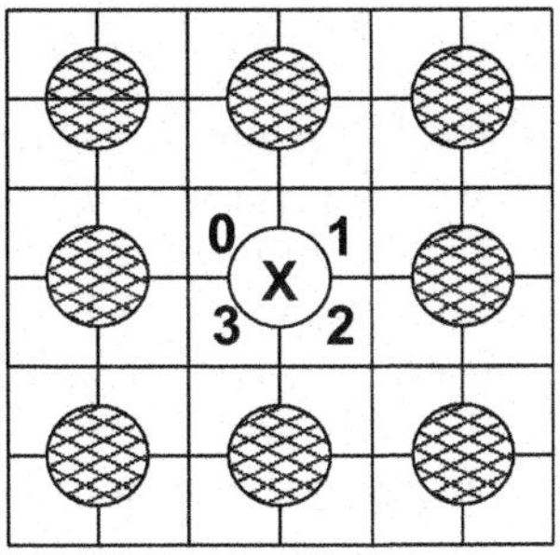

FIGURE 7.11 **Pixels used for determining the typical prediction status of a pixel.**

process begins with comparing each low-resolution pixel in a row, for example pixel X in Figure 7.11, to the eight neighboring low-resolution pixels (the shaded pixels in the figure) and the immediate neighboring high-resolution pixels (numbered 1,2,3 and 4). If the pixel being compared has the same color as the eight low-resolution pixels in the neighborhood but differs in color from one of the four high-resolution pixels, the pixel is labeled a nontypical pixel. If a line of low-resolution pixels contains any nontypical pixels, the line is declared to be nontypical. This information is conveyed to the decoder by inserting a fictitious pixel called the LNTP (line not typical) pixel prior to every other high-resolution line. If the corresponding low-resolution line contains pixels that are nontypical, then the value of LNTP is set to one, otherwise it is set to zero. For each low-resolution line for which the LNTP value is zero, the color of the high-resolution pixels corresponding to low-resolution pixels that have the same color as their eight neighbors is known, and, therefore, these high-resolution pixels are not encoded. In many documents, we can avoid encoding up to 95% of the high-resolution pixels [101] in this manner.

Typical prediction can also be used for sequentially encoded images or, equivalently, for the lowest resolution of progressively encoded images. In this case, the encoding makes use of the fact that in a bi-level image, a line of pixels is often identical to the line above. For typical prediction in this case, if the current line is the same as the line above, a bit flag called $LNTP_n$ is set to 0, and the line is not transmitted. If the line is not the same, the flag is set to 1, and the line is coded using the contexts currently used for the low-resolution layer in JBIG. The value of $LNTP_n$ is encoded by generating another bit, $SLNTP_n$, according to the rule

$$SLNTP_n = !(LNTP_n \oplus LNTP_{n-1})$$

which is treated as a virtual pixel to the left of each row. If the decoder decodes an *LNTP* value of 0, it copies the line above. If it decodes an *LNTP* value of 1, the following bits in the segment data are decoded using an arithmetic decoder and the contexts described previously.

Deterministic Prediction Deterministic prediction is only used in images encoded using a progressive mode. Because the resolution reduction is carried out using a table-based algorithm, it is sometimes possible to determine the exact value of a high-resolution pixel given the values of the pixels already encoded. The amount of reduction available through the use of deterministic prediction is about 7% [101].

Arithmetic Coding

Many bi-level images have a lot of local structure. Consider a digitized page of text. In large portions of the image we will encounter white pixels with a probability approaching 1. In other parts of the image, there will be a high probability of encountering a black pixel. We can make a reasonable guess of the situation for a particular pixel by looking at values of the pixels in the neighborhood of the pixel being encoded. For example, if the pixels in the neighborhood of the pixel being encoded are mostly white, then there is a high probability that the pixel to be encoded is also white. On the other hand, if most of the pixels in the neighborhood are black, there is a high probability that the pixel being encoded is also black. Each case gives us a skewed probability—a situation ideally suited for arithmetic coding. If we treat each case separately, using a different arithmetic coder for each of the two situations, we should be able to obtain improvement over the case where we use the same arithmetic coder for all pixels. Consider the following example.

Suppose the probability of encountering a black pixel is 0.2 and the probability of encountering a white pixel is 0.8. The entropy for this source is given by

$$H = -0.2\log_2 0.2 - 0.8\log_2 0.8 = 0.722. \tag{11}$$

If we use a single arithmetic coder to encode this source, we will get an average bit rate close to 0.722 bits per pixel. Now suppose, based on the neighborhood of the pixels, that we can divide the pixels into two sets, one comprising 80% of the pixels and the other 20%. In the first set, the probability of encountering a white pixel is 0.95, and in the second set the probability of encountering a black pixel is 0.7. The entropy of these sets is 0.286 and 0.881, respectively. If we used two different arithmetic coders for the two sets with frequency tables matched to the probabilities, we would get rates close to 0.286 bits per pixel about 80% of the time and close to 0.881 bits per pixel about 20% of the time. The average rate would be about 0.405 bits per pixel, which is almost half the rate required if we used a single arithmetic coder. If we used only those pixels in the neighborhood that had already been transmitted to the receiver to make our decision about which arithmetic coder to use, the decoder could keep track of which encoder was used to encode a particular pixel.

As we have mentioned before, the arithmetic coding approach is particularly amenable to the use of multiple coders. All coders use the same computational machinery, with each coder using a different set of probabilities. The JBIG algorithm makes full use of this feature of arithmetic coding. Instead of checking to see if most of the pixels in the neighborhood are white or black, the JBIG encoder uses the pattern of pixels in the neighborhood, or *context*, to decide which set of probabilities to use in encoding a particular pixel. If the neighborhood consists of 10 pixels with each pixel capable of taking on two different values, the number of possible patterns is 1024. The JBIG coder uses 1024 to 4096 coders, depending on whether a low- or high-resolution layer is being encoded.

For the low-resolution layer, the JBIG encoder uses one of the two different neighborhoods shown in Figure 7.12. The pixel to be coded is marked **X**, while the pixels to be used for templates are marked **A** or **O**. The **A** and **O** pixels are previously encoded pixels and are available to both the encoder and decoder. The **A** pixel can be thought of as a floating member of the neighborhood. Its placement is dependent on the input being encoded. Suppose the image has vertical lines 30 pixels apart. The **A** pixel would be placed 30 pixels to the left of

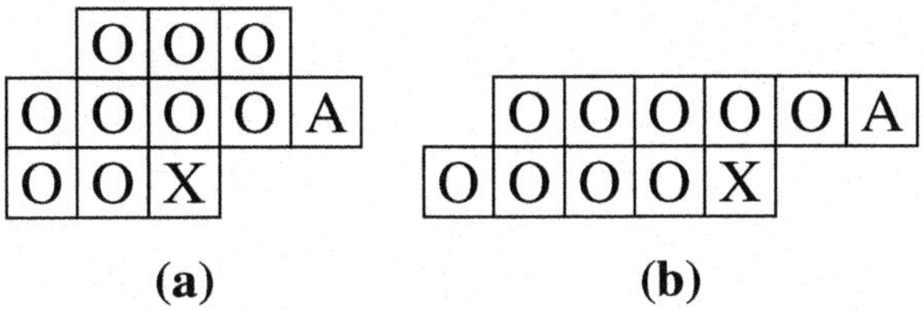

FIGURE 7.12 **(a) Three-line and (b) two-line neighborhoods.**

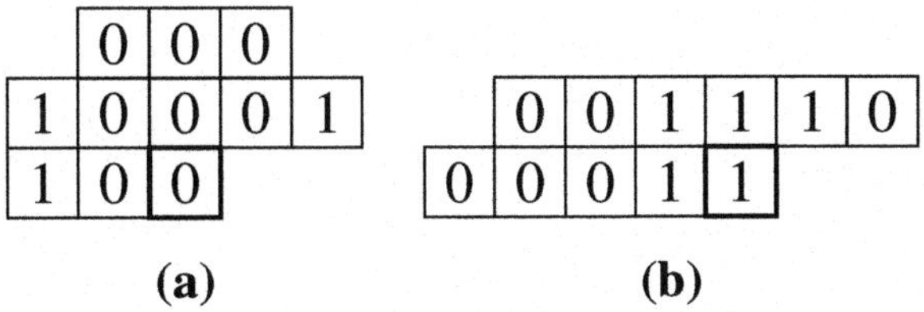

FIGURE 7.13 **(a) Three-line and (b) two-line contexts.**

the pixel being encoded. The **A** pixel can be moved around to capture any structure that might exist in the image. This is especially useful in halftone images in which the **A** pixels are used to capture the periodic structure. The location and movement of the **A** pixel are transmitted to the decoder as side information.

In Figure 7.13, the symbols in the neighborhoods have been replaced by 0s and 1s. We take 0 to correspond to white pixels, while 1 corresponds to black pixels. The pixel to be encoded is enclosed by the heavy box. The pattern of 0s and 1s is interpreted as a binary number, which is used as an index to the set of probabilities. The context in the case of the three-line neighborhood (reading left to right, top to bottom) is 0001000110, which corresponds to an index of 70. For the two-line neighborhood, the context is 0011100001, or 225. Since there are 10 bits in these templates, we will have 1024 different arithmetic coders.

In the JBIG standard, the 1024 arithmetic coders are a variation of the arithmetic coder known as the QM coder described in Chapter 4. The QM coder is a modification of an adaptive binary arithmetic coder called the Q coder [51, 52, 53], which in turn is an extension of another binary adaptive arithmetic coder called the skew coder [102].

As the lower-resolution layers are obtained from the higher-resolution images, we can use them when encoding the higher-resolution images. The JBIG specification makes use of the lower-resolution images when encoding the higher-resolution images by using the pixels of the lower-resolution images as part of the context for encoding the higher-resolution images. The contexts used for coding the lowest-resolution layer are those shown in Figure 7.12. The contexts used in coding the higher-resolution layer are shown in Figure 7.14.

Ten pixels are used in each context. If we include the 2 bits required to indicate which context template is being used, 12 bits will be used to indicate the context. This means that we can have 4096 different contexts.

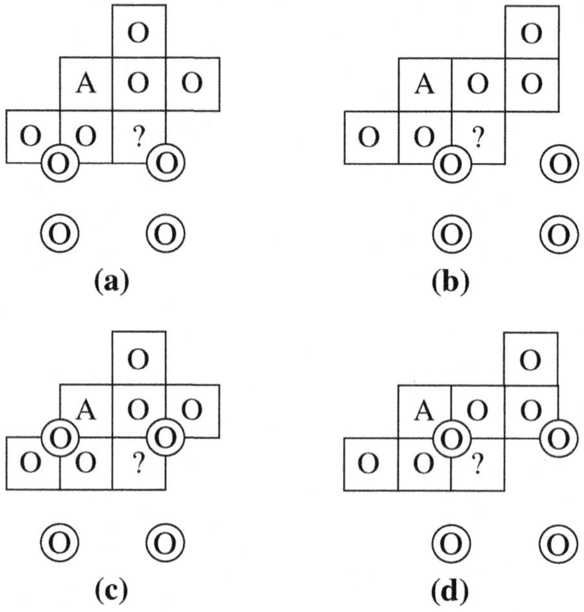

FIGURE 7.14 **Contexts used in the coding of higher-resolution layers.**

7.7.4 Comparison of MH, MR, MMR, and JBIG

Before we proceed to the more modern techniques found in T.88 and T.44, lets's compare the performance of the earliest of the modern techniques, JBIG, with the older facsimile coding algorithms, namely modified Huffman, modified READ, and modified modified READ. As we might expect, the JBIG algorithm performs better than the MMR algorithm, which performs better than the MR algorithm, which in turn performs better than the MH algorithm. The level of complexity also follows the same trend, although we could argue that MMR is actually less complex than MR.

A comparison of the schemes for some facsimile sources is shown in Table 7.5. The modified READ algorithm was used with $K = 4$, while the JBIG algorithm was used with an adaptive three-line template and adaptive arithmetic coder to obtain the results in this table. As we go from the one-dimensional MH coder to the two-dimensional MMR coder, we get a factor of two reduction in file size for the sparse text sources. We get even more reduction when we use an adaptive coder and an adaptive model, as is true for the JBIG coder. When we come to the dense text, the advantage of the two-dimensional MMR over the one-dimensional MH is not as significant, as the amount of two-dimensional correlation becomes substantially less.

The compression schemes specified in T.4 and T.6 break down when we try to use them to encode halftone images. In halftone images, gray levels are represented using binary pixel patterns. A gray level closer to black would be represented by a pattern that contains more black pixels, while a gray level closer to white would be represented by a pattern with fewer black pixels. Thus, the model that was used to develop the compression schemes specified in

TABLE 7.5 Comparison of binary image coding schemes. Data from [103].

Source Description	Original Size (pixels)	MH (bytes)	MR (bytes)	MMR (bytes)	JBIG (bytes)
Letter	4352×3072	20,605	14,290	8,531	6,682
Sparse text	4352×3072	26,155	16,676	9,956	7,696
Dense text	4352×3072	135,705	105,684	92,100	70,703

T.4 and T.6 is not valid for halftone images. The JBIG algorithm, with its adaptive model and coder, suffers from no such drawbacks and performs well for halftone images [103].

7.7.5 JBIG2-T.88

The JBIG2 standard was approved in February 2000. Besides facsimile transmission, the standard is also intended for document storage, archiving, wireless transmission, print spooling, and coding of images on the Web. The standard provides specifications only for the decoder, leaving the encoder design open. This means that the encoder design can be constantly refined, subject only to compatibility with the decoder specifications. This situation also allows for lossy compression, because the encoder can incorporate lossy transformations to the data that enhance the level of compression.

The compression algorithm in JBIG provides excellent compression of a generic bi-level image. The compression algorithm proposed for JBIG2 uses the same arithmetic coding scheme as JBIG. However, it takes advantage of the fact that a significant number of bi-level images contain structure that can be used to enhance compression performance. A large percentage of bi-level images consist of text on some background, while another significant percentage of bi-level images are or contain halftone images. The JBIG2 approach allows the encoder to select the compression technique that would provide the best performance for the type of data. To do so, the encoder divides the page to be compressed into three types of regions called *symbol regions*, *halftone regions*, and *generic regions*. The symbol regions are those containing text data, the halftone regions are those containing halftone images, and the generic regions are all the regions that do not fit into either category.

The partitioning information has to be supplied to the decoder. The decoder requires that all information provided to it be organized into *segments* that are made up of a segment header, a data header, and segment data. The page information segment contains information about the page including the size and resolution. The decoder uses this information to set up the page buffer. It then decodes the various regions using the appropriate decoding procedure and places the different regions in the appropriate location.

Generic Decoding Procedures

There are two procedures used for decoding the generic regions: the generic region decoding procedure and the generic refinement region decoding procedure. The generic region decoding procedure uses one of two procedures; the MMR technique used in the Group 3 and Group

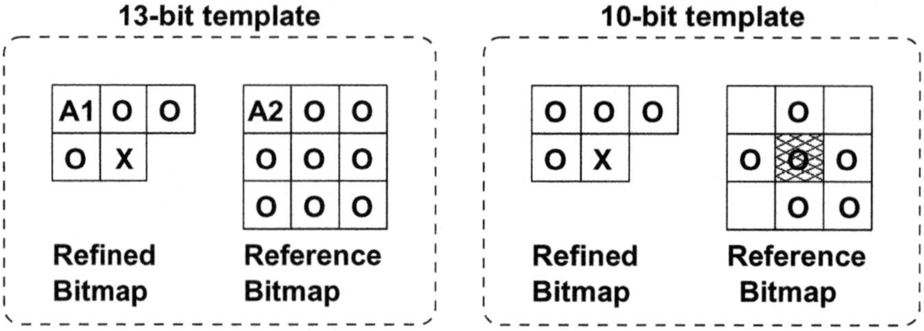

FIGURE 7.15 **Contexts used in the generic refinement decoding procedure.**

4 fax standards or a variation of the technique used to encode the lowest-resolution layer in the JBIG recommendation. The first procedure, MMR algorithm was described earlier in this chapter (Section 7.7.2). The second procedure, called *typical prediction*, is similar to the procedure used for JBIG as described earlier.

The adaptive arithmetic coding used in JBIG2 uses the same algorithm as that used in JBIG with four templates available for generating the context. There is a 16-bit template which contains four variable locations, a 13-bit template with one variable location, and the two 10-bit templates used for the encoding of the pixels in the JBIG bottom layer.

The generic refinement decoding procedure assumes the existence of a *reference* layer and decodes the segment data with reference to this layer. The standard leaves open the specification of the reference layer. The decoding of the *refined* layer is done in raster scan order using both the reference bitmap and the portions of the refined layer that have already been decoded. The decoding uses context adaptive arithmetic coding where the context consists of already decoded bits from the refined bitmap as well as bits from the reference bitmap. The two templates used in the refinement decoding procedure are shown in Figure 7.15. In Figure 7.15, the pixel marked X is the pixel being decoded while the shaded pixel in the reference bitmap is the pixel in the same spatial location as the pixel being decoded. The pixels marked $A1$ and $A2$ are the adaptive pixels. The $A1$ pixel location can be anywhere within 128 pixels of the pixel being decoded in the vertical and horizontal directions within the set of already decoded pixels. If we denote the X pixel location as (0,0) the $A2$ pixel location can be anywhere between (-128,-128) to (127,127) in the reference bitmap.

The generic refinement procedure can also use a typical prediction mode. In the generic refinement procedure a pixel is said to be a typical pixel if the 3×3 block of pixels in the reference bitmap centered on the location of the pixel being decoded all have the same value and the pixel being decoded also has that value. If every pixel in a line corresponding to a 3×3 block of identical valued pixels in the reference bitmap is a typical pixel, the typical pixel flag *LTP* is set to one. When the typical pixel flag is set to one, the pixels corresponding to 3×3 blocks of identically valued pixels are not encoded, and the decoder simply copies that value from the reference bitmap.

Symbol Region Decoding

The symbol region decoding procedure is a dictionary-based decoding procedure. The symbol region segment is decoded with the help of a symbol dictionary contained in the symbol dictionary segment. The data in the symbol region segment contains the location where a symbol is to be placed, as well as the index to an entry in the symbol dictionary. The symbol dictionary consists of a set of bitmaps and is decoded using the generic decoding procedures. Note that because JBIG2 allows for lossy compression, the symbols do not have to exactly match the symbols in the original document. This feature can significantly increase compression performance when the original document contains noise that may preclude exact matches with the symbols in the dictionary.

Halftone Region Decoding

The halftone region decoding procedure is also a dictionary-based decoding procedure. The halftone region segment is decoded with the help of a halftone dictionary contained in the halftone dictionary segment. The halftone dictionary segment is decoded using the generic decoding procedures. The data in the halftone region segment consists of the location of the halftone region and indices to the halftone dictionary. The dictionary is a set of fixed-size halftone patterns. As in the case of the symbol region, if lossy compression is allowed, the halftone patterns do not have to exactly match the patterns in the original document. By allowing for nonexact matches, the dictionary can be kept small, resulting in higher compression.

7.8 MRC-T.44

With the rapid advance of technology for document production, documents have changed in appearance. Where a document used to be a set of black and white printed pages, now documents contain multicolored text as well as color images. To deal with this new type of document, the ITU-T developed recommendation T.44 for Mixed Raster Content (MRC). This recommendation takes the approach of separating the document into elements that can be compressed using available techniques. Thus, it is more an approach of partitioning a document image than a compression technique. The compression strategies employed here are borrowed from previous standards such as JPEG (T.81), JBIG (T.82), and even T.6.

The T.44 recommendation divides a page into slices where the width of the slice is equal to the width of the entire page. The height of the slice is variable. In the base mode, each slice is represented by three layers: a background layer, a foreground layer, and a mask layer. These layers are used to effectively represent three basic data types: color images (which may be continuous tone or color mapped), bi-level data, and multilevel (multicolor) data. The multilevel image data is put in the background layer, and the mask and foreground layers are used to represent the bi-level and multilevel nonimage data. To work through the various definitions, let us use the document shown in Figure 7.16 as an example. We have divided the document into two slices. The top slice contains the picture of the cake and two lines

FIGURE 7.16 **Ruby's birthday invitation.**

FIGURE 7.17 **The background layer.**

of writing in two "colors." Notice that the heights of the two slices are not the same, and the complexity of the information contained in the two slices is not the same. The top slice contains multicolored text, and a continuous tone image whereas the bottom slice contains only bi-level text. Let us take the upper slice first, and see how to divide it into the three layers. We will discuss how to code these layers later. The background layer consists of the cake and nothing else. The default color for the background layer is white (though this can be changed). Therefore, we do not need to send the left half of this layer, which contains only white pixels (Figure 7.17).

The mask layer (Figure 7.18) consists of a bi-level representation of the textual information, while the foreground layer contains the colors used in the text. To reassemble the slice, we begin with the background layer. We then add to it pixels from the foreground layer using the mask layer as the guide. Wherever the mask layer pixel is black (1), we pick the corresponding pixel from the foreground layer. Wherever the mask pixel is white (0), we use the pixel from the background layer. Because of its role in selecting pixels, the mask layer is also known as the selector layer. During transmission, the mask layer is transmitted first, followed by the background and the foreground layers. During the rendering process, the background layer is rendered first (Figure 7.19).

When we look at the lower slice, we notice that it contains only bi-level information. In this case, we only need the mask layer because the other two layers would be superfluous. In order to deal with this kind of situation, the standard defines three different kinds of stripes. Three-layer stripes (3LS) contain all three layers and are useful when there is both image and

FIGURE 7.18 **The mask layer.**

FIGURE 7.19 **The foreground layer.**

textual data in the stripe. Two-layer stripes (2LS) only contain two layers, with the third set to a constant value. This kind of stripe would be useful when encoding a stripe with multicolored text and no images, or a stripe with images and bi-level text or line drawings. The third kind of stripe is a one-layer stripe (1LS), which would be used when a stripe contains only bi-level text or line art or only continuous tone images.

Once the document has been partitioned it can be compressed. Notice that the types of data we have after partitioning are continuous tone images, bi-level information, and multi-level regions. We already have efficient standards for compressing these types of data. For the mask layer containing bi-level information, the recommendation suggests that one of several approaches can be used, including modified Huffman or modified READ (as described in recomendation T.4), MMR (as described in recommendation T.6), or JBIG (recommendation T.82). The encoder includes information in the datastream about which algorithm has been used. For the continuous tone images and the multilevel regions contained in the foreground and background layers, the recommendation suggests the use of the JPEG standard (recommendation T.81) or the JBIG standard. The header for each slice contains information about which algorithm is used for compression-3.

7.9 Summary

In this chapter, we have examined a number of ways to compress images. All of these approaches exploit the fact that pixels in an image are generally highly correlated with their neighbors. This correlation can be used to predict the actual value of the current pixel. The prediction error can then be encoded and transmitted. Where the correlation is especially high, as in the case of bi-level images, long stretches of pixels can be encoded together using their similarity with previous rows. Finally, by identifying different components of an image that have common characteristics, an image can be partitioned and each partition encoded using the algorithm best suited to it.

Further Reading

1. A detailed survey of lossless image compression techniques can be found in "Lossless Image Compression" by K.P. Subbalakshmi. This chapter appears in the *Lossless Compression Handbook*, Academic Press, 2003 [268].
2. For a detailed description of the LOCO-I and JPEG-LS compression algorithm, see "The LOCO-I Lossless Image Compression Algorithm: Principles and Standardization into JPEG-LS," Hewlett-Packard Laboratories Technical Report HPL-98-193, November 1998 [87].
3. The JBIG and JBIG2 standards are described in a very accessible manner in "Lossless Bilevel Image Compression," by M.W. Hoffman. This chapter appears in the *Lossless Compression Handbook*, Academic Press, 2003 [269].
4. The area of lossless image compression is a very active one, and new schemes are being published all the time. These articles appear in a number of journals, including *Journal of Electronic Imaging*, *Optical Engineering*, *IEEE Transactions on Image Processing*, *IEEE Transactions on Communications*, *Communications of the ACM*, *IEEE Transactions on Computers*, and *Image Communication*, among others.

7.10 Projects and Problems

1. Encode the binary image shown in Figure 7.20 using the modified Huffman scheme.
2. Encode the binary image shown in Figure 7.20 using the modified READ scheme.

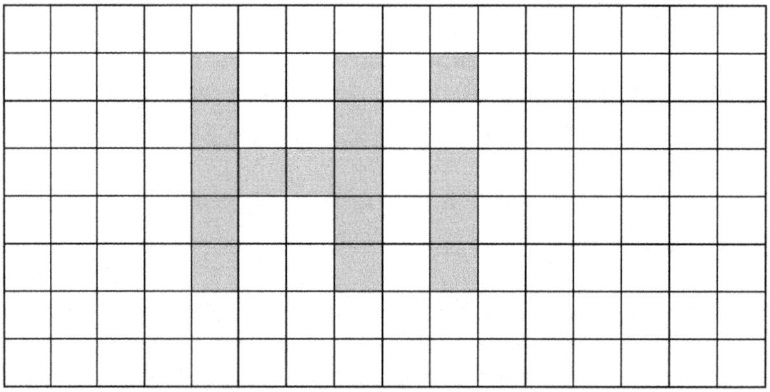

FIGURE 7.20 **An 8 × 16 binary image.**

3. Encode the binary image shown in Figure 7.20 using the modified modified READ scheme.

4. Suppose we want to transmit a 512×512, 8-bits-per-pixel image over a 9600 bits per second line.

(a) If we were to transmit this image using raster scan order, after 15 seconds how many rows of the image will the user have received? To what fraction of the image does this correspond?

(b) If we were to transmit the image using the method of Example 7.6.1, how long would it take the user to receive the first approximation? How long would it take to receive the first two approximations?

5. An implementation of the progressive transmission example (Example 7.6.1) is included in the programs accompanying this book. The program is called `prog_tran1.c`. Using this program as a template, experiment with different ways of generating approximations (you could use various types of weighted averages) and comment on the qualitative differences (or lack thereof) with using various schemes. Try different block sizes and comment on the practical effects in terms of quality and rate.

6. The program `jpegll_enc.c` generates the residual image for the different JPEG prediction modes, while the program `jpegll_dec.c` reconstructs the original image from the residual image. The output of the encoder program can be used as the input to the public domain arithmetic coding program mentioned in Chapter 4 and the Huffman coding programs mentioned in Chapter 3. Study the performance of different combinations of prediction mode and entropy coder using three images of your choice. Account for any differences you see.

7. Extend `jpegll_enc.c` and `jpegll_dec.c` with an additional prediction mode—be creative! Compare the performance of your predictor with the JPEG predictors.

8. Implement the portions of the CALIC algorithm described in this chapter. Encode the Sena image using your implementation.

8

Mathematical Preliminaries for Lossy Coding

8.1 Overview

Before we discussed lossless compression, we presented some of the mathematical background necessary for understanding and appreciating the compression schemes that followed. We will try to do the same here for lossy compression schemes. In lossless compression schemes, rate is the general concern. With lossy compression schemes, the loss of information associated with such schemes is also a concern. We will look at different ways of assessing the impact of the loss of information. We will also briefly revisit the subject of information theory, mainly to get an understanding of the part of the theory that deals with the trade-offs involved in reducing the rate, or number of bits per sample, at the expense of the introduction of distortion in the decoded information. This aspect of information theory is also known as rate distortion theory. We will also look at some of the models used in the development of lossy compression schemes.

8.2 Introduction

This chapter will provide some mathematical background that is necessary for discussing lossy compression techniques. Most of the material covered in this chapter is common to many of the compression techniques described in the later chapters. Material that is specific to a particular technique is described in the chapter in which the technique is presented.

Some of the material presented in this chapter is not essential for understanding the techniques described in this book. However, to follow some of the literature in this area, familiarity with these topics is necessary. We have marked these sections with a $\star$. If you are primarily interested in the techniques, you may wish to skip these sections, at least on first reading. On the other hand, if you wish to delve more deeply into these topics, we have included a list of resources at the end of this chapter that provide a more mathematically rigorous treatment of this material.

When we were looking at lossless compression, one thing we never had to worry about was how the reconstructed sequence would differ from the original sequence. By definition, the reconstruction of a losslessly constructed sequence is identical to the original sequence. However, there is only a limited amount of compression that can be obtained with lossless compression. There is a floor (a hard one) defined by the entropy of the source, below which we cannot drive the size of the compressed sequence. As long as we wish to preserve all of the information in the source, the entropy, like the speed of light, is a fundamental limit.

The limited amount of compression available from using lossless compression schemes may be acceptable in several circumstances. The storage or transmission resources available to us may be sufficient to handle our data requirements after lossless compression. Or the possible consequences of a loss of information may be much more expensive than the cost of additional storage and/or transmission resources. This would be the case with the storage and archiving of bank records; an error in the records could turn out to be much more expensive than the cost of buying additional storage media.

If neither of these conditions hold—that is, resources are limited and we do not require absolute integrity—we can improve the amount of compression by accepting a certain degree of loss during the compression process. Performance measures are necessary to determine the efficiency of our *lossy* compression schemes. For the lossless compression schemes, we essentially used only the rate as the performance measure. That would not be feasible for lossy compression. If rate were the only criterion for lossy compression schemes, where loss of information is permitted, the best lossy compression scheme would be simply to throw away all the data! Therefore, we need some additional performance measure, such as some measure of the difference between the original and reconstructed data, which we will refer to as the *distortion* in the reconstructed data. In the next section, we will look at some of the more well-known measures of difference and discuss their advantages and shortcomings.

In the best of all possible worlds, we would like to incur the minimum amount of distortion while compressing to the lowest rate possible. Obviously, there is a trade-off between minimizing the rate and keeping the distortion small. The extreme cases are when we transmit no information, in which case the rate is zero, or keep all the information, in which case the distortion is zero. The rate for a discrete source is simply the entropy. The study of the situations between these two extremes is called *rate distortion theory*. In this chapter we will take a brief look at some important concepts related to this theory.

We will need to expand the dictionary of models available for our use, for several reasons. First, because we are now able to introduce distortion, we need to determine how to add distortion intelligently. For this, we often need to look at the sources somewhat differently than we have done previously. Another reason is that we will be looking at compression schemes for sources that are analog in nature, even though we have treated them as discrete sources in the past. We need models that more precisely describe the true nature of these

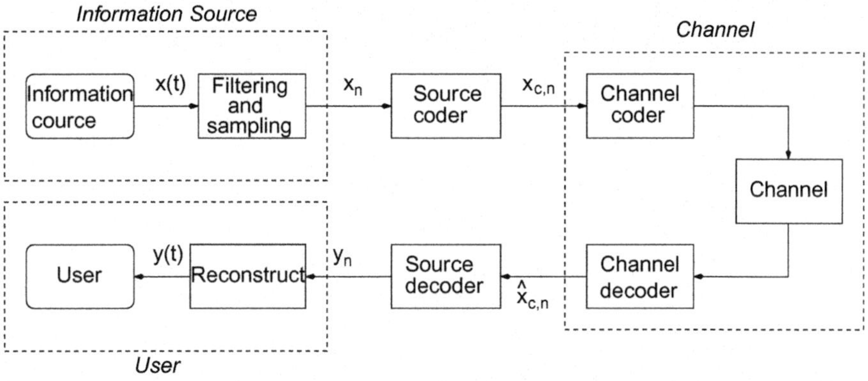

FIGURE 8.1 A simplified block diagram of a communication system.

sources. We will describe several different models that are widely used in the development of lossy compression algorithms.

Finally, let us briefly look at the context of the compression operation. A very simplified block diagram of a communication system is shown in Figure 8.1. An information source generates a signal (often an analog signal). In the figure we have represented the signal as a function of time. This would be the case if the signal was a speech or audio signal. However, the information signal may be a two-dimensional spatial signal such as an image, or a function of both space and time such as video. If this information is not in discrete form it is discretized to generate a sequence of values. Again we have depicted this signal as being a function of a single index as would be the case for a speech or audio signal. However, it could just as well be a function of two or three indices. This discrete sequence is the input to the compression algorithm, or in more general terms, the source coder, which attempts to reduce the average number of bits per sample used to encode the sequence. The process of source coding usually results in the removal of redundancy in the signal, making it vulnerable to noise. In order to protect against errors the output of the source coder is encoded using a channel encoder that introduces redundancy in a controlled fashion. The block labeled "*channel*" is usually thought of as the medium over which the information is transmitted from one location to another. In the case of something like a cell phone the locations may be spatially distinct—the information is transmitted from "here" to "there." However, in many cases, such as when we store the information to be examined later, the locations my be temporally distinct—a communication from "now" to "then." We will generally only deal with the discrete form of the data and then only the source coding aspect, so we tend to simplify this diagram even further by lumping together the blocks in the dashed boxes. Thus we have an information source generating a sequence of samples $\{x_n\}$ which are encoded by the source coder and transmitted, spatially or temporally, over the channel. We will assume in this book that the channel does not introduce any distortion so the source coder output is identical to the source decoder input. The source decoder output $\{y_n\}$ is then supplied to the user. The goal of the source coder is to use as few bits as possible to encode the information-bearing sequence $\{x_n\}$ while still keeping the reconstruction sequence $\{y_n\}$ as a close approximation. We are using very fuzzy language here when we

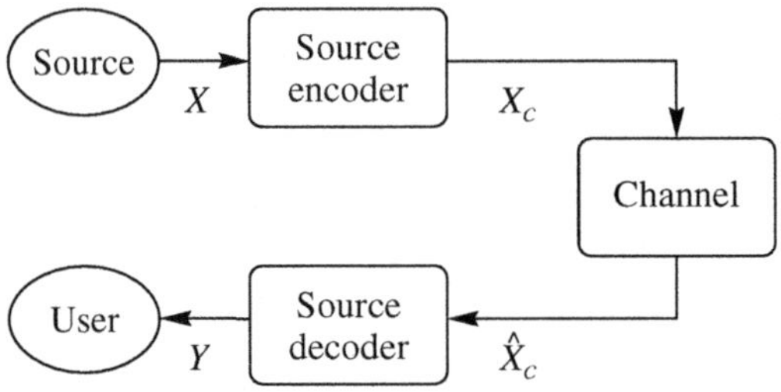

FIGURE 8.2 Block diagram of a generic compression scheme.

talk about "as few bits as possible," and "close approximation," and in the following sections we will introduce quantitative measures that will allow us to be more precise in our language.

The information-bearing sequence $\{x_n\}$ is generally highly correlated, with significant dependencies between samples. However, for many of the theoretical results we will assume that the sequence can be modeled as independent identically distributed random variables. Surprisingly, even with this gross level of simplification we can obtain results that provide significant insight into the source coding process. For these theoretical developments we will use the block diagram and notation shown in Figure 8.2. The output of the source is modeled as a random variable X. The *source coder* takes the source output and produces the compressed representation X_c. The channel block represents all transformations the compressed representation undergoes before the source is reconstructed. We will take the channel to be the identity mapping, which means $X_c = \widehat{X}_c$. The source decoder takes the compressed representation and produces a reconstruction of the source output for the user.

In the following section we will examine the more popular measures of fidelity, or closeness, between the information sequence $\{x_n\}$ and the reconstructed sequence $\{y_n\}$.

8.3 Distortion Criteria

How do we measure the closeness or fidelity of a reconstructed source sequence to the original? The answer frequently depends on what is being compressed and who is doing the answering. Suppose we were to compress and then reconstruct an image. If the image is a work of art and the resulting reconstruction is to be part of a book on art, the best way to find out how much distortion was introduced and in what manner is to ask a person familiar with the work to look at the image and provide an opinion. If the image is that of a house and is to be used in an advertisement, the best way to evaluate the quality of the reconstruction is probably to ask a real estate agent. However, if the image is from a satellite and is to be processed by a machine to obtain information about the objects in the image, the best measure of fidelity is to see how the distortion introduced affects the functioning of the machine. Similarly, if we were to compress and then reconstruct an audio segment, the judgment of how close the reconstructed sequence is to the original depends on the type of material being examined as well as the manner in which the judging is done. An audiophile is much more likely to perceive distortion in the reconstructed sequence, and distortion is much more likely to be noticed in a musical piece than in a politician's speech.

In the best of all worlds, we would always use the end user of a particular source output to assess quality and provide the feedback required for the design. In practice this is not often possible, especially when the end user is a human, because it is difficult to incorporate the human response into mathematical design procedures. Also, there is difficulty in objectively reporting the results. The people asked to assess one person's design may be more easygoing than the people who were asked to assess another person's design. Even though the reconstructed output using one person's design is rated "excellent" and the reconstructed output using the other person's design is only rated "acceptable," switching observers may change the ratings. We could reduce this kind of bias by recruiting a large number of observers in the hope that the various biases will cancel each other out. This is often the option used, especially in the final stages of the design of compression systems. However, the rather cumbersome nature of this process is limiting. We generally need a more practical method for looking at how close the reconstructed signal is to the original.

A natural thing to do when looking at the fidelity of a reconstructed sequence is to look at the differences between the original and reconstructed values—in other words, the distortion introduced in the compression process. Two popular measures of distortion or difference between the original and reconstructed sequences are the squared error measure and the absolute difference measure. These are called *difference distortion measures*. If $\{x_n\}$ is the source output and $\{y_n\}$ is the reconstructed sequence, then the squared error measure is given by

$$d(x_n, y_n) = (x_n - y_n)^2 \tag{1}$$

and the absolute difference measure is given by

$$d(x_n, y_n) = |x_n - y_n| \tag{2}$$

In general, it is difficult to examine the difference on a term-by-term basis. Therefore, a number of average measures are used to summarize the information in the difference sequence. The most often used average measure is the average of the squared error measure. This is called the *mean squared error* (mse). If we model the information sequence as a sequence of random variables $\{X_n\}$ and the reconstruction sequence as the sequence of random variables $\{Y_n\}$, for a sequence of length N we can define the mean squared error as

$$D = E\left\{\frac{1}{N}\sum_{n=1}^{N}(X_n - Y_n)^2\right\} = \frac{1}{N}\sum_{n=1}^{N} E\left[(X_n - Y_n)^2\right] \tag{3}$$

where the expectation is with respect to the joint distribution of X_n and Y_n. If we assume the sequences are independent and identically distributed this expression simplifies to

$$D = E\left[(X - Y)^2\right]$$

In practice we assume the information and reconstruction sequences are ergodic and replace the ensemble averages in Equation (3) with time averages to obtain

$$\sigma_d^2 = \frac{1}{N}\sum_{n=1}^{N}(x_n - y_n)^2 \tag{4}$$

Notice we have used the symbol σ_d^2 for the mean squared error, which implies that the variance of the distortion sequence $d(x_n, y_n)$ is equal to the second moment, or that the distortion sequence is zero mean. This would happen if the error introduced during source coding is additive. This idea that the error incurred during compression is additive is a common (though often unstated) assumption. If we are interested in the size of the error relative to the signal, we can find the ratio of the average squared value of the source output and the mse. This is called the *signal-to-noise ratio* (SNR):

$$\text{SNR} = \frac{\sigma_x^2}{\sigma_d^2} \tag{5}$$

where σ_x^2 is the average squared value of the source output, or signal, and σ_d^2 is the mse. The SNR is often measured on a logarithmic scale, and the units of measurement are *decibels* (abbreviated to dB):

$$\text{SNR(dB)} = 10 \log_{10} \frac{\sigma_x^2}{\sigma_d^2} \tag{6}$$

Sometimes we are more interested in the size of the error relative to the peak value of the signal x_{peak} than in the size of the error relative to the average squared value of the signal. This ratio is called the *peak-signal-to-noise-ratio* (PSNR) and is given by

$$\text{PSNR(dB)} = 10 \log_{10} \frac{x_{\text{peak}}^2}{\sigma_d^2} \tag{7}$$

Another difference distortion measure that is used quite often, although not as often as the mse, is the average of the absolute difference, or

$$d_1 = \frac{1}{N} \sum_{n=1}^{N} |x_n - y_n| \tag{8}$$

This measure seems especially useful for evaluating image compression algorithms.

In some applications, the distortion is not perceptible as long as it is below some threshold. In these situations, we might be interested in the maximum value of the error magnitude,

$$d_\infty = \max_n |x_n - y_n| \tag{9}$$

We have looked at two approaches to measuring the fidelity of a reconstruction. The first method involving humans may provide a very accurate measure of perceptible fidelity, but it is not practical or useful in mathematical design approaches. The second is mathematically tractable, but it usually does not provide a very accurate indication of the perceptible fidelity of the reconstruction. A middle ground is to find a mathematical model for human perception, transform both the source output and the reconstruction to this perceptual space, and then measure the difference in the perceptual space. For example, suppose we could find a transformation $\mathcal{V}$ that represented the actions performed by the human visual system (HVS) on the light intensity impinging on the retina before it is "perceived" by the cortex. We could then find $\mathcal{V}(x)$ and $\mathcal{V}(y)$ and examine the difference between them. There are two problems with this approach. First, the process of human perception is inferential and very difficult to model. Second, even if we could find a mathematical model for perception, the odds are that it would be so complex that it would be mathematically intractable.

In spite of these disheartening prospects, the study of perception mechanisms is still important from the perspective of design and analysis of compression systems. Even if we cannot obtain a transformation that accurately models perception, we can learn something about the properties of perception that may come in handy in the design of compression systems. In the following, we will look at some of the properties of the human visual system and the perception of sound. Our review will be far from thorough, but the intent here is to present some properties that will be useful in later chapters when we talk about compression of images, video, speech, and audio.

8.3.1 The Human Visual System

The eye is a globe-shaped object with a lens in the front that focuses objects onto the retina in the back of the eye. The retina contains two kinds of receptors, called *rods* and *cones*. The rods are more sensitive to light than cones, and in low light most of our vision is due to the operation of rods. There are three kinds of cones, each of which is most sensitive at a different wavelength of the visible spectrum. The peak sensitivities of the cones are in the red, blue, and green regions of the visible spectrum [104]. The cones are mostly concentrated in a very small area of the retina called the *fovea*. Although the rods are more numerous than the cones, the cones provide better resolution because they are more closely packed in the fovea. The muscles of the eye move the eyeball, positioning the image of the object on the fovea. This becomes a drawback in low light. One way to improve what you see in low light is to focus to one side of the object. This way the object is imaged on the rods, which are more sensitive to light.

The eye is sensitive to light over an enormously large range of intensities; the upper end of the range is about 10^{10} times the lower end of the range. However, at a given instant we cannot perceive the entire range of brightness. Instead, the eye adapts to an average brightness level. The range of brightness levels that the eye can perceive at any given instant is much smaller than the total range it is capable of perceiving.

If we illuminate a screen with a certain intensity I and shine a spot on it with different intensity, the spot becomes visible when the difference in intensity is ΔI. This is called the *just noticeable difference* (jnd). The ratio $\frac{\Delta I}{I}$ is known as the *Weber fraction* or *Weber ratio*. This ratio is known to be constant at about 0.02 over a wide range of intensities in the absence of background illumination. However, if the background illumination is changed, the range over which the Weber ratio remains constant becomes relatively small. The constant range is centered around the intensity level to which the eye adapts.

If $\frac{\Delta I}{I}$ is constant, then we can infer that the sensitivity of the eye to intensity is a logarithmic function ($d(\log I) = dI/I$). Thus, we can model the eye as a receptor whose output goes to a logarithmic nonlinearity. We also know that the eye acts as a spatial low-pass filter [105, 106]. Putting all of this information together, we can develop a model for monochromatic vision, shown in Figure 8.3.

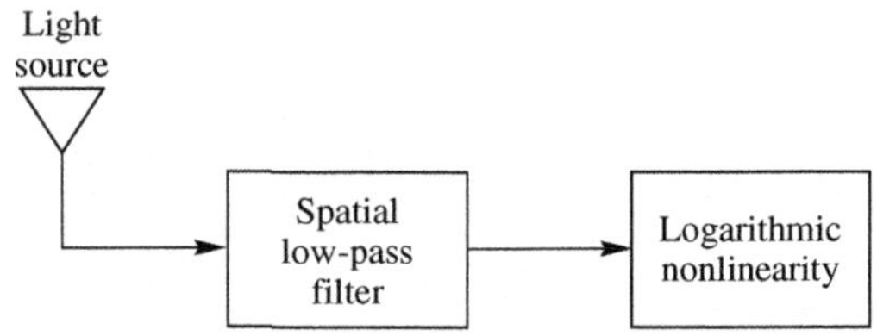

FIGURE 8.3 A model of monochromatic vision.

How does this description of the human visual system relate to coding schemes? Notice that the mind does not perceive everything the eye sees. We can use this knowledge to design compression systems such that the distortion introduced by our lossy compression scheme is not noticeable.

8.3.2 Auditory Perception

The ear is divided into three parts, creatively named the outer ear, the middle ear, and the inner ear. The outer ear consists of the structure that directs the sound waves, or pressure waves, to the *tympanic membrane*, or eardrum. This membrane separates the outer ear from the middle ear. The middle ear is an air-filled cavity containing three small bones that provide coupling between the tympanic membrane and the *oval window*, which leads into the inner ear. The tympanic membrane and the bones convert the pressure waves in the air to acoustical vibrations. The inner ear contains, among other things, a snail-shaped passage called the *cochlea* that contains the transducers that convert the acoustical vibrations to nerve impulses.

The human ear can hear sounds from approximately 20 Hz to 20 kHz, a 1000:1 range of frequencies. The range decreases with age; older people are usually unable to hear the higher frequencies. As in vision, auditory perception has several nonlinear components. One is that loudness is a function not only of the sound level, but also of the frequency. Thus, for example, a pure 1 kHz tone presented at a 20 dB intensity level will have the same apparent loudness as a 50 Hz tone presented at a 50 dB intensity level. By plotting the amplitude of tones at different frequencies that sound equally loud, we get a series of curves called the *Fletcher-Munson curves* [107].

Another very interesting audio phenomenon is that of *masking*, where one sound blocks out or masks the perception of another sound. The fact that one sound can drown out another seems reasonable. What is not so intuitive about masking is that if we were to try to mask a pure tone with noise, only the noise in a small frequency range around the tone being masked contributes to the masking. This range of frequencies is called the *critical band*. For most frequencies, when the noise just masks the tone, the ratio of the power of the tone divided by the power of the noise in the critical band is a constant [108]. The width of the critical band varies with frequency. This fact has led to the modeling of auditory perception as a bank of band-pass filters. There are a number of other, more complicated masking phenomena that also lend support to this theory (see [108, 109] for more information). The limitations of auditory perception play a major role in the design of audio compression algorithms. We will delve further into these limitations when we discuss audio compression in Chapter 17.

8.4 Information Theory Revisited ★

In order to study the trade-offs between rate and the distortion of lossy compression schemes, we would like to have rate defined explicitly as a function of the distortion for a given distortion measure. Unfortunately, this is generally not possible; and we have to go about it in a more roundabout way. Before we head down this path, we need a few more concepts from information theory.

In Chapter 2, when we talked about information, we were referring to letters from a single alphabet. In the case of lossy compression, we have to deal with two alphabets, the source alphabet and the reconstruction alphabet. These two alphabets are generally different from each other.

Example 8.4.1:

A simple lossy compression approach is to drop a certain number of the least significant bits from the source output. We might use such a scheme between a source that generates monochrome images at 8 bits per pixel and a user whose display facility can display only 64 different shades of gray. We could drop the two least significant bits from each pixel before transmitting the image to the user. There are other methods we can use in this situation that are much more effective, but this is certainly simple.

Suppose our source output consists of 4-bit words $\{0, 1, 2, \ldots, 15\}$. The source encoder encodes each value by shifting out the least significant bit. The output alphabet for the source coder is $\{0, 1, 2, \ldots, 7\}$. At the receiver we cannot recover the original value exactly. However, we can get an approximation by shifting in a 0 as the least significant bit or, in other words, multiplying the source encoder output by two. Thus, the reconstruction alphabet is $\{0, 2, 4, \ldots, 14\}$, and the source and reconstruction do not take values from the same alphabet. ♦

As the source and reconstruction alphabets can be distinct, we need to be able to talk about the information relationships between two random variables that take on values from two different alphabets.

8.4.1 Conditional Entropy

Let X be a random variable that takes values from the source alphabet $\mathcal{X} = \{x_0, x_1, \ldots, x_{N-1}\}$. Let Y be a random variable that takes on values from the reconstruction alphabet $\mathcal{Y} = \{y_0, y_1, \ldots, y_{M-1}\}$. From Chapter 2, we know that the entropy of the source and the reconstruction are given by

$$H(X) = -\sum_{i=0}^{N-1} P(x_i) \log_2 P(x_i)$$

and

$$H(Y) = -\sum_{j=0}^{M-1} P(y_j)\log_2 P(y_j)$$

A measure of the relationship between two random variables is the *conditional entropy* (the average value of the conditional self-information). Recall that the self-information for an event A is defined as

$$i(A) = \log \frac{1}{P(A)} = -\log P(A)$$

In a similar manner, the conditional self-information of an event A, given that another event B has occurred, can be defined as

$$i(A|B) = \log \frac{1}{P(A|B)} = -\log P(A|B)$$

Suppose B is the event "Frazer has not drunk anything in two days," and A is the event "Frazer is thirsty." Then $P(A|B)$ should be close to one, which means that the conditional self-information $i(A|B)$ would be close to zero. This makes sense from an intuitive point of view as well. If we know that Frazer has not drunk anything in two days, then the statement that Frazer is thirsty would not be at all surprising to us and would contain very little information.

As in the case of self-information, we are generally interested in the average value of the conditional self-information. This average value is called the conditional entropy. The conditional entropies of the source and reconstruction alphabets are given as

$$H(X|Y) = -\sum_{i=0}^{N-1}\sum_{j=0}^{M-1} P(x_i|y_j)P(y_j)\log_2 P(x_i|y_j) \tag{10}$$

and

$$H(Y|X) = -\sum_{i=0}^{N-1}\sum_{j=0}^{M-1} P(y_j|x_i)P(x_i)\log_2 P(y_j|x_i) \tag{11}$$

The conditional entropy $H(X|Y)$ can be interpreted as the amount of uncertainty remaining about the random variable X, or the source output, given that we know what value the reconstruction Y took. The additional knowledge of Y should reduce the uncertainty about X, and we can show that

$$H(X|Y) \leqslant H(X) \tag{12}$$

(see Problem 5 at the end of this chapter).

Example 8.4.2:

Suppose we have the 4-bits-per-symbol source and compression scheme described in Example 8.4.1. Assume that the source is equally likely to select any letter from its alphabet. Let us calculate the various entropies for this source and compression scheme.

As the source outputs are all equally likely, $P(X = i) = \frac{1}{16}$ for all $i \in \{0, 1, 2, \ldots, 15\}$, and, therefore,

$$H(X) = -\sum_i \frac{1}{16} \log \frac{1}{16} = \log 16 = 4 \text{ bits} \tag{13}$$

We can calculate the probabilities of the reconstruction alphabet:

$$P(Y = j) = P(X = j) + P(X = j + 1) = \frac{1}{16} + \frac{1}{16} = \frac{1}{8} \tag{14}$$

Therefore, $H(Y) = 3$ bits. To calculate the conditional entropy $H(X|Y)$, we need the conditional probabilities $\{P(x_i|y_j)\}$. From our construction of the source encoder, we see that

$$P(X = i|Y = j) = \begin{cases} \frac{1}{2} & \text{if } i = j \text{ or } i = j + 1, \text{ for } j = 0, 2, 4, \ldots, 14 \\ 0 & \text{otherwise} \end{cases} \tag{15}$$

Substituting this in the expression for $H(X|Y)$ in Equation (10), we get

$$\begin{aligned} H(X|Y) &= -\sum_i \sum_j P(X = i|Y = j)P(Y = j) \log P(X = i|Y = j) \\ &= -\sum_j \big[P(X = j|Y = j)P(Y = j) \log P(X = j|Y = j) \\ &\quad + P(X = j + 1|Y = j)P(Y = j) \log P(X = j + 1|Y = j)\big] \\ &\quad -8\left[\frac{1}{2} \cdot \frac{1}{8} \log \frac{1}{2} + \frac{1}{2} \cdot \frac{1}{8} \log \frac{1}{2}\right] && (16) \\ &= 1 && (17) \end{aligned}$$

Let us compare this answer to what we would have intuitively expected the uncertainty to be, based on our knowledge of the compression scheme. With the coding scheme described here, knowledge of Y means that we know the first 3 bits of the input X. The only thing about the input that we are uncertain about is the value of the last bit. In other words, if we know the value of the reconstruction, our uncertainty about the source output is 1 bit. Therefore, at least in this case, our intuition matches the mathematical definition.

To obtain $H(Y|X)$, we need the conditional probabilities $\{P(y_j|x_i)\}$. From our knowledge of the compression scheme, we see that

$$P(Y = j|X = i) = \begin{cases} 1 & \text{if } i = j \text{ or } i = j + 1, \text{ for } j = 0, 2, 4, \ldots, 14 \\ 0 & \text{otherwise} \end{cases} \tag{18}$$

If we substitute these values into Equation (11), we get $H(Y|X) = 0$ bits (note that $0 \log 0 = 0$).

This also makes sense. For the compression scheme described here, if we know the source output, we know 4 bits, the first 3 of which are the reconstruction. Therefore, in this example, knowledge of the source output at a specific time completely specifies the corresponding reconstruction. ♦

8.4.2 Average Mutual Information

We make use of one more quantity that relates the uncertainty or entropy of two random variables. This quantity is called *mutual information* and is defined as

$$i(x_k; y_j) = \log\left[\frac{P(x_k|y_j)}{P(x_k)}\right] \tag{19}$$

We will use the average value of this quantity, appropriately called *average mutual information*, which is given by

$$I(X;Y) = \sum_{i=0}^{N-1}\sum_{j=0}^{M-1} P(x_i, y_j)\log\left[\frac{P(x_i|y_j)}{P(x_i)}\right] \tag{20}$$

$$= \sum_{i=0}^{N-1}\sum_{j=0}^{M-1} P(x_i|y_j)P(y_j)\log\left[\frac{P(x_i|y_j)}{P(x_i)}\right] \tag{21}$$

We can write the average mutual information in terms of the entropy and the conditional entropy by expanding the argument of the logarithm in Equation (21):

$$I(X;Y) = \sum_{i=0}^{N-1}\sum_{j=0}^{M-1} P(x_i, y_j)\log\left[\frac{P(x_i|y_j)}{P(x_i)}\right] \tag{22}$$

$$= \sum_{i=0}^{N-1}\sum_{j=0}^{M-1} P(x_i, y_j)\log P(x_i|y_j) - \sum_{i=0}^{N-1}\sum_{j=0}^{M-1} P(x_i, y_j)\log P(x_i) \tag{23}$$

$$= H(X) - H(X|Y) \tag{24}$$

where the second term in Equation (23) is $H(X)$, and the first term is $-H(X|Y)$. Thus, the average mutual information is the entropy of the source minus the uncertainty that remains about the source output after the reconstructed value has been received. The average mutual information can also be written as

$$I(X;Y) = H(Y) - H(Y|X) = I(Y;X). \tag{25}$$

We can show this easily by using Bayes' theorem. According to Bayes' theorem

$$P(x_i|y_j) = \frac{P(y_j|x_i)P(x_i)}{P(y_j)}$$

Substituting this expression for $P(x_i|y_j)$ in Equation (20) we get

$$I(X;Y) = \sum_{i=0}^{N-1}\sum_{j=0}^{M-1} P(x_i, y_j)\log\left[\frac{P(y_j|x_i)P(x_i)}{P(y_j)P(x_i)}\right] \quad (26)$$

$$= \sum_{i=0}^{N-1}\sum_{j=0}^{M-1} P(y_j|x_i)P(x_i)\log\left[\frac{P(y_j|x_i)}{P(y_j)}\right] \quad (27)$$

Expanding the argument of the logarithm in Equation (27) we obtain Equation (25).

Example 8.4.3:

For the source coder of Example 8.4.2, $H(X) = 4$ bits, and $H(X|Y) = 1$ bit. Therefore, using Equation (24), the average mutual information $I(X;Y)$ is 3 bits. If we wish to use Equation (25) to compute $I(X;Y)$, we would need $H(Y)$ and $H(Y|X)$, which from Example 8.4.2 are 3 and 0, respectively. Thus, the value of $I(X;Y)$ still works out to be 3 bits. ♦

8.4.3 Differential Entropy

Up to this point, we have assumed that the source picks its outputs from a discrete alphabet. When we study lossy compression techniques, we will see that for many sources of interest to us this assumption is not true. In this section, we will extend some of the information theoretic concepts defined for discrete random variables to the case of random variables with continuous distributions.

Unfortunately, we run into trouble from the very beginning. Recall that the first quantity we defined was self-information, which was given by $\log \frac{1}{P(x_i)}$, where $P(x_i)$ is the probability that the random variable will take on the value x_i. For a random variable with a continuous distribution, this probability is zero. Therefore, if the random variable has a continuous distribution, the "self information" associated with any value is infinity.

If we do not have the concept of self-information, how do we go about defining entropy, which is the average value of the self-information? We know that many continuous functions can be written as limiting cases of their discretized version. We will try to take this route in order to define the entropy of a continuous random variable X with probability density function (*pdf*) $f_X(x)$.

While the random variable X cannot generally take on a particular value with nonzero probability, it can take on a value in an *interval* with nonzero probability. Therefore, let us divide the range of the random variable into intervals of size Δ. Then, by the mean value theorem, in each interval $[(i-1)\Delta, i\Delta)$, there exists a number x_i, such that

$$f_X(x_i)\Delta = \int_{(i-1)\Delta}^{i\Delta} f_X(x)\,dx \quad (28)$$

Let us define a discrete random variable X_d with *pdf*

$$P(X_d = x_i) = f_X(x_i)\Delta \quad (29)$$

Then we can obtain the entropy of this random variable as

$$H(X_d) = -\sum_{i=-\infty}^{\infty} P(x_i) \log P(x_i) \tag{30}$$

$$= -\sum_{i=-\infty}^{\infty} f_X(x_i)\Delta \log f_X(x_i)\Delta \tag{31}$$

$$= -\sum_{i=-\infty}^{\infty} f_X(x_i)\Delta \log f_X(x_i) - \sum_{i=-\infty}^{\infty} f_X(x_i)\Delta \log \Delta \tag{32}$$

$$= -\sum_{i=-\infty}^{\infty} \left[f_X(x_i) \log f_X(x_i)\right] \Delta - \log \Delta \tag{33}$$

Taking the limit as $\Delta \to 0$ of Equation (33), the first term goes to $-\int_{-\infty}^{\infty} f_X(x) \log f_X(x)\, dx$, which looks like the analog to our definition of entropy for discrete sources. However, the second term is $-\log \Delta$, which goes to plus infinity when Δ goes to zero. It seems there is not an analog to entropy as defined for discrete sources. However, the first term in the limit serves some functions similar to that served by entropy in the discrete case and is a useful function in its own right. We call this term the *differential entropy* of a continuous source and denote it by $h(X)$. In an analogous manner we can also define the conditional entropy $h(X|Y)$ as

$$h(X|Y) = -\int_{-\infty}^{\infty} f_{XY}(x, y) \log f_{X|Y}(x|y) dx dy$$

Example 8.4.4:

Suppose we have a random variable X that is uniformly distributed in the interval $[a, b)$. The differential entropy of this random variable is given by

$$h(X) = -\int_{-\infty}^{\infty} f_X(x) \log f_X(x) dx \tag{34}$$

$$= -\int_a^b \frac{1}{b-a} \log \frac{1}{b-a} dx \tag{35}$$

$$= \log(b-a) \tag{36}$$

Notice that when $b-a$ is less than one, the differential entropy becomes negative—in contrast to the entropy, which never takes on negative values. ♦

Later in this chapter, we will find particular use for the differential entropy of the Gaussian source.

Example 8.4.5:

Suppose we have a random variable X that has a Gaussian *pdf*,

$$f_X(x) = \frac{1}{\sqrt{2\pi\sigma^2}} \exp\frac{-(x-\mu)^2}{2\sigma^2} \tag{37}$$

The differential entropy is given by

$$h(X) = -\int_{-\infty}^{\infty} \frac{1}{\sqrt{2\pi\sigma^2}} \exp\frac{-(x-\mu)^2}{2\sigma^2} \log\left[\frac{1}{\sqrt{2\pi\sigma^2}} \exp\frac{-(x-\mu)^2}{2\sigma^2}\right] dx \tag{38}$$

$$= -\log\frac{1}{\sqrt{2\pi\sigma^2}} \int_{-\infty}^{\infty} f_X(x)dx + \int_{-\infty}^{\infty} \frac{(x-\mu)^2}{2\sigma^2} \log e f_X(x)dx \tag{39}$$

$$= \frac{1}{2}\log 2\pi\sigma^2 + \frac{1}{2}\log e \tag{40}$$

$$= \frac{1}{2}\log 2\pi e\sigma^2 \tag{41}$$

Thus, the differential entropy of a Gaussian random variable is an increasing function of its variance. ♦

The differential entropy for the Gaussian distribution has the added distinction that it is larger than the differential entropy for any other continuously distributed random variable with the same variance. That is, for any random variable X, with variance σ^2

$$h(X) \leqslant \frac{1}{2}\log 2\pi e\sigma^2 \tag{42}$$

The proof of this statement depends on the fact that for any two continuous distributions $f_X(X)$ and $g_X(X)$

$$-\int_{-\infty}^{\infty} f_X(x)\log f_X(x)dx \leqslant -\int_{-\infty}^{\infty} f_X(x)\log g_X(x)dx \tag{43}$$

We will not prove Equation (43) here, but you may refer to [110] for a simple proof. To obtain Equation (42), we substitute the expression for the Gaussian distribution for $g_X(x)$. Noting that the left-hand side of Equation (43) is simply the differential entropy of the random variable X, we have

$$\begin{aligned}
h(X) &\leqslant -\int_{-\infty}^{\infty} f_X(x)\log\frac{1}{\sqrt{2\pi\sigma^2}} \exp\frac{-(x-\mu)^2}{2\sigma^2} dx \\
&= \frac{1}{2}\log(2\pi\sigma^2) + \log e\int_{-\infty}^{\infty} f_X(x)\frac{(x-\mu)^2}{2\sigma^2} dx \\
&= \frac{1}{2}\log(2\pi\sigma^2) + \frac{\log e}{2\sigma^2}\int_{-\infty}^{\infty} f_X(x)(x-\mu)^2 dx \\
&= \frac{1}{2}\log(2\pi e\sigma^2)
\end{aligned} \tag{44}$$

We seem to be striking out with continuous random variables. There is no analog for self-information and really none for entropy either. However, the situation improves when we look for an analog for the average mutual information. Let us define the random variable Y_d in a manner similar to the random variable X_d, as the discretized version of a continuous valued random variable Y. Then we can show (see Problem 4 at the end of this chapter)

$$H(X_d|Y_d) = -\sum_{i=-\infty}^{\infty}\sum_{j=-\infty}^{\infty}\left[f_{X|Y}(x_i|y_j)f_Y(y_j)\log f_{X|Y}(x_i|y_j)\right]\Delta\Delta - \log\Delta \quad (45)$$

Therefore, the average mutual information for the discretized random variables is given by

$$I(X_d; Y_d) = H(X_d) - H(X_d|Y_d) \quad (46)$$

$$= -\sum_{i=-\infty}^{\infty} f_X(x_i)\Delta \log f_X(x_i) \quad (47)$$

$$-\sum_{i=-\infty}^{\infty}\left[\sum_{j=-\infty}^{\infty} f_{X|Y}(x_i|y_j)f_Y(y_j)\log f_{X|Y}(x_i|y_j)\Delta\right]\Delta \quad (48)$$

Notice that the two $\log\Delta$s in the expression for $H(X_d)$ and $H(X_d|Y_d)$ cancel each other out, and as long as $h(X)$ and $h(X|Y)$ are not equal to infinity, when we take the limit as $\Delta \to 0$ of $I(X_d; Y_d)$ we get

$$I(X; Y) = h(X) - h(X|Y) \quad (49)$$

The average mutual information in the continuous case can be obtained as a limiting case of the average mutual information for the discrete case and has the same physical significance.

We have gone through a lot of mathematics in this section. But the information will be used immediately to define the rate distortion function for a random source.

8.5 Rate Distortion Theory ★

Rate distortion theory is concerned with the trade-offs between distortion and rate in lossy compression schemes. Rate is defined as the average number of bits used to represent each sample value. One way of representing the trade-offs is via a *rate distortion function R(D)*. The rate distortion function $R(D)$ specifies the lowest rate at which the output of a source can be encoded while keeping the distortion less than or equal to D. The distortion D is generally taken to be the expected value of a single letter distortion measure $d(x_i, y_j)$, which is a measure of the difference between x_i and y_j:

$$D = E[d(X, Y)] = \sum_{i=0}^{N-1}\sum_{j=0}^{M-1} d(x_i, y_j)P(x_i, y_j)$$

On our way to mathematically defining the rate distortion function, let us look at the rate and distortion for some different lossy compression schemes.

In Example 8.4.2, knowledge of the value of the input at time k completely specifies the reconstructed value at time k. In this situation,

$$P(y_j|x_i) = \begin{cases} 1 \text{ for some } j = j_i \\ 0 \text{ otherwise} \end{cases} \tag{50}$$

Therefore,

$$D = \sum_{i=0}^{N-1} \sum_{j=0}^{M-1} P(y_j|x_i)P(x_i)d(x_i, y_j) \tag{51}$$

$$= \sum_{i=0}^{N-1} P(x_i)d(x_i, y_{j_i}) \tag{52}$$

where we used the fact that $P(x_i, y_j) = P(y_j|x_i)P(x_i)$ in Equation (51). The rate for this source coder is the output entropy $H(Y)$ of the source decoder. If this were always the case, the task of obtaining a rate distortion function would be relatively simple. Given a distortion constraint D^*, we could look at all encoders with distortion less than D^* and pick the one with the lowest output entropy. This entropy would be the rate corresponding to the distortion D^*. However, the requirement that knowledge of the input at time k completely specifies the reconstruction at time k is very restrictive, and there are many efficient compression techniques that would have to be excluded under this requirement. Consider the following example.

Example 8.5.1:

With a data sequence that consists of height and weight measurements, obviously height and weight are quite heavily correlated. In fact, after studying a long sequence of data, we find that if we plot the height along the x axis and the weight along the y axis, the data points cluster along the line $y = 2.5x$. In order to take advantage of this correlation, we devise the following compression scheme. For a given pair of height and weight measurements, we find the orthogonal projection on the $y = 2.5x$ line as shown in Figure 8.4. The point on this line can be represented as the distance to the nearest integer from the origin. Thus, we encode a pair of values into a single value. At the time of reconstruction, we simply map this value back into a pair of height and weight measurements.

For instance, suppose somebody is 72 inches tall and weighs 200 pounds (point A in Figure 8.4). This corresponds to a point at a distance of 212 along the $y = 2.5x$ line. The reconstructed values of the height and weight corresponding to this value are 79 and 197. Notice that the reconstructed values differ from the original values. Suppose we now have another individual who is also 72 inches tall but weighs 190 pounds (point B in Figure 8.4). The source coder output for this pair would be 203, and the reconstructed values for height and weight are 75 and 188, respectively. Notice that while the height value in both cases was the same, the reconstructed value of the height is different from the actual value. The reason for this is that the reconstructed value for the height depends on the weight. Thus, for this particular source coder, we do not have a conditional probability density function $\{P(y_j|x_i)\}$ of the form shown in Equation (50). ♦

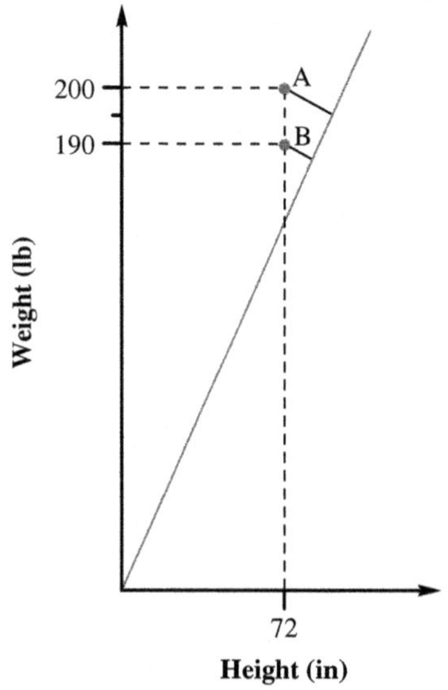

FIGURE 8.4 Compression scheme for encoding height-weight pairs.

Let us examine the distortion for this scheme a little more closely. As the conditional probability for this scheme is not of the form of Equation (50), we can no longer write the distortion in the form of Equation (52). Recall that the general form of the distortion is

$$D = \sum_{i=0}^{N-1} \sum_{j=0}^{M-1} d(x_i, y_j)P(x_i)P(y_j|x_i) \tag{53}$$

Each term in the summation consists of three factors: the distortion measure $d(x_i, y_j)$, the source density $P(x_i)$, and the conditional probability $P(y_j|x_i)$. The distortion measure is a measure of closeness of the original and reconstructed versions of the signal and is generally determined by the particular application. The source probabilities are solely determined by the source. The third factor, the set of conditional probabilities, can be seen as a description of the compression scheme.

Therefore, for a given source with some *pdf* $\{P(x_i)\}$ and a specified distortion measure $d(\cdot, \cdot)$, the distortion is a function of only the conditional probabilities $\{P(y_j|x_i)\}$; that is,

$$D = D(\{P(y_j|x_i)\}) \tag{54}$$

Therefore, we can write the constraint that the distortion D be less than some value D^* as a requirement that the conditional probabilities for the compression scheme belong to a set of conditional probabilities Γ that have the property that

$$\Gamma = \{\{P(y_j|x_i\} \text{ such that } D(\{P(y_j|x_i)\}) \leqslant D^*\} \tag{55}$$

Once we know the set of compression schemes to which we have to confine ourselves, we can start to look at the rate of these schemes. In Example 8.4.2, the rate was the entropy of Y. However, that was a result of the fact that the conditional probability describing that particular source coder took on only the values 0 and 1. Consider the following trivial situation.

Example 8.5.2:

Suppose we have the same source as in Example 8.4.2 and the same reconstruction alphabet. Suppose the distortion measure is

$$d(x_i, y_j) = (x_i - y_j)^2$$

and $D^* = 225$. One compression scheme that satisfies the distortion constraint randomly maps the input to any one of the outputs; that is,

$$P(y_j|x_i) = \frac{1}{8} \quad \text{for } i = 0, 1, \ldots, 15 \text{ and } j = 0, 2, \ldots, 14$$

We can see that this conditional probability assignment satisfies the distortion constraint:

$$\begin{aligned} D &= \sum_{i=0}^{N-1} \sum_{j=0}^{M-1} d(x_i, y_j) P(x_i) P(y_j|x_i) \\ &= \sum_{i=0}^{N-1} \sum_{j=0}^{M-1} (x_i - y_j)^2 \frac{1}{16} \frac{1}{8} \\ &= \frac{1}{128} \sum_{i=0}^{15} \sum_{j=0}^{7} (i - 2j)^2 \\ &= 42.5 \end{aligned}$$

As each of the eight reconstruction values is equally likely, $H(Y)$ is 3 bits. However, we are not transmitting *any* information. We could get exactly the same results by transmitting 0 bits and randomly picking Y at the receiver. ♦

Therefore, the entropy of the reconstruction $H(Y)$ cannot be a measure of the rate. In his 1959 paper on source coding [111], Shannon showed that the minimum rate for a given distortion is given by

$$R(D) = \min_{\{P(y_j|x_i)\} \in \Gamma} I(X; Y) \tag{56}$$

To prove this is beyond the scope of this book. (Further information can be found in [38,112].) However, we can at least convince ourselves that defining the rate as average mutual information gives sensible answers when used for the examples shown here. Consider Example 8.4.2. The average mutual information in this case is 3 bits as shown in Example 8.4.3, which is what we said the rate was. In fact, notice that whenever the conditional probabilities are constrained to be of the form of Equation (50),

$$H(Y|X) = 0$$

then

$$I(X; Y) = H(Y)$$

which had been our measure of rate.
In Example 8.5.2, the average mutual information is 0 bits, which accords with our intuitive feeling of what the rate should be. Again, whenever

$$H(Y|X) = H(Y)$$

that is, knowledge of the source gives us no knowledge of the reconstruction,

$$I(X; Y) = 0$$

which seems entirely reasonable. We should not have to transmit any bits when we are not sending any information.

At least for the examples here, it seems that the average mutual information does represent the rate. However, earlier we said that the average mutual information between the source output and the reconstruction is a measure of the information conveyed by the reconstruction about the source output. Then why are we looking for compression schemes that *minimize* this value? To understand this, we have to remember that the process of finding the performance of the optimum compression scheme had two parts. In the first part, we specified the desired distortion. The entire set of conditional probabilities over which the average mutual information is minimized satisfies the distortion constraint. Therefore, we can leave the question of distortion, or fidelity, aside and concentrate on minimizing the rate.

Finally, how do we find the rate distortion function? There are two ways. One is a computational approach developed by Arimoto [113] and Blahut [114]. While the derivation of the algorithm is beyond the scope of this book, the algorithm itself is relatively simple. The other approach is to find a lower bound for the average mutual information and then show that we can achieve this bound. We use this approach to find the rate distortion functions for two important sources.

Example 8.5.3: Rate distortion function for the binary source

Suppose we have a source alphabet $\{0, 1\}$, with $P(0) = p$. The reconstruction alphabet is also binary. Given the distortion measure

$$d(x_i, y_j) = x_i \oplus y_j, \tag{57}$$

where $\oplus$ is modulo 2 addition, let us find the rate distortion function. Assume for the moment that $p < \frac{1}{2}$. For $D > p$ an encoding scheme that would satisfy the distortion criterion would be not to transmit anything and fix $Y = 1$. So for $D \geqslant p$

$$R(D) = 0 \tag{58}$$

We will find the rate distortion function for the distortion range $0 \leqslant D < p$.

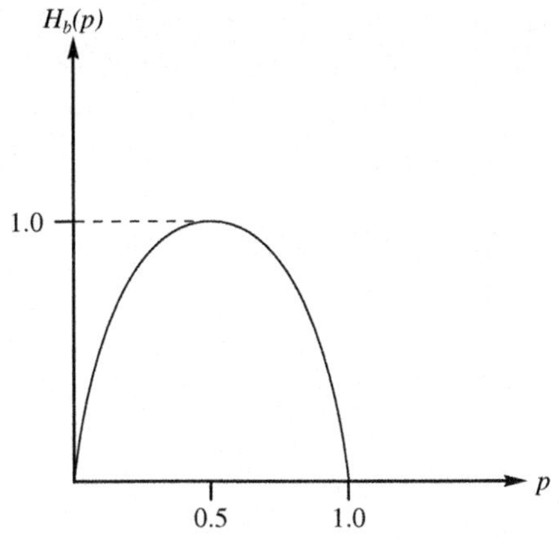

FIGURE 8.5 **The binary entropy function.**

Find a lower bound for the average mutual information:

$$I(X;Y) = H(X) - H(X|Y) \quad (59)$$
$$= H(X) - H(X \oplus Y|Y) \quad (60)$$
$$\geqslant H(X) - H(X \oplus Y) \quad \text{from Equation (12)} \quad (61)$$

In the second step, we have used the fact that if we know Y, then knowing X we can obtain $X \oplus Y$ and vice versa as $X \oplus Y \oplus Y = X$.

Let us look at the terms on the right-hand side of (12):

$$H(X) = -p \log_2 p - (1-p)\log_2(1-p) = H_b(p) \quad (62)$$

where $H_b(p)$ is called the *binary entropy function* and is plotted in Figure 8.5. Note that $H_b(p) = H_b(1-p)$.

Given that $H(X)$ is completely specified by the source probabilities, our task now is to find the conditional probabilities $\{P(x_i|y_j)\}$ such that $H(X \oplus Y)$ is maximized while the average distortion $E[d(x_i, y_j)] \leqslant D. H(X \oplus Y)$ is simply the binary entropy function $H_b(P(X \oplus Y = 1))$, where

$$P(X \oplus Y = 1) = P(X = 0, Y = 1) + P(X = 1, Y = 0) \quad (63)$$

Therefore, to maximize $H(X \oplus Y)$, we would want $P(X \oplus Y = 1)$ to be as close as possible to one-half. However, the selection of $P(X \oplus Y)$ also has to satisfy the distortion constraint. The distortion is given by

$$E[d(x_i, y_j)] = 0 \times P(X = 0, Y = 0) + 1 \times P(X = 0, Y = 1)$$
$$+1 \times P(X = 1, Y = 0) + 0 \times P(X = 1, Y = 1)$$
$$= P(X = 0, Y = 1) + P(X = 1, Y = 0)$$
$$= P(Y = 1|X = 0)p + P(Y = 0|X = 1)(1-p) \quad (64)$$

But this is simply the probability that $X \oplus Y = 1$. Therefore, the maximum value that $P(X \oplus Y = 1)$ can have is D. Our assumptions were that $D < p$ and $p \leqslant \frac{1}{2}$, which means that $D < \frac{1}{2}$. Therefore, $P(X \oplus Y = 1)$ is closest to $\frac{1}{2}$ while being less than or equal to D when $P(X \oplus Y = 1) = D$. Therefore,

$$I(X; Y) \geqslant H_b(p) - H_b(D) \tag{65}$$

We can show that for $P(X = 0|Y = 1) = P(X = 1|Y = 0) = D$, this bound is achieved. That is, if $P(X = 0|Y = 1) = P(X = 1|Y = 0) = D$, then

$$I(X; Y) = H_b(p) - H_b(D) \tag{66}$$

Therefore, for $D < p$ and $p \leqslant \frac{1}{2}$,

$$R(D) = H_b(p) - H_b(D) \tag{67}$$

Finally, if $p > \frac{1}{2}$, then we simply switch the roles of p and $1 - p$. Putting all this together, the rate distortion function for a binary source is

$$R(D) = \begin{cases} H_b(p) - H_b(D) & \text{for } D < \min\{p, 1 - p\} \\ 0 & \text{otherwise} \end{cases} \tag{68}$$

♦

Example 8.5.4: Rate Distortion Function for the Gaussian Source

Suppose we have a continuous amplitude source that has a zero mean Gaussian *pdf* with variance σ^2. If our distortion measure is given by

$$d(x, y) = (x - y)^2 \tag{69}$$

our distortion constraint is given by

$$E\left[(X - Y)^2\right] \leqslant D \tag{70}$$

Our approach to finding the rate distortion function will be the same as in the previous example; that is, find a lower bound for $I(X; Y)$ given a distortion constraint, and then show that this lower bound can be achieved.

First we find the rate distortion function for $D < \sigma^2$:

$$\begin{aligned} I(X; Y) &= h(X) - h(X|Y) & (71) \\ &= h(X) - h(X - Y|Y) & (72) \\ &\geqslant h(X) - h(X - Y) & (73) \end{aligned}$$

In order to minimize the right-hand side of Equation (73), we have to maximize the second term subject to the constraint given by Equation (70). This term is maximized if $X - Y$ is

Gaussian, and the constraint can be satisfied if $E\left[(X-Y)^2\right] = D$. Therefore, $h(X-Y)$ is the differential entropy of a Gaussian random variable with variance D, and the lower bound becomes

$$I(X;Y) \geqslant \frac{1}{2}\log(2\pi e\sigma^2) - \frac{1}{2}\log(2\pi eD) \tag{74}$$

$$= \frac{1}{2}\log\frac{\sigma^2}{D} \tag{75}$$

This average mutual information can be achieved if Y is zero mean Gaussian with variance $\sigma^2 - D$, and

$$f_{X|Y}(x|y) = \frac{1}{\sqrt{2\pi D}}\exp\frac{-x^2}{2D} \tag{76}$$

For $D > \sigma^2$, if we set $Y = 0$, then

$$I(X;Y) = 0 \tag{77}$$

and

$$E\left[(X-Y)^2\right] = \sigma^2 < D \tag{78}$$

Therefore, the rate distortion function for the Gaussian source can be written as

$$R(D) = \begin{cases} \frac{1}{2}\log\frac{\sigma^2}{D} & \text{for } D < \sigma^2 \\ 0 & \text{for } D > \sigma^2 \end{cases} \tag{79}$$

We plot the rate distortion function for $\sigma^2 = 1$ in Figure 8.6 ♦

Like the differential entropy for the Gaussian source, the rate distortion function for the Gaussian source also has the distinction of being larger than the rate distortion function for any other source with a continuous distribution and the same variance. This is especially valuable because for many sources it can be very difficult to calculate the rate distortion function. In these situations, it is helpful to have an upper bound for the rate distortion function. It would be very nice if we also had a lower bound for the rate distortion function of a continuous random variable. Shannon described such a bound in his 1948 paper [3], and it is appropriately called the *Shannon lower bound*. We will simply state the bound here without derivation (for more information, see [112]).

The Shannon lower bound for a random variable X and the magnitude error criterion

$$d(x,y) = |x - y| \tag{80}$$

is given by

$$R_{SLB}(D) = h(X) - \log(2eD) \tag{81}$$

If we use the squared error criterion, the Shannon lower bound is given by

$$R_{SLB}(D) = h(X) - \frac{1}{2}\log(2\pi eD). \tag{82}$$

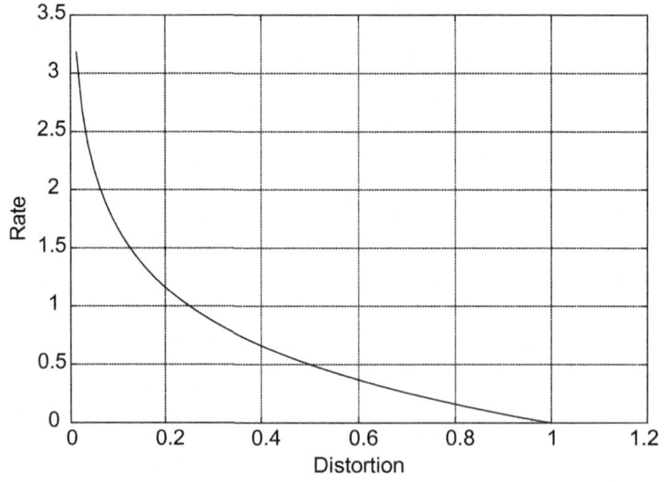

FIGURE 8.6 **The rate distortion function for a Gaussian random variable with variance 1.**

In this section, we have defined the rate distortion function and obtained the rate distortion function for two important sources. We have also obtained upper and lower bounds on the rate distortion function for an arbitrary *iid* source. These functions and bounds are especially useful when we want to know if it is possible to design compression schemes to provide a specified rate and distortion given a particular source. They are also useful in determining the amount of performance improvement that we could obtain by designing a better compression scheme. In these ways, the rate distortion function plays the same role for lossy compression that entropy plays for lossless compression.

8.6 Models

As in the case of lossless compression, models play an important role in the design of lossy compression algorithms; there are a variety of approaches available. The set of models we can draw on for lossy compression is much wider than the set of models we studied for lossless compression. We will look at some of these models in this section. What is presented here is by no means an exhaustive list of models. Our only intent is to describe those models that will be useful in the following chapters.

8.6.1 Probability Models

An important method for characterizing a particular source is through the use of probability models. As we shall see later, knowledge of the probability model is important for the design of a number of compression schemes.

Probability models used for the design and analysis of lossy compression schemes differ from those used in the design and analysis of lossless compression schemes. When developing models in the lossless case, we tried for an exact match. The probability of each symbol was estimated as part of the modeling process. When modeling sources in order to design or analyze lossy compression schemes, we look more to the general rather than exact correspondence. The reasons are more pragmatic than theoretical. Certain probability distribution functions are more analytically tractable than others, and we try to match the distribution of the source with one of these "nice" distributions.

The uniform, Gaussian, Laplacian, and gamma distributions are four probability models commonly used in the design and analysis of lossy compression systems:

- **Uniform Distribution:** As for lossless compression, this is again our ignorance model. If we do not know anything about the distribution of the source output, except possibly the range of values, we can use the uniform distribution to model the source. The probability density function for a random variable uniformly distributed between a and b is

$$f_X(x) = \begin{cases} \frac{1}{b-a} & \text{for } a \leqslant x \leqslant b \\ 0 & \text{otherwise} \end{cases} \tag{83}$$

- **Gaussian Distribution:** The Gaussian distribution is one of the most commonly used probability models for two reasons: it is mathematically tractable and, by virtue of the central limit theorem, it can be argued that in the limit the distribution of interest goes to a Gaussian distribution. The probability density function for a random variable with a Gaussian distribution and mean μ and variance σ^2 is

$$f_X(x) = \frac{1}{\sqrt{2\pi\sigma^2}} \exp -\frac{(x-\mu)^2}{2\sigma^2} \tag{84}$$

- **Laplacian Distribution:** Many sources that we deal with have distributions that have a sharp peak at zero. For example, speech consists mainly of silence. Therefore, samples of speech will be zero or close to zero with high probability. Image pixels themselves do not have any attraction to small values. However, there is a high degree of correlation among pixels. Therefore, a large number of the pixel-to-pixel differences will have values close to zero. In these situations, a Gaussian distribution is not a very close match to the data. A closer match is the Laplacian distribution, which is peaked at zero. The distribution function for a zero mean random variable with Laplacian distribution and variance σ^2 is

$$f_X(x) = \frac{1}{\sqrt{2\sigma^2}} \exp \frac{-\sqrt{2}\,|x|}{\sigma} \tag{85}$$

- **Gamma Distribution:** A distribution that is even more peaked, though considerably less tractable, than the Laplacian distribution is the gamma distribution. The distribution function for a gamma-distributed random variable with zero mean and variance σ^2 is

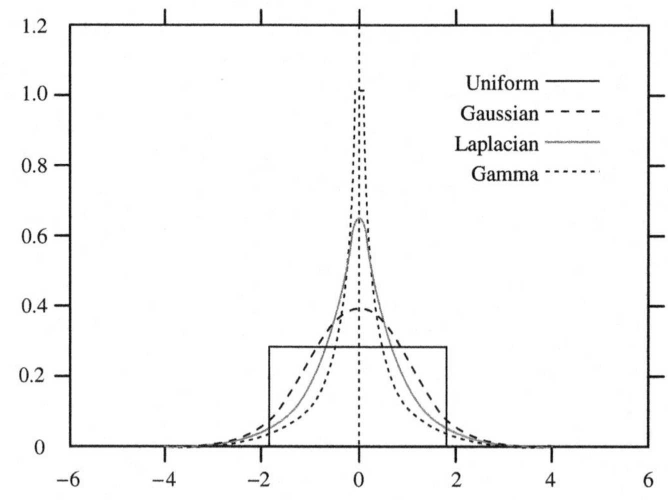

FIGURE 8.7 Uniform, Gaussian, Laplacian, and gamma distributions.

given by

$$f_X(x) = \frac{\sqrt[4]{3}}{\sqrt{8\pi\sigma\,|x|}} \exp\frac{-\sqrt{3}\,|x|}{2\sigma} \tag{86}$$

The shapes of these four distributions, assuming a mean of zero and a variance of one, are shown in Figure 8.7.

One way of obtaining the estimate of the distribution of a particular source is to divide the range of outputs into "bins" or intervals I_k. We can then find the number of values n_k that fall into each interval. A plot of $\frac{n_k}{n_T}$, where n_T is the total number of source outputs being considered, should give us some idea of what the input distribution looks like. Be aware that this is a rather crude method and can at times be misleading. For example, if we were not careful in our selection of the source output, we might end up modeling some local peculiarities of the source. If the bins are too large, we might effectively filter out some important properties of the source. If the bin sizes are too small, we may miss out on some of the gross behavior of the source.

Once we have decided on some candidate distributions, we can select between them using a number of sophisticated tests. These tests are beyond the scope of this book but are described in [115].

Many of the sources that we deal with when we design lossy compression schemes have a great deal of structure in the form of sample-to-sample dependencies. The probability models described here capture none of these dependencies. Fortunately, we have a lot of models that can capture most of this structure. We describe some of these models in the next section.

8.6.2 Linear System Models

In the previous sections we have assumed that the information sequence $\{x_n\}$ can be modeled by a sequence of *iid* random variables. In practice most information sequences derived from real sources such as speech will contain dependencies. In an ideal world we would characterize these dependencies using the joint *pdf* of the sequence elements. In practice such an approach is not feasible. Instead, we try to characterize the dependencies in terms of correlation between samples. An intuitive and useful way of modeling the correlation between samples is to view the information sequence as the output of a linear system governed by a difference equation with an *iid* input. The structure of the linear system as reflected in the parameters of the difference equation introduces the correlation we observe in the information sequence.

The information sequence can be modeled in the form of the following difference equation:

$$x_n = \sum_{i=1}^{N} a_i x_{n-i} + \sum_{j=1}^{M} b_j \epsilon_{n-j} + \epsilon_n \tag{87}$$

where $\{x_n\}$ are samples of the process we wish to model, and $\{\epsilon_n\}$ is a white noise sequence. We will assume throughout this book that we are dealing with real valued samples. Recall that a zero-mean wide-sense-stationary noise sequence $\{\epsilon_n\}$ is a sequence with an autocorrelation function

$$R_{\epsilon\epsilon}(k) = \begin{cases} \sigma_\epsilon^2 & \text{for } k = 0 \\ 0 & \text{otherwise} \end{cases} \tag{88}$$

In digital signal-processing terminology, Equation (87) represents the output of a linear discrete time invariant filter with N poles and M zeros. In the statistical literature, this model is called an autoregressive moving average model of order (N,M), or an ARMA (N,M) model. The autoregressive label is because of the first summation in Equation (87), while the second summation gives us the moving average portion of the name.

If all of the b_j were zero in Equation (87), only the autoregressive part of the ARMA model would remain:

$$x_n = \sum_{i=1}^{N} a_i x_{n-i} + \epsilon_n \tag{89}$$

This model is called an Nth-order autoregressive model and is denoted by AR(N). In digital signal-processing terminology, this is an *all pole filter*. The AR(N) model is the most popular of all the linear models, especially in speech compression, where it arises as a natural consequence of the speech production model. We will look at it a bit more closely.

First notice that for the AR(N) process, knowing all of the past history of the process gives no more information than knowing the last N samples of the process; that is,

$$P(x_n | x_{n-1}, x_{n-2}, \ldots) = P(x_n | x_{n-1}, x_{n-2}, \ldots, x_{n-N}) \tag{90}$$

which means that the AR(N) process is a Markov model of order N.

The autocorrelation function of a process can tell us a lot about the sample-to-sample behavior of a sequence. A slowly decaying autocorrelation function indicates a high sample-to-sample correlation, while a fast decaying autocorrelation denotes low sample-to-sample correlation. In the case of *no* sample-to-sample correlation, such as white noise, the autocorrelation function is zero for lags greater than zero, as seen in Equation (88). The autocorrelation function for the AR(N) process can be obtained as follows:

$$R_{xx}(k) = E\left[x_n x_{n-k}\right] \tag{91}$$

$$= E\left[\left(\sum_{i=1}^{N} a_i x_{n-i} + \epsilon_n\right)(x_{n-k})\right] \tag{92}$$

$$= E\left[\sum_{i=1}^{N} a_i x_{n-i} x_{n-k}\right] + E\left[\epsilon_n x_{n-k}\right] \tag{93}$$

$$= \begin{cases} \sum_{i=1}^{N} a_i R_{xx}(k-i) & \text{for } k > 0 \\ \sum_{i=1}^{N} a_i R_{xx}(i) + \sigma_\epsilon^2 & \text{for } k = 0 \end{cases} \tag{94}$$

where

$$E\left[\epsilon_n x_{n-k}\right] = \begin{cases} \sigma_\epsilon^2 & \text{for } k = 0 \\ 0 & \text{for } k > 0 \end{cases} \tag{95}$$

Example 8.6.1:

Suppose we have an AR(3) process. Let us write out the equations for the autocorrelation coefficient for lags 1, 2, and 3:

$$R_{xx}(1) = a_1 R_{xx}(0) + a_2 R_{xx}(1) + a_3 R_{xx}(2)$$
$$R_{xx}(2) = a_1 R_{xx}(1) + a_2 R_{xx}(0) + a_3 R_{xx}(1)$$
$$R_{xx}(3) = a_1 R_{xx}(2) + a_2 R_{xx}(1) + a_3 R_{xx}(0)$$

If we know the values of the autocorrelation function $R_{xx}(k)$, for $k = 0, 1, 2, 3$, we can use this set of equations to find the AR(3) coefficients $\{a_1, a_2, a_3\}$. On the other hand, if we know the model coefficients and σ_ϵ^2, we can use the above equations along with the equation for $R_{xx}(0)$ to find the first four autocorrelation coefficients. All of the other autocorrelation values can be obtained by using Equation (94). ♦

To see how the autocorrelation function is related to the temporal behavior of the sequence, let us look at the behavior of a simple AR(1) source.

Example 8.6.2:

An AR(1) source is defined by the equation

$$x_n = a_1 x_{n-1} + \epsilon_n \tag{96}$$

The autocorrelation function for this source (see Problem 8 at the end of this chapter) is given by

$$R_{xx}(k) = \frac{1}{1-a_1^2} a_1^k \sigma_\epsilon^2 \tag{97}$$

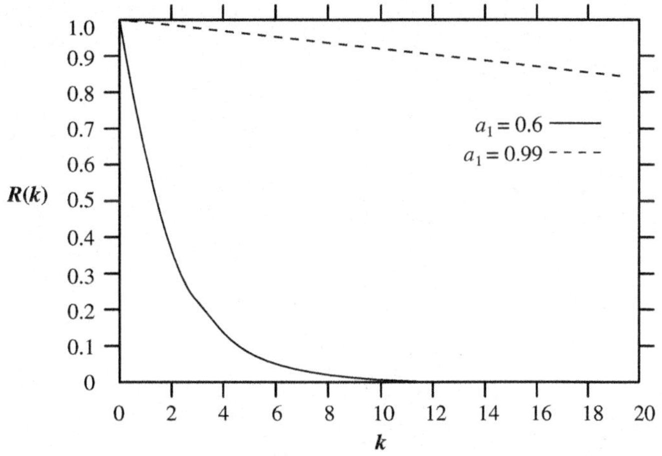

FIGURE 8.8 **Autocorrelation function of an AR(1) process with two values of a_1.**

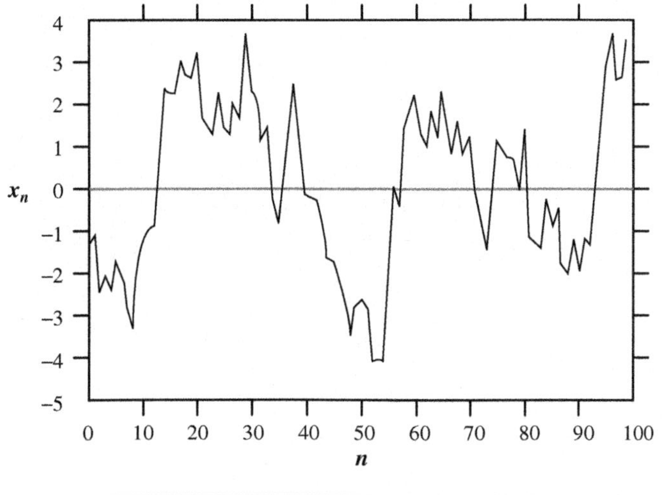

FIGURE 8.9 **Sample function of an AR(1) process with $a_1 = 0.99$.**

From this we can see that the autocorrelation will decay more slowly for larger values of a_1. Remember that the value of a_1 in this case is an indicator of how closely the current sample is related to the previous sample. The autocorrelation function is plotted for two values of a_1 in Figure 8.8. Notice that for a_1 close to 1, the autocorrelation function decays extremely slowly. As the value of a_1 moves farther away from 1, the autocorrelation function decays much faster.

Sample waveforms for $a_1 = 0.99$ and $a_1 = 0.6$ are shown in Figures 8.9 and 8.10. Notice the slower variations in the waveform for the process with a higher value of a_1. Because the waveform in Figure 8.9 varies more slowly than the waveform in Figure 8.10, samples of this waveform are much more likely to be close in value than the samples of the waveform of Figure 8.10.

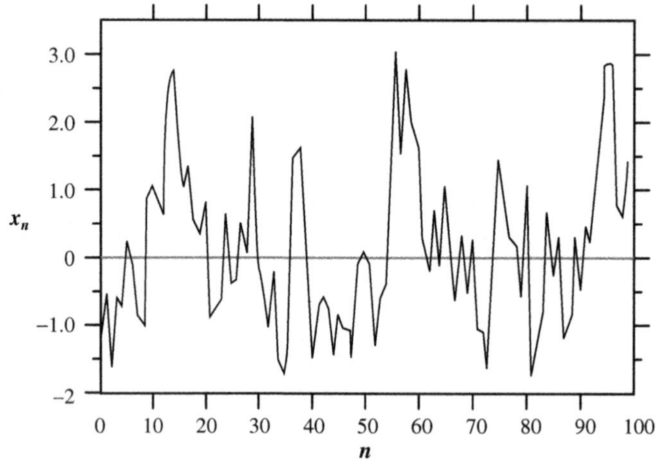

FIGURE 8.10 **Sample function of an AR(1) process with $a_1 = 0.6$.**

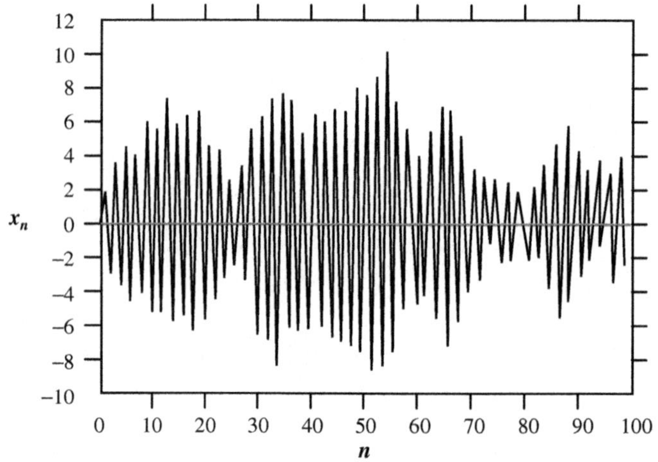

FIGURE 8.11 **Sample function of an AR(1) process with $a_1 = -0.99$.**

Let's look at what happens when the AR(1) coefficient is negative. The sample waveforms are plotted in Figures 8.11 and 8.12. The sample-to-sample variation in these waveforms is much higher than in the waveforms shown in Figures 8.9 and 8.10. However, if we were to look at the variation in magnitude, we can see that the higher value of a_1 results in magnitude values that are closer together.

This behavior is also reflected in the autocorrelation function, shown in Figure 8.13, as we might expect from looking at Equation (97). ♦

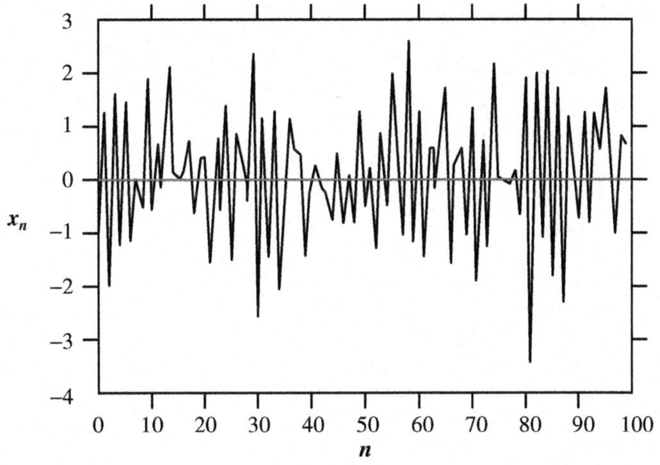

FIGURE 8.12 Sample function of an AR(1) process with $a_1 = -0.6$.

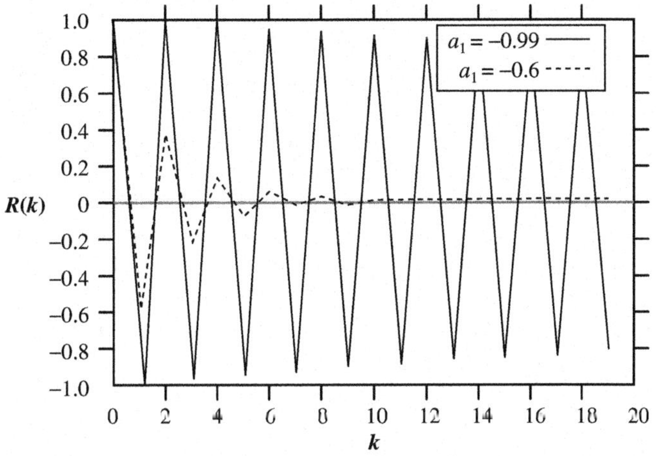

FIGURE 8.13 Autocorrelation function of an AR(1) process with two negative values of a_1.

In Equation (87), instead of setting all of the $\{b_j\}$ coefficients to zero, if we set all of the $\{a_i\}$ coefficients to zero, we will be left with the moving average part of the ARMA process:

$$x_n = \sum_{j=1}^{M} b_j \epsilon_{n-j} + \epsilon_n \tag{98}$$

This process is called an Mth-order moving average process. This is a weighted average of the current and M past samples. Because of the form of this process, it is most useful when modeling slowly varying processes.

8.6.3 Physical Models

Physical models are based on the physics of the source output production. The physics are generally complicated and not amenable to a reasonable mathematical approximation. An exception to this rule is speech generation.

Speech Production

There has been a significant amount of research conducted in the area of speech production [116], and volumes have been written about it. We will try to summarize some of the pertinent aspects in this section.

Speech is produced by forcing air through an elastic opening, the vocal cords, and then through cylindrical tubes with nonuniform diameter (the laryngeal, oral, nasal, and pharynx passages), and finally through cavities with changing boundaries, such as the mouth and the nasal cavity. Everything past the vocal cords is generally referred to as the *vocal tract.* The first action generates the sound, which is then modulated into speech as it traverses through the vocal tract.

We will often be talking about filters in the coming chapters. We will try to describe filters more precisely at that time. For our purposes at present, a filter is a system that has an input and an output and a rule for converting the input to the output, which we will call the *transfer function.* If we think of speech as the output of a filter, the sound generated by the air rushing past the vocal cords can be viewed as the input, while the rule for converting the input to the output is governed by the shape and physics of the vocal tract.

The output depends on the input and the transfer function. Let's look at each in turn. There are several different forms of input that can be generated by different conformations of the vocal cords and the associated cartilages. If the vocal cords are stretched shut and we force air through, the vocal cords vibrate, providing a periodic input. If a small aperture is left open, the input resembles white noise. By opening an aperture at different locations along the vocal cords, we can produce a white noise–like input with certain dominant frequencies that depend on the location of the opening. The vocal tract can be modeled as a series of tubes of unequal diameter. If we now examine how an acoustic wave travels through this series of tubes, we find that the mathematical model that best describes this process is an autoregressive model. We will often encounter the autoregressive model when we discuss speech compression algorithms.

8.7 Summary

In this chapter, we have looked at a variety of topics that will be useful to us when we study various lossy compression techniques, including distortion and its measurement, some new concepts from information theory, average mutual information and its connection to the rate of a compression scheme, and the rate distortion function. We have also briefly looked at some of the properties of the human visual and auditory systems—most importantly, visual and auditory masking. The masking phenomena allow us to incur distortion in such a way that the distortion is not perceptible to the human observer. We also presented a model for speech production.

Further Reading

There are a number of excellent books available that delve more deeply in the area of information theory:

1. *Information Theory*, by R.B. Ash [117].
2. *Information Transmission*, by R.M. Fano [20].
3. *Information Theory and Reliable Communication*, by R.G. Gallagher [7].
4. *Entropy and Information Theory*, by R.M. Gray [118].
5. *Elements of Information Theory*, by T.M. Cover and J.A. Thomas [38].

The subject of rate distortion theory is discussed in very clear terms in *Rate Distortion Theory*, by T. Berger [112].

For an introduction to the concepts behind speech perception, see *Voice and Speech Processing*, by T. Parsons [119].

8.8 Projects and Problems

1. Although SNR is a widely used measure of distortion, it often does not correlate with perceptual quality. In order to see this, we can conduct the following experiment. Using one of the images provided, generate two "reconstructed" images. For one of the reconstructions, add a value of 10 to each pixel. For the other reconstruction, randomly add either +10 or -10 to each pixel.

   (a) What is the SNR for each of the reconstructions? Do the relative values reflect the difference in the perceptual quality?

   (b) Devise a mathematical measure that will better reflect the difference in perceptual quality for this particular case.

2. Consider the following lossy compression scheme for binary sequences. We divide the binary sequence into blocks of size M. For each block we count the number of 0s. If this number is greater than or equal to $M/2$, we send a 0; otherwise, we send a 1.

   (a) If the sequence is random with $P(0) = 0.8$, compute the rate and distortion (use Equation (57)) for $M = 1, 2, 4, 8, 16$. Compare your results with the rate distortion function for binary sources.

   (b) Repeat assuming that the output of the encoder is encoded at a rate equal to the entropy of the output.

3. Write a program to implement the compression scheme described in the previous problem.

(a) Generate a random binary sequence with $P(0) = 0.8$, and compare your simulation results with the analytical results.
(b) Generate a binary first-order Markov sequence with $P(0|0) = 0.9$, and $P(1|1) = 0.9$. Encode it using your program. Discuss and comment on your results.

4. Show that

$$H(X_d|Y_d) = -\sum_{j=-\infty}^{\infty}\sum_{i=-\infty}^{\infty} f_{X|Y}(x_i|y_j) f_Y(y_j)\Delta\Delta \log f_{X|Y}(x_i|y_j) - \log \Delta \quad (99)$$

5. For two random variables X and Y, show that

$$H(X|Y) \leqslant H(X)$$

with equality if X is independent of Y.
Hint: $E[\log(f(x))] \leqslant \log\{E[f(x)]\}$ (Jensen's inequality).

6. Given two random variables X and Y, show that $I(X;Y) = I(Y;X)$.

7. For a binary source with $P(0) = p$, $P(X = 0|Y = 1) = P(X = 1|Y = 0) = D$, and distortion measure

$$d(x_i, y_j) = x_i \oplus y_j$$

show that

$$I(X;Y) = H_b(p) - H_b(D) \quad (100)$$

8. Find the autocorrelation function in terms of the model coefficients and σ_ϵ^2 for

(a) an AR(1) process,
(b) an MA(1) process, and
(c) an AR(2) process.

9

Scalar Quantization

9.1 Overview

In this chapter, we begin our study of quantization, one of the simplest and most general ideas in lossy compression. We will look at scalar quantization and continue with vector quantization in the next chapter. First, the general quantization problem is stated, then various solutions are examined, starting with the simpler solutions, which require the most assumptions, and proceeding to more complex solutions that require fewer assumptions. We describe uniform quantization with fixed-length codewords, first assuming a uniform source, then a source with a known probability density function (*pdf*) that is not necessarily uniform, and finally a source with unknown or changing statistics. We then look at *pdf*-optimized nonuniform quantization, followed by companded quantization. Finally, we return to the more general statement of the quantizer design problem and study entropy-coded quantization.

9.2 Introduction

In many lossy compression applications, we are required to represent each source output using one of a small number of codewords. The number of possible distinct source output values is generally much larger than the number of codewords available to represent them. The process of representing a large—possibly infinite—set of values with a much smaller set is called *quantization*.

Consider a source that generates numbers between -10.0 and 10.0. A simple quantization scheme would be to represent each output of the source with the integer value closest to it. (If the source output is equally close to two integers, we will randomly pick one of them.) For example, if the source output is 2.47, we would represent it as 2, and if the source output is 3.1415926, we would represent it as 3.

This approach reduces the size of the alphabet required to represent the source output; the infinite number of values between -10.0 and 10.0 are represented with a set that contains only 21 values ($\{-10, \ldots, 0, \ldots, 10\}$). At the same time we have also forever lost the original value of the source output. If we are told that the reconstruction value is 3, we cannot tell whether the source output was 2.95, 3.16, 3.057932, or any other of an infinite set of values. In other words, we have lost some information. This loss of information is the reason for the use of the word "lossy" in many lossy compression schemes.

The set of inputs and outputs of a quantizer can be scalars or vectors. If they are scalars, we call the quantizers *scalar quantizers*. If they are vectors, we call the quantizers *vector quantizers*. We will study scalar quantizers in this chapter and vector quantizers in Chapter 10.

9.3 The Quantization Problem

Quantization is a very simple process. However, the design of the quantizer has a significant impact on the amount of compression obtained and loss incurred in a lossy compression scheme. Therefore, we will devote a lot of attention to issues related to the design of quantizers.

In practice, the quantizer consists of two mappings: an encoder mapping and a decoder mapping. The encoder divides the range of values that the source generates into a number of intervals. Each interval is represented by a distinct codeword. The encoder represents all the source outputs that fall into a particular interval by the codeword representing that interval. As there could be many—possibly infinitely many—distinct sample values that can fall in any given interval, the encoder mapping is irreversible. Knowing the code only tells us the interval to which the sample value belongs. It does not tell us which of the many values in the interval is the actual sample value. When the sample value comes from an analog source, the encoder is called an analog-to-digital (A/D) converter.

The encoder mapping for a quantizer with eight reconstruction values is shown in Figure 9.1. For this encoder, all samples with values between -1 and 0 would be assigned the code 011. All

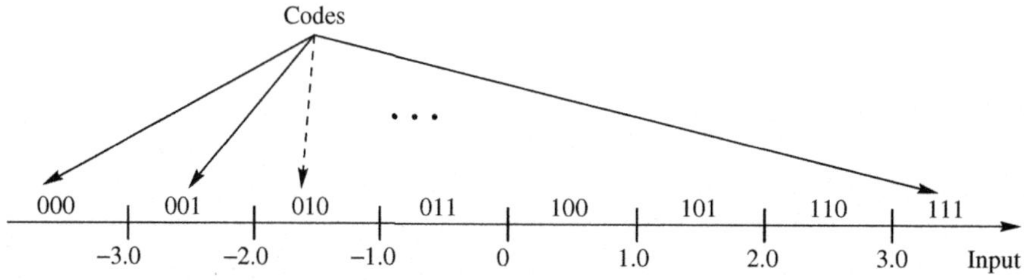

FIGURE 9.1 Mapping for a 3-bit encoder.

Input Codes	Output
000	−3.5
001	−2.5
010	−1.5
011	−0.5
100	0.5
101	1.5
110	2.5
111	3.5

FIGURE 9.2 Mapping for a 3-bit D/A converter.

values between 0 and 1.0 would be assigned the code 100, and so on. On the two boundaries, all inputs with values greater than 3 would be assigned the code 111, and all inputs with values less than −3.0 would be assigned the code 000. Thus, any input that we receive will be assigned a codeword depending on the interval in which it falls. As we are using 3 bits to represent each value, we refer to this quantizer as a 3-bit quantizer.

For every codeword generated by the encoder, the decoder generates a reconstruction value. Because a codeword represents an entire interval, and there is no way of knowing which value in the interval was actually generated by the source, the decoder puts out a value that, in some sense, best represents all the values in the interval. Later, we will see how to use information we may have about the distribution of the input in the interval to obtain a representative value. For now, we simply use the midpoint of the interval as the representative value generated by the decoder. If the reconstruction is analog, the decoder is often referred to as a digital-to-analog (D/A) converter. A decoder mapping corresponding to the 3-bit encoder shown in Figure 9.1 is shown in Figure 9.2.

Example 9.3.1:

Suppose a sinusoid $4\cos(2\pi t)$ was sampled every 0.05 second. The sample was digitized using the A/D mapping shown in Figure 9.1 and reconstructed using the D/A mapping shown in Figure 9.2. The first few inputs, codewords, and reconstruction values are given in Table 9.1. Notice the first two samples in Table 9.1. Although the two input values are distinct, they both fall into the same interval in the quantizer. The encoder, therefore, represents both inputs with the same codeword, which in turn leads to identical reconstruction values.

TABLE 9.1 Digitizing a sine wave.

t	$4\cos(2\pi t)$	A/D Output	D/A Output	Error
0.05	3.804	111	3.5	0.304
0.10	3.236	111	3.5	−0.264
0.15	2.351	110	2.5	−0.149
0.20	1.236	101	1.5	−0.264

♦

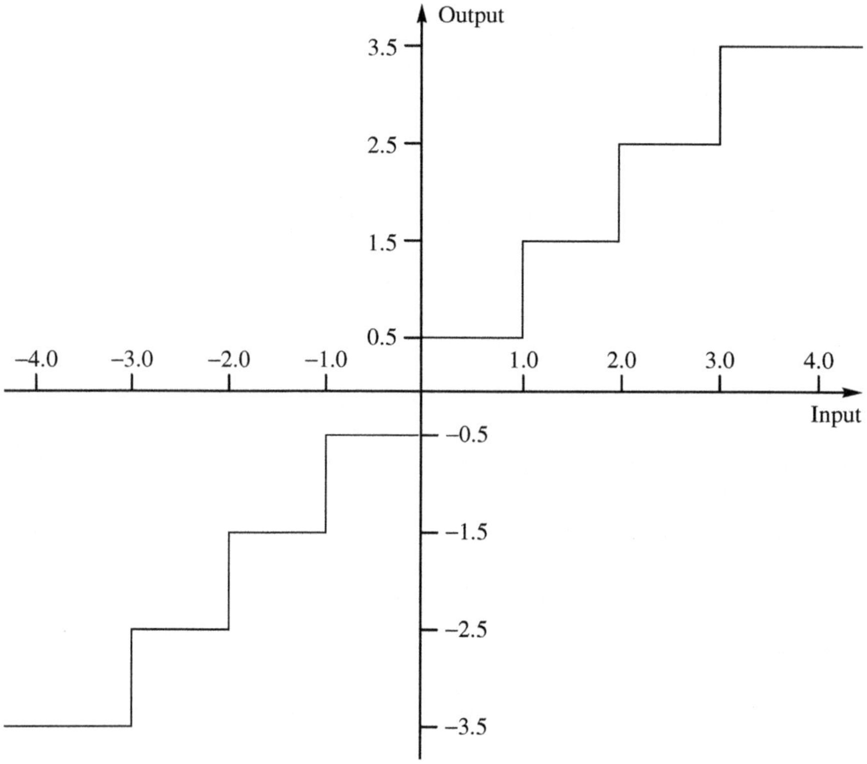

FIGURE 9.3 Quantizer input-output map.

Construction of the intervals (their location, etc.) can be viewed as part of the design of the encoder. Selection of reconstruction values is part of the design of the decoder. However, the fidelity of the reconstruction depends on both the intervals and the reconstruction values. Therefore, when designing or analyzing encoders and decoders, it is reasonable to view them as a pair. We call this encoder-decoder pair a *quantizer*. The quantizer mapping for the 3-bit encoder-decoder pair shown in Figures 9.1 and 9.2 can be represented by the input-output map shown in Figure 9.3. The quantizer accepts sample values, and depending on the interval in which the sample values fall, it provides an output codeword and a representation value. Using the map of Figure 9.3, we can see that an input to the quantizer of 1.7 will result in an output of 1.5, and an input of −0.3 will result in an output of −0.5.

From Figures 9.1, 9.2, 9.3 we can see that we need to know how to divide the input range into intervals, assign binary codes to these intervals, and find representation or output values for these intervals in order to specify a quantizer. We need to do all of this while satisfying distortion and rate criteria. In this chapter, we will define distortion to be the average squared difference between the quantizer input and output. We call this the mean squared quantization error (*msqe*) and denote it by σ_q^2. The rate of the quantizer is the average number of bits required to represent a single quantizer output. We would like to get the lowest distortion for a given rate or the lowest rate for a given distortion.

Let us pose the design problem in precise terms. Suppose we have an input modeled by a random variable X with *pdf* $f_X(x)$. If we wished to quantize this source using a quantizer with M intervals, we would have to specify $M + 1$ endpoints for the intervals and a representative value for each of the M intervals. The endpoints of the intervals are known as *decision boundaries*, while the representative values are called *reconstruction levels*. We will often model discrete sources with continuous distributions. For example, the difference between neighboring pixels is often modeled using a Laplacian distribution even though the differences can only take on a limited number of discrete values. Discrete processes are modeled with continuous distributions because these models can simplify the design process considerably, and the resulting designs perform well in spite of the incorrect assumption. Several of the continuous distributions used to model source outputs are unbounded—that is, the range of values is infinite. In these cases, the first and last endpoints are generally chosen to be $\pm\infty$.

Let us denote the decision boundaries by $\{b_i\}_{i=0}^{M}$, the reconstruction levels by $\{y_i\}_{i=1}^{M}$, and the quantization operation by $Q(\cdot)$. Then

$$Q(x) = y_i \quad \text{if} \quad b_{i-1} < x \leqslant b_i \tag{1}$$

The mean squared quantization error is then given by

$$\sigma_q^2 = \int_{-\infty}^{\infty} (x - Q(x))^2 f_X(x)dx \tag{2}$$

$$= \sum_{i=1}^{M} \int_{b_{i-1}}^{b_i} (x - y_i)^2 f_X(x)dx \tag{3}$$

The difference between the quantizer input x and output $y = Q(x)$, besides being referred to as the quantization error, is also called the *quantizer distortion* or *quantization noise*. But the word "noise" is somewhat of a misnomer. Generally, when we talk about noise we mean a process external to the source process. Because of the manner in which the quantization error is generated, it is dependent on the source process and, therefore, cannot be regarded as external to the source process. One reason for the use of the word "noise" in this context is that from time to time we will find it useful to model the quantization process as an additive noise process, as shown in Figure 9.4.

If we use fixed-length codewords to represent the quantizer output, then the size of the output alphabet immediately specifies the rate. If the number of quantizer outputs is M, then the rate is given by

$$R = \lceil \log_2 M \rceil \tag{4}$$

For example, if $M = 8$, then $R = 3$. In this case, we can pose the quantizer design problem as follows:

> Given an input *pdf* $f_X(x)$ and the number of levels M in the quantizer, find the decision boundaries $\{b_i\}$ and the reconstruction levels $\{y_i\}$ so as to minimize the mean squared quantization error given by Equation (3).

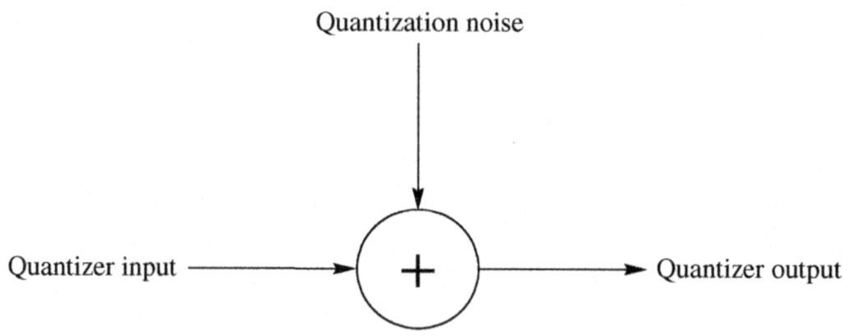

FIGURE 9.4 Additive noise model of a quantizer.

TABLE 9.2 Codeword assignment for an eight-level quantizer.

y_1	1110
y_2	1100
y_3	100
y_4	00
y_5	01
y_6	101
y_7	1101
y_8	1111

However, if we are allowed to use variable-length codes, such as Huffman codes or arithmetic codes, along with the size of the alphabet, the selection of the decision boundaries will also affect the rate of the quantizer. Consider the codeword assignment for the output of an eight-level quantizer shown in Table 9.2.

According to this codeword assignment, if the output y_4 occurs, we use 2 bits to encode it, while if the output y_1 occurs, we need 4 bits to encode it. Obviously, the rate will depend on how often we have to encode y_4 versus how often we have to encode y_1. In other words, the rate will depend on the probability of occurrence of the outputs. If l_i is the length of the codeword corresponding to the output y_i, and $P(y_i)$ is the probability of occurrence of y_i, then the rate is given by

$$R = \sum_{i=1}^{M} l_i P(y_i) \tag{5}$$

However, the probabilities $\{P(y_i)\}$ depend on the decision boundaries $\{b_i\}$. For example, the probability of y_i occurring is given by

$$P(y_i) = \int_{b_{i-1}}^{b_i} f_X(x)dx$$

Therefore, the rate R is a function of the decision boundaries and is given by the expression

$$R = \sum_{i=1}^{M} l_i \int_{b_{i-1}}^{b_i} f_X(x)dx \tag{6}$$

From this discussion and Equations (3) and (6), we see that for a given source input, the partitions we select and the representation for those partitions will determine the distortion incurred during the quantization process. The partitions we select and the binary codes for the partitions will determine the rate for the quantizer. Thus, the problem of finding the optimum partitions, codes, and representation levels are all linked. In light of this information, we can restate our problem statement:

> Given a distortion constraint
>
> $$\sigma_q^2 \leqslant D^* \tag{7}$$
>
> find the decision boundaries, reconstruction levels, and binary codes that minimize the rate given by Equation (6) while satisfying Equation (7).
> Or, given a rate constraint
>
> $$R \leqslant R^* \tag{8}$$
>
> find the decision boundaries, reconstruction levels, and binary codes that minimize the distortion given by Equation (3), while satisfying Equation (8).

This problem statement of quantizer design, while more general than our initial statement, is substantially more complex. Fortunately, in practice there are situations in which we can simplify the problem. We often use fixed-length codewords to encode the quantizer output. In this case, the rate is simply the number of bits used to encode each output, and we can use our initial statement of the quantizer design problem. We start our study of quantizer design by looking at this simpler version of the problem and later use what we have learned in this process to attack the more complex version.

9.4 Uniform Quantizer

The simplest type of quantizer is the uniform quantizer. All intervals are the same size in the uniform quantizer, except possibly for the two outer intervals. In other words, the decision boundaries are spaced evenly. The reconstruction values are also spaced evenly, with the same spacing as the decision boundaries; in the inner intervals, they are the midpoints of the intervals. This constant spacing is usually referred to as the step size and is denoted by Δ. The quantizer shown in Figure 9.3 is a uniform quantizer with $\Delta = 1$. It does not have zero as one of its representation levels. Such a quantizer is called a *midrise quantizer*. An alternative uniform quantizer could be the one shown in Figure 9.5. This is called a *midtread quantizer*. As the midtread quantizer has zero as one of its output levels, it is especially useful in situations where it is important that the zero value be represented—for example, control systems in which it is important to represent a zero value accurately, and audio coding schemes in which we need to represent

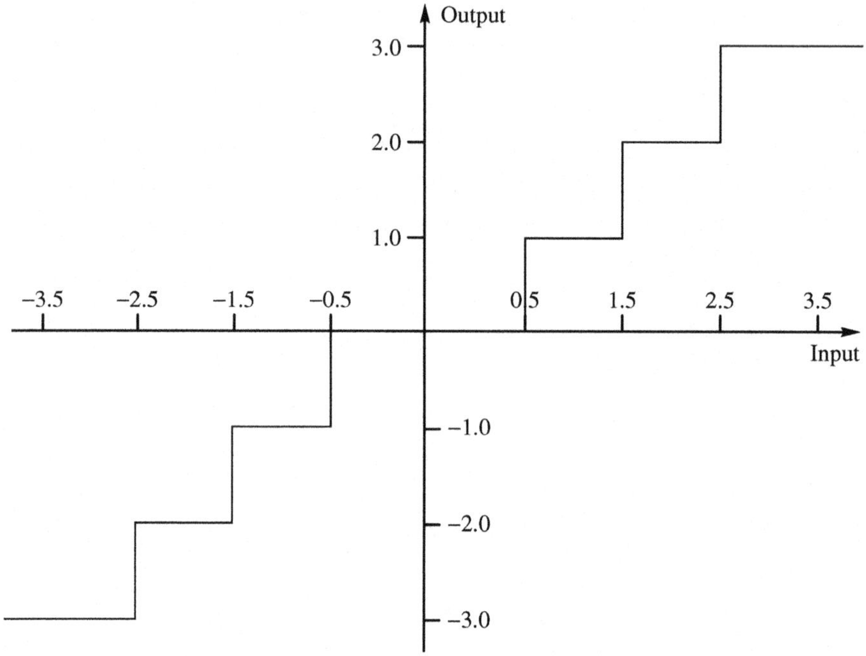

FIGURE 9.5 A midtread quantizer.

silence periods. Notice that the midtread quantizer has only seven intervals or levels. That means that if we were using a fixed-length 3-bit code, we would have one codeword left over.

Usually, we use a midrise quantizer if the number of levels is even and a midtread quantizer if the number of levels is odd. For the remainder of this chapter, unless we specifically mention otherwise, we will assume that we are dealing with midrise quantizers. We will also generally assume that the input distribution is symmetric around the origin and the quantizer is also symmetric. (The optimal minimum mean squared error quantizer for a symmetric distribution need not be symmetric [120].) Given all these assumptions, the design of a uniform quantizer consists of finding the step size Δ that minimizes the distortion for a given input process and number of decision levels.

Uniform Quantization of a Uniformly Distributed Source

We start our study of quantizer design with the simplest of all cases: design of a uniform quantizer for a uniformly distributed source. Suppose we want to design an M-level uniform quantizer for an input that is uniformly distributed in the interval $[-X_{\max}, X_{\max}]$. This means we need to divide the $[-X_{\max}, X_{\max}]$ interval into M equally sized intervals. In this case, the step size Δ is given by

$$\Delta = \frac{2X_{\max}}{M} \tag{9}$$

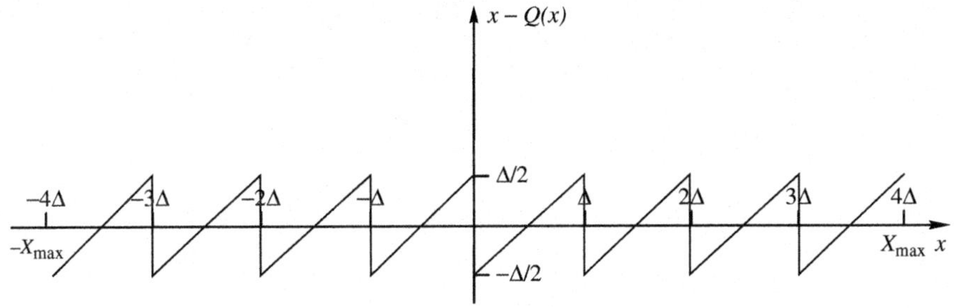

FIGURE 9.6 **Quantization error for a uniform midrise quantizer with a uniformly distributed input.**

The distortion in this case becomes

$$\sigma_q^2 = 2\sum_{i=1}^{\frac{M}{2}} \int_{(i-1)\Delta}^{i\Delta} \left(x - \frac{2i-1}{2}\Delta\right)^2 \frac{1}{2X_{\max}} dx \tag{10}$$

If we evaluate this integral (after some suffering), we find that the *msqe* is $\Delta^2/12$.

The same result can be more easily obtained if we examine the behavior of the quantization error q given by

$$q = x - Q(x) \tag{11}$$

In Figure 9.6, we plot the quantization error versus the input signal for an eight-level uniform quantizer with an input that lies in the interval $[-X_{\max}, X_{\max}]$. Notice that the quantization error lies in the interval $[-\frac{\Delta}{2}, \frac{\Delta}{2}]$. As the input is uniform, it is not difficult to establish that the quantization error is also uniform over this interval. Thus, the mean squared quantization error is the second moment of a random variable uniformly distributed in the interval $[-\frac{\Delta}{2}, \frac{\Delta}{2}]$:

$$\sigma_q^2 = \frac{1}{\Delta}\int_{-\frac{\Delta}{2}}^{\frac{\Delta}{2}} q^2 dq \tag{12}$$

$$= \frac{\Delta^2}{12} \tag{13}$$

Let us also calculate the signal-to-noise ratio for this case. The signal variance σ_s^2 for a uniform random variable that takes on values in the interval $[-X_{\max}, X_{\max}]$ is $\frac{(2X_{\max})^2}{12}$. The value of the step size Δ is related to $X_{\max}$ and the number of levels M by

$$\Delta = \frac{2X_{\max}}{M}$$

For the case where we use a fixed-length code, with each codeword being made up of n bits, the number of codewords or the number of reconstruction levels M is 2^n. Combining all this, we have

$$\text{SNR(dB)} = 10\log_{10}\left(\frac{\sigma_s^2}{\sigma_q^2}\right) \tag{14}$$

$$= 10\log_{10}\left(\frac{(2X_{\max})^2}{12}\cdot\frac{12}{\Delta^2}\right) \tag{15}$$

$$= 10\log_{10}\left(\frac{(2X_{\max})^2}{12}\frac{12}{(\frac{2X_{\max}}{M})^2}\right) \tag{16}$$

$$= 10\log_{10}(M^2)$$

$$= 20\log_{10}(2^n)$$

$$= 6.02n \text{ dB} \tag{17}$$

This equation says that for every additional bit in the quantizer, we get an increase in the signal-to-noise ratio of 6.02 dB. This is a well-known result and is often used to get an indication of the maximum gain available if we increase the rate. However, remember that we obtained this result under some assumptions about the input. If the assumptions are not true, this result will not hold true either.

Example 9.4.1: Image Compression

A probability model for the variations of pixels in an image is almost impossible to obtain because of the great variety of images available. A common approach is to declare the pixel values to be uniformly distributed between 0 and $2^b - 1$, where b is the number of bits per pixel. For most of the images we deal with, the number of bits per pixel is 8; therefore, the pixel values would be assumed to vary uniformly between 0 and 255. Let us quantize our test image Sena using a uniform quantizer.

If we wanted to use only 1 bit per pixel, we would divide the range [0, 255] into two intervals, [0, 127] and [128, 255]. The first interval would be represented by the value 64, the midpoint of the first interval; the pixels in the second interval would be represented by the pixel value 196, the midpoint of the second interval. In other words, the boundary values are {0, 128, 255}, while the reconstruction values are {64, 196}. The quantized image is shown in Figure 9.7. As expected, almost all the details in the image have disappeared. If we were to use a 2-bit quantizer, with boundary values {0, 64, 128, 196, 255} and reconstruction levels {32, 96, 160, 224}, we would get considerably more detail. The level of detail increases as the use of bits increases until at 6 bits per pixel, the reconstructed image is indistinguishable from the original, at least to a casual observer. The 1-, 2-, and 3-bit images are shown in Figure 9.7.

Looking at the lower-rate images, we notice a couple of things. First, the lower-rate images are darker than the original, and the lowest-rate reconstructions are the darkest. The reason for this is that the quantization process usually results in scaling down of the dynamic range of the input. For example, in the 1-bit-per-pixel reproduction, the highest pixel value is 196, as opposed to 255 for the original image. As higher gray values represent lighter shades, there is a corresponding darkening of the reconstruction. The other thing to notice in the low-rate reconstruction is that wherever there were smooth changes in gray values there are now abrupt

FIGURE 9.7 **Top left: original Sena image; top right: 1 bit/pixel image; bottom left: 2 bits/pixel; bottom right: 3 bits/pixel.**

transitions. This is especially evident in the face and neck area, where gradual shading has been transformed to blotchy regions of constant values. This is because a range of values is being mapped to the same value, as was the case for the first two samples of the sinusoid in Example 9.3.1. For obvious reasons, this effect is called *contouring*. The perceptual effect of contouring can be reduced by a procedure called *dithering* [121]. ♦

Uniform Quantization of Nonuniform Sources

Quite often the sources we deal with do not have a uniform distribution; however, we still want the simplicity of a uniform quantizer. In these cases, even if the sources are bounded, simply dividing the range of the input by the number of quantization levels does not produce a very good design.

Example 9.4.2:

Suppose our input fell within the interval $[-1, 1]$ with probability 0.95, and fell in the intervals $[-100, 1), (1, 100]$ with probability 0.05. Suppose we wanted to design an eight-level uniform quantizer. If we followed the procedure described in the previous section, the step size would be 25. This means that inputs in the $[-1, 0)$ interval would be represented by the value -12.5, and inputs in the interval $[0, 1)$ would be represented by the value 12.5. The maximum quantization error that can be incurred is 12.5. However, at least 95% of the time, the *minimum* error that will be incurred is 11.5. Obviously, this is not a very good design. A much better approach would be to use a smaller step size, which would result in better representation of the values in the $[-1, 1]$ interval, even if it meant a larger maximum error. Suppose we pick a step size of 0.3. In this case, the maximum quantization error goes from 12.5 to 98.95. However, 95% of the time the quantization error will be less than 0.15. Therefore, the average distortion, or *msqe*, for this quantizer would be substantially less than the *msqe* for the first quantizer. ♦

We can see that when the distribution is no longer uniform, it is not a good idea to obtain the step size by simply dividing the range of the input by the number of levels. This approach becomes totally impractical when we model our sources with distributions that are unbounded, such as the Gaussian distribution. Therefore, we include the *pdf* of the source in the design process.

Our objective is to find the step size that, for a given value of M, will minimize the distortion. The simplest way to do this is to write the distortion as a function of the step size, and then minimize this function. An expression for the distortion, or *msqe*, for an M-level uniform quantizer as a function of the step size can be found by replacing the b_is and y_is in Equation (3) with functions of Δ. As we are dealing with a symmetric condition, we need only compute the distortion for positive values of x; the distortion for negative values of x will be the same.

From Figure 9.8, we see that the decision boundaries are integral multiples of Δ, and the representation level for the interval $[(k-1)\Delta, k\Delta)$ is simply $\frac{2k-1}{2}\Delta$. Therefore, the expression for *msqe* becomes

$$\sigma_q^2 = 2\sum_{i=1}^{\frac{M}{2}-1}\int_{(i-1)\Delta}^{i\Delta}\left(x - \frac{2i-1}{2}\Delta\right)^2 f_X(x)dx$$
$$+2\int_{\left(\frac{M}{2}-1\right)\Delta}^{\infty}\left(x - \frac{M-1}{2}\Delta\right)^2 f_X(x)dx \qquad (18)$$

To find the optimal value of Δ, we simply take a derivative of this equation and set it equal to zero [122] (see Problem 1 at the end of this chapter):

$$\frac{d\sigma_q^2}{d\Delta} = -\sum_{i=1}^{\frac{M}{2}-1}(2i-1)\int_{(i-1)\Delta}^{i\Delta}(x - \frac{2i-1}{2}\Delta)f_X(x)dx$$
$$-(M-1)\int_{\left(\frac{M}{2}-1\right)\Delta}^{\infty}\left(x - \frac{M-1}{2}\Delta\right)f_X(x)dx = 0 \qquad (19)$$

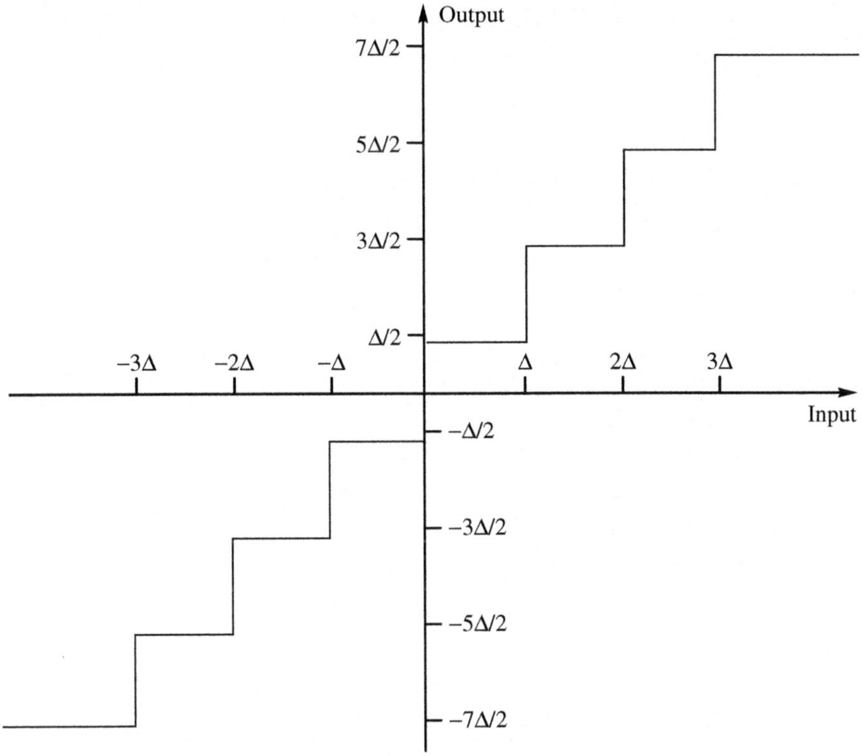

FIGURE 9.8 A uniform midrise quantizer.

This is a rather messy-looking expression, but given the *pdf* $f_X(x)$, it is easy to solve using any one of a number of numerical techniques (see Problem 2 at the end of the chapter). In Table 9.3, we list step sizes found by solving (19) for nine different alphabet sizes and three different distributions.

Before we discuss the results in Table 9.3, let's take a look at the quantization noise for the case of nonuniform sources. Nonuniform sources are often modeled by *pdfs* with unbounded support. That is, there is a nonzero probability of getting an unbounded input. In practical situations, we are not going to get inputs that are unbounded, but often it is very convenient to model the source process with an unbounded distribution. The classic example of this is measurement error, which is often modeled as having a Gaussian distribution, even when the measurement error is known to be bounded. If the input is unbounded, the quantization error is no longer bounded either. The quantization error as a function of input is shown in Figure 9.9. We can see that in the inner intervals the error is still bounded by $\frac{\Delta}{2}$; however, the quantization error in the outer intervals is unbounded. These two types of quantization errors are given different names. The bounded error is called *granular error* or *granular noise*, while the unbounded error is called *overload error* or *overload noise*. In the expression for the *msqe* in Equation (18), the first term represents the granular noise, while the second term represents

TABLE 9.3 **Optimum step size and SNR for uniform quantizers for different distributions and alphabet sizes. The variance for each distribution is 1.[122,123]**

Alphabet Size	Uniform		Gaussian		Laplacian	
	Step Size	SNR	Step Size	SNR	Step Size	SNR
2	1.732	6.02	1.596	4.40	1.414	3.00
4	0.866	12.04	0.9957	9.24	1.0873	7.05
6	0.577	15.58	0.7334	12.18	0.8707	9.56
8	0.433	18.06	0.5860	14.27	0.7309	11.39
10	0.346	20.02	0.4908	15.90	0.6334	12.81
12	0.289	21.60	0.4238	17.25	0.5613	13.98
14	0.247	22.94	0.3739	18.37	0.5055	14.98
16	0.217	24.08	0.3352	19.36	0.4609	15.84
32	0.108	30.10	0.1881	24.56	0.2799	20.46

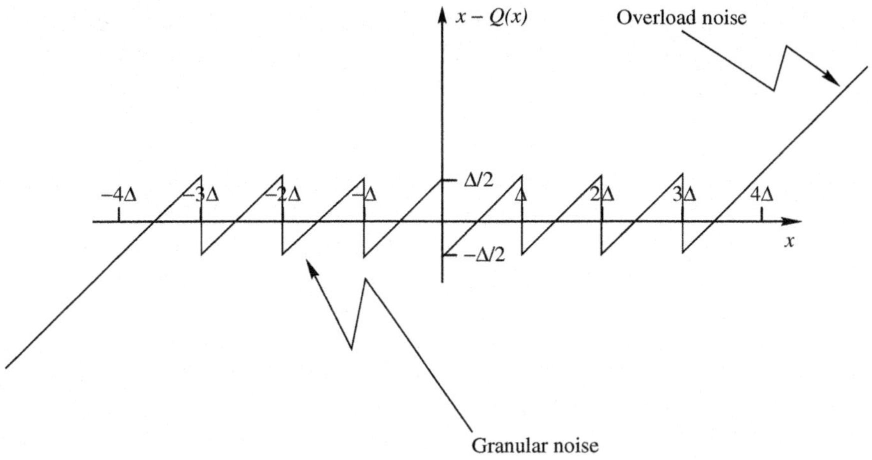

FIGURE 9.9 Quantization error for a uniform midrise quantizer.

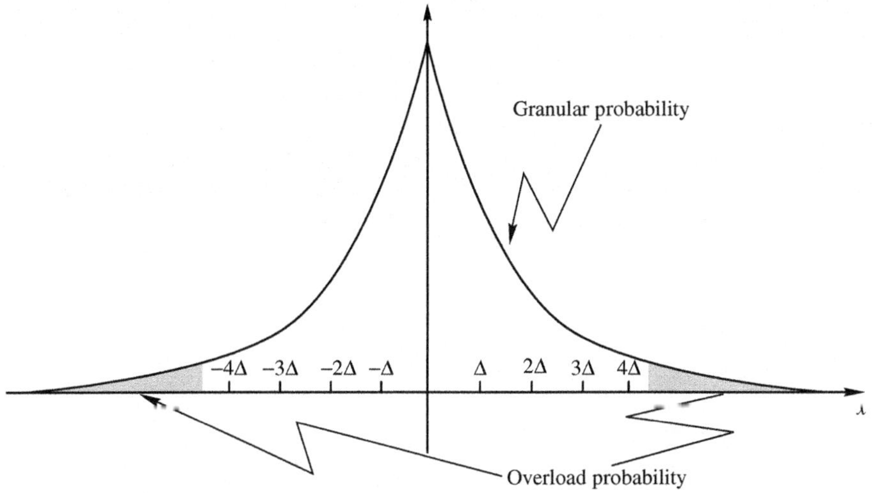

FIGURE 9.10 Overload and granular regions for a 3-bit uniform quantizer.

the overload noise. The probability that the input will fall into the overload region is called the *overload probability* (Figure 9.10).

The nonuniform sources we deal with have probability density functions that are generally peaked at zero and decay as we move away from the origin. Therefore, the overload probability is generally much smaller than the probability of the input falling in the granular region. As we see from Equation (19), an increase in the step size Δ will result in an increase in the value of $\left(\frac{M}{2}-1\right)\Delta$, which in turn will result in a decrease in the overload probability and the second term in Equation (19). However, an increase in the step size Δ will also increase the granular noise, which is the first term in Equation (19). The design process for the uniform quantizer is a balancing of these two effects. An important parameter that describes this trade-off is the loading factor f_l, defined as the ratio of the maximum value the input can take in the granular region to the standard deviation. A common value of the loading factor is 4. This is also referred to as 4σ *loading*.

Recall that when quantizing an input with a uniform distribution, the SNR and bit rate are related by Equation (17), which says that for each bit increase in the rate there is an increase of 6.02 dB in the SNR. In Table 9.3, along with the step sizes, we have also listed the SNR obtained when a million input values with the appropriate *pdf* are quantized using the indicated quantizer.

From this table, we can see that, although the SNR for the uniform distribution follows the rule of a 6.02 dB increase in the signal-to-noise ratio for each additional bit, this is not true for the other distributions. Remember that we made some assumptions when we obtained the $6.02n$ rule that are only valid for the uniform distribution. Notice that the more peaked a distribution is (that is, the further away from uniform it is), the more it seems to vary from the 6.02 dB rule.

We also said that the selection of Δ is a balance between the overload and granular errors. The Laplacian distribution has more of its probability mass away from the origin in its tails than the Gaussian distribution. This means that for the same step size and number of levels, there is a higher probability of being in the overload region if the input has a Laplacian distribution than if the input has a Gaussian distribution. The uniform distribution is the extreme case, where the overload probability is zero. If we increase the step size for the same number of levels, the size of the overload region (and hence the overload probability) is reduced at the expense of granular noise. Therefore, for a given number of levels, if we are picking the step size to balance the effects of the granular and overload noise, distributions that have heavier tails will tend to have larger step sizes. This effect can be seen in Table 9.3. For example, for eight levels, the step size for the uniform quantizer is 0.433. The step size for the Gaussian quantizer is larger (0.586), while the step size for the Laplacian quantizer is larger still (0.7309).

Mismatch Effects

We have seen that for a result to hold, the assumptions we used to obtain the result have to hold. When we obtain the optimum step size for a particular uniform quantizer using Equation (19), we make some assumptions about the statistics of the source. We assume a certain distribution and certain parameters of the distribution. What happens when our assumptions do not hold? Let's try to answer this question empirically.

We will look at two types of mismatches. The first is when the assumed distribution type matches the actual distribution type, but the variance of the input is different from the assumed variance. The second mismatch is when the actual distribution type is different from the distribution type assumed when obtaining the value of the step size. Throughout our discussion, we will assume that the mean of the input distribution is zero.

In Figure 9.11, we have plotted the signal-to-noise ratio as a function of the ratio of the actual to assumed variance of a 4-bit Gaussian uniform quantizer, with a Gaussian input. (To see the effect under different conditions, see Problem 5 at the end of this chapter.) Remember

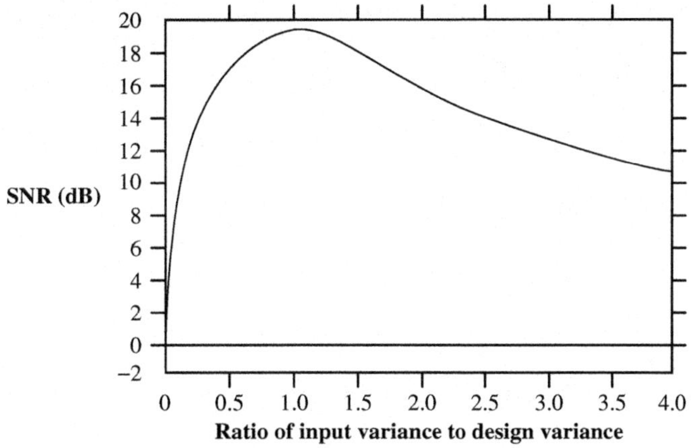

FIGURE 9.11 Effect of variance mismatch on the performance of a 4-bit uniform quantizer.

TABLE 9.4 Demonstration of the effect of mismatch using eight-level quantizers (dB). Each input distribution has unit variance.

Input Distribution	Uniform Quantizer	Gaussian Quantizer	Laplacian Quantizer	Gamma Quantizer
Uniform	18.06	15.56	13.29	12.41
Gaussian	12.40	14.27	13.37	12.73
Laplacian	8.80	10.79	11.39	11.28
Gamma	6.98	8.06	8.64	8.76

that for a distribution with zero mean, the variance is given by $\sigma_x^2 = E[X^2]$, which is also a measure of the power in the signal X. As we can see from the figure, the signal-to-noise ratio is maximum when the input signal variance matches the variance assumed when designing the quantizer. From the plot we also see that there is an asymmetry; the SNR is considerably worse when the input variance is lower than the assumed variance. This is because the SNR is a ratio of the input variance and the mean squared quantization error. When the input variance is smaller than the assumed variance, the mean squared quantization error actually drops because there is less overload noise. However, because the input variance is low, the ratio is small. When the input variance is higher than the assumed variance, the *msqe* increases substantially, but because the input power is also increasing, the ratio does not decrease as dramatically. To see this more clearly, we have plotted the mean squared error versus the signal variance separately in Figure 9.12. We can see from these figures that the decrease in signal-to-noise ratio does not always correlate directly with an increase in *msqe*.

The second kind of mismatch is where the input distribution does not match the distribution assumed when designing the quantizer. In Table 9.4 we have listed the SNR when inputs with different distributions are quantized using several different eight-level quantizers. The quantizers were designed assuming a particular input distribution.

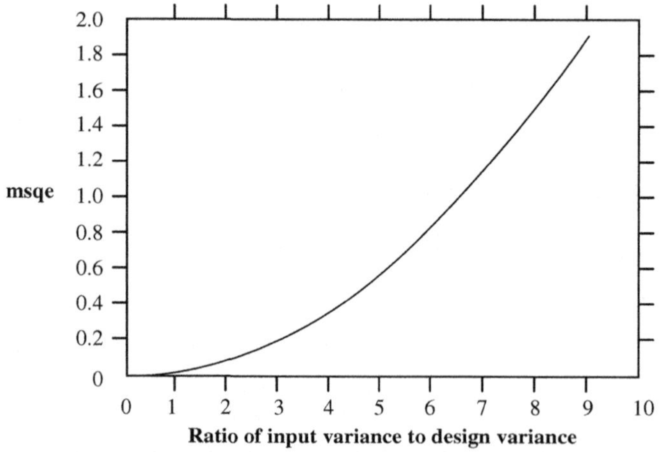

FIGURE 9.12 **The *msqe* as a function of variance mismatch with a 4-bit uniform quantizer.**

Notice that as we go from left to right in the table, the designed step size becomes progressively larger than the "correct" step size. This is similar to the situation where the input variance is smaller than the assumed variance. As we can see when we have a mismatch that results in a smaller step size relative to the optimum step size, there is a greater drop in performance than when the quantizer step size is larger than its optimum value.

9.5 Adaptive Quantization

One way to deal with the mismatch problem is to adapt the quantizer to the statistics of the input. Several things might change in the input relative to the assumed statistics, including the mean, the variance, and the *pdf*. The strategy for handling each of these variations can be different, though certainly not exclusive. If more than one aspect of the input statistics changes, it is possible to combine the strategies for handling each case separately. If the mean of the input is changing with time, the best strategy is to use some form of differential encoding (discussed in some detail in Chapter 11). For changes in the other statistics, the common approach is to adapt the quantizer parameters to the input statistics.

There are two main approaches to adapting the quantizer parameters: an *offline* or *forward adaptive* approach, and an *online* or *backward adaptive* approach. In forward adaptive quantization, the source output is divided into blocks of data. Each block is analyzed before quantization, and the quantizer parameters are set accordingly. The settings of the quantizer are then transmitted to the receiver as *side information*. In backward adaptive quantization, the adaptation is performed based on the quantizer output. As this is available to both transmitter and receiver, there is no need for side information.

9.5.1 Forward Adaptive Quantization

Let us first look at approaches for adapting to changes in input variance using the forward adaptive approach. This approach necessitates a delay of at least the amount of time required to process a block of data. The insertion of side information in the transmitted data stream may also require the resolution of some synchronization problems. The size of the block of data processed also affects a number of other things. If the size of the block is too large, then the adaptation process may not capture the changes taking place in the input statistics. Furthermore, large block sizes mean more delay, which may not be tolerable in certain applications. On the other hand, small block sizes mean that the side information has to be transmitted more often, which in turn means the amount of overhead per sample increases. The selection of the block size is a trade-off between the increase in side information necessitated by small block sizes and the loss of fidelity due to large block sizes (see Problem 7 at the end of this chapter).

The variance estimation procedure is rather simple. At time n we use a block of N future samples to compute an estimate of the variance:

$$\hat{\sigma}_q^2 = \frac{1}{N} \sum_{i=0}^{N-1} x_{n+i}^2 \tag{20}$$

Note that we are assuming that our input has a mean of zero. The variance information also needs to be quantized so that it can be transmitted to the receiver. Usually, the number of bits used to quantize the value of the variance is significantly larger than the number of bits used to quantize the sample values.

Example 9.5.1:

In Figure 9.13, we show a segment of speech quantized using a fixed 3-bit quantizer. The step size of the quantizer was adjusted based on the statistics of the entire sequence. The sequence was the `testm.raw` sequence from the sample data sets, consisting of about 4000 samples of a male speaker saying the word "test." The speech signal was sampled at 8000 samples per second and digitized using a 16-bit A/D.

We can see from the figure that, as in the case of the example of the sinusoid earlier in this chapter, there is a considerable loss in amplitude resolution. Sample values that are close together have been quantized to the same value.

The same sequence quantized with a forward adaptive quantizer is shown in Figure 9.14. For this example, we divided the input into blocks of 128 samples. Before quantizing the samples in a block, the standard deviation for the samples in the block was obtained. This value was quantized using an 8-bit quantizer and sent to both the transmitter and receiver. The samples in the block were then normalized using this value of the standard deviation. Notice that the reconstruction follows the input much more closely, though there seems to be room for improvement, especially in the latter half of the displayed samples. ♦

Example 9.5.2:

In Example 9.4.1, we used a uniform quantizer with the assumption that the input is uniformly distributed. Let us refine this source model a bit and say that while the source is uniformly

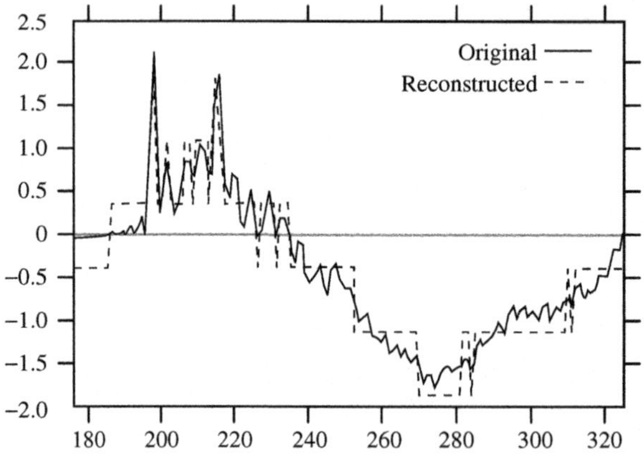

FIGURE 9.13 **Original 16-bit speech and compressed 3-bit speech sequences.**

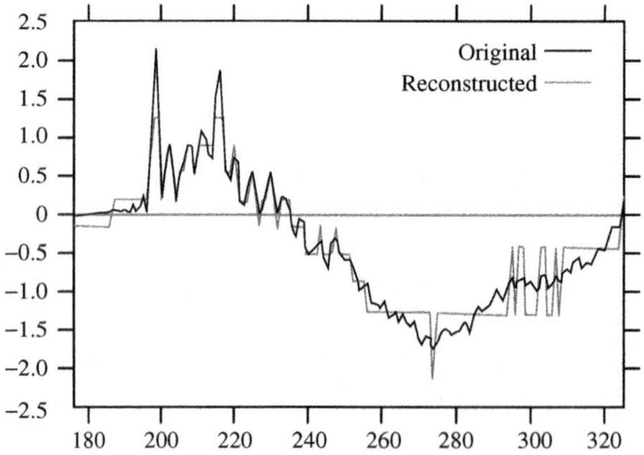

FIGURE 9.14 **Original 16-bit speech sequence and sequence obtained using an eight-level forward adaptive quantizer.**

distributed over different regions, the range of the input changes. In a forward adaptive quantization scheme, we would obtain the minimum and maximum values for each block of data, which would be transmitted as side information. In Figure 9.15, we see the Sena image quantized with a block size of 8×8 using 3-bit forward adaptive uniform quantization. The side information consists of the minimum and maximum values in each block, which require 8 bits each. Therefore, the overhead in this case is $\frac{16}{8\times8}$ or 0.25 bits per pixel, which is quite small compared to the number of bits per sample used by the quantizer.

The resulting image is hardly distinguishable from the original. Certainly at higher rates, forward adaptive quantization seems to be a very good alternative. ♦

FIGURE 9.15 Sena image quantized to 3.25 bits per pixel using forward adaptive quantization.

9.5.2 Backward Adaptive Quantization

In backward adaptive quantization, only the past quantized samples are available for use in adapting the quantizer. The values of the input are only known to the encoder; therefore, this information cannot be used to adapt the quantizer. How can we get information about mismatch simply by examining the output of the quantizer without knowing what the input was? If we studied the output of the quantizer for a long period of time, we could get some idea about mismatch from the distribution of output values. If the quantizer step size Δ is well matched to the input, the probability that an input to the quantizer would land in a particular interval would be consistent with the *pdf* assumed for the input. However, if the actual *pdf* differs from the assumed *pdf*, the number of times the input falls in the different quantization intervals will be inconsistent with the assumed *pdf*. If Δ is smaller than what it should be, the input will fall in the outer levels of the quantizer an excessive number of times. On the other hand, if Δ is larger than it should be for a particular source, the input will fall in the inner levels an excessive number of times. Therefore, it seems that we should observe the output of the quantizer for a long period of time, then expand the quantizer step size if the input falls in the outer levels an excessive number of times, and contract the step size if the input falls in the inner levels an excessive number of times.

Nuggehally S. Jayant at Bell Labs showed that we did not need to observe the quantizer output over a long period of time [124]. In fact, we could adjust the quantizer step size after observing a single output. Jayant named this quantization approach "quantization with one word memory." The quantizer is better known as the *Jayant quantizer*. The idea behind the Jayant quantizer is very simple. If the input falls in the outer levels, the step size needs to be expanded, and if the input falls in the inner quantizer levels, the step size needs to be reduced. The expansions and contractions should be done in such a way that once the quantizer is matched to the input, the product of the expansions and contractions is unity.

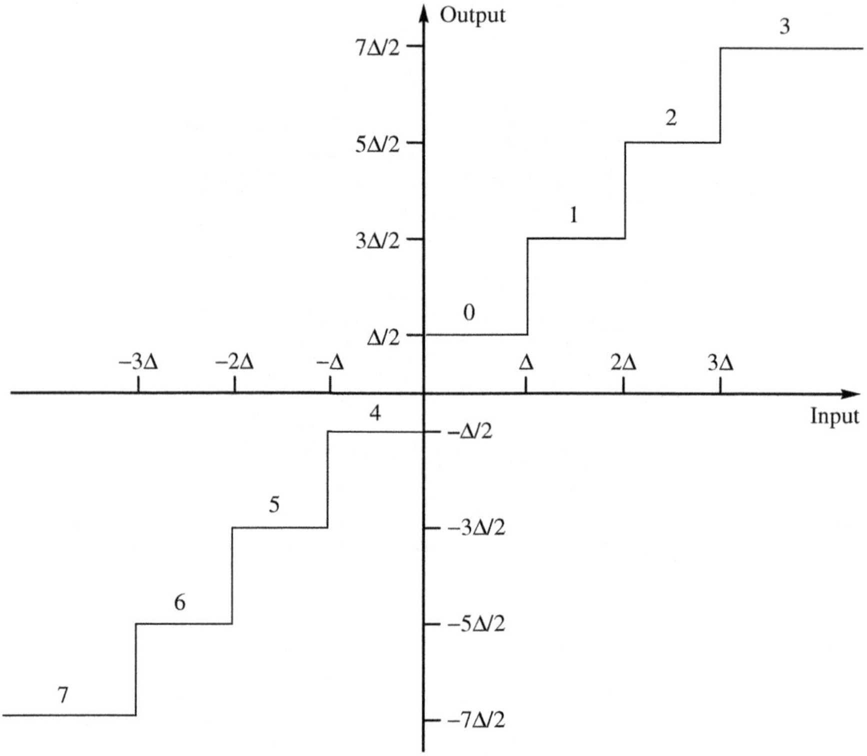

FIGURE 9.16 Output levels for the Jayant quantizer.

The expansion and contraction of the step size is accomplished in the Jayant quantizer by assigning a *multiplier* M_k to each interval. If the $(n-1)$th input falls in the kth interval, the step size to be used for the nth input is obtained by multiplying the step size used for the $(n-1)$th input with M_k. The multiplier values for the inner levels in the quantizer are less than one, and the multiplier values for the outer levels of the quantizer are greater than one. Therefore, if an input falls into the inner levels, the quantizer used to quantize the next input will have a smaller step size. Similarly, if an input falls into the outer levels, the step size will be multiplied with a value greater than one, and the next input will be quantized using a larger step size. Notice that the step size for the current input is modified based on the previous quantizer output. The previous quantizer output is available to both the transmitter and receiver, so there is no need to send any additional information to inform the receiver about the adaptation. Mathematically, the adaptation process can be represented as

$$\Delta_n = M_{l(n-1)}\Delta_{n-1} \tag{21}$$

where $l(n-1)$ is the quantization interval at time $n-1$.

In Figure 9.16 we show a 3-bit uniform quantizer. We have eight intervals represented by the different quantizer outputs. However, the multipliers for symmetric intervals are identical

TABLE 9.5 Operation of a Jayant quantizer.

n	Δ_n	Input	Output Level	Output	Error	Update Equation
0	0.5	0.1	0	0.25	0.15	$\Delta_1 = M_0 \times \Delta_0$
1	0.4	−0.2	4	−0.2	0.0	$\Delta_2 = M_4 \times \Delta_1$
2	0.32	0.2	0	0.16	0.04	$\Delta_3 = M_0 \times \Delta_2$
3	0.256	0.1	0	0.128	0.028	$\Delta_4 = M_0 \times \Delta_3$
4	0.2048	−0.3	5	−0.3072	−0.0072	$\Delta_5 = M_5 \times \Delta_4$
5	0.1843	0.1	0	0.0922	−0.0078	$\Delta_6 = M_0 \times \Delta_5$
6	0.1475	0.2	1	0.2212	0.0212	$\Delta_7 = M_1 \times \Delta_6$
7	0.1328	0.5	3	0.4646	−0.0354	$\Delta_8 = M_3 \times \Delta_7$
8	0.1594	0.9	3	0.5578	−0.3422	$\Delta_9 = M_3 \times \Delta_8$
9	0.1913	1.5	3	0.6696	−0.8304	$\Delta_{10} = M_3 \times \Delta_9$
10	0.2296	1.0	3	0.8036	0.1964	$\Delta_{11} = M_3 \times \Delta_{10}$
11	0.2755	0.9	3	0.9643	0.0643	$\Delta_{12} = M_3 \times \Delta_{11}$

because of symmetry:

$$M_0 = M_4 \quad M_1 = M_5 \quad M_2 = M_6 \quad M_3 = M_7$$

Therefore, we only need four multipliers. To see how the adaptation proceeds, let us work through a simple example using this quantizer.

Example 9.5.3: Jayant Quantizer

For the quantizer in Figure 9.16, suppose the multiplier values are $M_0 = M_4 = 0.8$, $M_1 = M_5 = 0.9$, $M_2 = M_6 = 1$, and $M_3 = M_7 = 1.2$; the initial value of the step size, Δ_0, is 0.5; and the sequence to be quantized is $0.1, -0.2, 0.2, 0.1, -0.3, 0.1, 0.2, 0.5, 0.9, 1.5, \ldots$. When the first input is received, the quantizer step size is 0.5. Therefore, the input falls into level 0, and the output value is 0.25, resulting in an error of 0.15. As this input fell into quantizer level 0, the new step size Δ_1 is $M_0 \times \Delta_0 = 0.8 \times 0.5 = 0.4$. The next input is −0.2, which falls into level 4. As the step size at this time is 0.4, the output is −0.2. To update, we multiply the current step size with M_4. Continuing in this fashion, we get the sequence of step sizes and outputs shown in Table 9.5.

Notice how the quantizer adapts to the input. In the beginning of the sequence, the input values are mostly small, and the quantizer step size becomes progressively smaller, providing better and better estimates of the input. At the end of the sample sequence, the input values are large, and the step size becomes progressively bigger. However, the size of the error is quite large during the transition. This means that if the input was changing rapidly, which would happen if we had a high-frequency input, such transition situations would be much more likely to occur, and the quantizer would not function very well. However, in cases where the statistics of the input change slowly, the quantizer could adapt to the input. As most natural sources such as speech and images tend to be correlated, their values do not change drastically from sample to sample. Even when some of this structure is removed through some transformation, the residual structure is generally enough for the Jayant quantizer (or some variation of it) to function quite effectively. ♦

The step size in the initial part of the sequence in this example is progressively getting smaller. We can easily conceive of situations where the input values would be small for a long period. Such a situation could occur during a silence period in speech-encoding systems or while encoding a dark background in image-encoding systems. If the step size continues to shrink for an extended period of time, it would result in a value of zero in a finite precision system. This would be catastrophic, effectively replacing the quantizer with a zero output device. Usually, a minimum value $\Delta_{\min}$ is defined, and the step size is not allowed to go below this value to prevent this from happening. Similarly, if we get a sequence of large values, the step size could increase to a point that, when we started getting smaller values, the quantizer would not be able to adapt fast enough. To prevent this from happening, a maximum value $\Delta_{\max}$ is defined, and the step size is not allowed to increase beyond this value.

The adaptivity of the Jayant quantizer depends on the values of the multipliers. The further the multiplier values are from unity, the more adaptive the quantizer. However, if the adaptation algorithm reacts too fast, this could lead to instability. So how do we go about selecting the multipliers?

First of all, we know that the multipliers correponding to the inner levels are less than one, and the multipliers for the outer levels are greater than one. If the input process is stationary and P_k represents the probability of being in quantizer interval k (generally estimated by using a fixed quantizer for the input data), then we can impose a stability criterion for the Jayant quantizer based on our requirement that once the quantizer is matched to the input, the product of the expansions and contractions are equal to unity. That is, if n_k is the number of times the input falls in the kth interval,

$$\prod_{k=0}^{M} M_k^{n_k} = 1 \tag{22}$$

Taking the Nth root of both sides (where N is the total number of inputs), we obtain

$$\prod_{k=0}^{M} M_k^{\frac{n_k}{N}} = 1$$

or

$$\prod_{k=0}^{M} M_k^{P_k} = 1 \tag{23}$$

where we have assumed that $P_k = n_k/N$.

There are an infinite number of multiplier values that would satisfy Equation (23). One way to restrict this number is to impose some structure on the multipliers by requiring them to be of the form

$$M_k = \gamma^{l_k} \tag{24}$$

where γ is a number greater than one and l_k takes on only integer values [125,126]. If we substitute this expression for M_k into Equation (23), we get

$$\prod_{k=0}^{M} \gamma^{l_k P_k} = 1 \tag{25}$$

which implies that

$$\sum_{k=0}^{M} l_k P_k = 0 \tag{26}$$

The final step is the selection of γ, which involves a significant amount of creativity. The value we pick for γ determines how fast the quantizer will respond to changing statistics. A large value of γ will result in faster adaptation, while a smaller value of γ will result in greater stability.

Example 9.5.4:

Suppose we have to obtain the multiplier functions for a 2-bit quantizer with input probabilities $P_0 = 0.8$, $P_1 = 0.2$. First, note that the multiplier value for the inner level has to be less than 1. Therefore, l_0 is less than 0. If we pick $l_0 = -1$ and $l_1 = 4$, this would satisfy Equation (26), while making M_0 less than 1 and M_1 greater than 1. Finally, we need to pick a value for γ.

In Figure 9.17, we see the effect of using different values of γ in a rather extreme example. The input is a square wave that switches between 0 and 1 every 30 samples. The input is quantized using a 2-bit Jayant quantizer. We have used $l_0 = -1$ and $l_1 = 2$. Notice what happens when the input switches from 0 to 1. At first the input falls in the outer level of the quantizer, and the step size increases. This process continues until Δ is just greater than 1. If γ is close to 1, Δ has been increasing quite slowly and should have a value close to 1 right before its value increases to greater than 1. Therefore, the output at this point is close to 1.5. When Δ becomes greater than 1, the input falls in the inner level, and if γ is close to 1, the output suddenly drops to about 0.5. The step size now decreases until it is just below 1, and the process repeats, causing the "ringing" seen in Figure 9.17. As γ increases, the quantizer adapts more rapidly, and the magnitude of the ringing effect decreases. The reason for the decrease is that right before the value of Δ increases above 1, its value is much smaller than 1, and subsequently the output value is much smaller than 1.5. When Δ increases beyond 1, it may increase by a significant amount, so the inner level may be much greater than 0.5. These two effects together compress the ringing phenomenon. Looking at this phenomenon, we can see that it may have been better to have two adaptive strategies, one for when the input is changing rapidly, as in the case of the transitions between 0 and 1, and one for when the input is constant, or nearly so. We will explore this approach further when we describe the quantizer used in CCITT standard G.726. ♦

When selecting multipliers for a Jayant quantizer, the best quantizers expand more rapidly than they contract. This makes sense when we consider that when the input falls into the outer levels of the quantizer, it is incurring overload error, which is essentially unbounded. This situation needs to be mitigated with dispatch. On the other hand, when the input falls in the inner levels, the noise incurred is granular noise, which is bounded and, therefore, may be more tolerable. Finally, the discussion of the Jayant quantizer was motivated by the need for robustness in the face of changing input statistics. Let us repeat the earlier experiment with changing input variance and distributions and see the performance of the Jayant quantizer

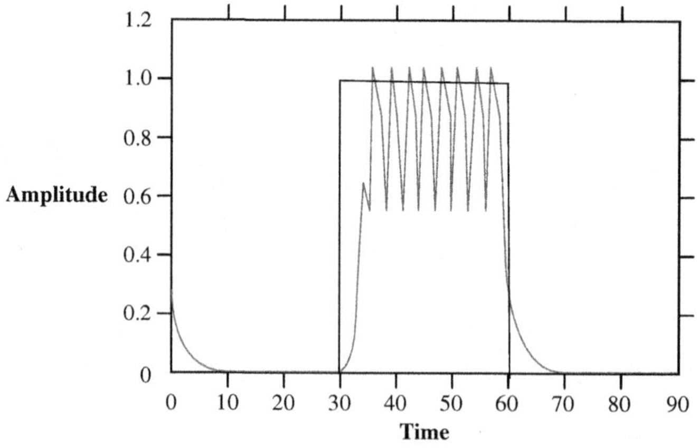

FIGURE 9.17 Effect of γ on the performance of the Jayant quantizer.

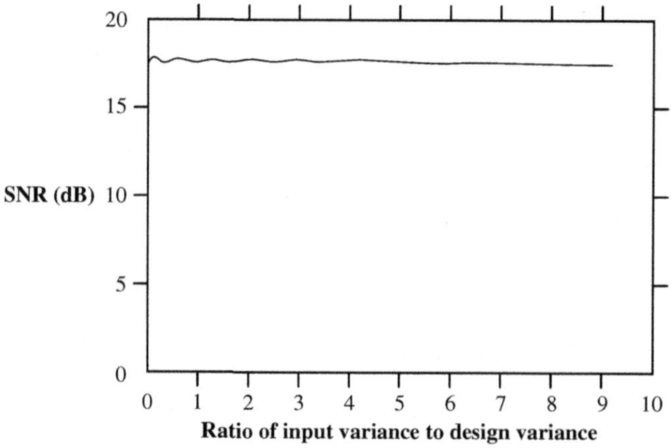

FIGURE 9.18 Performance of the Jayant quantizer for different input variances.

compared to the *pdf*-optimized quantizer. The results for these experiments are presented in Figure 9.18.

Notice how flat the performance curve is. Compare this with the performance curve shown in Figure 9.11, which shows the drastic effects of mismatch between the design and source variance. While the performance of the Jayant quantizer is much better than the nonadaptive uniform quantizer over a wide range of input variances, the performance of the nonadaptive quantizer is significantly better than the performance of the Jayant quantizer at the point where the input variance and design variance agree. This means that if we know the input statistics and we are reasonably certain that the input statistics will not change over time, it is better to design for those statistics than to design an adaptive system.

9.6 Nonuniform Quantization

As we can see from Figure 9.10, the input is more likely to fall in the inner levels of the quantizer if the input distribution has more mass near the origin. Recall that in lossless compression, in order to minimize the *average* number of bits per input symbol, we assigned shorter codewords to symbols that occurred with higher probability and longer codewords to symbols that occurred with lower probability. In an analogous fashion we can try to approximate the input better in regions of high probability, perhaps at the cost of worse approximations in regions of lower probability, in order to decrease the average distortion. We can do this by making the quantization intervals smaller in those regions that have more probability mass. If the source distribution is like the distribution shown in Figure 9.10, we would have smaller intervals near the origin. If we wanted to keep the number of intervals constant, this would mean we would have larger intervals away from the origin. A quantizer that has nonuniform intervals is called a *nonuniform quantizer*, an example of which is shown in Figure 9.19.

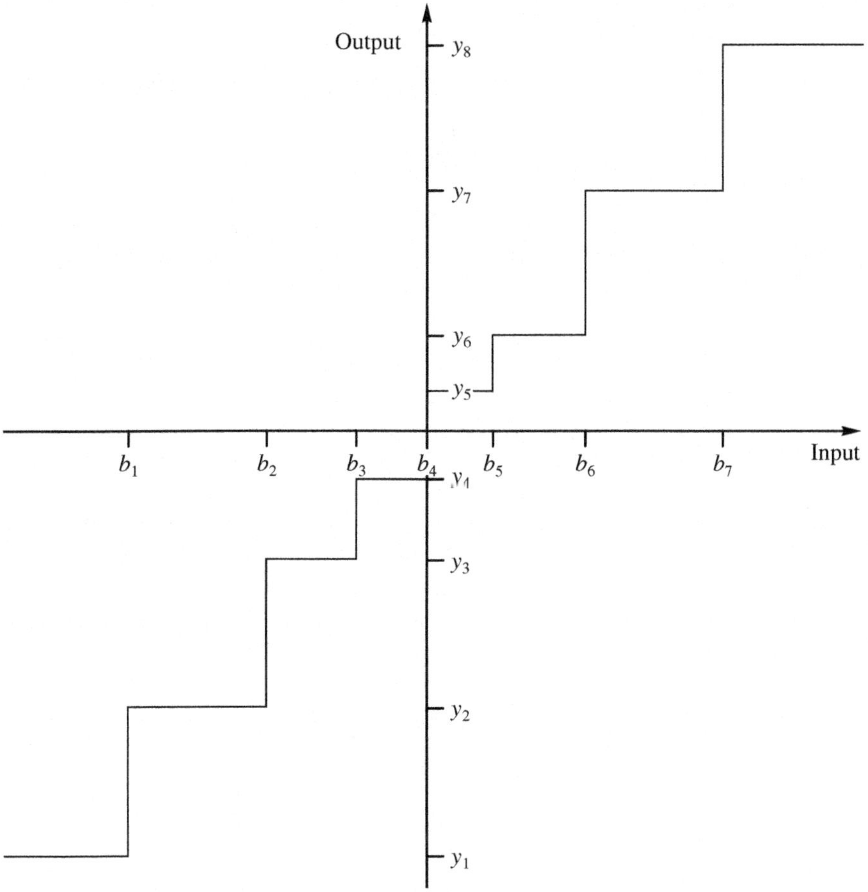

FIGURE 9.19 **A nonuniform midrise quantizer.**

Notice that the intervals closer to zero are smaller. Hence, the maximum value that the quantizer error can take on is also smaller, resulting in a better approximation. We pay for this improvement in accuracy at lower input levels by incurring larger errors when the input falls in the outer intervals. However, as the probability of getting smaller input values is much higher than getting larger signal values, on average the distortion will be lower than if we had a uniform quantizer. While a nonuniform quantizer provides lower average distortion, the design of nonuniform quantizers is also somewhat more complex. However, the basic idea is quite straightforward: find the decision boundaries and reconstruction levels that minimize the mean squared quantization error. We look at the design of nonuniform quantizers in more detail in the following sections.

9.6.1 *pdf*-Optimized Quantization

A direct approach for locating the best nonuniform quantizer, if we have a probability model for the source, is to find the $\{b_i\}$ and $\{y_i\}$ that minimize Equation (3). Setting the derivative of Equation (3) with respect to y_j to zero, and solving for y_j, we get

$$y_j = \frac{\int_{b_{j-1}}^{b_j} x f_X(x)dx}{\int_{b_{j-1}}^{b_j} f_X(x)dx} \tag{27}$$

The output point for each quantization interval is the centroid of the probability mass in that interval. Using the Leibniz integral rule, and taking the derivative of Equation (3) with respect to b_j and setting it equal to zero, we get an expression for b_j as

$$b_j = \frac{y_{j+1} + y_j}{2} \tag{28}$$

The decision boundary is simply the midpoint of the two neighboring reconstruction levels. Solving these two equations will give us the values for the reconstruction levels and decision boundaries that minimize the mean squared quantization error. Unfortunately, to solve for y_j, we need the values of b_j and b_{j-1}, and to solve for b_j, we need the values of y_{j+1} and y_j. In a 1960 paper, Joel Max [122] showed how to solve the two equations iteratively. The same approach was described by Stuart P. Lloyd in a 1957 internal Bell Labs memorandum. Generally, credit goes to whomever publishes first, but in this case, because much of the early work in quantization was done at Bell Labs, Lloyd's work was given due credit and the algorithm became known as the Lloyd-Max algorithm. However, the story does not end (begin?) there. Allen Gersho [127] points out that the same algorithm was published by Lukaszewicz and Steinhaus in a Polish journal in 1955 [128]! Lloyd's paper remained unpublished until 1982, when it was finally published in a special issue of the *IEEE Transactions on Information Theory* devoted to quantization [129].

To see how this algorithm works, let us apply it to a specific situation. Suppose we want to design an M-level symmetric midrise quantizer. To define our symbols, we will use Figure 9.20. From the figure, we see that we need to obtain the reconstruction levels $\{y_1, y_2, \ldots, y_{\frac{M}{2}}\}$ and the decision boundaries $\{b_1, b_2, \ldots, b_{\frac{M}{2}-1}\}$ to design this quantizer. The reconstruction levels

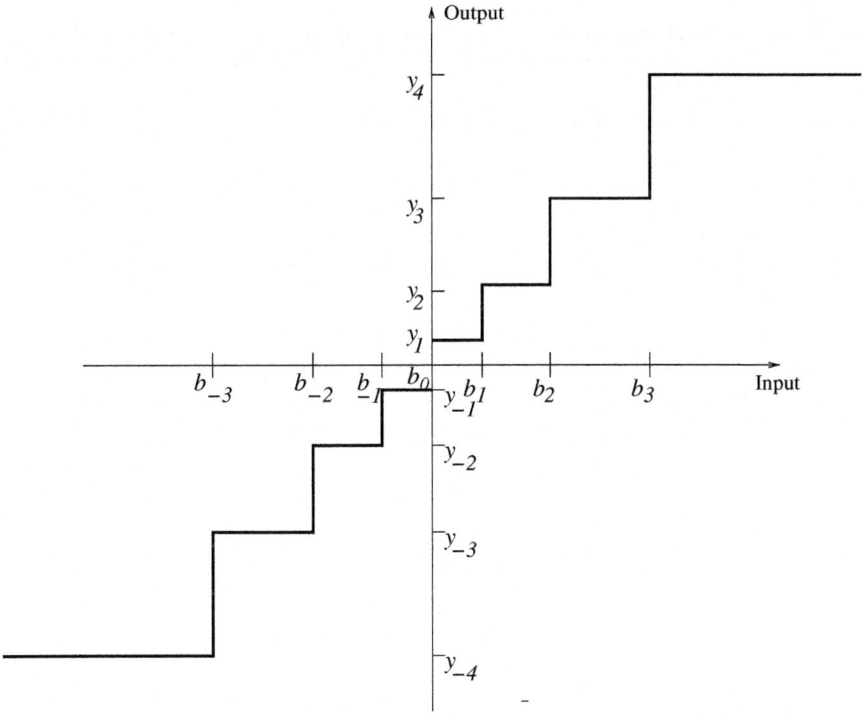

FIGURE 9.20 **A symmetric nonuniform midrise quantizer.**

$\{y_{-1}, y_{-2}, \ldots, y_{-\frac{M}{2}}\}$ and the decision boundaries $\{b_{-1}, b_{-2}, \ldots, b_{-(\frac{M}{2}-1)}\}$ can be obtained through symmetry, the decision boundary b_0 is zero, and the decision boundary $b_{\frac{M}{2}}$ is simply the largest value the input can take on (for unbounded inputs this would be ∞).

Let us set j equal to 1 in Equation (27):

$$y_1 = \frac{\int_{b_0}^{b_1} x f_X(x)dx}{\int_{b_0}^{b_1} f_X(x)dx} \tag{29}$$

As b_0 is known to be 0, we have two unknowns in this equation, b_1 and y_1. We make a guess at y_1, and later we will try to refine this guess. Using this guess in Equation (29), we numerically find the value of b_1 that satisfies Equation (29). Setting j equal to 1 in Equation (28) and rearranging things slightly, we get

$$y_2 = 2b_1 - y_1 \tag{30}$$

from which we can compute y_2. This value of y_2 can then be used in Equation (27) with $j = 2$ to find b_2, which in turn can be used to find y_3. We continue this process, until we obtain a value for $\{y_1, y_2, \ldots, y_{\frac{M}{2}}\}$ and $\{b_1, b_2, \ldots, b_{\frac{M}{2}-1}\}$. Note that the accuracy of all the values obtained to this point depends on the quality of our initial estimate of y_1. We can check this by

TABLE 9.6 Quantizer boundary and reconstruction levels for nonuniform Gaussian and Laplacian quantizers. The input distributions have unit variance.

	Gaussian			Laplacian		
Levels	b_i	y_i	SNR	b_i	y_i	SNR
4	0.0	0.4528		0.0	0.4196	
	0.9816	1.510	9.3 dB	1.1269	1.8340	7.54 dB
6	0.0	0.3177		0.0	0.2998	
	0.6589	1.0		0.7195	1.1393	
	1.447	1.894	12.41 dB	1.8464	2.5535	10.51 dB
8	0.0	0.2451		0.0	0.2334	
	0.5006	0.7550		0.5332	0.8330	
	1.050	1.3440		1.2527	1.6725	
	1.748	2.1520	14.62 dB	2.3796	3.0867	12.64 dB

noting that $y_{\frac{M}{2}}$ is the centroid of the probability mass of the interval $[b_{\frac{M}{2}-1}, b_{\frac{M}{2}}]$. We know $b_{\frac{M}{2}}$ from our knowledge of the data. Therefore, we can compute the integral

$$\frac{\int_{b_{\frac{M}{2}-1}}^{b_{\frac{M}{2}}} x f_X(x)dx}{\int_{b_{\frac{M}{2}-1}}^{b_{\frac{M}{2}}} f_X(x)dx} \tag{31}$$

and compare it with the previously computed value of $y_{\frac{M}{2}}$. If the difference is less than some tolerance threshold, we can stop. Otherwise, we adjust the estimate of y_1 in the direction indicated by the sign of the difference and repeat the procedure.

Decision boundaries and reconstruction levels for various distributions and number of levels generated using this procedure are shown in Table 9.6. Notice that the distributions that have heavier tails also have larger outer step sizes. However, these same quantizers have smaller inner step sizes because they are more heavily peaked. The SNR for these quantizers is also listed in the table. Comparing these values with those for the *pdf*-optimized uniform quantizers, we can see a significant improvement, especially for distributions further away from the uniform distribution. Both uniform and nonuniform *pdf*-optimized, or Lloyd-Max, quantizers have a number of interesting properties. We list these properties here (their proofs can be found in [270–272]):

- **Property 1:** The mean values of the input and output of a Lloyd-Max quantizer are equal.
- **Property 2:** For a given Lloyd-Max quantizer, the variance of the output is always less than or equal to the variance of the input.
- **Property 3:** The mean squared quantization error for a Lloyd-Max quantizer is given by

$$\sigma_q^2 = \sigma_x^2 - \sum_{j=1}^{M} y_j^2 P[b_{j-1} \leqslant X < b_j] \tag{32}$$

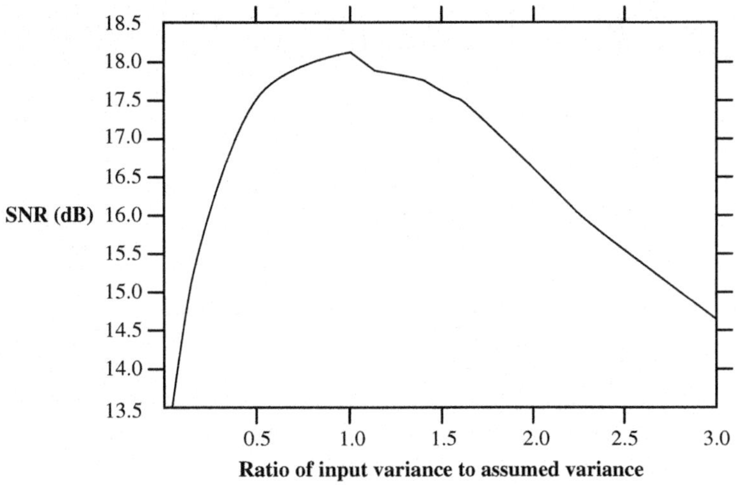

FIGURE 9.21 **Effect of mismatch on nonuniform quantization.**

where σ_x^2 is the variance of the quantizer input, and the second term on the right-hand side is the second moment of the output (or variance if the input is zero mean).

- **Property 4:** Let N be the random variable corresponding to the quantization error. Then for a given Lloyd-Max quantizer,

$$E[XN] = -\sigma_q^2 \tag{33}$$

- **Property 5:** For a given Lloyd-Max quantizer, the quantizer output and the quantization noise are orthogonal:

$$E[Q(X)N|b_0, b_1, \ldots, b_M] = 0 \tag{34}$$

Mismatch Effects

As in the case of uniform quantizers, the *pdf*-optimized nonuniform quantizers also have problems when the assumptions underlying their design are violated. In Figure 9.21 we show the effects of variance mismatch on a 4-bit Laplacian nonuniform quantizer.

This mismatch effect is a serious problem because in most communication systems, the input variance can change considerably over time. A common example of this is the telephone system. Different people speak with differing amounts of loudness into the telephone. The quantizer used in the telephone system needs to be quite robust to the wide range of input variances in order to provide satisfactory service.

One solution to this problem is the use of adaptive quantization to match the quantizer to the changing input characteristics. We have already looked at adaptive quantization for the uniform quantizer. Generalizing the uniform adaptive quantizer to the nonuniform case is relatively straightforward, and we leave that as a practice exercise (see Problem 8 at the end

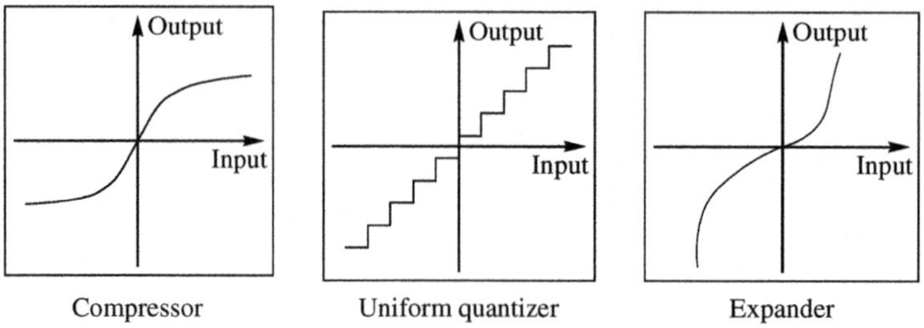

FIGURE 9.22 Block diagram for companded quantization.

of this chapter). A somewhat different approach is to use a nonlinear mapping to flatten the performance curve shown in Figure 9.21. In order to study this approach, we need to view the nonuniform quantizer in a slightly different manner.

9.6.2 Companded Quantization

Instead of making the step size small, we could make the interval in which the input lies with high probability large—that is, expand the region in which the input lands in proportion to the probability with which it lands there. This is the idea behind companded quantization. This quantization approach can be represented by the block diagram shown in Figure 9.22. The input is first mapped through a *compressor* function. This function "stretches" the high-probability regions close to the origin and correspondingly "compresses" the low-probability regions away from the origin. Thus, regions close to the origin in the input to the compressor occupy a greater fraction of the total region covered by the compressor. If the output of the compressor function is quantized using a uniform quantizer and the quantized value is transformed via an *expander* function, the overall effect is the same as using a nonuniform quantizer. To see this, we devise a simple compander and see how the process functions.

Example 9.6.1:

Suppose we have a source that can be modeled as a random variable taking values in the interval $[-4, 4]$ with more probability mass near the origin than away from it. We want to quantize this using the quantizer of Figure 9.3. Let us try to flatten out this distribution using the following compander and then compare the companded quantization with straightforward uniform quantization. The compressor characteristic we will use is given by the following equation:

$$c(x) = \begin{cases} 2x & \text{if } -1 \leqslant x \leqslant 1 \\ \frac{2x}{3} + \frac{4}{3} & x > 1 \\ \frac{2x}{3} - \frac{4}{3} & x < -1 \end{cases} \tag{35}$$

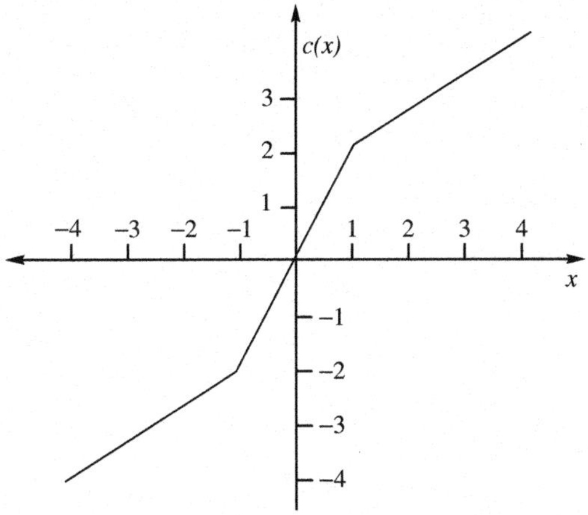

FIGURE 9.23 **Compressor mapping.**

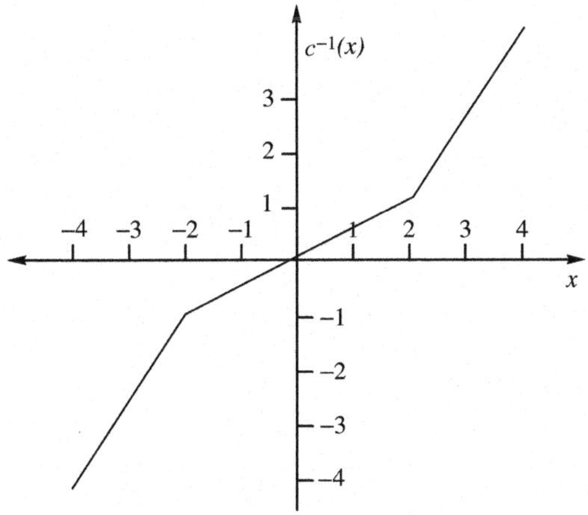

FIGURE 9.24 **Expander mapping.**

The mapping is shown graphically in Figure 9.23. The inverse mapping is given by

$$c^{-1}(x) = \begin{cases} \frac{x}{2} & \text{if } -2 \leqslant x \leqslant 2 \\ \frac{3x}{2} - 2 & x > 2 \\ \frac{3x}{2} + 2 & x < -2 \end{cases} \tag{36}$$

The inverse mapping is shown graphically in Figure 9.24.

Let's see how using these mappings affects the quantization error both near and far from the origin. Suppose we had an input of 0.9. If we quantize directly with the uniform quantizer, we get an output of 0.5, resulting in a quantization error of 0.4. If we use the companded quantizer, we first use the compressor mapping, mapping the input value of 0.9 to 1.8. Quantizing this

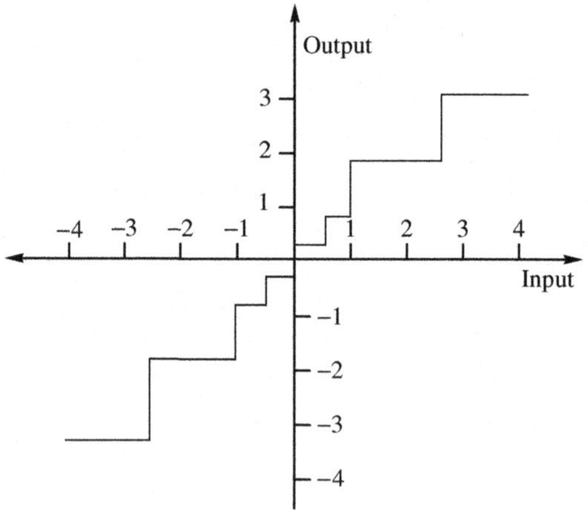

FIGURE 9.25 **Nonuniform companded quantizer.**

with the same uniform quantizer results in an output of 1.5, with an apparent error of 0.3. The expander then maps this to the final reconstruction value of 0.75, which is 0.15 away from the input. Comparing 0.15 with 0.4, we can see that relative to the input we get a substantial reduction in the quantization error. In fact, we will not get any increase in the quantization error for all values in the interval $[-1, 1]$, and for most values, we will get a decrease in the quantization error (see Problem 6 at the end of this chapter). Of course, this will not be true for values outside the $[-1, 1]$ interval. Suppose we have an input of 2.7. If we quantized this directly with the uniform quantizer, we would get an output of 2.5 with a corresponding error of 0.2. Applying the compressor mapping, the value of 2.7 would be mapped to 3.13 resulting in a quantized value of 3.5. Mapping this back through the expander, we get a reconstructed value of 3.25, which differs from the input by 0.55.

As we can see, the companded quantizer effectively works like a nonuniform quantizer with smaller quantization intervals in the interval $[-1, 1]$ and larger quantization intervals outside this interval. What is the effective input-output map of this quantizer? Notice that all inputs in the interval $[0, 0.5]$ get mapped into the interval $[0, 1]$, for which the quantizer output is 0.5, which in turn corresponds to the reconstruction value of 0.25. Essentially, all values in the interval $[0, 0.5]$ are represented by the value 0.25. Similarly, all values in the interval $[0.5, 1]$ are represented by the value 0.75 and so on. The effective quantizer input-output map is shown in Figure 9.25. ♦

If we bound the source output by some value $x_{\max}$, any nonuniform quantizer can always be represented as a companding quantizer. Let us see how we can use this fact to come up with quantizers that are robust to mismatch. First we need to look at some of the properties of high-rate quantizers, or quantizers with a large number of levels.

Define

$$\Delta_k = b_k - b_{k-1} \tag{37}$$

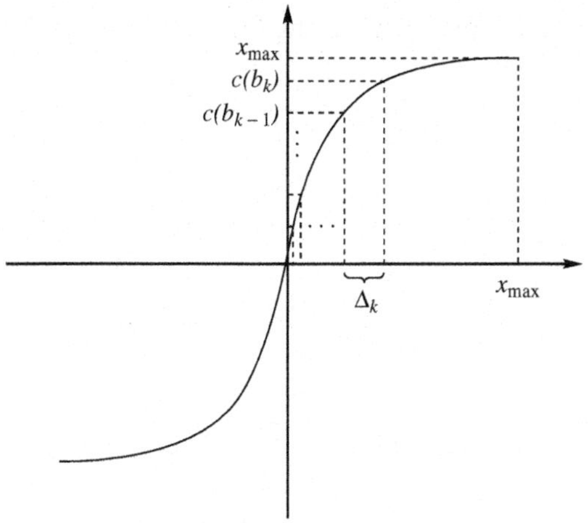

FIGURE 9.26 **A compressor function.**

If the number of levels is high, then the size of each quantization interval will be small, and we can assume that the *pdf* of the input $f_X(x)$ is essentially constant in each quantization interval. Then

$$f_X(x) = f_X(y_k) \quad \text{if} \quad b_{k-1} \leqslant x < b_k \tag{38}$$

Using this, we can rewrite Equation (3) as

$$\sigma_q^2 = \sum_{i=1}^{M} f_X(y_i) \int_{b_{i-1}}^{b_i} (x - y_i)^2 dx \tag{39}$$

$$= \frac{1}{12} \sum_{i=1}^{M} f_X(y_i) \Delta_i^3 \tag{40}$$

where we have used

$$y_i = \frac{b_i + b_{i-1}}{2}$$

Armed with this result, let us return to companded quantization. Let $c(x)$ be a companding characteristic for a symmetric quantizer, and let $c'(x)$ be the derivative of the compressor characteristic with respect to x. If the rate of the quantizer is high, that is, if there are a large number of levels, then within the kth interval, the compressor characteristic can be approximated by a straight line segment (see Figure 9.26). We can write

$$c'(y_k) = \frac{c(b_k) - c(b_{k-1})}{\Delta_k} \tag{41}$$

From Figure 9.26, we can also see that $c(b_k) - c(b_{k-1})$ is the step size of a uniform M-level quantizer. Therefore,

$$c(b_k) - c(b_{k-1}) = \frac{2x_{max}}{M} \tag{42}$$

Substituting this into Equation (41) and solving for Δ_k, we get

$$\Delta_k = \frac{2x_{\max}}{Mc'(y_k)} \tag{43}$$

Finally, substituting this expression for Δ_k into Equation (40), we get the following relationship between the quantizer distortion, the *pdf* of the input, and the compressor characteristic:

$$\begin{aligned}\sigma_q^2 &= \frac{1}{12}\sum_{i=1}^{M} f_X(y_i)\left(\frac{2x_{\max}}{Mc'(y_i)}\right)^3 \\ &= \frac{x_{\max}^2}{3M^2}\sum_{i=1}^{M}\frac{f_X(y_i)}{c'^2(y_i)}\cdot\frac{2x_{\max}}{Mc'(y_i)} \\ &= \frac{x_{\max}^2}{3M^2}\sum_{i=1}^{M}\frac{f_X(y_i)}{c'^2(y_i)}\Delta_i \end{aligned}\tag{44}$$

which for small Δ_i can be written as

$$\sigma_q^2 = \frac{x_{\max}^2}{3M^2}\int_{-x_{\max}}^{x_{\max}}\frac{f_X(x)}{(c'(x))^2}dx \tag{45}$$

This is a famous result, known as the Bennett integral after its discoverer, W.R. Bennett [130], and it has been widely used to analyze quantizers. We can see from this integral that the quantizer distortion is dependent on the *pdf* of the source sequence. However, it also tells us how to get rid of this dependence. Assume

$$c'(x) = \frac{x_{\max}}{\alpha\,|x|} \tag{46}$$

where α is a constant. From the Bennett integral we get

$$\sigma_q^2 = \frac{x_{\max}^2}{3M^2}\frac{\alpha^2}{x_{\max}^2}\int_{-x_{\max}}^{x_{\max}} x^2 f_X(x)dx \tag{47}$$

$$= \frac{\alpha^2}{3M^2}\sigma_x^2 \tag{48}$$

where

$$\sigma_x^2 = \int_{-x_{\max}}^{x_{\max}} x^2 f_X(x)dx \tag{49}$$

Substituting the expression for σ_q^2 into the expression for SNR, we get

$$\text{SNR} = 10\log_{10}\frac{\sigma_x^2}{\sigma_q^2} \tag{50}$$

$$= 10\log_{10}(3M^2) - 20\log_{10}\alpha \tag{51}$$

which is independent of the input *pdf*. This means that if we use a compressor characteristic whose derivative satisfies Equation (46), the signal-to-noise ratio will remain constant regardless of the input variance. This is an impressive result. However, we do need some caveats.

Notice that we are not saying that the mean squared quantization error is independent of the quantizer input. It is not, as is clear from Equation (48). Remember also that this result is valid as long as the underlying assumptions are valid. When the input variance is very small, our assumption about the *pdf* being constant over the quantization interval is no longer valid, and when the variance of the input is very large, our assumption about the input being bounded by x_{max} may no longer hold.

With fair warning, let us look at the resulting compressor characteristic. We can obtain the compressor characteristic by integrating Equation (46):

$$c(x) = x_{\max} + \beta \log \frac{|x|}{x_{\max}} \tag{52}$$

where β is a constant. The only problem with this compressor characteristic is that it becomes very large for small x. Therefore, in practice we approximate this characteristic with a function that is linear around the origin and logarithmic away from it.

Two companding characteristics that are widely used today are μ-law companding and A-law companding. The μ-law compressor function is given by

$$c(x) = x_{\max} \frac{\ln\left(1 + \mu \frac{|x|}{x_{\max}}\right)}{\ln(1+\mu)} \operatorname{sgn}(x) \tag{53}$$

The expander function is given by

$$c^{-1}(x) = \frac{x_{\max}}{\mu}\left[(1+\mu)^{\frac{|x|}{x_{\max}}} - 1\right]\operatorname{sgn}(x) \tag{54}$$

This companding characteristic with $\mu = 255$ is used in the telephone systems in North America and Japan. The rest of the world uses the A-law characteristic, which is given by

$$c(x) = \begin{cases} \dfrac{A\,|x|}{1+\ln A}\operatorname{sgn}(x) & 0 \leqslant \frac{|x|}{x_{\max}} \leqslant \frac{1}{A} \\[2ex] x_{\max} \dfrac{1+\ln \frac{A\,|x|}{x_{\max}}}{1+\ln A}\operatorname{sgn}(x) & \dfrac{1}{A} \leqslant \dfrac{|x|}{x_{\max}} \leqslant 1 \end{cases} \tag{55}$$

and

$$c^{-1}(x) = \begin{cases} \frac{|x|}{A}(1+\ln A) & 0 \leqslant \frac{|x|}{x_{\max}} \leqslant \frac{1}{1+\ln A} \\[2ex] \frac{x_{\max}}{A} \exp\left[\frac{|x|}{x_{\max}}(1+\ln A) - 1\right] & \frac{1}{1+\ln A} \leqslant \frac{|x|}{x_{\max}} \leqslant 1 \end{cases} \tag{56}$$

9.7 Entropy-Coded Quantization

In Section 9.3 we mentioned three tasks: selection of boundary values, selection of reconstruction levels, and selection of codewords. Up to this point we have talked about accomplishment of the first two tasks, with the performance measure being the mean squared quantization error. In this section we will look at accomplishing the third task, assigning codewords to the quantization interval. Recall that this becomes an issue when we use variable-length codes. In this section we will be looking at the latter situation, with the rate being the performance measure.

We can take two approaches to the variable-length coding of quantizer outputs. We can redesign the quantizer by taking into account the fact that the selection of the decision boundaries will affect the rate, or we can keep the design of the quantizer the same (i.e., Lloyd-Max quantization) and simply entropy-code the quantizer output. Since the latter approach is by far the simpler one, let's look at it first.

9.7.1 Entropy Coding of Lloyd-Max Quantizer Outputs

The process of trying to find the optimum quantizer for a given number of levels and rate is a rather difficult task. An easier approach to incorporating entropy coding is to design a quantizer that minimizes the *msqe*, that is, a Lloyd-Max quantizer, then entropy-code its output.

In Table 9.7 we list the output entropies of uniform and nonuniform Lloyd-Max quantizers. Notice that while the difference in rate for lower levels is relatively small, for a larger number of levels, there can be a substantial difference between the fixed-rate and entropy-coded cases. For example, for 32 levels a fixed-rate quantizer would require 5 bits per sample. However, the entropy of a 32-level uniform quantizer for the Laplacian case is 3.779 bits per sample, which is more than 1 bit less. Notice that the difference between the fixed rate and the uniform quantizer entropy is generally greater than the difference between the fixed rate and the entropy of the output of the nonuniform quantizer. This is because the nonuniform quantizers have smaller step sizes in high-probability regions and larger step sizes in low-probability regions. This brings the probability of an input falling into a low-probability region and the probability of an input falling in a high-probability region closer together. This, in turn, raises the output entropy of the nonuniform quantizer with respect to the uniform quantizer. Finally, the closer the distribution is to being uniform, the less difference in the rates. Thus, the difference in rates is much less for the quantizer for the Gaussian source than the quantizer for the Laplacian source.

TABLE 9.7 Output entropies in bits per sample for minimum mean squared error quantizers.

Number of Levels	Gaussian		Laplacian	
	Uniform	Nonuniform	Uniform	Nonuniform
4	1.904	1.911	1.751	1.728
6	2.409	2.442	2.127	2.207
8	2.759	2.824	2.394	2.479
16	3.602	3.765	3.063	3.473
32	4.449	4.730	3.779	4.427

9.7.2 Entropy-Constrained Quantization ★

Although entropy coding the Lloyd-Max quantizer output is certainly simple, it is easy to see that we could probably do better if we take a fresh look at the problem of quantizer design, this time with the entropy as a measure of rate rather than the alphabet size. The entropy of the quantizer output is given by

$$H(Q) = -\sum_{i=1}^{M} P_i \log_2 P_i \tag{57}$$

where P_i is the probability of the input to the quantizer falling in the ith quantization interval and is given by

$$P_i = \int_{b_{i-1}}^{b_i} f_X(x)dx \tag{58}$$

Notice that the selection of the representation values $\{y_j\}$ has no effect on the rate. This means that we can select the representation values solely to minimize the distortion. However, the selection of the boundary values affects both the rate and the distortion. Initially, we found the reconstruction levels and decision boundaries that minimized the distortion, while keeping the rate fixed by fixing the quantizer alphabet size and assuming fixed-rate coding. In an analogous fashion, we can now keep the entropy fixed and try to minimize the distortion. Or, more formally:

> For a given R_o, find the decision boundaries $\{b_j\}$ that minimize σ_q^2 given by Equation (3), subject to $H(Q) \leqslant R_o$.

The solution to this problem involves the solution of the following $M - 1$ nonlinear equations [131]:

$$\ln \frac{P_{l+1}}{P_l} = \lambda(y_{k+1} - y_k)(y_{k+1} + y_k - 2b_k) \tag{59}$$

where λ is adjusted to obtain the desired rate, and the reconstruction levels are obtained using Equation (27). A generalization of the method used to obtain the minimum mean squared error quantizers can be used to obtain solutions for this equation [132]. The process of finding optimum entropy-constrained quantizers looks complex. Fortunately, at higher rates we can show that the optimal quantizer is a uniform quantizer, simplifying the problem. Furthermore, while these results are derived for the high-rate case, it has been shown that the results also hold for lower rates [132].

9.7.3 High-Rate Optimum Quantization ★

At high rates, the design of optimum quantizers becomes simple, at least in theory. Gish and Pierce's work [133] says that at high rates the optimum entropy-coded quantizer is a uniform quantizer. Recall that any nonuniform quantizer can be represented by a compander and a uniform quantizer. Let us try to find the optimum compressor function at high rates that minimizes the entropy for a given distortion. Using the calculus of variations approach, we will construct the functional

$$J = H(Q) + \lambda\sigma_q^2 \tag{60}$$

then find the compressor characteristic to minimize it.

For the distortion σ_q^2, we will use the Bennett integral shown in Equation (45). The quantizer entropy is given by Equation (57). For high rates, we can assume (as we did before) that the *pdf* $f_X(x)$ is constant over each quantization interval Δ_i, and we can replace Equation (58) by

$$P_i = f_X(y_i)\Delta_i \tag{61}$$

Substituting this into Equation (57), we get

$$\begin{aligned} H(Q) &= -\sum f_X(y_i)\Delta_i \log[f_X(y_i)\Delta_i] & (62)\\ &= -\sum f_X(y_i)\log[f_X(y_i)]\Delta_i - \sum f_X(y_i)\log[\Delta_i]\Delta_i & (63)\\ &= -\sum f_X(y_i)\log[f_X(y_i)]\Delta_i - \sum f_X(y_i)\log\frac{2x_{\max}/M}{c'(y_i)}\Delta_i & (64) \end{aligned}$$

where we have used Equation (43) for Δ_i. For small Δ_i we can write this as

$$\begin{aligned} H(Q) &= -\int f_X(x)\log f_X(x)dx - \int f_X(x)\log\frac{2x_{\max}/M}{c'(x)}dx & (65)\\ &= -\int f_X(x)\log f_X(x)dx - \log\frac{2x_{\max}}{M} + \int f_X(x)\log c'(x)dx & (66) \end{aligned}$$

where the first term is the differential entropy of the source $h(X)$. Let's define $g = c'(x)$. Then substituting the value of $H(Q)$ into Equation (60) and differentiating with respect to g, we get

$$\int f_X(x)[g^{-1} - 2\lambda\frac{x_{\max}^2}{3M^2}g^{-3}]dx = 0 \tag{67}$$

This equation is satisfied if the integrand is zero, which gives us

$$g = \sqrt{\frac{2\lambda}{3}}\frac{x_{\max}}{M} = K\ (constant) \tag{68}$$

Therefore,

$$c'(x) = K \tag{69}$$

and

$$c(x) = Kx + \alpha \tag{70}$$

If we now use the boundary conditions $c(0) = 0$ and $c(x_{\max}) = x_{\max}$, we get $c(x) = x$, which is the compressor characteristic for a uniform quantizer. Thus, at high rates the optimum quantizer is a uniform quantizer.

Substituting this expression for the optimum compressor function in the Bennett integral, we get an expression for the distortion for the optimum quantizer:

$$\sigma_q^2 = \frac{x_{\max}^2}{3M^2} \tag{71}$$

Substituting the expression for $c(x)$ in Equation (66), we get the expression for the entropy of the optimum quantizer:

$$H(Q) = h(X) - \log \frac{2x_{\max}}{M} \tag{72}$$

Note that while this result provides us with an easy method for designing optimum quantizers, our derivation is only valid if the source *pdf* is entirely contained in the interval $[-x_{\max}, x_{\max}]$, and if the step size is small enough that we can reasonably assume the *pdf* to be constant over a quantization interval. Generally, these conditions can only be satisfied if we have an extremely large number of quantization intervals. While theoretically this is not much of a problem, most of these reconstruction levels will be rarely used. In practice, as mentioned in Chapter 3, entropy-coding a source with a large output alphabet is very problematic. One way we can get around this is through the use of a technique called *recursive indexing*.

Recursive indexing is a mapping of a countable set to a collection of sequences of symbols from another set with finite size [85]. Given a countable set $A = \{a_0, a_1, \ldots\}$ and a finite set $B = \{b_0, b_1, \ldots, b_M\}$ of size $M + 1$, we can represent any element in A by a sequence of elements in B in the following manner:

1. Take the index i of element a_i of A.
2. Find the quotient m and remainder r of the index i such that

$$i = mM + r$$

3. Generate the sequence $\underbrace{b_M b_M \cdots b_M}_{m \text{ times}} b_r$

B is called the representation set. We can see that given any element in A we will have a unique sequence from B representing it. Furthermore, no representative sequence is a prefix of any other sequence. Therefore, recursive indexing can be viewed as a trivial, uniquely decodable prefix code. The inverse mapping is given by

$$\underbrace{b_M b_M \cdots b_M}_{m \text{ times}} b_r \mapsto a_{mM+r}$$

Since the mapping is one-to-one, if it is used to convert the index sequence of the quantizer output into the sequence of the recursive indices, the former can be recovered without error from the latter. Furthermore, when the size $M + 1$ of the representation set B is chosen appropriately, in effect we can achieve a reduction in the size of the output alphabets that are used for entropy coding.

Example 9.7.1:

Suppose we want to represent the set of nonnegative integers $A = \{0, 1, 2, \ldots\}$ with the representation set $B = \{0, 1, 2, 3, 4, 5\}$. Then the value 12 would be represented by the sequence 5, 5, 2, and the value 16 would be represented by the sequence 5, 5, 5, 1. Whenever the decoder sees the value 5, it simply adds on the next value until the next value is smaller than 5. For example, the sequence 3, 5, 1, 2, 5, 5, 1, 5, 0 would be decoded as 3, 6, 2, 11, 5.♦

Recursive indexing is applicable to any representation of a large set by a small set. One way of applying recursive indexing to the problem of quantization is as follows: For a given step size $\Delta > 0$ and a positive integer K, define x_l and x_h as follows:

$$x_l = -\left\lfloor \frac{K-1}{2} \right\rfloor \Delta$$
$$x_h = x_l + (K-1)\Delta$$

where $\lfloor x \rfloor$ is the largest integer not exceeding x. We define a recursively indexed quantizer of size K to be a uniform quantizer with step size Δ and with x_l and x_h being its smallest and largest output levels. (Q defined this way also has 0 as its output level.) The quantization rule Q, for a given input value x, is as follows:

1. If x falls in the interval $(x_l + \frac{\Delta}{2}, x_h - \frac{\Delta}{2})$, then $Q(x)$ is the nearest output level.
2. If x is greater than $x_h - \frac{\Delta}{2}$, see if $x_1 = x - x_h \in (x_l + \frac{\Delta}{2}, x_h - \frac{\Delta}{2})$. If so, $Q(x) = (x_h, Q(x_1))$. If not, form $x_2 = x - 2x_h$ and do the same as for x_1. This process continues until for some m, $x_m = x - mx_h$ falls in $(x_l + \frac{\Delta}{2}, x_h - \frac{\Delta}{2})$, which will be quantized into

$$Q(x) = (\underbrace{x_h, x_h, \ldots, x_h}_{m \text{ times}}, Q(x_m)) \tag{73}$$

3. If x is smaller than $x_l + \frac{\Delta}{2}$, a similar procedure to the above is used; that is, form $x_m = x + mx_l$ so that it falls in $(x_l + \frac{\Delta}{2}, x_h - \frac{\Delta}{2})$, and quantize it to $(x_l, x_l, \ldots, x_l, Q(x_m))$.

In summary, the quantizer operates in two modes: one when the input falls in the range (x_l, x_h), the other when it falls outside of the specified range. The recursive nature of the second mode gives it the name.

We pay for the advantage of encoding a larger set by a smaller set in several ways. If we get a large input to our quantizer, the representation sequence may end up being intolerably large. We also get an increase in the rate. If $H(Q)$ is the entropy of the quantizer output, and γ is the average number of representation symbols per input symbol, then the minimum rate for the recursively indexed quantizer is $\gamma H(Q)$.

In practice, neither cost is too large. We can avoid the problem of intolerably large sequences by adopting some simple strategies for representing these sequences, and the value of γ is quite close to one for reasonable values of M. For Laplacian and Gaussian quantizers, a typical value for M would be 15 [85].

9.8 Summary

Quantization is a well-researched area and much is known about the subject. In this chapter, we looked at the design and performance of uniform and nonuniform quantizers for a variety of sources, and how the performance is affected when the assumptions used in the design process are not correct. When the source statistics are not well known or change with time, we can use an adaptive strategy. One of the more popular approaches to adaptive quantization is the Jayant quantizer. We also looked at the issues involved with entropy-coded quantization.

Further Reading

With an area as broad as quantization, we had to keep some of the coverage rather cursory. However, there is a wealth of information on quantization available in the published literature. The following sources are especially useful for a general understanding of the area:

1. A very thorough coverage of quantization can be found in *Digital Coding of Waveforms*, by N.S. Jayant and P. Noll [134].
2. The paper "Quantization," by A. Gersho, in *IEEE Communication Magazine*, September 1977 [127], provides excellent tutorial coverage of many of the topics listed here.
3. The original paper by J. Max, "Quantization for Minimum Distortion," *IRE Transactions on Information Theory* [122], contains a very accessible description of the design of *pdf*-optimized quantizers.
4. A thorough study of the effects of mismatch is provided by W. Mauersberger in [135].

9.9 Projects and Problems

1. Show that the derivative of the distortion expression in Equation (18) results in the expression in Equation (19). You will have to use a result called Leibnitz's rule and the idea of a telescoping series. Leibnitz's rule states that if $a(t)$ and $b(t)$ are monotonic, then

$$\frac{\delta}{\delta t}\int_{a(t)}^{b(t)} f(x,t)dx = \int_{a(t)}^{b(t)} \frac{\delta f(x,t)}{\delta t}dx + f(b(t),t)\frac{\delta b(t)}{\delta t} - f(a(t),t)\frac{\delta a(t)}{\delta t} \quad (74)$$

2. Use the program `falspos` to solve Equation (19) numerically for the Gaussian and Laplacian distributions. You may have to modify the function `func` in order to do this.
3. Design a 3-bit uniform quantizer (specify the decision boundaries and representation levels) for a source with a Laplacian *pdf*, with a mean of 3 and a variance of 4.
4. The pixel values in the Sena image are not really distributed uniformly. Obtain a histogram of the image (you can use the `hist_image` routine), and using the fact that the quantized image should be as good an approximation as possible for the original, design 1-, 2-, and 3-bit quantizers for this image. Compare these with the results displayed in Figure 9.7. (For better comparison, you can reproduce the results in the book using the program `uquan_img`.)
5. Use the program `misuquan` to study the effect of mismatch between the input and assumed variances. How do these effects change with the quantizer alphabet size and the distribution type?
6. For the companding quantizer of Example 9.6.1, what are the outputs for the following inputs: $-0.8, 1.2, 0.5, 0.6, 3.2, -0.3$? Compare your results with the case when the input is directly quantized with a uniform quantizer with the same number of levels. Comment on your results.

7. Use the test images Sena and Bookshelf1 to study the trade-offs involved in the selection of block sizes in the forward adaptive quantization scheme described in Example 9.5.2. Compare this with a more traditional forward adaptive scheme in which the variance is estimated and transmitted. The variance information should be transmitted using a uniform quantizer with differing number of bits.

8. Generalize the Jayant quantizer to the nonuniform case. Assume that the input is from a known distribution with unknown variance. Simulate the performance of this quantizer over the same range of ratio of variances as we have done for the uniform case. Compare your results to the fixed nonuniform quantizer and the adaptive uniform quantizer. To get a start on your program, you may wish to use `misnuq.c` and `juquan.c`.

9. Let's look at the rate distortion performance of the various quantizers.

(a) Plot the rate-distortion function $R(D)$ for a Gaussian source with mean zero and variance $\sigma_X^2 = 2$.

(b) Assuming fixed length codewords, compute the rate and distortion for 1-, 2-, and 3-bit *pdf*-optimized nonuniform quantizers. Also, assume that X is a Gaussian random variable with mean zero and $\sigma_X^2 = 2$. Plot these values on the same graph with **x**'s.

(c) For the 2- and 3-bit quantizers, compute the rate and distortion assuming that the quantizer outputs are entropy coded. Plot these on the graph with **o**'s.

10

Vector Quantization

10.1 Overview

By grouping source outputs together and encoding them as a single block, we can obtain efficient lossy as well as lossless compression algorithms. Many of the lossless compression algorithms that we looked at took advantage of this fact. We can do the same with quantization. In this chapter, several quantization techniques that operate on blocks of data are described. We can view these blocks as vectors, hence the name "vector quantization." We will describe several different approaches to vector quantization. We will explore how to design vector quantizers and how these quantizers can be used for compression.

10.2 Introduction

In the last chapter, we looked at different ways of quantizing the output of a source. In all cases the quantizer inputs were scalar values, and each quantizer codeword represented a single sample of the source output. In Chapter 2 we saw that, by taking longer and longer sequences of input samples, it is possible to extract the structure in the source coder output. In Chapter 4 we saw that, even when the input is random, encoding sequences of samples instead of encoding individual samples separately provides a more efficient code. Encoding sequences of samples is more advantageous in the lossy compression framework as well. By "advantageous" we mean a lower distortion for a given rate, or a lower rate for a given distortion. As in the previous chapter, by "rate" we mean the average number of bits per input sample, and the measures of distortion will generally be the mean squared error and the signal-to-noise ratio.

The idea that encoding sequences of outputs can provide an advantage over the encoding of individual samples was first put forward by Shannon, and the basic results in information theory were all proved by taking longer and longer sequences of inputs. This indicates that a quantization strategy that works with sequences or blocks of output would provide some improvement in performance over scalar quantization. In other words, we wish to generate a representative set of sequences. Given a source output sequence, we would represent it with one of the elements of the representative set.

In vector quantization we group the source output into blocks or vectors. For example, we can treat L consecutive samples of speech as the components of an L-dimensional vector. Or, we can take a block of L pixels from an image and treat each pixel value as a component of a vector of size or dimension L. This vector of source outputs forms the input to the vector quantizer. At both the encoder and decoder of the vector quantizer, we have a set of L-dimensional vectors called the *codebook* of the vector quantizer. The vectors in this codebook, known as *code-vectors*, are selected to be representative of the vectors we generate from the source output. Each code-vector is assigned a binary index. At the encoder, the input vector is compared to each code-vector in order to find the code-vector closest to the input vector. The elements of this code-vector are the quantized values of the source output. In order to inform the decoder about which code-vector was found to be the closest to the input vector, we transmit or store the binary index of the code-vector. Because the decoder has exactly the same codebook, it can retrieve the code-vector given its binary index. A pictorial representation of this process is shown in Figure 10.1.

Although the encoder may have to perform a considerable amount of computations in order to find the closest reproduction vector to the vector of source outputs, the decoding consists of a table lookup. This makes vector quantization a very attractive encoding scheme for applications in which the resources available for decoding are considerably less than the resources available for encoding. For example, in multimedia applications, considerable computational resources

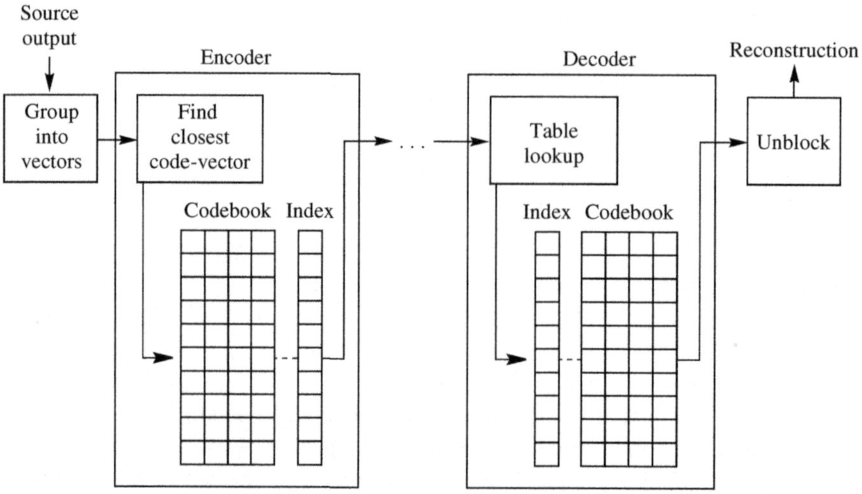

FIGURE 10.1 The vector quantization procedure.

may be available for the encoding operation. However, if the decoding is to be done in software, the amount of computational resources available to the decoder may be quite limited.

The subject of vector quantization is a broad one. In this chapter we will try to introduce you to as much of this fascinating area as we can. If your appetite is whetted by what is available here and you wish to explore further, there is an excellent book by Gersho and Gray [136] devoted to the subject of vector quantization.

Our approach in this chapter is as follows: First, we try to answer the question of why we would want to use vector quantization over scalar quantization. There are several answers to this question, each illustrated through examples. In our discussion, we assume that you are familiar with the material in Chapter 9. We will then turn to one of the most important elements in the design of a vector quantizer, the generation of the codebook. While there are a number of ways of obtaining the vector quantizer codebook, most of them are based on one particular approach, popularly known as the Linde-Buzo-Gray (LBG) algorithm. We devote a considerable amount of time to describing some of the details of this algorithm. Our intent here is to provide you with enough information so that you can write your own programs for design of vector quantizer codebooks. In the software accompanying this book, we have also included programs for designing codebooks that are based on the descriptions in this chapter. If you are not currently thinking of implementing vector quantization routines, you may wish to skip these sections (Sections 10.4.1 and 10.4.2). We follow our discussion of the LBG algorithm with some examples of image compression using codebooks designed with this algorithm, and then with a brief sampling of the many different kinds of vector quantizers. Finally, we describe another quantization strategy, called trellis-coded quantization (TCQ), which, though different in implementation from vector quantizers, also makes use of the advantage to be gained from operating on sequences.

Before we begin our discussion of vector quantization, let us define some of the terminology we will be using. The amount of compression will be described in terms of the rate, which will be measured in bits per sample. Suppose we have a codebook of size K, and the input vector is of dimension L. In order to inform the decoder of which code-vector was selected, we need to use $\lceil \log_2 K \rceil$ bits. For example, if the codebook contained 256 code-vectors, we would need 8 bits to specify which of the 256 code-vectors had been selected at the encoder. Thus, the number of bits *per vector* is $\lceil \log_2 K \rceil$ bits. As each code-vector contains the reconstruction values for L source output samples, the number of bits *per sample* would be $\frac{\lceil \log_2 K \rceil}{L}$. Thus, the rate for an L-dimensional vector quantizer with a codebook of size K is $\frac{\lceil \log_2 K \rceil}{L}$. As our measure of distortion we will use the mean squared error. When we say that in a codebook $\mathcal{C}$, containing the K code-vectors $\{\mathbf{y}_i\}$, the input vector $\mathbf{x}$ is closest to $\mathbf{y}_j$, we will mean that

$$\left\|\mathbf{x} - \mathbf{y}_j\right\|^2 \leqslant \|\mathbf{x} - \mathbf{y}_i\|^2 \quad \text{for all } \mathbf{y}_i \in \mathcal{C} \tag{1}$$

where $\mathbf{x} = (x_1 x_2 \cdots x_L)$ and

$$\|\mathbf{x}\|^2 = \sum_{i=1}^{L} x_i^2 \tag{2}$$

The term *sample* will always refer to a scalar value. Thus, when we are discussing compression of images, a sample refers to a single pixel. The output points of the quantizer are often referred to as *levels*. Thus, when we wish to refer to a quantizer with K output points or code-vectors, we may refer to it as a K-level quantizer. Finally, whenever we measure the performance of the quantizer by computing the distortion, it will always be on a per-sample basis.

10.3 Advantages of Vector Quantization over Scalar Quantization

For a given rate (in bits per sample), use of vector quantization results in a lower distortion than when scalar quantization is used at the same rate, for several reasons. In this section we will explore these reasons with examples (for a more theoretical explanation, see [38,112,118]). Note that whenever we compare performance it is always on a per-sample basis rather than on a per-vector basis.

If the source output is correlated, vectors of source output values will tend to fall in clusters. By selecting the quantizer output points to lie in these clusters, we have a more accurate representation of the source output. Consider the following example.

Example 10.3.1:

In Example 8.5.1, we introduced a source that generates the height and weight of individuals. Suppose the height of these individuals varies uniformly between 40 and 80 inches, and the weight varies uniformly between 40 and 240 pounds. Suppose we are allowed a total of 6 bits to represent each pair of values. We can use 3 bits to quantize the height and 3 bits to quantize the weight. Thus, the weight range between 40 and 240 pounds will be divided into eight intervals of equal width of 25, with reconstruction values $\{52, 77, \ldots, 227\}$. Similarly, the height range between 40 and 80 inches can be divided into eight intervals of width 5, with reconstruction levels $\{42, 47, \ldots, 77\}$. When we look at the representation of height and weight separately, this approach seems reasonable. But let's look at this quantization scheme in two dimensions. We will plot the height values along the x-axis and the weight values along the y-axis. Note that we are not changing anything in the quantization process. The height values are still being quantized to the same eight different values, as are the weight values. The two-dimensional representation of these two quantizers is shown in Figure 10.2.

From the figure we can see that we effectively have a quantizer output for a person who is 80 inches (6 feet 8 inches) tall and weighs 40 pounds, as well as a quantizer output for an individual who is 42 inches tall but weighs more than 200 pounds. Obviously, these outputs will never be used, as is the case for many of the other outputs. A more sensible approach would be to use a quantizer like the one shown in Figure 10.3, where we take account of the fact that the height and weight are correlated. This quantizer has exactly the same number of output points as the quantizer in Figure 10.2; however, the output points are clustered in the area occupied by the input. Using this quantizer, we can no longer quantize the height and weight separately. We have to consider them as the coordinates of a point in two dimensions in order to find the closest quantizer output point. However, this method provides a much finer quantization of the input.

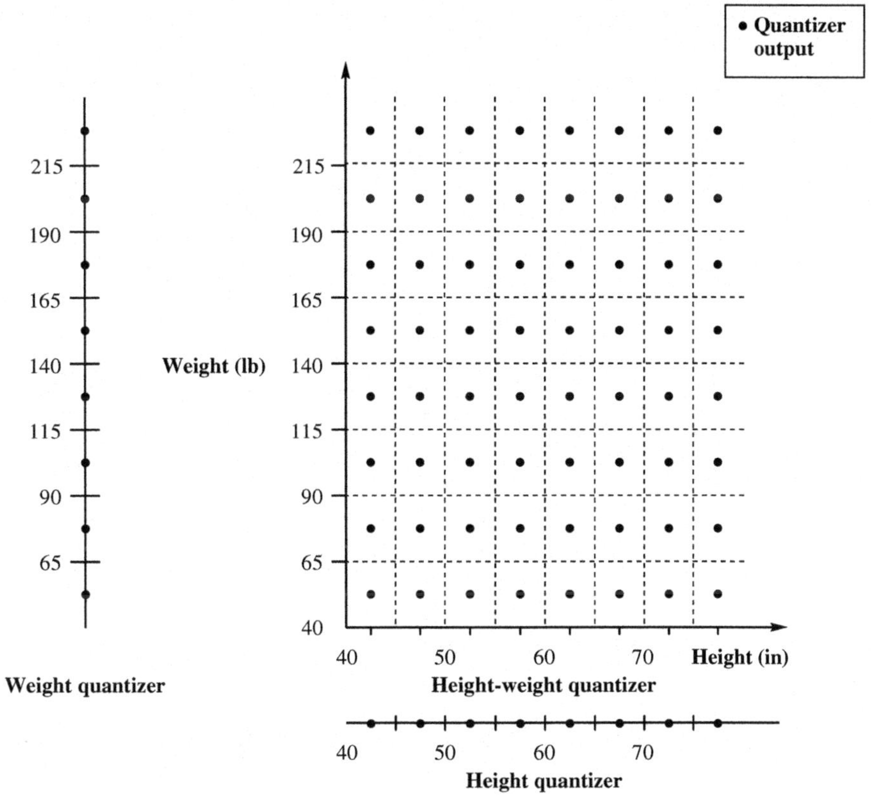

FIGURE 10.2 **The height/weight scalar quantizers when viewed in two dimensions.**

Note that we have not said how we would obtain the locations of the quantizer outputs shown in Figure 10.3. These output points make up the codebook of the vector quantizer, and we will be looking at codebook design in some detail later in this chapter. ♦

We can see from this example that, as in lossless compression, looking at longer sequences of inputs brings out the structure in the source output. This structure can then be used to provide more efficient representations.

We can easily see how structure in the form of correlation between source outputs can make it more efficient to look at sequences of source outputs rather than looking at each sample separately. However, the vector quantizer is also more efficient than the scalar quantizer when the source output values are not correlated. The reason for this is actually quite simple. As we look at longer and longer sequences of source outputs, we are afforded more flexibility in terms of our design. This flexibility in turn allows us to match the design of the quantizer to the source characteristics. Consider the following example.

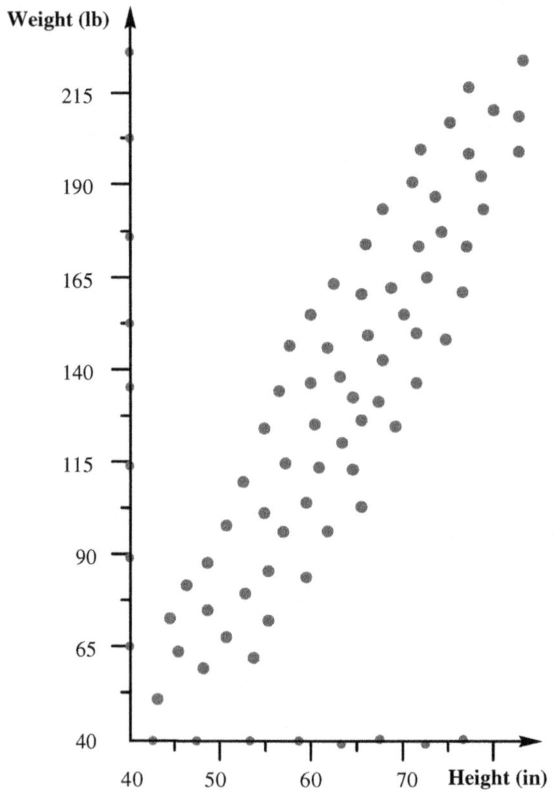

FIGURE 10.3 **The height-weight vector quantizer when height and weight are taken to be components of a vector.**

Example 10.3.2:

Suppose we have to design a uniform quantizer with eight output values for a Laplacian input. Using the information from Table 9.3 in Chapter 9, we would obtain the quantizer shown in Figure 10.4, where Δ is equal to 0.7309. As the input has a Laplacian distribution, the probability of the source output falling in the different quantization intervals is not the same. For example, the probability that the input will fall in the interval $[0, \Delta)$ is 0.3242, while the probability that a source output will fall in the interval $[3\Delta, \infty)$ is 0.0225. Let's look at how this quantizer will quantize two consecutive source outputs. As we did in the previous example, let's plot the first sample along the x-axis and the second sample along the y-axis. We can represent this two-dimensional view of the quantization process as shown in Figure 10.5. Note that, as in the previous example, we have not changed the quantization process; we are simply representing it differently. The first quantizer input, which we have represented in the figure as x_1, is quantized to the same eight possible output values as before. The same is true for the second quantizer input, which we have represented in the figure as x_2. This two-dimensional representation allows us to examine the quantization process in a slightly different manner.

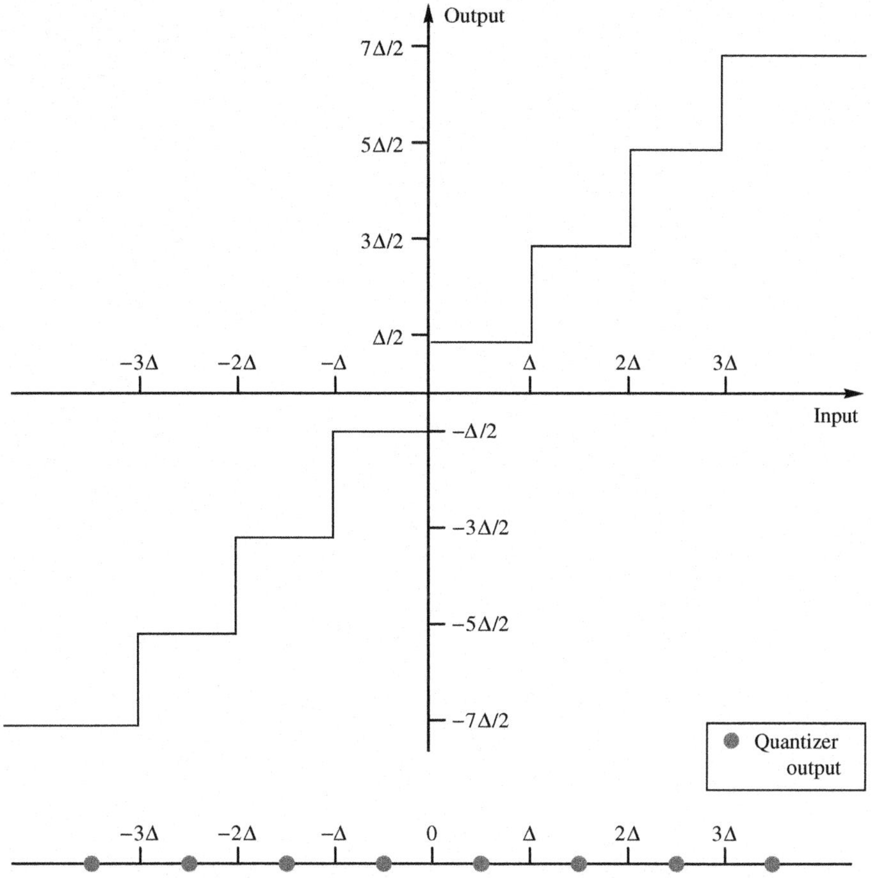

FIGURE 10.4 **Two representations of an eight-level scalar quantizer.**

Each filled-in circle in the figure represents a sequence of two quantizer outputs. For example, the top right-most circle represents the two quantizer outputs that would be obtained if we had two consecutive source outputs with a value greater than 3Δ. We computed the probability of a single source output greater than 3Δ to be 0.0225. The probability of two consecutive source outputs greater than 2.193 is simply $0.0225 \times 0.0225 = 0.0005$, which is quite small. Given that we do not use this output point very often, we could simply place it somewhere else where it would be of more use. Let us move this output point to the origin, as shown in Figure 10.6. We have now modified the quantization process. Now if we get two consecutive source outputs with values greater than 3Δ, the quantizer output corresponding to the second source output may not be the same as the first source output.

If we compare the rate distortion performance of the two vector quantizers, the SNR for the first vector quantizer is 11.44 dB, which agrees with the result in Chapter 9 for the uniform quantizer with a Laplacian input. The SNR for the modified vector quantizer, however, is 11.73 dB, an increase of about 0.3 dB. Recall that the SNR is a measure of the ratio of the average

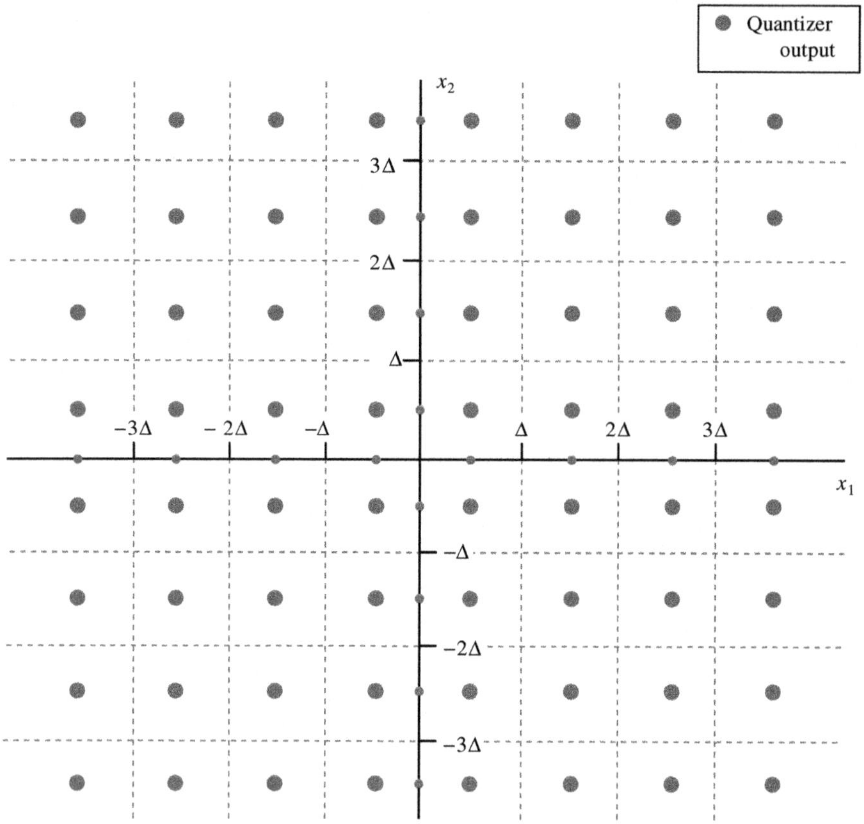

FIGURE 10.5 Input-output map for consecutive quantization of two inputs using an eight-level scalar quantizer.

squared value of the source output samples and the mean squared error. As the average squared value of the source output is the same in both cases, an increase in SNR means a decrease in the mean squared error. Whether this increase in SNR is significant will depend on the particular application. What is important here is that by treating the source output in groups of two we could effect a positive change with only a minor modification. We could argue that this modification is really not that minor since the uniform characteristic of the original quantizer has been destroyed. However, if we begin with a nonuniform quantizer and modify it in a similar way, we get similar results.

Could we do something similar with the scalar quantizer? If we move the output point at $\frac{7\Delta}{2}$ to the origin, the SNR *drops* from 11.44 dB to 10.8 dB. What is it that permits us to make modifications in the vector case, but not in the scalar case? This advantage is caused by the added flexibility we get by viewing the quantization process in higher dimensions. Consider the effect of moving the output point from $\frac{7\Delta}{2}$ to the origin in terms of two consecutive inputs. This one change in one dimension corresponds to moving 15 output points in two dimensions. Thus, modifications at the scalar quantizer level are gross modifications when viewed from

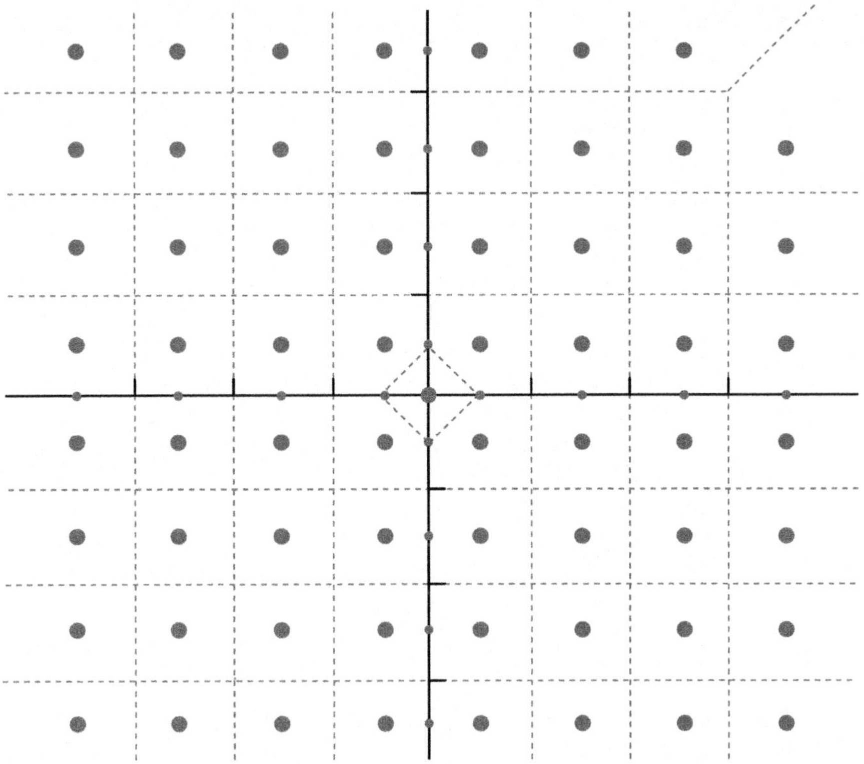

FIGURE 10.6 Modified two-dimensional vector quantizer.

the point of view of the vector quantizer. Remember that in this example we have only looked at two-dimensional vector quantizers. As we block the input into larger and larger blocks or vectors, these higher dimensions provide even greater flexibility and the promise of further gains to be made. ♦

In Figure 10.6, notice how the quantization regions have changed for the outputs around the origin, as well as for the two neighbors of the output point that were moved. The decision boundaries between the reconstruction levels can no longer be described as easily as in the case for the scalar quantizer. However, if we know the distortion measure, simply knowing the output points gives us sufficient information to implement the quantization process. Instead of defining the quantization rule in terms of the decision boundary, we can define the quantization rule as follows:

$$Q(\mathbf{x}) = \mathbf{y}_j \quad \text{iff} \quad d(\mathbf{x}, \mathbf{y}_j) < d(\mathbf{x}, \mathbf{y}_i) \ \forall i \neq j \tag{3}$$

For the case where the input $\mathbf{x}$ is equidistant from two output points, we can use a simple tie-breaking rule such as "use the output point with the smaller index." The quantization regions

V_j can then be defined as

$$V_j = \{\mathbf{x} : d(\mathbf{x}, \mathbf{y}_j) < d(\mathbf{x}, \mathbf{y}_i) \; \forall i \neq j\} \tag{4}$$

Thus, the quantizer is completely defined by the output points and a distortion measure.

From a multidimensional point of view, using a scalar quantizer for each input restricts the output points to a rectangular grid. Observing several source output values at once allows us to move the output points around. Another way of looking at this is that in one dimension the quantization intervals are restricted to be intervals, and the only parameter that we can manipulate is the size of these intervals. When we divide the input into vectors of some length n, the quantization regions are no longer restricted to be rectangles or squares. We have the freedom to divide the range of the inputs in an infinite number of ways.

These examples have shown two ways in which the vector quantizer can be used to improve performance. In the first case, we exploited the sample-to-sample dependence of the input. In the second case, there was no sample-to-sample dependence; the samples were independent. However, looking at two samples together still improved performance.

These two examples can be used to motivate two somewhat different approaches toward vector quantization. One approach is a pattern-matching approach, similar to the process used in Example 10.3.1, while the other approach deals with the quantization of random inputs. We will look at both of these approaches in this chapter.

10.4 The Linde-Buzo-Gray Algorithm

In Example 10.3.1 we saw that one way of exploiting the structure in the source output is to place the quantizer output points where the source output points (blocked into vectors) are most likely to congregate. The set of quantizer output points is called the *codebook* of the quantizer, and the process of placing these output points is often referred to as *codebook design*. When we group the source output in two-dimensional vectors, as in the case of Example 10.3.1, we might be able to obtain a good codebook design by plotting a representative set of source output points and then visually locating where the quantizer output points should be. However, this approach to codebook design breaks down when we design higher-dimensional vector quantizers. Consider designing the codebook for a 16-dimensional quantizer. Obviously, a visual placement approach will not work in this case. We need an automatic procedure for locating where the source outputs are clustered.

This is a familiar problem in the field of pattern recognition. It is no surprise, therefore, that the most popular approach to designing vector quantizers is a clustering procedure known as the k-means algorithm, which was developed for pattern recognition applications.

The k-means algorithm functions as follows: Given a large set of output vectors from the source, known as the *training set*, and an initial set of k representative patterns, assign each element of the training set to the closest representative pattern. After an element is assigned, the representative pattern is updated by computing the centroid of the training set vectors assigned to it. When the assignment process is complete, we will have k groups of vectors clustered around each of the output points.

Stuart Lloyd [129] used this approach to generate the *pdf*-optimized scalar quantizer, except that instead of using a training set, he assumed that the source distribution $f_X(x)$ was known. The Lloyd algorithm functions as follows:

1. Start with an initial set of reconstruction values $\left\{y_i^{(1)}\right\}_{i=1}^{M}$. Set $k = 1$, $D^{(0)} = 0$. Select threshold ϵ.
2. Find decision boundaries
$$b_j^{(k)} = \frac{y_{j+1}^{(k)} + y_j^{(k)}}{2} \qquad j = 1, 2, \ldots, M-1$$
3. Compute the distortion
$$D^{(k)} = \sum_{i=1}^{M} \int_{b_{i-1}^{(k)}}^{b_i^{(k)}} (x - y_i)^2 f_X(x) dx$$
4. If $D^{(k)} - D^{(k-1)} < \epsilon$, stop; otherwise, continue.
5. $k = k + 1$. Compute new reconstruction values
$$y_j^{(k)} = \frac{\int_{b_{j-1}^{(k-1)}}^{b_j^{(k-1)}} x f_X(x) dx}{\int_{b_{j-1}^{(k-1)}}^{b_j^{(k-1)}} f_X(x) dx}$$
Go to Step 2.

Linde, Buzo, and Gray generalized this algorithm to the case where the inputs are no longer scalars [137]. For the case where the distribution is known, the algorithm looks very much like the Lloyd algorithm described above.

1. Start with an initial set of reconstruction values $\left\{\mathbf{y}_i^{(1)}\right\}_{i=1}^{M}$. Set $k = 1$, $D^{(0)} = 0$. Select threshold ϵ.
2. Find quantization regions
$$V_i^{(k)} = \{\mathbf{x} : d(\mathbf{x}, \mathbf{y}_i) < d(\mathbf{x}, \mathbf{y}_j)\ \forall j \neq i\} \qquad j = 1, 2, \ldots, M$$
3. Compute the distortion
$$D^{(k)} = \sum_{i=1}^{M} \int_{V_i^{(k)}} \left\| \mathbf{x} - \mathbf{y}_i^{(k)} \right\|^2 f_X(\mathbf{x}) d\mathbf{x}$$
4. If $\frac{(D^{(k)} - D^{(k-1)})}{D^{(k)}} < \epsilon$, stop; otherwise, continue.
5. $k = k + 1$. Find new reconstruction values $\left\{\mathbf{y}_i^{(k)}\right\}_{i=1}^{M}$ that are the centroids of $\left\{V_i^{(k-1)}\right\}$. Go to Step 2.

This algorithm is not very practical because the integrals required to compute the distortions and centroids are over odd-shaped regions in n dimensions, where n is the dimension of the input vectors. Generally, these integrals are extremely difficult to compute, making this particular algorithm more of an academic interest.

Of more practical interest is the algorithm for the case where we have a training set available. In this case, the algorithm looks very much like the k-means algorithm.

1. Start with an initial set of reconstruction values $\left\{\mathbf{y}_i^{(1)}\right\}_{i=1}^{M}$ and a set of training vectors $\{\mathbf{x}_n\}_{n=1}^{N}$. Set $k = 1$, $D^{(0)} = 0$. Select threshold ϵ.
2. The quantization regions $\left\{V_i^{(k)}\right\}_{i=1}^{M}$ are given by

$$V_i^{(k)} = \{\mathbf{x}_n : d(\mathbf{x}_n, \mathbf{y}_i) < d(\mathbf{x}_n, \mathbf{y}_j)\ \forall j \neq i\} \qquad i = 1, 2, \ldots, M$$

   We assume that none of the quantization regions are empty. (Later we will deal with the case where $V_i^{(k)}$ is empty for some i and k.)
3. Compute the average distortion $D^{(k)}$ between the training vectors and the representative reconstruction value:

$$D^{(k)} = \frac{1}{N}\sum_{i=1}^{M}\sum_{\mathbf{x_n}\in V_i^{(k)}} ||\mathbf{x_n} - \mathbf{y}_i||^2$$

4. If $\frac{(D^{(k)} - D^{(k-1)})}{D^{(k)}} < \epsilon$, stop; otherwise, continue.
5. $k = k + 1$. Find new reconstruction values $\left\{\mathbf{y}_i^{(k)}\right\}_{i=1}^{M}$ that are the average value of the elements of each of the quantization regions $V_i^{(k-1)}$. Go to Step 2.

This algorithm forms the basis of most vector quantizer designs. It is popularly known as the Linde-Buzo-Gray or LBG algorithm, or the generalized Lloyd algorithm (GLA) [137]. Although the paper of Linde, Buzo, and Gray [137] is a starting point for most of the work on vector quantization, the latter algorithm had been used several years prior by Edward E. Hilbert at the NASA Jet Propulsion Laboratories in Pasadena, California. Hilbert's starting point was the idea of clustering, and although he arrived at the same algorithm as described above, he called it the *cluster compression algorithm* [138].

In order to see how this algorithm functions, consider the following example of a two-dimensional vector quantizer codebook design.

Example 10.4.1:

Suppose our training set consists of the height and weight values shown in Table 10.1. The initial set of output points is shown in Table 10.2. (For ease of presentation, we will always round the coordinates of the output points to the nearest integer.) The inputs, outputs, and quantization regions are shown in Figure 10.7.

The input (44, 41) has been assigned to the first output point; the inputs (56, 91), (57, 88), (59, 119), and (60, 110) have been assigned to the second output point; the inputs (62, 114), and (65, 120) have been assigned to the third output; and the five remaining vectors from the training

TABLE 10.1 Training set for designing vector quantizer codebook.

Height	Weight
72	180
65	120
59	119
64	150
65	162
57	88
72	175
44	41
62	114
60	110
56	91
70	172

TABLE 10.2 Initial set of output points for codebook design.

Height	Weight
45	50
75	117
45	117
80	180

set have been assigned to the fourth output. The distortion for this assignment is 387.25. We now find the new output points. There is only one vector in the first quantization region, so the first output point is (44, 41). The average of the four vectors in the second quantization region (rounded up) is the vector (58, 102), which is the new second output point. In a similar manner, we can compute the third and fourth output points as (64, 117) and (69, 168). The new output points and the corresponding quantization regions are shown in Figure 10.8. From Figure 10.8, we can see that, while the training vectors that were initially part of the first and fourth quantization regions are still in the same quantization regions, the training vectors (59,115) and (60,120), which were in quantization region 2, are now in quantization region 3. The distortion corresponding to this assignment of training vectors to quantization regions is 89, considerably less than the original 387.25. Given the new assignments, we can obtain a new set of output points. The first and fourth output points do not change because the training vectors in the corresponding regions have not changed. However, the training vectors in regions 2 and 3 have changed. Recomputing the output points for these regions, we get (57, 90) and (62, 116). The final form of the quantizer is shown in Figure 10.9. The distortion corresponding to the final assignments is 60.17. ♦

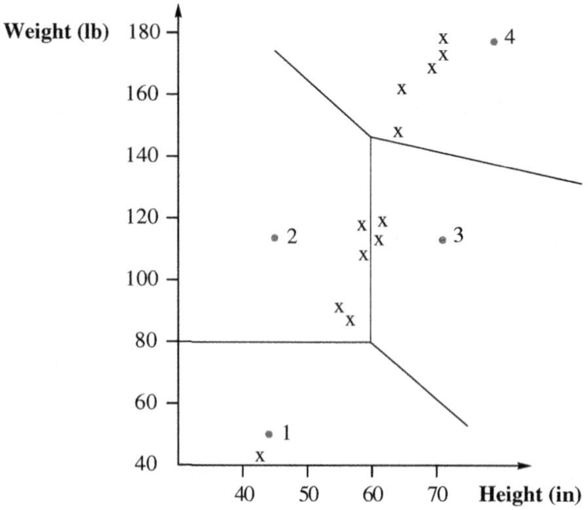

FIGURE 10.7 Initial state of the vector quantizer.

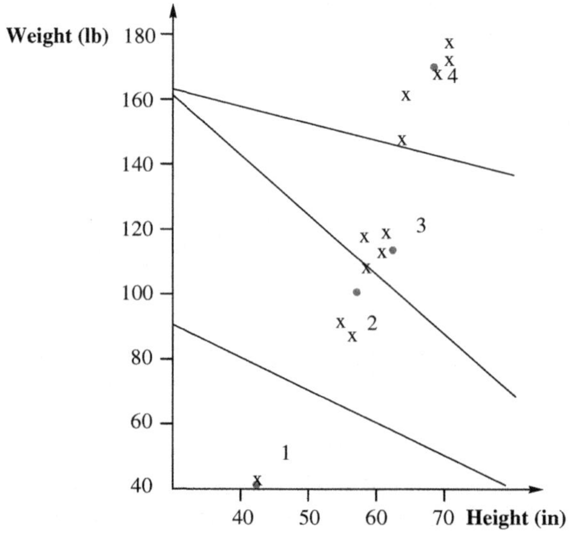

FIGURE 10.8 The vector quantizer after one iteration.

The LBG algorithm is conceptually simple, and as we shall see later, the resulting vector quantizer is remarkably effective in the compression of a wide variety of inputs, both by itself and in conjunction with other schemes. In the next two sections we will look at some of the details of the codebook design process. While these details are important to consider when designing codebooks, they are not necessary for the understanding of the quantization process. If you are not currently interested in these details, you may wish to proceed directly to Section 10.4.3.

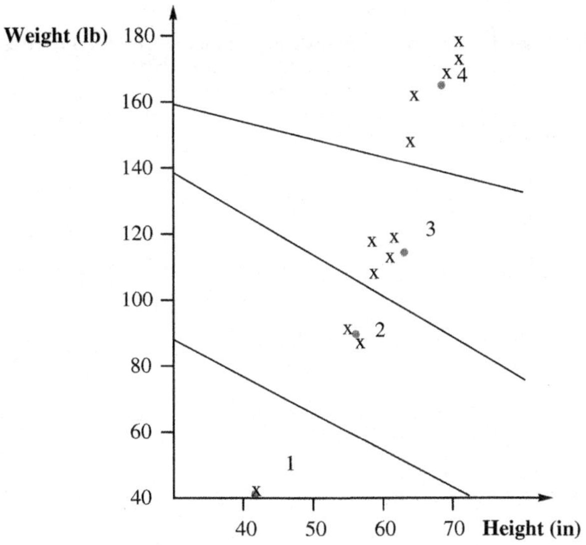

FIGURE 10.9 **Final state of the vector quantizer.**

TABLE 10.3 **An alternate initial set of output points.**

Height	Weight
75	50
75	117
75	127
80	180

10.4.1 Initializing the LBG Algorithm

The LBG algorithm guarantees that the distortion from one iteration to the next will not increase. However, there is no guarantee that the procedure will converge to the optimal solution. The solution to which the algorithm converges is heavily dependent on the initial conditions. For example, if our initial set of output points in Example 10.4.1 had been those shown in Table 10.3 instead of the set in Table 10.2, by using the LBG algorithm we would get the final codebook shown in Table 10.4.

The resulting quantization regions and their membership are shown in Figure 10.10. This is a very different quantizer than the one we previously obtained. Given this heavy dependence on initial conditions, the selection of the initial codebook is a matter of some importance. We will look at some of the better-known methods of initialization in the following section.

Linde, Buzo, and Gray described a technique in their original paper [137] called the *splitting technique* for initializing the design algorithm. In this technique, we begin by designing a vector quantizer with a single output point; in other words, a codebook of size one, or a one-level

TABLE 10.4 Final codebook obtained using the alternative initial codebook.

Height	Weight
44	41
60	107
64	150
70	172

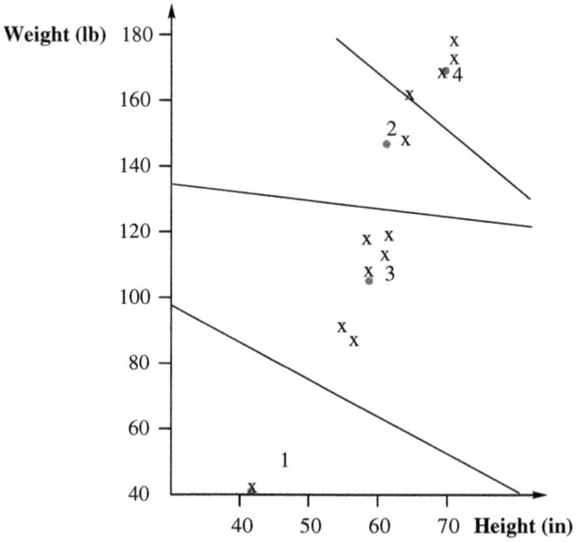

FIGURE 10.10 Final state of the vector quantizer using the alternative initial codebook.

vector quantizer. With a one-element codebook, the quantization region is the entire input space, and the output point is the average value of the entire training set. From this output point, the initial codebook for a two-level vector quantizer can be obtained by including the output point for the one-level quantizer and a second output point obtained by adding a fixed perturbation vector ϵ. We then use the LBG algorithm to obtain the two-level vector quantizer. Once the algorithm has converged, the two codebook vectors are used to obtain the initial codebook of a four-level vector quantizer. This initial four-level codebook consists of the two codebook vectors from the final codebook of the two-level vector quantizer and another two vectors obtained by adding ϵ to the two codebook vectors. The LBG algorithm can then be used until this four-level quantizer converges. In this manner we keep doubling the number of levels until we reach the desired number of levels. By including the final codebook of the previous stage at each "splitting," we guarantee that the codebook after splitting will be at least as good as the codebook prior to splitting.

TABLE 10.5 Progression of codebooks using splitting.

Codebook	Height	Weight
One-level	62	127
Initial two-level	62	127
	72	137
Final two-level	58	98
	69	168
Initial four-level	58	98
	68	108
	69	168
	79	178
Final four-level	52	73
	62	116
	65	156
	71	176

Example 10.4.2:

Let's revisit Example 10.4.1. This time, instead of using the initial codewords used in Example 10.4.1, we will use the splitting technique. For the perturbations, we will use a fixed vector $\epsilon = (10, 10)$. The perturbation vector is usually selected randomly; however, for purposes of explanation it is more useful to use a fixed perturbation vector.

We begin with a single-level codebook. The codeword is simply the average value of the training set. The progression of codebooks is shown in Table 10.5.

The perturbed vectors are used to initialize the LBG design of a two-level vector quantizer. The resulting two-level vector quantizer is shown in Figure 10.11. The resulting distortion is 468.58. These two vectors are perturbed to get the initial output points for the four-level design. Using the LBG algorithm, the final quantizer obtained is shown in Figure 10.12. The distortion is 156.17. The average distortion for the training set for this quantizer using the splitting algorithm is higher than the average distortion obtained previously. However, because the sample size used in this example is rather small, this is no indication of relative merit. ♦

If the desired number of levels is not a power of two, then in the last step, instead of generating two initial points from each of the output points of the vector quantizer designed previously, we can perturb as many vectors as necessary to obtain the desired number of vectors. For example, if we needed an eleven-level vector quantizer, we would generate a one-level vector quantizer first, then a two-level, then a four-level, and then an eight-level vector quantizer. At this stage, we would perturb only three of the eight vectors to get the eleven initial output points of the eleven-level vector quantizer. The three points should be those with the largest number of training set vectors, or the largest distortion.

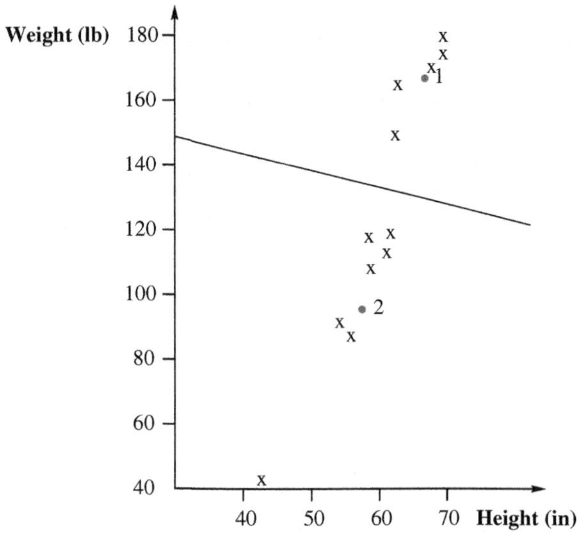

FIGURE 10.11 **Two-level vector quantizer using splitting approach.**

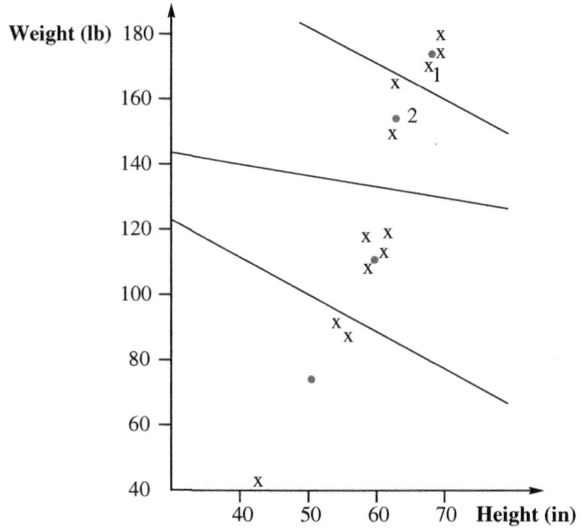

FIGURE 10.12 **Final design using the splitting approach.**

The approach used by Hilbert [138] to obtain the initial output points of the vector quantizer was to pick the output points randomly from the training set. This approach guarantees that, in the initial stages, there will always be at least one vector from the training set in each quantization region. However, we can still get different codebooks if we use different subsets of the training set as our initial codebook.

TABLE 10.6 Effect of using different subsets of the training sequence as the initial codebook.

Codebook	Height	Weight
Initial Codebook 1	72	180
	72	175
	65	120
	59	119
Final Codebook 1	71	176
	65	156
	62	116
	52	73
Initial Codebook 2	65	120
	44	41
	59	119
	57	88
Final Codebook 2	69	168
	44	41
	62	116
	57	90

Example 10.4.3:

Using the training set of Example 10.4.1, we selected different vectors of the training set as the initial codebook. The results are summarized in Table 10.6. If we pick the codebook labeled "Initial Codebook 1," we obtain the codebook labeled "Final Codebook 1." This codebook is identical to the one obtained using the split algorithm. The set labeled "Initial Codebook 2" results in the codebook labeled "Final Codebook 2." This codebook is identical to the quantizer we obtained in Example 10.4.1. In fact, most of the other selections result in one of these two quantizers. ♦

Notice that by picking different subsets of the input as our initial codebook, we can generate different vector quantizers. A good approach to codebook design is to initialize the codebook randomly several times, and pick the one that generates the least distortion in the training set from the resulting quantizers.

In 1989, Equitz [139] introduced a method for generating the initial codebook called the *pairwise nearest neighbor* (PNN) algorithm. In the PNN algorithm, we start with as many clusters as there are training vectors and end with the initial codebook. At each stage, we combine the two closest vectors into a single cluster and replace the two vectors by their mean. The idea is to merge those clusters that would result in the smallest increase in distortion. Equitz showed that when we combine two clusters C_i and C_j, the increase in distortion is

$$\frac{n_i n_j}{n_i + n_j} \left\| \mathbf{y}_i - \mathbf{y}_j \right\|^2 \tag{5}$$

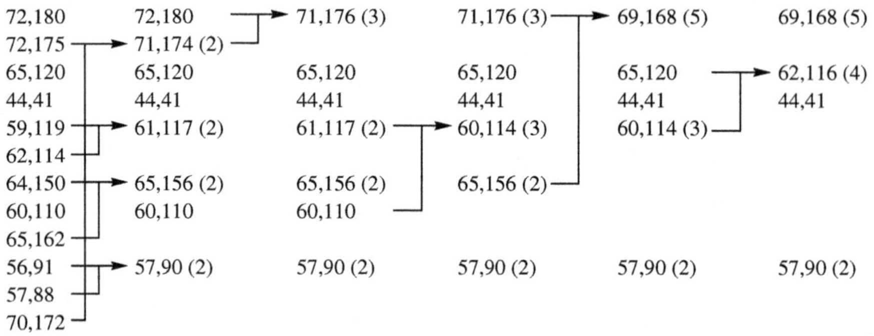

FIGURE 10.13 Obtaining initial output points using the PNN approach.

where n_i is the number of elements in the cluster C_i, and $\mathbf{y}_i$ is the corresponding output point. In the PNN algorithm, we combine clusters that cause the smallest increase in the distortion.

Example 10.4.4:

Using the PNN algorithm, we combine the elements in the training set as shown in Figure 10.13. At each step we combine the two clusters that are closest in the sense of Equation (5). If we use these values to initialize the LBG algorithm, we get a vector quantizer shown with output points (70, 172), (60, 107), (44, 41), and (64, 150), and a distortion of 104.08. ♦

Although it was a relatively easy task to generate the initial codebook using the PNN algorithm in Example 10.4.4, we can see that, as the size of the training set increases, this procedure becomes progressively more time-consuming. In order to avoid this cost, we can use a fast PNN algorithm that does not attempt to find the absolute smallest cost at each step (see [139] for details).

Finally, a simple initial codebook is the set of output points from the corresponding scalar quantizers. In the beginning of this chapter we saw how scalar quantization of a sequence of inputs can be viewed as vector quantization using a rectangular vector quantizer. We can use this rectangular vector quantizer as the initial set of outputs.

Example 10.4.5:

Return once again to the quantization of the height-weight data set. If we assume that the heights are uniformly distributed between 40 and 180, then a two-level scalar quantizer would have reconstruction values 75 and 145. Similarly, if we assume that the weights are uniformly distributed between 40 and 80, the reconstruction values would be 50 and 70. The initial reconstruction values for the vector quantizer are (50, 75), (50, 145), (70, 75), and (70, 145).

The final design for this initial set is the same as the one obtained in Example 10.4.1 with a distortion of 60.17. ♦

We have looked at four different ways of initializing the LBG algorithm. Each has its own advantages and drawbacks. The PNN initialization has been shown to result in better designs, producing a lower distortion for a given rate than the splitting approach [139]. However, the procedure for obtaining the initial codebook is much more involved and complex. We cannot make any general claims regarding the superiority of any one of these initialization techniques. Even the PNN approach cannot be proven to be optimal. In practice, if we are dealing with a wide variety of inputs, the effect of using different initialization techniques appears to be insignificant.

10.4.2 The Empty Cell Problem

Let's take a closer look at the progression of the design in Example 10.4.5. When we assign the inputs to the initial output points, no input point gets assigned to the output point at (70, 75). This is a problem because in order to update an output point, we need to take the average value of the input vectors. Obviously, some strategy is needed. The strategy that we actually used in Example 10.4.5 was not to update the output point if there were no inputs in the quantization region associated with it. This strategy seems to have worked in this particular example; however, there is a danger that we will end up with an output point that is never used. A common approach to avoid this is to remove an output point that has no inputs associated with it, and replace it with a point from the quantization region with the most output points. This can be done by selecting a point at random from the region with the highest population of training vectors, or the highest associated distortion. A more systematic approach is to design a two-level quantizer for the training vectors in the most heavily populated quantization region. This approach is computationally expensive and provides no significant improvement over the simpler approach. In the program accompanying this book, we have used the first approach. (To compare the two approaches, see Problem 3 at the end of this chapter.)

10.4.3 Use of LBG for Image Compression

One application for which the vector quantizer described in this section has been extremely popular is image compression. For image compression, the vector is formed by taking blocks of pixels of size $N \times M$ and treating them as an $L = NM$ dimensional vector. Generally, we take $N = M$. Instead of forming vectors in this manner, we could form the vector by taking L pixels in a row of the image. However, this does not allow us to take advantage of the two-dimensional correlations in the image. Recall that correlation between the samples provides the clustering of the input, and the LBG algorithm takes advantage of this clustering.

FIGURE 10.14 Original Sinan image.

Example 10.4.6:

Let us quantize the Sinan image shown in Figure 10.14 using a 16-dimensional quantizer. The input vectors are constructed using 4×4 blocks of pixels. The codebook was trained on the Sinan image.

The results of the quantization using codebooks of size 16, 64, 256, and 1024 are shown in Figure 10.15. The rates and compression ratios are summarized in Table 10.7. To see how these quantities were calculated, recall that if we have K vectors in a codebook, we need $\lceil \log_2 K \rceil$ bits to inform the receiver which of the K vectors is the quantizer output. This quantity is listed in the second column of Table 10.7 for the different values of K. If the vectors are of dimension L, this means that we have used $\lceil \log_2 K \rceil$ bits to send the quantized value of L pixels. Therefore, the rate in bits per pixel is $\frac{\lceil \log_2 K \rceil}{L}$. (We have assumed that the codebook is available to both transmitter and receiver, and therefore we do not have to use any bits to transmit the codebook from the transmitter to the receiver.) This quantity is listed in the third column of Table 10.7. Finally, the compression ratio, given in the last column of Table 10.7, is the ratio of the number of bits per pixel in the original image to the number of bits per pixel in the compressed image. The Sinan image was digitized using 8 bits per pixel. Using this information and the rate after compression, we can obtain the compression ratios.

Looking at the images, we see that reconstruction using a codebook of size 1024 is very close to the original. At the other end, the image obtained using a codebook with 16 reconstruction vectors contains a lot of visible artifacts. The utility of each reconstruction depends on the demands of the particular application. ♦

FIGURE 10.15 Top left: codebook size 16; top right: codebook size 64; bottom left: codebook size 256; bottom right: codebook size 1024.

In this example, we used codebooks trained on the image itself. Generally, this is not the preferred approach because the receiver has to have the same codebook in order to reconstruct the image. Either the codebook must be transmitted along with the image, or the receiver

TABLE 10.7 Summary of compression measures for image compression example.

Codebook Size (# of codewords)	Bits Needed to Select a Codeword	Bits per Pixel	Compression Ratio
16	4	0.25	32:1
64	6	0.375	21.33:1
256	8	0.50	16:1
1024	10	0.625	12.8:1

has the same training image so that it can generate an identical codebook. This is impractical because, if the receiver already has the image in question, much better compression can be obtained by simply sending the name of the image to the receiver. Sending the codebook with the image is not unreasonable. However, the transmission of the codebook is overhead that could be avoided if a more generic codebook, one that is available to both transmitter and receiver, were to be used.

In order to compute the overhead, we need to calculate the number of bits required to transmit the codebook to the receiver. If each codeword in the codebook is a vector with L elements and if we use B bits to represent each element, then in order to transmit the codebook of a K-level quantizer we need $B \times L \times K$ bits. In our example, $B = 8$ and $L = 16$. Therefore, we need $K \times 128$ bits to transmit the codebook. As our image consists of 256×256 pixels, the overhead in bits per pixel is $128K/65,536$. The overhead for different values of K is summarized in Table 10.8. We can see that while the overhead for a codebook of size 16 seems reasonable, the overhead for a codebook of size 1024 is over three times the rate required for quantization.

Given the excessive amount of overhead required for sending the codebook along with the vector quantized image, there has been substantial interest in the design of codebooks that are more generic in nature and, therefore, can be used to quantize a number of images. To investigate the issues that might arise, we quantized the Sinan image using four different codebooks generated by the Sena, Sensin, Earth, and Omaha images. The results are shown in Figure 10.16.

As expected, the reconstructed images from this approach are not of the same quality as when the codebook is generated from the image to be quantized. However, this is only true as long as the overhead required for storage or transmission of the codebook is ignored. If we include the extra rate required to encode and transmit the codebook of output points, using the codebook generated by the image to be quantized seems unrealistic. Although using the codebook generated by another image to perform the quantization may be realistic, the quality of the reconstructions is quite poor. Later in this chapter we will take a closer look at the subject of vector quantization of images and consider a variety of ways to improve this performance.

You may have noticed that the bit rates for the vector quantizers used in the examples are quite low. The reason is that the size of the codebook increases exponentially with the rate. Suppose we want to encode a source using R bits per sample; that is, the average number of bits per sample in the compressed source output is R. By "sample" we mean a scalar element of the source output sequence. If we wanted to use an L-dimensional quantizer, we would group

TABLE 10.8 Overhead in bits per pixel for codebooks of different sizes.

Codebook Size K	Overhead in Bits per Pixel
16	0.03125
64	0.125
256	0.50
1024	2.0

FIGURE 10.16 Sinan image quantized at the rate of 0.5 bits per pixel. The images used to obtain the codebook were (clockwise from top left) Sensin, Sena, Earth, and Omaha.

L samples together into vectors. This means that we would have RL bits available to represent each vector. With RL bits, we can represent 2^{RL} different output vectors. In other words, the size of the codebook for an L-dimensional R-bits-per-sample quantizer is 2^{RL}. From Table 10.7, we can see that when we quantize an image using 0.25 bits per pixel and 16-dimensional quantizers, we have $16 \times 0.25 = 4$ bits available to represent each vector. Hence, the size of the codebook is $2^4 = 16$. The quantity RL is often called the *rate dimension product*. Note that the size of the codebook grows exponentially with this product.

Consider the problems. The codebook size for a 16-dimensional, 2-bits-per-sample vector quantizer would be $2^{16\times 2}$! (If the source output was originally represented using 8 bits per sample, a rate of 2 bits per sample for the compressed source corresponds to a compression ratio of 4:1.) This large size causes problems both with storage and with the quantization process. To store 2^{32} 16-dimensional vectors, assuming that we can store each component of the vector in a single byte, requires $2^{32} \times 16$ bytes—approximately 64 gigabytes of storage. Furthermore, to quantize a single input vector would require over four billion vector comparisons to find the closest output point. Obviously, neither the storage requirements nor the computational requirements are realistic. Because of this problem, most vector quantization applications operate at low bit rates. In many applications, such as low-rate speech coding, we want to operate at very low rates; therefore, this is not a drawback. However, for applications such as high-quality video coding, which requires higher rates, this is definitely a problem.

There are several approaches to solving these problems. Each entails the introduction of some structure in the codebook and/or the quantization process. While the introduction of structure mitigates some of the storage and computational problems, there is generally a trade-off in terms of the distortion performance. We will look at some of these approaches in the following sections.

10.5 Tree-Structured Vector Quantizers

One way we can introduce structure is to organize our codebook in such a way that it is easy to pick which part contains the desired output vector. Consider the two-dimensional vector quantizer shown in Figure 10.17. Note that the output points in each quadrant are the mirror image of the output points in neighboring quadrants. Given an input to this vector quantizer, we can reduce the number of comparisons necessary for finding the closest output point by using the sign on the components of the input. The sign on the components of the input vector will tell us in which quadrant the input lies. Because all the quadrants are mirror images of the neighboring quadrants, the closest output point to a given input will lie in the same quadrant as the input itself. Therefore, we only need to compare the input to the output points that lie in the same quadrant, thus reducing the number of required comparisons by a factor of four. This approach can be extended to L dimensions, where the signs on the L components of the input vector can tell us in which of the 2^L hyperquadrants the input lies, which in turn would reduce the number of comparisons by 2^L.

This approach works well when the output points are distributed in a symmetrical manner. However, it breaks down as the distribution of the output points becomes less symmetrical.

Example 10.5.1:

Consider the vector quantizer shown in Figure 10.18. This is different from the output points in Figure 10.17; we have dropped the mirror image requirement of the previous example. The output points are shown as filled circles, and the input point is the X. It is obvious from the figure that while the input is in the first quadrant, the closest output point is in the fourth quadrant. However, the quantization approach described above will force the input to be represented by an output in the first quadrant.

The situation gets worse as we lose more and more of the symmetry. Consider the situation in Figure 10.19. In this quantizer, not only will we get an incorrect output point when the input is close to the boundaries of the first quadrant, but also there is no significant reduction in the amount of computation required.

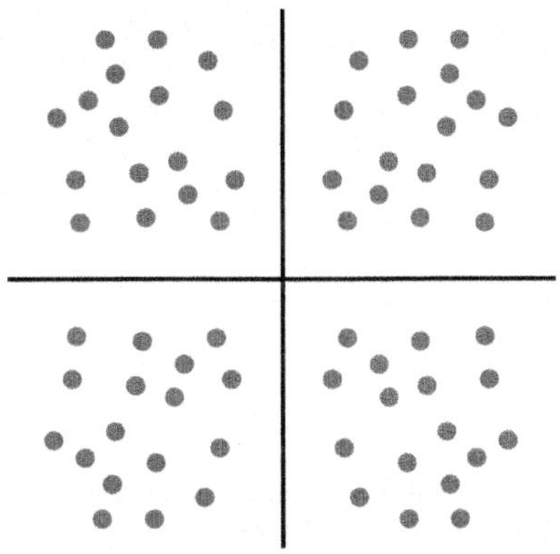

FIGURE 10.17 **A symmetrical vector quantizer in two dimensions.**

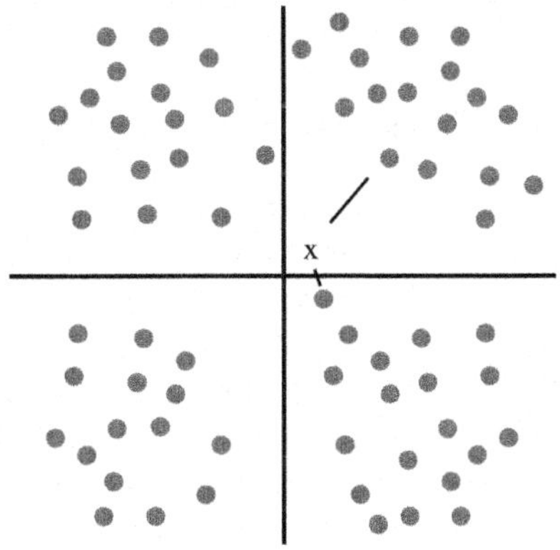

FIGURE 10.18 **Breakdown of the method using the quadrant approach.**

Most of the output points are in the first quadrant. Therefore, whenever the input falls in the first quadrant, which it will do quite often if the quantizer design is reflective of the distribution of the input, knowing that it is in the first quadrant does not lead to a great reduction in the number of comparisons. ♦

The idea of using the L-dimensional equivalents of quadrants to partition the output points in order to reduce the computational load can be extended to nonsymmetrical situations, like those shown in Figure 10.19, in the following manner. Divide the set of output points into two groups, *group0* and *group1*, and assign to each group a test vector such that output points in each group are closer to the test vector assigned to that group than to the test vector assigned to the other group (Figure 10.20). Label the two test vectors 0 and 1. When we get an input vector, we compare it against the test vectors. Depending on the outcome, the input is compared to the output points associated with the test vector closest to the input. After these two comparisons, we can discard half of the output points. Comparison with the test vectors takes the place of

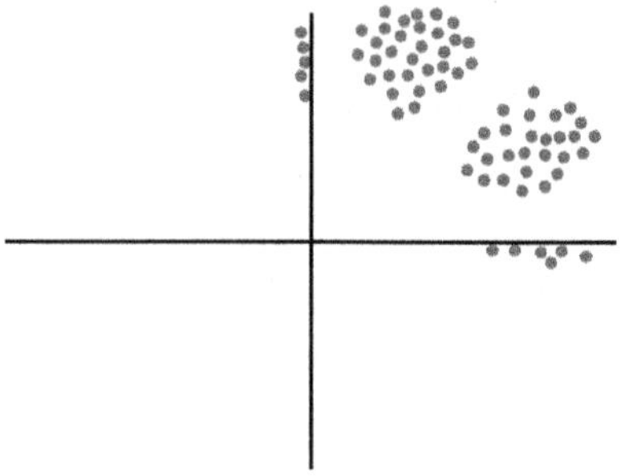

FIGURE 10.19 Failure of the quadrant approach when the data is not uniformly distributed amongst quadrants.

looking at the signs of the components to decide which set of output points to discard from contention. If the total number of output points is K, with this approach we have to make $\frac{K}{2}+2$ comparisons instead of K comparisons.

This process can be continued by splitting the output points in each group into two groups and assigning a test vector to the subgroups. So *group0* would be split into *group00* and *group01*, with associated test vectors labeled 00 and 01, and *group1* would be split into *group10* and *group11*, with associated test vectors labeled 10 and 11. Suppose the result of the first set of comparisons was that the output point would be searched for in *group1*. The input would be compared to the test vectors 10 and 11. If the input was closer to the test vector 10, then the output points in *group11* would be discarded, and the input would be compared to the output points in *group10*. We can continue the procedure by successively dividing each group of output points into two, until finally, if the number of output points is a power of two, the last set of groups would consist of single points. The number of comparisons required to obtain the final output point would be $2\log K$ instead of K. Thus, for a codebook of size 4096 we would need 24 vector comparisons instead of 4096 vector comparisons.

This is a remarkable decrease in computational complexity. However, we pay for this decrease in two ways. The first penalty is a possible increase in distortion. It is possible at some stage that the input is closer to one test vector while at the same time being closest to an

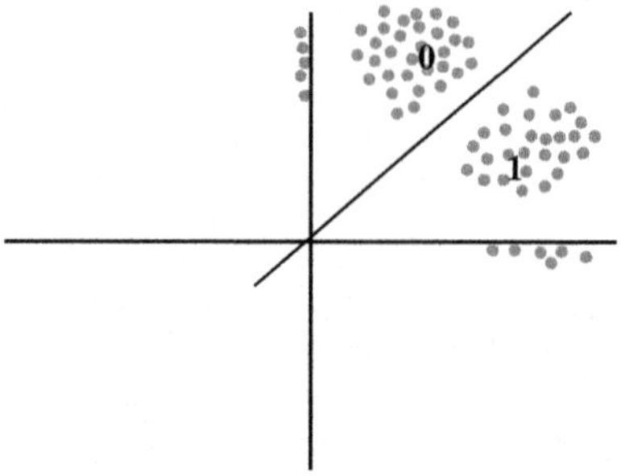

FIGURE 10.20 Division of output points into two groups.

output belonging to the rejected group. This is similar to the situation shown in Figure 10.18. The other penalty is an increase in storage requirements. Now we not only have to store the output points from the vector quantizer codebook, we also must store the test vectors. This means almost a doubling of the storage requirement.

The comparisons that must be made at each step are shown in Figure 10.21. The label inside each node is the label of the test vector that we compare the input against. This tree of decisions is what gives tree-structured vector quantizers (TSVQ) their name. Notice also that, as we are progressing down a tree, we are also building a binary string. As the leaves of the tree are the output points, by the time we reach a particular leaf or, in other words, select a particular output point, we have obtained the binary codeword corresponding to that output point.

This process of building the binary codeword as we progress through the series of decisions required to find the final output can result in some other interesting properties of tree-structured vector quantizers. For instance, even if a partial codeword is transmitted, we can still get an approximation of the input vector. In Figure 10.21, if the quantized value was the codebook vector 5, the binary codeword would be 011. However, if only the first two bits 01 were received by the decoder, the input can be approximated by the test vector labeled 01.

10.5.1 Design of Tree-Structured Vector Quantizers

In the last section we saw how we could reduce the computational complexity of the design process by imposing a tree structure on the vector quantizer. Rather than imposing this structure after the vector quantizer has been designed, it makes sense to design the vector quantizer within the framework of the tree structure. We can do this by a slight modification of the splitting design approach proposed by Linde et al. [137].

We start the design process in a manner identical to the splitting technique. First, obtain the average of all the training vectors, perturb it to obtain a second vector, and use these vectors to form a two-level vector quantizer. Let us label these two vectors 0 and 1, and the

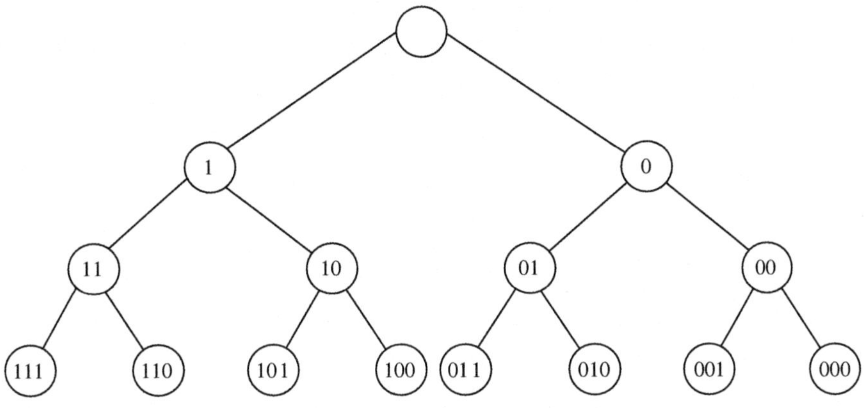

FIGURE 10.21 Decision tree for quantization.

groups of training set vectors that would be quantized to each of these two vectors *group 0* and *group1*. We will later use these vectors as test vectors. We perturb these output points to get the initial vectors for a four-level vector quantizer. At this point, the design procedure for the tree-structured vector quantizer deviates from the splitting technique. Instead of using the entire training set to design a four-level vector quantizer, we use the training set vectors in *group0* to design a two-level vector quantizer with output points labeled 00 and 01. We use the training set vectors in *group1* to design a two-level vector quantizer with output points labeled 10 and 11. We also split the training set vectors in *group0* and *group1* into two groups each. The vectors in *group0* are split, based on their proximity to the vectors labeled 00 and 01, into *group00* and *group01*, and the vectors in *group1* are divided in a like manner into the groups *group10* and *group11*. The vectors labeled 00, 01, 10, and 11 will act as test vectors at this level. To get an eight-level quantizer, we use the training set vectors in each of the four groups to obtain four two-level vector quantizers. We continue in this manner until we have the required number of output points. Notice that in the process of obtaining the output points, we have also obtained the test vectors required for the quantization process.

10.5.2 Pruned Tree-Structured Vector Quantizers

Once we have built a tree-structured codebook, we can sometimes improve its rate distortion performance by removing carefully selected subgroups. Removal of a subgroup, referred to as *pruning*, will reduce the size of the codebook and hence the rate. It may also result in an increase in distortion. Therefore, the objective of the pruning is to remove those subgroups that will result in the best trade-off of rate and distortion. Chou, Lookabaugh, and Gray [140] have developed an optimal pruning algorithm called the *generalized BFOS algorithm*. The name of the algorithm derives from the fact that it is an extension of an algorithm originally developed by Brieman, Freidman, Olshen, and Stone [141] for classification applications. (See [140, 136] for description and discussion of the algorithm.)

Pruning output points from the codebook has the unfortunate effect of removing the structure that was previously used to generate the binary codeword corresponding to the output points. If we used the structure to generate the binary codewords, the pruning would cause the codewords to be of variable length. As the variable-length codes would correspond to the leaves of a binary tree, this code would be a prefix code and, therefore, certainly usable. However, it would not require a large increase in complexity to assign fixed-length codewords to the output points using another method. This increase in complexity is generally offset by the improvement in performance that results from the pruning [142].

10.6 Structured Vector Quantizers

The tree-structured vector quantizer solves the complexity problem, but exacerbates the storage problem. We now take an entirely different tack and develop vector quantizers that do not have these storage problems; however, we pay for this relief in other ways.

Example 10.3.1 was our motivation for the quantizer obtained by the LBG algorithm. This example showed that the correlation between samples of the output of a source leads to clustering. This clustering is exploited by the LBG algorithm by placing output points at the location of these clusters. However, in Example 10.3.2, we saw that even when there is no correlation between samples, there is a kind of probabilistic structure that becomes more evident as we group the random inputs of a source into larger and larger blocks or vectors.

In Example 10.3.2, we changed the position of the output point in the top-right corner. All four corner points have the same probability, so we could have chosen any of these points. In the case of the two-dimensional Laplacian distribution in Example 10.3.2, all points that lie on the contour described by $|x| + |y| = constant$ have equal probability. These are called *contours of constant probability*. For spherically symmetrical distributions like the Gaussian distribution, the contours of constant probability are circles in two dimensions, spheres in three dimensions, and hyperspheres in higher dimensions.

We mentioned in Example 10.3.2 that the points away from the origin have very little probability mass associated with them. Based on what we have said about the contours of constant probability, we can be a little more specific and say that the points on constant probability contours farther away from the origin have very little probability mass associated with them. Therefore, we can get rid of all of the points outside some contour of constant probability without incurring much of a distortion penalty. In addition, as the number of reconstruction points is reduced, there is a decrease in rate, thus improving the rate distortion performance.

Example 10.6.1:

Let us design a two-dimensional uniform quantizer by keeping only the output points in the quantizer of Example 10.3.2 that lie on or within the contour of constant probability given by $|x_1| + |x_2| = 5\Delta$. If we count all the points that are retained, we get 60 points. This is close enough to 64 that we can compare it with the eight-level uniform scalar quantizer. If we simulate this quantization scheme with a Laplacian input, and the same step size as the scalar quantizer, that is, $\Delta = 0.7309$, we get an SNR of 12.22 dB. Comparing this to the 11.44 dB obtained with the scalar quantizer, we see that there is a definite improvement. We can get slightly more improvement in performance if we modify the step size. ♦

Notice that the improvement in the previous example is obtained only by restricting the outer boundary of the quantizer. Unlike Example 10.3.2, we did not change the shape of any of the inner quantization regions. This gain is referred to in the quantization literature as *boundary gain*. In terms of the description of quantization noise in Chapter 8, we reduced the overload error by reducing the overload probability, without a commensurate increase in the granular noise. In Figure 10.22, we have marked the 12 output points that belonged to the original 64-level quantizer, but do not belong to the 60-level quantizer, by drawing circles around them. Removal of these points results in an increase in overload probability. We also marked the eight output points that belong to the 60-level quantizer, but were not part of the original 64-level quantizer, by drawing squares around them. Adding these points results in a decrease in the overload probability. If we calculate the increases and decreases (see Problem 5 at the end of this chapter), we find that the net result is a decrease in overload probability. This overload probability is further reduced as the dimension of the vector is increased.

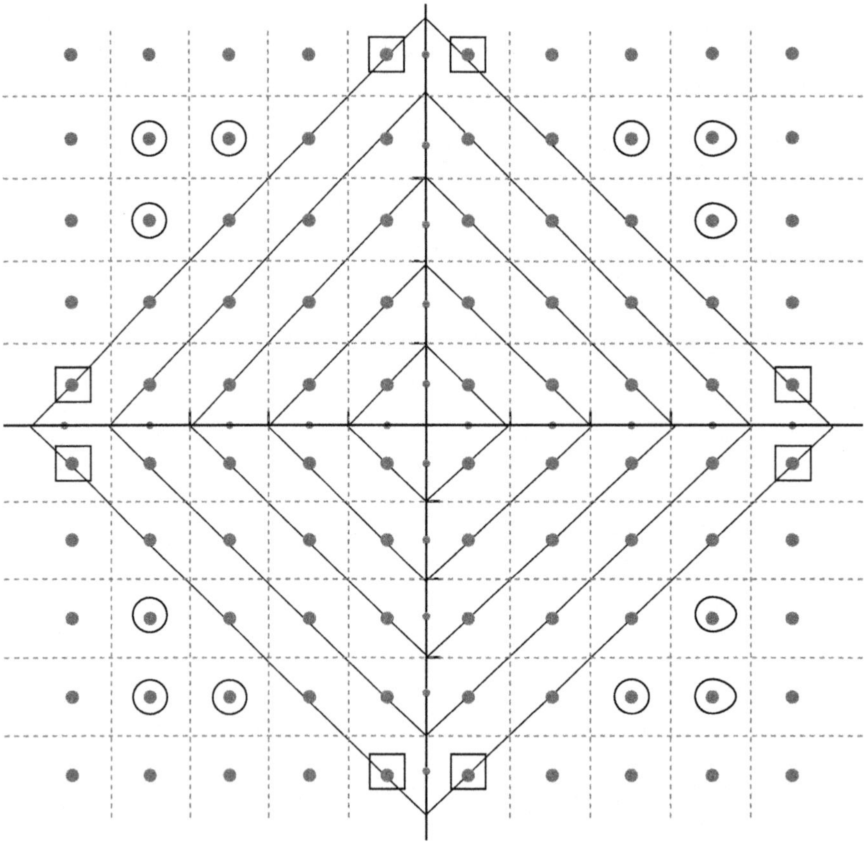

FIGURE 10.22 Contours of constant probability.

10.6.1 Pyramid Vector Quantization

As the dimension of the input vector increases, something interesting happens. Suppose we are quantizing a random variable X with *pdf* $f_X(X)$ and differential entropy $h(X)$. Suppose we block samples of this random variable into a random vector $\mathbf{X}$. A result of Shannon's, called the *asymptotic equipartition property* (AEP), states that for sufficiently large L and arbitrarily small ϵ

$$\left|\frac{\log f_{\mathbf{X}}(\mathbf{X})}{L} + h(X)\right| < \epsilon \tag{6}$$

for all but a set of vectors with a vanishingly small probability [3]. This means that almost all the L-dimensional vectors will lie on a contour of constant probability given by

$$\left|\frac{\log f_{\mathbf{X}}(\mathbf{X})}{L}\right| = -h(X) \tag{7}$$

Given that this is the case, Sakrison [143] suggested that an optimum manner to encode the source would be to distribute 2^{RL} points uniformly in this region. Fischer [144] used this insight to design a vector quantizer called the *pyramid vector quantizer* for the Laplacian source that looks quite similar to the quantizer described in Example 10.6.1. The vector quantizer consists of points of the rectangular quantizer that fall on the hyperpyramid given by

$$\sum_{i=1}^{L} |x_i| = C$$

where C is a constant depending on the variance of the input. Shannon's result is asymptotic, and for realistic values of L, the input vector is generally not localized to a single hyperpyramid.

For this case, Fischer first finds the distance

$$r = \sum_{i=1}^{L} |x_i|$$

This value is quantized and transmitted to the receiver. The input is normalized by this gain term and quantized using a single hyperpyramid. The quantization process for the shape term consists of two stages: finding the output point on the hyperpyramid closest to the scaled input, and finding a binary codeword for this output point. (See [144] for details about the quantization and coding process.) This approach is quite successful, and for a rate of 3 bits per sample and a vector dimension of 16, we get an SNR value of 16.32 dB. If we increase the vector dimension to 64, we get an SNR value of 17.03. Compared to the SNR obtained from using a nonuniform scalar quantizer, this is an improvement of more than 4 dB.

Notice that in this approach we separated the input vector into a *gain* term and a pattern or *shape* term. Quantizers of this form are called *gain-shape vector quantizers*, or *product code vector quantizers* [145].

10.6.2 Polar and Spherical Vector Quantizers

For the Gaussian distribution, the contours of constant probability are circles in two dimensions and spheres and hyperspheres in three and higher dimensions. In two dimensions, we can quantize the input vector by first transforming it into polar coordinates r and θ:

$$r = \sqrt{x_1^2 + x_2^2} \tag{8}$$

and

$$\theta = \tan^{-1} \frac{x_2}{x_1} \tag{9}$$

r and θ can then be either quantized independently [146], or we can use the quantized value of r as an index to a quantizer for θ [147]. The former is known as a polar quantizer; the latter, an unrestricted polar quantizer. The advantage to quantizing r and θ independently is one of simplicity. The quantizers for r and θ are independent scalar quantizers. However, the performance of the polar quantizers is not significantly higher than that of scalar quantization of the components of the two-dimensional vector. The unrestricted polar quantizer has a more complex implementation, as the quantization of θ depends on the quantization of r. However, the performance is also somewhat better than the polar quantizer. The polar quantizer can be extended to three or more dimensions [148].

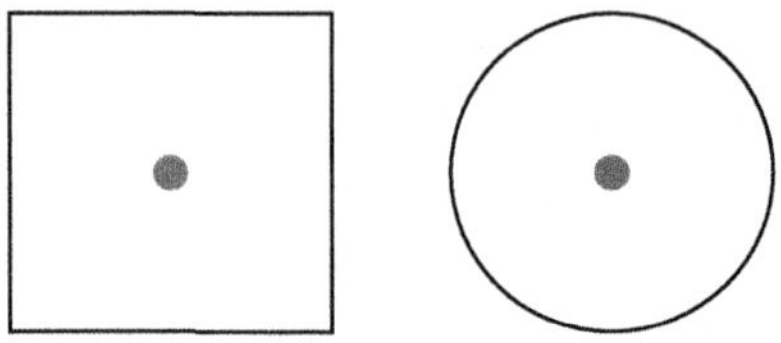

FIGURE 10.23 Possible quantization regions.

10.6.3 Lattice Vector Quantizers

Recall that quantization error is composed of two kinds of error, overload error and granular error. The overload error is determined by the location of the quantization regions furthest from the origin, or the boundary. We have seen how we can design vector quantizers to reduce the overload probability and thus the overload error. We called this the boundary gain of vector quantization. In scalar quantization, the granular error was determined by the size of the quantization interval. In vector quantization, the granular error is affected by the size and shape of the quantization interval.

Consider the square and circular quantization regions shown in Figure 10.23. We show only the quantization region at the origin. These quantization regions need to be distributed in a regular manner over the space of source outputs. However, for now, let us simply consider the quantization region at the origin. Let's assume they both have the same area so that we can compare them. This way it would require the same number of quantization regions to cover a given area. That is, we will be comparing two quantization regions of the same "size." To have an area of one, the square has to have sides of length one. As the area of a circle is given by πr^2, the radius of the circle is $\frac{1}{\sqrt{\pi}}$. The maximum quantization error possible with the square quantization region is when the input is at one of the four corners of the square. In this case, the error is $\frac{1}{\sqrt{2}}$, or about 0.707. For the circular quantization region, the maximum error occurs when the input falls on the boundary of the circle. In this case, the error is $\frac{1}{\sqrt{\pi}}$, or about 0.56. Thus, the maximum granular error is larger for the square region than the circular region.

In general, we are more concerned with the average squared error than the maximum error. If we compute the average squared error for the square region, we obtain

$$\int_{\text{Square}} \|\mathbf{x}\|^2 \, d\mathbf{x} = 0.166\bar{6}$$

For the circle, we obtain

$$\int_{\text{Circle}} \|\mathbf{x}\|^2 \, d\mathbf{x} = 0.159$$

Thus, the circular region would introduce less granular error than the square region.

Our choice seems to be clear; we will use the circle as the quantization region. Unfortunately, a basic requirement for the quantizer is that for every possible input vector there should be a unique output vector. In order to satisfy this requirement and have a quantizer with sufficient structure that can be used to reduce the storage space, a union of translates of the quantization region should cover the output space of the source. In other words, the quantization region should *tile* space. A two-dimensional region can be tiled by squares, but it cannot be tiled by circles. If we tried to tile the space with circles, we would either get overlaps or holes.

Apart from squares, other shapes that tile space include rectangles and hexagons. It turns out that the best shape to pick for a quantization region in two dimensions is a hexagon [149].

In two dimensions, it is relatively easy to find the shapes that tile space, then select the one that gives the smallest amount of granular error. However, when we start looking at higher dimensions, it is difficult, if not impossible, to visualize different shapes, let alone find which ones tile space. An easy way out of this dilemma is to remember that a quantizer can be completely defined by its output points. In order for this quantizer to possess structure, these points should be spaced in some regular manner.

Regular arrangements of output points in space are called *lattices*. Mathematically, we can define a lattice as follows:

Let $\{\mathbf{a_1}, \mathbf{a_2}, \ldots, \mathbf{a_L}\}$be L independent L-dimensional vectors. Then the set

$$\mathcal{L} = \left\{ \mathbf{x} : \mathbf{x} = \sum_{i=1}^{L} u_i \mathbf{a}_i \right\} \tag{10}$$

is a lattice if $\{u_i\}$ are all integers.

When a subset of lattice points is used as the output points of a vector quantizer, the quantizer is known as a *lattice vector quantizer*. From this definition, the pyramid vector quantizer described earlier can be viewed as a lattice vector quantizer. Basing a quantizer on a lattice solves the storage problem. As any lattice point can be regenerated if we know the basis set, there is no need to store the output points. Further, the highly structured nature of lattices makes finding the closest output point to an input relatively simple. Note that what we give up when we use lattice vector quantizers is the clustering property of LBG quantizers.

Let's take a look at a few examples of lattices in two dimensions. If we pick $a_1 = (1, 0)$ and $a_2 = (0, 1)$, we obtain the integer lattice—the lattice that contains all points in two dimensions whose coordinates are integers.

If we pick $a_1 = (1, 1)$ and $a_2 = (1, -1)$, we get the lattice shown in Figure 10.24. This lattice has a rather interesting property. Any point in the lattice is given by $na_1 + ma_2$, where n and m are integers. But

$$na_1 + ma_2 = \begin{bmatrix} n + m \\ n - m \end{bmatrix}$$

and the sum of the coefficients is $n + m + n - m = 2n$, which is even for all n. Therefore, all points in this lattice have an even coordinate sum. Lattices with these properties are called *D lattices*.

Finally, if $a_1 = (1, 0)$ and $a_2 = \left(-\frac{1}{2}, \frac{\sqrt{3}}{2}\right)$, we get the hexagonal lattice shown in Figure 10.25. This is an example of an *A lattice*.

There are a large number of lattices that can be used to obtain lattice vector quantizers. In fact, given a dimension L, there are an infinite number of possible sets of L independent vectors. Among these, we would like to pick the lattice that produces the greatest reduction in

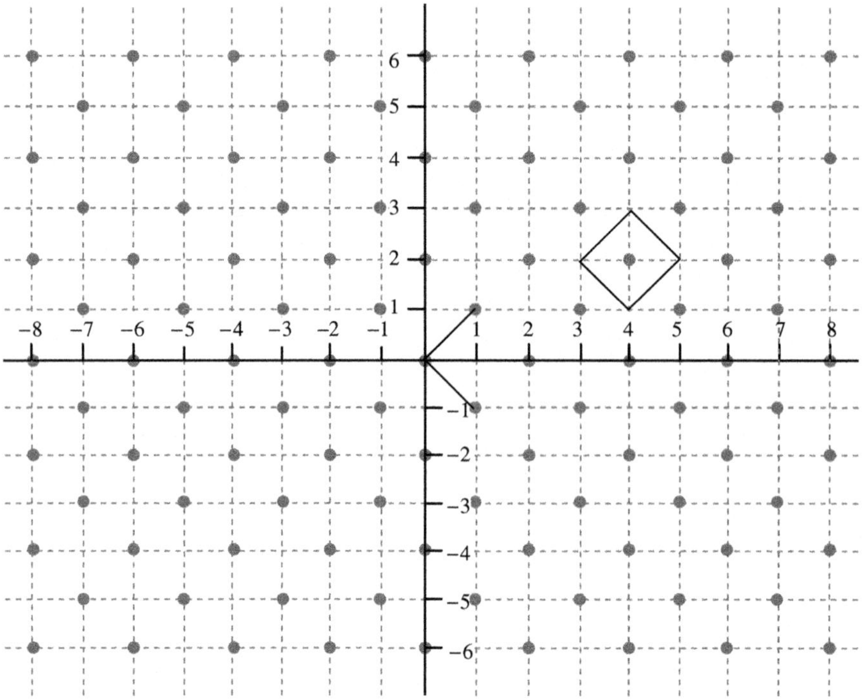

FIGURE 10.24 **The D_2 lattice.**

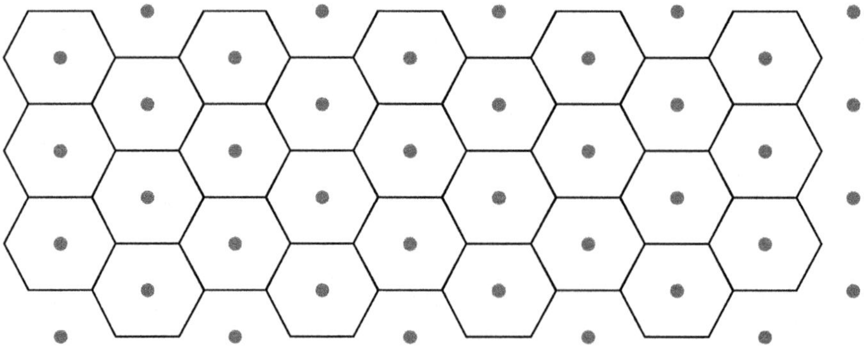

FIGURE 10.25 **The A_2 lattice.**

granular noise. When comparing the square and circle as candidates for quantization regions, we used the integral over the shape of $\|\mathbf{x}\|^2$. This is simply the second moment of the shape. The shape with the smallest second moment for a given volume is known to be the circle in two dimensions and the sphere and hypersphere in higher dimensions [150]. Unfortunately, circles and spheres cannot tile space; either there will be overlap or there will be holes. As the ideal case is unattainable, we can try to approximate it. We can look for ways of arranging spheres so that they cover space with minimal overlap [151], or look for ways of packing spheres with the least amount of space left over [150]. The centers of these spheres can then be used as the output points. The quantization regions will not be spheres, but they may be close approximations to spheres.

The problems of sphere covering and sphere packing are widely studied in a number of different areas. Lattices discovered in these studies have also been useful as vector quantizers [150]. Some of these lattices, such as the A_2 and D_2 lattices described earlier, are based on the root systems of Lie algebras [152]. The study of Lie algebras is beyond the scope of this book; however, we have included a brief discussion of the root systems and how to obtain the corresponding lattices in Appendix C.

One of the nice things about root lattices is that we can use their structural properties to obtain fast quantization algorithms. For example, consider building a quantizer based on the D_2 lattice. Because of the way in which we described the D_2 lattice, the size of the lattice is fixed. We can change the size by picking the basis vectors as (Δ, Δ) and $(\Delta, -\Delta)$, instead of (1,1) and (1,−1). We can have exactly the same effect by dividing each input by Δ before quantization, and then multiplying the reconstruction values by Δ. Suppose we pick the latter approach and divide the components of the input vector by Δ. If we want to find the closest lattice point to the input, all we need to do is find the closest integer to each coordinate of the scaled input. If the sum of these integers is even, we have a lattice point. If not, find the coordinate that incurred the largest distortion during conversion to an integer and then find the next closest integer. The sum of coordinates of this new vector differs from the sum of coordinates of the previous vector by one. Therefore, if the sum of the coordinates of the previous vector was odd, the sum of the coordinates of the current vector will be even, and we have the closest lattice point to the input.

Example 10.6.2:

Suppose the input vector is given by (2.3, 1.9). Rounding each coefficient to the nearest integer, we get the vector (2, 2). The sum of the coordinates is even; therefore, this is the closest lattice point to the input.

Suppose the input is (3.4, 1.8). Rounding the components to the nearest integer, we get (3, 2). The sum of the components is 5, which is odd. The differences between the components of the input vector and the nearest integer are 0.4 and 0.2. The largest difference was incurred by the first component, so we round it up to the next closest integer, and the resulting vector is (4, 2). The sum of the coordinates is 6, which is even; therefore, this is the closest lattice point. ♦

Many of the lattices have similar properties that can be used to develop fast algorithms for finding the closest output point to a given input [153,152].

To review our coverage of lattice vector quantization, overload error can be reduced by careful selection of the boundary, and we can reduce the granular noise by selection of the lattice. The lattice also provides us with a way to avoid storage problems. Finally, we can use the structural properties of the lattice to find the closest lattice point to a given input.

Now we need two things: to know how to find the closest *output* point (remember, not all lattice points are output points), and to find a way of assigning a binary codeword to the output point and recovering the output point from the binary codeword. This can be done by again making use of the specific structures of the lattices. While the procedures necessary are simple, explanations of the procedures are lengthy and involved (see [154,152] for details).

10.7 Variations on the Theme

Because of its capability to provide high compression with relatively low distortion, vector quantization has been one of the more popular lossy compression techniques over the last decade in such diverse areas as video compression and low-rate speech compression. During this period, several people have come up with variations on the basic vector quantization approach. We briefly look at a few of the more well-known variations here, but this is by no means an exhaustive list. For more information, see [136,155].

10.7.1 Gain-Shape Vector Quantization

In some applications such as speech, the dynamic range of the input is quite large. One effect of this is that, in order to be able to represent the various vectors from the source, we need a very large codebook. This requirement can be reduced by normalizing the source output vectors, then quantizing the normalized vector and the normalization factor separately [156,145]. In this way, the variation due to the dynamic range is represented by the normalization factor or *gain*, while the vector quantizer is free to do what it does best, which is to capture the structure in the source output. Vector quantizers that function in this manner are called *gain-shape vector quantizers*. The pyramid quantizer discussed earlier is an example of a gain-shape vector quantizer.

10.7.2 Mean-Removed Vector Quantization

If we were to generate a codebook from an image, differing amounts of background illumination would result in vastly different codebooks. This effect can be significantly reduced if we remove the mean from each vector before quantization. The mean and the mean-removed vector can then be quantized separately. The mean can be quantized using a scalar quantization scheme, while the mean-removed vector can be quantized using a vector quantizer. Of course, if this strategy is used, the vector quantizer should be designed using mean-removed vectors as well.

Example 10.7.1:

Let us encode the Sinan image using a codebook generated by the Sena image, as we did in Figure 10.16 However, this time we will use a mean-removed vector quantizer. The result is

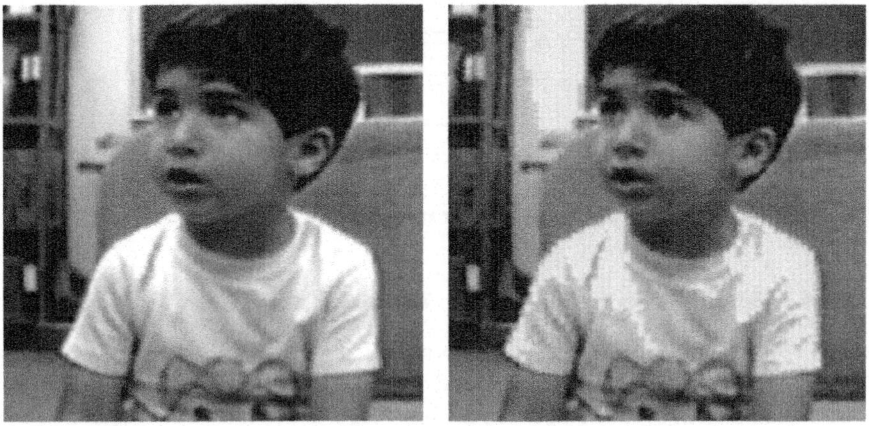

FIGURE 10.26 Left: Reconstructed image using mean-removed vector quantization and the Sena image as the training set. Right: LBG vector quantization with the Sena image as the training set.

shown in Figure 10.26. For comparison we have also included the reconstructed image from Figure 10.16. Notice the annoying blotches on the shoulder have disappeared. However, the reconstructed image also suffers from more blockiness. The blockiness increases because adding the mean back into each block accentuates the discontinuity at the block boundaries.

Each approach has its advantages and disadvantages. Which approach we use in a particular application depends very much on the application. ♦

10.7.3 Classified Vector Quantization

We can sometimes divide the source output into separate classes with different spatial properties. In these cases, it can be very beneficial to design separate vector quantizers for the different classes. This approach, referred to as *classified vector quantization*, is especially useful in image compression, where edges and nonedge regions form two distinct classes. We can separate the training set into vectors that contain edges and vectors that do not. A separate vector quantizer can be developed for each class. During the encoding process, the vector is first tested to see if it contains an edge. A simple way to do this is to check the variance of the pixels in the vector. A large variance will indicate the presence of an edge. More sophisticated techniques for edge detection can also be used. Once the vector is classified, the corresponding codebook can be used to quantize the vector. The encoder transmits both the label for the codebook used and the label for the vector in the codebook [157].

A slight variation of this strategy is to use different kinds of quantizers for the different classes of vectors. For example, if certain classes of source outputs require quantization at a higher rate than is possible using LBG vector quantizers, we can use lattice vector quantizers. An example of this approach can be found in [158].

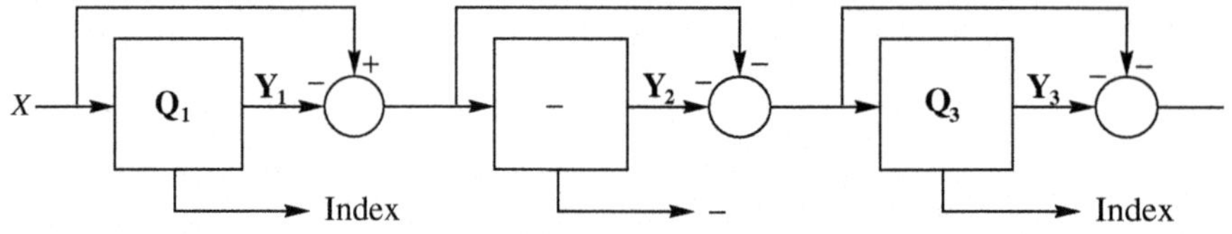

FIGURE 10.27 **A three-stage vector quantizer.**

10.7.4 Multistage Vector Quantization

Multistage vector quantization [159] is an approach that reduces both the encoding complexity and the memory requirements for vector quantization, especially at high rates. In this approach, the input is quantized in several stages. In the first stage, a low-rate vector quantizer is used to generate a coarse approximation of the input. This coarse approximation, in the form of the label of the output point of the vector quantizer, is transmitted to the receiver. The error between the original input and the coarse representation is quantized by the second-stage quantizer, and the label of the output point is transmitted to the receiver. In this manner, the input to the nth-stage vector quantizer is the difference between the original input and the reconstruction obtained from the outputs of the preceding $n-1$ stages. The difference between the input to a quantizer and the reconstruction value is often called the *residual*, and the multistage vector quantizers are also known as *residual vector quantizers* [160]. The reconstructed vector is the sum of the output points of each of the stages. Suppose we have a three-stage vector quantizer, with the three quantizers represented by $\mathbf{Q}_1$, $\mathbf{Q}_2$, and $\mathbf{Q}_3$. Then for a given input $\mathbf{x}$, we find

$$\begin{aligned} \mathbf{y}_1 &= \mathbf{Q}_1(\mathbf{x}) \\ \mathbf{y}_2 &= \mathbf{Q}_2(\mathbf{x} - \mathbf{Q}_1(\mathbf{x})) \\ \mathbf{y}_3 &= \mathbf{Q}_3(\mathbf{x} - \mathbf{Q}_1(\mathbf{x}) - \mathbf{Q}_2(\mathbf{x} - \mathbf{Q}_1(\mathbf{x}))) \end{aligned} \tag{11}$$

The reconstruction $\hat{\mathbf{x}}$ is given by

$$\hat{\mathbf{x}} = \mathbf{y}_1 + \mathbf{y}_2 + \mathbf{y}_3 \tag{12}$$

This process is shown in Figure 10.27.

If we have K stages, and the codebook size of the nth-stage vector quantizer is L_n, then the effective size of the overall codebook is $L_1 \times L_2 \times \cdots \times L_K$. However, we need to store only $L_1 + L_2 + \cdots + L_K$ vectors, which is also the number of comparisons required. Suppose we have a five-stage vector quantizer, each with a codebook size of 32, meaning that we would have to store 160 codewords. This would provide an effective codebook size of $32^5 = 33,554,432$. The computational savings are also of the same order.

This approach allows us to use vector quantization at much higher rates than we could otherwise. However, at rates at which it is feasible to use LBG vector quantizers, the performance of the multistage vector quantizers is generally lower than the LBG vector quantizers [136]. The reason for this is that after the first few stages, much of the structure used by the vector quantizer has been removed, and the vector quantization advantage that depends on this structure is not available. Details on the design of residual vector quantizers can be found in [160,161].

There may be some vector inputs that can be well represented by fewer stages than others. A multistage vector quantizer with a variable number of stages can be implemented by extending the idea of recursively indexed scalar quantization to vectors. It is not possible to do this directly because there are some fundamental differences between scalar and vector quantizers. The input to a scalar quantizer is assumed to be *iid*. On the other hand, the vector quantizer can be viewed as a pattern-matching algorithm [162]. The input is assumed to be one of a number of different patterns. The scalar quantizer is used after the redundancy has been removed from the source sequence, while the vector quantizer takes advantage of the redundancy in the data.

With these differences in mind, the recursively indexed vector quantizer (RIVQ) can be described as a two-stage process. The first stage performs the normal pattern-matching function, while the second stage recursively quantizes the residual if the magnitude of the residual is greater than some prespecified threshold. The codebook of the second stage is ordered so that the magnitude of the codebook entries is a nondecreasing function of its index. We then choose an index I that will determine the mode in which the RIVQ operates.

The quantization rule Q, for a given input value $\mathbf{x}$, is as follows:

- Quantize $\mathbf{x}$ with the first-stage quantizer Q_1.
- If the residual $\|\mathbf{x} - \mathbf{Q}_1(\mathbf{x})\|$ is below a specified threshold, then $\mathbf{Q}_1(\mathbf{x})$ is the nearest output level.
- Otherwise, generate $\mathbf{x}_1 = \mathbf{x} - \mathbf{Q}_1(\mathbf{x})$ and quantize using the second-stage quantizer $\mathbf{Q}_2$. Check if the index J_1 of the output is below the index I. If so,

  $$\mathbf{Q}(\mathbf{x}) = \mathbf{Q}_1(\mathbf{x}) + \mathbf{Q}_2(\mathbf{x}_1)$$

  If not, form

  $$\mathbf{x}_2 = \mathbf{x}_1 - \mathbf{Q}(\mathbf{x}_1)$$

  and do the same for $\mathbf{x}_2$ as we did for $\mathbf{x}_1$.

This process is repeated until for some m, the index J_m falls below the index I, in which case $\mathbf{x}$ will be quantized to

$$\mathbf{Q}(\mathbf{x}) = \mathbf{Q}_1(\mathbf{x}) + \mathbf{Q}_2(\mathbf{x}_1) + \cdots + \mathbf{Q}_2(\mathbf{x}_M)$$

Thus, the RIVQ operates in two modes: when the index J of the quantized input falls below a given index I and when the index J falls above the index I.

Details on the design and performance of the recursively indexed vector quantizer can be found in [163,164].

10.7.5 Adaptive Vector Quantization

While LBG vector quantizers function by using the structure in the source output, this reliance on the use of the structure can also be a drawback when the characteristics of the source change over time. For situations like these, we would like to have the quantizer adapt to the changes in the source output.

For mean-removed and gain-shape vector quantizers, we can adapt the scalar aspect of the quantizer, that is, the quantization of the mean or the gain, using the techniques discussed in the previous chapter. In this section, we look at a few approaches to adapting the codebook of the vector quantizer to changes in the characteristics of the input.

One way of adapting the codebook to changing input characteristics is to start with a very large codebook designed to accommodate a wide range of source characteristics [165]. This large codebook can be ordered in some manner known to both transmitter and receiver. Given a sequence of input vectors to be quantized, the encoder can select a subset of the larger codebook to be used. Information about which vectors from the large codebook were used can be transmitted as a binary string. For example, if the large codebook contained 10 vectors, and the encoder was to use the second, third, fifth, and ninth vectors, we would send the binary string 0110100010, with a 1 representing the position of the codeword used in the large codebook. This approach permits the use of a small codebook that is matched to the local behavior of the source.

This approach can be used with particular effectiveness with the recursively indexed vector quantizer [163]. Recall that in the recursively indexed vector quantizer, the quantized output is always within a prescribed distance of the inputs, determined by the index I. This means that the set of output values of the RIVQ can be viewed as an accurate representation of the inputs and their statistics. Therefore, we can treat a subset of the output set of the previous intervals as our large codebook. We can then use the method described in [165] to inform the receiver of which elements of the previous outputs form the codebook for the next interval. This method (while not the most efficient) is quite simple. Suppose an output set, in order of first appearance, is $\{p, a, q, s, l, t, r\}$, and the desired codebook for the interval to be encoded is $\{a, q, l, r\}$. Then we would transmit the binary string 0110101 to the receiver. The 1s correspond to the letters in the output set, which would be elements of the desired codebook. We select the subset for the current interval by finding the closest vectors from our collection of past outputs to the input vectors of the current set. This means that there is an inherent delay of one interval imposed by this approach. The overhead required to send the codebook selection is M/N, where M is the number of vectors in the output set and N is the interval size.

Another approach to updating the codebook is to check the distortion incurred while quantizing each input vector. Whenever this distortion is above some specified threshold, a different higher-rate mechanism is used to encode the input. The higher-rate mechanism might be the scalar quantization of each component, or the use of a high-rate lattice vector quantizer. This quantized representation of the input is transmitted to the receiver and, at the same time, added to both the encoder and decoder codebooks. In order to keep the size of the codebook the same, an entry must be discarded when a new vector is added to the codebook. Selecting an entry to discard is handled in a number of different ways. Variations of this approach have been used for speech coding, image coding, and video coding (see [154, 155, 156, 157, 158] for more details).

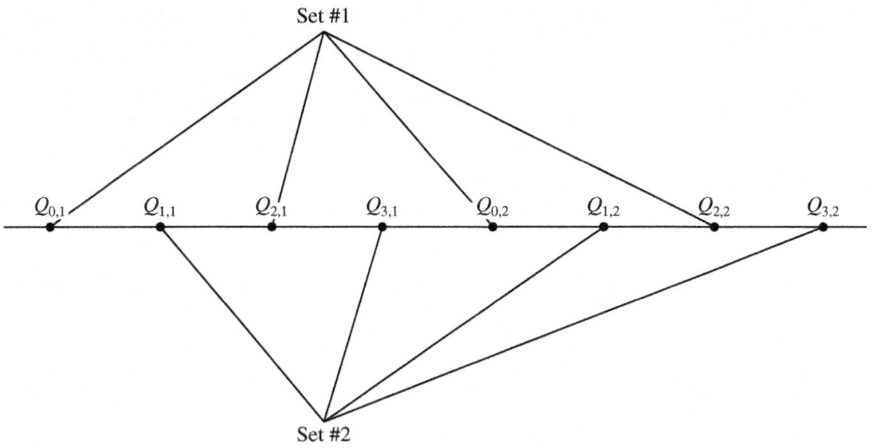

FIGURE 10.28 Reconstruction levels for a 2-bit trellis-coded quantizer.

10.8 Trellis-Coded Quantization

Finally, we look at a quantization scheme that appears to be somewhat different from other vector quantization schemes. In fact, some may argue that it is not a vector quantizer at all. However, the trellis-coded quantization (TCQ) algorithm gets its performance advantage by exploiting the statistical structure exploited by the lattice vector quantizer. Therefore, we can argue that it should be classified as a vector quantizer.

The trellis-coded quantization algorithm was inspired by the appearance of a revolutionary concept in modulation called trellis-coded modulation (TCM). The TCQ algorithm and its entropy-constrained variants provide some of the best performance when encoding random sources. This quantizer can be viewed as a vector quantizer with very large dimension, but a restricted set of values for the components of the vectors.

Like a vector quantizer, the TCQ quantizes sequences of source outputs. Each element of a sequence is quantized using 2^R reconstruction levels selected from a set of 2^{R+1} reconstruction levels, where R is the number of bits per sample used by a trellis-coded quantizer. The 2^R element subsets are predefined; which particular subset is used is based on the reconstruction level used to quantize the previous quantizer input. However, the TCQ algorithm allows us to postpone a decision on which reconstruction level to use until we can look at a sequence of decisions. This way we can select the sequence of decisions that gives us the lowest amount of average distortion.

Let's take the case of a 2-bit quantizer. As described above, this means that we will need 2^3, or 8, reconstruction levels. Let's label these reconstruction levels as shown in Figure 10.28. The set of reconstruction levels is partitioned into two subsets: one consisting of the reconstruction values labeled $Q_{0,i}$ and $Q_{2,i}$, and the remainder comprising the second set. We use the first set to perform the quantization if the previous quantization level was one labeled $Q_{0,i}$ or $Q_{1,i}$; otherwise, we use the second set. Because the current reconstructed value defines the subset that can be used to perform the quantization on the next input, sometimes it may be advantageous to actually accept more distortion than necessary for the current sample in order to have less distortion in the next quantization step. In fact, at times it may be advantageous to accept poor quantization for several samples so that several samples down the line the quantization can result in less distortion. If you have followed this reasoning, you can see how we might be able to get lower overall distortion by looking at the quantization of an entire sequence of source outputs. The problem with delaying a decision is that the number of choices increases exponentially with each sample. In the 2-bit example, for the first sample we have four choices; for each of these four choices we have four choices for the second sample. For each of these 16 choices we have four choices for the third sample, and so on. Luckily, there is a technique that can be used to keep this explosive growth of choices under control. The technique, called the *Viterbi algorithm* [166], is widely used in error control coding.

In order to explain how the Viterbi algorithm works, we will need to formalize some of what we have been discussing. The sequence of choices can be viewed in terms of a state diagram. Let's suppose we have four states: S_0, S_1, S_2, and S_3. We will say we are in state S_k if we use the reconstruction levels $Q_{k,1}$ or $Q_{k,2}$. Thus, if we use the reconstruction levels $Q_{0,i}$, we are in state S_0. We have said that we use the elements of Set #1 if the previous quantization levels were $Q_{0,i}$ or $Q_{1,i}$. As Set #1 consists of the quantization levels $Q_{0,i}$ and $Q_{2,i}$, this means that we can go from state S_0 and S_1 to states S_0 and S_2. Similarly, from states S_2 and S_3 we can only go to states S_1 and S_3. The state diagram can be drawn as shown in Figure 10.29.

Let's suppose we go through two sequences of choices that converge to the same state, after which both sequences are identical. This means that the sequence of choices that had incurred a higher distortion at the time the two sequences converged will have a higher distortion from then on. In the end we will select the sequence of choices that results in the lowest distortion; therefore, there is no point in continuing to keep track of a sequence that we will discard anyway. This means that whenever two sequences of choices converge, we can discard one of them. How often does this happen? In order to see this, let's introduce time into our state diagram. The state diagram with the element of time introduced into it is called a *trellis diagram*. The trellis for this particular example is shown in Figure 10.30. At each time instant, we can go from one state to two other states. And, at each step we have two sequences that converge to each state. If we discard one of the two sequences that converge to each state, we can see that, no matter how long a sequence of decisions we use, we will always end up with four sequences.

Notice that, assuming the initial state is known to the decoder, any path through this particular trellis can be described to the decoder using 1 bit per sample. From each state we can only go to two other states. In Figure 10.31, we have marked the branches with the bits used to signal that transition. Given that each state corresponds to two quantization levels, specifying the quantization level for each sample would require an additional bit, resulting in a total of 2 bits per sample. Let's see how all this works together in an example.

Example 10.8.1:

Using the quantizer whose quantization levels are shown in Figure 10.32, we will quantize the sequence of values 0.2, 1.6, 2.3. For the distortion measure we will use the sum of absolute differences. If we simply used the quantization levels marked as Set #1 in Figure 10.28, we would quantize 0.2 to the reconstruction value 0.5, for a distortion of 0.3. The second sample

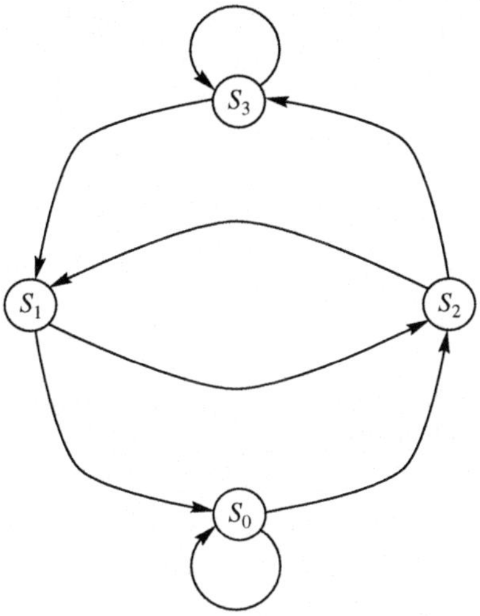

FIGURE 10.29 State diagram for the selection process.

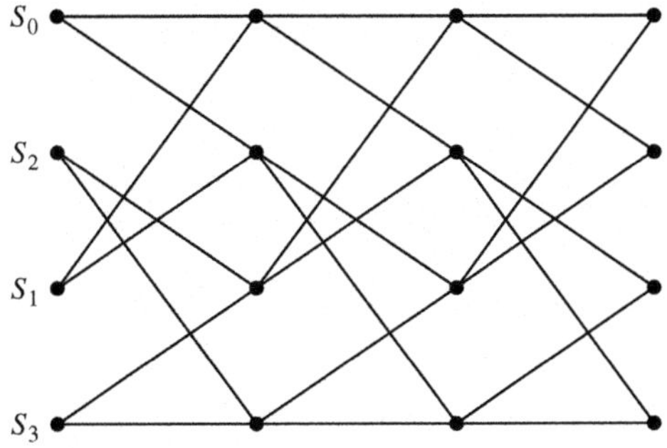

FIGURE 10.30 Trellis diagram for the selection process.

value of 1.6 would be quantized to 2.5, and the third sample value of 2.3 would also be quantized to 2.5, resulting in a total distortion of 1.4. If we used Set #2 to quantize these values, we would end up with a total distortion of 1.6. Let's see how much distortion results when using the TCQ algorithm.

We start by quantizing the first sample using the two quantization levels $Q_{0,1}$ and $Q_{0,2}$. The reconstruction level $Q_{0,2}$, or 0.5, is closer and results in an absolute difference of 0.3. We mark this on the first node corresponding to S_0. We then quantize the first sample using $Q_{1,1}$ and $Q_{1,2}$. The closest reconstruction value is $Q_{1,2}$, or 1.5, which results in a distortion value of 1.3. We mark the first node corresponding to S_1. Continuing in this manner, we get a distortion value of 1.7 when we use the reconstruction levels corresponding to state S_2 and a distortion value of 0.7 when we use the reconstruction levels corresponding to state S_3. At this point the trellis looks like Figure 10.33. Now we move on to the second sample. Let's first quantize the second sample value of 1.6 using the quantization levels associated with state S_0.

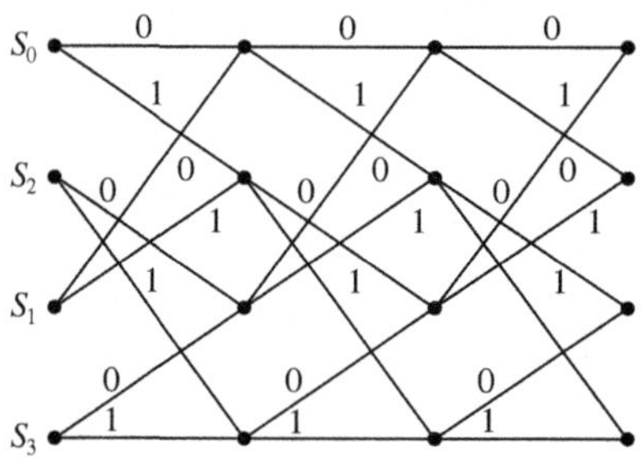

FIGURE 10.31 **Trellis diagram for the selection process with binary labels for the state transitions.**

The reconstruction levels associated with state S_0 are -3.5 and 0.5. The closest value to 1.6 is 0.5. This results in an absolute difference for the second sample of 1.1. We can reach S_0 from S_0 and from S_1. If we accept the first sample reconstruction corresponding to S_0, we will end up with an accumulated distortion of 1.4. If we accept the reconstruction corresponding to state S_1, we get an accumulated distortion of 2.4. Since the accumulated distortion is less if we accept the transition from state S_0, we do so and discard the transition from state S_1. Continuing in this fashion for the remaining states, we end up with the situation depicted in Figure 10.34. The sequence of decisions that have been terminated are shown by an X on the branch corresponding to the particular transition. The accumulated distortion is listed at each node. Repeating this procedure for the third sample value of 2.3, we obtain the trellis shown in Figure 10.35. If we want to terminate the algorithm at this time, we can pick the sequence of decisions with the smallest accumulated distortion. In this particular example, the sequence would be S_3, S_1, S_2. The accumulated distortion is 1.0, which is less than what we would have obtained using either Set #1 or Set #2. ♦

10.9 Summary

In this chapter we introduced the technique of vector quantization. We have seen how we can make use of the structure exhibited by groups, or vectors, of values to obtain compression. Because there are different kinds of structure in different kinds of data, there are a number of different ways to design vector quantizers. Because data from many sources, when viewed as vectors, tend to form clusters, we can design quantizers that essentially consist of representations of these clusters. We also described aspects of the design of vector quantizers and looked

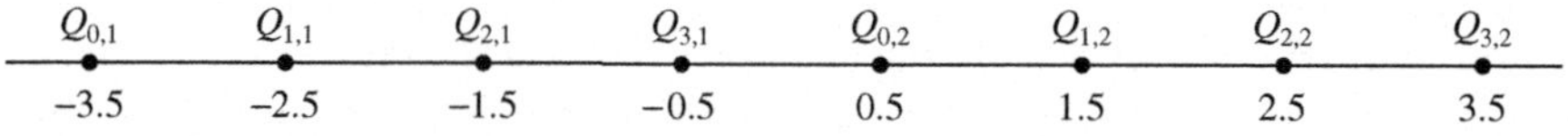

FIGURE 10.32 **Reconstruction levels for a 2-bit trellis-coded quantizer.**

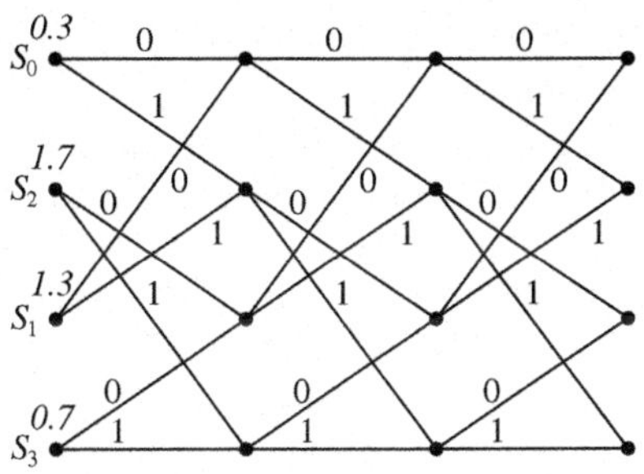

FIGURE 10.33 **Quantizing the first sample.**

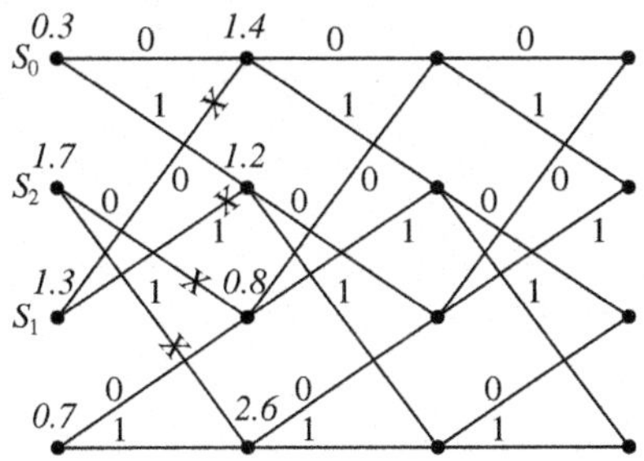

FIGURE 10.34 **Quantizing the second sample.**

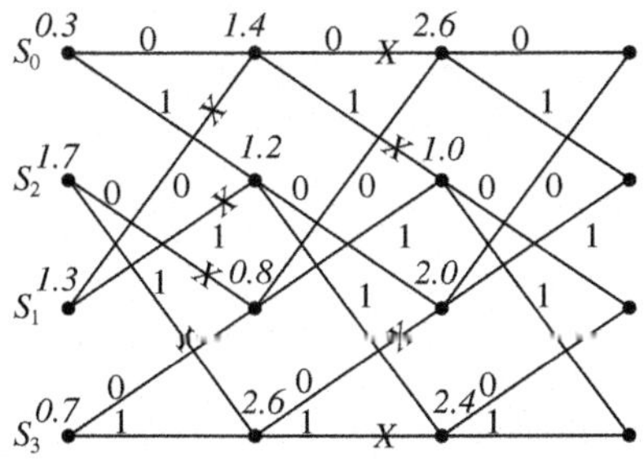

FIGURE 10.35 **Quantizing the third sample.**

at some applications. Recent literature in this area is substantial, and we have barely skimmed the surface of the large number of interesting variations of this technique.

Further Reading The subject of vector quantization is dealt with extensively in the book *Vector Quantization and Signal Compression*, by A. Gersho and R.M. Gray [136]. There is also an excellent collection of papers called *Vector Quantization*, edited by H. Abut and published by IEEE Press [155].

There are a number of excellent tutorial articles on this subject:

1. "Vector Quantization," by R.M. Gray, in the April 1984 issue of *IEEE Acoustics, Speech, and Signal Processing Magazine* [273].
2. "Vector Quantization: A Pattern Matching Technique for Speech Coding," by A. Gersho and V. Cuperman, in the December 1983 issue of *IEEE Communications Magazine* [162].
3. "Vector Quantization in Speech Coding," by J. Makhoul, S. Roucos, and H. Gish, in the November 1985 issue of the *Proceedings of the IEEE* [167].
4. "Vector Quantization," by P.F. Swaszek, in *Communications and Networks*, edited by I.F. Blake and H.V. Poor [274].
5. A survey of various image-coding applications of vector quantization can be found in "Image Coding Using Vector Quantization: A Review," by N.M. Nasrabadi and R.A. King, in the August 1988 issue of the *IEEE Transactions on Communications* [168].
6. A thorough review of lattice vector quantization can be found in "Lattice Quantization," by J.D. Gibson and K. Sayood, in *Advances in Electronics and Electron Physics* [152].

The area of vector quantization is an active one, and new techniques that use vector quantization are continually being developed. The journals that report work in this area include *IEEE Transactions on Information Theory*, *IEEE Transactions on Communications*, *IEEE Transactions on Signal Processing*, and *IEEE Transactions on Image Processing*, among others.

10.10 Projects and Problems

1. In Example 10.3.2 we increased the SNR by about 0.3 dB by moving the top-left output point to the origin. What would happen if we moved the output points at the four corners to the positions $(\pm\Delta, 0)$, $(0, \pm\Delta)$? As in the example, assume the input has a Laplacian distribution with mean zero and variance one, and $\Delta = 0.7309$. You can obtain the answer analytically or through simulation.
2. For the quantizer of the previous problem, rather than moving the output points to $(\pm\Delta, 0)$ and $(0, \pm\Delta)$, we could have moved them to other positions that might have provided a larger increase in SNR. Write a program to test different (reasonable) possibilities and report on the best and worst cases.
3. In the program `trainvq.c` the empty cell problem is resolved by replacing the vector with no associated training set vectors with a training set vector from the quantization region with the largest number of vectors. In this problem, we will investigate some possible alternatives.
   Generate a sequence of pseudorandom numbers with a triangular distribution between 0 and 2. (You can obtain a random number with a triangular distribution by adding two uniformly distributed random numbers.) Design an eight-level, two-dimensional vector quantizer with the initial codebook shown in Table 10.9.
   (a) Use the `trainvq` program to generate a codebook with 10,000 random numbers as the training set. Comment on the final codebook you obtain. Plot the elements of the codebook and discuss why they ended up where they did.

TABLE 10.9 Initial codebook for Problem 3.

1	1
1	2
1	0.5
0.5	1
0.5	0.5
1.5	1
2	5
3	3

(b) Modify the program so that the empty cell vector is replaced with a vector from the quantization region with the largest distortion. Comment on any changes in the distortion (or lack of change). Is the final codebook different from the one you obtained earlier?

(c) Modify the program so that whenever an empty cell problem arises, a two-level quantizer is designed for the quantization region with the largest number of output points. Comment on any differences in the codebook and distortion from the previous two cases.

4. Generate a 16-dimensional codebook of size 64 for the Sena image. Construct the vector as a 4×4 block of pixels, an 8×2 block of pixels, and a 16×1 block of pixels. Comment on the differences in the mean squared errors and the quality of the reconstructed images. You can use the program `trvqsp_img` to obtain the codebooks.

5. In Example10.6.1 we designed a 60-level two-dimensional quantizer by taking the two-dimensional representation of an 8-level scalar quantizer, removing 12 output points from the 64 output points, and adding 8 points in other locations. Assume the input is Laplacian with zero mean and unit variance, and $\Delta = 0.7309$.

(a) Calculate the increase in the probability of overload by the removal of the 12 points from the original 64.

(b) Calculate the decrease in overload probability when we added the 8 new points to the remaining 52 points.

6. In this problem we will compare the performance of a 16-dimensional pyramid vector quantizer and a 16-dimensional LBG vector quantizer for two different sources. In each case the codebook for the pyramid vector quantizer consists of 272 elements:

- 32 vectors with 1 element equal to $\pm\Delta$, and the other 15 equal to zero, and
- 240 vectors with 2 elements equal to $\pm\Delta$ and the other 14 equal to zero.

The value of Δ should be adjusted to give the best performance. The codebook for the LBG vector quantizer will be obtained by using the program `trvqsp_img` on the source output. You will have to modify `trvqsp_img` slightly to give you a codebook that is not a power of two.

(a) Use the two quantizers to quantize a sequence of 10,000 zero mean unit variance Laplacian random numbers. Using either the mean squared error or the SNR as a measure of performance, compare the performance of the two quantizers.

(b) Use the two quantizers to quantize the Sinan image. Compare the two quantizers using either the mean squared error or the SNR and the reconstructed image. Compare the difference between the performance of the two quantizers with the difference when the input was random.

7. Design one-, two-, three-, four-, and eight-dimensional quantizers at the rate of 1 bit/sample using the LGB algorithm (you can use *trvqsplit*), for the uniform, Gaussian, and Laplacian distributions. Study how your design algorithm performs and suggest ways of improving it.

Use the random number generator programs to design and test the quantizers (`rangen.c`).

8. The purpose of this project is to have you familiarize yourself with vector quantization, in particular with vector quantization of images. The programs you will use include `trvqsp_img`, `vqimg_enc`, and `vqimg_dec`. The `trvqsp_img` program designs vector quantization codebooks for images using the splitting approach. The programs `vqimg_enc` and `vqimg_dec` are the vector quantization encoder and decoder for images. The indices generated by the encoder are coded using fixed-length codes.

Experiment with different ways of selecting vectors and examine the effect of the training set used for obtaining a codebook on the VQ performance. Suggest some ways of resolving the mismatch problem between the image being encoded and the codebook.

11

Differential Encoding

11.1 Overview

Sources such as speech and images have a great deal of correlation from sample to sample. We can use this fact to predict each sample based on its past and only encode and transmit the differences between the prediction and the sample value. Differential encoding schemes are built around this premise. Because the prediction techniques are rather simple, these schemes are much easier to implement than other compression schemes. In this chapter, we will look at various components of differential encoding schemes and study how they are used to encode sources—in particular, speech. We will also look at a widely used international differential encoding standard for speech encoding.

11.2 Introduction

In the last chapter we looked at vector quantization—a rather complex scheme requiring a significant amount of computational resources—as one way of taking advantage of the structure in the data to perform lossy compression. In this chapter, we look at a different approach that uses the structure in the source output in a slightly different manner, resulting in a significantly less complex system.

When we design a quantizer for a given source, the size of the quantization interval depends on the variance of the input. If we assume the input is uniformly distributed, the variance depends on the dynamic range of the input. In turn, the size of the quantization interval determines the amount of quantization noise incurred during the quantization process.

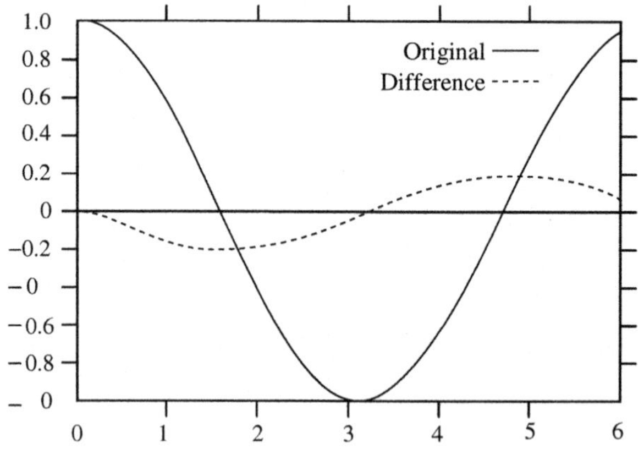

FIGURE 11.1 Sinusoid and sample-to-sample differences.

In many sources of interest, the sampled source output $\{x_n\}$ does not change a great deal from one sample to the next. This means that both the dynamic range and the variance of the sequence of differences $\{d_n = x_n - x_{n-1}\}$ are significantly smaller than that of the source output sequence. Furthermore, for correlated sources the distribution of d_n is highly peaked at zero. We made use of this skew, and resulting loss in entropy, for the lossless compression of images in Chapter 7. Given the relationship between the variance of the quantizer input and the incurred quantization error, it is also useful, in terms of lossy compression, to look at ways to encode the difference from one sample to the next rather than encoding the actual sample value. Techniques that transmit information by encoding differences are called *differential encoding techniques*.

Example 11.2.1:

Consider the half cycle of a sinusoid shown in Figure 11.1 that has been sampled at the rate of 30 samples per cycle. The value of the sinusoid ranges between 1 and -1. If we wanted to quantize the sinusoid using a uniform four-level quantizer, we would use a step size of 0.5, which would result in quantization errors in the range $[-0.25, 0.25]$. If we take the sample-to-sample differences (excluding the first sample), the differences lie in the range $[-0.2, 0.2]$. To quantize this range of values with a four-level quantizer requires a step size of 0.1, which results in quantization noise in the range $[-0.05, 0.05]$. ♦

The sinusoidal signal in the previous example is somewhat contrived. However, if we look at some of the real-world sources that we want to encode, we see that the dynamic range that contains most of the differences is significantly smaller than the dynamic range of the source output.

Example 11.2.2:

Figure 11.2 is the histogram of the Sinan image. Notice that the pixel values vary over almost the entire range of 0 to 255. To represent these values exactly, we need 8 bits per pixel. To represent these values in a lossy manner to within an error in the least significant bit, we need 7 bits per pixel. Figure 11.3 is the histogram of the differences.

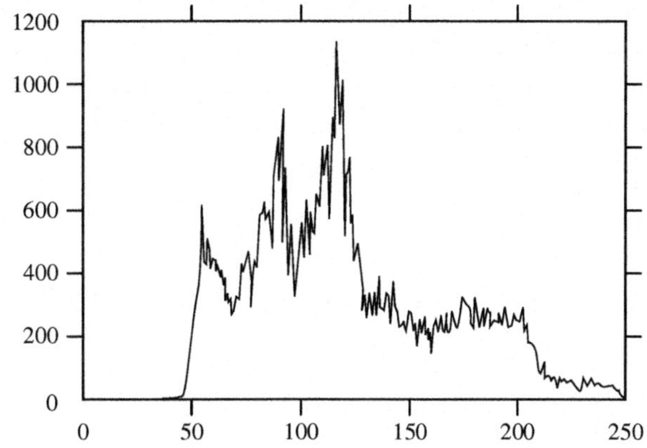

FIGURE 11.2 **Histogram of the Sinan image.**

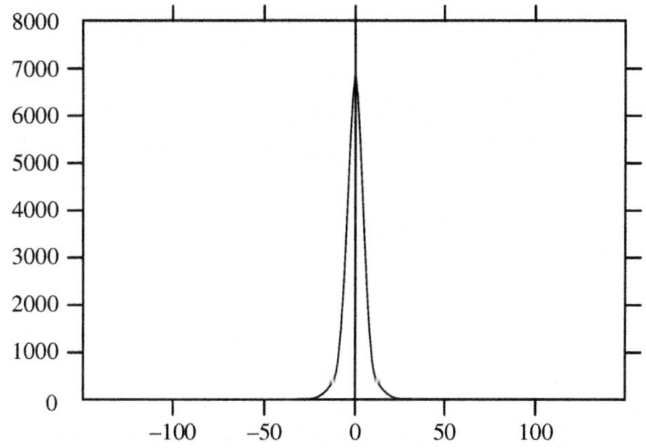

FIGURE 11.3 **Histogram of pixel-to-pixel differences of the Sinan image.**

More than 99% of the pixel values lie in the range −31 to 31. Therefore, if we are willing to accept distortion in the least significant bit, for more than 99% of the difference values we need 5 bits per pixel rather than 7. In fact, if we are willing to have a small percentage of the differences with a larger error, we could get by with 4 bits for each difference value. ♦

In both examples, we have shown that the dynamic range of the differences between samples is substantially less than the dynamic range of the source output. In the following sections we describe encoding schemes that take advantage of this fact to provide improved compression performance.

11.3 The Basic Algorithm

Although it takes fewer bits to encode differences than it takes to encode the original pixel, we have not said whether it is possible to recover an acceptable reproduction of the original sequence from the quantized difference value. When we were looking at lossless compression schemes, we found that if we encoded and transmitted the first value of a sequence, followed by the encoding of the differences between samples, we could losslessly recover the original sequence. Unfortunately, a strictly analogous situation does not exist for lossy compression.

Example 11.3.1:

Suppose a source puts out the sequence

$$6.2\ 9.7\ 13.2\ 5.9\ 8\ 7.4\ 4.2\ 1.8$$

We could generate the following sequence by taking the difference between samples (assume that the first sample value is zero):

$$6.2\ 3.5\ 3.5\ -7.3\ 2.1\ -0.6\ -3.2\ -2.4$$

If we losslessly encoded these values, we could recover the original sequence at the receiver by adding back the difference values. For example, to obtain the second reconstructed value, we add the difference 3.5 to the first received value 6.2 to obtain a value of 9.7. The third reconstructed value can be obtained by adding the received difference value of 3.5 to the second reconstructed value of 9.7, resulting in a value of 13.2, which is the same as the third value in the original sequence. Thus, by adding the nth received difference value to the $(n-1)$th reconstruction value, we can recover the original sequence exactly.

Now let us look at what happens if these difference values are encoded using a lossy scheme. Suppose we had a seven-level quantizer with output values $-6, -4, -2, 0, 2, 4, 6$. The quantized sequence would be

$$6\ 4\ 4\ -6\ 2\ 0\ -4\ -2$$

If we follow the same procedure for reconstruction as we did for the lossless compression scheme, we get the sequence

$$6\ 10\ 14\ 8\ 10\ 10\ 6\ 4$$

The difference or error between the original sequence and the reconstructed sequence is

$$0.2\ -0.3\ -0.8\ -2.1\ -2\ -2.6\ -1.8\ -2.2$$

Notice that initially the magnitudes of the error are quite small (0.2, 0.3). As the reconstruction progresses, the magnitudes of the error become significantly larger (2.6, 1.8, 2.2). ♦

To see what is happening, consider a sequence $\{x_n\}$. A difference sequence $\{d_n\}$ is generated by taking the differences $x_n - x_{n-1}$. This difference sequence is quantized to obtain the sequence $\{\hat{d}_n\}$:

$$\hat{d}_n = Q[d_n] = d_n + q_n$$

where q_n is the quantization error. At the receiver, the reconstructed sequence $\{\hat{x}_n\}$ is obtained by adding $\hat{d}_n$ to the previous reconstructed value $\hat{x}_{n-1}$:

$$\hat{x}_n = \hat{x}_{n-1} + \hat{d}_n$$

Let us assume that both transmitter and receiver start with the same value x_0, that is, $\hat{x}_0 = x_0$. Follow the quantization and reconstruction process for the first few samples:

$$d_1 = x_1 - x_0 \quad (1)$$
$$\hat{d}_1 = Q[d_1] = d_1 + q_1 \quad (2)$$
$$\hat{x}_1 = x_0 + \hat{d}_1 = x_0 + d_1 + q_1 = x_1 + q_1 \quad (3)$$
$$d_2 = x_2 - x_1 \quad (4)$$
$$\hat{d}_2 = Q[d_2] = d_2 + q_2 \quad (5)$$
$$hatx_2 = \hat{x}_1 + \hat{d}_2 = x_1 + q_1 + d_2 + q_2 \quad (6)$$
$$= x_2 + q_1 + q_2 \quad (7)$$

Continuing this process, at the nth iteration we get

$$\hat{x}_n = x_n + \sum_{k=1}^{n} q_k \quad (8)$$

We can see that the quantization error accumulates as the process continues. Theoretically, if the quantization error process is zero mean, the errors will cancel each other out in the long run. In practice, often long before that can happen, the finite precision of the machines causes the reconstructed value to overflow.

Notice that the encoder and decoder are operating with different pieces of information. The encoder generates the difference sequence based on the original sample values, while the decoder adds back the quantized difference onto a distorted version of the original signal. We can solve this problem by forcing both encoder and decoder to use the same information during the differencing and reconstruction operations. The only information available to the receiver about the sequence $\{x_n\}$ is the reconstructed sequence $\{\hat{x}_n\}$. As this information is also available to the transmitter, we can modify the differencing operation to use the reconstructed value of the previous sample, instead of the previous sample itself, that is,

$$d_n = x_n - \hat{x}_{n-1} \quad (9)$$

Using this new differencing operation, let's repeat our examination of the quantization and reconstruction process. We again assume that $\hat{x}_0 = x_0$.

$$d_1 = x_1 - x_0 \quad (10)$$
$$\hat{d}_1 = Q[d_1] = d_1 + q_1 \quad (11)$$
$$\hat{x}_1 = x_0 + \hat{d}_1 = x_0 + d_1 + q_1 = x_1 + q_1 \quad (12)$$
$$d_2 = x_2 - \hat{x}_1 \quad (13)$$
$$\hat{d}_2 = Q[d_2] = d_2 + q_2 \quad (14)$$
$$\hat{x}_2 = \hat{x}_1 + \hat{d}_2 = \hat{x}_1 + d_2 + q_2 \quad (15)$$
$$= x_2 + q_2 \quad (16)$$

At the nth iteration we have

$$\hat{x}_n = x_n + q_n \quad (17)$$

and there is no accumulation of the quantization noise. In fact, the quantization noise in the nth reconstructed sequence is the quantization noise incurred by the quantization of the nth difference. The quantization error for the difference sequence is substantially less than the quantization error for the original sequence. Therefore, this procedure leads to an overall reduction of the quantization error. If we are satisfied with the quantization error for a given number of bits per sample, then we can use fewer bits with a differential encoding procedure to attain the same distortion.

Example 11.3.2:

Let us try to quantize and then reconstruct the sinusoid of Example 11.2.1 using the two different differencing approaches. Using the first approach, we get a dynamic range of differences from −0.2 to 0.2. Therefore, we use a quantizer step size of 0.1. In the second approach, the differences lie in the range [−0.4, 0.4]. In order to cover this range, we use a step size in the quantizer of 0.2. The reconstructed signals are shown in Figure 11.4.

Notice in the first case that the reconstruction diverges from the signal as we process more and more of the signal. Although the second differencing approach uses a larger step size, this approach provides a more accurate representation of the input. ♦

A block diagram of the differential encoding system as we have described it to this point is shown in Figure 11.5. We have drawn a dotted box around the portion of the encoder that mimics the decoder. The encoder must mimic the decoder in order to obtain a copy of the reconstructed sample used to generate the next difference.

We would like our difference value to be as small as possible. For this to happen, given the system we have described to this point, $\hat{x}_{n-1}$ should be as close to x_n as possible. However, $\hat{x}_{n-1}$ is the reconstructed value of x_{n-1}; therefore, we would like $\hat{x}_{n-1}$ to be close to x_{n-1}. Unless x_{n-1} is always very close to x_n, some function of past values of the reconstructed sequence can often provide a better prediction of x_n. We will look at some of these *predictor* functions later in this chapter. For now, let's modify Figure 11.5 and replace the delay block with a predictor block to obtain our basic differential encoding system as shown in Figure 11.6. The output of the predictor is the prediction sequence $\{p_n\}$ given by

$$p_n = f(\hat{x}_{n-1}, \hat{x}_{n-2}, \ldots, \hat{x}_0) \quad (18)$$

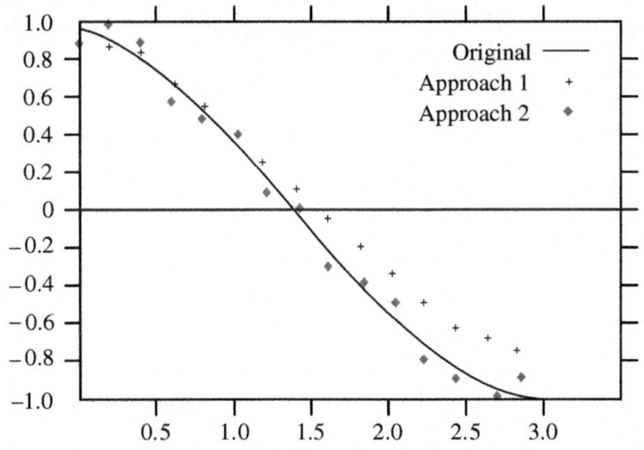

FIGURE 11.4 Sinusoid and reconstructions.

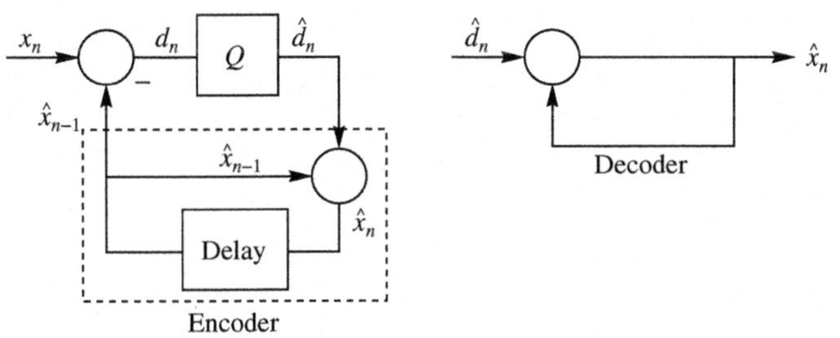

FIGURE 11.5 A simple differential encoding system.

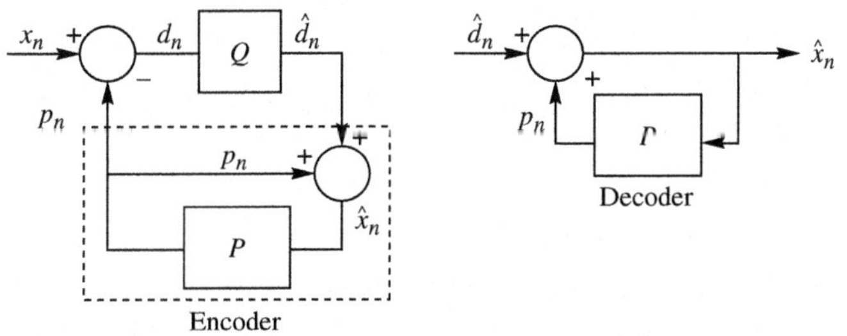

FIGURE 11.6 The basic algorithm.

This basic differential encoding system is known as the differential pulse code modulation (DPCM) system. The DPCM system was developed at Bell Laboratories a few years after World War II [169]. It is most popular as a speech-encoding system and is widely used in telephone communications.

As we can see from Figure 11.6, the DPCM system consists of two major components, the predictor and the quantizer. The study of DPCM is basically the study of these two components.

In the following sections, we will look at various predictor and quantizer designs and see how they function together in a differential encoding system.

11.4 Prediction in DPCM

Differential encoding systems like DPCM gain their advantage by the reduction in the variance and dynamic range of the difference sequence. How much the variance is reduced depends on how well the predictor can predict the next symbol based on the past reconstructed symbols. In this section we will mathematically formulate the prediction problem. The analytical solution to this problem will give us one of the more widely used approaches to the design of the predictor. In order to follow this development, some familiarity with the mathematical concepts of expectation and correlation is needed. These concepts are described in Appendix A.

Define σ_d^2, the variance of the difference sequence, as

$$\sigma_d^2 = E[(x_n - p_n)^2] \tag{19}$$

where $E[]$ is the expectation operator. As the predictor outputs p_n are given by (18), the design of a good predictor is essentially the selection of the function $f(\cdot)$ that minimizes σ_d^2. One problem with this formulation is that $\hat{x}_n$ is given by

$$\hat{x}_n = x_n + q_n$$

and q_n depends on the variance of d_n. Thus, by picking $f(\cdot)$, we affect σ_d^2, which in turn affects the reconstruction $\hat{x}_n$, which then affects the selection of $f(\cdot)$. This coupling makes an explicit solution extremely difficult for even the most well-behaved source [170]. As most real sources are far from well behaved, the problem becomes computationally intractable in most applications.

We can avoid this problem by making an assumption known as the *fine quantization assumption*. We assume that quantizer step sizes are so small that we can replace $\hat{x}_n$ by x_n, and therefore

$$p_n = f(x_{n-1}, x_{n-2}, \ldots, x_0) \tag{20}$$

Once the function $f(\cdot)$ has been found, we can use it with the reconstructed values $\hat{x}_n$ to obtain p_n. If we now assume that the output of the source is a stationary process, from the study of random processes [171] we know that the function that minimizes σ_d^2 is the conditional expectation $E[x_n | x_{n-1}, x_{n-2}, \ldots, x_0]$. Unfortunately, the assumption of stationarity is generally not true, and even if it were, finding this conditional expectation requires the knowledge of nth-order conditional probabilities, which would generally not be available.

Given the difficulty of finding the best solution, in many applications we simplify the problem by restricting the predictor function to be linear. That is, the prediction p_n is given by

$$p_n = \sum_{i=1}^{N} a_i \hat{x}_{n-i} \tag{21}$$

The value of N specifies the order of the predictor. Using the fine quantization assumption, we can now write the predictor design problem as follows: Find the $\{a_i\}$ so as to minimize σ_d^2:

$$\sigma_d^2 = E\left[\left(x_n - \sum_{i=1}^{N} a_i x_{n-i}\right)^2\right] \tag{22}$$

where we assume that the source sequence is a realization of a real-valued wide-sense stationary process. Take the derivative of σ_d^2 with respect to each of the a_i and set this equal to zero. We get N equations and N unknowns:

$$\frac{\partial \sigma_d^2}{\partial a_1} = -2E\left[\left(x_n - \sum_{i=1}^{N} a_i x_{n-i}\right) x_{n-1}\right] = 0 \tag{23}$$

$$\frac{\partial \sigma_d^2}{\partial a_2} = -2E\left[\left(x_n - \sum_{i=1}^{N} a_i x_{n-i}\right) x_{n-2}\right] = 0 \tag{24}$$

$$\vdots \quad \vdots$$

$$\frac{\partial \sigma_d^2}{\partial a_N} = -2E\left[\left(x_n - \sum_{i=1}^{N} a_i x_{n-i}\right) x_{n-N}\right] = 0 \tag{25}$$

Taking the expectations, we can rewrite these equations as

$$\sum_{i=1}^{N} a_i R_{xx}(i-1) = R_{xx}(1) \tag{26}$$

$$\sum_{i=1}^{N} a_i R_{xx}(i-2) = R_{xx}(2) \tag{27}$$

$$\vdots \quad \vdots$$

$$\sum_{i=1}^{N} a_i R_{xx}(i-N) = R_{xx}(N) \tag{28}$$

where $R_{xx}(k)$ is the autocorrelation function of x_n:

$$R_{xx}(k) = E[x_n x_{n+k}] \tag{29}$$

We can write these equations in matrix form as

$$\mathbf{Ra} = \mathbf{p} \tag{30}$$

where

$$\mathbf{R} = \begin{bmatrix} R_{xx}(0) & R_{xx}(1) & R_{xx}(2) & \cdots & R_{xx}(N-1) \\ R_{xx}(1) & R_{xx}(0) & R_{xx}(1) & \cdots & R_{xx}(N-2) \\ R_{xx}(2) & R_{xx}(1) & R_{xx}(0) & \cdots & R_{xx}(N-3) \\ \vdots & \vdots & & & \vdots \\ R_{xx}(N-1) & R_{xx}(N-2) & R_{xx}(N-3) & \cdots & R_{xx}(0) \end{bmatrix} \tag{31}$$

$$\mathbf{a} = \begin{bmatrix} a_1 \\ a_2 \\ a_3 \\ \vdots \\ a_N \end{bmatrix} \tag{32}$$

$$\mathbf{p} = \begin{bmatrix} R_{xx}(1) \\ R_{xx}(2) \\ R_{xx}(3) \\ \vdots \\ R_{xx}(N) \end{bmatrix} \tag{33}$$

where we have used the fact that $R_{xx}(-k) = R_{xx}(k)$ for real-valued wide-sense stationary processes. These equations are referred to as the discrete form of the Wiener-Hopf equations. If we know the autocorrelation values $\{R_{xx}(k)\}$ for $k = 0, 1, \ldots, N$, then we can find the predictor coefficients as

$$\mathbf{a} = \mathbf{R}^{-1}\mathbf{p} \tag{34}$$

Example 11.4.1:

For the speech sequence shown in Figure 11.7, let us find predictors of orders one, two, and three and examine their performance. We begin by estimating the autocorrelation values from the data. Given M data points, we use the following average to find the value for $R_{xx}(k)$:

$$R_{xx}(k) = \frac{1}{M-k} \sum_{i=1}^{M-k} x_i x_{i+k} \tag{35}$$

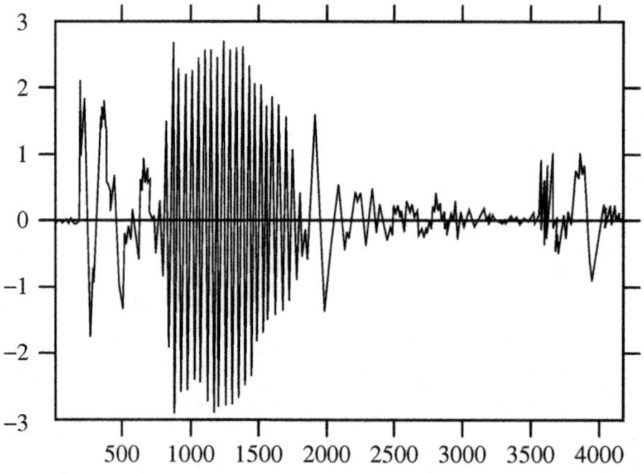

FIGURE 11.7 **A segment of speech: a male speaker saying the word "test."**

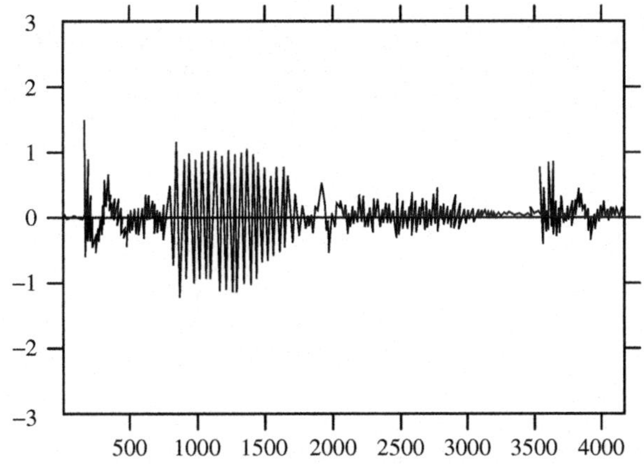

FIGURE 11.8 The residual sequence using a third-order predictor.

Using these autocorrelation values, we obtain the following coefficients for the three different predictors. For $N = 1$, the predictor coefficient is $a_1 = 0.66$; for $N = 2$, the coefficients are $a_1 = 0.596$ and $a_2 = 0.096$; and for $N = 3$, the coefficients are $a_1 = 0.577$, $a_2 = -0.025$, and $a_3 = 0.204$. We used these coefficients to generate the residual sequence. In order to see the reduction in variance, we computed the ratio of the source output variance to the variance of the residual sequence. For comparison, we also computed this ratio for the case where the residual sequence is obtained by taking the difference of neighboring samples. The sample-to-sample differences resulted in a ratio of 1.63. Compared to this, the ratio of the input variance to the variance of the residuals from the first-order predictor was 2.04. With a second-order predictor, this ratio rose to 3.37, and with a third-order predictor, the ratio was 6.28.

The residual sequence for the third-order predictor is shown in Figure 11.8. Notice that although there has been a reduction in the dynamic range, there is still substantial structure in the residual sequence, especially in the range of samples from about the 700th sample to the 2000th sample. We will look at ways of removing this structure when we discuss speech coding.

Let us now introduce a quantizer into the loop and look at the performance of the DPCM system. For simplicity, we will use a uniform quantizer. If we look at the histogram of the residual sequence, we find that it is highly peaked. Therefore, we will assume that the input to the quantizer will be Laplacian. We will also adjust the step size of the quantizer based on the variance of the residual. The step sizes provided in Chapter 9 are based on the assumption that the quantizer input has a unit variance. It is easy to show that when the variance differs from unity, the optimal step size can be obtained by multiplying the step size for a variance of one with the standard deviation of the input. Using this approach for a four-level Laplacian quantizer, we obtain step sizes of 0.75, 0.59, and 0.43 for the first-, second-, and third-order predictors, and step sizes of 0.3, 0.4, and 0.5 for an eight-level Laplacian quantizer. We measure the performance using two different measures, the signal-to-noise ratio (SNR) and

TABLE 11.1 Performance of DPCM system with different predictors and quantizers.

Quantizer	Predictor Order	SNR (dB)	SPER (dB)
Four-level	None	2.43	0
	1	3.37	2.65
	2	8.35	5.9
	3	8.74	6.1
Eight-level	None	3.65	0
	1	3.87	2.74
	2	9.81	6.37
	3	10.16	6.71

the signal-to-prediction error ratio. These are defined as follows:

$$\text{SNR(dB)} = 10\log_{10}\left(\frac{\sum_{i=1}^{M} x_i^2}{\sum_{i=1}^{M}(x_i - \hat{x}_i)^2}\right) \tag{36}$$

$$\text{SPER(dB)} = 10\log_{10}\left(\frac{\sum_{i=1}^{M} x_i^2}{\sum_{i=1}^{M}(x_i - p_i)^2}\right) \tag{37}$$

The results are tabulated in Table 11.1. For comparison we have also included the results when no prediction is used; that is, we directly quantize the input. Notice the large difference between using a first-order predictor and a second-order predictor, and then the relatively minor increase when going from a second-order predictor to a third-order predictor. This is fairly typical when using a fixed quantizer.

Finally, let's take a look at the reconstructed speech signal. The speech coded using a third-order predictor and an eight-level quantizer is shown in Figure 11.9. Although the reconstructed sequence looks like the original, notice that there is significant distortion in areas where the source output values are small. This is because in these regions the input to the quantizer is close to zero. Because the quantizer does not have a zero output level, the output of the quantizer flips between the two inner levels. If we listened to this signal, we would hear a hissing sound in the reconstructed signal.

The speech signal used to generate this example is contained among the data sets accompanying this book in the file `testm.raw`. The function `readau.c` can be used to read the file. You are encouraged to reproduce the results in this example and listen to the resulting reconstructions. ♦

If we look at the speech sequence in Figure 11.7, we can see that there are several distinct segments of speech. Between sample number 700 and sample number 2000, the speech looks periodic. Between sample number 2200 and sample number 3500, the speech is low amplitude and noiselike. Given the distinctly different characteristics in these two regions, it would make sense to use different approaches to encode these segments. Some approaches to dealing with

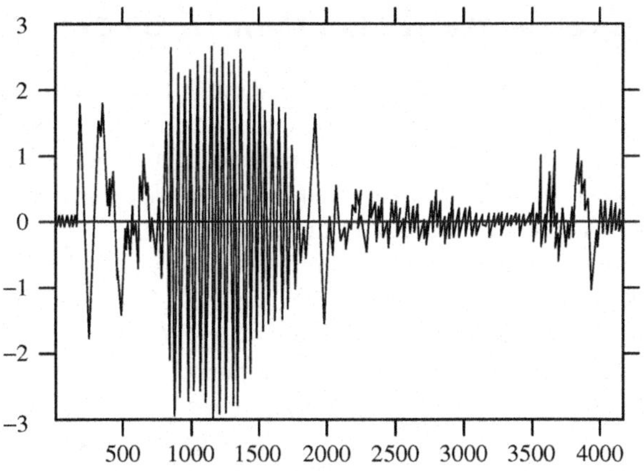

FIGURE 11.9 **The reconstructed sequence using a third-order predictor and an eight-level uniform quantizer.**

these issues are specific to speech coding, and we will encounter them when we specifically discuss encoding speech using DPCM. However, the problem is also much more widespread than when encoding speech. A general response to the nonstationarity of the input is the use of adaptation in prediction. We will look at some of these approaches in the next section.

11.5 Adaptive DPCM

As DPCM consists of two main components, the quantizer and the predictor, making DPCM adaptive means making the quantizer and the predictor adaptive. Recall that we can adapt a system based on its input or output. The former approach is called forward adaptation; the latter, backward adaptation. In the case of forward adaptation, the parameters of the system are updated based on the input to the encoder, which is not available to the decoder. Therefore, the updated parameters have to be sent to the decoder as side information. In the case of backward adaptation, the adaptation is based on the output of the encoder. As this output is also available to the decoder, there is no need for transmission of side information.

In cases where the predictor is adaptive, especially when it is backward adaptive, we generally use adaptive quantizers (forward or backward). The reason for this is that the backward adaptive predictor is adapted based on the quantized outputs. If for some reason the predictor does not adapt properly at some point, this results in predictions that are far from the input, and the residuals will be large. In a fixed quantizer, these large residuals will tend to fall in the overload regions with consequently unbounded quantization errors. The reconstructed values with these large errors will then be used to adapt the predictor, which will result in the predictor moving further and further from the input.

The same constraint is not present for quantization, and we can have adaptive quantization with fixed predictors.

11.5.1 Adaptive Quantization in DPCM

In forward adaptive quantization, the input is divided into blocks. The quantizer parameters are estimated for each block. These parameters are transmitted to the receiver as side information. In DPCM, the quantizer is in a feedback loop, which means that the input to the quantizer is not conveniently available in a form that can be used for forward adaptive quantization. Therefore, most DPCM systems use backward adaptive quantization.

The backward adaptive quantization used in DPCM systems is basically a variation of the backward adaptive Jayant quantizer described in Chapter 9. In Chapter 9, the Jayant algorithm was used to adapt the quantizer to a stationary input. In DPCM, the algorithm is used to adapt the quantizer to the local behavior of nonstationary inputs. Consider the speech segment shown in Figure 11.7 and the residual sequence shown in Figure 11.8. Obviously, the quantizer used around the 3000th sample should not be the same quantizer that was used around the 1000th sample. The Jayant algorithm provides an effective approach to adapting the quantizer to the variations in the input characteristics.

Example 11.5.1:

Let's encode the speech sample shown in Figure 11.7 using a DPCM system with a backward adaptive quantizer. We will use a third-order predictor and an eight-level quantizer. We will also use the following multipliers [124]:

$$M_0 = 0.90 \quad M_1 = 0.90 \quad M_2 = 1.25 \quad M_3 = 1.75$$

The results are shown in Figure 11.10. Notice the region at the beginning of the speech sample and between the 3000th and 3500th sample, where the DPCM system with the fixed quantizer had problems. Because the step size of the adaptive quantizer can become quite small, these regions have been nicely reproduced. However, right after this region, the speech output has a larger spike than the reconstructed waveform. This is an indication that the quantizer is not expanding rapidly enough. This can be remedied by increasing the value of M_3. The program used to generate this example is `dpcm_aqb`. You can use this program to study the behavior of the system for different configurations. ♦

11.5.2 Adaptive Prediction in DPCM

The equations used to obtain the predictor coefficients were derived based on the assumption of stationarity. However, we see from Figure 11.7 that this assumption is not true. In the speech segment shown in Figure 11.7, different segments have different characteristics. This is true for most sources we deal with; while the source output may be locally stationary over any significant length of the output, the statistics may vary considerably. In this situation, it is better to adapt the predictor to match the local statistics. This adaptation can be forward adaptive or backward adaptive.

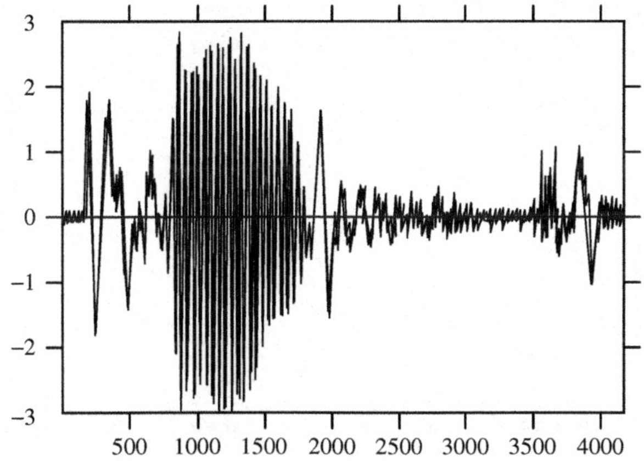

FIGURE 11.10 **The reconstructed sequence using a third-order predictor and an eight-level Jayant quantizer.**

DPCM with Forward Adaptive Prediction (DPCM-APF)

In forward adaptive prediction, the input is divided into segments or blocks. In speech coding this block usually consists of about 16 ms of speech. At a sampling rate of 8000 samples per second, this corresponds to 128 samples per block [134,172]. In image coding, we use an 8×8 block [173].

The autocorrelation coefficients are computed for each block. The predictor coefficients are obtained from the autocorrelation coefficients and quantized using a relatively high-rate quantizer. If the coefficient values are to be quantized directly, we need to use at least 12 bits per coefficient [134]. This number can be reduced considerably if we represent the predictor coefficients in terms of *parcor coefficients*; we will describe how to obtain the parcor coefficients in Chapter 17. For now, let's assume that the coefficients can be transmitted with an expenditure of about 6 bits per coefficient.

In order to estimate the autocorrelation for each block, we generally assume that the sample values outside each block are zero. Therefore, for a block length of M, the autocorrelation function for the lth block would be estimated by

$$R_{xx}^{(l)}(k) = \frac{1}{M-k} \sum_{i=(l-1)M+1}^{lM-k} x_i x_{i+k} \tag{38}$$

for k positive, or

$$R_{xx}^{(l)}(k) = \frac{1}{M+k} \sum_{i=(l-1)M+1-k}^{lM} x_i x_{i+k} \tag{39}$$

for k negative. Notice that $R_{xx}^{(l)}(k) = R_{xx}^{(l)}(-k)$, which agrees with our initial assumption.

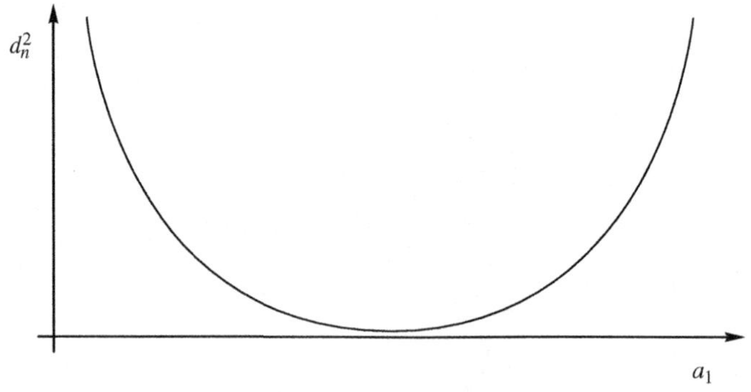

FIGURE 11.11 A plot of the residual squared versus the predictor coefficient.

DPCM with Backward Adaptive Prediction (DPCM-APB)

Forward adaptive prediction requires that we buffer the input. This introduces delay in the transmission of the speech. As the amount of buffering is small, the use of forward adaptive prediction when there is only one encoder and decoder is not a big problem. However, in the case of speech, the connection between two parties may be several links, each of which may consist of a DPCM encoder and decoder. In such tandem links, the amount of delay can become large enough to be a nuisance. Furthermore, the need to transmit side information makes the system more complex. In order to avoid these problems, we can adapt the predictor based on the output of the encoder, which is also available to the decoder. The adaptation is done in a sequential manner [172, 174].

In our derivation of the optimum predictor coefficients, we took the derivative of the statistical average of the squared prediction error or residual sequence. In order to do this, we had to assume that the input process was stationary. Let us now remove that assumption and try to figure out how to adapt the predictor to the input algebraically. To keep matters simple, we will start with a first-order predictor and then generalize the result to higher orders.

For a first-order predictor, the value of the residual squared at time n would be given by

$$d_n^2 = (x_n - a_1 \hat{x}_{n-1})^2 \tag{40}$$

If we could plot the value of d_n^2 against a_1, we would get a graph similar to the one shown in Figure 11.11. Let's take a look at the derivative of d_n^2 as a function of whether the current value of a_1 is to the left or right of the optimal value of a_1—that is, the value of a_1 for which d_n^2 is minimum. When a_1 is to the left of the optimal value, the derivative is negative. Furthermore, the derivative will have a larger magnitude when a_1 is further away from the optimal value. If we were asked to adapt a_1, we would add to the current value of a_1. The amount to add would be large if a_1 was far from the optimal value, and small if a_1 was close to the optimal value. If the current value was to the right of the optimal value, the derivative would be positive, and we would subtract some amount from a_1 to adapt it. The amount to subtract would be larger if we were further from the optimal, and as before, the derivative would have a larger magnitude if a_1 were further from the optimal value.

At any given time, in order to adapt the coefficient at time $n+1$, we add an amount proportional to the magnitude of the derivative with a sign that is opposite to that of the derivative of d_n^2 at time n:

$$a_1^{(n+1)} = a_1^{(n)} - \alpha \frac{\partial d_n^2}{\partial a_1} \tag{41}$$

where α is some positive proportionality constant.

$$\frac{\partial d_n^2}{\partial a_1} = -2(x_n - a_1 \hat{x}_{n-1})\hat{x}_{n-1} \tag{42}$$

$$= -2 d_n \hat{x}_{n-1}. \tag{43}$$

Substituting this into (41), we get

$$a_1^{(n+1)} = a_1^{(n)} + \alpha d_n \hat{x}_{n-1} \tag{44}$$

where we have absorbed the 2 into α. The residual value d_n is available only to the encoder. Therefore, in order for both the encoder and decoder to use the same algorithm, we replace d_n by $\hat{d}_n$ in (44) to obtain

$$a_1^{(n+1)} = a_1^{(n)} + \alpha \hat{d}_n \hat{x}_{n-1} \tag{45}$$

Extending this adaptation equation for a first-order predictor to an Nth-order predictor is relatively easy. The equation for the squared prediction error is given by

$$d_n^2 = \left(x_n - \sum_{i=1}^{N} a_i \hat{x}_{n-i} \right)^2 \tag{46}$$

Taking the derivative with respect to a_j will give us the adaptation equation for the jth predictor coefficient:

$$a_j^{(n+1)} = a_j^{(n)} + \alpha \hat{d}_n \hat{x}_{n-j} \tag{47}$$

We can combine all N equations in vector form to get

$$\mathbf{A}^{(n+1)} = \mathbf{A}^{(n)} + \alpha \hat{d}_n \widehat{X}_{n-1} \tag{48}$$

where

$$\widehat{X}_n = \begin{bmatrix} \hat{x}_n \\ \hat{x}_{n-1} \\ \vdots \\ \hat{x}_{n-N+1} \end{bmatrix} \tag{49}$$

This particular adaptation algorithm is called the least mean squared (LMS) algorithm [175].

11.6 Delta Modulation

A very simple form of DPCM that has been widely used in a number of speech-coding applications is the delta modulator (DM). The DM can be viewed as a DPCM system with a 1-bit

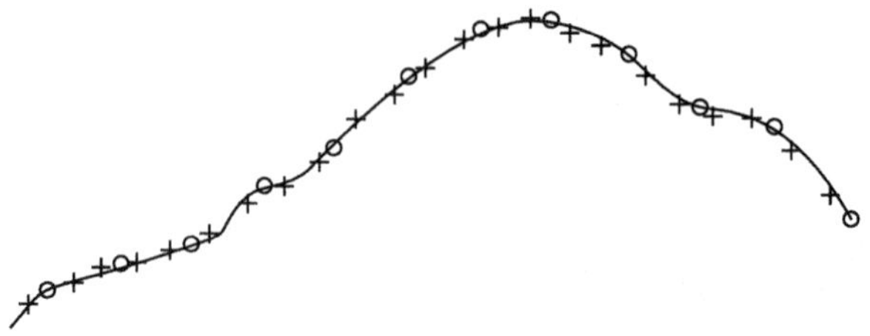

FIGURE 11.12 A signal sampled at two different rates.

(two-level) quantizer. With a two-level quantizer with output values $\pm\Delta$, we can only represent a sample-to-sample difference of Δ. If, for a given source sequence, the sample-to-sample difference is often very different from Δ, then we may incur substantial distortion. One way to limit the difference is to sample more often. In Figure 11.12 we see a signal that has been sampled at two different rates. The lower-rate samples are shown by open circles, while the higher-rate samples are represented by $+$. It is apparent that the lower-rate samples are not only further apart in time, they are also further apart in value.

The rate at which a signal is sampled is governed by the highest frequency component of a signal. If the highest frequency component in a signal is W, then in order to obtain an exact reconstruction of the signal, we need to sample it at least at twice the highest frequency, or $2W$ (see Section 12.7). In systems that use delta modulation, we usually sample the signal at much more than twice the highest frequency. If F_s is the sampling frequency, then the ratio of F_s to $2W$ can range from almost 1 to almost 100 [134]. The higher sampling rates are used for high-quality A/D converters, while the lower rates are more common for low-rate speech coders.

If we look at a block diagram of a delta modulation system, we see that, while the block diagram of the encoder is identical to that of the DPCM system, the standard DPCM decoder is followed by a filter. The reason for the existence of the filter is evident from Figure 11.13, where we show a source output and the unfiltered reconstruction. The samples of the source output are represented by the filled circles. As the source is sampled at several times the highest frequency, the staircase shape of the reconstructed signal results in distortion in frequency bands outside the band of frequencies occupied by the signal. The filter can be used to remove these spurious frequencies.

The reconstruction shown in Figure 11.13 was obtained with a delta modulator using a fixed quantizer. Delta modulation systems that use a fixed step size are often referred to as linear delta modulators. Notice that the reconstructed signal shows one of two behaviors. In regions where the source output is relatively constant, the output alternates up or down by Δ; these regions are called the *granular regions*. In the regions where the source output rises or falls fast, the reconstructed output cannot keep up; these regions are called the *slope overload regions*. If we want to reduce the granular error, we need to make the step size Δ small. However, this will make it more difficult for the reconstruction to follow rapid changes in the input. In other words, it will result in an increase in the overload error. To avoid the overload condition, we need to make the step size large so that the reconstruction can quickly catch up with rapid changes in the input. However, this will increase the granular error.

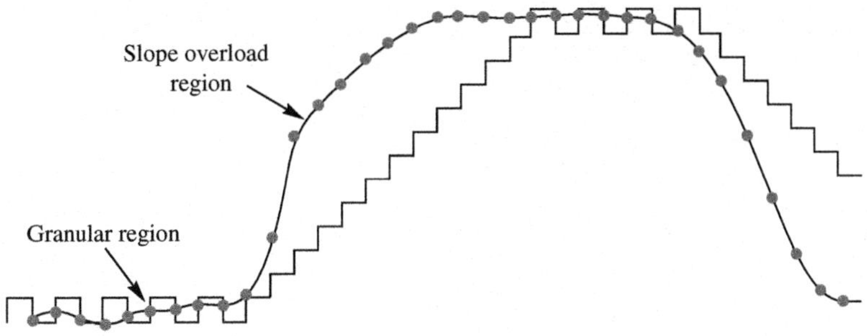

FIGURE 11.13 **A source output sampled and coded using delta modulation.**

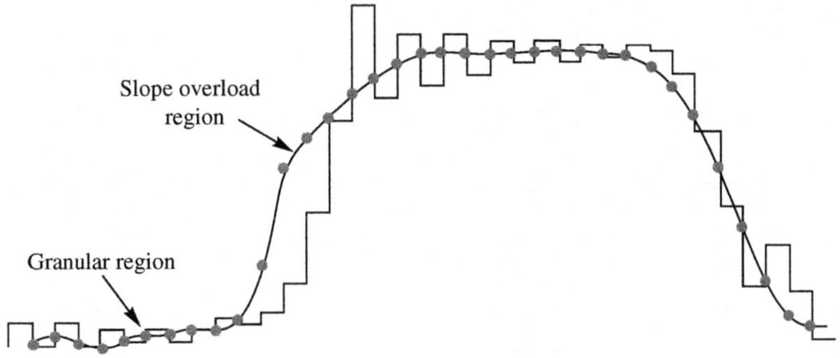

FIGURE 11.14 **A source output sampled and coded using adaptive delta modulation.**

One way to avoid this impasse is to adapt the step size to the characteristics of the input, as shown in Figure 11.14. In quasi-constant regions, make the step size small in order to reduce the granular error. In regions of rapid change, increase the step size in order to reduce overload error. There are various ways of adapting the delta modulator to the local characteristics of the source output. We describe two of the more popular ways here.

11.6.1 Constant Factor Adaptive Delta Modulation (CFDM)

The objective of adaptive delta modulation is clear: increase the step size in overload regions and decrease it in granular regions. The problem lies in knowing when the system is in each of these regions. Looking at Figure 11.13, we see that in the granular region the output of the quantizer changes sign with almost every input sample; in the overload region, the sign of the quantizer output is the same for a string of input samples. Therefore, we can define an overload or granular condition based on whether the output of the quantizer has been changing signs. A very simple system [176] uses a history of one sample to decide whether the system is in overload or granular condition and whether to expand or contract the step size. If s_n denotes the sign of the quantizer output $\hat{d}_n$,

$$s_n = \begin{cases} 1 & if\ \hat{d}_n > 0 \\ -1 & if\ \hat{d}_n < 0 \end{cases} \tag{50}$$

the adaptation logic is given by

$$\Delta_n = \begin{cases} M_1 \Delta_{n-1} & s_n = s_{n-1} \\ M_2 \Delta_{n-1} & s_n \neq s_{n-1} \end{cases} \tag{51}$$

where $M_1 = \frac{1}{M_2} = M > 1$. In general, $M < 2$.

By increasing the memory, we can improve the response of the CFDM system. For example, if we looked at two past samples, we could decide that the system was moving from overload to granular condition if the sign had been the same for the past two samples and then changed with the current sample:

$$s_n \neq s_{n-1} = s_{n-2} \tag{52}$$

In this case it would be reasonable to assume that the step size had been expanding previously and, therefore, needed a sharp contraction. If

$$s_n = s_{n-1} \neq s_{n-2} \tag{53}$$

then it would mean that the system was probably entering the overload region, while

$$s_n = s_{n-1} = s_{n-2} \tag{54}$$

would mean the system was in overload and the step size should be expanded rapidly.

For the encoding of speech, the following multipliers M_i are recommended by [177] for a CFDM system with two-sample memory:

$$s_n \neq s_{n-1} s_{n-2} \qquad M_1 = 0.4 \tag{55}$$

$$s_n \neq s_{n-1} \neq s_{n-2} \qquad M_2 = 0.9 \tag{56}$$

$$s_n = s_{n-1} \neq s_{n-2} \qquad M_3 = 1.5 \tag{57}$$

$$s_n = s_{n-1} = s_{n-2} \qquad M_4 = 2.0 \tag{58}$$

The amount of memory can be increased further with a concurrent increase in complexity. The space shuttle used a delta modulator with a memory of seven [178].

11.6.2 Continuously Variable Slope Delta Modulation

The CFDM systems described use a rapid adaptation scheme. For low-rate speech coding, it is more pleasing if the adaptation is over a longer period of time. This slower adaptation results in a decrease in the granular error and generally an increase in overload error. Delta modulation systems that adapt over longer periods of time are referred to as *syllabically* companded. A popular class of syllabically companded delta modulation systems are continuously variable slope delta modulation systems.

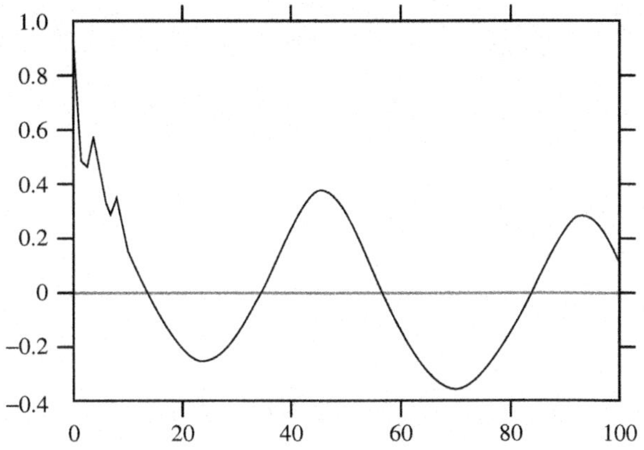

FIGURE 11.15 **Autocorrelation function for** `test.snd`.

The adaptation logic used in CVSD systems is as follows [134]:

$$\Delta_n = \beta \Delta_{n-1} + \alpha_n \Delta_0 \tag{59}$$

where β is a number less than but close to one, and α_n is equal to one if J of the last K quantizer outputs were of the same sign. That is, we look in a window of length K to obtain the behavior of the source output. If this condition is not satisfied, then α_n is equal to zero. Standard values for J and K are $J = 3$ and $K = 3$.

11.7 Speech Coding

Differential encoding schemes are immensely popular for speech encoding. They are used in the telephone system, voice messaging, and multimedia applications, among others. Adaptive DPCM is a part of several international standards (ITU-T G.721, ITU G.723, ITU G.726, ITU-T G.722), which we will look at here and in later chapters.

Before we do that, let's take a look at one issue specific to speech coding. In Figure 11.7, we see that there is a segment of speech that looks highly periodic. We can see this periodicity if we plot the autocorrelation function of the speech segment (Figure 11.15).

The autocorrelation peaks at a lag value of 47 and multiples of 47. This indicates a periodicity of 47 samples. This period is called the *pitch period*. The predictor we originally designed did not take advantage of this periodicity, as the largest predictor was a third-order predictor, and this periodic structure takes 47 samples to show up. We can take advantage of this periodicity by constructing an outer prediction loop around the basic DPCM structure as shown in Figure 11.16. This can be a simple single coefficient predictor of the form $b\hat{x}_{n-\tau}$, where τ is the pitch period. Using this system on `testm.raw`, we get the residual sequence shown in Figure 11.17. Notice the decrease in amplitude in the periodic portion of the speech.

Finally, remember that we have been using mean squared error as the distortion measure in all of our discussions. However, perceptual tests do not always correlate with the mean squared

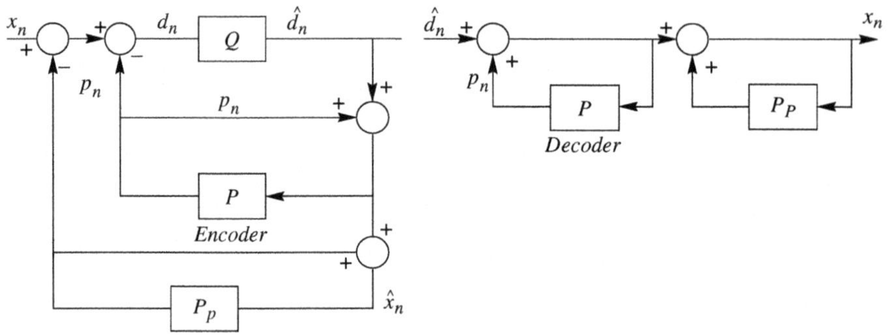

FIGURE 11.16 **The DPCM structure with a pitch predictor.**

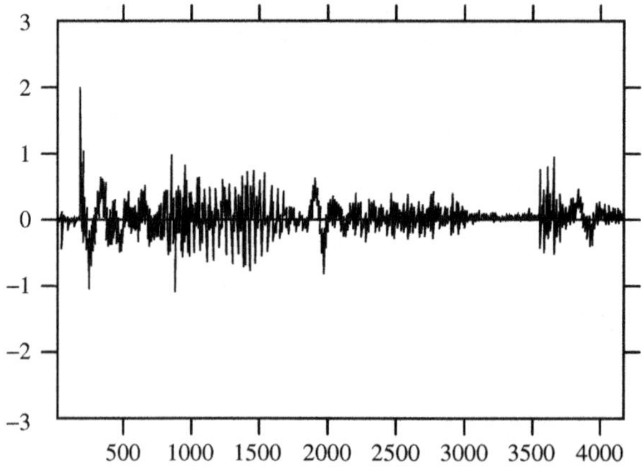

FIGURE 11.17 **The residual sequence using the DPCM system with a pitch predictor.**

error. The level of distortion we perceive is often related to the level of the speech signal. In regions where the speech signal is of higher amplitude, we have a harder time perceiving the distortion, but the same amount of distortion in a different frequency band, where the speech is of lower amplitude, might be very perceptible. We can take advantage of this by shaping the quantization error so that most of the error lies in the region where the signal has a higher amplitude. This variation of DPCM is called *noise feedback coding* (NFC) (see [134] for details).

11.7.1 G.726

The International Telecommunications Union has published several recommendations for a standard ADPCM system, including recommendations G.721, G.723, and G.726. G.726 supersedes G.721 and G.723. In this section we will describe the G.726 recommendation for ADPCM systems at rates of 40, 32, 24, and 16 kbits.

TABLE 11.2 Recommended input-output characteristics of the quantizer for 24-kbits-per-second operation.

Input Range	Label	Output
$\log_2 \frac{d_k}{\alpha_k}$	$\lvert I_k \rvert$	$\log_2 \frac{d_k}{\alpha_k}$
$[2.58, \infty)$	3	2.91
$[1.70, 2.58)$	2	2.13
$[0.06, 1.70)$	1	1.05
$(-\infty, -0.06)$	0	$-\infty$

The Quantizer

The recommendation assumes that the speech output is sampled at the rate of 8000 samples per second, so the rates of 40, 32, 24, and 16 kbits per second translate 5 bits per sample, 4 bits per sample, 3 bits per sample, and 2 bits per sample. Comparing this to the PCM rate of 8 bits per sample, this would mean compression ratios of 1.6:1, 2:1, 2.67:1, and 4:1. Except for the 16 kbits per second system, the number of levels in the quantizer are $2^{n_b} - 1$, where n_b is the number of bits per sample. Thus, the number of levels in the quantizer is odd, which means that for the higher rates we use a midtread quantizer.

The quantizer is a backward adaptive quantizer with an adaptation algorithm that is similar to the Jayant quantizer. The recommendation describes the adaptation of the quantization interval in terms of the adaptation of a scale factor. The input d_k is normalized by a scale factor α_k. This normalized value is quantized, and the normalization removed by multiplying with α_k. In this way the quantizer is kept fixed and α_k is adapted to the input. Therefore, for example, instead of expanding the step size, we would increase the value of α_k.

The fixed quantizer is a nonuniform midtread quantizer. The recommendation describes the quantization boundaries and reconstruction values in terms of the log of the scaled input. The input-output characteristics for the 24 kbit system are shown in Table 11.2. An output value of $-\infty$ in the table corresponds to a reconstruction value of 0.

The adaptation algorithm is described in terms of the logarithm of the scale factor:

$$y(k) = \log_2 \alpha_k \tag{60}$$

The adaptation of the scale factor α or its log $y(k)$ depends on whether the input is speech or speechlike, where the sample-to-sample difference can fluctuate considerably, or whether the input is voice-band data, which might be generated by a modem, where the sample-to-sample fluctuation is quite small. In order to handle both these situations, the scale factor is composed of two values, a *locked* slow scale factor for when the sample-to-sample differences are quite small, and an *unlocked* value for when the input is more dynamic:

$$y(k) = a_l(k)y_u(k-1) + (1 - a_l(k))y_l(k-1) \tag{61}$$

The value of $a_l(k)$ depends on the variance of the input. It will be close to one for speech inputs and close to zero for tones and voice-band data.

The unlocked scale factor is adapted using the Jayant algorithm with one slight modification. If we were to use the Jayant algorithm, the unlocked scale factor could be adapted as

$$\alpha_u(k) = \alpha_{k-1} M[I_{k-1}] \tag{62}$$

where $M[\cdot]$ is the multiplier. In terms of logarithms, this becomes

$$y_u(k) = y(k-1) + \log M[I_{k-1}] \tag{63}$$

The modification consists of introducing some memory into the adaptive process so that the encoder and decoder converge following transmission errors:

$$y_u(k) = (1-\epsilon)y(k-1) + \epsilon W[I_{k-1}] \tag{64}$$

where $W[\cdot] = \log M[\cdot]$, and $\epsilon = 2^{-5}$.

The locked scale factor is obtained from the unlocked scale factor through

$$y_l(k) = (1-\gamma)y_l(k-1) + \gamma y_u(k), \qquad \gamma = 2^{-6} \tag{65}$$

The Predictor

The recommended predictor is a backward adaptive predictor that uses a linear combination of the past two reconstructed values as well as the six past quantized differences to generate the prediction:

$$p_k = \sum_{i=1}^{2} a_i^{(k-1)} \hat{x}_{k-i} + \sum_{i=1}^{6} b_i^{(k-1)} \hat{d}_{k-i} \tag{66}$$

The set of predictor coefficients is updated using a simplified form of the LMS algorithm:

$$a_1^{(k)} = (1-2^{-8})a_1^{(k-1)} + 3 \times 2^{-8}\text{sgn}[z(k)]\text{sgn}[z(k-1)] \tag{67}$$

$$a_2^{(k)} = (1-2^{-7})a_2^{(k-1)} + 2^{-7}(\text{sgn}[z(k)]\text{sgn}[z(k-2)] - f\left(a_1^{(k-1)}\text{sgn}[z(k)]\text{sgn}[z(k-1)]\right)) \tag{68}$$

where

$$z(k) = \hat{d}_k + \sum_{i=1}^{6} b_i^{(k-1)} \hat{d}_{k-i} \tag{69}$$

$$f(\beta) = \begin{cases} 4\beta & |\beta| \leqslant \frac{1}{2} \\ 2\text{sgn}(\beta) & |\beta| > \frac{1}{2} \end{cases} \tag{70}$$

The coefficients $\{b_i\}$ are updated using the following equation:

$$b_i^{(k)} = (1-2^{-8})b_i^{(k-1)} + 2^{-7}\text{sgn}[\hat{d}_k]\text{sgn}[\hat{d}_{k-i}] \tag{71}$$

Notice that in the adaptive algorithms we have replaced products of reconstructed values and products of quantizer outputs with products of their signs. This is computationally much simpler and does not lead to any significant degradation of the adaptation process. Furthermore, the values of the coefficients are selected such that multiplication with these coefficients can be accomplished using shifts and adds. The predictor coefficients are all set to zero when the input moves from tones to speech.

11.8 Image Coding

We saw in Chapter 7 that differential encoding provided an efficient approach to the lossless compression of images. The case for using differential encoding in the lossy compression of images has not been made as clearly. In the early days of image compression, both differential encoding and transform coding were popular forms of lossy image compression. At the current time differential encoding has a much more restricted role as part of other compression strategies. Several currently popular approaches to image compression decompose the image into lower and higher frequency components. As low-frequency signals have high sample-to-sample correlation, several schemes use differential encoding to compress the low-frequency components. We will see this use of differential encoding when we look at subband- and wavelet-based compression schemes and, to a lesser extent, when we study transform coding.

For now let us look at the performance of a couple of stand-alone differential image compression schemes. We will compare the performance of these schemes with the performance of the JPEG compression standard.

Consider a simple differential encoding scheme in which the predictor $p[j,k]$ for the pixel in the jth row and the kth column is given by

$$p[j,k] = \begin{cases} \hat{x}[j,k-1] & \text{for } k > 0 \\ \hat{x}[j-1,k] & \text{for } k = 0 \text{ and } j > 0 \\ 128 & \text{for } j = 0 \text{ and } k = 0 \end{cases}$$

where $\hat{x}[j,k]$ is the reconstructed pixel in the jth row and kth column. We use this predictor in conjunction with a fixed four-level uniform quantizer and code the quantizer output using an arithmetic coder. The coding rate for the compressed image is approximately 1 bit per pixel. We compare this reconstructed image with a JPEG-coded image at the same rate in Figure 11.18. The signal-to-noise ratio for the differentially encoded image is 22.33 dB (PSNR 31.42 dB), while for the JPEG-encoded image it is 32.52 dB (PSNR 41.60 dB), a difference of more than 10 dB!

However, this is an extremely simple system compared to the JPEG standard, which has been fine-tuned for encoding images. Let's make our differential encoding system slightly more complicated by replacing the uniform quantizer with a recursively indexed quantizer and by using a somewhat more complex predictor. For each pixel (except for the boundary pixels) we compute the following three values:

$$\begin{aligned} p_1 &= 0.5 \times \hat{x}[j-1,k] + 0.5 \times \hat{x}[j,k-1] \\ p_2 &= 0.5 \times \hat{x}[j-1,k-1] + 0.5 \times \hat{x}[j,k-1] \\ p_3 &= 0.5 \times \hat{x}[j-1,k-1] + 0.5 \times \hat{x}[j-1,k] \end{aligned} \tag{72}$$

then obtain the predicted value as

$$p[j,k] = \text{median}\{p_1, p_2, p_3\}$$

FIGURE 11.18 Left: Reconstructed image using differential encoding at 1 bit per pixel. Right: Reconstructed image using JPEG at 1 bit per pixel.

FIGURE 11.19 Left: Reconstructed image using differential encoding at 1 bit per pixel using median predictor and recursively indexed quantizer. Right: Reconstructed image using JPEG at 1 bit per pixel.

For the boundary pixels we use the simple prediction scheme. At a coding rate of 1 bit per pixel, we obtain the image shown in Figure 11.19. For reference we show it next to the JPEG-coded image at the same rate. The signal-to-noise ratio for this reconstruction is 29.20 dB (PSNR 38.28 dB). We have made up two-thirds of the difference using some relatively minor modifications. We can see that it might be feasible to develop differential encoding schemes that are competitive with other image compression techniques. Therefore, it makes sense not to dismiss differential encoding out of hand when we need to develop image compression systems.

11.9 Summary

In this chapter we described some of the more well-known differential encoding techniques. Although differential encoding does not provide compression as high as vector quantization, it is very simple to implement. This approach is especially suited to the encoding of speech, where it has found broad application. The DPCM system consists of two main components, the quantizer and the predictor. We spent a considerable amount of time discussing the quantizer in Chapter 9, so most of the discussion in this chapter focused on the predictor. We have seen different ways of making the predictor adaptive, and looked at some of the improvements to be obtained from source-specific modifications to the predictor design.

Further Reading

1. *Digital Coding of Waveforms*, by N.S. Jayant and P. Noll [134], contains some very detailed and highly informative chapters on differential encoding.
2. "Adaptive Prediction in Speech Differential Encoding Systems," by J.D. Gibson [172], is a comprehensive treatment of the subject of adaptive prediction.
3. A real-time video coding system based on DPCM has been developed by NASA. Details can be found in [179].

11.10 Projects and Problems

1. Generate an AR(1) process using the relationship

$$x_n = 0.9 \times x_{n-1} + \epsilon_n$$

   where ϵ_n is the output of a Gaussian random number generator (this is option 2 in `rangen`).

   (a) Encode this sequence using a DPCM system with a one-tap predictor with predictor coefficient 0.9 and a three-level Gaussian quantizer. Compute the variance of the prediction error. How does this compare with the variance of the input? How does the variance of the prediction error compare with the variance of the $\{\epsilon_n\}$ sequence?

   (b) Repeat using predictor coefficient values of 0.5, 0.6, 0.7, 0.8, and 1.0. Comment on the results.

2. Generate an AR(5) process using the following coefficients: 1.381, 0.6, 0.367, −0.7, 0.359.

   (a) Encode this with a DPCM system with a 3-bit Gaussian nonuniform quantizer and a first-, second-, third-, fourth-, and fifth-order predictor. Obtain these predictors by solving (30). For each case compute the variance of the prediction error and the SNR in dB. Comment on your results.

   (b) Repeat using a 3-bit Jayant quantizer.

3. DPCM can also be used for encoding images. Encode the Sinan image using a one-tap predictor of the form

$$\hat{x}_{i,j} = a \times x_{i,j-1}$$

and a 2-bit quantizer. Experiment with quantizers designed for different distributions. Comment on your results.

4. Repeat the image-coding experiment of the previous problem using a Jayant quantizer.

5. DPCM-encode the Sinan, Elif, and Bookshelf1 images using a one-tap predictor and a four-level quantizer followed by a Huffman coder. Repeat using a five-level quantizer. Compute the SNR for each case, and compare the rate distortion performances.

6. We want to DPCM-encode images using a two-tap predictor of the form

$$\hat{x}_{i,j} = a \times x_{i,j-1} + b \times x_{i-1,j}$$

and a four-level quantizer followed by a Huffman coder. Find the equations we need to solve to obtain coefficients a and b that minimize the mean squared error.

7. **(a)** DPCM-encode the Sinan, Elif, and Bookshelf1 images using a two-tap predictor and a four-level quantizer followed by a Huffman coder.

(b) Repeat using a five-level quantizer. Compute the SNR and rate (in bits per pixel) for each case.

(c) Compare the rate distortion performances with the one-tap case.

(d) Repeat using a five-level quantizer. Compute the SNR for each case, and compare the rate distortion performances using a one-tap and two-tap predictor.

12

Mathematical Preliminaries for Transforms, Subbands, and Wavelets

12.1 Overview

In this chapter we will review some of the mathematical background necessary for the study of transforms, subbands, and wavelets. The topics include Fourier series, Fourier transforms, and their discrete counterparts. We will also look at sampling and briefly review some linear system concepts.

12.2 Introduction

The roots of many of the techniques we will study can be found in the mathematical literature. Therefore, in order to understand the techniques, we will need some mathematical background. Our approach in general will be to introduce the mathematical tools just prior to when they are needed. However, there is a certain amount of background that is required for most of what we will be looking at. In this chapter we will present only that material that is a common background to all the techniques we will be studying. Our approach will be rather utilitarian; more sophisticated coverage of these topics can be found in [180]. We will be introducing a rather large number of concepts, many of which depend on each other. In order to make it easier for you to find a particular concept, we will identify the paragraph in which the concept is first introduced.

We will begin our coverage with a brief introduction to the concept of vector spaces, and in particular the concept of the inner product. We will use these concepts in our description of Fourier series and Fourier transforms. Next is a brief overview of linear systems, then a look at the issues involved in sampling a function. Finally, we will revisit the Fourier concepts in the context of sampled functions and provide a brief introduction to Z-transforms. Throughout, we will try to get a physical feel for the various concepts.

12.3 Vector Spaces

The techniques we will be using to obtain compression will involve manipulations and decompositions of (sampled) functions of time. In order to do this we need some sort of mathematical framework. This framework is provided through the concept of vector spaces.

We are very familiar with vectors in two- or three-dimensional space. An example of a vector in two-dimensional space is shown in Figure12.1. This vector can be represented in a number of different ways: we can represent it in terms of its magnitude and direction, or we can represent it as a weighted sum of the unit vectors in the x and y directions, or we can represent it as an array whose components are the coefficients of the unit vectors. Thus, the vector $\mathbf{v}$ in Figure12.1 has a magnitude of 5 and an angle of 36.86 degrees,

$$\mathbf{v} = 4u_x + 3u_y$$

and

$$\mathbf{v} = \begin{bmatrix} 4 \\ 3 \end{bmatrix}$$

We can view the second representation as a decomposition of V into simpler building blocks, namely, the *basis vectors*. The nice thing about this is that any vector in two dimensions can be decomposed in exactly the same way. Given a particular vector $\mathbf{A}$ and a basis set (more

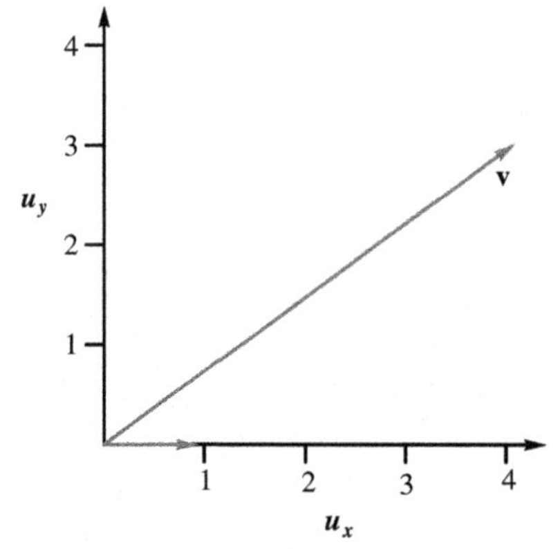

FIGURE 12.1 **A vector.**

on this later), decomposition means finding the coefficients with which to weight the unit vectors of the basis set. In our simple example it is easy to see what these coefficients should be. However, we will encounter situations where it is not a trivial task to find the coefficients that constitute the decomposition of the vector. We therefore need some machinery to extract these coefficients. The particular machinery we will use here is called the *dot product* or the *inner product*.

12.3.1 Dot or Inner Product

Given two vectors **a** and **b** such that

$$\mathbf{a} = \begin{bmatrix} a_1 \\ a_2 \end{bmatrix}, \quad \mathbf{b} = \begin{bmatrix} b_1 \\ b_2 \end{bmatrix}$$

the inner product between **a** and **b** is defined as

$$\mathbf{a} \cdot \mathbf{b} = a_1 b_1 + a_2 b_2$$

Two vectors are said to be *orthogonal* if their inner product is zero. A set of vectors is said to be orthogonal if each vector in the set is orthogonal to every other vector in the set. The inner product between a vector and a unit vector from an orthogonal basis set will give us the coefficient corresponding to that unit vector. It is easy to see that this is indeed so. We can write u_x and u_y as

$$u_x = \begin{bmatrix} 1 \\ 0 \end{bmatrix}, \quad u_y = \begin{bmatrix} 0 \\ 1 \end{bmatrix}$$

These are obviously orthogonal. Therefore, the coefficient a_1 can be obtained by

$$\mathbf{a} \cdot u_x = a_1 \times 1 + a_2 \times 0 = a_1$$

and the coefficient of u_y can be obtained by

$$\mathbf{a} \cdot u_y = a_1 \times 0 + a_2 \times 1 = a_2$$

The inner product between two vectors is in some sense a measure of how "similar" they are, but we have to be a bit careful in how we define "similarity." For example, consider the vectors in Figure 12.2. The vector **a** is closer to u_x than to u_y. Therefore $\mathbf{a} \cdot u_x$ will be greater than $\mathbf{a} \cdot u_y$. The reverse is true for **b**.

12.3.2 Vector Space

In order to handle not just two- or three-dimensional vectors but general sequences and functions of interest to us, we need to generalize these concepts. Let us begin with a more general definition of vectors and the concept of a vector space.

A *vector space* consists of a set of elements called vectors that have the operations of vector addition and scalar multiplication defined on them. Furthermore, the results of these operations are also elements of the vector space.

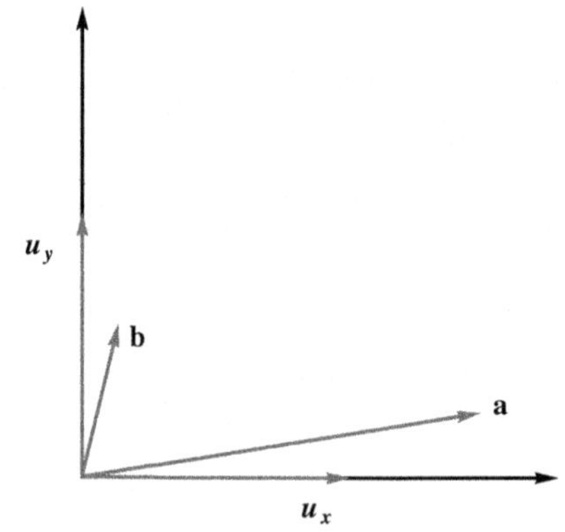

FIGURE 12.2 Example of different vectors.

By *vector addition* of two vectors, we mean the vector obtained by the pointwise addition of the components of the two vectors. For example, given two vectors **a** and **b**:

$$\mathbf{a} = \begin{bmatrix} a_1 \\ a_2 \\ a_3 \end{bmatrix}, \quad \mathbf{b} = \begin{bmatrix} b_1 \\ b_2 \\ b_3 \end{bmatrix} \tag{1}$$

the vector addition of these two vectors is given as

$$\mathbf{a} + \mathbf{b} = \begin{bmatrix} a_1 + b_1 \\ a_2 + b_2 \\ a_3 + b_3 \end{bmatrix} \tag{2}$$

By *scalar multiplication*, we mean the multiplication of a vector with a real or complex number. For this set of elements to be a vector space it has to satisfy certain axioms.

Suppose V is a vector space; $\mathbf{x}, \mathbf{y}$, and $\mathbf{z}$ are vectors; and α and β are scalars. Then the following axioms are satisfied:

1. $\mathbf{x} + \mathbf{y} = \mathbf{y} + \mathbf{x}$ (commutativity).
2. $(\mathbf{x} + \mathbf{y}) + \mathbf{z} = \mathbf{x} + (\mathbf{y} + \mathbf{z})$ and $(\alpha\beta)\mathbf{x} = \alpha(\beta\mathbf{x})$ (associativity).
3. There exists an element θ in V such that $\mathbf{x} + \theta = \mathbf{x}$ for all $\mathbf{x}$ in V. θ is called the additive identity.
4. $\alpha(\mathbf{x} + \mathbf{y}) = \alpha\mathbf{x} + \alpha\mathbf{y}$, and $(\alpha + \beta)\mathbf{x} = \alpha\mathbf{x} + \beta\mathbf{x}$ (distributivity).
5. $1 \cdot \mathbf{x} = \mathbf{x}$, and $0 \cdot \mathbf{x} = \theta$.
6. For every $\mathbf{x}$ in V, there exists a $(-\mathbf{x})$ such that $\mathbf{x} + (-\mathbf{x}) = \theta$.

A simple example of a vector space is the set of real numbers. In this set zero is the additive identity. We can easily verify that the set of real numbers with the standard operations of addition and multiplication obey the axioms stated above. See if you can verify that the set of real numbers is a vector space. One of the advantages of this exercise is to emphasize the fact that a vector is more than a line with an arrow at its end.

Example 12.3.1:

Another example of a vector space that is of more practical interest to us is the set of all functions $f(t)$ with finite energy. That is,

$$\int_{-\infty}^{\infty} |f(t)|^2 \, dt < \infty \tag{3}$$

Let's see if this set constitutes a vector space. If we define additions as pointwise addition and scalar multiplication in the usual manner, the set of functions $f(t)$ obviously satisfies axioms 1, 2, and 4.

- If $f(t)$ and $g(t)$ are functions with finite energy, and α is a scalar, then the functions $f(t) + g(t)$ and $\alpha f(t)$ also have finite energy.
- If $f(t)$ and $g(t)$ are functions with finite energy, then $f(t) + g(t) = g(t) + f(t)$ (axiom 1).
- If $f(t)$, $g(t)$, and $h(t)$ are functions with finite energy, and α and β are scalars, then $(f(t) + g(t)) + h(t) = f(t) + (g(t) + h(t))$ and $(\alpha\beta)f(t) = \alpha(\beta f(t))$ (axiom 2).
- If $f(t)$, $g(t)$, and $h(t)$ are functions with finite energy, and α is a scalar, then $\alpha(f(t) + g(t)) = \alpha f(t) + \alpha g(t)$ and $(\alpha + \beta)f(t) = \alpha f(t) + \beta f(t)$ (axiom 4).

Let us define the additive identity function $\theta(t)$ as the function that is identically zero for all t. This function satisfies the requirement of finite energy, and we can see that axioms 3 and 5 are also satisfied. Finally, if a function $f(t)$ has finite energy, then from Equation (3), the function $-f(t)$ also has finite energy, and axiom 6 is satisfied. Therefore, the set of all functions with finite energy constitutes a vector space. This space is denoted by $L_2(f)$, or simply L_2. ♦

12.3.3 Subspace

A *subspace* S of a vector space V is a subset of V whose members satisfy all the axioms of the vector space. It has the additional property that if $\mathbf{x}$ and $\mathbf{y}$ are in S, and α is a scalar, then $\mathbf{x} + \mathbf{y}$ and $\alpha\mathbf{x}$ are also in S.

Example 12.3.2:

Consider the set S of continuous bounded functions on the interval $[0, 1]$. Then S is a subspace of the vector space L_2. ♦

12.3.4 Basis

One way we can generate a subspace is by taking linear combinations of a set of vectors. If this set of vectors is *linearly independent*, then the set is called a *basis* for the subspace.

> A set of vectors $\{\mathbf{x}_1, \mathbf{x}_2, \ldots\}$ is said to be linearly independent if no vector of the set can be written as a linear combination of the other vectors in the set.

A direct consequence of this definition is the following theorem:

Theorem 12.1 *A set of vectors* $\mathbf{X} = \{\mathbf{x}_1, \mathbf{x}_2, \ldots, \mathbf{x}_N\}$ *is linearly independent if and only if the expression* $\sum_{i=1}^{N} \alpha_i \mathbf{x}_i = \theta$ *implies that* $\alpha_i = 0$ *for all* $i = 1, 2, \ldots, N$.

Proof The proof of this theorem can be found in most books on linear algebra [180].

The set of vectors formed by all possible linear combinations of vectors from a linearly independent set **X** forms a vector space (see Problem 1 at the end of this chapter). The set **X** is said to be the *basis* for this vector space. The basis set contains the smallest number of linearly independent vectors required to represent each element of the vector space. More than one set can be the basis for a given space.

Example 12.3.3:

Consider the vector space consisting of vectors $[a\ b]^T$, where a and b are real numbers, and T denotes transpose (the transpose of a vector involves writing rows as columns and columns as rows). Then the set

$$\mathbf{X} = \left\{ \begin{bmatrix} 1 \\ 0 \end{bmatrix}, \begin{bmatrix} 0 \\ 1 \end{bmatrix} \right\}$$

forms a basis for this space, as does the set

$$\mathbf{X} = \left\{ \begin{bmatrix} 1 \\ 1 \end{bmatrix}, \begin{bmatrix} 1 \\ 0 \end{bmatrix} \right\}$$

In fact, any two vectors that are not scalar multiples of each other form a basis for this space. ◆

The number of basis vectors required to generate the space is called the *dimension* of the vector space. In the previous example the dimension of the vector space is two. The dimension of the space of all continuous functions on the interval [0, 1] is infinity.

Given a particular basis, we can find a representation with respect to this basis for any vector in the space.

Example 12.3.4:

If $\mathbf{a} = [3\ 4]^T$, then

$$\mathbf{a} = 3 \begin{bmatrix} 1 \\ 0 \end{bmatrix} + 4 \begin{bmatrix} 0 \\ 1 \end{bmatrix}$$

and

$$\mathbf{a} = 4\begin{bmatrix}1\\1\end{bmatrix} + (-1)\begin{bmatrix}1\\0\end{bmatrix}$$

so the representation of **a** with respect to the first basis set is (3, 4), and the representation of **a** with respect to the second basis set is (4, -1). ♦

In the beginning of this section we had described a mathematical machinery for finding the components of a vector that involved taking the dot product or inner product of the vector to be decomposed with basis vectors. In order to use the same machinery in more abstract vector spaces we need to generalize the notion of inner product.

12.3.5 Inner Product—Formal Definition

An inner product between two vectors $\mathbf{x}$ and $\mathbf{y}$, denoted by $\langle \mathbf{x}, \mathbf{y} \rangle$, associates a scalar value with each pair of vectors. The inner product satisfies the following axioms:

1. $\langle \mathbf{x}, \mathbf{y} \rangle = \langle \mathbf{y}, \mathbf{x} \rangle^*$, where * denotes complex conjugate.
2. $\langle \mathbf{x} + \mathbf{y}, \mathbf{z} \rangle = \langle \mathbf{x}, \mathbf{z} \rangle + \langle \mathbf{y}, \mathbf{z} \rangle$.
3. $\langle \alpha \mathbf{x}, \mathbf{y} \rangle = \alpha \langle \mathbf{x}, \mathbf{y} \rangle$, where α can be a real or complex number.
4. $\langle \mathbf{x}, \mathbf{x} \rangle \geqslant 0$, with equality if and only if $\mathbf{x} = \theta$. The quantity $\sqrt{\langle \mathbf{x}, \mathbf{x} \rangle}$ denoted by $\|\mathbf{x}\|$ is called the *norm* of $\mathbf{x}$ and is analogous to our usual concept of distance.

12.3.6 Orthogonal and Orthonormal Sets

As in the case of Euclidean space, two vectors are said to be *orthogonal* if their inner product is zero. If we select our basis set to be orthogonal (that is, each vector is orthogonal to every other vector in the set) and further require that the norm of each vector be one (that is, the basis vectors are unit vectors), such a basis set is called an *orthonormal basis set*. Given an orthonormal basis, it is easy to find the representation of any vector in the space in terms of the basis vectors using the inner product. Suppose we have a vector space S_N with an orthonormal basis set $\{\mathbf{x}_i\}_{i=1}^{N}$. Given a vector $\mathbf{y}$ in the space S_N, by definition of the basis set we can write $\mathbf{y}$ as a linear combination of the vectors $\mathbf{x}_i$:

$$\mathbf{y} = \sum_{i=1}^{N} \alpha_i \mathbf{x}_i$$

To find the coefficient α_k, we find the inner product of both sides of this equation with $\mathbf{x}_k$:

$$\langle \mathbf{y}, \mathbf{x}_k \rangle = \sum_{i=1}^{N} \alpha_i \langle \mathbf{x}_i, \mathbf{x}_k \rangle$$

Because of orthonormality,

$$\langle \mathbf{x}_i, \mathbf{x}_k \rangle = \begin{cases} 1 & i = k \\ 0 & i \neq k \end{cases}$$

and

$$\langle \mathbf{y}, \mathbf{x}_k \rangle = \alpha_k$$

By repeating this with each $\mathbf{x}_i$, we can get all the coefficients α_i. Note that in order to use this machinery, the basis set has to be orthonormal.

We now have sufficient information in hand to begin looking at some of the well-known techniques for representing functions of time. This was somewhat of a crash course in vector spaces, and you might, with some justification, be feeling somewhat dazed. Basically, the important ideas that we would like you to remember are the following:

- Vectors are not simply points in two- or three-dimensional space. In fact, functions of time can be viewed as elements in a vector space.
- Collections of vectors that satisfy certain axioms make up a vector space.
- All members of a vector space can be represented as linear, or weighted, combinations of the basis vectors (keep in mind that you can have many different basis sets for the same space). If the basis vectors have unit magnitude and are orthogonal, they are known as an *orthonormal basis set.*
- If a basis set is orthonormal, the weights, or coefficients, can be obtained by taking the inner product of the vector with the corresponding basis vector.

In the next section we use these concepts to show how we can represent periodic functions as linear combinations of sines and cosines.

12.4 Fourier Series

The representation of periodic functions in terms of a series of sines and cosines was used by Jean Baptiste Joseph Fourier to solve equations describing heat diffusion. This approach has since become indispensable in the analysis and design of systems. The work was awarded the grand prize for mathematics in 1812 and has been called one of the most revolutionary contributions of the last century. A very readable account of the life of Fourier and the impact of his discovery can be found in [181].

Fourier showed that any periodic function, no matter how awkward looking, could be represented as the sum of smooth, well-behaved sines and cosines. Given a periodic function $f(t)$ with period T,

$$f(t) = f(t + nT) \qquad n = \pm 1, \pm 2, \ldots$$

we can write $f(t)$ as

$$f(t) = a_0 + \sum_{n=1}^{\infty} a_n \cos n w_0 t + \sum_{n=1}^{\infty} b_n \sin n w_0 t, \qquad w_0 = \frac{2\pi}{T} \tag{4}$$

This form is called the *trigonometric Fourier series representation* of $f(t)$.

A more useful form of the Fourier series representation from our point of view is the exponential form of the Fourier series:

$$f(t) = \sum_{n=-\infty}^{\infty} c_n e^{jnw_0 t} \tag{5}$$

We can easily move between the exponential and trigonometric representations by using Euler's identity

$$e^{j\phi} = \cos\phi + j\sin\phi$$

where $j = \sqrt{-1}$.

In the terminology of the previous section, all periodic functions with period T form a vector space. The complex exponential functions $\{e^{jn\omega_0 t}\}$ constitute a basis for this space. The parameters $\{c_n\}_{n=-\infty}^{\infty}$ are the representations of a given function $f(t)$ with respect to this basis set. Therefore, by using different values of $\{c_n\}_{n=-\infty}^{\infty}$, we can build different periodic functions. If we wanted to inform somebody what a particular periodic function looked like, we could send the values of $\{c_n\}_{n=-\infty}^{\infty}$ and they could synthesize the function.

We would like to see if this basis set is orthonormal. If it is, we want to be able to obtain the coefficients that make up the Fourier representation using the approach described in the previous section. In order to do all this, we need a definition of the inner product on this vector space. If $f(t)$ and $g(t)$ are elements of this vector space, the inner product is defined as

$$\langle f(t), g(t)\rangle = \frac{1}{T}\int_{t_0}^{t_0+T} f(t)g(t)^* dt \tag{6}$$

where t_0 is an arbitrary constant and * denotes complex conjugate. For convenience we will take t_0 to be zero.

Using this inner product definition, let us check to see if the basis set is orthonormal.

$$\langle e^{jn\omega_0 t}, e^{jm\omega_0 t}\rangle = \frac{1}{T}\int_0^T e^{jn\omega_0 t} e^{-jm\omega_0 t} dt \tag{7}$$

$$= \frac{1}{T}\int_0^T e^{j(n-m)\omega_0 t} dt \tag{8}$$

When $n = m$, Equation (7) becomes the norm of the basis vector, which is clearly one. When $n \neq m$, let us define $k = n - m$. Then

$$\langle e^{jn\omega_0 t}, e^{jm\omega_0 t}\rangle = \frac{1}{T}\int_0^T e^{jk\omega_0 t} dt \tag{9}$$

$$= \frac{1}{jk\omega_0 T}(e^{jk\omega_0 T} - 1) \tag{10}$$

$$= \frac{1}{jk\omega_0 T}(e^{jk2\pi} - 1) \tag{11}$$

$$= 0 \tag{12}$$

where we have used the facts that $\omega_0 = \frac{2\pi}{T}$ and

$$e^{jk2\pi} = \cos(2k\pi) + j\sin(2k\pi) = 1$$

Thus, the basis set is orthonormal.

Using this fact, we can find the coefficient c_n by taking the inner product of $f(t)$ with the basis vector $e^{jn\omega_0 t}$:

$$c_n = \langle f(t), e^{jn\omega_0 t}\rangle = \frac{1}{T}\int_0^T f(t)e^{-jn\omega_0 t}\,dt \tag{13}$$

What do we gain from obtaining the Fourier representation $\{c_n\}_{n=-\infty}^{\infty}$ of a function $f(t)$? Before we answer this question, let us examine the context in which we generally use Fourier analysis. We start with some signal generated by a source. If we wish to look at how this signal changes its amplitude over a period of time (or space), we represent it as a function of time $f(t)$ (or a function of space $f(x)$). Thus, $f(t)$ (or $f(x)$) is a representation of the signal that brings out how this signal varies in time (or space). The sequence $\{c_n\}_{n=-\infty}^{\infty}$ is a different representation of the same signal which brings out a different aspect of the signal. The basis functions are sinusoids that differ from each other in how fast they fluctuate in a given time interval. The basis vector $e^{2j\omega_0 t}$ fluctuates twice as fast as the basis vector $e^{j\omega_0 t}$. The coefficients of the basis vectors $\{c_n\}_{n=-\infty}^{\infty}$ give us a measure of the different amounts of fluctuation present in the signal. Fluctuation of this sort is usually measured in terms of frequency. A frequency of 1 Hz denotes the completion of one period in one second, a frequency of 2 Hz denotes the completion of two cycles in one second, and so on. Thus, the coefficients $\{c_n\}_{n=-\infty}^{\infty}$ provide us with a frequency profile of the signal: how much of the signal changes at the rate of $\frac{\omega_0}{2\pi}$ Hz, how much of the signal changes at the rate of $\frac{2\omega_0}{2\pi}$ Hz, and so on. This information cannot be obtained by looking at the time representation $f(t)$. Note that the use of the $\{c_n\}_{n=-\infty}^{\infty}$ representation tells us little about how the signal changes with time. Each representation emphasizes a different aspect of the signal. The ability to view the same signal in different ways helps us to better understand the nature of the signal, and thus develop tools for manipulation of the signal. Later, when we talk about wavelets, we will look at representations that provide information about both the time profile and the frequency profile of the signal.

The Fourier series provides us with a frequency representation of *periodic* signals. However, many of the signals we will be dealing with are not periodic. Fortunately, the Fourier series concepts can be extended to nonperiodic signals.

12.5 Fourier Transform

Consider the function $f(t)$ shown in Figure 12.3. Let us define a function $f_P(t)$ as

$$f_P(t) = \sum_{n=-\infty}^{\infty} f(t - nT) \tag{14}$$

where $T > t_1$. This function, which is obviously periodic ($f_P(t+T) = f_P(t)$), is called the *periodic extension* of the function $f(t)$. Because the function $f_P(t)$ is periodic, we can define

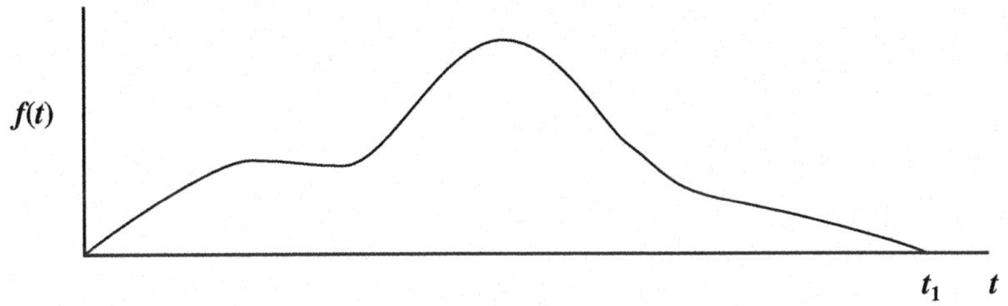

FIGURE 12.3 **A function of time.**

a Fourier series expansion for it:

$$c_n = \frac{1}{T}\int_{-\frac{T}{2}}^{\frac{T}{2}} f_P(t)e^{-jn\omega_0 t}dt \quad (15)$$

$$f_P(t) = \sum_{n=-\infty}^{\infty} c_n e^{jn\omega_0 t} \quad (16)$$

Define

$$C(n, T) = c_n T$$

and

$$\Delta\omega = \omega_0$$

and let us slightly rewrite the Fourier series equations:

$$C(n, T) = \int_{-\frac{T}{2}}^{\frac{T}{2}} f_P(t)e^{-jn\Delta\omega t}dt \quad (17)$$

$$f_P(t) = \sum_{n=-\infty}^{\infty} \frac{C(n, T)}{T} e^{jn\Delta\omega t} \quad (18)$$

We can recover $f(t)$ from $f_P(t)$ by taking the limit of $f_P(t)$ as T goes to infinity. Because $\Delta\omega = \omega_0 = \frac{2\pi}{T}$, this is the same as taking the limit as $\Delta\omega$ goes to zero. As $\Delta\omega$ goes to zero, $n\Delta\omega$ goes to a continuous variable ω. Therefore,

$$\lim_{T\to\infty \Delta\omega\to 0} \int_{-\frac{T}{2}}^{\frac{T}{2}} f_P(t)e^{-jn\Delta\omega t}dt = \int_{-\infty}^{\infty} f(t)e^{-j\omega t}dt \quad (19)$$

From the right-hand side, we can see that the resulting function is a function only of ω. We call this function the Fourier transform of $f(t)$, and we will denote it by $F(\omega)$. To recover $f(t)$ from $F(w)$, we apply the same limits to Equation (18):

$$f(t) = \lim_{T\to\infty} f_P(t) = \lim_{T\to\infty \Delta\omega\to 0} \sum_{n=-\infty}^{\infty} C(n, T)\frac{\Delta\omega}{2\pi} e^{jn\Delta\omega t} \quad (20)$$

$$= \frac{1}{2\pi}\int_{-\infty}^{\infty} F(\omega)e^{j\omega t}d\omega \quad (21)$$

The equation

$$F(\omega) = \int_{-\infty}^{\infty} f(t)e^{-j\omega t} dt \tag{22}$$

is generally called the *Fourier transform*. The function $F(\omega)$ tells us how the signal fluctuates at different frequencies. The equation

$$f(t) = \frac{1}{2\pi} \int_{-\infty}^{\infty} F(w)e^{j\omega t} d\omega \tag{23}$$

is called the *inverse Fourier transform*, and it shows us how we can construct a signal using components that fluctuate at different frequencies. We will denote the operation of the Fourier transform by the symbol $\mathcal{F}$. Thus, in the preceding, $F(\omega) = \mathcal{F}[f(t)]$.

There are several important properties of the Fourier transform, three of which will be of particular use to us. We state them here and leave the proof to the problems (problems 2, 3, and 4 at the end of this chapter).

12.5.1 Parseval's Theorem

The Fourier transform is an energy-preserving transform; that is, the total energy when we look at the time representation of the signal is the same as the total energy when we look at the frequency representation of the signal. This makes sense because the total energy is a physical property of the signal and should not change when we look at it using different representations. Mathematically, this is stated as

$$\int_{-\infty}^{\infty} |f(t)|^2 = \frac{1}{2\pi} \int_{-\infty}^{\infty} |F(\omega)|^2 d\omega \tag{24}$$

The $\frac{1}{2\pi}$ factor is a result of using units of radians (ω) for frequency instead of Hertz (f). If we substitute $\omega = 2\pi f$ in Equation (24), the 2π factor will go away. This property applies to any vector space representation obtained using an orthonormal basis set.

12.5.2 Modulation Property

If $f(t)$ has the Fourier transform $F(\omega)$, then the Fourier transform of $f(t)e^{j\omega_0 t}$ is $F(w - w_0)$. That is, multiplication with a complex exponential in the time domain corresponds to a shift in the frequency domain. As a sinusoid can be written as a sum of complex exponentials, multiplication of $f(t)$ by a sinusoid will also correspond to shifts of $F(\omega)$. For example,

$$\cos(\omega_0 t) = \frac{e^{j\omega_0 t} + e^{-j\omega_0 t}}{2}$$

Therefore,

$$\mathcal{F}[f(t)\cos(\omega_0 t)] = \frac{1}{2}\left(F(\omega - \omega_0) + F(\omega + \omega_0)\right)$$

12.5.3 Convolution Theorem

When we examine the relationships between the input and output of linear systems, we will encounter integrals of the following forms:

$$f(t) = \int_{-\infty}^{\infty} f_1(\tau) f_2(t - \tau) d\tau$$

or

$$f(t) = \int_{-\infty}^{\infty} f_1(t - \tau) f_2(\tau) d\tau$$

These are called convolution integrals. The convolution operation is often denoted as

$$f(t) = f_1(t) \otimes f_2(t)$$

The convolution theorem states that if $F(\omega) = \mathcal{F}[f(t)] = \mathcal{F}[f_1(t) \otimes f_2(t)]$, $F_1(\omega) = \mathcal{F}[f_1(t)]$, and $F_2(\omega) = \mathcal{F}[f_2(t)]$, then

$$F(\omega) = F_1(\omega) F_2(\omega)$$

We can also go in the other direction. If

$$F(\omega) = F_1(\omega) \otimes F_2(\omega) = \int F_1(\sigma) F_2(\omega - \sigma) d\sigma$$

then

$$f(t) = f_1(t) f_2(t)$$

As mentioned earlier, this property of the Fourier transform is important because the convolution integral relates the input and output of linear systems, which brings us to one of the major reasons for the popularity of the Fourier transform. We have claimed that the Fourier series and Fourier transform provide us with an alternative frequency profile of a signal. Although sinusoids are not the only basis set that can provide us with a frequency profile, they do have an important property that helps us study linear systems, which we describe in the next section.

12.6 Linear Systems

A linear system is a system that has the following two properties:

- **Homogeneity:** Suppose we have a linear system L with input $f(t)$ and output $g(t)$:

  $$g(t) = L[f(t)]$$

  If we have two inputs, $f_1(t)$ and $f_2(t)$, with corresponding outputs, $g_1(t)$ and $g_2(t)$, then the output of the sum of the two inputs is simply the sum of the two outputs:

  $$L[f_1(t) + f_2(t)] = g_1(t) + g_2(t)$$

- **Scaling:** Given a linear system L with input $f(t)$ and output $g(t)$, if we multiply the input with a scalar α, then the output will be multiplied by the same scalar:

$$L[\alpha f(t)] = \alpha L[f(t)] = \alpha g(t)$$

The two properties together are referred to as *superposition*.

12.6.1 Time Invariance

Of specific interest to us are linear systems that are *time invariant*. A time-invariant system has the property that the shape of the response of the system does not depend on the time at which the input was applied. If the response of a linear system L to an input $f(t)$ is $g(t)$,

$$L[f(t)] = g(t)$$

and we delay the input by some interval t_0, then if L is a time-invariant system, the output will be $g(t)$ delayed by the same amount:

$$L[f(t - t_0)] = g(t - t_0) \tag{25}$$

12.6.2 Transfer Function

Linear time-invariant systems have a very interesting (and useful) response when the input is a sinusoid. If the input to a linear system is a sinusoid of a certain frequency ω_0, then the output is also a sinusoid of the same frequency that has been scaled and delayed; that is,

$$L[\cos(\omega_0 t)] = \alpha \cos(\omega_0 (t - t_d))$$

or in terms of the complex exponential

$$L[e^{j\omega_0 t}] = \alpha e^{j\omega_0 (t - t_d)}$$

Thus, given a linear system, we can characterize its response to sinusoids of a particular frequency by a pair of parameters, the gain α and the delay t_d. In general, we use the phase $\phi = \omega_0 t_d$ in place of the delay. The parameters α and ϕ will generally be a function of the frequency, so in order to characterize the system for all frequencies, we will need a pair of functions $\alpha(\omega)$ and $\phi(\omega)$. As the Fourier transform allows us to express the signal as coefficients of sinusoids, given an input $f(t)$, all we need to do is, for each frequency ω, multiply the Fourier transform of $f(t)$ with some $\alpha(\omega)e^{j\phi(\omega)}$, where $\alpha(\omega)$ and $\phi(\omega)$ are the gain and phase terms of the linear system for that particular frequency.

This pair of functions $\alpha(\omega)$ and $\phi(\omega)$ constitute the *transfer function* of the linear time-invariant system $H(\omega)$:

$$H(\omega) = |H(\omega)|\, e^{j\phi(\omega)}$$

where $|H(\omega)| = \alpha(\omega)$.

Because of the specific way in which a linear system responds to a sinusoidal input, given a linear system with transfer function $H(\omega)$, input $f(t)$, and output $g(t)$, the Fourier transforms of the input and output $F(\omega)$ and $G(\omega)$ are related by

$$G(w) = H(\omega)F(\omega)$$

Using the convolution theorem, $f(t)$ and $g(t)$ are related by

$$g(t) = \int_{-\infty}^{\infty} f(\tau)h(t-\tau)d\tau$$

or

$$g(t) = \int_{-\infty}^{\infty} f(t-\tau)h(\tau)d\tau$$

where $H(\omega)$ is the Fourier transform of $h(t)$.

12.6.3 Impulse Response

To see what $h(t)$ is, let us look at the input-output relationship of a linear time-invariant system from a different point of view. Let us suppose we have a linear system L with input $f(t)$. We can obtain a staircase approximation $f_S(t)$ to the function $f(t)$, as shown in Figure 12.4:

$$f_S(t) = \sum f(n\Delta t)\text{rect}\left(\frac{t-n\Delta t}{\Delta t}\right) \tag{26}$$

where

$$\text{rect}\left(\frac{t}{T}\right) = \begin{cases} 1 & |t| < \frac{T}{2} \\ 0 & \text{otherwise} \end{cases} \tag{27}$$

The response of the linear system can be written as

$$L[f_S(t)] = L\left[\sum f(n\Delta t)\text{rect}\left(\frac{t-n\Delta t}{\Delta t}\right)\right] \tag{28}$$

$$= L\left[\sum f(n\Delta t)\frac{\text{rect}\left(\frac{t-n\Delta t}{\Delta t}\right)}{\Delta t}\Delta t\right] \tag{29}$$

For a given value of Δt, we can use the superposition property of linear systems to obtain

$$L[f_S(t)] = \sum f(n\Delta t)L\left[\frac{\text{rect}(\frac{t-n\Delta t}{\Delta t})}{\Delta t}\right]\Delta t \tag{30}$$

If we now take the limit as Δt goes to zero in this equation, on the left-hand side $f_S(t)$ will go to $f(t)$. To see what happens on the right-hand side of the equation, first let's look at the effect of this limit on the function $\text{rect}(\frac{t}{\Delta t})/\Delta t$. As Δt goes to zero, this function becomes narrower and taller. However, at all times the integral of this function is equal to one. The

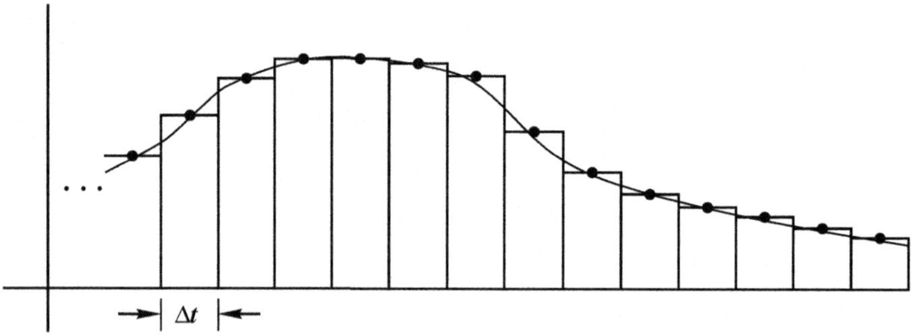

FIGURE 12.4 **A function of time.**

limit of this function as Δt goes to zero is called the *Dirac delta function*, or *impulse function*, and is denoted by $\delta(t)$:

$$\lim_{\Delta t \to 0} \frac{\text{rect}(\frac{t}{\Delta t})}{\Delta t} = \delta(t) \tag{31}$$

Therefore,

$$L[f(t)] = \lim_{\Delta t \to 0} L[f_S(t)] = \int f(\tau)L[\delta(t-\tau)]d\tau \tag{32}$$

where $n\Delta t$ goes to τ as Δt goes to zero. Denote the response of the system L to an impulse, or the *impulse response*, by $h(t)$:

$$h(t) = L[\delta(t)] \tag{33}$$

Then, if the system is also time invariant,

$$L[f(t)] = \int f(\tau)h(t-\tau)d\tau \tag{34}$$

Using the convolution theorem, we can see that the Fourier transform of the impulse response $h(t)$ is the transfer function $H(\omega)$.

The Dirac delta function is an interesting function. In fact, it is not clear that it is a function at all. It has an integral that is clearly one, but at the only point where it is not zero, it is undefined! One property of the delta function that makes it very useful is the *sifting* property:

$$\int_{t_1}^{t_2} f(t)\delta(t-t_0)dt = \begin{cases} f(t_0) & t_1 \leqslant t_0 \leqslant t_2 \\ 0 & \text{otherwise} \end{cases} \tag{35}$$

12.6.4 Filter

The linear systems of most interest to us will be systems that permit certain frequency components of the signal to pass through, while attenuating all other components of the signal. Such systems are called *filters*. If the filter allows only frequency components below a certain frequency W Hz to pass through, the filter is called a *low-pass filter*. The transfer function of an ideal low-pass filter is given by

$$H(\omega) = \begin{cases} e^{-j\omega} & |\omega| < 2\pi W \\ 0 & \text{otherwise} \end{cases} \quad (36)$$

This filter is said to have a *bandwidth* of W Hz. The magnitude of this filter is shown in Figure 12.5. A low-pass filter will produce a smoothed version of the signal by blocking higher-frequency components that correspond to fast variations in the signal.

A filter that attenuates the frequency components below a certain frequency W and allows the frequency components above this frequency to pass through is called a *high-pass filter*. A high-pass filter will remove slowly changing trends from the signal. Finally, a signal that lets through a range of frequencies between two specified frequencies, say, W_1 and W_2, is called a *band-pass filter*. The bandwidth of this filter is said to be $W_2 - W_1$ Hz. The magnitude of the transfer functions of an ideal high-pass filter and an ideal band-pass filter with bandwidth W are shown in Figure 12.6. In all the ideal filter characteristics, there is a sharp transition between the *passband* of the filter (the range of frequencies that are not attenuated) and the *stopband* of the filter (those frequency intervals where the signal is completely attenuated).

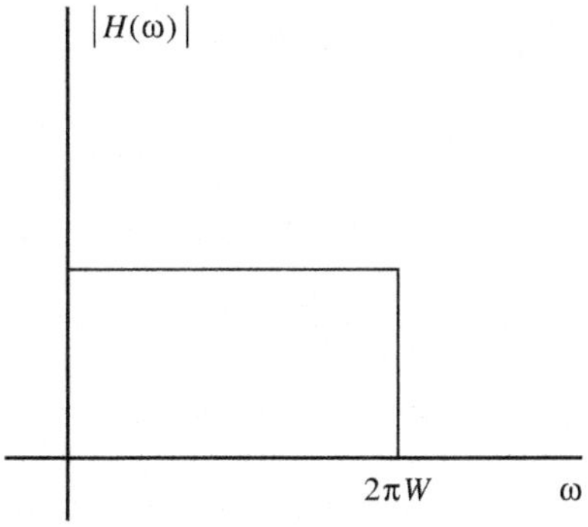

FIGURE 12.5 **Magnitude of the transfer function of an ideal low-pass filter.**

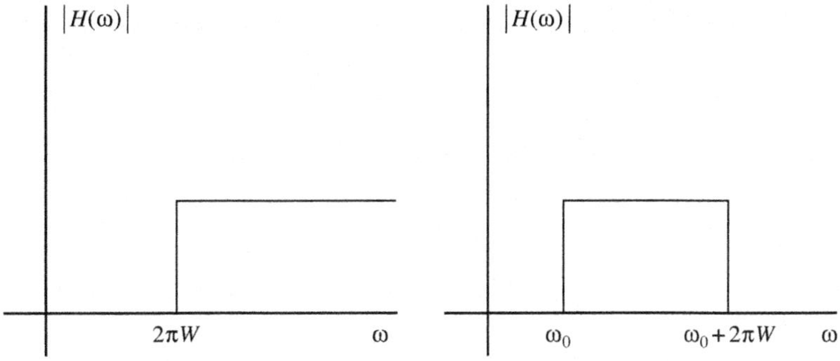

FIGURE 12.6 **Magnitudes of the transfer functions of ideal high-pass (left) and ideal band-pass (right) filters.**

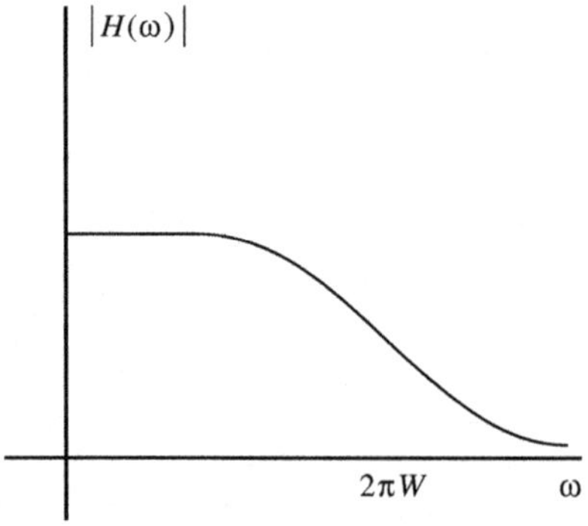

FIGURE 12.7 Magnitude of the transfer functions of a realistic low-pass filter.

Real filters do not have such sharp transitions, or *cutoffs*. The magnitude characteristics of a more realistic low-pass filter are shown in Figure 12.7. Notice the more gentle rolloff. But when the cutoff between stopband and passband is not sharp, how do we define the bandwidth? There are several different ways of defining the bandwidth. The most common way is to define the frequency at which the magnitude of the transfer function is $1/\sqrt{2}$ of its maximum value (or the magnitude squared is 1/2 of its maximum value) as the cutoff frequency.

12.7 Sampling

In 1928 Harry Nyquist at Bell Laboratories showed that if we have a signal whose Fourier transform is zero above some frequency W Hz, it can be accurately represented using $2W$ equally spaced samples per second. This very important result, known as the *sampling theorem*, is at the heart of our ability to transmit analog waveforms such as speech and video using digital means. There are several ways to prove this result. We will use the results presented in the previous section to do so.

12.7.1 Ideal Sampling-Frequency Domain View

Let us suppose we have a function $f(t)$ with Fourier transform $F(\omega)$, shown in Figure 12.8, which is zero for ω greater than $2\pi W$. Define the periodic extension of $F(\omega)$ as

$$F_P(\omega) = \sum_{n=-\infty}^{\infty} F(\omega - n\sigma_0), \qquad \sigma_0 = 4\pi W \tag{37}$$

The periodic extension is shown in Figure 12.9. As $F_P(\omega)$ is periodic, we can express it in terms of a Fourier series expansion:

$$F_P(\omega) = \sum_{n=-\infty}^{\infty} c_n e^{jn\frac{1}{2W}\omega} \tag{38}$$

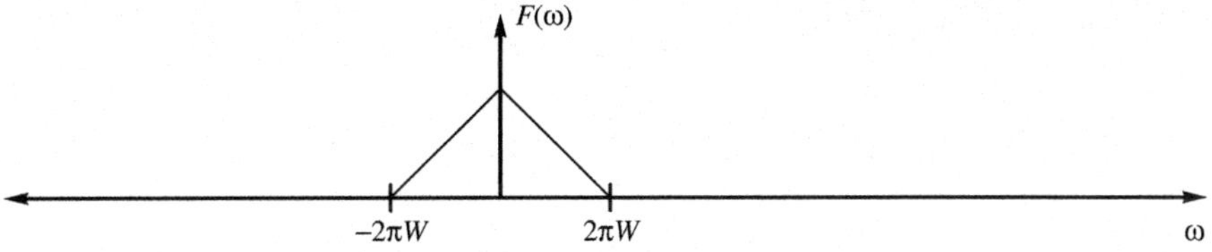

FIGURE 12.8 **A function $F(\omega)$.**

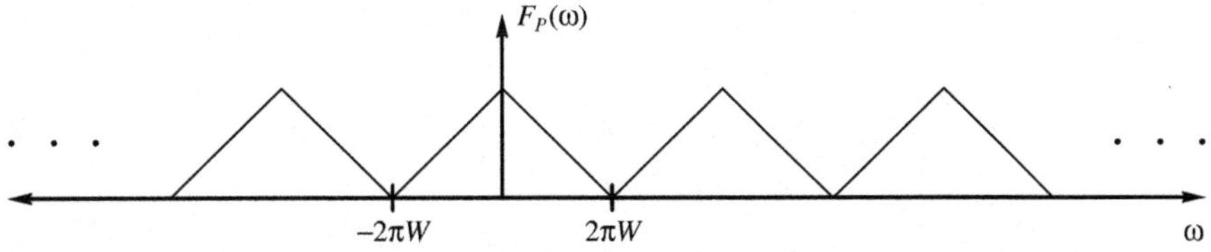

FIGURE 12.9 **The periodic extension $F_P(\omega)$.**

The coefficients of the expansion $\{c_n\}_{n=-\infty}^{\infty}$ are then given by

$$c_n = \frac{1}{4\pi W}\int_{-2\pi W}^{2\pi W} F_P(\omega)e^{-jn\frac{1}{2W}\omega}d\omega \tag{39}$$

However, in the interval $(-2\pi W, 2\pi W)$, $F(\omega)$ is identical to $F_P(\omega)$; therefore,

$$c_n = \frac{1}{4\pi W}\int_{-2\pi W}^{2\pi W} F(\omega)e^{-jn\frac{1}{2W}\omega}d\omega \tag{40}$$

The function $F(\omega)$ is zero outside the interval $(-2\pi W, 2\pi W)$, so we can extend the limits to infinity without changing the result.

$$c_n = \frac{1}{2W}\left[\frac{1}{2\pi}\int_{-\infty}^{\infty} F(\omega)e^{-jn\frac{1}{2W}\omega}d\omega\right] \tag{41}$$

The expression in brackets is simply the inverse Fourier transform evaluated at $t = \frac{n}{2W}$; therefore,

$$c_n = \frac{1}{2W}f\left(\frac{n}{2W}\right) \tag{42}$$

Knowing $\{c_n\}_{n=-\infty}^{\infty}$ and the value of W, we can reconstruct $F_P(\omega)$. Because $F_P(\omega)$ and $F(\omega)$ are identical in the interval $(-2\pi W, 2\pi W)$, we can also reconstruct $F(\omega)$ in this interval. But $\{c_n\}_{n=-\infty}^{\infty}$ are simply the samples of $f(t)$ every $\frac{1}{2W}$ seconds, and $F(\omega)$ is zero outside this interval. Therefore, given the samples of a function $f(t)$ obtained at a rate of $2W$ samples per second, we should be able to exactly reconstruct the function $f(t)$.

Let us see how we can do this:

$$f(t) = \frac{1}{2\pi}\int_{-\infty}^{\infty} F(\omega)e^{-j\omega t}d\omega \tag{43}$$

$$= \frac{1}{2\pi}\int_{-2\pi W}^{2\pi W} F(\omega)e^{-j\omega t}d\omega \tag{44}$$

$$= \frac{1}{2\pi}\int_{-2\pi W}^{2\pi W} F_P(\omega)e^{-j\omega t}d\omega \tag{45}$$

$$= \frac{1}{2\pi}\int_{-2\pi W}^{2\pi W} \sum_{n=-\infty}^{\infty} c_n e^{jn\frac{1}{2W}\omega}e^{-j\omega t}d\omega \tag{46}$$

$$= \frac{1}{2\pi}\sum_{n=-\infty}^{\infty} c_n \int_{-2\pi W}^{2\pi W} e^{jw(t-\frac{n}{2W})}d\omega \tag{47}$$

Evaluating the integral and substituting for c_n from Equation (42), we obtain

$$f(t) = \sum_{n=-\infty}^{\infty} f\left(\frac{n}{2W}\right) \text{Sinc}\left[2W\left(t - \frac{n}{2W}\right)\right] \tag{48}$$

where

$$\text{Sinc}[x] = \frac{\sin(\pi x)}{\pi x} \tag{49}$$

Thus, given samples of $f(t)$ taken every $\frac{1}{2W}$ seconds, or, in other words, samples of $f(t)$ obtained at a rate of $2W$ samples per second, we can reconstruct $f(t)$ by interpolating between the samples using the Sinc function.

12.7.2 Ideal Sampling—Time Domain View

Let us look at this process from a slightly different point of view, starting with the sampling operation. Mathematically, we can represent the sampling operation by multiplying the function $f(t)$ with a train of impulses to obtain the sampled function $f_S(t)$:

$$f_S(t) = f(t)\sum_{n=-\infty}^{\infty} \delta(t - nT), \qquad T < \frac{1}{2W} \tag{50}$$

To obtain the Fourier transform of the sampled function, we use the convolution theorem:

$$\mathcal{F}\left[f(t)\sum_{n=-\infty}^{\infty} \delta(t - nT)\right] = \mathcal{F}[f(t)] \otimes \mathcal{F}\left[\sum_{n=-\infty}^{\infty} \delta(t - nT)\right] \tag{51}$$

Let us denote the Fourier transform of $f(t)$ by $F(\omega)$. The Fourier transform of a train of impulses in the time domain is a train of impulses in the frequency domain (see Problem 5 at the end of this chapter):

$$\mathcal{F}\left[\sum_{n=-\infty}^{\infty} \delta(t - nT)\right] = \sigma_0 \sum_{n=-\infty}^{\infty} \delta(w - n\sigma_0) \qquad \sigma_0 = \frac{2\pi}{T} \tag{52}$$

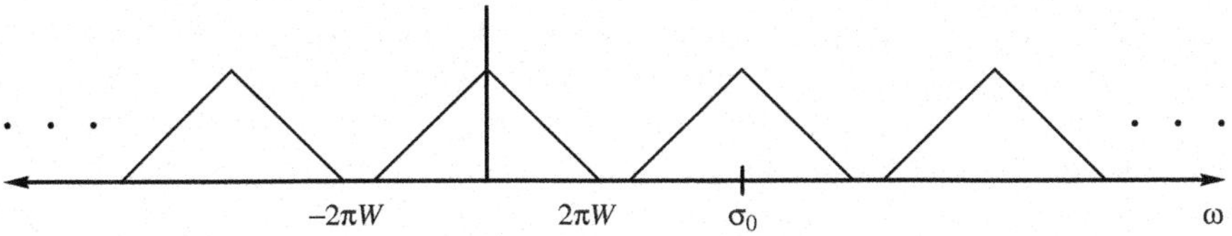

FIGURE 12.10 **Fourier transform of the sampled function.**

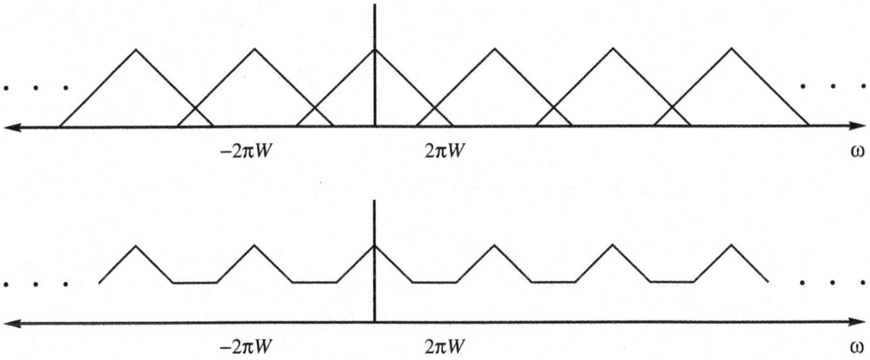

FIGURE 12.11 **Effect of sampling at a rate less than *2W* samples per second.**

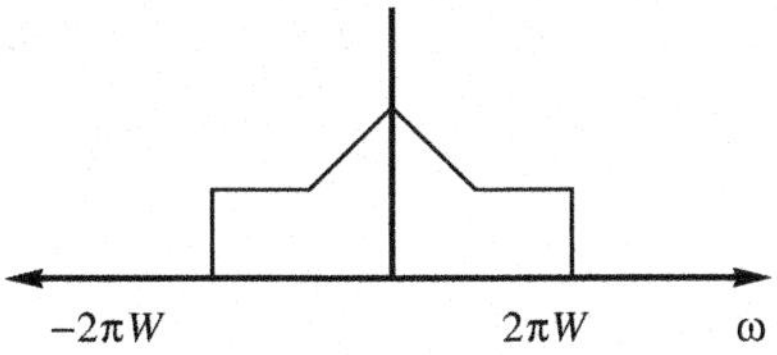

FIGURE 12.12 **Aliased reconstruction.**

Thus, the Fourier transform of $f_S(t)$ is

$$F_S(\omega) = F(\omega) \otimes \sum_{n=-\infty}^{\infty} \delta(w - n\sigma_0) \tag{53}$$

$$= \sum_{n=-\infty}^{\infty} F(\omega) \otimes \delta(w - n\sigma_0) \tag{54}$$

$$= \sum_{n=-\infty}^{\infty} F(\omega - n\sigma_0) \tag{55}$$

where the last equality is due to the sifting property of the delta function.

Pictorially, for $F(\omega)$ as shown in Figure 12.8, $F_S(\omega)$ is shown in Figure 12.10. Note that if T is less than $\frac{1}{2W}$, σ_0 is greater than $4\pi W$, and as long as σ_0 is greater than $4\pi W$, we can recover $F(\omega)$ by passing $F_S(\omega)$ through an ideal low-pass filter with bandwidth W Hz ($2\pi W$ radians).

What happens if we do sample at a rate less than $2W$ samples per second (that is, σ_0 is less than $4\pi W$)? Again we can see the results most easily in a pictorial fashion. The result for σ_0 equal to $3\pi W$ is shown in Figure 12.11. Filtering this signal through an ideal low-pass filter, we get the distorted signal shown in Figure 12.12. Therefore, if σ_0 is less than $4\pi W$, we can no longer recover the signal $f(t)$ from its samples. This distortion is known as *aliasing*. In order to prevent aliasing, it is useful to filter the signal prior to sampling using a low-pass filter with a bandwidth less than half the sampling frequency.

Once we have the samples of a signal, sometimes the actual times they were sampled at are not important. In these situations we can normalize the sampling frequency to unity. This means that the highest frequency component in the signal is at 0.5 Hz, or π radians. Thus, when dealing with sampled signals, we will often talk about frequency ranges of $-\pi$ to π.

12.8 Discrete Fourier Transform

The procedures that we gave for obtaining the Fourier series and transform were based on the assumption that the signal we were examining could be represented as a continuous function of time. However, for the applications that we will be interested in, we will primarily be dealing with samples of a signal. To obtain the Fourier transform of nonperiodic signals, we started from the Fourier series and modified it to take into account the nonperiodic nature of the signal. To obtain the discrete Fourier transform (DFT), we again start from the Fourier series. We begin with the Fourier series representation of a sampled function, the discrete Fourier series.

Recall that the Fourier series coefficients of a periodic function $f(t)$ with period T are given by

$$c_k = \frac{1}{T}\int_0^T f(t)e^{-jkw_0 t}\,dt \tag{56}$$

Suppose instead of a continuous function, we have a function sampled N times during each period T. We can obtain the coefficients of the Fourier series representation of this sampled function as

$$F_k = \frac{1}{T}\int_0^T f(t)\sum_{n=0}^{N-1}\delta\left(t - \frac{n}{N}T\right)e^{-jkw_0 t}\,dt \tag{57}$$

$$= \frac{1}{T}\sum_{n=0}^{N-1} f\left(\frac{n}{N}T\right)e^{-j\frac{2\pi kn}{N}} \tag{58}$$

where we have used the fact that $w_0 = \frac{2\pi}{T}$, and we have replaced c_k by F_k. Taking $T = 1$ for convenience and defining

$$f_n = f\left(\frac{n}{N}\right)$$

we get the coefficients for the discrete Fourier series (DFS) representation:

$$F_k = \sum_{n=0}^{N-1} f_n e^{-j\frac{2\pi kn}{N}} \tag{59}$$

Notice that the sequence of coefficients $\{F_k\}$ is periodic with period N.

The Fourier series representation was given by

$$f(t) = \sum_{k=-\infty}^{\infty} c_k e^{jn\omega_0 t} \tag{60}$$

Evaluating this for $t = \frac{n}{N}T$, we get

$$f_n = f(\frac{n}{N}T) = \sum_{k=-\infty}^{\infty} c_k e^{j\frac{2\pi kn}{N}} \tag{61}$$

Let us write this in a slightly different form:

$$f_n = \sum_{k=0}^{N-1} \sum_{l=-\infty}^{\infty} c_{k+lN} e^{j\frac{2\pi n(k+lN)}{N}} \tag{62}$$

but

$$e^{j\frac{2\pi n(k+lN)}{N}} = e^{j\frac{2\pi kn}{N}} e^{j2\pi nl} \tag{63}$$

$$= e^{j\frac{2\pi kn}{N}} \tag{64}$$

Therefore,

$$f_n = \sum_{k=0}^{N-1} e^{j\frac{2\pi kn}{N}} \sum_{l=-\infty}^{\infty} c_{k+lN} \tag{65}$$

Define

$$\bar{c}_k = \sum_{l=-\infty}^{\infty} c_{k+lN} \tag{66}$$

Clearly, $\bar{c}_k$ is periodic with period N. In fact, we can show that $\bar{c}_k = \frac{1}{N}F_k$ and

$$f_n = \frac{1}{N} \sum_{k=0}^{N-1} F_k e^{j\frac{2\pi kn}{N}} \tag{67}$$

Obtaining the discrete Fourier transform from the discrete Fourier series is simply a matter of interpretation. We are generally interested in the discrete Fourier transform of a finite-length sequence. If we assume that the finite-length sequence is one period of a periodic sequence, then we can use the DFS equations to represent this sequence. The only difference is that the expressions are only valid for one "period" of the "periodic" sequence.

The DFT is an especially powerful tool because of the existence of a fast algorithm, appropriately called the *fast Fourier transform* (FFT), that can be used to compute it.

12.9 Z-Transform

In the previous section we saw how to extend the Fourier series for use with sampled functions. We can also do the same with the Fourier transform. Recall that the Fourier transform was given by the equation

$$F(\omega) = \int_{-\infty}^{\infty} f(t)e^{-j\omega t}dt \tag{68}$$

Replacing $f(t)$ with its sampled version, we get

$$F(\omega) = \int_{-\infty}^{\infty} f(t)\sum_{n=-\infty}^{\infty} \delta(t - nT)e^{-j\omega t}dt \tag{69}$$

$$= \sum_{n=-\infty}^{\infty} f_n e^{-j\omega nT} \tag{70}$$

where $f_n = f(nT)$. This is called the discrete time Fourier transform. The Z-transform of the sequence $\{f_n\}$ is a generalization of the discrete time Fourier transform and is given by

$$F(z) = \sum_{n=-\infty}^{\infty} f_n z^{-n} \tag{71}$$

where

$$z = e^{\sigma T + jwT} \tag{72}$$

Notice that if we let σ equal zero, we get the original expression for the Fourier transform of a discrete time sequence. We denote the Z-transform of a sequence by

$$F(z) = \mathcal{Z}[f_n]$$

We can express this another way. Notice that the magnitude of z is given by

$$|z| = e^{\sigma T}$$

Thus, when σ equals zero, the magnitude of z is one. Because z is a complex number, the magnitude of z is equal to one on the unit circle in the complex plane. Therefore, we can say that the Fourier transform of a sequence can be obtained by evaluating the Z-transform of the sequence on the unit circle. Notice that the Fourier transform thus obtained will be periodic, which we expect because we are dealing with a sampled function. Further, if we assume T to be one, ω varies from $-\pi$ to π, which corresponds to a frequency range of -0.5 to 0.5 Hz. This makes sense because, by the sampling theorem, if the sampling rate is one sample per second, the highest frequency component that can be recovered is 0.5 Hz.

For the Z-transform to exist—in other words, for the power series to converge—we need to have

$$\sum_{n=-\infty}^{\infty} \left| f_n z^{-n} \right| < \infty$$

Whether this inequality holds will depend on the sequence itself and the value of z. The values of z for which the series converges are called the *region of convergence* of the Z-transform. From our earlier discussion, we can see that for the Fourier transform of the sequence to exist, the region of convergence should include the unit circle. Let us look at a simple example.

Example 12.9.1:

Given the sequence

$$f_n = a^n u[n]$$

where $u[n]$ is the unit step function

$$u[n] = \begin{cases} 1 & n \geqslant 0 \\ 0 & n < 0 \end{cases} \tag{73}$$

the Z-transform is given by

$$F(z) = \sum_{n=0}^{\infty} a^n z^{-n} \tag{74}$$

$$= \sum_{n=0}^{\infty} (az^{-1})^n \tag{75}$$

This is simply the sum of a geometric series. As we confront this kind of sum quite often, let us briefly digress and obtain the formula for the sum of a geometric series.

Suppose we have a sum

$$S_{mn} = \sum_{k=m}^{n} x^k = x^m + x^{m+1} + \cdots + x^n \tag{76}$$

Then

$$xS_{mn} = x^{m+1} + x^{m+2} + \cdots + x^{n+1} \tag{77}$$

Subtracting Equation (77) from Equation (76), we get

$$(1 - x)S_{mn} = x^m - x^{n+1}$$

and

$$S_{mn} = \frac{x^m - x^{n+1}}{1 - x}$$

If the upper limit of the sum is infinity, we take the limit as n goes to infinity. This limit exists only when $|x| < 1$.

Using this formula, we get the Z-transform of the $\{f_n\}$ sequence as

$$F(z) = \frac{1}{1 - az^{-1}}, \qquad \left|az^{-1}\right| < 1 \tag{78}$$

$$= \frac{z}{z - a}, \qquad |z| > |a|. \tag{79}$$

♦

In this example the region of convergence is the region $|z| > a$. For the Fourier transform to exist, we need to include the unit circle in the region of convergence. In order for this to happen, a has to be less than one.

Using this example, we can get some other Z-transforms that will be useful to us.

Example 12.9.2:

In the previous example we found that

$$\sum_{n=0}^{\infty} a^n z^{-n} = \frac{z}{z-a}, \qquad |z| > |a| \tag{80}$$

If we take the derivative of both sides of the equation with respect to a, we get

$$\sum_{n=0}^{\infty} n a^{n-1} z^{-n} = \frac{z}{(z-a)^2}, \qquad |z| > |a| \tag{81}$$

Thus,

$$\mathcal{Z}[n a^{n-1} u[n]] = \frac{z}{(z-a)^2}, \qquad |z| > |a|$$

If we differentiate Equation (80) m times, we get

$$\sum_{n=0}^{\infty} n(n-1)\cdots(n-m+1) a^{n-m} z^{-n} = \frac{m!z}{(z-a)^{m+1}}$$

In other words,

$$\mathcal{Z}\left[\binom{n}{m} a^{n-m} u[n]\right] = \frac{z}{(z-a)^{m+1}} \tag{82}$$

♦

In these examples the Z-transform is a ratio of polynomials in z. For sequences of interest to us, this will generally be the case, and the Z-transform will be of the form

$$F(z) = \frac{N(z)}{D(z)}$$

The values of z for which $F(z)$ is zero are called the *zeros* of $F(z)$; the values for which $F(z)$ is infinity are called the *poles* of $F(z)$. For finite values of z, the poles will occur at the roots of the polynomial $D(z)$.

The inverse Z-transform is formally given by the contour integral [?]

$$\frac{1}{2\pi j} \oint_C F(z) z^{n-1} dz$$

where the integral is over the counterclockwise contour C, and C lies in the region of convergence. This integral can be difficult to evaluate directly; therefore, in most cases we use alternative methods for finding the inverse Z-transform.

TABLE 12.1 **Some Z-transform pairs.**

$\{f_n\}$	$F(z)$
$a^n u[n]$	$\frac{z}{z-a}$
$nTu[n]$	$\frac{Tz^{-1}}{(1-z^{-1})^2}$
$\sin(\alpha nT)$	$\frac{(\sin \alpha nT)z^{-1}}{1-2\cos(\alpha T)z^{-1}+z^{-2}}$
$\cos(\alpha nT)$	$\frac{(\cos \alpha nT)z^{-1}}{1-2\cos(\alpha T)z^{-1}+z^{-2}}$
δ_n	1

12.9.1 Tabular Method

The inverse Z-transform has been tabulated for a number of interesting cases (see Table. 12.1). If we can write $F(z)$ as a sum of these functions

$$F(z) = \sum \alpha_i F_i(z)$$

then the inverse Z-transform is given by

$$f_n = \sum \alpha_i f_{i,n}$$

where $F_i(z) = \mathcal{Z}[\{f_{i,n}\}]$.

Example 12.9.3:

$$F(z) = \frac{z}{z-0.5} + \frac{2z}{z-0.3}$$

From our earlier example we know the inverse Z-transform of $z/(z-a)$. Using that, the inverse Z-transform of $F(z)$ is

$$f_n = 0.5^n u[n] + 2(0.3)^n u[n]$$

♦

12.9.2 Partial Fraction Expansion

In order to use the tabular method, we need to be able to decompose the function of interest to us as a sum of simpler terms. The partial fraction expansion approach does exactly that when the function is a ratio of polynomials in z.

Suppose $F(z)$ can be written as a ratio of polynomials $N(z)$ and $D(z)$. For the moment let us assume that the degree of $D(z)$ is greater than the degree of $N(z)$, and that all the roots of $D(z)$ are distinct (distinct roots are referred to as simple roots); that is,

$$F(z) = \frac{N(z)}{(z-z_1)(z-z_2)\cdots(z-z_L)} \quad (83)$$

Then we can write $F(z)/z$ as

$$\frac{F(z)}{z} = \sum_{i=1}^{L} \frac{A_i}{z - z_i} \tag{84}$$

If we can find the coefficients A_i, then we can write $F(z)$ as

$$F(z) = \sum_{i=1}^{L} \frac{A_i z}{z - z_i}$$

and the inverse Z-transform will be given by

$$f_n = \sum_{i=1}^{L} A_i z_i^n u[n]$$

The question then becomes one of finding the value of the coefficients A_i. This can be simply done as follows: Suppose we want to find the coefficient A_k. Multiply both sides of Equation (84) by $(z - z_k)$. Simplifying this we obtain

$$\frac{F(z)(z - z_k)}{z} = \sum_{i=1}^{L} \frac{A_i(z - z_k)}{z - z_i} \tag{85}$$

$$= A_k + \sum_{\substack{i=1 \\ i \neq k}}^{L} \frac{A_i(z - z_k)}{z - z_i} \tag{86}$$

Evaluating this equation at $z = z_k$, all the terms in the summation go to zero and

$$A_k = \left. \frac{F(z)(z - z_k)}{z} \right|_{z=z_k} \tag{87}$$

Example 12.9.4:

Let us use the partial fraction expansion method to find the inverse Z-transform of

$$F(z) = \frac{6z^2 - 9z}{z^2 - 2.5z + 1}$$

Then

$$\frac{F(z)}{z} = \frac{1}{z} \frac{6z^2 - 9z}{z^2 - 2.5z + 1} \tag{88}$$

$$= \frac{6z - 9}{(z - 0.5)(z - 2)} \tag{89}$$

We want to write $F(z)/z$ in the form

$$\frac{F(z)}{z} = \frac{A_1}{z - 0.5} + \frac{A_2}{z - 2}$$

Using the approach described above, we obtain

$$A_1 = \left.\frac{(6z-9)(z-0.5)}{(z-0.5)(z-2)}\right|_{z=0.5} \tag{90}$$

$$= 4 \tag{91}$$

$$A_2 = \left.\frac{(6z-9)(z-2)}{(z-0.5)(z-2)}\right|_{z=2} \tag{92}$$

$$= 2 \tag{93}$$

Therefore,

$$F(z) = \frac{4z}{z-0.5} + \frac{2z}{z-2}$$

and

$$f_n = [4(0.5)^n + 2(2)^n]u[n]$$

♦

The procedure becomes slightly more complicated when we have repeated roots of $D(z)$. Suppose we have a function

$$F(z) = \frac{N(z)}{(z-z_1)(z-z_2)^2}$$

The partial fraction expansion of this function is

$$\frac{F(z)}{z} = \frac{A_1}{z-z_1} + \frac{A_2}{z-z_2} + \frac{A_3}{(z-z_2)^2}$$

The values of A_1 and A_3 can be found as shown previously:

$$A_1 = \left.\frac{F(z)(z-z_1)}{z}\right|_{z=z_1} \tag{94}$$

$$A_3 = \left.\frac{F(z)(z-z_2)^2}{z}\right|_{z=z_2} \tag{95}$$

However, we run into problems when we try to evaluate A_2. Let's see what happens when we multiply both sides by $(z - z_2)$:

$$\frac{F(z)(z-z_2)}{z} = \frac{A_1(z-z_2)}{z-z_1} + A_2 + \frac{A_3}{z-z_2} \tag{96}$$

If we now evaluate this equation at $z = z_2$, the third term on the right-hand side becomes undefined. In order to avoid this problem, we first multiply both sides by $(z - z_2)^2$ and take the derivative with respect to z prior to evaluating the equation at $z = z_2$:

$$\frac{F(z)(z-z_2)^2}{z} = \frac{A_1(z-z_2)^2}{z-z_1} + A_2(z-z_2) + A_3 \tag{97}$$

Taking the derivative of both sides with respect to z, we get

$$\frac{d}{dz}\frac{F(z)(z-z_2)^2}{z} = \frac{2A_1(z-z_2)(z-z_1) - A_1(z-z_2)^2}{(z-z_1)^2} + A_2 \tag{98}$$

If we now evaluate the expression at $z = z_2$, we get

$$A_2 = \left.\frac{d}{dz}\frac{F(z)(z-z_2)^2}{z}\right|_{z=z_2} \tag{99}$$

Generalizing this approach, we can show that if $D(z)$ has a root of order m at some z_k, that portion of the partial fraction expansion can be written as

$$\frac{F(z)}{z} = \frac{A_1}{z-z_k} + \frac{A_2}{(z-z_k)^2} + \cdots + \frac{A_m}{(z-z_k)^m} \tag{100}$$

and the lth coefficient can be obtained as

$$A_l = \left.\frac{1}{(m-l)!}\frac{d^{(m-l)}}{dz^{(m-l)}}\frac{F(z)(z-z_k)^m}{z}\right|_{z=z_k} \tag{101}$$

Finally, let us drop the requirement that the degree of $D(z)$ be greater or equal to the degree of $N(z)$. When the degree of $N(z)$ is greater than the degree of $D(z)$, we can simply divide $N(z)$ by $D(z)$ to obtain

$$F(z) = \frac{N(z)}{D(z)} = Q(z) + \frac{R(z)}{D(z)} \tag{102}$$

where $Q(z)$ is the quotient and $R(z)$ is the remainder of the division operation. Clearly, $R(z)$ will have degree less than $D(z)$.

To see how all this works together, consider the following example.

Example 12.9.5:

Let us find the inverse Z-transform of the function

$$F(z) = \frac{2z^4+1}{2z^3-5z^2+4z-1} \tag{103}$$

The degree of the numerator is greater than the degree of the denominator, so we divide once to obtain

$$F(z) = z + \frac{5z^3-4z^2+z+1}{2z^3-5z^2+4z-1} \tag{104}$$

The inverse Z-transform of z is δ_{n-1}, where δ_n is the discrete delta function defined as

$$\delta_n = \begin{cases} 1 & n = 0 \\ 0 & \text{otherwise} \end{cases} \tag{105}$$

Let us call the remaining ratio of polynomials $F_1(z)$. We find the roots of the denominator of $F_1(z)$ as

$$F_1(z) = \frac{5z^3 - 4z^2 + z + 1}{2(z - 0.5)(z - 1)^2} \tag{106}$$

Then

$$\frac{F_1(z)}{z} = \frac{5z^3 - 4z^2 + z + 1}{2z(z - 0.5)(z - 1)^2} \tag{107}$$

$$= \frac{A_1}{z} + \frac{A_2}{z - 0.5} + \frac{A_3}{z - 1} + \frac{A_4}{(z - 1)^2} \tag{108}$$

Thus

$$A_1 = \left.\frac{5z^3 - 4z^2 + z + 1}{2(z - 0.5)(z - 1)^2}\right|_{z=0} = -1 \tag{109}$$

$$A_2 = \left.\frac{5z^3 - 4z^2 + z + 1}{2z(z - 1)^2}\right|_{z=0.5} = 4.5 \tag{110}$$

$$A_4 = \left.\frac{5z^3 - 4z^2 + z + 1}{2z(z - 0.5)}\right|_{z=1} = 3 \tag{111}$$

To find A_3, we take the derivative with respect to z, then set $z = 1$:

$$A_3 = \left.\frac{d}{dz}\left[\frac{5z^3 - 4z^2 + 2z + 1}{2z(z - 0.5)}\right]\right|_{z=1} = -3 \tag{112}$$

Therefore,

$$F_1(z) = -1 + \frac{4.5z}{z - 0.5} - \frac{3z}{z - 1} + \frac{3z}{(z - 1)^2} \tag{113}$$

and

$$f_{1,n} = -\delta_n + 4.5(0.5)^n u[n] - 3u[n] + 3nu[n] \tag{114}$$

and

$$f_n = \delta_{n-1} - \delta_n + 4.5(0.5)^n u[n] - (3 - 3n)u[n] \tag{115}$$

♦

12.9.3 Long Division

If we could write $F(z)$ as a power series, then from the Z-transform expression the coefficients of z^{-n} would be the sequence values f_n.

Example 12.9.6:

Let's find the inverse z-transform of

$$F(z) = \frac{z}{z - a}$$

Dividing the numerator by the denominator we get the following:

$$
\begin{array}{r|lclcl}
 & 1 & + & az^{-1} & + & a^2z^{-2} \quad \cdots \\
\hline
z-a & z & & & & \\
 & z & - & a & & \\
\hline
 & & & a & & \\
 & & & a & - & a^2z^{-1} \\
\hline
 & & & & & a^2z^{-1}
\end{array}
$$

Thus, the quotient is

$$1 + az^{-1} + a^2z^{-2} + \cdots = \sum_{n=0}^{\infty} a^n z^{-n}$$

We can easily see that the sequence for which $F(z)$ is the Z-transform is

$$f_n = a^n u[n]$$

♦

12.9.4 Z-Transform Properties

Analogous to the continuous linear systems, we can define the transfer function of a discrete linear system as a function of z that relates the Z-transform of the input to the Z-transform of the output. Let $\{f_n\}_{n=-\infty}^{\infty}$ be the input to a discrete linear time-invariant system, and $\{g_n\}_{n=-\infty}^{\infty}$ be the output. If $F(z)$ is the Z-transform of the input sequence, and $G(z)$ is the Z-transform of the output sequence, then these are related to each other by

$$G(z) = H(z)F(z) \tag{116}$$

and $H(z)$ is the transfer function of the discrete linear time-invariant system.

If the input sequence $\{f_n\}_{n=-\infty}^{\infty}$ had a Z-transform of one, then $G(z)$ would be equal to $H(z)$. It is an easy matter to find the requisite sequence:

$$F(z) = \sum_{n=-\infty}^{\infty} f_n z^{-n} = 1 \Rightarrow f_n = \begin{cases} 1 & n = 0 \\ 0 & \text{otherwise} \end{cases} \tag{117}$$

This particular sequence is called the *discrete delta function*. The response of the system to the discrete delta function is called the impulse response of the system. Obviously, the transfer function $H(z)$ is the Z-transform of the impulse response.

12.9.5 Discrete Convolution

In the continuous time case, the output of the linear time-invariant system was a convolution of the input with the impulse response. Does the analogy hold in the discrete case? We can check this out easily by explicitly writing out the Z-transforms in Equation (116). For simplicity

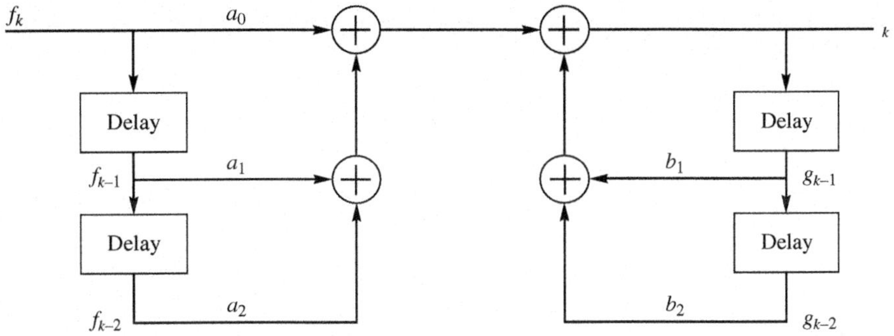

FIGURE 12.13 **A discrete system.**

let us assume the sequences are all one-sided; that is, they are only nonzero for nonnegative values of the subscript:

$$\sum_{n=0}^{\infty} g_n z^{-n} = \sum_{n=0}^{\infty} h_n z^{-n} \sum_{m=0}^{\infty} f_m z^{-m} \tag{118}$$

Equating like powers of z, we obtain:

$$\begin{aligned} g_0 &= h_0 f_0 \\ g_1 &= f_0 h_1 + f_1 h_0 \\ g_2 &= f_0 h_2 + f_1 h_1 + f_2 h_0 \\ &\vdots \\ g_n &= \sum_{m=0}^{n} f_m h_{n-m} \end{aligned}$$

Thus, the output sequence is a result of the discrete convolution of the input sequence with the impulse response.

Most of the discrete linear systems we will be dealing with will be made up of delay elements, and their input-output relations can be written as constant coefficient difference equations. For example, for the system shown in Figure 12.13, the input-output relationship can be written in the form of the following difference equation:

$$g_k = a_0 f_k + a_1 f_{k-1} + a_2 f_{k-2} + b_1 g_{k-1} + b_2 g_{k-2} \tag{119}$$

The transfer function of this system can be easily found by using the *shifting theorem.* The shifting theorem states that if the Z-transform of a sequence $\{f_n\}$ is $F(z)$, then the Z-transform of the sequence shifted by some integer number of samples n_0 is $z^{-n_0}F(z)$.

The theorem is easy to prove. Suppose we have a sequence $\{f_n\}$ with Z-transform $F(z)$. Let us look at the Z-transform of the sequence $\{f_{n-n_0}\}$:

$$\mathcal{Z}[\{f_{n-n_0}\}] = \sum_{n=-\infty}^{\infty} f_{n-n_0} z^{-n} \tag{120}$$

$$= \sum_{m=-\infty}^{\infty} f_m z^{-m-n_0} \tag{121}$$

$$= z^{-n_0} \sum_{m=-\infty}^{\infty} f_m z^{-m} \tag{122}$$

$$= z^{-n_0} F(z) \tag{123}$$

Assuming $G(z)$ is the Z-transform of $\{g_n\}$ and $F(z)$ is the Z-transform of $\{f_n\}$, we can take the Z-transform of both sides of the difference Equation (119):

$$G(z) = a_0 F(z) + a_1 z^{-1} F(z) + a_2 z^{-2} F(z) + b_1 z^{-1} G(z) + b_2 z^{-2} G(z) \tag{124}$$

From this we obtain the relationship between $G(z)$ and $F(z)$ as

$$G(z) = \frac{a_0 + a_1 z^{-1} + a_2 z^{-2}}{1 - b_1 z^{-1} - b_2 z^{-2}} F(z) \tag{125}$$

By definition the transfer function $H(z)$ is therefore

$$H(z) = \frac{G(z)}{F(z)} \tag{126}$$

$$= \frac{a_0 + a_1 z^{-1} + a_2 z^{-2}}{1 - b_1 z^{-1} - b_2 z^{-2}} \tag{127}$$

12.10 Summary

In this chapter we have reviewed some of the mathematical tools we will be using throughout the remainder of this book. We started with a review of vector space concepts, followed by a look at a number of ways we can represent a signal, including the Fourier series, the Fourier transform, the discrete Fourier series, the discrete Fourier transform, and the Z-transform. We also looked at the operation of sampling and the conditions necessary for the recovery of the continuous representation of the signal from its samples.

Further Reading

1. There are a large number of books that provide a much more detailed look at the concepts described in this chapter. A nice one is *Signal Processing and Linear Systems*, by B.P. Lathi [275].

2. For a through treatment of the fast Fourier transform (FFT), see *Numerical Recipes in C*, by W.H. Press, S.A. Teukolsky, W.T. Vetterling, and B.J. Flannery [182].

12.11 Projects and Problems

1. Let X be a set of N linearly independent vectors, and let V be the collection of vectors obtained using all linear combinations of the vectors in X.

(a) Show that given any two vectors in V, the sum of these vectors is also an element of V.
(b) Show that V contains an additive identity.
(c) Show that for every $\mathbf{x}$ in V, there exists a $(-\mathbf{x})$= in V such that their sum is the additive identity.

2. Prove Parseval's theorem for the Fourier transform.

3. Prove the modulation property of the Fourier transform.

4. Prove the convolution theorem for the Fourier transform.

5. Show that the Fourier transform of a train of impulses in the time domain is a train of impulses in the frequency domain:

$$\mathcal{F}\left[\sum_{n=-\infty}^{\infty} \delta(t - nT)\right] = \sigma_0 \sum_{n=-\infty}^{\infty} \delta(w - n\sigma_0) \qquad \sigma_0 = \frac{2\pi}{T} \tag{128}$$

6. Find the Z-transform for the following sequences:

(a) $h_n = 2^{-n}u[n]$, where $u[n]$ is the unit step function
(b) $h_n = (n^2 \quad n)3^{-n}u[n]$
(c) $h_n = (n2^{-n} + (0.6)^n)u[n]$

7. Consider the following input-output relationship:

$$y_n = 0.6y_{n-1} + 0.5x_n + 0.2x_{n-1}$$

(a) Find the transfer function $H(z)$.
(b) Find the impulse response $\{h_n\}$.

8. Find the inverse Z-transform of the following:

(a) $H(z) = \frac{5}{z-2}$
(b) $H(z) = \frac{z}{z^2-0.25}$
(c) $H(z) = \frac{z}{z-0.5}$

13

Transform Coding

13.1 Overview

In this chapter we will describe a technique in which the source output is decomposed, or transformed, into components that are then coded according to their individual characteristics. We will then look at a number of different transforms, including the popular discrete cosine transform, and discuss the issues of quantization and coding of the transformed coefficients. This chapter concludes with a description of the baseline sequential JPEG image-coding algorithm and some of the issues involved with transform coding of audio signals.

13.2 Introduction

In the previous chapter, we developed a number of tools that can be used to transform a given sequence into different representations. If we take a sequence of inputs and transform them into another sequence in which most of the information is contained in only a few elements, we can then encode and transmit those elements, along with their location in the new sequence, resulting in data compression. In our discussion, we will use the terms "variance" and "information" interchangeably. The justification for this is shown in the results in Chapter 7. For example, recall that for a Gaussian source the differential entropy is given as $\frac{1}{2}\log 2\pi e\sigma^2$. Thus, an increase in the variance results in an increase in the entropy, which is a measure of the information contained in the source output.

To begin our discussion of transform coding, consider the following example.

TABLE 13.1 Original sequence.

Height	Weight
65	170
75	188
60	150
70	170
56	130
80	203
68	160
50	110
40	80
50	153
69	148
62	140
76	164
64	120

Example 13.2.1:

Let's revisit Example 8.5.1. In Example 8.5.1, we studied the encoding of the output of a source that consisted of a sequence of pairs of numbers. Each pair of numbers corresponds to the height and weight of an individual. In particular, let's look at the sequence of outputs shown in Table 13.1.

If we look at the height and weight as the coordinates of a point in two-dimensional space, the sequence can be shown graphically as in Figure 13.1. Notice that the output values tend to cluster around the line $y = 2.5x$. We can rotate this set of values by the transformation

$$\theta = \mathbf{A}\mathbf{x} \tag{1}$$

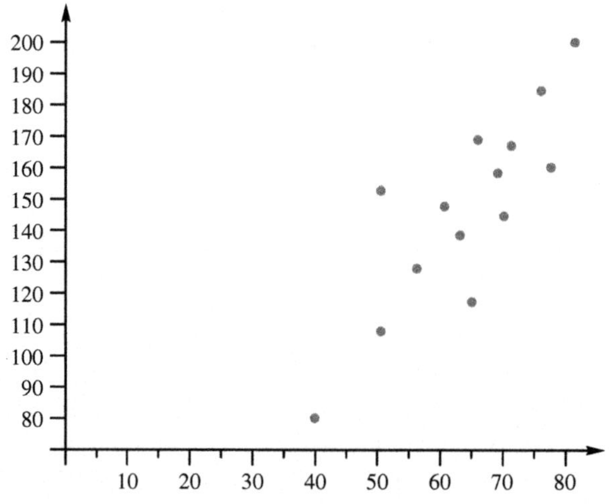

FIGURE 13.1 Source output sequence.

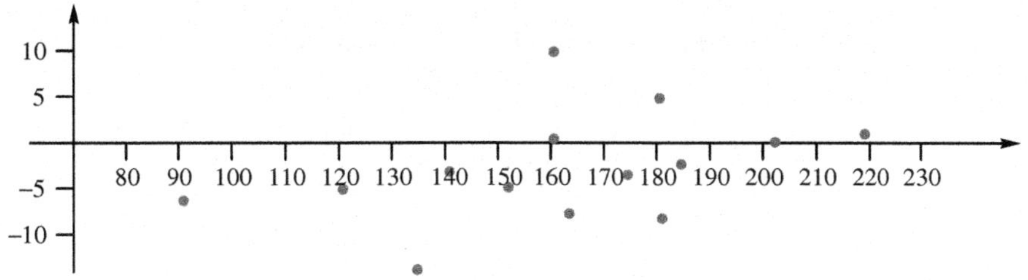

FIGURE 13.2 The transformed sequence.

where $\mathbf{x}$ is the two-dimensional source output vector

$$\mathbf{x} = \begin{bmatrix} x_0 \\ x_1 \end{bmatrix} \tag{2}$$

x_0 corresponds to height and x_1 corresponds to weight, A is the rotation matrix

$$\mathbf{A} = \begin{bmatrix} \cos\phi & \sin\phi \\ -\sin\phi & \cos\phi \end{bmatrix} \tag{3}$$

ϕ is the angle between the x-axis and the $y = 2.5x$ line, and

$$\theta = \begin{bmatrix} \theta_0 \\ \theta_1 \end{bmatrix} \tag{4}$$

is the rotated or transformed set of values. For this particular case, the matrix $\mathbf{A}$ is

$$\mathbf{A} = \begin{bmatrix} 0.37139068 & 0.92847669 \\ -0.92847669 & 0.37139068 \end{bmatrix} \tag{5}$$

and the transformed sequence (rounded to the nearest integer) is shown in Table 13.2. (For a brief review of matrix concepts, see Appendix B.)

Notice that for each pair of values, almost all the energy is compacted into the first element of the pair, while the second element of the pair is significantly smaller. If we plot this sequence in pairs, we get the result shown in Figure 13.2. Note that we have rotated the original values by an angle of approximately 68 degrees (arctan 2.5).

Suppose we set all of the second elements of the transformation to zero, that is, the second coordinates of the sequence shown in Table 13.2. This reduces the number of elements that need to be encoded by half. What is the effect of throwing away half the elements of the sequence? We can find that out by taking the inverse transform of the reduced sequence. The inverse transform consists of reversing the rotation. We can do this by multiplying the blocks of two of the transformed sequences with the second element in each block set to zero with the matrix

$$\mathbf{A}^{-1} = \begin{bmatrix} \cos\phi & -\sin\phi \\ \sin\phi & \cos\phi \end{bmatrix} \tag{6}$$

We obtain the reconstructed sequence shown in Table 13.3. Comparing this to the original sequence in Table 13.1, we see that, even though we transmitted only half the number of

TABLE 13.2 Transformed sequence.

First Coordinate	Second Coordinate
182	3
202	0
162	0
184	−2
141	−4
218	1
174	−4
121	−6
90	−7
161	10
163	−9
153	−6
181	−9
135	−15

TABLE 13.3 Reconstructed sequence.

Height	Weight
68	169
75	188
60	150
68	171
53	131
81	203
65	162
45	112
34	84
60	150
61	151
57	142
67	168
50	125

elements present in the original sequence, this "reconstructed" sequence is very close to the original. The reason there is so little error introduced in the sequence $\{x_n\}$ is that for this particular transformation the error introduced into the $\{x_n\}$ sequence is equal to the error introduced into the $\{\theta_n\}$ sequence. That is,

$$\sum_{i=0}^{N-1}(x_i - \hat{x}_i)^2 = \sum_{i=0}^{N-1}(\theta_i - \hat{\theta}_i)^2 \tag{7}$$

where $\{\hat{x}_n\}$ is the reconstructed sequence, and

$$\hat{\theta}_i = \begin{cases} \theta_i & i = 0, 2, 4, \ldots \\ 0 & \text{otherwise} \end{cases} \tag{8}$$

(see Problem 1 at the end of this chapter). The error introduced in the $\{\theta_n\}$ sequence is the sum of squares of the θ_ns that are set to zero. The magnitudes of these elements are quite small and, therefore, the total error introduced into the reconstructed sequence is quite small also. ♦

We could reduce the number of samples we need to code because most of the information contained in each pair of values is put into one element of each pair. As the other element of the pair contains very little information, we could discard it without a significant effect on the fidelity of the reconstructed sequence. The transform in this case acts on pairs of values; therefore, the maximum reduction in the number of significant samples is a factor of two. We can extend this idea to longer blocks of data. By compacting most of the information in a source output sequence into a few elements of the transformed sequence using a reversible transform, and then discarding the elements of the sequence that do not contain much information, we can get a large amount of compression. This is the basic idea behind transform coding.

In Example 13.2.1 we present a geometric view of the transform process. We can also examine the transform process in terms of the changes in statistics between the original and transformed sequences. It can be shown that we can get the maximum amount of compaction if we use a transform that decorrelates the input sequence; that is, the sample-to-sample correlation of the transformed sequence is zero. The first transform to provide decorrelation for discrete data was presented by Hotelling [183] in the *Journal of Educational Psychology* in 1933. He called his approach the *method of principal components*. The analogous transform for continuous functions was obtained by Karhunen [184] and Loéve [185]. This decorrelation approach was first utilized for compression, in what we now call transform coding, by Kramer and Mathews [186], and Huang and Schultheiss [187].

Transform coding consists of three steps. First, the data sequence $\{x_n\}$ is divided into blocks of size N. Each block is mapped into a transform sequence $\{\theta_n\}$ using a reversible mapping in a manner similar to that described in Example 13.2.1. As shown in the example, different elements of each block of the transformed sequence generally have different statistical properties. In Example 13.2.1, most of the energy of the block of two input values was contained in the first element of the block of two transformed values, while very little of the energy was contained in the second element. This meant that the second element of each block of the transformed sequence would have a small magnitude, while the magnitude of the first element could vary considerably depending on the magnitude of the elements in the input block. The second step consists of quantizing the transformed sequence. The quantization strategy used will depend on three main factors: the desired average bit rate, the statistics of the various elements of the transformed sequence, and the effect of distortion in the transformed coefficients on the reconstructed sequence. In Example 13.2.1, we could take all the bits available to us and use them to quantize the first coefficient. In more complex situations, the strategy used may be very different. In fact, we may use different techniques, such as differential encoding and vector quantization [151], to encode the different coefficients.

Finally, the quantized value needs to be encoded using some binary encoding technique. The binary coding may be as simple as using a fixed-length code or as complex as a combination of run-length coding and Huffman or arithmetic coding. We will see an example of the latter when we describe the JPEG algorithm.

The various quantization and binary coding techniques have been described at some length in previous chapters, so we will spend the next section describing various transforms. We will then discuss quantization and coding strategies in the context of these transforms.

13.3 The Transform

All of the transforms we deal with will be linear transforms; that is, we can get the sequence $\{\theta_n\}$ from the sequence $\{x_n\}$ as

$$\theta_n = \sum_{i=0}^{N-1} x_i a_{n,i} \tag{9}$$

This is referred to as the *forward transform*. For the transforms that we will be considering, a major difference between the transformed sequence $\{\theta_n\}$ and the original sequence $\{x_n\}$ is that the characteristics of the elements of the θ sequence are determined by their position within the sequence. For instance, in Example 13.2.1 the first element of each pair of the transformed sequence is more likely to have a large magnitude compared to the second element. In general, we cannot make such statements about the source output sequence $\{x_n\}$. A measure of the differing characteristics of the different elements of the transformed sequence $\{\theta_n\}$ is the variance σ_n^2 of each element. These variances will strongly influence how we encode the transformed sequence. The size of the block N is dictated by practical considerations. In general, the complexity of the transform grows more than linearly with N. Therefore, beyond a certain value of N, the computational costs overwhelm any marginal improvements that might be obtained by increasing N. Furthermore, in most real sources, the statistical characteristics of the source output can change abruptly. For example, when we go from a silence period to a voiced period in speech, the statistics change drastically. Similarly, in images, the statistical characteristics of a smooth region of the image can be very different from the statistical characteristics of a busy region of the image. If N is large, the probability that the statistical characteristics change significantly within a block increases. This generally results in a larger number of transform coefficients with large values, which in turn leads to a reduction in the compression ratio.

The original sequence $\{x_n\}$ can be recovered from the transformed sequence $\{\theta_n\}$ via the *inverse transform*:

$$x_n = \sum_{i=0}^{N-1} \theta_i b_{n,i} \tag{10}$$

The transforms can be written in matrix form as

$$\theta = \mathbf{A}\mathbf{x} \tag{11}$$

$$\mathbf{x} = \mathbf{B}\theta \tag{12}$$

where $\mathbf{A}$ and $\mathbf{B}$ are $N \times N$ matrices, and the (i, j)th element of the matrices is given by

$$[\mathbf{A}]_{i,j} = a_{i,j} \tag{13}$$

$$[\mathbf{B}]_{i,j} = b_{i,j} \tag{14}$$

The forward and inverse transform matrices **A** and **B** are inverses of each other; that is, $\mathbf{AB} = \mathbf{BA} = \mathbf{I}$, where **I** is the identity matrix.

Equations (9) and (10) deal with the transform coding of one-dimensional sequences, such as sampled speech and audio sequences. However, transform coding is one of the most popular methods used for image compression. In order to take advantage of the two-dimensional nature of dependencies in images, we need to look at two-dimensional transforms.

Let $X_{i,j}$ be the (i, j)th pixel in an image. A general linear, two-dimensional transform for a block of size $N \times N$ is given as

$$\Theta_{k,l} = \sum_{i=0}^{N-1} \sum_{j=0}^{N-1} X_{i,j} a_{i,j,k,l} \tag{15}$$

All two-dimensional transforms in use today are *separable* transforms; that is, we can transform a two-dimensional block by first taking the transform along one dimension, then repeating the operation along the other direction. In terms of matrices, this involves first taking the (one-dimensional) transform of the rows and then taking the column-by-column transform of the resulting matrix. We can also reverse the order of the operations, first taking the transform of the columns, and then taking the row-by-row transform of the resulting matrix. The transform operation can be represented as

$$\Theta_{k,l} = \sum_{i=0}^{N-1} \sum_{j=0}^{N-1} a_{k,i} X_{i,j} a_{j,l} \tag{16}$$

which in matrix terminology would be given by

$$\Theta = \mathbf{A}\mathbf{X}\mathbf{A}^T \tag{17}$$

The inverse transform is given as

$$\mathbf{X} = \mathbf{B}\Theta\mathbf{B}^T \tag{18}$$

All of the transforms we deal with will be *orthonormal transforms*. An orthonormal transform has the property that the inverse of the transform matrix is simply its transpose because the rows of the transform matrix form an orthonormal basis set:

$$\mathbf{B} = \mathbf{A}^{-1} = \mathbf{A}^T \tag{19}$$

For an orthonormal transform, the inverse transform will be given as

$$\mathbf{X} = \mathbf{A}^T \Theta \mathbf{A} \tag{20}$$

Orthonormal transforms are energy preserving; that is, the sum of the squares of the transformed sequence is the same as the sum of the squares of the original sequence. We can see this most easily in the case of the one-dimensional transform:

$$\sum_{i=0}^{N-1} \theta_i^2 = \theta^T \theta \tag{21}$$

$$= (\mathbf{A}\mathbf{x})^T \mathbf{A}\mathbf{x} \tag{22}$$

$$= \mathbf{x}^T \mathbf{A}^T \mathbf{A}\mathbf{x} \tag{23}$$

If $\mathbf{A}$ is an orthonormal transform, $\mathbf{A}^T\mathbf{A} = \mathbf{A}^{-1}\mathbf{A} = \mathbf{I}$, then

$$\mathbf{x}^T\mathbf{A}^T\mathbf{A}\mathbf{x} = \mathbf{x}^T\mathbf{x} \tag{24}$$

$$= \sum_{n=0}^{N-1} x_n^2 \tag{25}$$

and

$$\sum_{i=0}^{N-1} \theta_i^2 = \sum_{n=0}^{N-1} x_n^2 \tag{26}$$

The efficacy of a transform depends on how much energy compaction is provided by the transform. One way of measuring the amount of energy compaction afforded by a particular orthonormal transform is to take a ratio of the arithmetic mean of the variances of the transform coefficient to their geometric means [134]. This ratio is also referred to as the transform coding gain G_{TC}:

$$G_{TC} = \frac{\frac{1}{N}\sum_{i=0}^{N-1} \sigma_i^2}{(\prod_{i=0}^{N-1} \sigma_i^2)^{\frac{1}{N}}} \tag{27}$$

where σ_i^2 is the variance of the ith coefficient θ_i.

Transforms can be interpreted in several ways. We have already mentioned a geometric interpretation and a statistical interpretation. We can also interpret them as a decomposition of the signal in terms of a basis set. For example, suppose we have a two-dimensional orthonormal transform $\mathbf{A}$. The inverse transform can be written as

$$\begin{bmatrix} x_0 \\ x_1 \end{bmatrix} = \begin{bmatrix} a_{00} & a_{10} \\ a_{01} & a_{11} \end{bmatrix} \begin{bmatrix} \theta_0 \\ \theta_1 \end{bmatrix} = \theta_0 \begin{bmatrix} a_{00} \\ a_{01} \end{bmatrix} + \theta_1 \begin{bmatrix} a_{10} \\ a_{11} \end{bmatrix} \tag{28}$$

We can see that the transformed values are actually the coefficients of an expansion of the input sequence in terms of the rows of the transform matrix. The rows of the transform matrix are often referred to as the basis vectors for the transform because they form an orthonormal basis set, and the elements of the transformed sequence are often called the transform coefficients. By characterizing the basis vectors in physical terms, we can get a physical interpretation of the transform coefficients.

Example 13.3.1:

Consider the following transform matrix:

$$\mathbf{A} = \frac{1}{\sqrt{2}} \begin{bmatrix} 1 & 1 \\ 1 & -1 \end{bmatrix} \tag{29}$$

We can verify that this is indeed an orthonormal transform.

Notice that the first row of the matrix would correspond to a "low-pass" signal (no change from one component to the next), while the second row would correspond to a "high-pass" signal. Thus, if we tried to express a sequence in which each element has the same value in terms of these two rows, the second coefficient should be zero. Suppose the original sequence is (α, α). Then

$$\begin{bmatrix} \theta_0 \\ \theta_1 \end{bmatrix} = \frac{1}{\sqrt{2}} \begin{bmatrix} 1 & 1 \\ 1 & -1 \end{bmatrix} \begin{bmatrix} \alpha \\ \alpha \end{bmatrix} = \begin{bmatrix} \sqrt{2}\alpha \\ 0 \end{bmatrix} \tag{30}$$

The "low-pass" coefficient has a value of $\sqrt{2}\alpha$, while the "high-pass" coefficient has a value of 0. The "low-pass" and "high-pass" coefficients are generally referred to as the low-frequency and high-frequency coefficients.

Let us take two sequences in which the components are not the same and the degree of variation is different. Consider the two sequences (3, 1) and (3, −1). In the first sequence, the second element differs from the first by 2; in the second sequence, the magnitude of the difference is 4. We could say that the second sequence is more "high pass" than the first sequence. The transform coefficients for the two sequences are $(2\sqrt{2}, \sqrt{2})$ and $(\sqrt{2}, 2\sqrt{2})$, respectively. Notice that the high-frequency coefficient for the sequence in which we see a larger change is twice that of the high-frequency coefficient for the sequence with less change. Thus, the two coefficients do seem to behave like the outputs of a low-pass filter and a high-pass filter.

Finally, notice that in every case the sum of the squares of the original sequence is the same as the sum of the squares of the transform coefficients; that is, the transform is energy preserving, as it must be, since **A** is orthonormal. ♦

We can interpret one-dimensional transforms as an expansion in terms of the rows of the transform matrix. Similarly, we can interpret two-dimensional transforms as expansions in terms of matrices that are formed by the outer product of the rows of the transform matrix. Recall that the outer product is given by

$$\mathbf{x}\mathbf{x}^T = \begin{bmatrix} x_0x_0 & x_0x_1 & \cdots & x_0x_{N-1} \\ x_1x_0 & x_1x_1 & \cdots & x_1x_{N-1} \\ \vdots & \vdots & & \vdots \\ x_{N-1}x_0 & x_{N-1}x_1 & \cdots & x_{N-1}x_{N-1} \end{bmatrix} \tag{31}$$

To see this more clearly, let us use the transform introduced in Example 13.3.1 for a two-dimensional transform.

Example 13.3.2:

For an $N \times N$ transform **A**, let $\alpha_{i,j}$ be the outer product of the ith and jth rows:

$$\alpha_{i,j} = \begin{bmatrix} a_{i0} \\ a_{i1} \\ \vdots \\ a_{iN-1} \end{bmatrix} \begin{bmatrix} a_{j0} & a_{j1} & \cdots & a_{jN-1} \end{bmatrix} \tag{32}$$

$$= \begin{bmatrix} a_{i0}a_{j0} & a_{i0}a_{j1} & \cdots & a_{i0}a_{jN-1} \\ a_{i1}a_{j0} & a_{i1}a_{j1} & \cdots & a_{i1}a_{jN-1} \\ \vdots & \vdots & & \vdots \\ a_{iN-1}a_{j0} & a_{iN-1}a_{j1} & \cdots & a_{iN-1}a_{jN-1} \end{bmatrix} \tag{33}$$

For the transform of Example 13.3.1, the outer products are

$$\alpha_{0,0} = \frac{1}{2}\begin{bmatrix} 1 & 1 \\ 1 & 1 \end{bmatrix} \qquad \alpha_{0,1} = \frac{1}{2}\begin{bmatrix} 1 & -1 \\ 1 & -1 \end{bmatrix} \tag{34}$$

$$\alpha_{1,0} = \frac{1}{2}\begin{bmatrix} 1 & 1 \\ -1 & -1 \end{bmatrix} \qquad \alpha_{1,1} = \frac{1}{2}\begin{bmatrix} 1 & -1 \\ -1 & 1 \end{bmatrix} \tag{35}$$

From (20), the inverse transform is given by

$$\begin{bmatrix} x_{01} & x_{01} \\ x_{10} & x_{11} \end{bmatrix} = \frac{1}{2}\begin{bmatrix} 1 & 1 \\ 1 & -1 \end{bmatrix}\begin{bmatrix} \theta_{00} & \theta_{01} \\ \theta_{10} & \theta_{11} \end{bmatrix}\begin{bmatrix} 1 & 1 \\ 1 & -1 \end{bmatrix} \tag{36}$$

$$= \frac{1}{2}\begin{bmatrix} \theta_{00} + \theta_{01} + \theta_{10} + \theta_{11} & \theta_{00} - \theta_{01} + \theta_{10} - \theta_{11} \\ \theta_{00} + \theta_{01} - \theta_{10} - \theta_{11} & \theta_{00} - \theta_{01} - \theta_{10} + \theta_{11} \end{bmatrix} \tag{37}$$

$$= \theta_{00}\alpha_{0,0} + \theta_{01}\alpha_{0,1} + \theta_{10}\alpha_{1,0} + \theta_{11}\alpha_{1,1} \tag{38}$$

The transform values θ_{ij} can be viewed as the coefficients of the expansion of $\mathbf{x}$ in terms of the matrices $\alpha_{i,j}$. The matrices $\alpha_{i,j}$ are known as the *basis* matrices.

For historical reasons, the coefficient θ_{00}, corresponding to the basis matrix $\alpha_{0,0}$, is called the DC coefficient, while the coefficients corresponding to the other basis matrices are called AC coefficients. DC stands for direct current, which is current that does not change with time. AC stands for alternating current, which does change with time. Notice that all of the elements of the basis matrix $\alpha_{0,0}$ are the same, hence the DC designation. ♦

In the following section, we will look at some of the variety of transforms available to us, then at some of the issues involved in quantization and coding. Finally, we will describe in detail two applications, one for image coding and one for audio coding.

13.4 Transforms of Interest

In Example 13.2.1, we construct a transform that is specific to the data. In practice, it is generally not feasible to construct a transform for the specific situation for several reasons. Unless the characteristics of the source output are stationary over a long interval, the transform needs to be recomputed often, and it is generally burdensome to compute a transform for every different set of data. Furthermore, the overhead required to transmit the transform itself might negate any compression gains. Both of these problems become especially acute when the size of the transform is large. However, there are times when we want to find out the best we can do with transform coding. In these situations, we can use data-dependent transforms to obtain an idea of the best performance available. The best-known data-dependent transform is the discrete Karhunen-Loéve transform (KLT). We will describe this transform in the next section.

13.4.1 Karhunen-Loéve Transform

The rows of the discrete Karhunen-Loéve transform [189], also known as the Hotelling transform, consist of the eigenvectors of the autocorrelation matrix. The autocorrelation matrix for a random process X is a matrix whose (i, j)th element $[R]_{i,j}$ is given by

$$[R]_{i,j} = E[X_n X_{n+|i-j|}] \tag{39}$$

We can show [134] that a transform constructed in this manner will minimize the geometric mean of the variance of the transform coefficients. Hence, the Karhunen-Loéve transform provides the largest transform coding gain of any transform coding method.

If the source output being compressed is nonstationary, the autocorrelation function will change with time. Thus, the autocorrelation matrix will change with time, and the KLT will have to be recomputed. For a transform of any reasonable size, this is a significant amount of computation. Furthermore, as the autocorrelation is computed based on the source output, it is not available to the receiver. Therefore, either the autocorrelation or the transform itself has to be sent to the receiver. The overhead can be significant and remove any advantages to using the optimum transform. However, in applications where the statistics change slowly and the transform size can be kept small, the KLT can be of practical use [190].

Example 13.4.1:

Let us see how to obtain the KLT transform of size two for an arbitrary input sequence. The autocorrelation matrix of size two for a stationary process is

$$\mathbf{R} = \begin{bmatrix} R_{xx}(0) & R_{xx}(1) \\ R_{xx}(1) & R_{xx}(0) \end{bmatrix} \tag{40}$$

Solving the equation $|\lambda\mathbf{I} - \mathbf{R}| = 0$, we get the two eigenvalues $\lambda_1 = R_{xx}(0) + R_{xx}(1)$ and $\lambda_2 = R_{xx}(0) - R_{xx}(1)$. The corresponding eigenvectors are

$$V_1 = \begin{bmatrix} \alpha \\ \alpha \end{bmatrix} \quad V_2 = \begin{bmatrix} \beta \\ -\beta \end{bmatrix} \tag{41}$$

where α and β are arbitrary constants. If we now impose the orthonormality condition, which requires the vectors to have a magnitude of 1, we get

$$\alpha = \beta = \frac{1}{\sqrt{2}}$$

and the transform matrix $\mathbf{K}$ is

$$\mathbf{K} = \frac{1}{\sqrt{2}} \begin{bmatrix} 1 & 1 \\ 1 & -1 \end{bmatrix} \tag{42}$$

Notice that this matrix is not dependent on the values of $R_{xx}(0)$ and $R_{xx}(1)$. This is only true of the 2×2 KLT. The transform matrices of higher order are functions of the autocorrelation values. ♦

Although the Karhunen-Loéve transform maximizes the transform coding gain as defined by (27), it is not practical in most circumstances. Therefore, we need transforms that do not depend on the data being transformed. We describe some of the more popular transforms in the following sections.

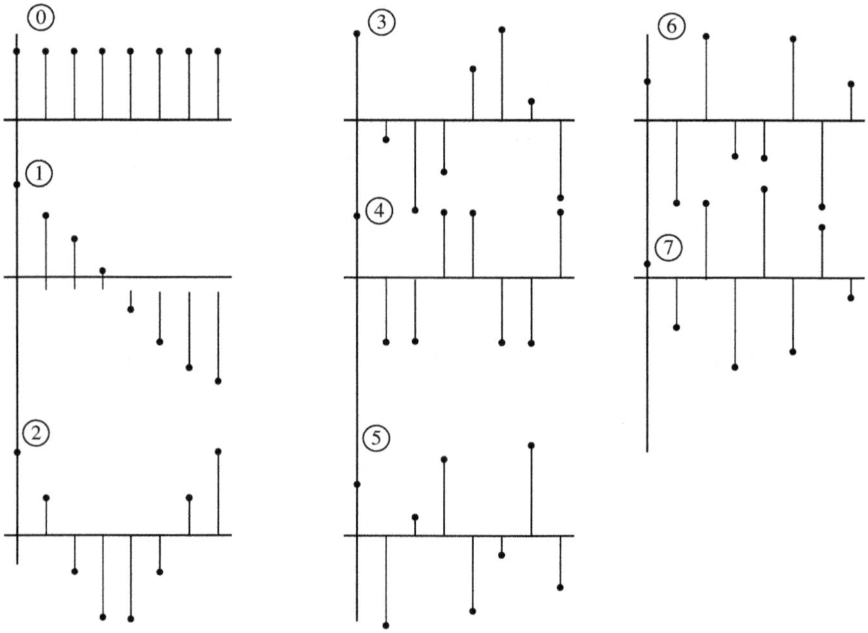

FIGURE 13.3 **Basis set for the discrete cosine transform. The numbers in the circles correspond to the row of the transform matrix.**

13.4.2 Discrete Cosine Transform

The discrete cosine transform (DCT) gets its name from the fact that the rows of the $N \times N$ transform matrix $\mathbf{C}$ are obtained as a function of cosines:

$$[\mathbf{C}]_{i,j} = \begin{cases} \sqrt{\frac{1}{N}} \cos \frac{(2j+1)i\pi}{2N} & i = 0,\ j = 0, 1, \ldots, N-1 \\ \sqrt{\frac{2}{N}} \cos \frac{(2j+1)i\pi}{2N} & i = 1, 2, \ldots, N-1,\ j = 0, 1, \ldots, N-1 \end{cases} \tag{43}$$

The rows of the transform matrix are shown in graphical form in Figure 13.3. Notice how the amount of variation increases as we progress down the rows; that is, the frequency of the rows increases as we go from top to bottom.

The outer products of the rows are shown in Figure 13.4. Notice that the basis matrices show increased variation as we go from the top-left matrix, corresponding to the θ_{00} coefficient, to the bottom-right matrix, corresponding to the $\theta_{(N-1)(N-1)}$ coefficient.

The DCT is closely related to the discrete Fourier transform (DFT) mentioned in Chapter 11 and, in fact, can be obtained from the DFT. However, in terms of compression, the DCT performs better than the DFT.

To see why, recall that when we find the Fourier coefficients for a sequence of length N, we assume that the sequence is periodic with period N. If the original sequence is as shown in Figure 13.5(a), the DFT assumes that the sequence outside the interval of interest behaves in the manner shown in Figure 13.5(b). This introduces sharp discontinuities at the beginning and the

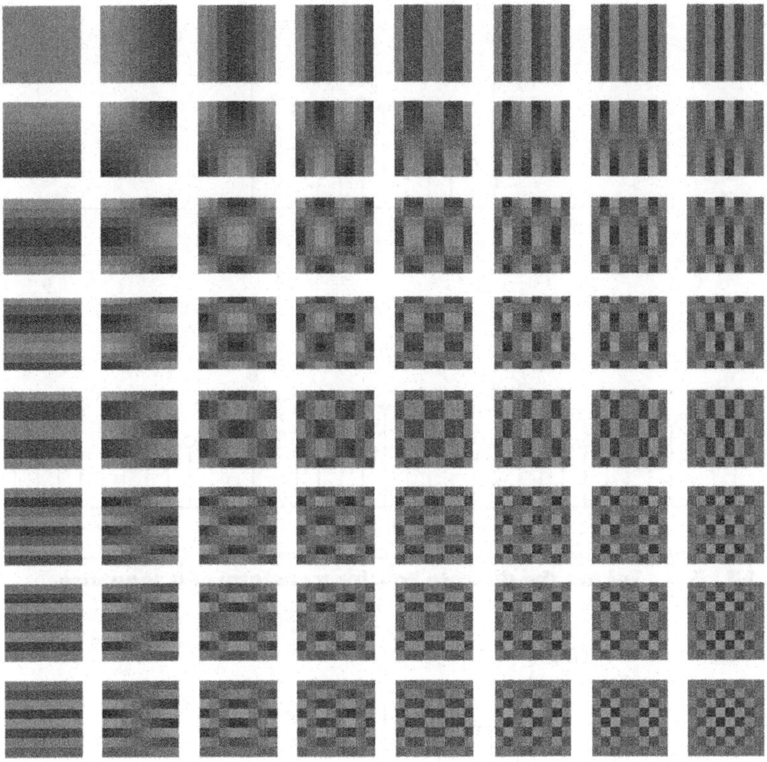

FIGURE 13.4 The basis matrices for the DCT.

end of the sequence. In order to represent these sharp discontinuities, the DFT needs nonzero coefficients for the high-frequency components. Because these components are needed only at the two endpoints of the sequence, their effect needs to be canceled out at other points in the sequence. Thus, the DFT adjusts other coefficients accordingly. When we discard the high-frequency coefficients (which should not have been there anyway) during the compression process, the coefficients that were canceling out the high-frequency effect in other parts of the sequence result in the introduction of additional distortion.

The DCT can be obtained using the DFT by mirroring the original N-point sequence to obtain a $2N$-point sequence, as shown in Figure 13.6(b). The DCT is simply the first N points of the resulting $2N$-point DFT. When we take the DFT of the $2N$-point mirrored sequence, we again have to assume periodicity. However, as we can see from Figure 13.6(c), this does not introduce any sharp discontinuities at the edges.

The DCT is substantially better at energy compaction for most correlated sources when compared to the DFT [134]. In fact, for Markov sources with high correlation coefficient ρ,

$$\rho = \frac{E[x_n x_{n+1}]}{E[x_n^2]} \tag{44}$$

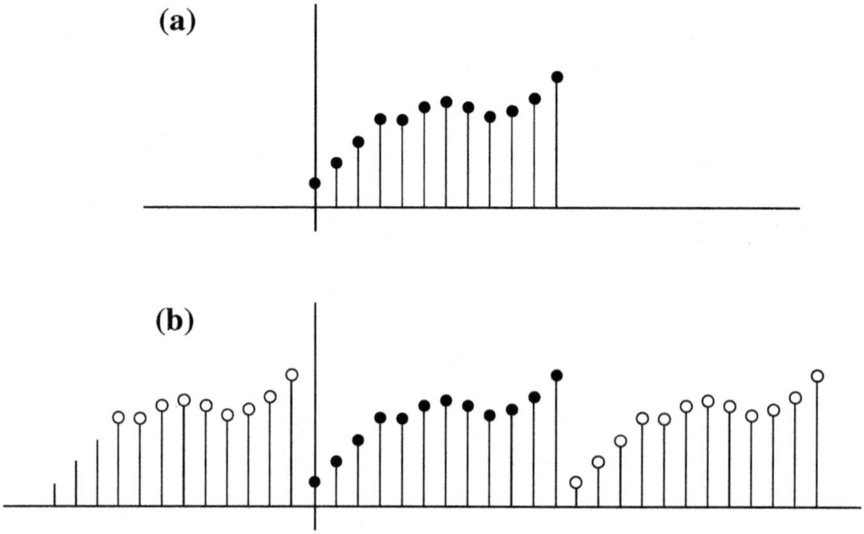

FIGURE 13.5 **Taking the discrete Fourier transform of a sequence.**

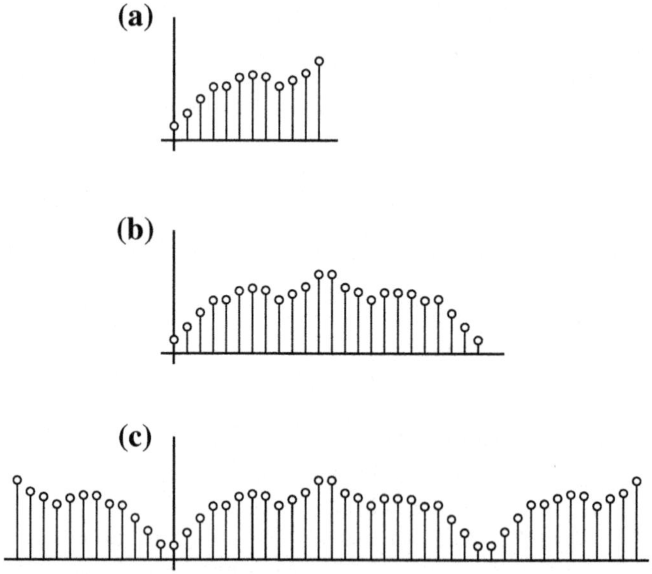

FIGURE 13.6 **Taking the discrete cosine transform of a sequence.**

the compaction ability of the DCT is very close to that of the KLT. As many sources can be modeled as Markov sources with high values for ρ, this superior compaction ability has made the DCT the most popular transform. It is a part of many international standards, including JPEG, MPEG, and CCITT H.261, among others.

13.4.3 Discrete Sine Transform

The discrete sine transform (DST) is a complementary transform to the DCT. Where the DCT provides performance close to the optimum KLT when the correlation coefficient ρ is large, the DST performs close to the optimum KLT in terms of compaction when the magnitude of ρ is small. Because of this property, it is often used as the complementary transform to DCT in image [191] and audio [192] coding applications.

The elements of the transform matrix for an $N \times N$ DST are

$$[\mathbf{S}]_{ij} = \sqrt{\frac{2}{N+1}} \sin \frac{\pi(i+1)(j+1)}{N+1} \qquad i, j = 0, 1, \ldots, N-1 \tag{45}$$

13.4.4 Discrete Walsh-Hadamard Transform

A transform that is especially simple to implement is the discrete Walsh-Hadamard transform (DWHT). The DWHT transform matrices are rearrangements of discrete Hadamard matrices, which are of particular importance in coding theory [193]. A Hadamard matrix of order N is defined as an $N \times N$ matrix H, with the property that $HH^T = NI$, where I is the $N \times N$ identity matrix. Hadamard matrices whose dimensions are a power of two can be constructed in the following manner:

$$H_{2N} = \begin{bmatrix} H_N & H_N \\ H_N & -H_N \end{bmatrix} \tag{46}$$

with $H_1 = [1]$. Therefore,

$$H_2 = \begin{bmatrix} H_1 & H_1 \\ H_1 & -H_1 \end{bmatrix} = \begin{bmatrix} 1 & 1 \\ 1 & -1 \end{bmatrix} \tag{47}$$

$$H_4 = \begin{bmatrix} H_2 & H_2 \\ H_2 & -H_2 \end{bmatrix} = \begin{bmatrix} 1 & 1 & 1 & 1 \\ 1 & -1 & 1 & -1 \\ 1 & 1 & -1 & -1 \\ 1 & -1 & -1 & 1 \end{bmatrix} \tag{48}$$

$$H_8 = \begin{bmatrix} H_4 & H_4 \\ H_4 & -H_4 \end{bmatrix} = \begin{bmatrix} 1 & 1 & 1 & 1 & 1 & 1 & 1 & 1 \\ 1 & -1 & 1 & -1 & 1 & -1 & 1 & -1 \\ 1 & 1 & -1 & -1 & 1 & 1 & -1 & -1 \\ 1 & -1 & -1 & 1 & 1 & -1 & -1 & 1 \\ 1 & 1 & 1 & 1 & -1 & -1 & -1 & -1 \\ 1 & -1 & 1 & -1 & -1 & 1 & -1 & 1 \\ 1 & 1 & -1 & -1 & -1 & -1 & 1 & 1 \\ 1 & -1 & -1 & 1 & -1 & 1 & 1 & -1 \end{bmatrix} \tag{49}$$

The DWHT transform matrix H can be obtained from the Hadamard matrix by multiplying it by a normalizing factor so that $HH^T = I$ instead of NI and by reordering the rows in increasing order of *sequency*. The sequency of a row is half the number of sign changes in that row. In H_8, the first row has a sequency of 0, the second row has a sequency of $7/2$, the third row has a sequency of $3/2$, and so on. Normalization involves multiplying the matrix by $\frac{1}{\sqrt{N}}$. Reordering the H_8 matrix in increasing order of sequency, we get

$$H = \frac{1}{\sqrt{8}} \begin{bmatrix} 1 & 1 & 1 & 1 & 1 & 1 & 1 & 1 \\ 1 & 1 & 1 & 1 & -1 & -1 & -1 & -1 \\ 1 & 1 & -1 & -1 & -1 & -1 & 1 & 1 \\ 1 & 1 & -1 & -1 & 1 & 1 & -1 & -1 \\ 1 & -1 & -1 & 1 & 1 & -1 & -1 & 1 \\ 1 & -1 & -1 & 1 & -1 & 1 & 1 & -1 \\ 1 & -1 & 1 & -1 & -1 & 1 & -1 & 1 \\ 1 & -1 & 1 & -1 & 1 & -1 & 1 & -1 \end{bmatrix} \quad (50)$$

Because the matrix without the scaling factor consists of ± 1, the transform operation consists simply of addition and subtraction. For this reason, this transform is useful in situations where minimizing the amount of computations is very important. However, the amount of energy compaction obtained with this transform is substantially less than the compaction obtained by the use of the DCT. Therefore, where sufficient computational power is available, DCT is the transform of choice.

13.5 Quantization and Coding of Transform Coefficients

If the amount of information conveyed by each coefficient is different, it makes sense to assign differing numbers of bits to the different coefficients. There are two approaches to assigning bits. One approach relies on the average properties of the transform coefficients, while the other approach assigns bits as needed by individual transform coefficients.

In the first approach, we first obtain an estimate of the variances of the transform coefficients. These estimates can be used by one of two algorithms to assign the number of bits used to quantize each of the coefficients. We assume that the relative variance of the coefficients corresponds to the amount of information contained in each coefficient. Thus, coefficients with higher variance are assigned more bits than coefficients with smaller variance.

Let us find an expression for the distortion, then find the bit allocation that minimizes the distortion. To perform the minimization, we will use the method of Lagrange [194]. If the average number of bits per sample to be used by the transform coding system is R and the average number of bits per sample used by the kth coefficient is R_k, then

$$R = \frac{1}{M} \sum_{k=1}^{M} R_k \quad (51)$$

where M is the number of transform coefficients. The reconstruction error variance for the kth quantizer $\sigma_{r_k}^2$ is related to the kth quantizer input variance $\sigma_{\theta_k}^2$ by the following:

$$\sigma_{r_k}^2 = \alpha_k 2^{-2R_k} \sigma_{\theta_k}^2 \quad (52)$$

where α_k is a factor that depends on the input distribution and the quantizer.

The total reconstruction error is given by

$$\sigma_r^2 = \sum_{k=1}^{M} \alpha_k 2^{-2R_k} \sigma_{\theta_k}^2 \tag{53}$$

The objective of the bit allocation procedure is to find R_k to minimize (53) subject to the constraint of (51). If we assume that α_k is a constant α for all k, we can set up the minimization problem in terms of Lagrange multipliers as

$$J = \alpha \sum_{k=1}^{M} 2^{-2R_k} \sigma_{\theta_k}^2 - \lambda \left(R - \frac{1}{M} \sum_{k=1}^{M} R_k \right) \tag{54}$$

Taking the derivative of J with respect to R_k and setting it equal to zero, we can obtain this expression for R_k:

$$R_k = \frac{1}{2} \log_2 \left(2\alpha \ln 2\sigma_{\theta_k}^2 \right) - \frac{1}{2} \log_2 \left(\frac{\lambda}{M} \right) \tag{55}$$

Substituting this expression for R_k in (51), we get a value for λ/M:

$$\frac{\lambda}{M} = \prod_{k=1}^{M} \left(2\alpha \ln 2\sigma_{\theta_k}^2 \right)^{\frac{1}{M}} 2^{-2R} \tag{56}$$

Substituting this expression for λ/M in (55), we finally obtain the individual bit allocations:

$$R_k = R + \frac{1}{2} \log_2 \frac{\sigma_{\theta_k}^2}{\prod_{k=1}^{M} (\sigma_{\theta_k}^2)^{\frac{1}{M}}} \tag{57}$$

Although these values of R_k will minimize (53), they are not guaranteed to be integers or even positive. The standard approach at this point is to set the negative R_ks to zero. This will increase the average bit rate above the constraint. Therefore, the nonzero R_ks are uniformly reduced until the average rate is equal to R.

If we substitute the value of R_k from Equation (57) into the expression for the reconstruction error variance for the kth quantizer shown in Equation (52), we obtain

$$\begin{aligned}
\sigma_{r_k}^2 &= \alpha_k \sigma_{\theta_k}^2 2^{-2(R + \frac{1}{2} \log_2 \frac{\sigma_{\theta_k}^2}{\prod_{k=1}^{M} (\sigma_{\theta_k}^2)^{\frac{1}{M}}})} \\
&= \alpha_k \sigma_{\theta_k}^2 2^{-2R} 2^{-\log_2 \frac{\sigma_{\theta_k}^2}{\prod_{k=1}^{M} (\sigma_{\theta_k}^2)^{\frac{1}{M}}}} \\
&= \frac{\alpha_k \sigma_{\theta_k}^2 \prod_{k=1}^{M} (\sigma_{\theta_k}^2)^{\frac{1}{M}} 2^{-2R}}{\sigma_{\theta_k}^2} \\
&= \alpha_k \prod_{k=1}^{M} (\sigma_{\theta_k}^2)^{\frac{1}{M}} 2^{-2R}
\end{aligned} \tag{58}$$

TABLE 13.4 Bit allocation map for an 8 × 8 transform.

8	7	5	3	1	1	0	0
7	5	3	2	1	0	0	0
4	3	2	1	1	0	0	0
3	3	2	1	1	0	0	0
2	1	1	1	0	0	0	0
1	1	0	0	0	0	0	0
1	0	0	0	0	0	0	0
0	0	0	0	0	0	0	0

Defining

$$\gamma = \prod_{k=1}^{M} (\sigma_{\theta_k}^2)^{\frac{1}{M}} \tag{59}$$

and assuming that α_k is a constant α for all k, we obtain the reconstruction error variance for the kth quantizer to be

$$\sigma_{r_k}^2 = \alpha\gamma 2^{-2R}$$

Note that the expression of the reconstruction error variance is independent of k. That is, each of the quantizers in this idealized situation, where the reconstruction error variance is given by Equation (52) with $\alpha_k = \alpha$ and rates R_k are allowed to be nonintegers, contributes equally to the total reconstruction error variance! We can use this insight to develop a recursive bit allocation algorithm. Notice from Equation (52) that an increase in one bit for the value of R_k will result in a reduction by half in the standard deviation of the reconstruction error σ_{r_k}.

Combining these two insights, we obtain the following recursive bit allocation algorithm.

1. Compute σ_{r_k} for each coefficient. As the number of bits allocated to each coefficient is zero, initially, $\sigma_{r_k} = \sigma_{\theta_k}$.
2. Set $R_k = 0$ for all k, and set $R_b = MR$, where R_b is the total number of bits available for distribution.
3. Sort the standard deviations $\{\sigma_{r_k}\}$. Suppose σ_{r_l} is the maximum.
4. Increment R_l by 1, and divide σ_{r_l} by 2.
5. Decrement R_b by 1. If $R_b = 0$, then stop; otherwise, go to Step 3.

If we follow this procedure, we end up allocating more bits to the coefficients with higher variance. Furthermore, we tend to equalize the amount of distortion contributed by each coefficient.

This form of bit allocation is called *zonal sampling*. The reason for this name can be seen from the example of a bit allocation map for the 8 × 8 DCT of an image shown in Table 13.4. Notice that there is a zone of coefficients that roughly comprises the right lower diagonal of the bit map that has been assigned zero bits. In other words, these coefficients are to be discarded. The advantage to this approach is its simplicity. Once the bit allocation has been obtained, every coefficient at a particular location is always quantized using the same number of bits. The disadvantage is that, because the bit allocations are performed based on average value, variations that occur on the local level are not reconstructed properly. For example, consider an image of an object with sharp edges in front of a relatively plain background. The number of pixels that occur on edges is quite small compared to the total number of pixels. Therefore, if we allocate bits based on average variances, the coefficients that are important for representing edges (the high-frequency coefficients) will get few or no bits assigned to them. This means that the reconstructed image will not contain a very good representation of the edges.

This problem can be avoided by using a different approach to bit allocation known as *threshold coding* [195, 104, 196]. In this approach, which coefficient to keep and which to discard is not decided a priori. In the simplest form of threshold coding, we specify a threshold value. Coefficients with magnitude below this threshold are discarded, while the other coefficients are quantized and transmitted. The information about which coefficients have been retained is sent to the receiver as side information. A simple approach, described by Pratt [104], is to code the first coefficient on each line regardless of the magnitude. After this, when we encounter a coefficient with a magnitude above the threshold value, we send two codewords: one for the quantized value of the coefficient and one for the count of the number of coefficients since the last coefficient with a magnitude greater than the threshold. For the two-dimensional case, the block size is usually small, and each "line" of the transform is very short. Thus, this approach would be quite expensive. Chen and Pratt [196] suggest scanning the block of transformed coefficients in a zigzag fashion, as shown in Figure 13.7. If we scan an 8×8 block of quantized transform coefficients in this manner, we will find that, in general, a large section of the tail end of the scan will consist of zeros. This is because, generally, the higher-order coefficients have smaller amplitude. This is reflected in the bit allocation table shown in Table 13.4. As we

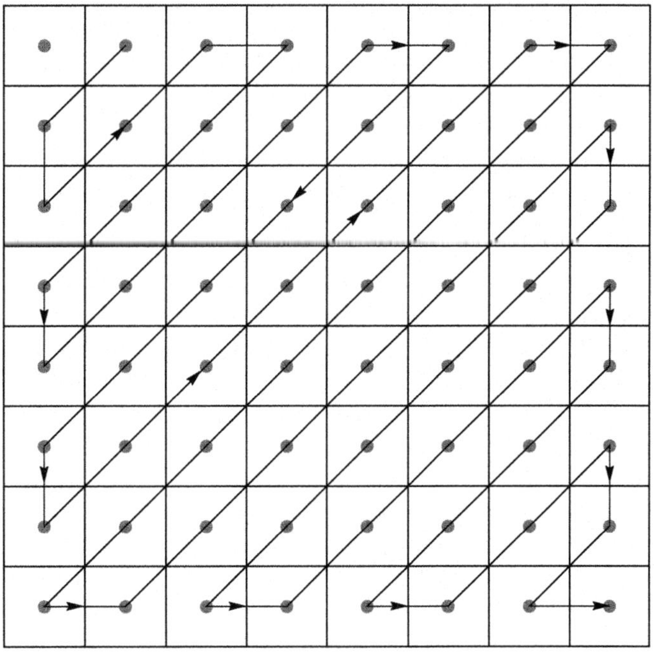

FIGURE 13.7 The zigzag scanning pattern for an 8×8 transform.

shall see later, if we use midtread quantizers (quantizers with a zero output level), combined with the fact that the step sizes for the higher-order coefficients are generally chosen to be quite large, many of these coefficients will be quantized to zero. Therefore, there is a high probability that after a few coefficients along the zigzag scan, all coefficients will be zero. In this situation, Chen and Pratt suggest the transmission of a special *end-of-block* (EOB) symbol. Upon reception of the EOB signal, the receiver would automatically set all remaining coefficients along the zigzag scan to zero.

The algorithm developed by the Joint Photographic Experts Group (JPEG), described in Section 13.6, uses a rather clever variation of this approach.

The bit allocation schemes described above assume knowledge about both the source and the quantizer performance that may not be accurate. An approach that attempts to limit the assumptions needed for bit allocation is called operational rate-distortion bit allocation. We briefly describe the operational approach in the next section.

13.5.1 Operational Rate-Distortion Bit Allocation

The bit allocation algorithms considered in the previous section assume that a quantizer performance function is available to the user. For example, in the previous section, we assumed that the reconstruction error variance could be obtained using Equation (52). We further idealized the situation by assuming that the α_k factor, which depends on the input distribution and the quantizer, is identical for all coefficients and quantizers. In practice, none of these assumptions may be correct. The alternative to using theoretical performance characteristics for quantizers is to use empirically obtained performance characteristics for bit allocation.

Let $B = \{b_1, b_2, \ldots, b_m\}$ be a set of bit allocations available for quantizing a particular coefficient θ_i. For each bit rate, we can find the quantizer that gives the minimum distortion when used for quantizing θ_i. We can find the best quantizer for θ_i by using a model for the distribution of θ_i or by using a training set of representative values. The bit rates and corresponding distortions can be organized into an *operational rate-distortion function* or an *operational distortion-rate function*. The rate-distortion and distortion-rate functions differ only in whether we plot the distortion on the ordinate and the rate on the abscissa, or vice versa. We will use the rate-distortion and distortion-rate functions interchangeably depending on whichever is more convenient. An example of an operational rate-distortion function is shown in Figure 13.8. In the figure, the circles represent rate-distortion coordinates for different quantizers. The empty circles denote the best quantizers for this particular source. The dashed line is the operational rate-distortion function while the solid line is the convex hull of the operational points. The operational rate-distortion function specifies the distortion and rate associated with quantizers that can feasibly be used in the compression process. We have also plotted the convex hull of the operational points. The convex hull is the convex function that comes closest in a Euclidean sense to the operational rate-distortion function. A usual way of visualizing the convex hull of a set of points in two dimensions is to visualize the points as pins stuck into a board and the convex hull as a rubber band enclosing the pins.

Given an operational rate-distortion function for each coefficient we can use a number of brute force algorithms to obtain the bit allocation. For example, we can initially assign zero bits to each coefficient. We can then sort the distortion corresponding to each coefficient and

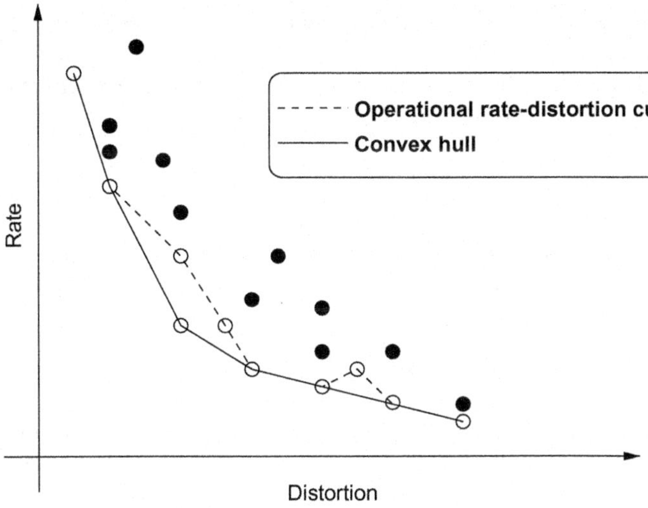

FIGURE 13.8 Operational rate-distortion function.

assign an additional bit to the coefficient contributing the highest distortion. The reduction in distortion can then be obtained using the operational rate-distortion function or through a simulation process using a training set. This process can be repeated until the bit budget is exhausted. This is a greedy approach in that at each step we try to maximize the impact of one bit—the bit that is being allocated. However, that might not be the best use for the bit when we look at the overall impact. Maybe it would have been better in terms of reducing the overall distortion if instead of assigning one bit to the coefficient with the highest distortion we assigned two bits to the coefficient with the next highest distortion. Exploring all the possibilities available can be computationally expensive. To obtain a computationally tractable solution, the problem of bit allocation can be posed as a dynamic programming problem.

A different approach was pioneered in the work of Shoham and Gersho [197]. Suppose we have a bit allocation B, the corresponding average distortion $D(B)$, and rate $R(B)$. The problem of bit allocation can then be posed as that of finding the allocation B^* that minimizes $D(B)$ while satisfying the constraint that $R(B) \leqslant R_b$. Shoham and Gersho examined the simpler unconstrained minimization of a function of the form

$$J = D(B) + \lambda R(B)$$

where the minimization is over all B in a specified set S. They noted that for any $\lambda \geqslant 0$, the solution $B^*(\lambda)$ to the unconstrained problem of minimizing J over all $B \in S$ is also the solution of the problem of minimizing $D(B)$ over all $B \in S$ while satisfying the constraint that $R(B) \leqslant R(B^*(\lambda))$. This latter constrained problem seems to be exactly the problem we are interested in solving if we take B to be bit allocation, $D(B)$ and $R(B)$ to be the corresponding distortion and rate, and $R(B^*(\lambda))$ to be our rate constraint R_b. Of course, there is no indication that for a given value of λ, $R(B^*(\lambda))$ is equal to, or even close to, R_b. However, if we can find a relationship between λ, the distortion, and the rate, we can develop an algorithm to find the λ that will result in the desired rate.

Let's take another look at the unconstrained problem proposed by Shoham and Gersho in the context of our bit allocation problem. Define

$$D(B) = \sum_i D_i(R_i) \text{ and } R(B) = \sum_i R_i$$

where R_i is the number of bits allocated to the coefficient θ_i and $D_i(R_i)$ is the corresponding distortion. The function for the unconstrained minimization problem then becomes

$$J = \sum_i D_i(R_i) + \lambda \sum_i R_i$$

Taking the derivative of this function with respect to the bit allocation for the jth coefficient R_j for $j = 1, 2, \ldots N$, we obtain

$$\frac{\partial J}{\partial R_j} = \frac{\partial D_j(R_j)}{\partial R_j} + \lambda \quad j = 1, 2, \ldots N$$

Setting this equal to zero for all j, we obtain

$$\frac{\partial D_j(R_j)}{\partial R_j} = -\lambda \quad j = 1, 2, \ldots N \tag{60}$$

This is an interesting result with several consequences. It provides us with an explanation for what λ represents. The left side of Equation (60) is the derivative of the distortion-rate function for the jth coefficient. Thus λ is simply the negative of the slope of the distortion-rate function for the jth coefficient. This equation also indicates that for an optimal bit allocation, this slope of $-\lambda$ is identical for each coefficient. Thus for the optimum bit allocation for each coefficient θ_j we should pick a rate R_j at which point the slope of the rate-distortion function for θ_j is exactly the same as the slope for the distortion-rate function for all other coefficients $\theta_i, i \neq j$.

We can easily show this latter result that the slopes of the distortion-rate functions at the operating rates for an optimum bit allocation have to be the same for each coefficient. Consider the case of a transform with two coefficients. Let's suppose that the optimum bit assignment (R_1^*, R_2^*) for a rate constraint R_b results in distortions D_1^* and D_2^* and, contrary to our assertion above, that the slopes $-\lambda_1$ and $-\lambda_2$ at (R_1^*, D_1^*) and (R_2^*, D_2^*) are not the same. Without loss of generality, we can assume $\lambda_1 < \lambda_2$, and we can define $\lambda_1 = \lambda_2 - \Delta\lambda$. Because these are the optimum bit assignments, any change in these assignments should lead to a decrease in the total distortion. Let's check to see if this is actually true. Let's increase R_1^* by an arbitrarily small amount $\epsilon > 0$. This will result in a decrease in the distortion. As ϵ is very small, we can approximate the distortion-rate function with the first term of the Taylor series expansion of the distortion-rate function and represent the decrease by $D_1^* - \lambda_1\epsilon$. If we increase R_1^* by ϵ, the total rate $R_1^* + R_2^*$ increases by ϵ. In order to keep the total number of bits the same, the bit assignment R_2^* has to be decreased by ϵ. This will increase the distortion for the second coefficient by an amount equal to $\lambda_2\epsilon$. Thus the new total distortion will be

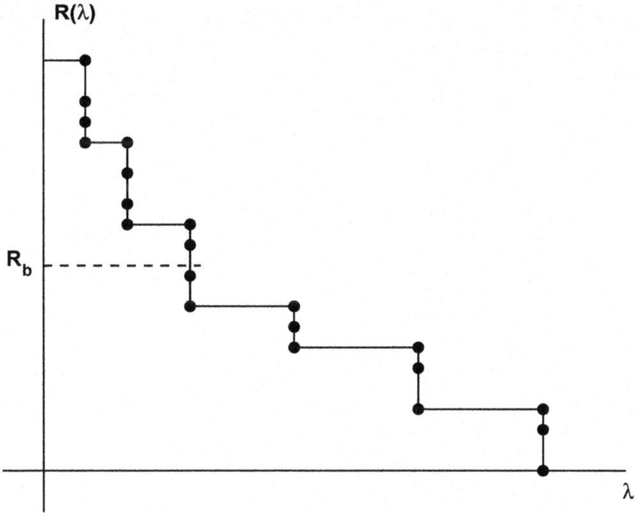

FIGURE 13.9 **A typical relationship between λ and the corresponding rate $R^*(\lambda)$ [197].**

$D_1^* - \lambda_1\epsilon + D_2^* + \lambda_2\epsilon$. But as $\lambda_1 = \lambda_2 - \Delta\lambda$, the new distortion is $D_1^* - (\lambda_2 - \Delta\lambda)\epsilon + D_2^* + \lambda_2\epsilon$, which is equal to $D_1^* + D_2^* + \Delta\lambda\epsilon$. But this is an increase in distortion that contradicts our assumption that the allocation was optimum. Thus, it is not possible to pick an optimum bit allocation in a manner that will result in differing slopes at the operating points of the respective distortion-rate functions.

We have reduced the problem of bit allocation to the problem of finding a value for λ that will result in rates that satisfy the rate constraint. Once we have this value, we can use the individual distortion-rate functions for the different coefficients to obtain the bit allocation and the corresponding distortion. In order to do this we need a relationship between λ and the rate.

Shoham and Gersho [197] show that while for a given value of λ there may be several rates that achieve the optimum value, the optimal rate for each λ is monotonically nonincreasing with λ. A plot of the typical dependency between the value of λ and the corresponding optimal rate is shown in Figure 13.9. In the figure, the filled-in circles correspond to multiple solutions for those values of λ. These multiple solutions cause the relationship between $R^*(\lambda)$ and λ to have a stair-step appearance.

Given this relationship, in order to find the value of λ needed for an optimal solution, we need only find a value λ_1 that results in a rate below the target rate and a value λ_2 that results in a rate above the target rate. We can then search between these two values of λ to find the rate that comes closest to satisfying our constraint. Note that it is possible that no set of quantizers exist that will exactly satisfy our rate constraint.

This approach will only allow us to select quantizers corresponding to the points on the convex hull. Points interior to the convex hull are unreachable using this approach. There might be situations where the points on the convex hull are not the best practical solution. For example, we might have a situation in which because of the allocations to other coefficients certain high rate points on the convex hull are no longer allowed. In such situations one could use the points on the convex hull to initiate a search for a point on the operational rate-distortion curve interior to the convex hull that would satisfy the operational constraint.

13.6 Application to Image Compression—JPEG

The JPEG standard is one of the most widely known standards for lossy image compression. It is a result of the collaboration of the International Standards Organization (ISO), which is a private organization, and what was the CCITT (now ITU-T), a part of the United Nations. The approach recommended by JPEG is a transform coding approach using the DCT. The approach is a modification of the scheme proposed by Chen and Pratt [196]. In this section, we will briefly describe the baseline JPEG algorithm. In order to illustrate the various components of the algorithm, we will use an 8×8 block of the Sena image, shown in Table 13.5. For more details, see [6].

13.6.1 The Transform

The transform used in the JPEG scheme is the DCT transform described earlier. The input image is first "level shifted" by 2^{P-1}; that is, we subtract 2^{P-1} from each pixel value, where P is the number of bits used to represent each pixel. Thus, if we are dealing with 8-bit images whose pixels take on values between 0 and 255, we would subtract 128 from each pixel so that the value of the pixel varies between -128 and 127. The image is divided into blocks of size 8×8, which are then transformed using an 8×8 forward DCT. If any dimension of the image is not a multiple of eight, the encoder replicates the last column or row until the final size is a multiple of eight. These additional rows or columns are removed during the decoding process. If we take the 8×8 block of pixels shown in Table 13.5, subtract 128 from it, and take the DCT of this level-shifted block, we obtain the DCT coefficients shown in Table 13.6. Notice that the lower-frequency coefficients in the top-left corner of the table have larger values than the higher-frequency coefficients. This is generally the case, except for situations in which there is substantial activity in the image block.

13.6.2 Quantization

The JPEG algorithm uses uniform midtread quantization to quantize the various coefficients. The quantizer step sizes are organized in a table called the *quantization table* and can be viewed as the fixed part of the quantization. An example of a quantization table from the JPEG recommendation [6] is shown in Table 13.7. Each quantized value is represented by a label. The label corresponding to the quantized value of the transform coefficient θ_{ij} is obtained as

$$l_{ij} = \left\lfloor \frac{\theta_{ij}}{Q_{ij}} + 0.5 \right\rfloor \tag{61}$$

TABLE 13.5 An 8 × 8 block from the Sena image.

124	125	122	120	122	119	117	118
121	121	120	119	119	120	120	118
126	124	123	122	121	121	120	120
124	124	125	125	126	125	124	124
127	127	128	129	130	128	127	125
143	142	143	142	140	139	139	139
150	148	152	152	152	152	150	151
156	159	158	155	158	158	157	156

TABLE 13.6 The DCT coefficients corresponding to the block of data from the Sena image after level shift.

39.88	6.56	−2.24	1.22	−0.37	−1.08	0.79	1.13
−102.43	4.56	2.26	1.12	0.35	−0.63	−1.05	−0.48
37.77	1.31	1.77	0.25	−1.50	−2.21	−0.10	0.23
−5.67	2.24	−1.32	−0.81	1.41	0.22	−0.13	0.17
−3.37	−0.74	−1.75	0.77	−0.62	−2.65	−1.30	0.76
5.98	−0.13	−0.45	−0.77	1.99	−0.26	1.46	0.00
3.97	5.52	2.39	−0.55	−0.051	−0.84	−0.52	−0.13
−3.43	0.51	−1.07	0.87	0.96	0.09	0.33	0.01

where Q_{ij} is the (i, j)th element of the quantization table, and $\lfloor x \rfloor$ is the largest integer smaller than x. Consider the θ_{00} coefficient from Table 13.6. The value of θ_{00} is 39.88. From Table 13.7, Q_{00} is 16. Therefore,

$$l_{00} = \left\lfloor \frac{39.88}{16} + 0.5 \right\rfloor = \lfloor 2.9925 \rfloor = 2 \tag{62}$$

The reconstructed value is obtained from the label by multiplying the label with the corresponding entry in the quantization table. Therefore, the reconstructed value of θ_{00} would be $l_{00} \times Q_{00}$, which is $2 \times 16 = 32$. The quantization error in this case is $39.88 - 32 = 7.88$. Similarly, from Tables 13.6 and 13.7, θ_{01} is 6.56 and Q_{01} is 11. Therefore,

$$l_{01} = \left\lfloor \frac{6.56}{11} + 0.5 \right\rfloor = \lfloor 1.096 \rfloor = 1 \tag{63}$$

The reconstructed value is 11, and the quantization error is $11 - 6.56 = 4.44$. Continuing in this fashion, we obtain the labels shown in Table 13.8.

From the sample quantization table shown in Table 13.7, we can see that the step size generally increases as we move from the DC coefficient to the higher-order coefficients. Because the quantization error is an increasing function of the step size, more quantization error will be introduced in the higher-frequency coefficients than in the lower-frequency coefficients. The decision on the relative size of the step sizes is based on how errors in these coefficients will

TABLE 13.7 Sample quantization table.

16	11	10	16	24	40	51	61
12	12	14	19	26	58	60	55
14	13	16	24	40	57	69	56
14	17	22	29	51	87	80	62
18	22	37	56	68	109	103	77
24	35	55	64	81	104	113	92
49	64	78	87	103	121	120	101
72	92	95	98	112	100	103	99

TABLE 13.8 The quantizer labels obtained by using the quantization table on the coefficients.

2	1	0	0	0	0	0	0
−9	0	0	0	0	0	0	0
3	0	0	0	0	0	0	0
0	0	0	0	0	0	0	0
0	0	0	0	0	0	0	0
0	0	0	0	0	0	0	0
0	0	0	0	0	0	0	0
0	0	0	0	0	0	0	0

be perceived by the human visual system. Different coefficients in the transform have widely different perceptual importance. Quantization errors in the DC and lower AC coefficients are more easily detectable than quantization errors in the higher AC coefficients. Therefore, we use larger step sizes for perceptually less important coefficients.

Because the quantizers are all midtread quantizers (that is, they all have a zero output level), the quantization process also functions as the thresholding operation. All coefficients with magnitudes less than half the corresponding step size will be set to zero. Because the step sizes at the tail end of the zigzag scan are larger, the probability of finding a long run of zeros increases at the end of the scan. This is the case for the 8×8 block of labels shown in Table 13.8. The entire run of zeros at the tail end of the scan can be coded with an EOB code after the last nonzero label, resulting in substantial compression.

Furthermore, this effect also provides us with a method to vary the rate. By making the step sizes larger, we can reduce the number of nonzero values that need to be transmitted, which translates to a reduction in the number of bits that need to be transmitted.

13.6.3 Coding

Chen and Pratt [196] used separate Huffman codes for encoding the label for each coefficient and the number of coefficients since the last nonzero label. The JPEG approach is somewhat more complex but results in higher compression. In the JPEG approach, the labels for the DC and AC coefficients are coded differently.

TABLE 13.9 Coding of the differences of the DC labels.

0			0			
1			−1	1		
2		−3	−2	2	3	
3	−7	···	−4	4	···	7
4	−15	···	−8	8	···	15
5	−31	···	−16	16	···	31
6	−63	···	−32	32	···	63
7	−127	···	−64	64	···	127
8	−255	···	−128	128	···	255
9	−511	···	−256	256	···	511
10	−1,023	···	−512	512	···	1,023
11	−2,047	···	−1,024	1,024	···	2,047
12	−4,095	···	−2,048	2,048	···	4,095
13	−8,191	···	−4,096	4,096	···	8,191
14	−16,383	···	−8,192	8,192	···	16,383
15	−32,767	···	−16,384	16,384	···	32,767
16			32,768			

From Figure 13.4, we can see that the basis matrix corresponding to the DC coefficient is a constant matrix. Thus, the DC coefficient is some multiple of the average value in the 8×8 block. The average pixel value in any 8×8 block will not differ substantially from the average value in the neighboring 8×8 block; therefore, the DC coefficient values will be quite close. Given that the labels are obtained by dividing the coefficients with the corresponding entry in the quantization table, the labels corresponding to these coefficients will be closer still. Therefore, it makes sense to encode the differences between neighboring labels rather than to encode the labels themselves.

Depending on the number of bits used to encode the pixel values, the number of values that the labels and, hence, the differences can take on may become quite large. A Huffman code for such a large alphabet would be quite unmanageable. The JPEG recommendation resolves this problem by partitioning the possible values that the differences can take on into categories. The size of these categories grows as a power of two. Thus, Category 0 has only one member (0), Category 1 has two members (−1 and 1), Category 2 has four members (−3, −2, 2, 3), and so on. The category numbers are then Huffman coded. The number of codewords in the Huffman code is equal to the base two logarithm of the number of possible values that the label differences can take on. If the differences can take on 4,096 possible values, the size of the Huffman code is $\log_2 4096 = 12$. The elements within each category are specified by tacking on extra bits to the end of the Huffman code for that category. As the categories are different sizes, we need a differing number of bits to identify the value in each category. For example, because Category 0 contains only one element, we need no additional bits to specify the value. Category 1 contains two elements, so we need 1 bit tacked on to the end of the Huffman code for Category 1 to specify the particular element in that category. Similarly, we need 2 bits to specify the element in Category 2, 3 bits for Category 3, and n bits for Category n.

TABLE 13.10 Sample table for obtaining the Huffman code for a given label value and run length. The values of *Z* are represented in hexadecimal.

Z/C	Codeword	Z/C	Codeword	⋯	Z/C	Codeword
0/0 (EOB)	1010			⋯	F/0 (ZRL)	11111111001
0/1	00	1/1	1100	⋯	F/1	1111111111110101
0/2	01	1/2	11011	⋯	F/2	1111111111110110
0/3	100	1/3	1111001	⋯	F/3	1111111111110111
0/4	1011	1/4	111110110	⋯	F/4	1111111111111000
0/5	11010	1/5	11111110110	⋯	F/5	1111111111111001
⋮	⋮	⋮	⋮		⋮	

The categories and the corresponding difference values are shown in Table 13.9. For example, if the difference between two labels was 6, we would send the Huffman code for Category 3. As Category 3 contains the eight values $\{-7, -6, -5, -4, 4, 5, 6, 7\}$, the Huffman code for Category 3 would be followed by 3 bits that would specify which of the eight values in Category 3 was being transmitted.

The binary code for the AC coefficients is generated in a slightly different manner. The category C that a nonzero label falls in and the number of zero-valued labels Z since the last nonzero label form a pointer to a specific Huffman code, as shown in Table 13.10. Thus, if the label being encoded falls in Category 3, and there have been 15 zero-valued labels prior to this nonzero label in the zigzag scan, then we form the pointer $F/3$, which points to the codeword 1111111111110111. Because the label falls in Category 3, we follow this codeword with 3 bits that indicate which of the eight possible values in Category 3 the label takes on.

There are two special codes shown in Table 13.10. The first is for the end-of-block (EOB). This is used in the same way as in the Chen and Pratt [196] algorithm; that is, if a particular label value is the last nonzero value along the zigzag scan, the code for it is immediately followed by the EOB code. The other code is the ZRL code, which is used when the number of consecutive zero values along the zigzag scan exceeds 15.

To see how all of this fits together, let's encode the labels in Table 13.8. The label corresponding to the DC coefficient is coded by first taking the difference between the value of the quantized label in this block and the quantized label in the previous block. If we assume that the corresponding label in the previous block was -1, then the difference would be 3. From Table 13.9, we can see that this value falls in Category 2. Therefore, we would send the Huffman code for Category 2 followed by the 2-bit sequence 11 to indicate that the value in Category 2 being encoded was 3, and not -3, -2, or 2. To encode the AC coefficients, we first order them using the zigzag scan. We obtain the sequence

$$1\ -9\ 3\ 0\ 0\ 0 \cdots 0$$

The first value, 1, belongs to Category 1. Because there are no zeros preceding it, we transmit the Huffman code corresponding to 0/1, which from Table 13.10 is 00. We then follow this by a single bit 1 to indicate that the value being transmitted is 1 and not -1. Similarly, -9

TABLE 13.11 The quantized values of the coefficients.

32	11	0	0	0	0	0	0
−108	0	0	0	0	0	0	0
42	0	0	0	0	0	0	0
0	0	0	0	0	0	0	0
0	0	0	0	0	0	0	0
0	0	0	0	0	0	0	0
0	0	0	0	0	0	0	0
0	0	0	0	0	0	0	0

TABLE 13.12 The reconstructed block.

123	122	122	121	120	120	119	119
121	121	121	120	119	118	118	118
121	121	120	119	119	118	117	117
124	124	123	122	122	121	120	120
130	130	129	129	128	128	128	127
141	141	140	140	139	138	138	137
152	152	151	151	150	149	149	148
159	159	158	157	157	156	155	155

is the seventh element in Category 4. Therefore, we send the binary string 1011, which is the Huffman code for 0/4, followed by 0110 to indicate that −9 is the seventh element in Category 4. The next label is 3, which belongs to Category 2, so we send the Huffman code 01 corresponding to 0/2, followed by the 2 bits 11. All the labels after this point are 0, so we send the EOB Huffman code, which in this case is 1010. If we assume that the Huffman code for the DC coefficient is 2 bits long, we have sent a grand total of 21 bits to represent this 8×8 block. This translates to an average of $\frac{21}{64}$ bits per pixel.

To obtain a reconstruction of the original block, we perform the dequantization, which simply consists of multiplying the labels in Table 13.8 with the corresponding values in Table 13.7. Taking the inverse transform of the quantized coefficients shown in Table 13.11 and adding 128, we get the reconstructed block shown in Table 13.12. We can see that in spite of going from 8 bits per pixel to $\frac{9}{32}$ bits per pixel, the reproduction is remarkably close to the original.

If we wanted an even more accurate reproduction, we could do so at the cost of an increased bit rate by multiplying the step sizes in the quantization table by one-half and using these values as the new step sizes. Using the same assumptions as before, we can show that this will result in an increase in the number of bits transmitted. We can go in the other direction by multiplying the step sizes with a number greater than one. This will result in a reduction in bit rate at the cost of increased distortion.

13.6.4 Format—JFIF

The most common format used for storing JPEG encoded images is the JPEG File Interchange Format (JFIF). It uses markers to partition the bitstream. The first byte of each marker is FF and the second byte is constrained to be neither FF nor 0. Markers can be stand-alone or they can be followed by a payload. Examples of standalone markers are the SOI (start of image) (FFD8) markers, which initiate all JFIF files, and the EOI (end of image) (FFD9) markers, which terminate all JFIF files. In the markers which act as headers for a payload, the first two bytes encode the length of the payload.

The application-specific APP0 marker (FFE0), which immediately follows the SOI marker has the following format:

Field	Size	Content
Length	2 bytes	Length of the header
Identifier	5 bytes	0x4A 0x46 0x49 0x46 0x00 (JFIF0)
Version	2 bytes	MSB for major revision, LSB for minor revisions
Units	1 byte	0: no units, 1: dots per inch, 2: dots per cm
Xdensity	2 bytes	Horizontal pixel density
Ydensity	2 bytes	Vertical pixel density
Xthumbnail	1 byte	Thumbnail horizontal pixel count
Ythumbnail	1 byte	Thumbnail vertical pixel count
$(\text{RGB})_n$	$3n$ bytes	24-bit RGB values for thumbnail
		$n = \text{Xthumbnail} \times \text{Ythumbnail}$

The structure of the JFIF file is shown in Figure 13.10 in which the representation optional elements are shown as dashed boxes. Optional elements include table markers such as DHT (define Huffman table) (FFC4) and DQT (define quantization table) (FFDB). The syntax of the DHT marker is shown in Figure 13.11. The first two bytes contain the length of the payload. The four most significant bits of the next byte indicate whether the Huffman code is for the DC coefficient or the AC coefficients. The number of Huffman codewords of lengths from 1 bit through 16 bits are stored in the next 16 bytes. This is followed by a variable number of bytes that describe the Huffman codes in terms of their lengths.

The syntax of the quantization table marker is shown in Figure 13.12. Again, the first two bytes contain the payload length. The most significant four bits of the next byte specify whether the quantization table values are stored in one or two bytes. A value of 0 indicates a single byte is used for each quantization value while a value of 1 indicates 2 bytes are used for each quantization value. The next 64 or 128 bytes contain the 64 quantization table values stored in zigzag order.

Optional table marker segments are followed by the frame header, which indicates the start of a frame and in turn can be followed by a number of scans. Each new scan begins with a marker segment, the DNL marker segment (FFDC), which defines the number of lines in the next scan. This is used to reset the dimension of the image set in the frame header. The scan can begin with optional tables followed by the scan header. The first two bytes in the scan header are the start of scan (SOS) marker followed by the length, number of components, and component parameters. The scan header is followed by the entropy-coded data interleaved with restart markers (DRI). These markers provide a degree of robustness to the JPEG datastream.

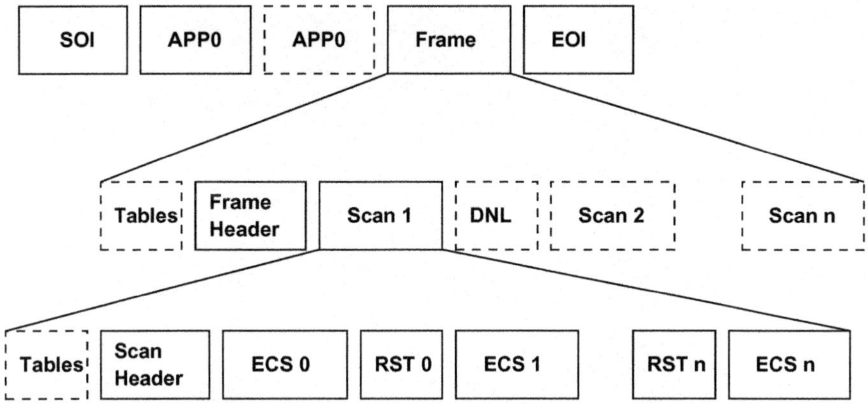

FIGURE 13.10 **Syntax of the JFIF file.**

FIGURE 13.11 **Syntax of the Huffman table marker.**

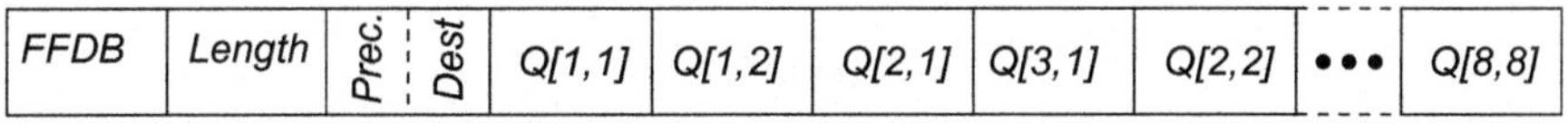

FIGURE 13.12 **Syntax of the quantization table marker.**

One of the drawbacks of using variable length codes is that bit errors tend to propagate. The DRI markers prevent the error from propagating beyond a restart interval. The file is terminated with an end of image (EOI) marker.

Finally, we present some examples of JPEG-coded images in Figures 13.8 and 13.9. These were coded using shareware generated by the Independent JPEG Group (organizer, Dr. Thomas G. Lane). Notice the high degree of "blockiness" in the lower-rate image (Figure 13.8). This is a standard problem of most block-based techniques and specifically of the transform coding approach. A number of solutions have been suggested for removing this blockiness, including postfiltering at the block edges as well as transforms that overlap the block boundaries. Each approach has its own drawbacks. The filtering approaches tend to reduce the resolution of the reconstructions, while the overlapped approaches increase the complexity. One particular overlapped approach that is widely used in audio compression is the modified DCT (MDCT), which is described in the next section (see Figs. 13.13 and 13.14).

FIGURE 13.13 Sinan image coded at 0.5 bits per pixel using the JPEG algorithm.

13.7 Application to Audio Compression—The MDCT

As mentioned in the previous section, the use of the block-based transform has the unfortunate effect of causing distortion at the block boundaries at low rates. A number of techniques that use overlapping blocks have been developed over the years [198]. One that has gained wide acceptance in audio compression is a transform based on the discrete cosine transform called the modified discrete cosine transform (MDCT). It is used in almost all popular audio coding standards from MP3 and AAC to Ogg Vorbis.

The MDCT used in these algorithms uses 50% overlap. That is, each block overlaps half of the previous block and half of the next block of data. Consequently, each audio sample is part of two blocks. If we were to keep all the frequency coefficients, we would end up with twice as many coefficients as samples. Reducing the number of frequency coefficients results in the introduction of distortion in the inverse transform. The distortion is referred to as time domain aliasing [199]. The reason for the name is evident if we consider that the distortion is being introduced by subsampling in the frequency domain. Recall that sampling at less than the Nyquist frequency in the time domain leads to an overlap of replicas of the frequency spectrum, or frequency aliasing. The lapped transforms are successful because they are constructed in such a way that while the inverse transform of each block results in time-domain aliasing, the aliasing in consecutive blocks cancels each other out.

Consider the scenario shown in Figure 13.15. Let's look at the coding for block i and block $i+1$. The inverse transform of the coefficients resulting from both these blocks results in the audio samples in the subblock q. We assume that the blocksize is N, and, therefore, the subblock size is $N/2$. The forward transform can be represented by an $N/2 \times N$ matrix P.

FIGURE 13.14 **Sinan image coded at 0.25 bits per pixel using the JPEG algorithm.**

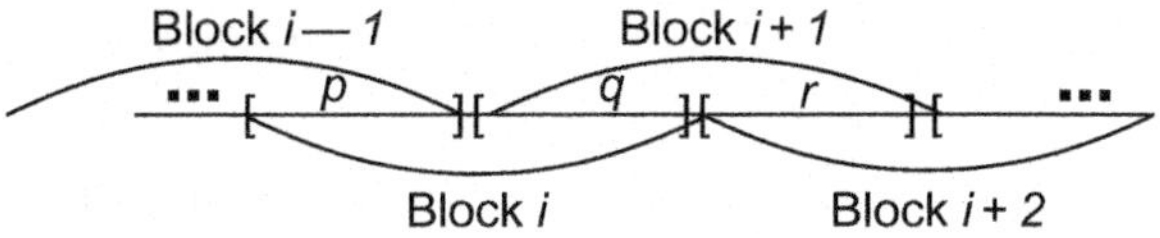

FIGURE 13.15 **Source output sequence.**

Let us partition the matrix into two $N/2 \times N/2$ blocks, A and B. Thus

$$P = [A|B]$$

Let $x_i = [p|q]$, then the forward transform Px_i can be written in terms of the subblocks as

$$X_i = [A|B]\begin{bmatrix} p \\ q \end{bmatrix}$$

The inverse transform matrix Q can be represented by an $N \times N/2$ matrix, which can be partitioned into two $N/2 \times N/2$ blocks, C and D.

$$Q = \begin{bmatrix} C \\ D \end{bmatrix}$$

Applying the inverse transform, we get the reconstruction values $\hat{x}$:

$$\hat{x}_i = QX_i = QPx_i = \begin{bmatrix} C \\ D \end{bmatrix}[A|B]\begin{bmatrix} p \\ q \end{bmatrix} = \begin{bmatrix} CAp + CBq \\ DAp + DBq \end{bmatrix}$$

Repeating the process for block $i + 1$ we get

$$\hat{x}_{i+1} = QX_{i+1} = QPx_{i+1} = \begin{bmatrix} C \\ D \end{bmatrix}[A|B]\begin{bmatrix} q \\ r \end{bmatrix} = \begin{bmatrix} CAq + CBr \\ DAq + DBr \end{bmatrix}$$

To cancel out the aliasing in the second half of the block, we need

$$CAq + CBr + DAp + DBq = q$$

From this we can get the requirements for the transform:

$$CB = 0 \tag{64}$$

$$DA = 0 \tag{65}$$

$$CA + DB = I \tag{66}$$

Note that the same requirements will help cancel the aliasing in the first half of block i by using the second half of the inverse transform of block $i - 1$. One selection that satisfies the last condition is

$$CA = \frac{1}{2}(I - J) \tag{67}$$

$$DB = \frac{1}{2}(I + J) \tag{68}$$

where J is the counteridentity matrix.

The forward modified discrete transform is given by the following equation:

$$X_k = \sum_{n=0}^{N-1} x_n \cos\left(\frac{2\pi}{N}(k + \frac{1}{2})(n + \frac{1}{2} + \frac{N}{4})\right) \tag{69}$$

where x_n are the audio samples and X_k are the frequency coefficients. The inverse MDCT is given by

$$y_n = \frac{2}{N}\sum_{k=0}^{\frac{N}{2}-1} X_k \cos\left(\frac{2\pi}{N}(k + \frac{1}{2})(n + \frac{1}{2} + \frac{N}{4})\right) \tag{70}$$

or in terms of our matrix notation,

$$[P]_{i,j} = \cos\left(\frac{2\pi}{N}(i + \frac{1}{2})(j + \frac{1}{2} + \frac{N}{4})\right) \tag{71}$$

$$[Q]_{i,j} = \frac{2}{N}\cos\left(\frac{2\pi}{N}(i + \frac{1}{2})(j + \frac{1}{2} + \frac{N}{4})\right) \tag{72}$$

It is easy to verify that, given a value of N, these matrices satisfy the conditions for alias cancellation.

Thus, while the inverse transform for any one block will contain aliasing, by using the inverse transform of neighboring blocks, the aliasing can be canceled. What about blocks that do not have neighbors—that is, the first and last blocks? One way to resolve this problem is to pad the sampled audio sequence with $N/2$ zeros at the beginning and end of the sequence. In practice, this is not necessary, because the data to be transformed is windowed prior to the transform. For the first and last blocks, we use a special window that has the same effect as introducing zeros. For information on the design of windows for the MDCT, see [200]. For more on how the MDCT is used in audio compression techniques, see Chapter 16.

13.8 Summary

In this chapter, we have described the concept of transform coding and provided some of the details needed for the investigation of this compression scheme. The basic encoding scheme works as follows:

- Divide the source output into blocks. In the case of speech or audio data, they will be one-dimensional blocks. In the case of images, they will be two-dimensional blocks. In image coding, a typical block size is 8×8. In audio coding, the blocks are generally overlapped by 50%.
- Take the transform of this block. In the case of one-dimensional data, this involves premultiplying the $N \times 1$ vector of source output samples by the transform matrix. In the case of image data, for the transforms we have looked at, this involves premultiplying the $N \times N$ block by the transform matrix and post-multiplying the result with the transpose of the transform matrix. Fast algorithms exist for performing the transforms described in this chapter (see [201]).
- Quantize the coefficients. Various techniques exist for the quantization of these coefficients. We have described the approach used by JPEG. In Chapter 16 we describe the quantization techniques used in various audio coding algorithms.
- Encode the quantized value. The quantized value can be encoded using a fixed-length code or any of the different variable-length codes described in earlier chapters. We have described the approach taken by JPEG.

The decoding scheme is the inverse of the encoding scheme for image compression. For the overlapped transform used in audio coding, the decoder adds the overlapped portions of the inverse transform to cancel aliasing.

The basic approach can be modified depending on the particular characteristics of the data. We have described some of the modifications used by various commercial algorithms for transform coding of audio signals.

Further Reading

1. For detailed information about the JPEG standard, *JPEG Still Image Data Compression Standard*, by W.B. Pennebaker and J.L. Mitchell [6], is an invaluable reference. This book also contains the entire text of the official draft JPEG recommendation, ISO DIS 10918-1 and ISO DIS 10918-2.
2. For a detailed discussion of the MDCT and how it is used in audio coding, an excellent source is *Introduction to Digital Audio Coding Standards*, by M. Bosi and R.E. Goldberg [200].
3. Chapter 11 in *Digital Coding of Waveforms*, by N.S. Jayant and P. Noll [134], provides a more mathematical treatment of the subject of transform coding.
4. A good source for information about transforms is *Fundamentals of Digital Image Processing*, by A.K. Jain [202]. Another one is *Digital Image Processing*, by R.C. Gonzales

and R.E. Wood [96]. This book has an especially nice discussion of the Hotelling transform.

5. The bit allocation problem and its solutions are described in *Vector Quantization and Signal Compression*, by A. Gersho and R.M. Gray [136].

6. A very readable description of transform coding of images is presented in *Digital Image Compression Techniques*, by M. Rabbani and P.W. Jones [91].

7. *The Data Compression Book*, by M. Nelson and J.-L. Gailly [69], provides a very readable discussion of the JPEG algorithm.

13.9 Projects and Problems

1. A square matrix $\mathbf{A}$ has the property that $\mathbf{A}^T\mathbf{A} = \mathbf{A}\mathbf{A}^T = \mathbf{I}$, where $\mathbf{I}$ is the identity matrix. If X_1 and X_2 are two N-dimensional vectors and

$$\Theta_1 = \mathbf{A}X_1$$
$$\Theta_2 = \mathbf{A}X_2$$

then show that

$$|X_1 - X_2|^2 = |\Theta_1 - \Theta_2|^2 \tag{73}$$

2. Consider the following sequence of values:

10	11	12	11	12	13	12	11
10	−10	8	−7	8	−8	7	−7

(a) Transform each row separately using an eight-point DCT. Plot the resulting 16 transform coefficients.

(b) Combine all 16 numbers into a single vector, and transform it using a 16-point DCT. Plot the 16 transform coefficients.

(c) Compare the results of (a) and (b). For this particular case would you suggest a block size of 8 or 16 for greater compression? Justify your answer.

3. Consider the following "image":

4	3	2	1
3	2	1	1
2	1	1	1
1	1	1	1

(a) Obtain the two-dimensional DWHT transform by first taking the one-dimensional transform of the rows, then taking the column-by-column transform of the resulting matrix.

(b) Obtain the two-dimensional DWHT transform by first taking the one-dimensional transform of the columns, then taking the row-by-row transform of the resulting matrix.

(c) Compare and comment on the results of (a) and (b).

4. (This problem was suggested by P.F. Swaszek.) Let us compare the energy compaction properties of the DCT and the DWHT transforms.

(a) For the Sena image, compute the mean squared value of each of the 64 coefficients using the DCT. Plot these values.

(b) For the Sena image, compute the mean squared value of each of the 64 coefficients using the DWHT. Plot these values.

(c) Compare the results of (a) and (b). Which transform provides more energy compaction? Justify your answer.

5. Implement the transform and quantization portions of the JPEG standard. For coding the labels, use an arithmetic coder instead of the modified Huffman code described in this chapter.

(a) Encode the Sena image using this transform coder at rates of (approximately) 0.25, 0.5, and 0.75 bits per pixel. Compute the mean squared error at each rate and plot the rate versus the mse.

(b) Repeat part (a) using one of the public domain implementations of JPEG.

(c) Compare the plots obtained using the two coders and comment on the relative performance of the coders.

6. One of the extensions to the JPEG standard allows for the use of multiple quantization matrices. Investigate the issues involved in designing a set of quantization matrices. Should the quantization matrices be similar or dissimilar? How would you measure their similarity? Given a particular block, do you need to quantize it with each quantization matrix to select the best? Or is there a computationally more efficient approach? Describe your findings in a report.

7. Assume we use the following transform coding scheme on the source sequence of *iid* Xs. First we form a vector with two consecutive X values. Then we take a one-dimensional transform using the following transform matrix:

$$A = \frac{1}{\sqrt{2}} \begin{bmatrix} 1 & 1 \\ 1 & -1 \end{bmatrix} \quad (74)$$

We discard the second coefficient θ_2 and only transmit the first coefficient θ_1 (assume θ_1 is transmitted after being quantized using a *pdf*-optimized nonuniform quantizer). The two consecutive samples of X are then reconstructed using the inverse transform. Starting with

$$2D = E\left[(X_1 - \hat{X}_1)^2\right] + E\left[(X_2 - \hat{X}_2)^2\right] \quad (75)$$

show that

$$D = \frac{\sigma_X^2}{2} + \sigma_q^2 \quad (76)$$

where σ_X^2 represents the variance of each input random variable and σ_q^2 denotes the variance of the quantization error.

8. Write a program simulating an 8×8 transform coder without quantization. The program should read the basis set from a file, perform the transform, retain specified coefficients only, and perform the inverse transform and reconstruction.

(a) Use this program to compare the discrete cosine transform and the discrete Walsh-Hadamard transform. Compare the transforms based on objective measures such as geometric mean and correlation among the coefficients, and by comparing the quality of the reconstructed image when only the first three and first six coefficients are retained.

(b) Finally examine the importance of the DC coefficient by looking at a reconstruction in which the DC coefficient has been discarded.

14

Subband Coding

14.1 Overview

In this chapter we present the second of three approaches to compression in which the source output is decomposed into constituent parts. Each constituent part is encoded using one or more of the methods that have been described previously. The approach described in this chapter, known as subband coding, relies on separating the source output into different bands of frequencies using digital filters. We provide a general description of the subband coding system and, for those readers with some knowledge of Z-transforms, a more mathematical analysis of the system. The sections containing the mathematical analysis are not essential to understanding the rest of the chapter and are marked with a $\star$. If you are not interested in the mathematical analysis, you should skip these sections. This is followed by a description of a popular approach to bit allocation. We conclude the chapter with applications to audio and image compression.

14.2 Introduction

In previous chapters we looked at a number of different compression schemes. Each of these schemes is most efficient when the data have certain characteristics. A vector quantization scheme is most effective if blocks of the source output show a high degree of clustering. A differential encoding scheme is most effective when the sample-to-sample difference is small. If the source output is truly random, it is best to use scalar quantization or lattice vector quantization. Thus, if a source exhibits certain well-defined characteristics, we can choose a compression scheme most suited to that characteristic. Unfortunately, most source outputs exhibit a combination of characteristics, which makes it difficult to select a compression scheme exactly suited to the source output.

In the last chapter we looked at techniques for decomposing the source output into different frequency bands using block transforms. The transform coefficients had differing statistics and differing perceptual importance. We made use of these differences in allocating bits for encoding the different coefficients. This variable bit allocation resulted in a decrease in the average number of bits required to encode the source output. One of the drawbacks of transform coding is the artificial division of the source output into blocks, which results in the generation of coding artifacts at the block edges, or blocking. One approach to avoiding this blocking is the lapped orthogonal transform (LOT) [198]. In this chapter we look at a popular approach to decomposing the image into different frequency bands without the imposition of an arbitrary block structure. After the input has been decomposed into its constituents, we can use the coding technique best suited to each constituent to improve compression performance. Furthermore, each component of the source output may have different perceptual characteristics. For example, quantization error that is perceptually objectionable in one component may be acceptable in a different component of the source output. Therefore, a coarser quantizer that uses fewer bits can be used to encode the component that is perceptually less important.

Consider the sequence $\{x_n\}$ plotted in Figure 14.1. We can see that, while there is a significant amount of sample-to-sample variation, there is also an underlying long-term trend shown by the dotted line that varies slowly.

One way to extract this trend is to average the sample values in a moving window. The averaging operation smooths out the rapid variations, making the slow variations more evident. Let's pick a window of size two and generate a new sequence $\{y_n\}$ by averaging neighboring values of x_n:

$$y_n = \frac{x_n + x_{n-1}}{2} \tag{1}$$

The consecutive values of y_n will be closer to each other than the consecutive values of x_n. Therefore, the sequence $\{y_n\}$ can be coded more efficiently using differential encoding than the sequence $\{x_n\}$. However, we want to encode the sequence $\{x_n\}$, not the sequence $\{y_n\}$. Therefore, we follow the encoding of the averaged sequence $\{y_n\}$ by the difference sequence $\{z_n\}$:

$$z_n = x_n - y_n = x_n - \frac{x_n + x_{n-1}}{2} = \frac{x_n - x_{n-1}}{2} \tag{2}$$

The sequences $\{y_n\}$ and $\{z_n\}$ can be coded independently of each other. This way we can use the compression schemes that are best suited for each sequence.

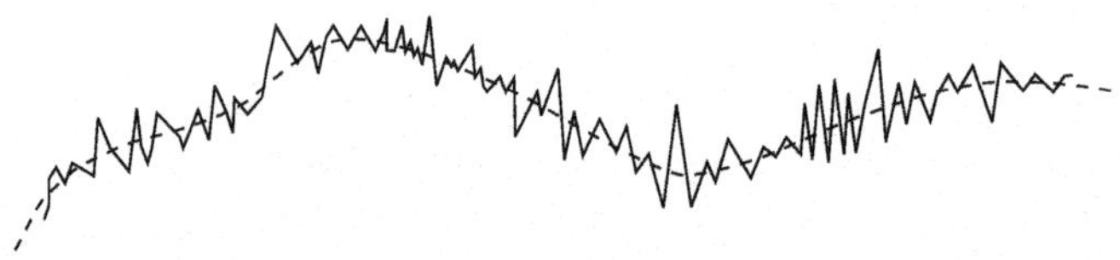

FIGURE 14.1 A rapidly changing source output that contains a long-term component with slow variations.

Example 14.2.1:

Suppose we want to encode the following sequence of values $\{x_n\}$:

$$10 \quad 14 \quad 10 \quad 12 \quad 14 \quad 8 \quad 14 \quad 12 \quad 10 \quad 8 \quad 10 \quad 12$$

There is a significant amount of sample-to-sample correlation, so we might consider using a DPCM scheme to compress this sequence. In order to get an idea of the requirements on the quantizer in a DPCM scheme, let us take a look at the sample-to-sample differences $x_n - x_{n-1}$:

$$10 \quad 4 \quad -4 \quad 2 \quad 2 \quad -6 \quad 6 \quad -2 \quad -2 \quad -2 \quad 2 \quad 2$$

Ignoring the first value, the dynamic range of the differences is from -6 to 6. Suppose we want to quantize these values using m bits per sample. This means we could use a quantizer with $M = 2^m$ levels or reconstruction values. If we choose a uniform quantizer, the size of each quantization interval, Δ, is the range of possible input values divided by the total number of reconstruction values. Therefore,

$$\Delta = \frac{12}{M}$$

which would give us a maximum quantization error of $\frac{\Delta}{2}$ or $\frac{6}{M}$.

Now let's generate two new sequences $\{y_n\}$ and $\{z_n\}$ according to (1) and (2). All three sequences are plotted in Figure 14.2. Notice that given y_n and z_n, we can always recover x_n:

$$x_n = y_n + z_n \tag{3}$$

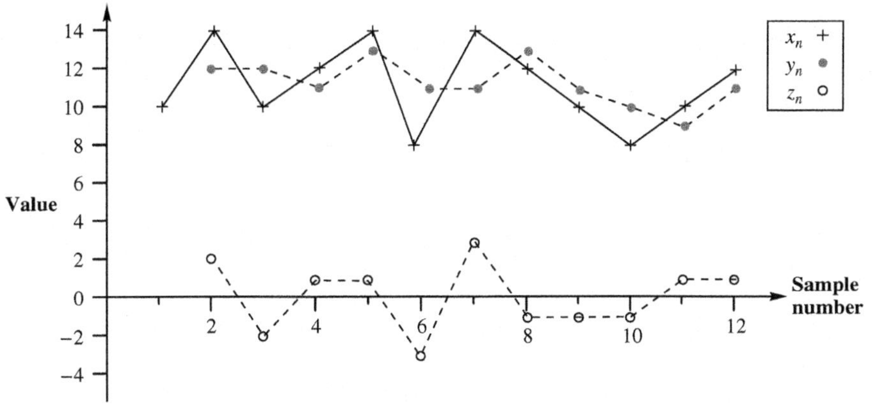

FIGURE 14.2 Original set of samples and the two components.

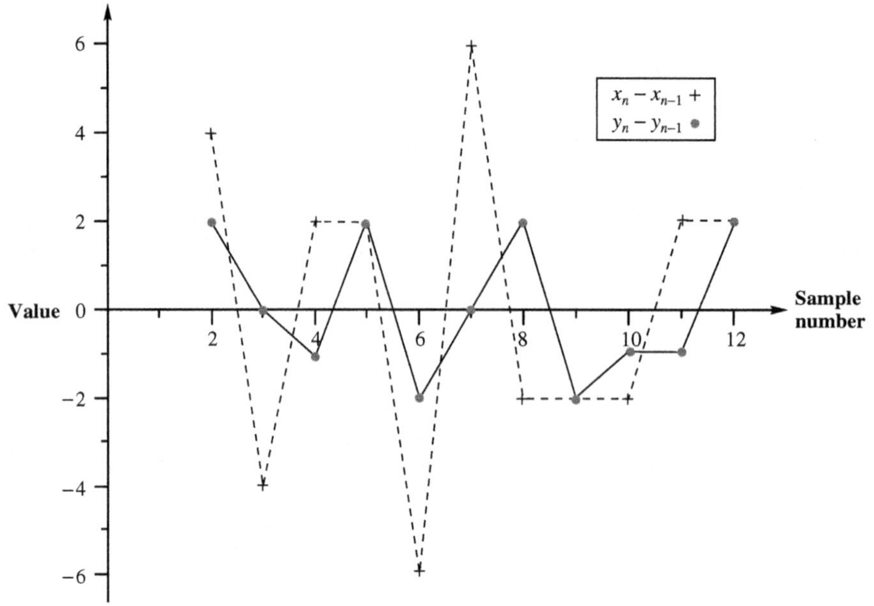

FIGURE 14.3 **Difference sequences generated from the original and averaged sequences.**

Let's try to encode each of these sequences. The sequence $\{y_n\}$ is

10 12 12 11 13 11 11 13 11 10 9 11

Notice that the $\{y_n\}$ sequence is "smoother" than the $\{x_n\}$ sequence—the sample-to-sample variation is much smaller. This becomes evident when we look at the sample-to-sample differences:

10 2 0 −1 2 −2 0 2 −2 −1 −1 2

The difference sequences $\{x_n - x_{n-1}\}$ and $\{y_n - y_{n-1}\}$ are plotted in Figure 14.3. Again, ignoring the first difference, the dynamic range of the differences $y_n - y_{n-1}$ is 4. If we take the dynamic range of these differences as a measure of the range of the quantizer, then for an M-level quantizer, the step size of the quantizer is $\frac{4}{M}$ and the maximum quantization error is $\frac{2}{M}$. This maximum quantization error is one-third the maximum quantization error incurred when the $\{x_n\}$ sequence is quantized using an M-level quantizer. However, in order to reconstruct $\{x_n\}$, we also need to transmit $\{z_n\}$. The $\{z_n\}$ sequence is

$$0 \quad 2 \quad -2 \quad 1 \quad 1 \quad -3 \quad 3 \quad -1 \quad -1 \quad -1 \quad 1 \quad 1$$

The dynamic range for z_n is 6, half the dynamic range of the difference sequence for $\{x_n\}$. (We could have inferred this directly from the definition of z_n.) The sample-to-sample difference varies more than the actual values. Therefore, instead of differentially encoding this sequence, we quantize each individual sample. For an M-level quantizer, the required step size would be $\frac{6}{M}$, giving a maximum quantization error of $\frac{3}{M}$.

For the same number of bits per sample, we can code both y_n and z_n and incur less distortion. At the receiver, we add y_n and z_n to get the original sequence x_n back. The maximum possible quantization error in the reconstructed sequence would be $\frac{5}{M}$, which is less than the maximum error we would incur if we encoded the $\{x_n\}$ sequence directly.

Although we use the same number of bits for each value of y_n and z_n, the number of elements in each of the $\{y_n\}$ and $\{z_n\}$ sequences is the same as the number of elements in the original $\{x_n\}$ sequence. Although we are using the same number of bits per sample, we are transmitting twice as many samples and, in effect, doubling the bit rate.

We can avoid this by sending every other value of y_n and z_n. Let's divide the sequence $\{y_n\}$ into subsequences $\{y_{2n}\}$ and $\{y_{2n-1}\}$—that is, a subsequence containing only the odd-numbered elements $\{y_1, y_3, \ldots\}$, and a subsequence containing only the even-numbered elements $\{y_2, y_4, \ldots\}$. Similarly, we divide the $\{z_n\}$ sequence into subsequences $\{z_{2n}\}$ and $\{z_{2n-1}\}$. If we transmit either the even-numbered subsequences or the odd-numbered subsequences, we would transmit only as many elements as in the original sequence. To see how we recover the sequence $\{x_n\}$ from these subsequences, suppose we only transmitted the subsequences $\{y_{2n}\}$ and $\{z_{2n}\}$:

$$y_{2n} = \frac{x_{2n} + x_{2n-1}}{2}$$

$$z_{2n} = \frac{x_{2n} - x_{2n-1}}{2}$$

To recover the even-numbered elements of the $\{x_n\}$ sequence, we add the two subsequences. In order to obtain the odd-numbered members of the $\{x_n\}$ sequence, we take the difference:

$$y_{2n} + z_{2n} = x_{2n} \tag{4}$$

$$y_{2n} - z_{2n} = x_{2n-1} \tag{5}$$

Thus, we can recover the entire original sequence $\{x_n\}$, sending only as many bits as required to transmit the original sequence while incurring less distortion.

Is the last part of the previous statement still true? In our original scheme we proposed to transmit the sequence $\{y_n\}$ by transmitting the differences $y_n - y_{n-1}$. As we now need to transmit the subsequence $\{y_{2n}\}$, we will be transmitting the differences $y_{2n} - y_{2n-2}$ instead. In order for our original statement about reduction in distortion to hold, the dynamic range of this new sequence of differences should be less than or equal to the dynamic range of the original difference. A quick check of the $\{y_n\}$ shows us that the dynamic range of the new differences is still 4, and our claim of incurring less distortion still holds. ♦

There are several things we can see from this example. First, the number of different values that we transmit is the same, whether we send the original sequence $\{x_n\}$ or the two subsequences $\{y_n\}$ and $\{z_n\}$. Decomposing the $\{x_n\}$ sequence into subsequences did not result in any increase in the number of values that we need to transmit. Second, the two subsequences had distinctly different characteristics, which led to our use of different techniques to encode the different sequences. If we had not split the $\{x_n\}$ sequence, we would have been using essentially the same approach to compress both subsequences. Finally, we could have used the same decomposition approach to decompose the two constituent sequences, which then could be decomposed further still.

While this example was specific to a particular set of values, we can see that decomposing a signal can lead to different ways of looking at the problem of compression. This added flexibility can lead to improved compression performance.

Before we leave this example let us formalize the process of decomposing or *analysis*, and recomposing or *synthesis*. In our example, we decomposed the input sequence $\{x_n\}$ into two subsequences $\{y_n\}$ and $\{z_n\}$ by the operations

$$y_n = \frac{x_n + x_{n-1}}{2} \tag{6}$$

$$z_n = \frac{x_n - x_{n-1}}{2} \tag{7}$$

We can implement these operations using discrete time filters. We briefly considered discrete time filters in Chapter 11. We take a slightly more detailed look at filters in the next section.

14.3 Filters

A system that isolates certain frequency components is called a *filter*. The analogy here with mechanical filters such as coffee filters is obvious. A coffee filter or a filter in a water purification system blocks coarse particles and allows only the finer-grained components of the input to pass through. The analogy is not complete, however, because mechanical filters always block the coarser components of the input, while the filters we are discussing can selectively let through or block any range of frequencies. Filters that only let through components below a certain frequency f_0 are called low-pass filters; filters that block all frequency components below a certain value f_0 are called high-pass filters. The frequency f_0 is called the *cutoff frequency*. Filters that let through components that have frequency content above some frequency f_1 but below frequency f_2 are called band-pass filters.

One way to characterize filters is by their *magnitude transfer function*—the ratio of the magnitude of the input and output of the filter as a function of frequency. In Figure 14.4 we show the magnitude transfer function for an ideal low-pass filter and a more realistic low-pass filter, both with a cutoff frequency of f_0. In the ideal case, all components of the input signal with frequencies below f_0 are unaffected except for a constant amount of amplification. All frequencies above f_0 are blocked. In other words, the cutoff is sharp. In the case of the more realistic filter, the cutoff is more gradual. Also, the amplification for the components with

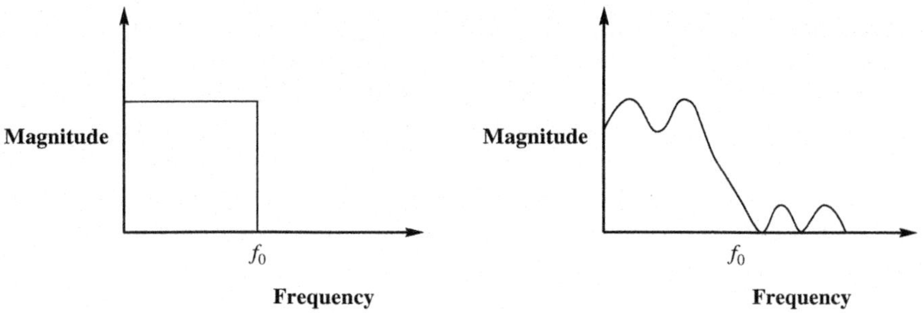

FIGURE 14.4 **Ideal and realistic low-pass filter characteristics.**

frequency less than f_0 is not constant, and components with frequencies above f_0 are not totally blocked. This phenomenon is referred to as *ripple* in the passband and stopband.

The filters we will discuss are digital filters, which operate on a sequence of numbers that are usually samples of a continuously varying signal. We discussed sampling in Chapter 12. For those of you who skipped that chapter, let us take a brief look at the sampling operation.

How often does a signal have to be sampled in order to reconstruct the signal from the samples? If one signal changes more rapidly than another, it is reasonable to assume that we would need to sample the more rapidly varying signal more often than the slowly varying signal in order to achieve an accurate representation. In fact, it can be shown mathematically that if the highest frequency component of a signal is f_0, then we need to sample the signal at more than $2f_0$ times per second. This result is known as the *Nyquist theorem* or *Nyquist rule* after Harry Nyquist, a famous mathematician from Bell Laboratories. His pioneering work laid the groundwork for much of digital communication. The Nyquist rule can also be extended to signals that only have frequency components between two frequencies f_1 and f_2. If f_1 and f_2 satisfy certain criteria, then we can show that in order to recover the signal exactly, we need to sample the signal at a rate of at least $2(f_2 - f_1)$ samples per second [134].

What would happen if we violated the Nyquist rule and sampled at less than twice the highest frequency? In Chapter 12 we showed that it would be impossible to recover the original signal from the sample. Components with frequencies higher than half the sampling rate show up at lower frequencies. This process is called *aliasing*. In order to prevent aliasing, most systems that require sampling will contain an "anti-aliasing filter" that restricts the input to the sampler to be less than half the sampling frequency. If the signal contains components at more than half the sampling frequency, we will introduce distortion by filtering out these components. However, the distortion due to aliasing is generally more severe than the distortion we introduce due to filtering.

Digital filtering involves taking a weighted sum of current and past inputs to the filter and, in some cases, the past outputs of the filter. The general form of the input-output relationships of the filter is given by

$$y_n = \sum_{i=0}^{N} a_i x_{n-i} + \sum_{i=1}^{M} b_i y_{n-i} \tag{8}$$

where the sequence $\{x_n\}$ is the input to the filter, the sequence $\{y_n\}$ is the output from the filter, and the values $\{a_i\}$ and $\{b_i\}$ are called the *filter coefficients*.

If the input sequence is a single 1 followed by all 0s, the output sequence is called the impulse response of the filter. Notice that if the b_i are all 0, then the impulse response will die out after N samples. These filters are called *finite impulse response* (FIR) filters. The number N is sometimes called the number of *taps* in the filter. If any of the b_i have nonzero values, the impulse response can, in theory, continue forever. Filters with nonzero values for some of the b_i are called *infinite impulse response* (IIR) filters.

Example 14.3.1:

Suppose we have a filter with $a_0 = 1.25$ and $a_1 = 0.5$. If the input sequence $\{x_n\}$ is given by

$$x_n = \begin{cases} 1 & n = 0 \\ 0 & n \neq 0 \end{cases} \tag{9}$$

then the output is given by

$$y_0 = a_0 x_0 + a_1 x_{-1} = 1.25$$

$$y_1 = a_0 x_1 + a_1 x_0 = 0.5$$

$$y_n = 0 \qquad n < 0 \text{ or } n > 1$$

This output is called the impulse response of the filter. The impulse response sequence is usually represented by $\{h_n\}$. Therefore, for this filter we would say that

$$h_n = \begin{cases} 1.25 & n = 0 \\ 0.5 & n = 1 \\ 0 & \text{otherwise} \end{cases} \tag{10}$$

Notice that if we know the impulse response we also know the values of a_i. Knowledge of the impulse response completely specifies the filter. Furthermore, because the impulse response goes to zero after a finite number of samples (two in this case), the filter is an FIR filter.

The filters we used in Example 14.2.1 are both two-tap FIR filters with impulse responses

$$h_n = \begin{cases} \frac{1}{2} & n = 0 \\ \frac{1}{2} & n = 1 \\ 0 & \text{otherwise} \end{cases} \tag{11}$$

for the "averaging" or low-pass filter, and

$$h_n = \begin{cases} \frac{1}{2} & n = 0 \\ -\frac{1}{2} & n = 1 \\ 0 & \text{otherwise} \end{cases} \tag{12}$$

for the "difference" or high-pass filter.

Now let's consider a different filter with $a_0 = 1$ and $b_1 = 0.9$. For the same input as above, the output is given by

$$y_0 = a_0x_0 + b_1y_{-1} = 1(1) + 0.9(0) = 1 \tag{13}$$

$$y_1 = a_0x_1 + b_1y_0 = 1(0) + 0.9(1) = 0.9 \tag{14}$$

$$y_2 = a_0x_2 + b_1y_1 = 1(0) + 0.9(0.9) = 0.81 \tag{15}$$

$$\vdots \quad \vdots$$

$$y_n = (0.9)^n \tag{16}$$

The impulse response can be written more compactly as

$$h_n = \begin{cases} 0 & n < 0 \\ (0.9)^n & n \geqslant 0 \end{cases} \tag{17}$$

Notice that the impulse response is nonzero for all $n \geqslant 0$, which makes this an IIR filter. ♦

Although it is not as clear in the IIR case as it was in the FIR case, the impulse response completely specifies the filter. Once we know the impulse response of the filter, we know the relationship between the input and output of the filter. If $\{x_n\}$ and $\{y_n\}$ are the input and output, respectively, of a filter with impulse response $\{h_n\}_{n=0}^{M}$, then $\{y_n\}$ can be obtained from $\{x_n\}$ and $\{h_n\}$ via the following relationship:

$$y_n = \sum_{k=0}^{M} h_k x_{n-k} \tag{18}$$

where M is finite for an FIR filter and infinite for an IIR filter. The relationship shown in (18) is known as *convolution* and can be easily obtained through the use of the properties of linearity and shift invariance (see Problem 1 at the end of this chapter).

Because FIR filters are simply weighted averages, they are always stable. When we say a filter is stable we mean that as long as the input is bounded, the output will also be bounded. This is not true of IIR filters. Certain IIR filters can give an unbounded output even when the input is bounded.

Example 14.3.2:

Consider a filter with $a_0 = 1$ and $b_1 = 2$. Suppose the input sequence is a single 1 followed by 0s. Then the output is

$$y_0 = a_0 x_0 + b_1 y_{-1} = 1(1) + 2(0) = 1 \quad (19)$$
$$y_1 = a_0 x_0 + b_1 y_0 = 1(0) + 2(1) = 2 \quad (20)$$
$$y_2 = a_0 x_1 + b_1 y_1 = 1(0) + 2(2) = 4 \quad (21)$$
$$\vdots \quad \vdots$$
$$y_n = 2^n \quad (22)$$

Even though the input contained a single 1, the output at time $n = 30$ is 2^{30}, or more than a billion! ♦

Although IIR filters can become unstable, they can also provide better performance, in terms of sharper cutoffs and less ripple in the passband and stopband for a fewer number of coefficients.

The study of design and analysis of digital filters is a fascinating and important subject. We provide some of the details in Sections 14.5, 14.6, 14.7, 14.8. If you are not interested in these topics, you can take a more utilitarian approach and make use of the literature to select the necessary filters rather than design them. In the following section we briefly describe some of the families of filters used to generate the examples in this chapter. We also provide filter coefficients that you can use for experiment.

14.3.1 Some Filters Used in Subband Coding

The most frequently used filter banks in subband coding consist of a cascade of stages, where each stage consists of a low-pass filter and a high-pass filter, as shown in Figure 14.5. The most popular among these filters are the *quadrature mirror filters* (QMF), which were first proposed by Crosier, Esteban, and Galand [203]. These filters have the property that if the impulse response of the low-pass filter is given by $\{h_n\}$, then the high-pass impulse response is given by $\{(-1)^n h_{N-1-n}\}$. The quadrature mirror filters designed by Johnston [204] are widely used in a number of applications. The filter coefficients for 8-, 16-, and 32-tap filters are given in Tables 14.1, 14.2, 14.3. Notice that the filters are symmetric; that is,

$$h_{N-1-n} = h_n \qquad n = 0, 1, \ldots, \frac{N}{2} - 1 \quad (23)$$

As we shall see later, the filters with fewer taps are less efficient in their decomposition than the filters with more taps. However, from Equation (18) we can see that the number of taps dictates the number of multiply-add operations necessary to generate the filter outputs. Thus, if we want to obtain more efficient decompositions, we do so by increasing the amount of computation.

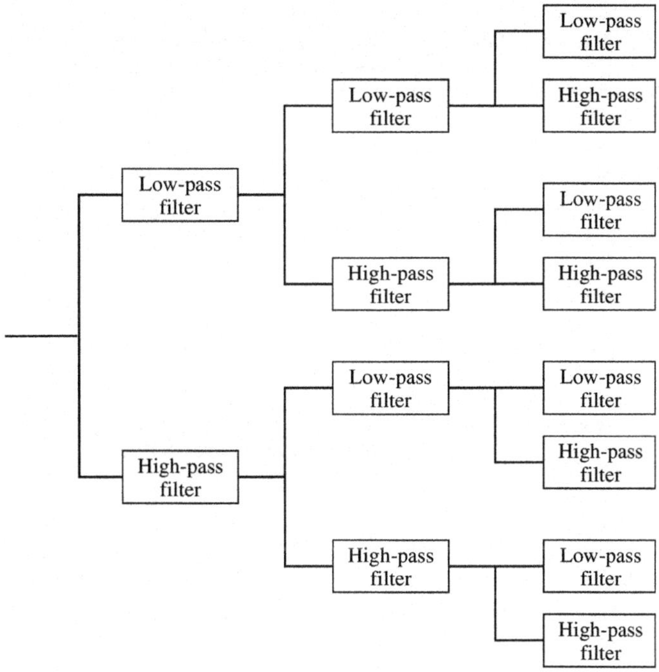

FIGURE 14.5 An eight-band filter bank.

TABLE 14.1 Coefficients for the eight-tap Johnston low-pass filter.

h_0, h_7	0.00938715
h_1, h_6	0.06942827
h_2, h_5	−0.07065183
h_3, h_4	0.48998080

Another popular set of filters are the Smith-Barnwell filters [205], some of which are shown in Tables 14.4 and 14.5.

These families of filters differ in a number of ways. For example, consider the Johnston eight-tap filter and the Smith-Barnwell eight-tap filter. The magnitude transfer functions for these two filters are plotted in Figure 14.6. Notice that the cutoff for the Smith-Barnwell filter is much sharper than the cutoff for the Johnston filter. This means that the separation provided by the eight-tap Johnston filter is not as good as that provided by the eight-tap Smith-Barnwell filter. We will see the effect of this when we look at image compression later in this chapter.

These filters are examples of some of the more popular filters. Many more filters exist in the literature, and more are being discovered.

TABLE 14.2 Coefficients for the 16-tap Johnston low-pass filter.

h_0, h_{15}	0.002898163
h_1, h_{14}	−0.009972252
h_2, h_{13}	−0.001920936
h_3, h_{12}	0.03596853
h_4, h_{11}	−0.01611869
h_5, h_{10}	−0.09530234
h_6, h_9	0.1067987
h_7, h_8	0.4773469

TABLE 14.3 Coefficients for the 32-tap Johnston low-pass filter.

h_0, h_{31}	0.0022551390
h_1, h_{30}	−0.0039715520
h_2, h_{29}	−0.0019696720
h_3, h_{28}	0.0081819410
h_4, h_{27}	0.00084268330
h_5, h_{26}	−0.014228990
h_6, h_{25}	0.0020694700
h_7, h_{24}	0.022704150
h_8, h_{23}	−0.0079617310
h_9, h_{22}	−0.034964400
h_{10}, h_{21}	0.019472180
h_{11}, h_{20}	0.054812130
h_{12}, h_{19}	−0.044524230
h_{13}, h_{18}	−0.099338590
h_{14}, h_{17}	0.13297250
h_{15}, h_{16}	0.46367410

TABLE 14.4 Coefficients for the eight-tap Smith-Barnwell low-pass filter.

h_0	0.0348975582178515
h_1	−0.01098301946252854
h_2	−0.06286453934951963
h_3	0.223907720892568
h_4	0.556856993531445
h_5	0.357976304997285
h_6	−0.02390027056113145
h_7	−0.07594096379188282

TABLE 14.5 **Coefficients for the 16-tap Smith-Barnwell low-pass filter.**

h_0	0.02193598203004352
h_1	0.001578616497663704
h_2	−0.06025449102875281
h_3	−0.0118906596205391
h_4	0.137537915636625
h_5	0.05745450056390939
h_6	−0.321670296165893
h_7	−0.528720271545339
h_8	−0.295779674500919
h_9	0.0002043110845170894
h_{10}	0.02906699789446796
h_{11}	−0.03533486088708146
h_{12}	−0.006821045322743358
h_{13}	0.02606678468264118
h_{14}	0.001033363491944126
h_{15}	−0.01435930957477529

14.4 The Basic Subband Coding Algorithm

The basic subband coding system is shown in Figure 14.7.

14.4.1 Analysis

The source output is passed through a bank of filters, called the analysis filter bank, which covers the range of frequencies that make up the source output. The passbands of the filters can be nonoverlapping or overlapping. Nonoverlapping and overlapping filter banks are shown in Figure 14.8. The outputs of the filters are then subsampled.

The justification for the subsampling is the Nyquist rule and its generalization, which tells us that we only need twice as many samples per second as the range of frequencies. This means that we can reduce the number of samples at the output of the filter because the range of frequencies at the output of the filter is less than the range of frequencies at the input to the filter. This process of reducing the number of samples is called *decimation*,[1] or *downsampling*. The amount of decimation depends on the ratio of the bandwidth of the filter output to the filter input. If the bandwidth at the output of the filter is $1/M$ of the bandwidth at the input to the filter, we would decimate the output by a factor of M by keeping every Mth sample. The symbol $M \downarrow$ is used to denote this decimation.

[1] The word *decimation* has a rather bloody origin. During the time of the Roman empire, if a legion broke ranks and ran during battle, its members were lined up and every tenth person was killed. This process was called decimation.

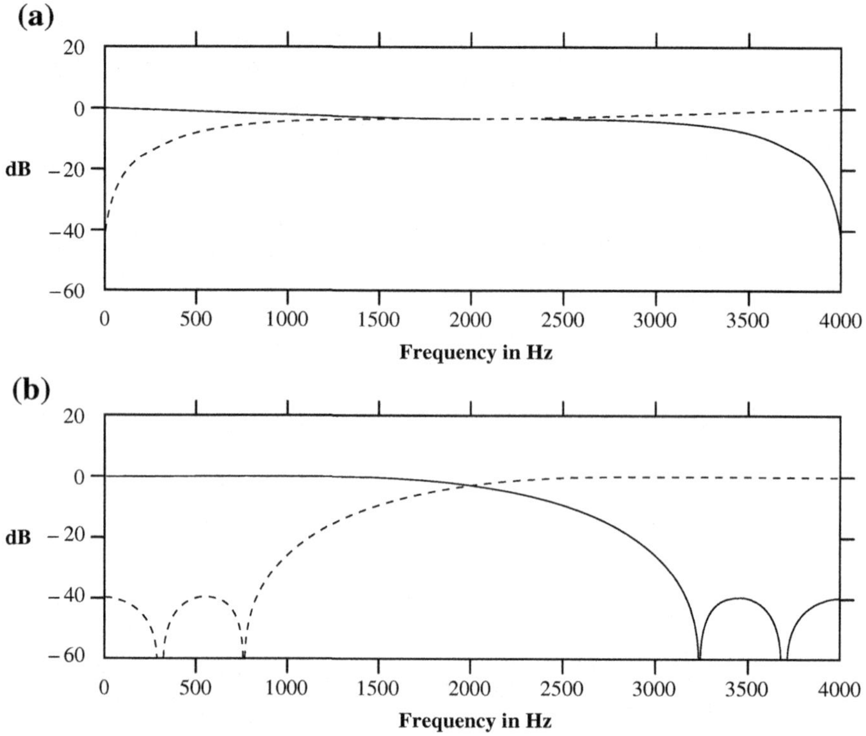

FIGURE 14.6 **Magnitude transfer functions of the (a) eight-tap Johnston and (b) eight-tap Smith-Barnwell filters.**

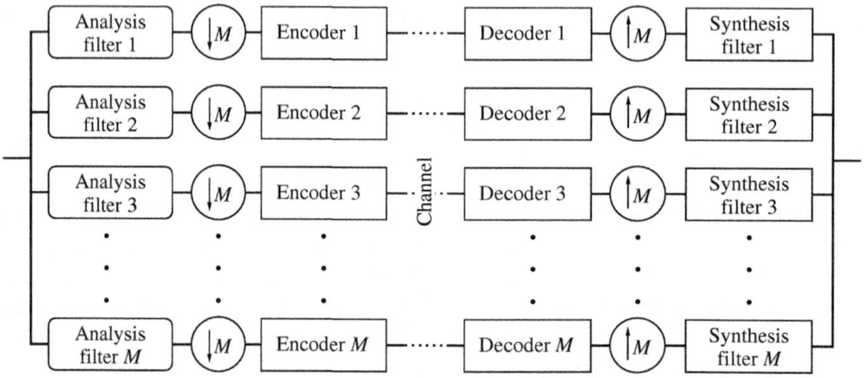

FIGURE 14.7 **Block diagram of the subband coding system.**

Once the output of the filters has been decimated, the output is encoded using one of several encoding schemes, including ADPCM, PCM, and vector quantization.

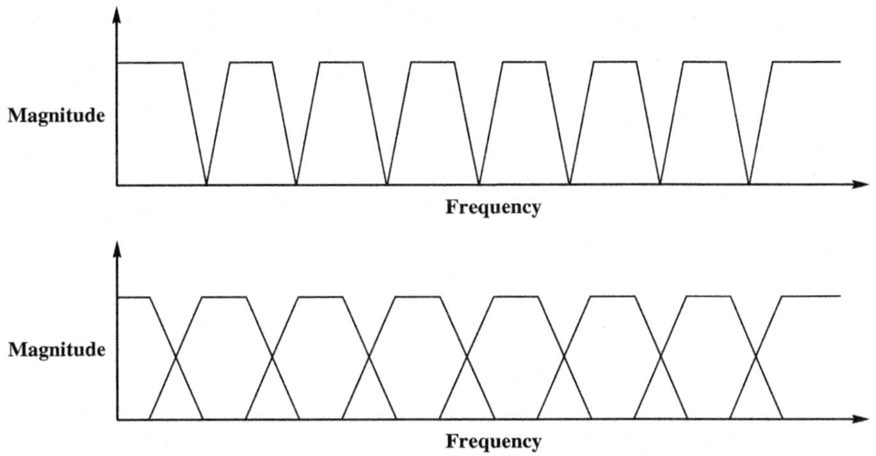

FIGURE 14.8 Nonoverlapping and overlapping filter banks.

14.4.2 Quantization and Coding

Along with the selection of the compression scheme, the allocation of bits between the subbands is an important design parameter. Different subbands contain differing amounts of information. Therefore, we need to allocate the available bits among the subbands according to some measure of the information content. There are a number of different ways we can distribute the available bits. For example, suppose we are decomposing the source output into four bands and we want a coding rate of 1 bit per sample. We can accomplish this by using 1 bit per sample for each of the four bands (recall that a 1/4 decimation occurs in each band). On the other hand, we can simply discard the output of two of the bands and use 2 bits per sample for the two remaining bands. Or, we can discard the output of three of the four filters and use 4 bits per sample to encode the output of the remaining filter.

This *bit allocation* procedure can have a significant impact on the quality of the final reconstruction, especially when the information content of different bands is very different.

If we use the variance of the output of each filter as a measure of information, and assume that the compression scheme is scalar quantization, we can arrive at several simple bit allocation schemes (see Section 13.5). If we use a slightly more sophisticated model for the outputs of the filters, we can arrive at significantly better bit allocation procedures (see Section 14.9).

14.4.3 Synthesis

The quantized and coded coefficients are used to reconstruct a representation of the original signal at the decoder. First, the encoded samples from each subband are decoded at the receiver. These decoded values are then upsampled by inserting an appropriate number of 0s between samples. Once the number of samples per second has been brought back to the original rate, the upsampled signals are passed through a bank of reconstruction filters. The outputs of the reconstruction filters are added to give the final reconstructed outputs.

We can see that the basic subband system is simple. The three major components of this system are the *analysis and synthesis filters*, the *bit allocation* scheme, and the *encoding* scheme. A substantial amount of research has focused on each of these components. Various filter bank structures have been studied in order to find filters that are simple to implement and provide good separation between the frequency bands. In the next section we briefly look at some of the techniques used in the design of filter banks, but our descriptions are necessarily limited. For a (much) more detailed look, see the excellent book by P.P. Vaidyanathan [206].

The bit allocation procedures have also been extensively studied in the contexts of subband coding, wavelet-based coding, and transform coding. We have already described some bit allocation schemes in Section 13.5, and we describe a different approach in Section 14.9. There are also some bit allocation procedures that have been developed in the context of wavelets, which we describe in the next chapter.

The separation of the source output according to frequency also opens up the possibility for innovative ways to use compression algorithms. The decomposition of the source output in this manner provides inputs for the compression algorithms, each of which has more clearly defined characteristics than the original source output. We can use these characteristics to select separate compression schemes appropriate to each of the different inputs.

Human perception of audio and video inputs is frequency dependent. We can use this fact to design our compression schemes so that the frequency bands that are most important to perception are reconstructed most accurately. Whatever distortion there has to be is introduced in the frequency bands to which humans are least sensitive. We describe some applications to the coding of speech, audio, and images later in this chapter.

Before we proceed to bit allocation procedures and implementations, we provide a more mathematical analysis of the subband coding system. We also look at some approaches to the design of filter banks for subband coding. The analysis relies heavily on the Z-transform concepts introduced in Chapter 11 and will primarily be of interest to readers with an electrical engineering background. The material is not essential to understanding the rest of the chapter; if you are not interested in these details, you should skip these sections and go directly to Section 14.9.

14.5 Design of Filter Banks ★

In this section we will take a closer look at the analysis, downsampling, upsampling, and synthesis operations. Our approach follows that of [207]. We assume familiarity with the Z-transform concepts of Chapter 12. We begin with some notation. Suppose we have a sequence $x_0, x_1, x_2, \ldots$. We can divide this sequence into two subsequences: $x_0, x_2, x_4, \ldots$ and $x_1, x_3, x_5, \ldots$ using the scheme shown in Figure 14.9, where z^{-1} corresponds to a delay of one sample and $\downarrow M$ denotes a subsampling by a factor of M. This subsampling process is called *downsampling* or *decimation*.

The original sequence can be recovered from the two downsampled sequences by inserting 0s between consecutive samples of the subsequences, delaying the top branch by one sample and adding the two together. Adding 0s between consecutive samples is called *upsampling* and is denoted by $\uparrow M$. The reconstruction process is shown in Figure 14.10.

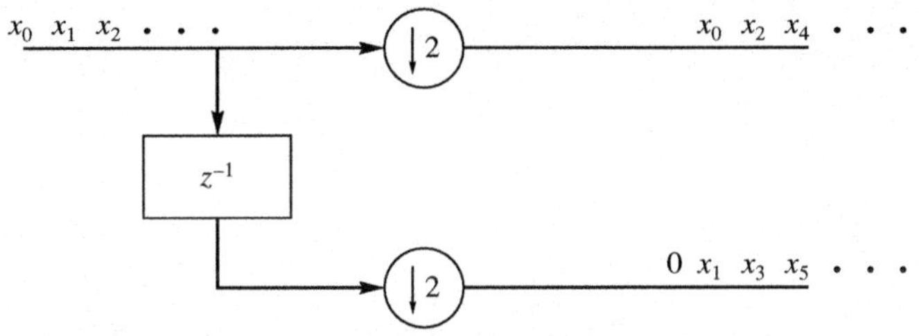

FIGURE 14.9 Decomposition of an input sequence into its odd and even components.

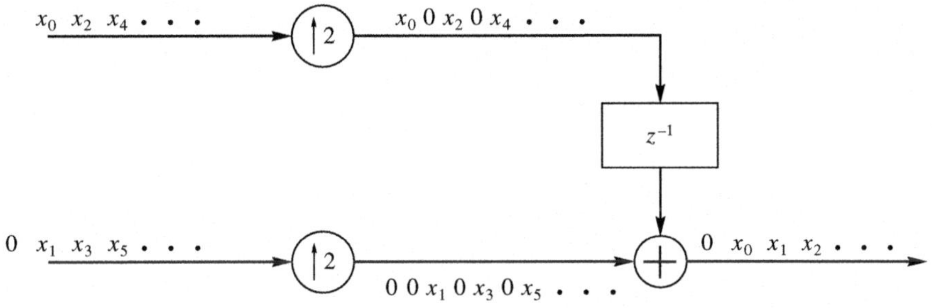

FIGURE 14.10 Reconstructing the input sequence from its odd and even components.

While we have decomposed the source output sequence into two subsequences, there is no reason for the statistical and spectral properties of these subsequences to be different. As our objective is to decompose the source output sequences into subsequences with differing characteristics, there is much more yet to be done.

Generalizing this, we obtain the system shown in Figure 14.11. The source output sequence is fed to an ideal low-pass filter and an ideal high-pass filter, each with a bandwidth of $\pi/2$. We assume that the source output sequence had a bandwidth of π. If the original source signal was sampled at the Nyquist rate, as the output of the two filters have bandwidths half that of the original sequence, the filter outputs are actually oversampled by a factor of two. We can, therefore, subsample these signals by a factor of two without any loss of information. The two bands now have different characteristics and can be encoded differently. For the moment let's assume that the encoding is performed in a lossless manner so that the reconstructed sequence exactly matches the source output sequence.

Let us look at how this system operates in the frequency domain. We begin by looking at the downsampling operation.

14.5.1 Downsampling ★

To see the effects of downsampling, we will obtain the Z-transform of the downsampled sequence in terms of the original source sequence. Because it is easier to understand what is going on if we can visualize the process, we will use the example of a source sequence that has

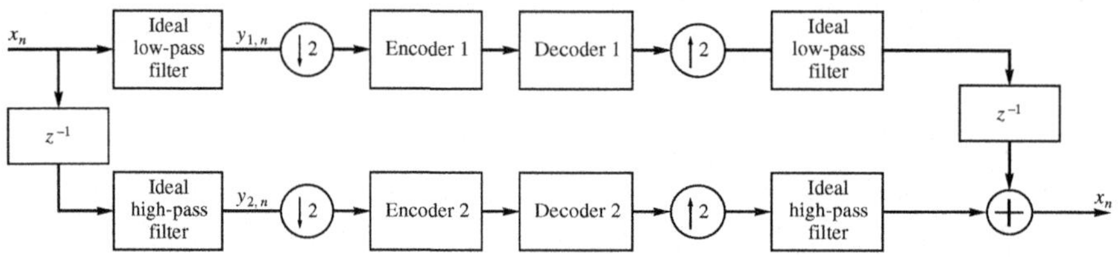

FIGURE 14.11 Decomposition into two bands using ideal filters.

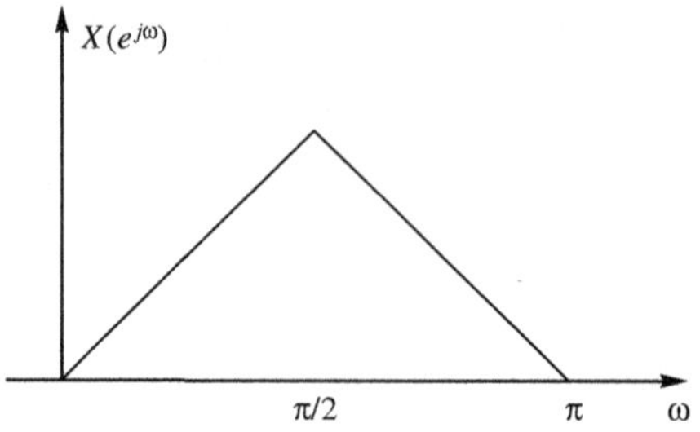

FIGURE 14.12 Spectrum of the source output.

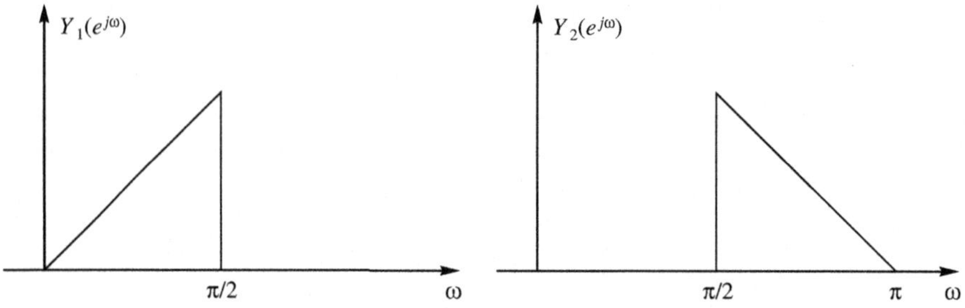

FIGURE 14.13 Spectrum of the outputs of the ideal filters.

the frequency profile shown in Figure 14.12. For this sequence the output of the ideal filters will have the shape shown in Figure 14.13.

Let's represent the downsampled sequence as $\{w_{i,n}\}$. The Z-transform $W_1(z)$ of the downsampled sequence $w_{1,n}$ is

$$W_1(z) = \sum w_{1,n} z^{-n} \tag{24}$$

The downsampling operation means that

$$w_{1,n} = y_{1,2n} \tag{25}$$

In order to find the Z-transform of this sequence, we go through a two-step process. Define the sequence

$$y'_{1,n} = \frac{1}{2}(1 + e^{jn\pi})y_{1,n} \tag{26}$$

$$= \begin{cases} y_{1,n} & n \text{ even} \\ 0 & \text{otherwise} \end{cases} \tag{27}$$

We could also have written Equation (26) as

$$y'_{1,n} = \frac{1}{2}(1 + (-1)^n)y_{1,n}$$

however, writing the relationship as in Equation (26) makes it easier to extend this development to the case where we divide the source output into more than two bands.

The Z-transform of $y'_{1,n}$ is given as

$$Y'_1(z) = \sum_{n=-\infty}^{\infty} \frac{1}{2}(1 + e^{jn\pi})y_{1,n}z^{-n} \tag{28}$$

Assuming all summations converge,

$$Y'_1(z) = \frac{1}{2}\sum_{n=-\infty}^{\infty} y_{1,n}z^{-n} + \frac{1}{2}\sum_{n=-\infty}^{\infty} y_{1,n}(ze^{-j\pi})^{-n} \tag{29}$$

$$= \frac{1}{2}Y_1(z) + \frac{1}{2}Y_1(-z) \tag{30}$$

where we have used the fact that

$$e^{-j\pi} = \cos(\pi) - j\sin\pi = -1$$

Note that

$$w_{1,n} = y'_{1,2n} \tag{31}$$

$$W_1(z) = \sum_{n=-\infty}^{\infty} w_{1,n}z^{-n} = \sum_{-\infty}^{\infty} y'_{1,2n}z^{-n} \tag{32}$$

Substituting $m = 2n$,

$$W_1(z) = \sum_{-\infty}^{\infty} y'_{1,m}z^{\frac{-m}{2}} \tag{33}$$

$$= Y'_1\left(z^{\frac{1}{2}}\right) \tag{34}$$

$$= \frac{1}{2}Y_1\left(z^{\frac{1}{2}}\right) + \frac{1}{2}Y_1\left(-z^{\frac{1}{2}}\right) \tag{35}$$

Why didn't we simply write the Z-transform of $w_{1,n}$ directly in terms of $y_{1,n}$ and use the substitution $m = 2n$? If we had, the equivalent equation to (33) would contain the odd indexed terms of $y_{1,n}$, which we know do not appear at the output of the downsampler. In

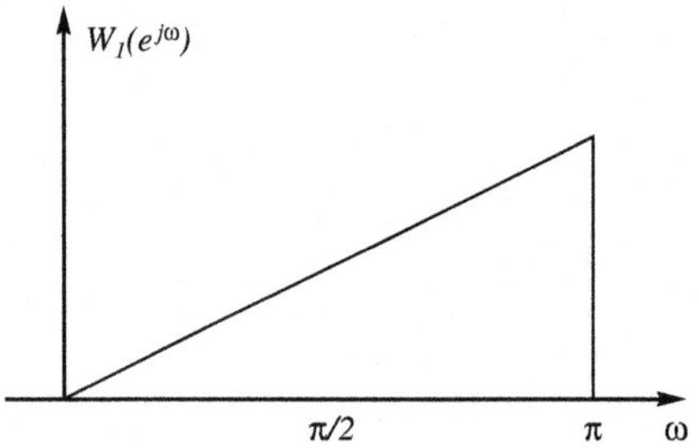

FIGURE 14.14 **Spectrum of the downsampled low-pass filter output.**

Equation (33), we also get the odd indexed terms of $y'_{1,n}$; however, as these terms are all zero (see Equation (26)), they do not contribute to the Z-transform.

Substituting $z = e^{j\omega}$ we get

$$W_1(e^{j\omega}) = \frac{1}{2}Y_1(e^{j\frac{\omega}{2}}) + \frac{1}{2}Y_1(-e^{j\frac{\omega}{2}}) \tag{36}$$

Plotting this for the $Y_1(e^{j\omega})$ of Figure 14.13, we get the spectral shape shown in Figure 14.14; that is, the spectral shape of the downsampled signal is a stretched version of the spectral shape of the original signal. A similar situation exists for the downsampled signal $w_{2,n}$.

14.5.2 Upsampling ★

Let's take a look now at what happens after the upsampling. The upsampled sequence $v_{1,n}$ can be written as

$$v_{1,n} = \begin{cases} w_{1,\frac{n}{2}} & n \text{ even} \\ 0 & n \text{ odd} \end{cases} \tag{37}$$

The Z-transform $V_1(z)$ is thus

$$\begin{aligned} V_1(z) &= \sum_{n=-\infty}^{\infty} v_{1,n} z^{-n} && (38) \\ &= \sum_{n=-\infty}^{\infty} w_{1,\frac{n}{2}} z^{-n} \quad n \text{ even} && (39) \\ &= \sum_{m=-\infty}^{\infty} w_{1,m} z^{-2m} && (40) \\ &= W_1(z^2) && (41) \end{aligned}$$

The spectrum is sketched in Figure 14.15. The "stretching" of the sequence in the time domain has led to a compression in the frequency domain. This compression has also resulted in a replication of the spectrum in the $[0, \pi]$ interval. This replication effect is called *imaging*. We remove the images by using an ideal low-pass filter in the top branch and an ideal high-pass filter in the bottom branch.

Because the use of the filters prior to sampling reduces the bandwidth, which in turn allows the downsampling operation to proceed without aliasing, these filters are called *anti-aliasing*

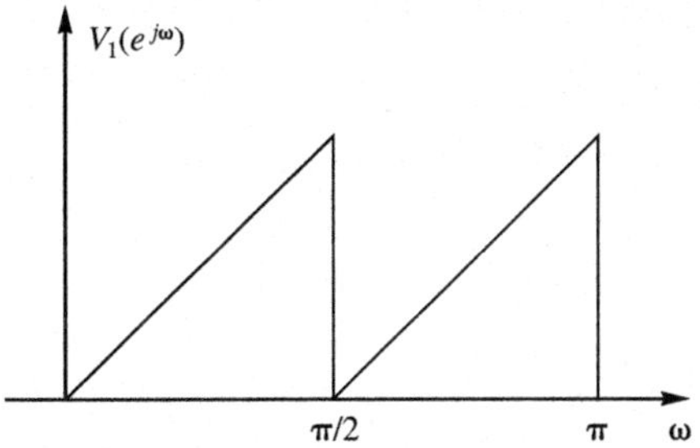

FIGURE 14.15 **Spectrum of the upsampled signal.**

filters. Because they decompose the source output into components, they are also called *analysis* filters. The filters after the upsampling operation are used to recompose the original signal; therefore, they are called *synthesis* filters. We can also view these filters as interpolating between nonzero values to recover the signal at the point that we have inserted zeros. Therefore, these filters are also called *interpolation* filters.

Although the use of ideal filters would give us perfect reconstruction of the source output, in practice we do not have ideal filters available. When we use more realistic filters in place of the ideal filters, we end up introducing distortion. In the next section we look at this situation and discuss how we can reduce or remove this distortion.

14.6 Perfect Reconstruction Using Two-Channel Filter Banks ★

Suppose we replace the ideal low-pass filter in Figure 14.11 with a more realistic filter with the magnitude response shown in Figure 14.4. The spectrum of the output of the low-pass filter is shown in Figure 14.16. Notice that we now have nonzero values for frequencies above $\frac{\pi}{2}$. If we now subsample by two, we will end up sampling at *less* than twice the highest frequency, or in other words, we will be sampling at below the Nyquist rate. This will result in the introduction of aliasing distortion, which will show up in the reconstruction. A similar situation will occur when we replace the ideal high-pass filter with a realistic high-pass filter.

In order to get perfect reconstruction after synthesis, we need to somehow get rid of the aliasing and imaging effects. Let us look at the conditions we need to impose upon the filters $H_1(z)$, $H_2(z)$, $K_1(z)$, and $K_2(z)$ in order to accomplish this. These conditions are called *perfect reconstruction* (PR) conditions.

Consider Figure 14.17. Let's obtain an expression for $\hat{X}(z)$ in terms of $H_1(z)$, $H_2(z)$, $K_1(z)$, and $K_2(z)$. We start with the reconstruction:

$$\hat{X}(z) = U_1(z) + U_2(z) \tag{42}$$

$$= V_1(z)K_1(z) + V_2(z)K_2(z) \tag{43}$$

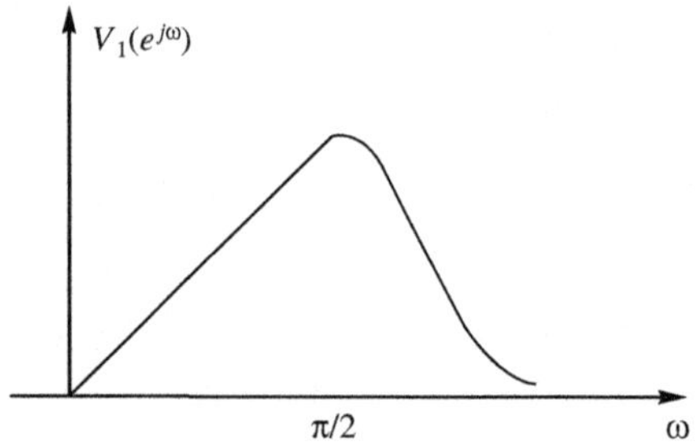

FIGURE 14.16 Output of the low-pass filter.

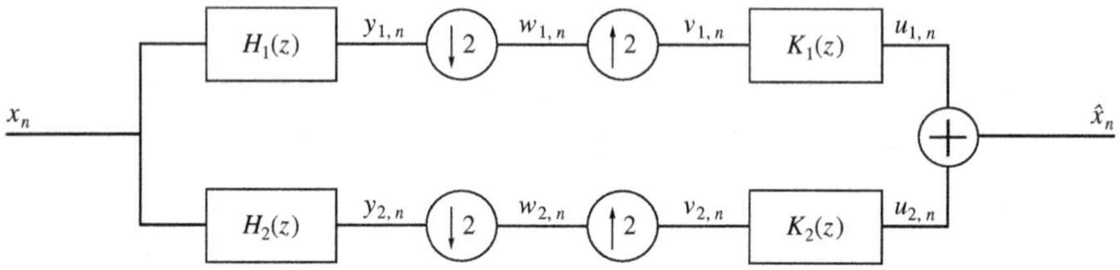

FIGURE 14.17 Two-channel subband decimation and interpolation.

Therefore, we need to find $V_1(z)$ and $V_2(z)$. The sequence $v_{1,n}$ is obtained by upsampling $w_{1,n}$. Therefore, from Equation (41),

$$V_1(z) = W_1(z^2) \tag{44}$$

The sequence $w_{1,n}$ is obtained by downsampling $y_{1,n}$,

$$Y_1(z) = X(z)H_1(z)$$

Therefore, from Equation (35),

$$W_1(z) = \frac{1}{2}\left[X(z^{\frac{1}{2}})H_1(z^{\frac{1}{2}}) + X(-z^{\frac{1}{2}})H_1(-z^{\frac{1}{2}})\right] \tag{45}$$

and

$$V_1(z) = \frac{1}{2}\left[X(z)H_1(z) + X(-z)H_1(-z)\right] \tag{46}$$

Similarly, we can also show that

$$V_2(z) = \frac{1}{2}\left[X(z)H_2(z) + X(-z)H_2(-z)\right] \tag{47}$$

Substituting the expressions for $V_1(z)$ and $V_2(z)$ into Equation (43) we obtain

$$\begin{aligned}\widehat{X}(z) &= \frac{1}{2}\left[H_1(z)K_1(z) + H_2(z)K_2(z)\right]X(z)\\ &\quad+\frac{1}{2}\left[H_1(-z)K_1(z) + H_2(-z)K_2(z)\right]X(-z)\end{aligned} \tag{48}$$

For perfect reconstruction we would like $\widehat{X}(z)$ to be a delayed and perhaps amplitude-scaled version of $X(z)$; that is,

$$\widehat{X}(z) = cX(z)z^{-n_0} \tag{49}$$

In order for this to be true, we need to impose conditions on $H_1(z)$, $H_2(z)$, $K_1(z)$, and $K_2(z)$. There are several ways we can do this, with each approach providing a different solution. One approach involves writing Equation (48) in matrix form as

$$\widehat{X}(z) = \frac{1}{2}\begin{bmatrix} K_1(z) & K_2(z) \end{bmatrix}\begin{bmatrix} H_1(z) & H_1(-z) \\ H_2(z) & H_2(-z) \end{bmatrix}\begin{bmatrix} X(z) \\ X(-z) \end{bmatrix} \tag{50}$$

For perfect reconstruction, we need

$$\begin{bmatrix} K_1(z) & K_2(z) \end{bmatrix}\begin{bmatrix} H_1(z) & H_1(-z) \\ H_2(z) & H_2(-z) \end{bmatrix} = \begin{bmatrix} cz^{-n_0} & 0 \end{bmatrix} \tag{51}$$

where we have absorbed the factor of $\frac{1}{2}$ into the constant c. This means that the synthesis filters $K_1(z)$ and $K_2(z)$ satisfy

$$\begin{bmatrix} K_1(z) & K_2(z) \end{bmatrix} = \frac{cz^{-n_0}}{\det[\mathcal{H}(z)]}\begin{bmatrix} H_2(-z) & -H_1(-z) \end{bmatrix} \tag{52}$$

where

$$\mathcal{H}(z) = \begin{bmatrix} H_1(z) & H_1(-z) \\ H_2(z) & H_2(-z) \end{bmatrix} \tag{53}$$

If $H_1(z)$ and/or $H_2(z)$ are IIR filters, the reconstruction filters can become quite complex. Therefore, we would like to have both the analysis and synthesis filters be FIR filters. If we select the analysis filters to be FIR, then in order to guarantee that the synthesis filters are also FIR we need

$$\det[\mathcal{H}(z)] = \gamma z^{-n_1}$$

where γ is a constant. Examining $\det[\mathcal{H}(z)]$,

$$\begin{aligned} \det[\mathcal{H}(z)] &= H_1(z)H_2(-z) - H_1(-z)H_2(z) \\ &= P(z) - P(-z) = \gamma z^{-n_1} \end{aligned} \tag{54}$$

where $P(z) = H_1(z)H_2(-z)$. If we examine Equation (54), we can see that n_1 has to be odd because all terms containing even powers of z in $P(z)$ will be canceled out by the corresponding terms in $P(-z)$. Thus, $P(z)$ can have an arbitrary number of even-indexed coefficients (as they will get canceled out), but there must be only one nonzero coefficient of an odd power of z. By choosing any valid factorization of the form

$$P(z) = P_1(z)P_2(z) \tag{55}$$

we can obtain many possible solutions of perfect reconstruction FIR filter banks with

$$H_1(z) = P_1(z) \tag{56}$$

and

$$H_2(z) = P_2(-z) \tag{57}$$

Although these filters are perfect reconstruction filters, for applications in data compression they suffer from one significant drawback. Because these filters may be of unequal bandwidth, the output of the larger bandwidth filter suffers from severe aliasing. If the output of both bands is available to the receiver, this is not a problem because the aliasing is canceled out in the reconstruction process. However, in many compression applications we discard the subband containing the least amount of energy, which will generally be the output of the filter with the smaller bandwidth. In this case the reconstruction will contain a large amount of aliasing distortion. In order to avoid this problem for compression applications, we generally wish to minimize the amount of aliasing in each subband. *Quadrature mirror filters* (QMF) are a class of filters that are useful in this situation. We look at these filters in the next section.

14.6.1 Two-Channel PR Quadrature Mirror Filters ★

Before we introduce quadrature mirror filters, let's rewrite Equation (48) as

$$\hat{X}(z) = T(z)X(z) + S(z)X(-z) \tag{58}$$

where

$$T(z) = \frac{1}{2}[H_1(z)K_1(z) + H_2(z)K_2(z)] \tag{59}$$

$$S(z) = \frac{1}{2}[H_1(-z)K_1(z) + H_2(-z)K_2(z)] \tag{60}$$

In order for the reconstruction of the input sequence $\{x_n\}$ to be a delayed, and perhaps scaled, version of $\{x_n\}$, we need to get rid of the aliasing term $X(-z)$ and have $T(z)$ be a pure delay. To get rid of the aliasing term, we need

$$S(z) = 0, \qquad \forall z$$

From Equation (60), this will happen if

$$K_1(z) = H_2(-z) \tag{61}$$

$$K_2(z) = -H_1(-z) \tag{62}$$

After removing the aliasing distortion, a delayed version of the input will be available at the output if

$$T(z) = cz^{-n_0} \qquad c \text{ is a constant} \tag{63}$$

Replacing z by $e^{j\omega}$, this means that we want

$$\left|T(e^{j\omega})\right| = \text{constant} \tag{64}$$

$$\arg(T(e^{j\omega})) = Kw \qquad K \text{ constant} \tag{65}$$

The first requirement eliminates amplitude distortion, while the second, the linear phase requirement, is necessary to eliminate phase distortion. If these requirements are satisfied,

$$\hat{x}(n) = cx(n - n_0) \tag{66}$$

That is, the reconstructed signal is a delayed version of input signal $x(n)$. However, meeting both requirements simultaneously is not a trivial task.

Consider the problem of designing $T(z)$ to have linear phase. Substituting (61) and (62) into Equation (59), we obtain

$$T(z) = \frac{1}{2}[H_1(z)H_2(-z) - H_1(-z)H_2(z)] \tag{67}$$

Therefore, if we choose $H_1(z)$ and $H_2(z)$ to be linear phase FIR, $T(z)$ will also be a linear phase FIR filter. In the QMF approach, we first select the low-pass filter $H_1(z)$, then define the high-pass filter $H_2(z)$ to be a mirror image of the low-pass filter:

$$H_2(z) = H_1(-z) \tag{68}$$

This is referred to as a *mirror* condition and is the original reason for the term quadrature mirror filter [206]. We can see that this condition will force both filters to have equal bandwidth.

Given the mirror condition and $H_1(z)$, a linear phase FIR filter, we will have linear phase and

$$T(z) = \frac{1}{2}[H_1^2(z) - H_1^2(-z)] \tag{69}$$

It is not clear that $|T(e^{j\omega})|$ is a constant. In fact, we will show in Section 14.8 that a linear phase two-channel FIR QMF bank with the filters chosen as in Equation (68) can have the PR property if and only if $H_1(z)$ is in the simple two-tap form

$$H_1(z) = h_0 z^{-2k_0} + h_1 z^{-(2k_1+1)} \tag{70}$$

Then $T(z)$ is given by

$$T(z) = 2h_0h_1z^{-(2k_0+2k_1+1)} \tag{71}$$

which is of the desired form cz^{-n_0}. However, if we look at the magnitude characteristics of the two filters, we see that they have poor cutoff characteristics. The magnitude of the low-pass filter is given by

$$\left|H_1(e^{j\omega})\right|^2 = h_0^2 + h_1^2 + 2h_0h_1\cos(2k_0 - 2k_1 - 1)\omega \tag{72}$$

and the high-pass filter is given by

$$\left|H_2(e^{j\omega})\right|^2 = h_0^2 + h_1^2 - 2h_0h_1\cos(2k_0 - 2k_1 - 1)\omega \tag{73}$$

For $h_0 = h_1 = k_0 = k_1 = 1$, the magnitude responses are plotted in Figure 14.18. Notice the poor cutoff characteristics of these two filters.

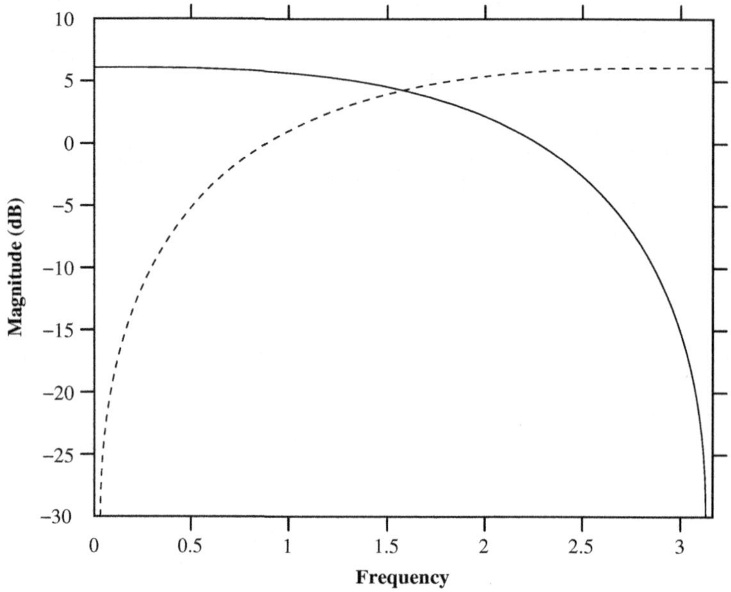

FIGURE 14.18 **Magnitude characteristics of the two-tap PR filters.**

Thus, for perfect reconstruction with no aliasing and no amplitude or phase distortion, the mirror condition does not seem like such a good idea. However, if we slightly relax these rather strict conditions, we can obtain some very nice designs. For example, instead of attempting to eliminate all phase and amplitude distortion, we could elect to eliminate only the phase distortion and *minimize* the amplitude distortion. We can optimize the coefficients of $H_1(z)$ such that $\left|T(e^{j\omega})\right|$ is made as close to a constant as possible, while minimizing the stopband energy of $H_1(z)$ in order to have a good low-pass characteristic. Such an optimization has been suggested by Johnston [204] and Jain and Crochiere [208]. They construct the objective function

$$J = \alpha \int_{\omega_s}^{\pi} \left|H_1(e^{j\omega})\right|^2 d\omega + (1-\alpha)\int_0^{\pi} (1 - \left|T(e^{j\omega})\right|^2) d\omega \tag{74}$$

which has to be minimized to obtain $H_1(z)$ and $T_1(z)$, where ω_s is the cutoff frequency of the filter.

We can also go the other way and eliminate the amplitude distortion, then attempt to minimize the phase distortion. A review of these approaches can be found in [207,206].

14.6.2 Power Symmetric FIR Filters ★

Another approach, independently discovered by Smith and Barnwell [205] and Mintzer [209], can be used to design a two-channel filter bank in which aliasing, amplitude distortion, and phase distortion can be completely eliminated. As discussed earlier, choosing

$$K_1(z) = -H_2(-z)$$
$$K_2(z) = H_1(-z) \tag{75}$$

eliminates aliasing. This leaves us with

$$T(z) = \frac{1}{2}[H_1(-z)H_2(z) - H_1(z)H_2(-z)]$$

In the approach due to Smith and Barnwell [205] and Mintzer [209], with N an odd integer, we select

$$H_2(z) = z^{-N} H_1(-z^{-1}) \tag{76}$$

so that

$$T(z) = \frac{1}{2} z^{-N} [H_1(z)H_1(z^{-1}) + H_1(-z)H_1(-z^{-1})] \tag{77}$$

Therefore, the perfect reconstruction requirement reduces to finding a prototype low-pass filter $H(z) = H_1(z)$ such that

$$Q(z) = H(z)H(z^{-1}) + H(-z)H(-z^{-1}) = \text{constant} \tag{78}$$

Defining

$$R(z) = H(z)H(z^{-1}) \tag{79}$$

the perfect reconstruction requirement becomes

$$Q(z) = R(z) + R(-z) = \text{constant} \tag{80}$$

But $R(z)$ is simply the Z-transform of the autocorrelation sequence of $h(n)$. The autocorrelation sequence $\rho(n)$ is given by

$$\rho(n) = \sum_{k=0}^{N} h_k h_{k+n} \tag{81}$$

The Z-transform of $\rho(n)$ is given by

$$R(z) = \mathcal{Z}[\rho(n)] = \mathcal{Z}\left[\sum_{k=0}^{N} h_k h_{k+n}\right] \tag{82}$$

We can express the sum $\sum_{k=0}^{N} h_k h_{k+n}$ as a convolution:

$$h_n \otimes h_{-n} = \sum_{k=0}^{N} h_k h_{k+n} \tag{83}$$

Using the fact that the Z-transform of a convolution of two sequences is the product of the Z-transforms of the individual sequences, we obtain

$$R(z) = \mathcal{Z}[h_n]\mathcal{Z}[h_{-n}] = H(z)H(z^{-1}) \tag{84}$$

Writing out $R(z)$ as the Z-transform of the sequence $\{\rho(n)\}$ we obtain

$$R(z) = \rho(N)z^N + \rho(N-1)z^{N-1} + \cdots + \rho(0) + \cdots + \rho(N-1)z^{-N-1} + \rho(N)z^{-N} \quad (85)$$

Then $R(-z)$ is

$$R(-z) = -\rho(N)z^N + \rho(N-1)z^{N-1} - \cdots + \rho(0) - \cdots + \rho(N-1)z^{-N-1} - \rho(N)z^{-N} \quad (86)$$

Adding $R(z)$ and $R(-z)$, we obtain $Q(z)$ as

$$Q(z) = 2\rho(N-1)z^{N-1} + 2\rho(N-3)z^{N-3} + \cdots + \rho(0) + \cdots + 2\rho(N-1)z^{-N-1} \quad (87)$$

Notice that the terms containing the odd powers of z got canceled out. Thus, for $Q(z)$ to be a constant all we need is that for even values of the lag n (except for $n = 0$), $\rho(n)$ be zero. In other words

$$\rho(2n) = \sum_{k=0}^{N} h_k h_{k+2n} = 0, \qquad n \neq 0 \quad (88)$$

Writing this requirement in terms of the impulse response,

$$\sum_{k=0}^{N} h_k h_{k+2n} = \begin{cases} 0 & n \neq 0 \\ \rho(0) & n = 0 \end{cases} \quad (89)$$

If we now normalize the impulse response,

$$\sum_{k=0}^{N} |h_k|^2 = 1 \quad (90)$$

we obtain the perfect reconstruction requirement

$$\sum_{k=0}^{N} h_k h_{k+2n} = \delta_n \quad (91)$$

In other words, for perfect reconstruction, the impulse response of the prototype filter is orthogonal to the twice-shifted version of itself.

14.7 M-Band Quadrature Mirror Filter Banks ★

We have looked at how we can decompose an input signal into two bands. In many applications it is necessary to divide the input into multiple bands. We can do this by using a recursive two-band splitting as shown in Figure 14.19, or we can obtain banks of filters that directly split the input into multiple bands. Given that we have good filters that provide two-band splitting, it would seem that using a recursive splitting, as shown in Figure 14.19, would be an efficient way of obtaining an M-band split. Unfortunately, even when the spectral characteristics of the

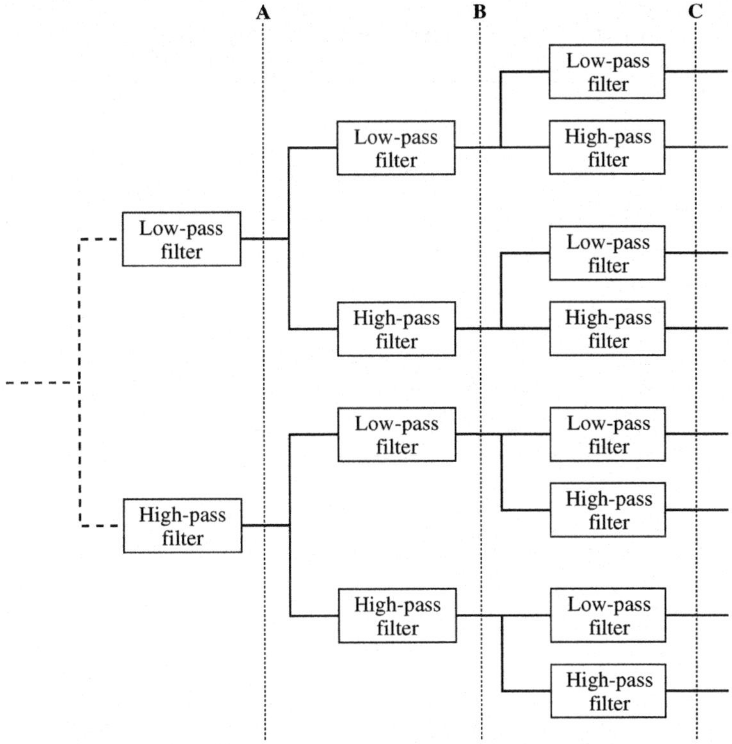

FIGURE 14.19 **Decomposition of an input sequence into multiple bands by recursively using a two-band split.**

filters used for the two-band split are quite good, when we employ them in the tree structure shown in Figure 14.19, the spectral characteristics may not be very good. For example, consider the four-tap filter with filter coefficients shown in Table 14.6. In Figure 14.20 we show what happens to the spectral characteristics when we look at the two-band split (at point **A** in Figure 14.19), the four-band split (at point **B** in Figure 14.19), and the eight-band split (at point **C** in Figure 14.19). For a two-band split the magnitude characteristic is flat, with some aliasing. When we employ these same filters to obtain a four-band split from the two-band split, there is an increase in the aliasing. When we go one step further to obtain an eight-band split, the magnitude characteristic deteriorates substantially, as evidenced by Figure 14.20. The various bands are no longer clearly distinct. There is significant overlap between the bands, and hence there will be a significant amount of aliasing in each band.

In order to see why there is an increase in distortion, let us follow the top branch of the tree. The path followed by the signal is shown in Figure 14.21(a). As we will show later (Section 14.8), the three filters and downsamplers can be replaced by a single filter and downsampler as shown in Figure 14.21(b), where

$$\mathbf{A}(z) = H_L(z)H_L(z^2)H_L(z^4) \tag{92}$$

TABLE 14.6 Coefficients for the four-tap Daubechies low-pass filter.

h_0	0.4829629131445341
h_1	0.8365163037378079
h_2	0.2241438680420134
h_3	−0.1294095225512604

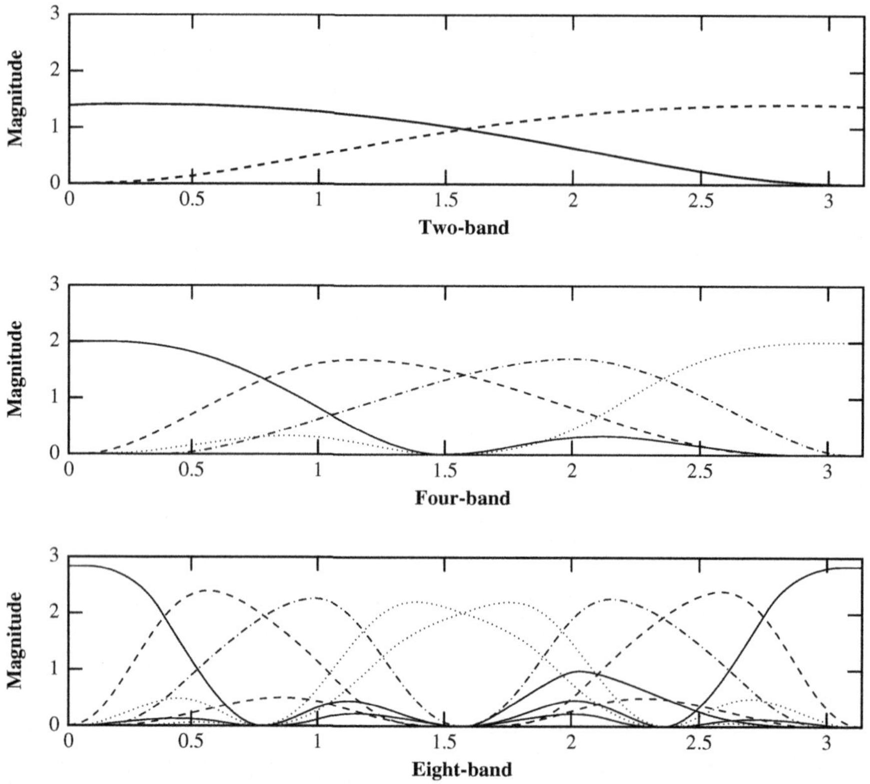

FIGURE 14.20 Spectral characteristics at points A, B, and C.

If $H_L(z)$ corresponds to a 4-tap filter, then $\mathbf{A}(z)$ corresponds to a 22-tap filter! However, this is a severely constrained filter because it was generated using only four coefficients. If we had set out to design a 22-tap filter from scratch, we would have had significantly more freedom in selecting the coefficients. This is a strong motivation for designing filters directly for the M-band case.

An M-band filter bank has two sets of filters that are arranged as shown in Figure 14.7. The input signal $x(n)$ is split into M frequency bands using an analysis bank of M filters of bandwidth π/M. The signal in any of these M channels is then downsampled by a factor L.

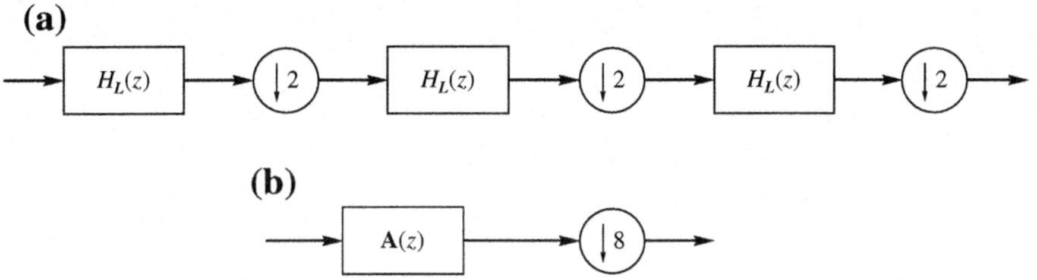

FIGURE 14.21 **Equivalent structures for recursive filtering using a two-band split.**

This constitutes the analysis bank. The subband signals $y_k(n)$ are encoded and transmitted. At the synthesis stage the subband signals are then decoded, upsampled by a factor of L by interlacing adjacent samples with $L - 1$ zeros, and then passed through the synthesis or interpolation filters. The output of all these synthesis filters is added together to obtain the reconstructed signal. This constitutes the synthesis filter bank. Thus, the analysis and synthesis filter banks together take an input signal $x(n)$ and produce an output signal $\hat{x}(n)$. These filters could be any combination of FIR and IIR filters.

Depending on whether M is less than, equal to, or greater than L, the filter bank is called an *underdecimated, critically (maximally) decimated*, or *overdecimated* filter bank. For most practical applications, maximal decimation or "critical subsampling" is used.

A detailed study of M-band filters is beyond the scope of this chapter. Suffice it to say that in broad outline much of what we said about two-band filters can be generalized to M-band filters. (For more on this subject, see [206].)

14.8 The Polyphase Decomposition ★

A major problem with representing the combination of filters and downsamplers is the time-varying nature of the up- and downsamplers. An elegant way of solving this problem is with the use of *polyphase decomposition*. In order to demonstrate this concept, let us first consider the simple case of two-band splitting. We will first consider the analysis portion of the system shown in Figure 14.22. Suppose the analysis filter $H_1(z)$ is given by

$$H_1(z) = h_0 + h_1 z^{-1} + h_2 z^{-2} + h_3 z^{-3} + \cdots \qquad (93)$$

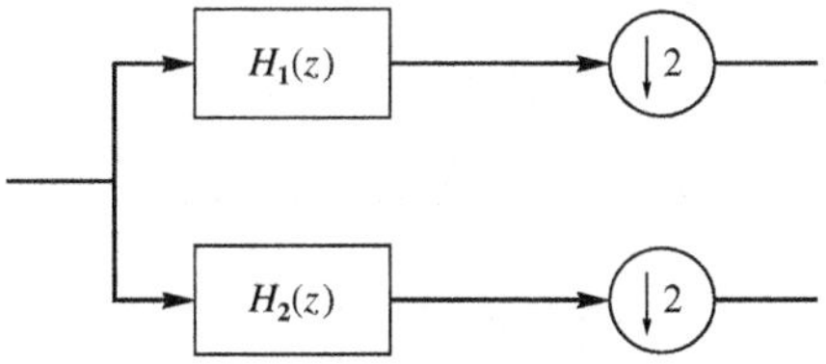

FIGURE 14.22 **Analysis portion of a two-band subband coder.**

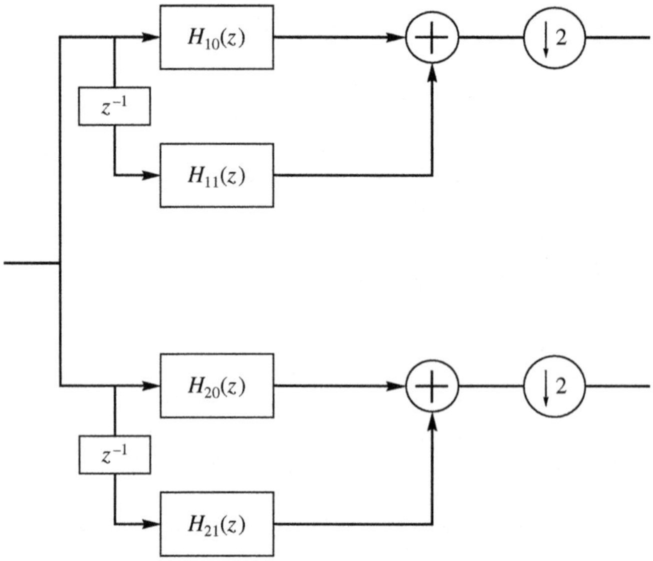

FIGURE 14.23 **Alternative representation of the analysis portion of a two-band subband coder.**

By grouping the odd and even terms together, we can write this as

$$H_1(z) = (h_0 + h_2z^{-2} + h_4z^{-4} + \cdots) + z^{-1}(h_1 + h_3z^{-2} + h_5z^{-4} + \cdots) \quad (94)$$

Define

$$H_{10}(z) = h_0 + h_2z^{-1} + h_4z^{-2} + \cdots \quad (95)$$
$$H_{11}(z) = h_1 + h_3z^{-1} + h_5z^{-2} + \cdots \quad (96)$$

Then $H_1(z) = H_{10}(z^2) + z^{-1}H_{11}(z^2)$. Similarly, we can decompose the filter $H_2(z)$ into components $H_{20}(z)$ and $H_{21}(z)$, and we can represent the system of Figure 14.22 as shown in Figure 14.23. The filters $H_{10}(z)$, $H_{11}(z)$ and $H_{20}(z)$, $H_{21}(z)$ are called the polyphase components of $H_1(z)$ and $H_2(z)$.

Let's take the inverse Z-transform of the polyphase components of $H_1(z)$:

$$h_{10}(n) = h_{2n} \qquad n = 0, 1, \ldots \quad (97)$$
$$h_{11}(n) = h_{2n+1} \qquad n = 0, 1, \ldots \quad (98)$$

Thus, $h_{10}(n)$ and $h_{11}(n)$ are simply the impulse response h_n downsampled by two. Consider the output of the downsampler for a given input $X(z)$. The input to the downsampler is $X(z)H_1(z)$; thus, the output from Equation (35) is

$$Y_1(z) = \frac{1}{2}X\left(z^{\frac{1}{2}}\right)H_1\left(z^{\frac{1}{2}}\right) + \frac{1}{2}X\left(-z^{\frac{1}{2}}\right)H_1\left(-z^{\frac{1}{2}}\right) \quad (99)$$

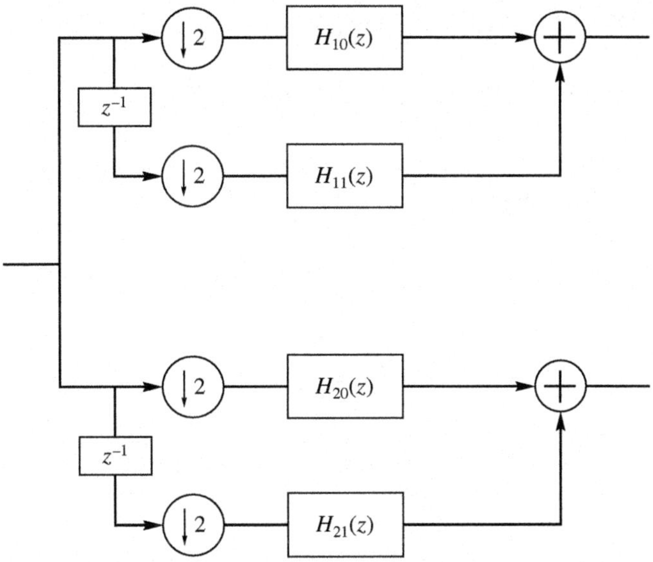

FIGURE 14.24 **Polyphase representation of the analysis portion of a two-band subband coder.**

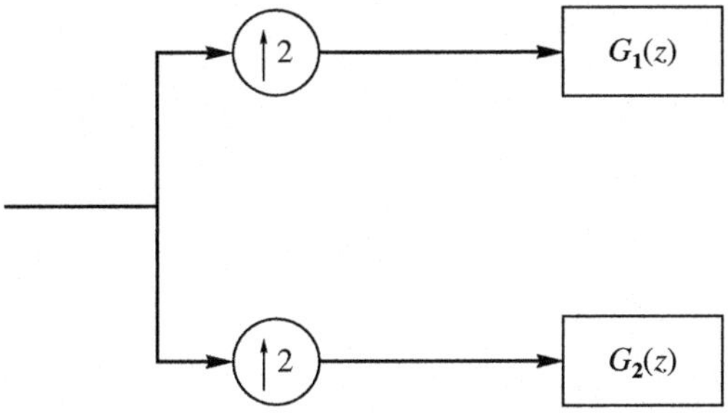

FIGURE 14.25 **The synthesis portion of a two-band subband coder.**

Replacing $H_1(z)$ with its polyphase representation, we get

$$Y_1(z) = \frac{1}{2}X\left(z^{\frac{1}{2}}\right)\left[H_{10}(z) + z^{-\frac{1}{2}}H_{11}(z)\right] + \frac{1}{2}X\left(-z^{\frac{1}{2}}\right)\left[H_{10}(z) - z^{-\frac{1}{2}}H_{11}(z)\right] \tag{100}$$

$$= H_{10}(z)\left[\frac{1}{2}X\left(z^{\frac{1}{2}}\right)+\frac{1}{2}X\left(-z^{\frac{1}{2}}\right)\right]+H_{11}(z)\left[\frac{1}{2}z^{-\frac{1}{2}}X\left(z^{\frac{1}{2}}\right)-\frac{1}{2}z^{-\frac{1}{2}}X\left(-z^{\frac{1}{2}}\right)\right] \tag{101}$$

Note that the first expression in square brackets is the output of a downsampler whose input is $X(z)$, while the quantity in the second set of square brackets is the output of a downsampler whose input is $z^{-1}X(z)$. Therefore, we could implement this system as shown in Figure 14.24.

Now let us consider the synthesis portion of the two-band system shown in Figure 14.25. As in the case of the analysis portion, we can write the transfer functions in terms of their polyphase representation. Thus,

$$G_1(z) = G_{10}(z^2) + z^{-1}G_{11}(z^2) \tag{102}$$
$$G_2(z) = G_{20}(z^2) + z^{-1}G_{21}(z^2) \tag{103}$$

Consider the output of the synthesis filter $G_1(z)$ given an input $Y_1(z)$. From Equation (41), the output of the upsampler is

$$U_1(z) = Y_1(z^2) \tag{104}$$

and the output of $G_1(z)$ is

$$V_1(z) = Y_1(z^2)G_1(z) \tag{105}$$
$$= Y_1(z^2)G_{10}(z^2) + z^{-1}Y_1(z^2)G_{11}(z^2) \tag{106}$$

The first term in the equation above is the output of an upsampler that *follows* a filter with transfer function $G_{10}(z)$ with input $Y(z)$. Similarly, $Y_1(z^2)G_{11}(z^2)$ is the output of an upsampler that follows a filter with transfer function $G_{11}(z)$ with input $Y(z)$. Thus, this system can be represented as shown in Figure 14.26.

Putting the polyphase representations of the analysis and synthesis portions together, we get the system shown in Figure 14.27. Looking at the portion in the dashed box, we can see that this is a completely linear time-invariant system.

The polyphase representation can be a very useful tool for the design and analysis of filters. While many of its uses are beyond the scope of this chapter, we can use this representation to prove our statement about the two-band perfect reconstruction quadrature mirror filters.

Recall that we want

$$T(z) = \frac{1}{2}[H_1(z)H_2(-z) - H_1(-z)H_2(z)] = cz^{-n_0}$$

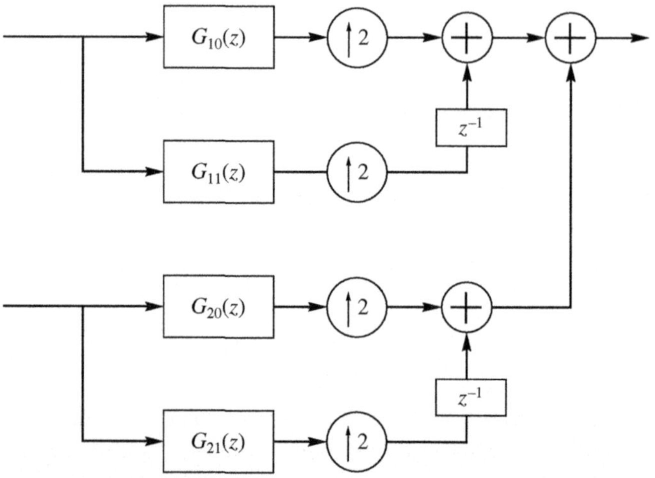

FIGURE 14.26 **Polyphase representation of the synthesis portion of a two-band subband coder.**

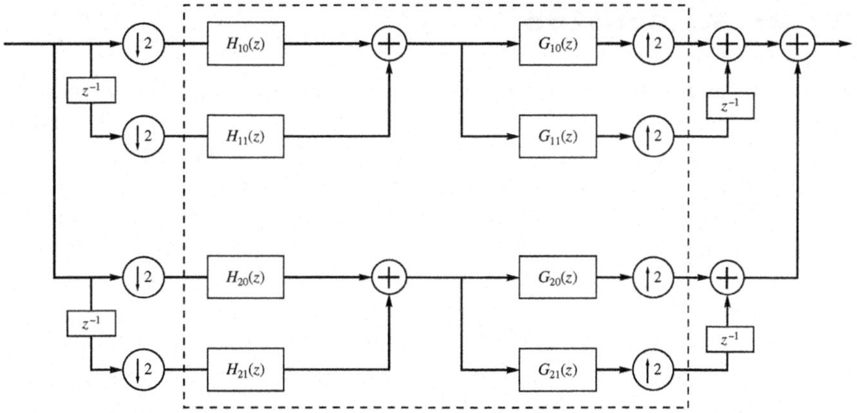

FIGURE 14.27 **Polyphase representation of the two-band subband coder.**

If we impose the mirror condition $H_2(z) = H_1(-z)$, $T(z)$ becomes

$$T(z) = \frac{1}{2}\left[H_1^2(z) - H_1^2(-z)\right] \tag{107}$$

The polyphase decomposition of $H_1(z)$ is

$$H_1(z) = H_{10}(z^2) + z^{-1}H_{11}(z^2)$$

Substituting this into Equation (107) for $H_1(z)$ and

$$H_1(-z) = H_{10}(z^2) - z^{-1}H_{11}(z^2)$$

for $H_1(-z)$, we obtain

$$T(z) = 2z^{-1}H_{10}(z^2)H_{11}(z^2) \tag{108}$$

Clearly, the only way $T(z)$ can have the form cz^{-n_0} is if both $H_{10}(z)$ and $H_{11}(z)$ are simple delays; that is,

$$H_{10}(z) = h_0 z^{-k_0} \tag{109}$$

$$H_{11}(z) = h_1 z^{-k_1} \tag{110}$$

This results in

$$T(z) = 2h_0h_1z^{-(2k_0+2k_1+1)} \tag{111}$$

which is of the form cz^{-n_0} as desired. The resulting filters have the transfer functions

$$H_1(z) = h_0 z^{-2k_0} + h_1 z^{-(2k_1+1)} \tag{112}$$

$$H_2(z) = h_0 z^{-2k_0} - h_1 z^{-(2k_1+1)} \tag{113}$$

14.9 Bit Allocation

Once we have separated the source output into the constituent sequences, we need to decide how much of the coding resource should be used to encode the output of each synthesis filter. In other words, we need to allocate the available bits between the subband sequences. In the previous chapter we described a bit allocation procedure that uses the variances of the transform coefficient. In this section we describe a bit allocation approach that attempts to use as much information about the subbands as possible to distribute the bits.

Let's begin with some notation. We have a total of B_T bits that we need to distribute among M subbands. Suppose R corresponds to the average rate in bits per sample for the overall system, and R_k is the average rate for subband k. Let's begin with the case where the input is decomposed into M equal bands, each of which is decimated by a factor of M. Finally, let's assume that we know the rate distortion function for each band. (If you recall from Chapter 8, this is a rather strong assumption and we will relax it shortly.) We also assume that the distortion measure is such that the total distortion is the sum of the distortion contribution of each band.

We want to find the bit allocation R_k such that

$$R = \frac{1}{M}\sum_{k=1}^{M} R_k \tag{114}$$

and the reconstruction error is minimized. Each value of R_k corresponds to a point on the rate distortion curve. The question is where should we operate on the rate distortion curve for each subband to minimize the average distortion. There is a trade-off between rate and distortion. If we decrease the rate (that is, move down the rate distortion curve), we will increase the distortion. Similarly, if we want to move to the left on the rate distortion curve and minimize the distortion, we end up increasing the rate. We need a formulation that incorporates both rate and distortion and the trade-off involved. The formulation we use is based on a landmark paper in 1988 by Yaacov Shoham and Allen Gersho [197]. Let's define a functional J_k:

$$J_k = D_k + \lambda R_k \tag{115}$$

where D_k is the distortion contribution from the kth subband and λ is a Lagrangian parameter. This is the quantity we wish to minimize. In this expression the parameter λ in some sense specifies the trade-off. If we are primarily interested in minimizing the distortion, we can set λ to a small value. If our primary interest is in minimizing the rate, we keep the value of λ large. We can show that the values of D_k and R_k that minimize J_k occur where the slope of the rate distortion curve is λ. Thus, given a value of λ and the rate distortion function, we can immediately identify the values of R_k and D_k. So what should the value of λ be, and how should it vary between subbands?

Let's take the second question first. We would like to allocate bits in such a way that any increase in any of the rates will have the same impact on the distortion. This will happen when we pick R_k in such a way that the slopes of the rate distortion functions for the different subbands are the same; that is, we want to use the same λ for each subband. Let's see what

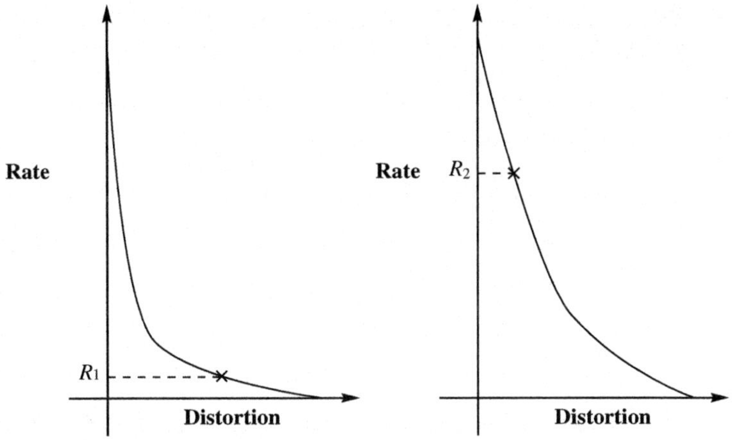

FIGURE 14.28 Two rate distortion functions.

happens if we do not. Consider the two rate distortion functions shown in Figure 14.28. Suppose the points marked **x** on the rate distortion functions correspond to the selected rates. Obviously, the slopes, and hence the values of λ, are different in the two cases. Because of the differences in the slope, an increase by ΔR in the rate R_1 will result in a much larger decrease in the distortion than the increase in distortion if we decreased R_2 by ΔR. Because the total distortion is the sum of the individual distortions, we can therefore reduce the overall distortions by increasing R_1 and decreasing R_2. We will be able to keep doing this until the slope corresponding to the rates is the same in both cases. Thus, the answer to our second question is that we want to use the same value of λ for all the subbands.

Given a set of rate distortion functions and a value of λ, we automatically get a set of rates R_k. We can then compute the average and check if it satisfies our constraint on the total number of bits we can spend. If it does not, we modify the value of λ until we get a set of rates that satisfies our rate constraint.

However, generally we do not have rate distortion functions available. In these cases we use whatever is available. For some cases we might have *operational* rate distortion curves available. By "operational" we mean performance curves for particular types of encoders operating on specific types of sources. For example, if we know we are going to be using *pdf*-optimized nonuniform quantizers with entropy coding, we can estimate the distribution of the subband and use the performance curve for *pdf*-optimized nonuniform quantizers for that distribution. We might only have the performance of the particular encoding scheme for a limited number of rates. In this case we need to have some way of obtaining the slope from a few points. We could estimate this numerically from these points. Or we could fit the points to a curve and estimate the slope from the curve. In these cases we might not be able to get exactly the average rate we wanted.

Finally, we have been talking about a situation where the number of samples in each subband is exactly the same, and therefore the total rate is simply the sum of the individual rates. If this is not true, we need to weight the rates of the individual subbands. The functional to be minimized becomes

$$J = \sum D_k + \lambda \sum \beta_k R_k \tag{116}$$

where β_k is the weight reflecting the relative length of the sequence generated by the kth filter. The distortion contribution from each subband might not be equally relevant, perhaps because of the filter construction or because of the perceptual weight attached to those frequencies [210]. In these cases we can modify our functional still further to include the unequal weighting of the distortion:

$$J = \sum w_k D_k + \lambda \sum \beta_k R_k \tag{117}$$

14.10 Application to Speech Coding—G.722

ITU-T recommendation G.722 provides a technique for wideband coding of speech signals that is based on subband coding. The basic objective of this recommendation is to provide high-quality speech at 64 kbits per second (kbps). The recommendation also contains two other modes that encode the input at 56 and 48 kbps. These two modes are used when an auxiliary channel is needed. These two modes provide for auxiliary channels of 8 and 16 kbps, respectively.

The speech output or audio signal is filtered to 7 kHz to prevent aliasing, then sampled at 16,000 samples per second. Notice that the cutoff frequency for the anti-aliasing filter is 7 kHz, not 8 kHz, even though we are sampling at 16,000 samples per second. One reason for this is that the cutoff for the anti-aliasing filter is not going to be sharp like that of the ideal low-pass filter. Therefore, the highest frequency component in the filter output will be greater than 7 kHz. Each sample is encoded using a 14-bit uniform quantizer. This 14-bit input is passed through a bank of two 24-coefficient FIR filters. The coefficients of the low-pass quadrature mirror filter are shown in Table 14.7.

The coefficients for the high-pass quadrature mirror filter can be obtained by the relationship

$$h_{HP,n} = (-1)^n h_{LP,n} \tag{118}$$

The low-pass filter passes all frequency components in the range of 0 to 4 kHz, while the high-pass filter passes all remaining frequencies. The output of the filters is downsampled by a factor of two. The downsampled sequences are encoded using adaptive differential PCM (ADPCM) systems.

The ADPCM system that encodes the downsampled output of the low-frequency filter uses 6 bits per sample, with the option of dropping 1 or 2 least significant bits in order to provide room for the auxiliary channel. The output of the high-pass filter is encoded using 2 bits per sample. Because the 2 least significant bits of the quantizer output of the low-pass ADPCM system could be dropped and thus not be available to the receiver, the adaptation and prediction at both the transmitter and receiver are performed using only the 4 most significant bits of the quantizer output.

TABLE 14.7 Transmit and receive QMF coefficient values.

h_0, h_{23}	3.66211×10^{-4}
h_1, h_{22}	-1.34277×10^{-3}
h_2, h_{21}	-1.34277×10^{-3}
h_3, h_{20}	6.46973×10^{-3}
h_4, h_{19}	1.46484×10^{-3}
h_5, h_{18}	-1.90430×10^{-2}
h_6, h_{17}	3.90625×10^{-3}
h_7, h_{16}	4.41895×10^{-2}
h_8, h_{15}	-2.56348×10^{-2}
h_9, h_{14}	-9.82666×10^{-2}
h_{10}, h_{13}	1.16089×10^{-1}
h_{11}, h_{12}	4.73145×10^{-1}

If all 6 bits are used in the encoding of the low-frequency subband, we end up with a rate of 48 kbps for the low band. Since the high band is encoded at 2 bits per sample, the output rate for the high subband is 16 kbps. Therefore, the total output rate for the subband-ADPCM system is 64 kbps.

The quantizer is adapted using a variation of the Jayant algorithm [124]. Both ADPCM systems use the past two reconstructed values and the past six quantizer outputs to predict the next sample, in the same way as the predictor for recommendation G.726 described in Chapter 11. The predictor is adapted in the same manner as the predictor used in the G.726 algorithm.

At the receiver, after being decoded by the ADPCM decoder, each output signal is upsampled by the insertion of a zero after each sample. The upsampled signals are passed through the reconstruction filters. These filters are identical to the filters used for decomposing the signal. The low-pass reconstruction filter coefficients are given in Table 14.7, and the coefficients for the high-pass filter can be obtained using Equation (118).

14.11 Application to Audio Coding—MPEG Audio

The Moving Picture Experts Group (MPEG) has proposed an audio coding scheme that is based in part on subband coding. Actually, MPEG has proposed three coding schemes, called Layer I, Layer II, and Layer III coding. Each is more complex than the previous and provides higher compression. The coders are also "upward" compatible; a Layer N decoder is able to decode the bitstream generated by the Layer $N-1$ encoder.

The Layer I and Layer II coders both use a bank of 32 filters, splitting the input into 32 bands, each with a bandwidth of $f_s/64$, where f_s is the sampling frequency. Allowable sampling frequencies are 32,000 samples per second, 44,100 samples per second, and 48,000 samples per second. Details of these coders are provided in Chapter 17.

14.12 Application to Image Compression

We have discussed how to separate a sequence into its components. However, all the examples we have used are one-dimensional sequences. What do we do when the sequences contain two-dimensional dependencies such as images? The obvious answer is that we need two-dimensional filters that separate the source output into components based on both the horizontal and vertical frequencies. Fortunately, in most cases, this two-dimensional filter can be implemented as two one-dimensional filters, which can be applied first in one dimension, then in the other. Filters that have this property are called *separable* filters. Two-dimensional nonseparable filters do exist [211]; however, the gains are offset by the increase in complexity.

Generally, for subband coding of images we filter each row of the image separately using a high-pass and low-pass filter. The output of the filters is decimated by a factor of two. Assume that the images were of size $N \times N$. After this first stage, we will have two images of size $N \times \frac{N}{2}$. We then filter each column of the two subimages, decimating the outputs of the filters again by a factor of two. This results in four images of size $\frac{N}{2} \times \frac{N}{2}$. We can stop at this point or continue the decomposition process with one or more of the four subimages, resulting in 7, 10, 13, or 16 images. Generally, of the four original subimages, only one or two are further decomposed. The reason for not decomposing the other subimages is that many of the pixel values in the high-frequency subimages are close to zero. Thus, there is little reason to spend computational power to decompose these subimages.

Example 14.12.1:

Let's take the "image" in Table 14.8 and decompose it using the low-pass and high-pass filters of Example 14.2.1. After filtering each row with the low-pass filter, the output is decimated by a factor of two. Each output from the filter depends on the current input and the past input. For the very first input (that is, the pixels at the left edge of the image), we will assume that the past values of the input were zero. The decimated output of the low-pass and high-pass filters is shown in Table 14.9.

We take each of these subimages and filter them column by column using the low-pass and high-pass filters and decimate the outputs by two. In this case, the first input to the filters is the top element in each row. We assume that there is a zero row of pixels right above this

TABLE 14.8 A sample "image."

10	14	10	12	14	8	14	12
10	12	8	12	10	6	10	12
12	10	8	6	8	10	12	14
8	6	4	6	4	6	8	10
14	12	10	8	6	4	6	8
12	8	12	10	6	6	6	6
12	10	6	6	6	6	6	6
6	6	6	6	6	6	6	6

TABLE 14.9 Filtered and decimated output.

Decimated Low-Pass Output					Decimated High-Pass Output		
5	12	13	11	5	−2	1	3
5	10	11	8	5	−2	−1	2
6	9	7	11	6	−1	1	1
4	5	5	7	4	−1	−1	1
7	11	7	5	7	−1	−1	1
6	10	8	6	6	2	−2	0
6	8	6	6	6	−2	0	0
3	6	6	6	3	0	0	0

TABLE 14.10 Four subimages.

Low-Low Image				Low-High Image			
2.5	6	6.5	5.5	2.5	6	6.5	5.5
5.5	9.5	9	9.5	0.5	−0.5	−2	1.5
5.5	8	6	6	1.5	3	1	−1
6	9	7	6	0	−1	−1	0
High-Low Image				**High-High Image**			
2.5	−1	0.5	1.5	2.5	−1	0.5	1.5
5.5	−1.5	0	1.5	0.5	0.5	1	−0.5
5.5	−1	−1	1	1.5	0	0	0
6	0	−1	0	0	−2	1	0

row in order to provide the filter with "past" values. After filtering and decimation, we get four subimages (Table 14.10). The subimage obtained by low-pass filtering of the columns of the subimage (which was the output of the row low-pass filtering) is called the low-low (LL) image. Similarly, the other images are called the low-high (LH), high-low (HL), and high-high (HH) images. ♦

If we look closely at the final set of subimages in the previous example, we notice that there is a difference in the characteristics of the values in the left or top row and the interiors of some of the subimages. For example, in the high-low subimage, the values in the first column are significantly larger than the other values in the subimage. Similarly, in the low-high subimage, the values in the first row are generally very different than the other values in the subimage. The reason for this variance is our assumption that the "past" of the image above the first row and to the left of the column is zero. The difference between zero and the image values is much larger than the normal pixel-to-pixel differences. Therefore, we end up adding some spurious structure to the image reflected in the subimages. Generally, this is undesirable because it is easier to select appropriate compression schemes when the characteristics of the subimages

TABLE 14.11 Alternate four subimages.

Low-Low Image				Low-High Image			
10	12	13	11	0	0	−0.5	−0.5
11	9.5	9	9.5	1	−0.5	−2	1.5
11	8	6	6	3	3	1	−1
12	9	7	6	0	−1	−1	0
High-Low Image				High-High Image			
0	−2	1	3	0	0	0	0
0	−1.5	0	1.5	0	0.5	1	−0.5
0	−1	−1	1	0	0	0	0
0	0	−1	0	0	−2	1	0

are as uniform as possible. For example, if we did not have the relatively large values in the first column of the high-low subimage, we could choose a quantizer with a smaller step size.

In this example, this effect is limited to a single row or column because the filters use a single past value. However, most filters use a substantially larger number of past values in the filtering operation, and a larger portion of the subimage is affected.

We can avoid this problem by assuming a different "past." There are a number of ways this can be done. A simple method that works well is to reflect the values of the pixels at the boundary. For example, for the sequence 6 9 5 4 7 2 · · ·, which is to be filtered with a three-tap filter, we assume the past as $\boxed{9\ 6}$ 6 9 5 4 7 2 · · ·. If we use this approach for the image in Example 14.12.1, the four subimages would be as shown in Table 14.11.

Notice how much sparser each image is, except for the low-low image. Most of the energy in the original image has been compacted into the low-low image. Since the other subimages have very few values that need to be encoded, we can devote most of our resources to the low-low subimage.

14.12.1 Decomposing an Image

Earlier a set of filters was provided for one-dimensional subband coding. We can use those same filters to decompose an image into its subbands.

Example 14.12.2:

Let's use the eight-tap Johnston filter to decompose the Sinan image into four subbands. The results of the decomposition are shown in Figure 14.29. Notice that, as in the case of the image in Example 14.12.1, most of the signal energy is concentrated in the low-low subimage. However, there remains a substantial amount of energy in the higher bands. To see this more clearly, let's look at the decomposition using the 16-tap Johnston filter. The results are shown in Figure 14.30. Notice how much less energy there is in the higher subbands. In fact, the high-high subband seems completely empty. As we shall see later, this difference in *energy compaction* can have a drastic effect on the reconstruction.

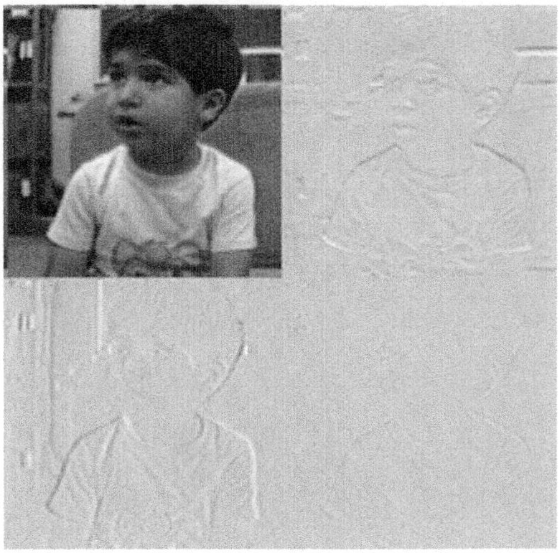

FIGURE 14.29 **Decomposition of Sinan image using the eight-tap Johnston filter.**

FIGURE 14.30 **Decomposition of Sinan image using the 16-tap Johnston filter.**

Increasing the size of the filter is not necessarily the only way of improving the energy compaction. Figure 14.31 shows the decomposition obtained using the eight-tap Smith-Barnwell filter. The results are almost identical to the 16-tap Johnston filter. Therefore, rather than increase the computational load by going to a 16-tap filter, we can keep the same computational load and simply use a different filter. ♦

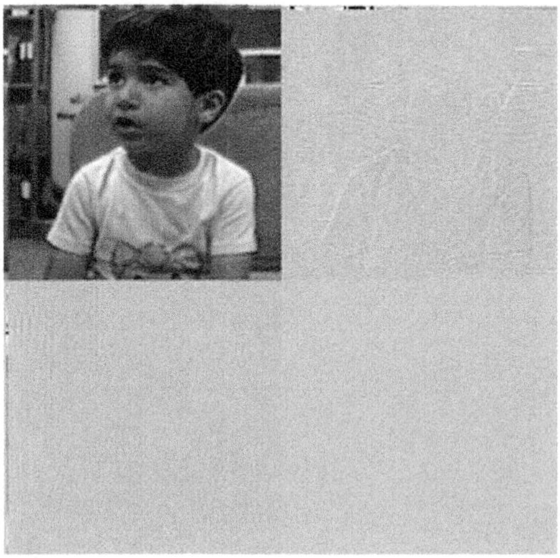

FIGURE 14.31 **Decomposition of Sinan image using the the eight-tap Smith-Barnwell filter.**

14.12.2 Coding the Subbands

Once we have decomposed an image into subbands, we need to find the best encoding scheme to use with each subband. The coding schemes we have studied to date are scalar quantization, vector quantization, and differential encoding. Let us encode some of the decomposed images from the previous section using two of the coding schemes we have studied earlier, scalar quantization and differential encoding.

Example 14.12.3:

In the previous example we noted the fact that the eight-tap Johnston filter did not compact the energy as well as the 16-tap Johnston filter or the eight-tap Smith-Barnwell filter. Let's see how this affects the encoding of the decomposed images.

When we encode these images at an average rate of 0.5 bits per pixel, there are $4 \times 0.5 = 2$ bits available to encode four values, one value from each of the four subbands. If we use the recursive bit allocation procedure on the eight-tap Johnston filter outputs, we end up allocating 1 bit to the low-low band and 1 bit to the high-low band. As the pixel-to-pixel difference in the low-low band is quite small, we use a DPCM encoder for the low-low band. The high-low band does not show this behavior, which means we can simply use scalar quantization for the high-low band. As there are no bits available to encode the other two bands, these bands can be discarded. This results in the image shown in Figure 14.32, which is far from pleasing. However, if we use the same compression approach with the image decomposed using the eight-tap Smith-Barnwell filter, the result is Figure 14.33, which is much more pleasing.

To understand why we get such different results from using the two filters, we need to look at the way the bits were allocated to the different bands. In this implementation, we used the

FIGURE 14.32 Sinan image coded at 0.5 bits per pixel using the eight-tap Johnston filter.

FIGURE 14.33 Sinan image coded at 0.5 bits per pixel using the eight-tap Smith-Barnwell filter.

recursive bit allocation algorithm. In the image decomposed using the Johnston filter, there was significant energy in the high-low band. The algorithm allocated 1 bit to the low-low band and 1 bit to the high-low band. This resulted in poor encoding for both, and subsequently poor reconstruction. There was very little signal content in any of the bands other than the low-low band for the image decomposed using the Smith-Barnwell filter. Therefore, the bit allocation algorithm assigned both bits to the low-low band, which provided a reasonable reconstruction.

If the problem with the encoding of the image decomposed by the Johnston filter is an insufficient number of bits for encoding the low-low band, why not simply assign both bits to the low-low band? The problem is that the bit allocation scheme assigned a bit to the high-low band because there was a significant amount of information in that band. If both bits were assigned to the low-low band, we would have no bits left for use in encoding the high-low band, and we would end up throwing away information necessary for the reconstruction. ♦

The issue of energy compaction becomes a very important factor in reconstruction quality. Filters that allow for more energy compaction permit the allocation of bits to a smaller number of subbands. This in turn results in a better reconstruction.

The coding schemes used in this example were DPCM and scalar quantization, the techniques generally preferred in subband coding. The advantage provided by subband coding is readily apparent if we compare the result shown in Figure 14.33 to results in the previous chapters where we used either DPCM or scalar quantization without prior decomposition.

It would appear that the subband approach lends itself naturally to vector quantization. After decomposing an image into subbands, we could design separate codebooks for each subband to reflect the characteristics of that particular subband. The only problem with this idea is that the low-low subband generally requires a large number of bits per pixel. As we mentioned in Chapter 10, it is generally not feasible to operate the nonstructured vector quantizers at high rates. Therefore, when vector quantizers are used, they are generally used only for encoding the higher frequency bands. This may change as vector quantization algorithms that operate at higher rates are developed.

14.13 Summary

In this chapter we introduced another approach to the decomposition of signals. In subband coding we decompose the source output into components. Each of these components can then be encoded using one of the techniques described in the previous chapters. The general subband encoding procedure can be summarized as follows:

- Select a set of filters for decomposing the source. We have provided a number of filters in this chapter. Many more filters can be obtained from the published literature (we give some references below).
- Using the filters, obtain the subband signals $\{y_{k,n}\}$:

$$y_{k,n} = \sum_{i=0}^{N-1} h_{k,i} x_{n-i} \qquad (119)$$

where $\{h_{k,n}\}$ are the coefficients of the kth filter.

- Decimate the output of the filters.
- Encode the decimated output.

The decoding procedure is the inverse of the encoding procedure. When encoding images the filtering and decimation operations have to be performed twice, once along the rows and once along the columns. Care should be taken to avoid problems at edges, as described in Section 14.12.

Further Reading

1. *Handbook for Digital Signal Processing*, edited by S.K. Mitra and J.F. Kaiser, is an excellent source of information about digital filters [276].
2. *Multirate Systems and Filter Banks*, by P.P. Vaidyanathan [206], provides detailed information on quadrature mirror filters, as well as the relationship between wavelets and filter banks and much more.
3. The topic of subband coding is also covered in *Digital Coding of Waveforms*, by N.S. Jayant and P. Noll [134].
4. The MPEG-1 audio coding algorithm is described in "ISO-MPEG-1 Audio: A Generic Standard for Coding of High-Quality Digital Audio," by K. Brandenburg and G. Stoll , in the October 1994 issue of the *Journal of the Audio Engineering Society* [277].
5. A review of the rate distortion method of bit allocation is provided in "Rate Distortion Methods for Image and Video Compression," by A. Ortega and K. Ramachandran, in the November 1998 issue of *IEEE Signal Processing Magazine* [278].

14.14 Projects and Problems

1. A linear shift invariant system has the following properties:

   - If for a given input sequence $\{x_n\}$ the output of the system is the sequence $\{y_n\}$, then if we delay the input sequence by k units to obtain the sequence $\{x_{n-k}\}$, the corresponding output will be the sequence $\{y_n\}$ delayed by k units.
   - If the output corresponding to the sequence $\{x_n^{(1)}\}$ is $\{y_n^{(1)}\}$, and the output corresponding to the sequence $\{x_n^{(2)}\}$ is $\{y_n^{(2)}\}$, then the output corresponding to the sequence $\{\alpha x_n^{(1)} + \beta x_n^{(2)}\}$ is $\{\alpha y_n^{(1)} + \beta y_n^{(2)}\}$.

   Use these two properties to show the convolution property given in Equation (18).

2. Let's design a set of simple four-tap filters that satisfies the perfect reconstruction condition.

(a) We begin with the low-pass filter. Assume that the impulse response of the filter is given by $\{h_{1,k}\}_{k=0}^{k=3}$. Further assume that

$$\left|h_{1,k}\right| = \left|h_{1,j}\right| \qquad \forall j, k$$

Find a set of values for $\{h_{i,j}\}$ that satisfies Equation (91).

(b) Plot the magnitude of the transfer function $H_1(z)$.

(c) Using Equation (23), find the high-pass filter coefficients $\{h_{2,k}\}$.

(d) Find the magnitude of the transfer function $H_2(z)$.

3. Consider the input sequence

$$x_n = \begin{cases} (-1)^n & n = 0, 1, 2, \ldots \\ 0 & \text{otherwise} \end{cases}$$

(a) Find the output sequence y_n if the filter impulse response is

$$h_n = \begin{cases} \frac{1}{\sqrt{2}} & n = 0, 1 \\ 0 & \text{otherwise} \end{cases}$$

(b) Find the output sequence w_n if the impulse response of the filter is

$$h_n = \begin{cases} \frac{1}{\sqrt{2}} & n = 0 \\ -\frac{1}{\sqrt{2}} & n = 1 \\ 0 & \text{otherwise} \end{cases}$$

(c) Looking at the sequences y_n and w_n, what can you say about the sequence x_n?

4. Consider the input sequence

$$x_n = \begin{cases} 1 & n = 0, 1, 2, \ldots \\ 0 & \text{otherwise} \end{cases}$$

(a) Find the output sequence y_n if the filter impulse response is

$$h_n = \begin{cases} \frac{1}{\sqrt{2}} & n = 0, 1 \\ 0 & \text{otherwise} \end{cases}$$

(b) Find the output sequence w_n if the impulse response of the filter is

$$h_n = \begin{cases} \frac{1}{\sqrt{2}} & n = 0 \\ -\frac{1}{\sqrt{2}} & n = 1 \\ 0 & \text{otherwise} \end{cases}$$

(c) Looking at the sequences y_n and w_n, what can you say about the sequence x_n?

5. Write a program to perform the analysis and downsampling operations and another to perform the upsampling and synthesis operations for an image compression application. The programs should read the filter parameters from a file. The synthesis program should read the output of the analysis program and write out the reconstructed images. The analysis program should also write out the subimages scaled so that they can be displayed. Test your program using the Johnston eight-tap filter and the Sena image.

6. In this problem we look at some of the many ways we can encode the subimages obtained after subsampling. Use the eight-tap Johnston filter to decompose the Sena image into four subimages.

(a) Encode the low-low band using an adptive delta modulator (CFDM or CVSD). Encode all other bands using a 1-bit scalar quantizer.

(b) Encode the low-low band using a 2-bit adaptive DPCM system. Encode the low-high and high-low bands using a 1-bit scalar quantizer.

(c) Encode the low-low band using a 3-bit adaptive DPCM system. Encode the low-high and high-low band using a 0.5 bit/pixel vector quantizer.

(d) Compare the reconstructions obtained using the different schemes.

15

Wavelets

15.1 Overview

In this chapter, we introduce the concept of wavelets and describe how to use wavelet-based decompositions in compression schemes. We begin with an introduction to wavelets and multiresolution analysis and then describe how we can implement a wavelet decomposition using filters. We then examine the implementations of several wavelet-based compression schemes.

15.2 Introduction

In the previous two chapters, we looked at a number of ways to decompose a signal. In this chapter, we look at another approach to decompose a signal that has become increasingly popular: the use of wavelets. Wavelets are being used in a number of different applications. Depending on the application, different aspects of wavelets can be emphasized. As our particular application is compression, we will emphasize those aspects of wavelets that are important in the design of compression algorithms. You should be aware that there is much more to wavelets than is presented in this chapter. At the end of the chapter we suggest options if you want to delve more deeply into this subject.

The practical implementation of wavelet compression schemes is very similar to that of subband coding schemes. As in the case of subband coding, we decompose the signal using analysis filter banks. The outputs of the filter banks are downsampled, quantized, and encoded. The decoder decodes the coded representations, upsamples, and recomposes the signal using a synthesis filter bank.

In the next several sections, we will briefly examine the construction of wavelets and describe how we can obtain a decomposition of a signal using multiresolution analysis. We will then describe some of the currently popular schemes for image compression. If you are primarily interested at this time in implementation of wavelet-based compression schemes, you should skip the next few sections and go directly to Section 15.5.

In the last two chapters, we have described several ways of decomposing signals. Why do we need another one? To answer this question, let's begin with our standard tool for analysis, the Fourier transform. Given a function $f(t)$, we can find the Fourier transform $F(\omega)$ as

$$F(\omega) = \int_{-\infty}^{\infty} f(t)e^{-j\omega t}dt$$

Integration is an averaging operation; therefore, the analysis we obtain, using the Fourier transform, is in some sense an "average" analysis, where the averaging interval is all of time. Thus, by looking at a particular Fourier transform, we can say, for example, that there is a large component of frequency 10 kHz in a signal, but we cannot tell when in time this component occurred. In other words, Fourier analysis provides excellent localization in frequency and none in time. The converse is true for the time function $f(t)$, which provides exact information about the value of the function at each instant of time but does not directly provide spectral information. It should be noted that both $f(t)$ and $F(\omega)$ represent the same function, and all the information is present in each representation. However, each representation makes different kinds of information easily accessible.

If we have a very nonstationary signal, like the one shown in Figure 15.1, we would like to know not only the frequency components but when in time the particular frequency components occurred. One way to obtain this information is via the *short-term Fourier transform* (STFT). With the STFT, we break the time signal $f(t)$ into pieces of length T and apply Fourier analysis to each piece. This way we can say, for example, that a component at 10 kHz occurred in the third piece—that is, between time $2T$ and time $3T$. Thus, we obtain an analysis that is a function of both time and frequency. If we simply chopped the function into pieces, we could get distortion in the form of boundary effects (see Problem 1 at the end of this chapter). In order to reduce the boundary effects, we *window* each piece before we take the Fourier transform. If the window shape is given by $g(t)$, the STFT is formally given by

$$F(\omega, \tau) = \int_{-\infty}^{\infty} f(t)g^*(t-\tau)e^{j\omega t}dt \tag{1}$$

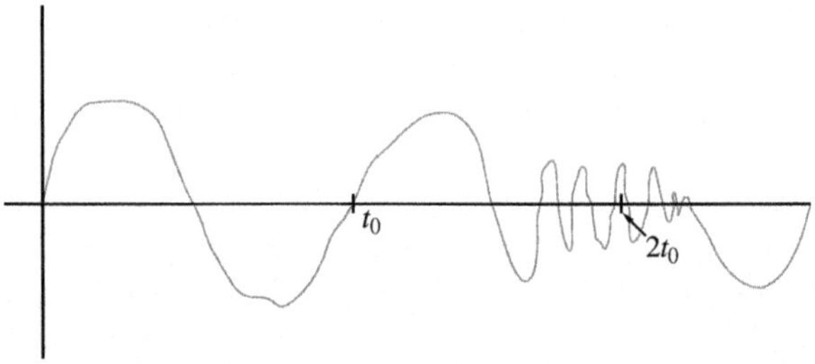

FIGURE 15.1 A nonstationary signal.

If the window function $g(t)$ is a Gaussian, the STFT is called the *Gabor transform*.

The problem with the STFT is the fixed window size. Consider Figure 15.1. In order to obtain the low-pass component at the beginning of the function, the window size should be at least t_0 so that the window will contain at least one cycle of the low-frequency component. However, a window size of t_0 or greater means that we will not be able to accurately localize the high-frequency spurt. A large window in the time domain corresponds to a narrow filter in the frequency domain, which is what we want for the low-frequency components—and what we do not want for the high-frequency components. This dilemma is formalized in the uncertainty principle, which states that for a given window $g(t)$, the product of the time spread σ_t^2 and the frequency spread σ_ω^2 is lower bounded by $\sqrt{1/2}$, where

$$\sigma_t^2 = \frac{\int t^2 \left|g(t)\right|^2 dt}{\int \left|g(t)\right|^2 dt} \tag{2}$$

$$\sigma_\omega^2 = \frac{\int \omega^2 \left|G(\omega)\right|^2 d\omega}{\int \left|G(\omega)\right|^2 d\omega} \tag{3}$$

Thus, if we wish to have finer resolution in time, that is, reduce σ_t^2, we end up with an increase in σ_ω^2 or a lower resolution in the frequency domain. How do we get around this problem?

Let's take a look at the discrete STFT in terms of basis expansion, and, for the moment, let's look at just one interval:

$$F(m, 0) = \int_{-\infty}^{\infty} f(t) g^*(t) e^{-jm\omega_0 t} dt \tag{4}$$

The basis functions are $g(t)$, $g(t)e^{j\omega_0 t}$, $g(t)e^{j2\omega_0 t}$, and so on. The first three basis functions are shown in Figure 15.2. We can see that we have a window with constant size, and within this window, we have sinusoids with an increasing number of cycles. Let's conjure up a different set of functions in which the number of cycles is constant, but the size of the window keeps changing, as shown in Figure 15.3. Notice that although the number of cycles of the sinusoid in each window is the same, as the size of the window gets smaller, these cycles occur in a smaller time interval; that is, the frequency of the sinusoid increases as the size of the window gets smaller. The lower frequency functions cover a longer time interval, while the higher frequency functions cover a shorter time interval, thus avoiding the problem that we had with the STFT. If we can write our function in terms of these functions and their translates, we have a representation that gives us time and frequency localization and can provide high frequency resolution at low frequencies (longer time window) and high time resolution at high frequencies (shorter time window). This, crudely speaking, is the basic idea behind wavelets.

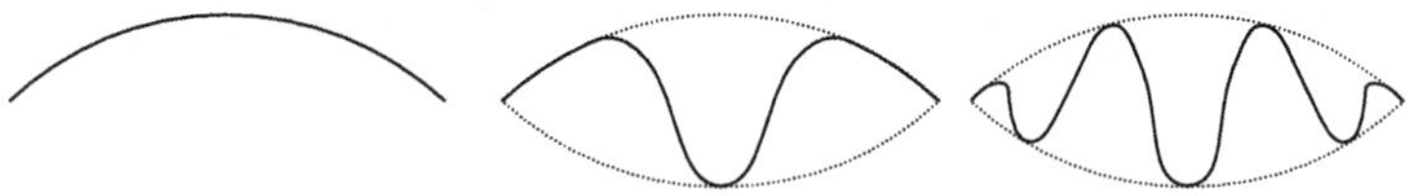

FIGURE 15.2 The first three STFT basis functions for the first time interval.

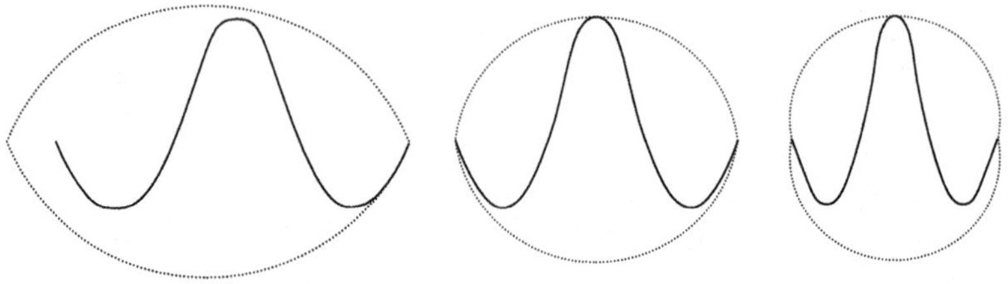

FIGURE 15.3 **Three wavelet basis functions.**

In the following section, we will formalize the concept of wavelets. Then we will discuss how to get from a wavelet basis set to an implementation. If you wish to move directly to implementation issues, you should skip to Section 15.5.

15.3 Wavelets

In the example at the end of the previous section, we started out with a single function. All other functions were obtained by changing the size of the function or *scaling* and translating this single function. This function is called the *mother wavelet*. Mathematically, we can scale a function $f(t)$ by replacing t with t/a, where the parameter a governs the amount of scaling. For example, consider the function

$$f(t) = \begin{cases} \cos(\pi t) & -1 \leqslant t \leqslant 1 \\ 0 & \text{otherwise} \end{cases}$$

We have plotted this function in Figure 15.4. To scale this function by 0.5, we replace t by $t/0.5$:

$$\begin{aligned} f\left(\frac{t}{0.5}\right) &= \begin{cases} \cos(\pi \frac{t}{0.5}) & -1 \leqslant \frac{t}{0.5} \leqslant 1 \\ 0 & \text{otherwise} \end{cases} \\ &= \begin{cases} \cos(2\pi t) & -\frac{1}{2} \leqslant t \leqslant \frac{1}{2} \\ 0 & \text{otherwise} \end{cases} \end{aligned}$$

We have plotted the scaled function in Figure 15.5. If we define the norm of a function $f(t)$ by

$$\|f(t)\|^2 = \int_{-\infty}^{\infty} f^2(t)dt$$

scaling obviously changes the norm of the function:

$$\begin{aligned} \left\|f\left(\frac{t}{a}\right)\right\|^2 &= \int_{-\infty}^{\infty} f^2\left(\frac{t}{a}\right)dt \\ &= a\int_{-\infty}^{\infty} f^2(x)dx \end{aligned}$$

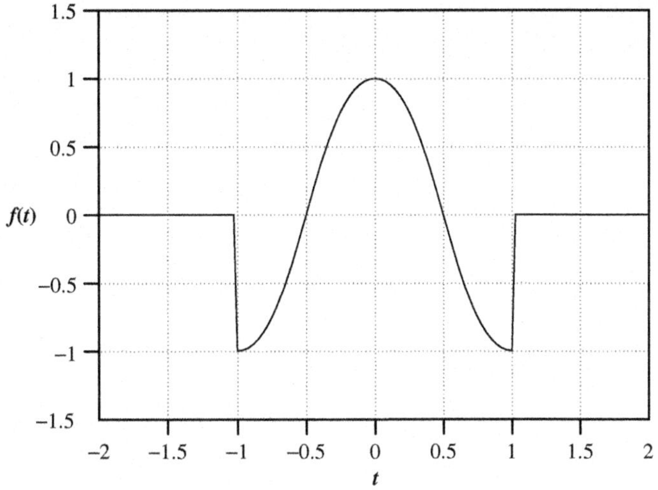

FIGURE 15.4 **The function f(t).**

where we have used the substitution $x = t/a$. Thus,

$$\left\| f\left(\frac{t}{a}\right) \right\|^2 = a \, \|f(t)\|^2$$

If we want the scaled function to have the same norm as the original function, we need to multiply it by $1/\sqrt{a}$. Transformations that change the size of an object but not its shape are often referred to as *dilation*, which is a term we will use interchangeably with scaling in our discussion.

Mathematically, we can represent the translation of a function to the right or left by an amount b by replacing t by $t - b$ or $t + b$. For example, if we want to translate the scaled function shown in Figure 15.5 by one, we have

$$f\left(\frac{t-1}{0.5}\right) = \begin{cases} \cos(2\pi(t-1)) & -\frac{1}{2} \leqslant t - 1 \leqslant \frac{1}{2} \\ 0 & \text{otherwise} \end{cases}$$
$$= \begin{cases} \cos(2\pi(t-1)) & \frac{1}{2} \leqslant t \leqslant \frac{3}{2} \\ 0 & \text{otherwise} \end{cases}$$

The scaled and translated function is shown in Figure 15.6.

We can use the translation and scaling operations to generate a family of functions from a single function. Thus, given a *mother* function $\psi(t)$, we can generate a whole family of functions $\{\psi_{a,b}(t)\}$ with

$$\psi_{a,b}(t) = \frac{1}{\sqrt{a}} \psi\left(\frac{t-b}{a}\right) \tag{5}$$

and Fourier transforms

$$\Psi(\omega) = \mathcal{F}[\psi(t)]$$
$$\Psi_{a,b}(\omega) = \mathcal{F}[\psi_{a,b}(t)] \tag{6}$$

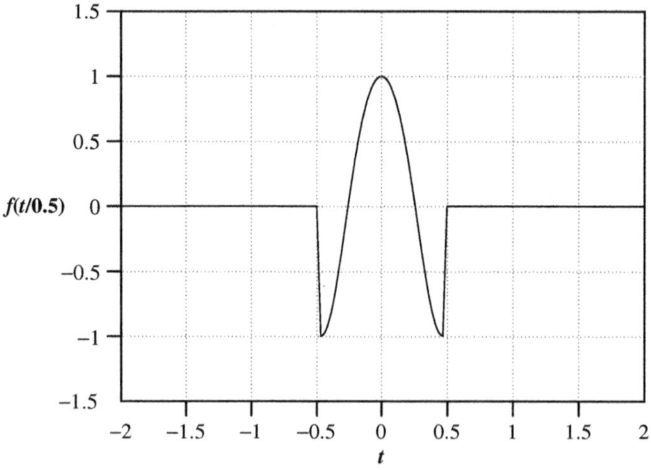

FIGURE 15.5 **The function $f(\frac{t}{0.5})$.**

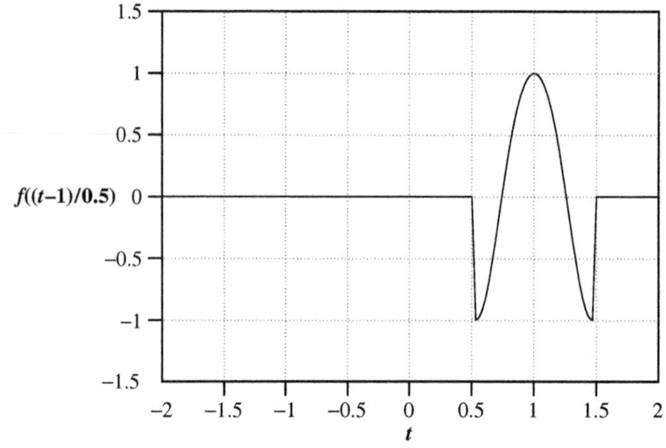

FIGURE 15.6 **A scaled and translated function.**

If this family of functions satisfies certain properties we can use them to represent signals of interest in a manner similar to the Fourier basis functions. Unlike the Fourier basis functions, because these functions are localized in time and frequency we can use them without giving up all the temporal information contained in the original function. Because of their limited spatial extent these functions were originally called *ondelettes*, which is French for small waves, or *wavelets*.

Our expansion using coefficients with respect to these functions is obtained from the inner product of $f(t)$ with the wavelet functions:

$$w_{a,b} = \langle \psi_{a,b}(t), f(t) \rangle = \int_{-\infty}^{\infty} \psi_{a,b}(t) f(t) dt \tag{7}$$

We can recover the function $f(t)$ from the $w_{a,b}$ by

$$f(t) = \frac{1}{C_\psi} \int_{-\infty}^{\infty} \int_{-\infty}^{\infty} w_{a,b} \psi_{a,b}(t) \frac{dadb}{a^2} \tag{8}$$

where

$$C_\psi = \int_0^\infty \frac{|\Psi(\omega)|^2}{\omega} d\omega \tag{9}$$

For integral (8) to exist, we need C_ψ to be finite. For C_ψ to be finite, we need $\Psi(0) = 0$. Otherwise, we have a singularity in the integrand of (9). Note that $\Psi(0)$ is the average value of $\psi(t)$; therefore, the mother wavelet must have zero mean. The condition that C_ψ be finite is often called the *admissibility condition*. We would also like the wavelets to have finite energy; that is, we want the wavelets to belong to the vector space L_2 (see Example 12.3.1). Using Parseval's relationship, we can write this requirement as

$$\int_{-\infty}^{\infty} |\Psi(\omega)|^2 d\omega < \infty$$

For this to happen, $|\Psi(\omega)|^2$ has to decay as ω goes to infinity. These requirements mean that the energy in $\Psi(\omega)$ is concentrated in a narrow frequency band, which gives the wavelet its frequency localization capability.

If a and b are continuous, then $w_{a,b}$ is called the *continuous wavelet transform* (CWT). Just as with other transforms, we will be more interested in the discrete version of this transform. However, we first obtain a series representation where the basis functions are continuous functions of time with discrete scaling and translating parameters a and b. The discrete versions of the scaling and translating parameters have to be related to each other, because if the scale is such that the basis functions are narrow, the translation step should be correspondingly small and vice versa. There are a number of ways we can choose these parameters. The most popular approach is to select a and b according to

$$a = a_0^{-m}, \qquad b = nb_0 a_0^{-m} \tag{10}$$

where m and n are integers, a_0 is selected to be 2, and b_0 has a value of 1. This gives us the wavelet set

$$\psi_{m,n}(t) = a_0^{m/2} \psi(a_0^m t - nb_0), \qquad m, n \in Z \tag{11}$$

For $a_0 = 2$ and $b_0 = 1$, we have

$$\psi_{m,n}(t) = 2^{m/2} \psi(2^m t - n) \tag{12}$$

(Note that these are the most commonly used choices, but they are not the only choices.) If this set is *complete*, then $\{\psi_{m,n}(t)\}$ are called *affine* wavelets. The wavelet coefficients are given by

$$w_{m,n} = \langle f(t), \psi_{m,n}(t)\rangle \tag{13}$$

$$= a_0^{m/2} \int f(t)\psi(a_0^m t - nb_0)dt \tag{14}$$

The function $f(t)$ can be reconstructed from the wavelet coefficients by

$$f(t) = \sum_m \sum_n w_{m,n}\psi_{m,n}(t) \tag{15}$$

Wavelets come in many shapes. We will look at some of the more popular ones later in this chapter. One of the simplest wavelets is the Haar wavelet, which we will use to explore the various aspects of wavelets. The Haar wavelet is given by

$$\psi(t) = \begin{cases} 1 & 0 \leqslant t < \frac{1}{2} \\ -1 & \frac{1}{2} \leqslant t < 1 \end{cases} \tag{16}$$

By translating and scaling this mother wavelet, we can synthesize a variety of functions.

This version of the transform, where $f(t)$ is a continuous function while the transform consists of discrete values, is a wavelet series analogous to the Fourier series. It is also called the *discrete time wavelet transform* (DTWT). We have moved from the continuous wavelet transform, where both the time function $f(t)$ and its transform $w_{a,b}$ were continuous functions of their arguments, to the wavelet series, where the time function is continuous but the time-scale wavelet representation is discrete. Given that in data compression we are generally dealing with sampled functions that are discrete in time, we would like both the time and frequency representations to be discrete. This is called the *discrete wavelet transform* (DWT). However, before we get to that, let's look into one additional concept—multiresolution analysis.

15.4 Multiresolution Analysis and the Scaling Function

The idea behind multiresolution analysis is fairly simple. Let's define a function $\phi(t)$ that we call a *scaling* function. We will later see that the scaling function is closely related to the mother wavelet. By taking linear combinations of the scaling function and its translates, we can generate a large number of functions

$$f(t) = \sum_k a_k\phi(t - k) \tag{17}$$

The scaling function has the property that a function that can be represented by the scaling function can also be represented by the dilated versions of the scaling function.

For example, one of the simplest scaling functions is the Haar scaling function:

$$\phi(t) = \begin{cases} 1 & 0 \leqslant t < 1 \\ 0 & \text{otherwise} \end{cases} \tag{18}$$

Then $f(t)$ can be any piecewise continuous function that is constant in the interval $[k, k+1)$ for all k.

Let's define

$$\phi_k(t) = \phi(t-k) \tag{19}$$

The set of all functions that can be obtained using a linear combination of the set $\{\phi_k(t)\}$, represented by

$$f(t) = \sum_k a_k \phi_k(t) \tag{20}$$

is called the *span* of the set $\{\phi_k(t)\}$, or Span$\{\phi_k(t)\}$. If we now add all functions that are limits of sequences of functions in Span$\{\phi_k(t)\}$, this is referred to as the closure of Span$\{\phi_k(t)\}$ and denoted by $\overline{\text{Span}\{\phi_k(t)\}}$. Let's call this set V_0.

If we want to generate functions at a higher resolution, say functions that are required to be constant over only half a unit interval, we can use a dilated version of the "mother" scaling function. In fact, we can obtain scaling functions at different resolutions in a manner similar to the procedure used for wavelets:

$$\phi_{j,k}(t) = 2^{j/2}\phi(2^j t - k) \tag{21}$$

The indexing scheme is the same as that used for wavelets, with the first index referring to the resolution while the second index denotes the translation. For the Haar example,

$$\phi_{1,0}(t) = \begin{cases} \sqrt{2} & 0 \leqslant t < \frac{1}{2} \\ 0 & \text{otherwise} \end{cases} \tag{22}$$

We can use translates of $\phi_{1,0}(t)$ to represent all functions that are constant over intervals $[k/2, (k+1)/2)$ for all k. Notice that in general any function that can be represented by the translates of $\phi(t)$ can also be represented by a linear combination of translates of $\phi_{1,0}(t)$. The converse, however, is not true. Defining

$$V_1 = \overline{\text{Span}\{\phi_{1,k}(t)\}} \tag{23}$$

we can see that $V_0 \subset V_1$. Similarly, we can show that $V_1 \subset V_2$ and so on.

Example 15.4.1:

Consider the function shown in Figure 15.7. We can approximate this function using translates of the Haar scaling function $\phi(t)$. The approximation is shown in Figure 15.8(a). If we call this approximation $\phi_f^{(0)}(t)$, then

$$\phi_f^{(0)}(t) = \sum_k c_{0,k}\phi_k(t) \tag{24}$$

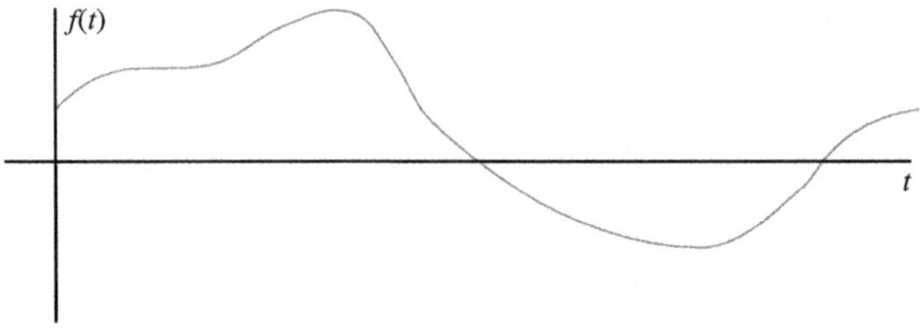

FIGURE 15.7 **A sample function.**

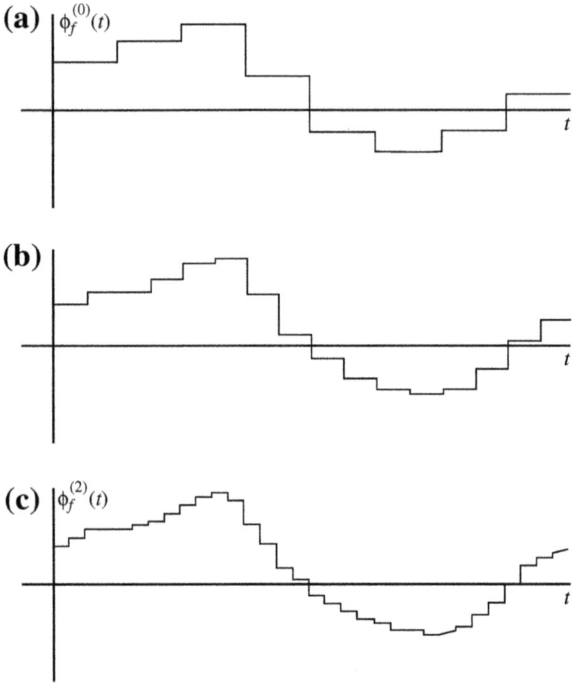

FIGURE 15.8 **Approximations of the function shown in Figure 15.7.**

where

$$c_{0,k} = \int_k^{k+1} f(t)\phi_k(t)dt \tag{25}$$

We can obtain a more refined approximation, or an approximation at a higher resolution, $\phi_f^{(1)}(t)$, shown in Figure 15.8(b), if we use the set $\{\phi_{1,k}(t)\}$:

$$\phi_f^{(1)}(t) = \sum_k c_{1,k}\phi_{1,k}(t) \tag{26}$$

Notice that we need twice as many coefficients at this resolution compared to the previous resolution. The coefficients at the two resolutions are related by

$$c_{0,k} = \frac{1}{\sqrt{2}}(c_{1,2k} + c_{1,2k+1}) \tag{27}$$

Continuing in this manner (Figure 15.8(c)), we can get higher and higher resolution approximations of $f(t)$ with

$$\phi_f^{(m)}(t) = \sum_k c_{m,k}\phi_{m,k}(t) \tag{28}$$

Recall that, according to the Nyquist rule, if the highest frequency component of a signal is at f_0 Hz, we need $2f_0$ samples per second to accurately represent it. Therefore, we could obtain an accurate representation of $f(t)$ using the set of translates $\{\phi_{j,k}(t)\}$, where $2^{-j} < \frac{1}{2f_0}$. As

$$c_{j,k} = 2^{j/2}\int_{\frac{k}{2^j}}^{\frac{k+1}{2^j}} f(t)dt \tag{29}$$

by the mean value theorem of calculus, $c_{j,k}$ is equal to a sample value of $f(t)$ (within a constant scale factor) in the interval $[k2^{-j}, (k+1)2^{-j})$. Therefore, the function $\phi_f^{(j)}(t)$ would represent more than $2f_0$ samples per second of $f(t)$. ♦

We said earlier that a scaling function has the property that any function that can be represented exactly by an expansion at some resolution j can also be represented by dilations of the scaling function at resolution $j+1$. In particular, this means that the scaling function itself can be represented by its dilations at a higher resolution:

$$\phi(t) = \sum_k h_k\phi_{1,k}(t) \tag{30}$$

Substituting $\phi_{1,k}(t) = \sqrt{2}\phi(2t-k)$, we obtain the *multiresolution analysis* (MRA) equation:

$$\phi(t) = \sum_k h_k\sqrt{2}\phi(2t-k) \tag{31}$$

This equation will be of great importance to us when we begin looking at ways of implementing the wavelet transform.

Example 15.4.2:

Consider the Haar scaling function. Picking

$$h_0 = h_1 = \frac{1}{\sqrt{2}}$$

and

$$h_k = 0 \qquad \text{for } k > 1$$

satisfies the recursion equation. ♦

Example 15.4.3:

Consider the triangle scaling function shown in Figure 15.9. We can write this as

$$\phi_\Delta(t) = \begin{cases} t & 0 \leqslant t \leqslant 1 \\ 2-t & 1 \leqslant t \leqslant 2 \\ 0 & t > 2 \end{cases}$$

We can see from Figure 15.9 that we can decompose $\phi_\Delta(t)$ as

$$\begin{aligned} \phi_\Delta(t) &= \frac{1}{2}\phi_\Delta(2t) + \phi_\Delta(2t-1) + \frac{1}{2}\phi_\Delta(2t-2) \\ &= \frac{1}{2\sqrt{2}}\sqrt{2}\phi_\Delta(2t-0) + \frac{1}{\sqrt{2}}\sqrt{2}\phi_\Delta(2t-1) + \frac{1}{2\sqrt{2}}\sqrt{2}\phi_\Delta(2t-2) \end{aligned}$$

Thus,

$$h_0 = \frac{1}{2\sqrt{2}}, \quad h_1 = \frac{1}{\sqrt{2}}, \quad h_2 = \frac{1}{2\sqrt{2}}$$

satisfies the recursion equation.

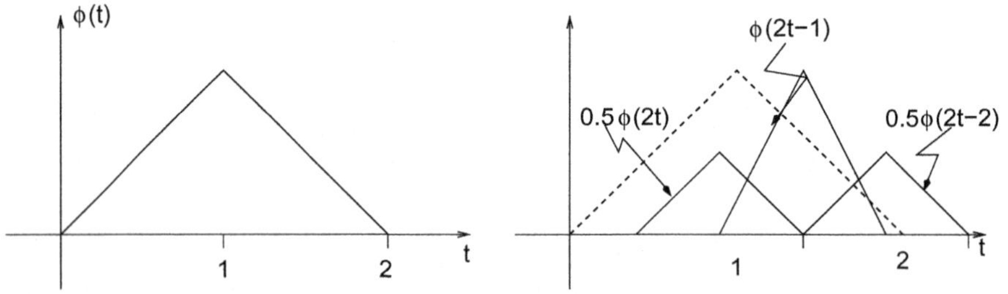

FIGURE 15.9 Triangular scaling function. ♦

While both the Haar scaling function and the triangle scaling functions are valid scaling functions, there is an important difference between the two. The Haar function is orthogonal to its translates; that is,

$$\int \phi(t)\phi(t-m)dt = \delta_m$$

This is obviously not true of the triangle function. In this chapter we will be principally concerned with scaling functions that are orthogonal because they give rise to orthonormal transforms that, as we have previously seen, are very useful in compression.

How about the Haar wavelet? Can it be used as a scaling function? Some reflection will show that we cannot obtain the Haar wavelet from a linear combination of its dilated versions.

So, where do wavelets come into the picture? Let's continue with our example using the Haar scaling function. Let us assume for the moment that there is a function $g(t)$ that can be exactly represented by $\phi_g^{(1)}(t)$; that is, $g(t)$ is a function in the set V_1. We can decompose

$\phi_g^{(1)}(t)$ into the sum of a lower-resolution version of itself, namely, $\phi_g^{(0)}(t)$, and the difference $\phi_g^{(1)}(t) - \phi_g^{(0)}(t)$. Let's examine this difference over an arbitrary unit interval $[k, k+1)$:

$$\phi_g^{(1)}(t) - \phi_g^{(0)}(t) = \begin{cases} c_{0,k} - \sqrt{2}c_{1,2k} & k \leqslant t < k + \frac{1}{2} \\ c_{0,k} - \sqrt{2}c_{1,2k+1} & k + \frac{1}{2} \leqslant t < k+1 \end{cases} \tag{32}$$

Substituting for $c_{0,k}$ from (27), we obtain

$$\phi_g^{(1)}(t) - \phi_g^{(0)}(t) = \begin{cases} -\frac{1}{\sqrt{2}}c_{1,2k} + \frac{1}{\sqrt{2}}c_{1,2k+1} & k \leqslant t < k + \frac{1}{2} \\ \frac{1}{\sqrt{2}}c_{1,2k} - \frac{1}{\sqrt{2}}c_{1,2k+1} & k + \frac{1}{2} \leqslant t < k+1 \end{cases} \tag{33}$$

Defining

$$d_{0,k} = -\frac{1}{\sqrt{2}}c_{1,2k} + \frac{1}{\sqrt{2}}c_{1,2k+1}$$

over the arbitrary interval $[k, k+1)$,

$$\phi_g^{(1)}(t) - \phi_g^{(0)}(t) = d_{0,k}\psi_{0,k}(t) \tag{34}$$

where

$$\psi_{0,k}(t) = \begin{cases} 1 & k \leqslant t < k + \frac{1}{2} \\ -1 & k + \frac{1}{2} \leqslant t < k+1 \end{cases} \tag{35}$$

But this is simply the kth translate of the Haar wavelet. Thus, for this particular case the function can be represented as the sum of a scaling function and a wavelet at the same resolution:

$$\phi_g^{(1)}(t) = \sum_k c_{0,k}\phi_{0,k}(t) + \sum_k d_{0,k}\psi_{0,k}(t) \tag{36}$$

In fact, we can show that this decomposition is not limited to this particular example. A function in V_1 can be decomposed into a function in V_0—that is, a function that is a linear combination of the scaling function at resolution 0 and a function that is a linear combination of translates of a mother wavelet. Denoting the set of functions that can be obtained by a linear combination of the translates of the mother wavelet as W_0, we can write this symbolically as

$$V_1 = V_0 \oplus W_0 \tag{37}$$

In other words, any function in V_1 can be represented using functions in V_0 and W_0.

Obviously, once a scaling function is selected, the choice of the wavelet function cannot be arbitrary. The wavelet that generates the set W_0 and the scaling function that generates the sets V_0 and V_1 are intrinsically related. In fact, from (37), $W_0 \subset V_1$, and therefore any function in W_0 can be represented by a linear combination of $\{\phi_{1,k}\}$. In particular, we can write the mother wavelet $\psi(t)$ as

$$\psi(t) = \sum_k w_k\phi_{1,k}(t) \tag{38}$$

or

$$\psi(t) = \sum_k w_k\sqrt{2}\phi(2t-k) \tag{39}$$

This is the counterpart of the multiresolution analysis equation for the wavelet function and will be of primary importance in the implementation of the decomposition.

All of this development has been for a function in V_1. What if the function can only be accurately represented at resolution $j+1$? If we define W_j as the closure of the span of $\psi_{j,k}(t)$, we can show that

$$V_{j+1} = V_j \oplus W_j \tag{40}$$

But, as j is arbitrary,

$$V_j = V_{j-1} \oplus W_{j-1} \tag{41}$$

and

$$V_{j+1} = V_{j-1} \oplus W_{j-1} \oplus W_j \tag{42}$$

Continuing in this manner, we can see that for any $k \leqslant j$

$$V_{j+1} = V_k \oplus W_k \oplus W_{k+1} \oplus \cdots \oplus W_j \tag{43}$$

In other words, if we have a function that belongs to V_{j+1} (i.e., that can be exactly represented by the scaling function at resolution $j+1$), we can decompose it into a sum of functions starting with a lower-resolution approximation followed by a sequence of functions generated by dilations of the wavelet that represent the leftover details. This is very much like what we did in subband coding. A major difference is that, while the subband decomposition is in terms of sines and cosines, the decomposition in this case can use a variety of scaling functions and wavelets. Thus, we can adapt the decomposition to the signal being decomposed by selecting the scaling function and wavelet.

15.5 Implementation Using Filters

One of the most popular approaches to implementing the decomposition discussed in the previous section is using a hierarchical filter structure similar to the one used in subband coding. In this section, we will look at how to obtain the structure and the filter coefficients.

We start with the MRA equation

$$\phi(t) = \sum_k h_k \sqrt{2}\phi(2t-k) \tag{44}$$

Substituting $t = 2^j t - m$, we obtain the equation for an arbitrary dilation and translation:

$$\begin{aligned}\phi(2^j t - m) &= \sum_k h_k \sqrt{2}\phi(2(2^j t - m) - k) && (45)\\ &= \sum_k h_k \sqrt{2}\phi(2^{j+1} t - 2m - k) && (46)\\ &= \sum_l h_{l-2m} \sqrt{2}\phi(2^{j+1} t - l) && (47)\end{aligned}$$

where in the last equation we have used the substitution $l = 2m+k$. Suppose we have a function $f(t)$ that can be accurately represented at resolution $j+1$ by some scaling function $\phi(t)$. We assume that the scaling function and its dilations and translations form an orthonormal set. The coefficients c_{j+1} can be obtained by

$$c_{j+1,k} = \int f(t)\phi_{j+1,k}(t)dt \tag{48}$$

If we can represent $f(t)$ accurately at resolution $j+1$ with a linear combination of $\phi_{j+1,k}(t)$, then from the previous section we can decompose it into two functions: one in terms of $\phi_{j,k}(t)$ and one in terms of the jth dilation of the corresponding wavelet $\{\psi_{j,k}(t)\}$. The coefficients $c_{j,k}$ are given by

$$c_{j,k} = \int f(t)\phi_{j,k}(t)dt \tag{49}$$

$$= \int f(t)2^{\frac{j}{2}}\phi(2^j t - k)dt \tag{50}$$

Substituting for $\phi(2^j t - k)$ from (47), we get

$$c_{j,k} = \int f(t)2^{\frac{j}{2}} \sum_l h_{l-2k}\sqrt{2}\phi(2^{j+1}t - l)dt \tag{51}$$

Interchanging the order of summation and integration, we get

$$c_{j,k} = \sum_l h_{l-2k} \int f(t)2^{\frac{j}{2}}\sqrt{2}\phi(2^{j+1}t - l)dt \tag{52}$$

But the integral is simply $c_{j+1,l}$. Therefore,

$$c_{j,k} = \sum_l h_{l-2k}c_{j+1,l} \tag{53}$$

We have encountered this relationship before in the context of the Haar function. Equation (27) provides the relationship between coefficients of the Haar expansion at two resolution levels. In a more general setting, the coefficients $\{h_j\}$ provide a link between the coefficients $\{c_{j,k}\}$ at different resolutions. Thus, given the coefficients at resolution level $j+1$, we can obtain the coefficients at all other resolution levels. But how do we start the process? Recall that $f(t)$ can be accurately represented at resolution $j+1$. Therefore, we can replace $c_{j+1,k}$ by the samples of $f(t)$. Let's represent these samples by x_k. Then the coefficients of the low-resolution expansion are given by

$$c_{j,m} = \sum_k h_{k-2m}x_k \tag{54}$$

In Chapter 12, we introduced the input-output relationship of a linear filter as

$$y_m = \sum_k h_k x_{m-k} = \sum_k h_{m-k}x_k \tag{55}$$

Replacing m by $2m$, we get every other sample of the output:

$$y_{2m} = \sum_k h_{2m-k} x_k \tag{56}$$

Comparing (56) with (54), we can see that the coefficients of the low-resolution approximation are every other output of a linear filter whose impulse response is h_{-k}. Recall that $\{h_k\}$ are the coefficients that satisfy the MRA equation. Using the terminology of subband coding, the coefficients $c_{j,k}$ are the downsampled output of the linear filter with impulse response $\{h_{-k}\}$.

The detail portion of the representation is obtained in a similar manner. Again we start from the recursion relationship. This time we use the recursion relationship for the wavelet function as our starting point:

$$\psi(t) = \sum_k w_k \sqrt{2}\phi(2t-k) \tag{57}$$

Again substituting $t = 2^j t - m$ and using the same simplifications, we get

$$\psi(2^j t - m) = \sum_k w_{k-2m} \sqrt{2}\phi(2^{j+1}t-k) \tag{58}$$

Using the fact that the dilated and translated wavelets form an orthonormal basis, we can obtain the detail coefficients $d_{j,k}$ by

$$d_{j,k} = \int f(t)\psi_{j,k}(t)dt \tag{59}$$

$$= \int f(t)2^{\frac{j}{2}}\psi(2^j t-k)dt \tag{60}$$

$$= \int f(t)2^{\frac{j}{2}} \sum_l w_{l-2k}\sqrt{2}\phi(2^{j+1}t-l)dt \tag{61}$$

$$= \sum_l w_{l-2k} \int f(t)2^{\frac{j+1}{2}}\phi(2^{j+1}t-l)dt \tag{62}$$

$$= \sum_l w_{l-2k} c_{j+1,l} \tag{63}$$

Thus, the detail coefficients are the decimated outputs of a filter with impulse response $\{w_{-k}\}$.

At this point, we can use exactly the same arguments to further decompose the coefficients $\{c_j\}$. A block diagram of a three-level decomposition is shown in Figure 15.10.

In order to retrieve $\{c_{j+1,k}\}$ from $\{c_{j,k}\}$ and $\{d_{j,k}\}$, we upsample the lower resolution coefficients and use filters with impulse response $\{h_k\}$ and $\{w_k\}$:

$$c_{j+1,k} = \sum_l c_{j,l} h_{k-2l} + \sum_l d_{j,l} w_{k-2l}$$

Assuming the three-level decomposition described above, the block diagram for the synthesis process is shown in Figure 15.11.

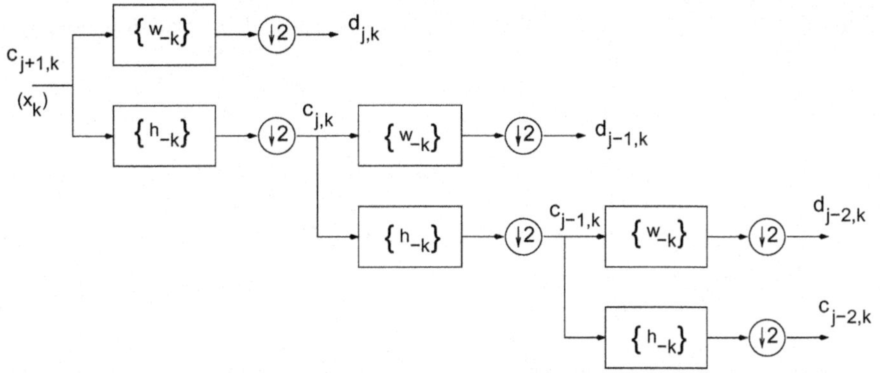

FIGURE 15.10 **Block diagram of a three-level wavelet decomposition.**

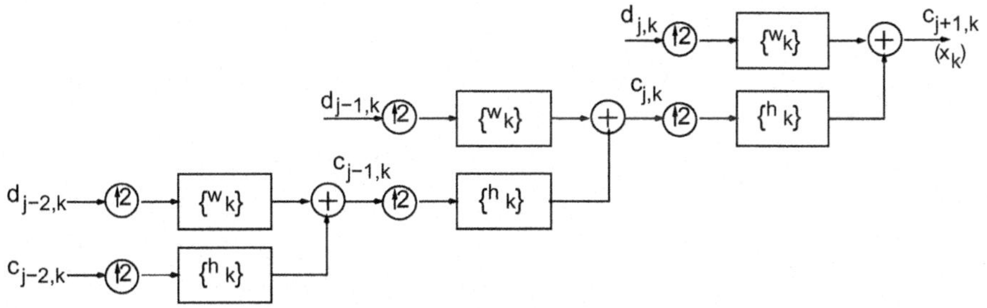

FIGURE 15.11 **Block diagram of a three-level wavelet reconstruction.**

15.5.1 Scaling and Wavelet Coefficients

In order to implement the wavelet decomposition, the coefficients $\{h_k\}$ and $\{w_k\}$ are of primary importance. In this section, we look at some of the properties of these coefficients that will help us in finding different decompositions.

We start with the MRA equation. Integrating both sides of the equation over all t, we obtain

$$\int_{-\infty}^{\infty} \phi(t)dt = \int_{-\infty}^{\infty} \sum_k h_k \sqrt{2}\phi(2t-k)dt \tag{64}$$

Interchanging the summation and integration on the right-hand side of the equation, we get

$$\int_{-\infty}^{\infty} \phi(t)dt = \sum_k h_k \sqrt{2} \int_{-\infty}^{\infty} \phi(2t-k)dt \tag{65}$$

Substituting $x = 2t - k$ with $dx = 2dt$ in the right-hand side of the equation, we get

$$\int_{-\infty}^{\infty} \phi(t)dt = \sum_k h_k \sqrt{2} \int_{-\infty}^{\infty} \phi(x) \frac{1}{2} dx \tag{66}$$

$$= \sum_k h_k \frac{1}{\sqrt{2}} \int_{-\infty}^{\infty} \phi(x)dx \tag{67}$$

Assuming that the average value of the scaling function is not zero, we can divide both sides by the integral to obtain

$$\sum_k h_k = \sqrt{2} \tag{68}$$

If we normalize the scaling function to have a magnitude of one, we can use the orthogonality condition on the scaling function to get another condition on $\{h_k\}$:

$$\int |\phi(t)|^2 \, dt = \int \sum_k h_k \sqrt{2} \phi(2t - k) \sum_m h_m \sqrt{2} \phi(2t - m)dt \tag{69}$$

$$= \sum_k \sum_m h_k h_m 2 \int \phi(2t - k)\phi(2t - m)dt \tag{70}$$

$$= \sum_k \sum_m h_k h_m \int \phi(x - k)\phi(x - m)dx \tag{71}$$

where in the last equation we have used the substitution $x = 2t$. The integral on the right-hand side is zero except when $k = m$. When $k = m$, the integral is unity and we obtain

$$\sum_k h_k^2 = 1 \tag{72}$$

We can actually elicit a more general property by using the orthogonality of the translates of the scaling function

$$\int \phi(t)\phi(t - m)dt = \delta_m \tag{73}$$

Rewriting this using the MRA equation to substitute for $\phi(t)$ and $\phi(t - m)$, we obtain

$$\int \left[\sum_k h_k \sqrt{2} \phi(2t - k) \right] \left[\sum_l h_l \sqrt{2} \phi(2t - 2m - l) \right] dt$$

$$= \sum_k \sum_l h_k h_l 2 \int \phi(2t - k)\phi(2t - 2m - l)dt \tag{74}$$

Substituting $x = 2t$, we obtain

$$\int \phi(t)\phi(t - m)dt = \sum_k \sum_l h_k h_l \int \phi(x - k)\phi(x - 2m - l)dx \tag{75}$$

$$= \sum_k \sum_l h_k h_l \delta_{k-(2m+l)} \tag{76}$$

$$= \sum_k h_k h_{k-2m} \tag{77}$$

Therefore, we have

$$\sum_k h_k h_{k-2m} = \delta_m \tag{78}$$

Notice that this is the same relationship we had to satisfy for perfect reconstruction in the previous chapter.

Using these relationships, we can generate scaling coefficients for filters of various lengths.

Example 15.5.1:

For $k = 2$, we have from (68) and (72)

$$h_0 + h_1 = \sqrt{2} \tag{79}$$

$$h_0^2 + h_1^2 = 1 \tag{80}$$

These equations are uniquely satisfied by

$$h_0 = h_1 = \frac{1}{\sqrt{2}}$$

which is the Haar scaling function. ♦

An orthogonal expansion does not exist for all lengths. In the following example, we consider the case of $k = 3$.

Example 15.5.2:

For $k = 3$, from the three conditions (68), (72), and (78), we have

$$h_0 + h_1 + h_2 = \sqrt{2} \tag{81}$$

$$h_0^2 + h_1^2 + h_2^2 = 1 \tag{82}$$

$$h_0 h_2 = 0 \tag{83}$$

The last condition can only be satisfied if $h_0 = 0$ or $h_2 = 0$. In either case we will be left with the two-coefficient filter for the Haar scaling function. ♦

In fact, we can see that for k odd, we will always end up with a condition that will force one of the coefficients to zero, thus leaving an even number of coefficients. When the number of coefficients gets larger than the number of conditions, we end up with an infinite number of solutions.

Example 15.5.3:

Consider the case when $k = 4$. The three conditions give us the following three equations:

$$h_0 + h_1 + h_2 + h_3 = \sqrt{2} \quad (84)$$

$$h_0^2 + h_1^2 + h_2^2 + h_3^2 = 1 \quad (85)$$

$$h_0 h_2 + h_1 h_3 = 0 \quad (86)$$

We have three equations and four unknowns; that is, we have one degree of freedom. We can use this degree of freedom to impose further conditions on the solution. The solutions to these equations include the Daubechies four-tap solution:

$$h_0 = \frac{1+\sqrt{3}}{4\sqrt{2}}, \quad h_1 = \frac{3+\sqrt{3}}{4\sqrt{2}}, \quad h_2 = \frac{3-\sqrt{3}}{4\sqrt{2}}, \quad h_3 = \frac{1-\sqrt{3}}{4\sqrt{2}}$$ ♦

Given the close relationship between the scaling function and the wavelet, it seems reasonable that we should be able to obtain the coefficients for the wavelet filter from the coefficients of the scaling filter. In fact, if the wavelet function is orthogonal to the scaling function at the same scale

$$\int \phi(t-k)\psi(t-m)dt = 0 \quad (87)$$

then

$$w_k = \pm(-1)^k h_{N-k} \quad (88)$$

and

$$\sum_k h_k w_{n-2k} = 0 \quad (89)$$

Furthermore,

$$\sum_k w_k = 0 \quad (90)$$

The proof of these relationships is somewhat involved [212].

15.5.2 Families of Wavelets

We have said that there is an infinite number of possible wavelets. Which one is best depends on the application. In this section, we list different wavelets and their corresponding filters. You are encouraged to experiment with these to find those best suited to your application.

The 4-tap, 12-tap, and 20-tap Daubechies filters are shown in Tables 15.1, 15.2, 15.3. The 6-tap, 12-tap, and 18-tap Coiflet filters are shown in Tables 15.4, 15.5, 15.6.

15.6 Biorthogonal Wavelets

We have focused on orthogonal wavelets in order to develop various concepts regarding wavelet decomposition. However, in various image compression schemes, which will be the focus of

TABLE 15.1 Coefficients for the 4-tap Daubechies low-pass filter.

h_0	0.4829629131445341
h_1	0.8365163037378079
h_2	0.2241438680420134
h_3	−0.1294095225512604

TABLE 15.2 Coefficients for the 12-tap Daubechies low-pass filter.

h_0	0.111540743350
h_1	0.494623890398
h_2	0.751133908021
h_3	0.315250351709
h_4	−0.226264693965
h_5	−0.129766867567
h_6	0.097501605587
h_7	0.027522865530
h_8	−0.031582039318
h_9	0.000553842201
h_{10}	0.004777257511
h_{11}	−0.001077301085

the next chapter, the type of wavelets used in practice are biorthogonal wavelets. In this section, we will describe what biorthogonal wavelets are and why we would want to use them rather than orthogonal wavelets in image compression applications. In order to do that, we first need to make a brief detour and discuss some implementation issues with the use of discrete time filters.

Until now in our development we have ignored the length of the sequence being decomposed. This length becomes important when we examine wavelet decomposition of images. Suppose we wish to filter a sequence of length N, $\{x_k\}_{k=0}^{N-1}$, using a filter with M coefficients $\{h_k\}_{k=0}^{M-1}$ (Figure 15.12). The output of the filter is given by the convolution equation

$$y_k = \sum_{m=0}^{N-1} h_{k-m} x_m$$

where the limits on the summation are because x_m is nonzero only for $m = 0, \cdots, N-1$. If we plot x_m and h_{k-m} for different values of k as shown in Figure 15.13, we can see that the product of h_{k-m} and x_m is nonzero only for specific values of k. When $k < 0$, the nonzero values of h_{k-m} and x_m do not overlap, and, therefore, the product, and hence y_k, is zero. When $k \geqslant 0$ and $k - M + 1 \leqslant 0$, the output y_k is given by

TABLE 15.3 Coefficients for the 20-tap Daubechies low-pass filter.

h_0	0.026670057901
h_1	0.188176800078
h_2	0.527201188932
h_3	0.688459039454
h_4	0.281172343661
h_5	−0.249846424327
h_6	−0.195946274377
h_7	0.127369340336
h_8	0.093057364604
h_9	−0.071394147166
h_{10}	−0.029457536822
h_{11}	0.033212674059
h_{12}	0.003606553567
h_{13}	−0.010733175483
h_{14}	0.001395351747
h_{15}	0.001992405295
h_{16}	−0.000685856695
h_{17}	−0.000116466855
h_{18}	0.000093588670
h_{19}	−0.000013264203

TABLE 15.4 Coefficients for the 6-tap Coiflet low-pass filter.

h_0	−0.051429728471
h_1	0.238929728471
h_2	0.602859456942
h_3	0.272140543058
h_4	−0.051429972847
h_5	−0.011070271529

$$y_k = \sum_{m=0}^{k} h_{k-m} x_m$$

When $k - M + 1 \geqslant 0$ and $k \leqslant N - 1$, the output y_k is given by

$$y_k = \sum_{m=k-M+1}^{k} h_{k-m} x_m$$

TABLE 15.5 **Coefficients for the 12-tap Coiflet low-pass filter.**

h_0	0.011587596739
h_1	−0.029320137980
h_2	−0.047639590310
h_3	0.273021046535
h_4	0.574682393857
h_5	0.294867193696
h_6	−0.054085607092
h_7	−0.042026480461
h_8	0.016744410163
h_9	0.003967883613
h_{10}	−0.001289203356
h_{11}	−0.000509505539

TABLE 15.6 **Coefficients for the 18-tap Coiflet low-pass filter.**

h_0	−0.002682418671
h_1	0.005503126709
h_2	0.016583560479
h_3	−0.046507764479
h_4	−0.043220763560
h_5	0.286503335274
h_6	0.561285256870
h_7	0.302983571773
h_8	−0.050770140755
h_9	−0.058196250762
h_{10}	0.024434094321
h_{11}	0.011229240962
h_{12}	0.006369601011
h_{13}	−0.001820458916
h_{14}	0.000790205101
h_{15}	0.000329665174
h_{16}	−0.000050192775
h_{17}	−0.000024465734

When $N-1 < k \leqslant N+M-2$ the output y_k is given by

$$y_k = \sum_{m=k-M+1}^{N-1} h_{k-m} x_m$$

Finally, when $k > N+M-2$, the nonzero portions of the two sequences no longer overlap and the output is 0. Thus for an input of length N, the output is of length $N+M-1$ where M

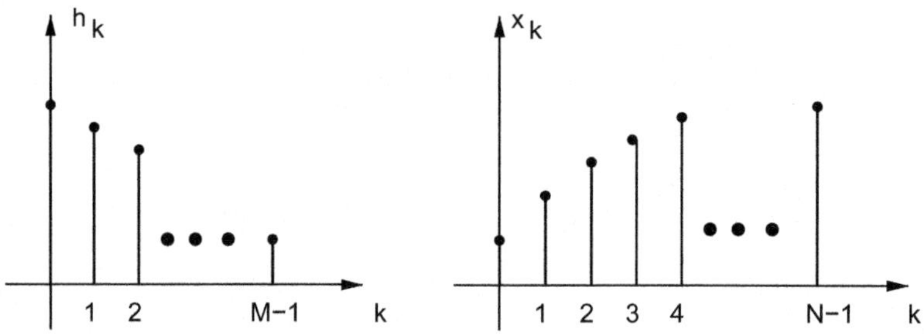

FIGURE 15.12 **An example filter of length M and an input sequence of length N.**

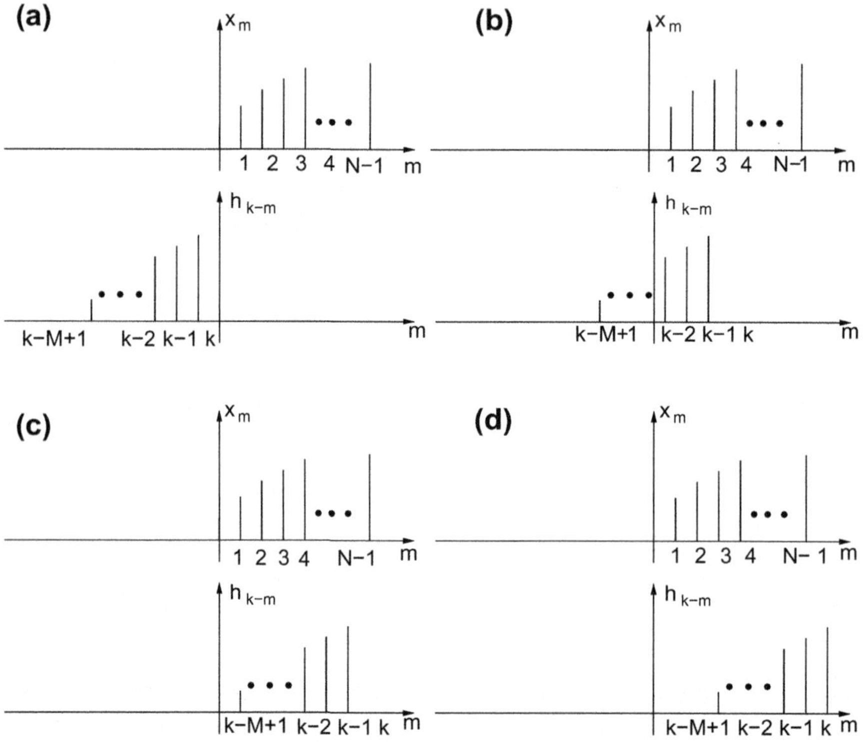

FIGURE 15.13 **A pictorial representation of the components in the convolution sum for different values of k: (a) for $k < 0$; (b) for $0 \leqslant k \leqslant M+1$; (c) for $M+1 < k \leqslant N-1$; and (d) for $N-1 < k \leqslant N+M-2$.**

is the length of the filter. We have increased the number of coefficients that need to be encoded. Clearly this is not a desirable feature for a data compression system. The problem becomes worse as we go through several levels of filtering operations, each resulting in an increase in the number of coefficients.

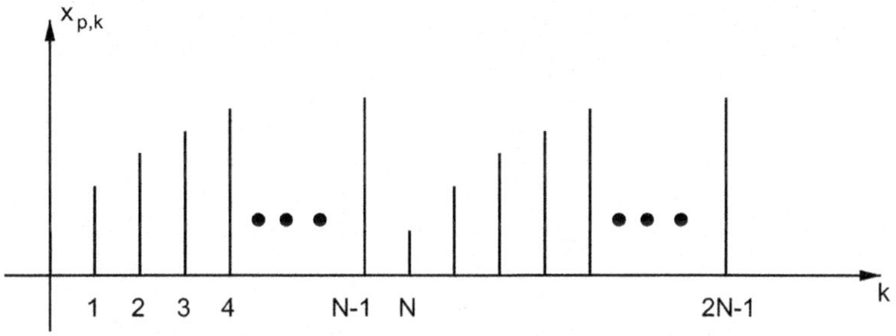

FIGURE 15.14 Periodic extension of the input sequence.

There are several approaches to resolving this problem. One is to generate a periodic extension of the input signal, as shown in Figure 15.14. At first sight this seems counterintuitive as we are increasing the length of the input sequence. However, for a linear time-invariant filter, a periodic input results in a periodic output. As the period is N, for an input sequence of length N, we need to retain only N coefficients. The problem with this approach is identical to the problem we discussed when describing the use of the discrete cosine transform in Chapter 13. Because the initial sample in the sequence and the final sample in the sequence can be very different, the periodic extension requires a significant change in the sample values at the boundaries, for example from the sample at $k = N - 1$ to the sample at $k = N$ in Figure 15.14. As in the case of the Fourier transform discussed in Chapter 13, this results in an increase in the magnitude of the high-frequency coefficients. These coefficients will need to be encoded, resulting in an expenditure of resources. If they are discarded, the result will be distortion in the reconstructed image.

Instead of generating a periodic extension, we can mimic the approach we used for the discrete cosine transform and generate a symmetric extension, as shown in Figure 15.15. In this case, the sequence being filtered is periodic with period $2N$ so we cannot use the fact that linear time-invariant filters preserve periodicity to reduce the number of output samples to N. However, if the output of the filter is also symmetric, we can use the symmetry property to again retain only N samples. This way we can retain the advantage of using the periodic extension without the cost involved with the introduction of spurious high frequencies. Fortunately, there are filters which will provide a symmetric output for a symmetric input. However, for an FIR filter to provide a symmetric output it must itself be symmetric, which in turn means that the filter has to have linear phase. Unfortunately, the only orthogonal wavelet that generates a linear phase FIR filter is the Haar wavelet. If we examine our development of wavelet decomposition we can see that we keep piling more and more requirements on the mother wavelet. Both the analysis filters and both the synthesis filters are based on the same wavelet. Now, in addition to the four filters, we want to add the requirement that the filters also have linear phase. It seems reasonable that the wavelet cannot satisfy this additional constraint. In order to generate linear-phase filters we need to relax some other constraint. The constraint we relax is the requirement of orthogonality.

We relax the condition of orthogonality on a wavelet and its translates by, in a sense, spreading this constraint over a broader set. Instead of requiring that the wavelet or scaling functions be orthogonal to their translates, we require they be orthogonal to the translates of a different separate set of wavelets.

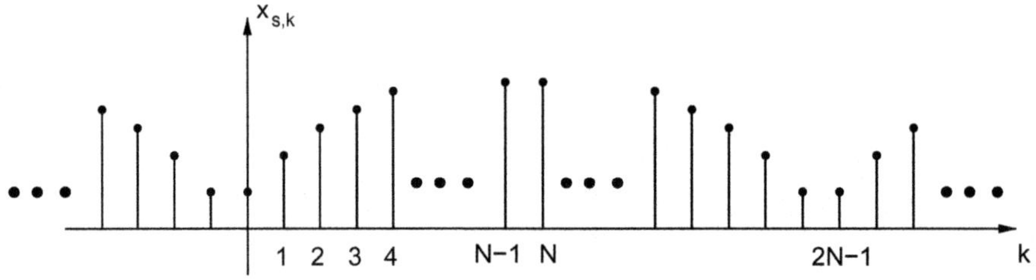

FIGURE 15.15 Symmetric extension of the input sequence.

Let's suppose we have a set of functions $\{g_k(t)\}$ that can be used to represent all functions $f(t)$ in some space $\mathcal{V}$ as

$$f(t) = \sum_k a_k g_k(t) \tag{91}$$

If the set $\{g_k(t)\}$ is orthonormal, we can obtain the coefficients a_j by taking the inner product of both sides of Equation (91) with $g_j(t)$:

$$\langle f(t), g_j(t)\rangle = \langle \sum_k a_k g_k(t), g_j(t)\rangle$$

Also,

$$\langle g_k(t), g_j(t)\rangle = \delta_{k,j}$$

and

$$\langle f(t), g_j(t)\rangle = a_j$$

This is the standard result we have used for finding various orthogonal representations. However, suppose we cannot impose an orthogonality condition on $\{g_k(t)\}$. If we can find another set of functions $\{\tilde{g}_k(t)\}$ such that

$$\langle g_k(t), \tilde{g}_j(t)\rangle = \delta_{k,j}$$

then taking the inner product of both sides of Equation (91) with $\tilde{g}_j(t)$ we can find the coefficient a_j as

$$a_j = \langle f(t), \tilde{g}_j(t)\rangle \tag{92}$$

and we have a *biorthogonal* expansion instead of an orthogonal expansion. In the case of the orthogonal expansion, the same set is used to provide both the analysis and the synthesis of the function. If $\{g_k(t)\}$ is an orthogonal basis set, then the analysis equation for a function $f(t)$ is given by

$$a_j = \langle f(t), g_j(t)\rangle \tag{93}$$

and the synthesis equation is given by

$$f(t) = \sum_k a_k g_k(t)$$

That is, both the analysis and synthesis functions are carried out using the same basis set. In this case of the biorthogonal expansion, analysis and synthesis are carried out by dual basis sets. In the development above, the synthesis is carried out using the set $\{g_k(t)\}$ as in Equation (91) while the analysis is carried out using the set $\{\tilde{g}_j(t)\}$ as in Equation (92).

In terms of wavelets, if we have a scaling function that satisfies the multiresolution analysis equation

$$\phi(t) = \sum_k h_k \sqrt{2}\phi(2t - k)$$

and a dual set that also satisfies the MRA

$$\tilde{\phi}(t) = \sum_k \tilde{h}_k \sqrt{2}\tilde{\phi}(2t - k)$$

and the two sets $\{\phi(t)\}$ and $\{\tilde{\phi}(t)\}$ are orthogonal, that is

$$\langle \phi(t), \tilde{\phi}(t - k) \rangle = \delta(k)$$

then we can obtain analysis filters from one set and synthesis filters from the other [212].

Because we do not have to obtain all four filters from the same mother wavelet or scaling function, we have much more freedom in picking wavelets to use. In particular, we can find scaling functions and wavelets with linear phase implementations.

15.7 Lifting

We have discussed the idea of wavelet decomposition mainly in terms of the separation of signals in the frequency domain. The input signal, through appropriate use of filters based on wavelets, gets decomposed into a low-frequency signal and a high-frequency signal. The low- and high-frequency signals are then downsampled, encoded, and transmitted. The received signal is decoded and upsampled, and the low- and high-frequency signals are recovered using low- and high-frequency reconstruction filters. An alternative way of looking at the decomposition is to view it on spatial, or temporal, terms. Frequency can be thought of as a way to measure how fast a signal changes either in space or in time, and as we saw in Chapter 11, the slowly changing components of a signal can be extracted through the use of prediction. Thus, high frequencies correspond to less predictable components of the signal and low frequencies correspond to more predictable components of the signal where the prediction is based on spatial or temporal correlation. This fact is utilized in an efficient implementation of wavelets developed by Wim Sweldens called *lifting* [213].

We began Chapter 14 on subband coding with a simple motivating example where we separated a sequence into its even-indexed and odd-indexed components. We return to that example again using different terminology to motivate our discussion of lifting. Consider the system shown in Figure 15.16 where the box labeled *Split* breaks the incoming sequence into the even- and odd-indexed sequences. At the decoder, the box *Merge* performs the opposite function. If the sequence is relatively slow changing, we can reduce the data rate by predicting

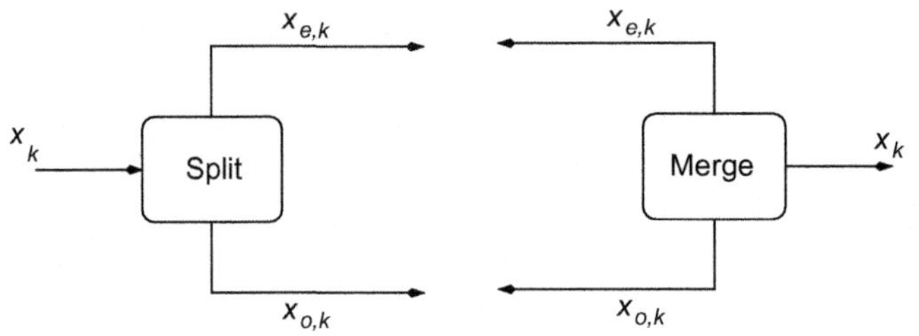

FIGURE 15.16 **Decomposing a sequence into its odd- and even-indexed components.**

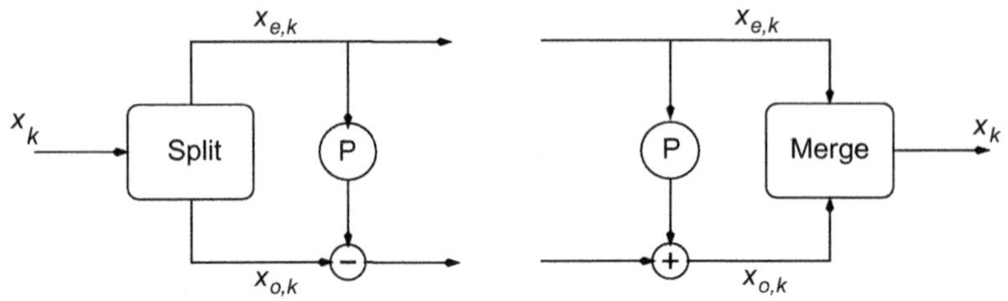

FIGURE 15.17 **Generating the high-frequency difference sequence.**

one sequence using the other. In this case, let's use the even sequence to predict the odd sequence as shown in Figure 15.17:

$$d_k = x_{o,k} - P(\{x_{e,k}\})$$

As long as the values of the sequences are preserved, the original sequence can be reconstructed by reversing the operation at the encoder. If we have a good model for the signal, the prediction residual $\{d_n\}$ will have a much smaller dynamic range than the original sequence $\{x_{o,n}\}$. Let's suppose the input sequence is highly correlated. In this case, we can predict an odd-indexed value by using the even-indexed neighbor,

$$\begin{aligned} d_k &= x_{o,k} - x_{e,k} \\ &= x_{2k+1} - x_{2k} \end{aligned}$$

The magnitude of the $\{d_k\}$ values will depend on the linear correlation between the samples of the original signal. We can see that the sequence $\{d_k\}$ corresponds, in some sense, to the high-frequency component of the original sequence. The sequence $\{x_{e,k}\}$, on the other hand, is simply the sampled version of the original signal. Depending on the frequency content of the original signal, this subsampling could lead to significant aliasing. We can try and reduce this aliasing effect by attempting to preserve some of the statistical properties of the original signal. The simplest thing is to preserve the average value of the sequence. We can do this by using the odd values, or rather the residual of the odd-indexed values, to update the even-indexed values as shown in Figure 15.18, where

$$s_k = x_{e,k} + U(\{x_{o,k}\})$$

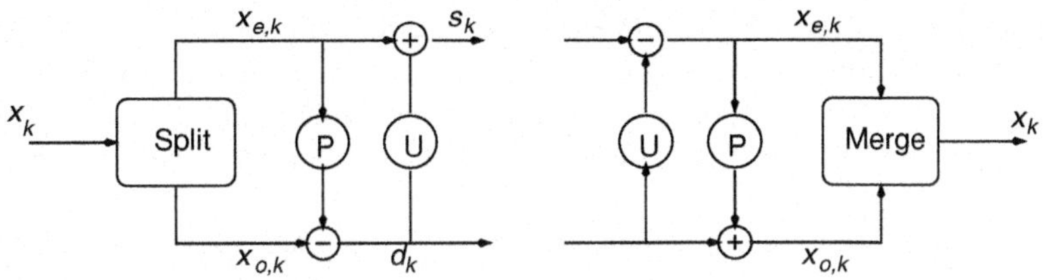

FIGURE 15.18 Decomposing a sequence using prediction and updating operations.

Corresponding to the simple prediction operation we can create a simple update equation using the weighted sum of the neighboring residual values:

$$s_k = x_{e,k} + \alpha d_k$$

We would like the sequence $\{s_k\}$ to have the same average value as the sequence $\{x_k\}$. As there are only half as many elements in the sequence $\{s_k\}$ as in the sequence $\{x_k\}$, this means that

$$\sum_k s_k = \frac{1}{2}\sum_k x_k \tag{94}$$

where the summation range for the right-hand sum is twice that of the left-hand sum. Substituting for s_k we obtain

$$\begin{aligned}\sum_k s_k &= \sum_k (x_{e,k} + \alpha d_k) \\ &= \sum_k (x_{2k} + \alpha(x_{2k+1} - x_{2k})) \\ &= (1-\alpha)\sum_k x_{2k} + \alpha \sum_k x_{2k+1}\end{aligned}$$

where we have ignored boundary effects. One way we can get this sum to satisfy Equation (94) is to assume that the sum of the even-indexed components and the odd-indexed components is the same. Therefore,

$$1 - \alpha = \alpha = \frac{1}{2}$$

Thus,

$$s_k = x_{e,k} + \frac{d_k}{2} \tag{95}$$

$$= x_{e,k} + \frac{x_{2k+1} - x_{2k}}{2} \tag{96}$$

$$= \frac{x_{2k+1} + x_{2k}}{2} \tag{97}$$

But this (within a scale factor) is simply the Haar wavelet expansion! The Haar wavelet expansion is such a simple expansion that this might seem like a rather small return for all this work, except for two factors. We developed this expansion without using any of the machinery of wavelets, in fact without explicitly using the idea of the frequency domain. All the development was in the spatial domain using spatial concepts like prediction. Secondly, we have used very simple (the simplest) predictor and update operators. By using more complicated predictors and update operators, we can get more sophisticated wavelet implementations.

Let's use slightly more complicated prediction and update operators. Instead of using single neighbor lets predict an odd-indexed value by using the the average of two even-indexed neighbors:

$$\begin{aligned} d_k &= x_{o,k} - \frac{x_{e,k} + x_{e,k+1}}{2} \\ &= x_{2k+1} - \frac{x_{2k} + x_{2k+2}}{2} \end{aligned}$$

Notice that if the input sequence corresponds to a linear or constant function, in other words a polynomial of degree less than 2, the detail coefficients will all be zero.

For the corresponding update operator, let's use the weighted sum of the residual neighbors:

$$s_k = x_{e,k} + \alpha(d_{k-1} + d_k)$$

To preserve the average value, we need to satisfy Equation (94):

$$\begin{aligned} \sum_k s_k &= \sum_k \left(x_{e,k} + \alpha(d_{k-1} + d_k)\right) \\ &= \sum_k \left(x_{2k} + \alpha\left(x_{2k-1} - \frac{x_{2k-2} + x_{2k}}{2} + x_{2k+1} - \frac{x_{2k} + x_{2k+2}}{2}\right)\right) \\ &= (1 - 2\alpha)\sum_k x_{2k} + 2\alpha \sum_k x_{2k+1} \end{aligned}$$

Again, assuming that the sum of the even-indexed components and the odd-indexed components are the same, we get

$$1 - 2\alpha = 2\alpha = \frac{1}{2}$$

which implies that α equals $1/4$. Thus,

$$s_k = x_{e,k} + \frac{(d_{k-1} + d_k)}{4}$$

This is an implementation of the Cohen-Daubechies-Feaveau (2,2) (CDF(2,2)) wavelet [213]. The numbers in parenthesis indicate the number of moments that are preserved in the update step and the order of the input polynomial below which the detail coefficients will be zero. In both of these examples we have used one prediction and one update step. In practice this is not a restriction. We can use any number of update and prediction steps to implement more and more sophisticated wavelet decompositions. An example that will be especially useful to us in the next chapter is CDF(4,4) biorthogonal wavelets. This requires two prediction and two

TABLE 15.7 Coefficients for the lifting implementation of the CDF(4,4) biorthogonal wavelet decomposition [214].

α	−1.58613432
β	−0.052980118
γ	0.882911075
δ	0.443506852

update steps:

$$\begin{aligned} d_k &\leftarrow x_{o,k} + \alpha(x_{e,k} + x_{e,k+1}) \\ s_k &\leftarrow x_{e,k} + \beta(d_{k-1} + d_k) \\ d_k &\leftarrow d_k + \gamma(s_k + s_{k+1}) \\ s_k &\leftarrow s_k + \delta(d_{k-1} + d_k) \end{aligned}$$

Here, we have used $\leftarrow$ instead of the equality sign to indicate that the computations are in-place. The values of the coefficients needed to satisfy the moment preservation constraints are given in Table 15.7. Unlike the previous case, in this example we also need to scale the two sets of coefficients where the scaling factor is $\kappa = 1.230174105$. The detail coefficients are divided by κ while the low-frequency coefficients are multiplied by κ.

As the decoder is a simple reversal of whatever is done at the encoder all we need to do is to focus on either the analysis or synthesis steps. Daubechies and Swelden [214] provide a step-by-step implementation of a general wavelet decomposition using a factorization of the polyphase matrix described in Chapter 14.

Apart from a mechanism for generating new kinds of wavelets based on some very simple steps, there are a number of benefits to lifting implementations. Lifting operations can provide significant reductions in implementation complexity. Also, we can see that the lifting operations are being performed "in-place." We do not need to store the values that occur earlier in the lifting pipeline. This can result in a significant reduction in the amount of memory needed for implementation. Finally, the use of the lifting approach provides us with a very concrete way of looking at wavelet decompositions, which can lead to better insight and better designs.

15.8 Summary

In this chapter, we have introduced the concepts of wavelets and multiresolution analysis; and we have seen how we can use wavelets to provide an efficient decomposition of signals prior to compression. We have introduced the concept of biorthogonal wavelets, which significantly enrich the choices we have for developing efficient wavelet decomposition approaches. We have also described several ways of implementing wavelets, including the lifting approach, which gives us a more physical feel for the process of wavelet decomposition. Wavelets and their applications are currently areas of intensive research.

Further Reading

1. There are a number of excellent introductory books on wavelets. One of the most accessible is *Introduction to Wavelets and Wavelet Transforms—A Primer*, by C.S. Burrus, R.A. Gopinath, and H. Guo [212].
2. Probably the best mathematical source on wavelets is the book *Ten Lectures on Wavelets*, by I. Daubechies [279].
3. There are a number of tutorials on wavelets available on the Internet. The best source for all matters related to wavelets (and more) on the Internet is "The Wavelet Digest" (<http://www.wavelet.org>). This site includes pointers to many other interesting and useful sites dealing with different applications of wavelets.

15.9 Projects and Problems

1. In this problem we consider the boundary effects encountered when using the short-term Fourier transform. Consider the signal

$$f(t) = \sin(2t)$$

   (a) Find the Fourier transform $F(\omega)$ of $f(t)$.

   (b) Find the STFT $F_1(\omega)$ of $f(t)$ using a rectangular window

$$g(t) = \begin{cases} 1 & -2 \leqslant t \leqslant 2 \\ 0 & \text{otherwise} \end{cases}$$

   for the interval $[-2, 2]$.

   (c) Find the STFT $F_2(\omega)$ of $f(t)$ using a window

$$g(t) = \begin{cases} 1 + \cos(\frac{\pi}{2}t) & -2 \leqslant t \leqslant 2 \\ 0 & \text{otherwise} \end{cases}$$

   (d) Plot $|F(\omega)|$, $|F_1(\omega)|$, and $|F_2(\omega)|$. Comment on the effect of using different window functions.

2. For the function

$$f(t) = \begin{cases} 1 + \sin(2t) & 0 \leqslant t \leqslant 1 \\ \sin(2t) & \text{otherwise} \end{cases}$$

   use the Haar wavelet to find and plot the coefficients $\{c_{j,k}\}$, $j = 0, 1, 2$; $k = 0, \ldots, 10$.

16

Wavelet-Based Image Compression

16.1 Overview

In this chapter we examine the implementations of several wavelet-based image compression schemes. We begin with the earliest, the embedded zerotree wavelet coder, and then look at an image coder based on set partitioning, and finally, we look at JPEG2000—the latest image compression standard.

16.2 Introduction

One of the most popular applications of wavelets has been image compression. The JPEG 2000 standard, which is increasingly becoming the compression method of choice for a wide range of applications from compression of movies to compression of medical images, uses wavelets instead of the discrete cosine transform (DCT) to perform decomposition of the image. During our discussion, we have always referred to the signal to be decomposed as a one-dimensional signal; however, images are two-dimensional signals. There are two approaches to the subband decomposition of two-dimensional signals: using two-dimensional filters, or using separable transforms that can be implemented using one-dimensional filters on the rows first and then on the columns (or vice versa). Most approaches, including JPEG 2000, use the second approach.

In Figure 16.1 we show how an image can be decomposed using subband decomposition. We begin with an $N \times M$ image. We filter each row and then downsample to obtain two $N \times \frac{M}{2}$ images. We then filter each column and subsample the filter output to obtain four $\frac{N}{2} \times \frac{M}{2}$ images. Of the four subimages, the one obtained by low-pass filtering of the rows

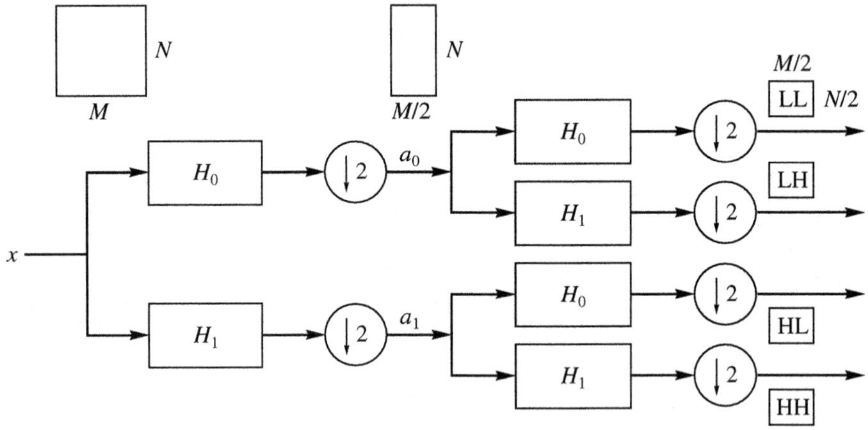

FIGURE 16.1 Subband decomposition of an $N \times M$ image.

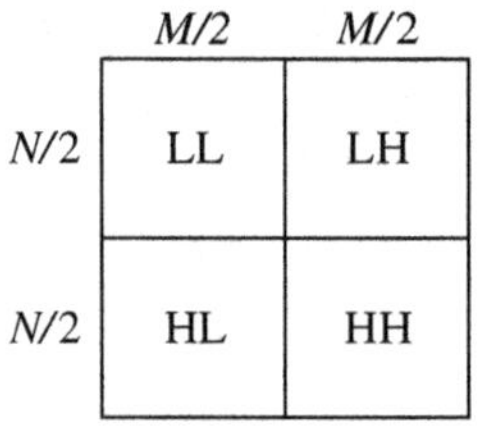

FIGURE 16.2 First-level decomposition.

and columns is referred to as the LL image; the one obtained by low-pass filtering of the rows and high-pass filtering of the columns is referred to as the LH image; the one obtained by high-pass filtering of the rows and low-pass filtering of the columns is called the HL image; and the subimage obtained by high-pass filtering of both the rows and columns is referred to as the HH image. This decomposition is sometimes represented as shown in Figure 16.2. Each of the subimages obtained in this fashion can then be filtered and subsampled to obtain four more subimages. This process can be continued until the desired subband structure is obtained. Three popular structures are shown in Figure 16.3. In the structure in Figure 16.3(a), in each step the LL subimage is further decomposed into four more subimages, resulting in a total of thirteen subimages. This is one of the more popular decompositions.

Example 16.2.1:

Let's use the Daubechies wavelet filter to repeat what we did in Examples 14.12.2 and 14.12.3 using the Johnston and the Smith-Barnwell filters. If we use the 4-tap Daubechies filter, we obtain the decomposition shown in Figure 16.4. Notice that even though we are only using a 4-tap filter, we get results comparable to the 16-tap Johnston filter and the 8-tap Smith-Barnwell filters.

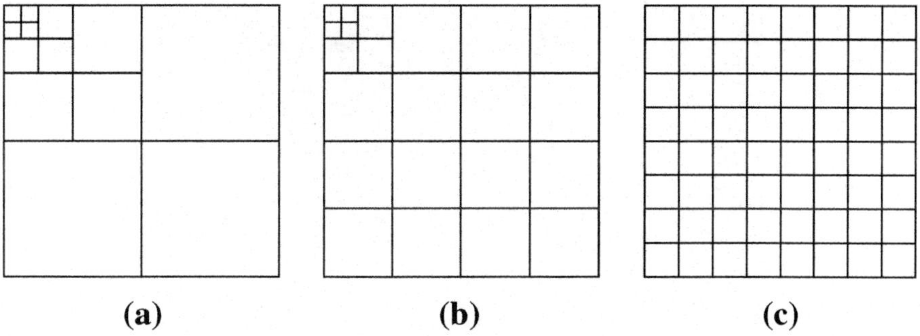

FIGURE 16.3 Three popular subband structures.

FIGURE 16.4 Decomposition of Sinan image using the four-tap Daubechies filter.

If we now encode this image at the rate of 0.5 bits per pixel, we get the reconstructed image shown in Figure 16.5. Notice that the quality is comparable to that obtained using filters requiring two or four times as much computation. ♦

In this example, we used a simple scalar quantizer for quantization of the coefficients. However, if we use strategies that are motivated by the properties of the coefficients themselves, we can obtain significant performance improvements. In the next sections, we examine two popular quantization strategies developed specifically for wavelets.

FIGURE 16.5 **Reconstruction of Sinan image encoded using 0.5 bits per pixel and the four-tap Daubechies filter.**

16.3 Embedded Zerotree Coder

The embedded zerotree wavelet (EZW) coder was introduced by Shapiro [215]. It is a quantization and coding strategy that incorporates some characteristics of the wavelet decomposition. Just as the quantization and coding approaches used in the JPEG standard, which were motivated by the characteristics of the coefficients, were superior to the generic zonal coding algorithms, the EZW approach and its descendants significantly outperform some of the generic approaches. The particular characteristic used by the EZW algorithm is that there are wavelet coefficients in different subbands that represent the same spatial location in the image. If the decomposition is such that the size of the different subbands is different (the first two decompositions in Figure 16.3), then a single coefficient in the smaller subband may represent the same spatial location as multiple coefficients in the other subbands.

In order to put our discussion on more solid ground, consider the ten-band decomposition shown in Figure 16.6. The coefficient a in the upper-left corner of Band I represents the same spatial location as coefficients a_1 in Band II, a_2 in Band III, and a_3 in Band IV. In turn, the coefficient a_1 represents the same spatial location as coefficients a_{11}, a_{12}, a_{13}, and a_{14} in Band V. Each of these pixels represents the same spatial location as four pixels in Band VIII, and so on. In fact, we can visualize the relationships of these coefficients in the form of a tree. The coefficient a forms the root of the tree with three descendants a_1, a_2, and a_3. The coefficient a_1 has descendants a_{11}, a_{12}, a_{13}, and a_{14}. The coefficient a_2 has descendants a_{21}, a_{22}, a_{23}, and a_{24}, and the coefficient a_3 has descendants a_{31}, a_{32}, a_{33}, and a_{34}. Each of these coefficients in turn has four descendants, making a total of 64 coefficients in this tree. A pictorial representation of the tree is shown in Figure 16.7.

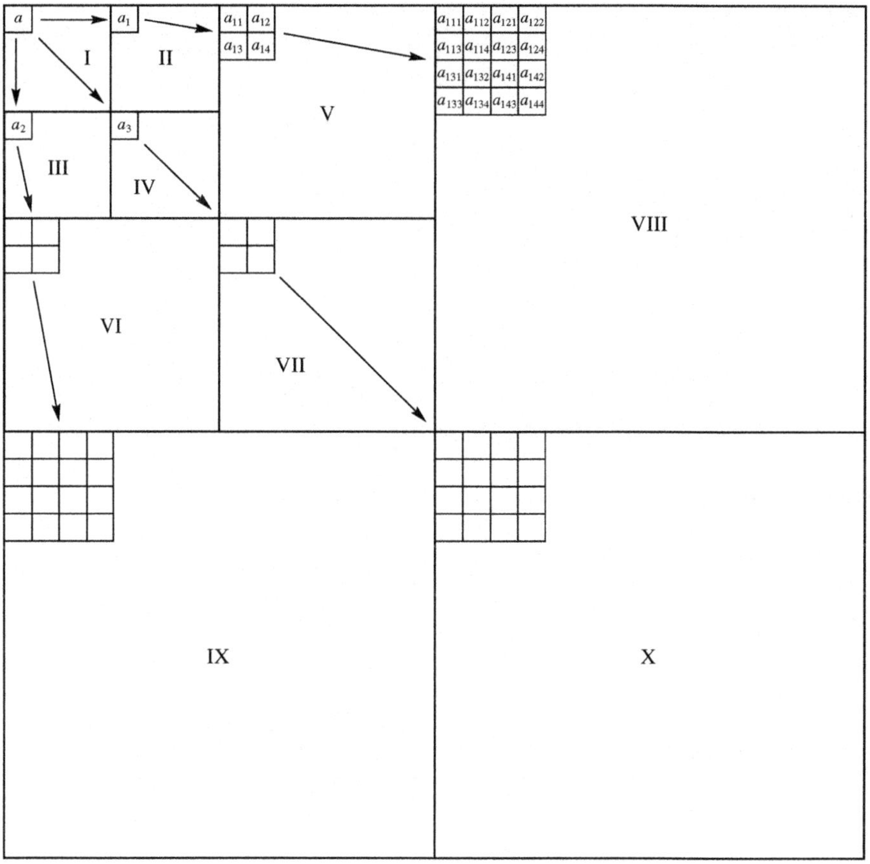

FIGURE 16.6 **A ten-band wavelet decomposition.**

Recall that when natural images are decomposed in this manner, most of the energy is compacted into the lower bands. Thus, in many cases, the coefficients closer to the root of the tree have higher magnitudes than coefficients further away from the root. This means that often if a coefficient has a magnitude less than a given threshold, all its descendants will have magnitudes less than that threshold. In a scalar quantizer, the outer levels of the quantizer correspond to larger magnitudes. Consider the 3-bit quantizer shown in Figure 16.8. If we determine that all coefficients arising from a particular root have magnitudes smaller than T_0 and we inform the decoder of this situation, then for all coefficients in that tree we need only use 2 bits per sample, while getting the same performance as we would have obtained using the 3-bit quantizer. If the binary coding scheme used in Figure 16.8 is used, in which the first bit is the sign bit and the next bit is the most significant bit of the magnitude, then the information that a set of coefficients has value less than T_0 is the same as saying that the most significant bit of the magnitude is 0. If there are N coefficients in the tree, this is a savings of N bits minus however many bits are needed to inform the decoder of this situation.

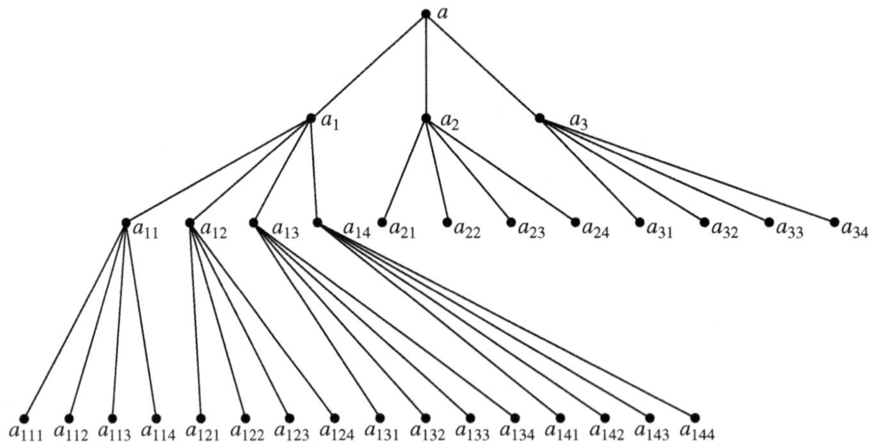

FIGURE 16.7 **Data structure used in the EZW coder.**

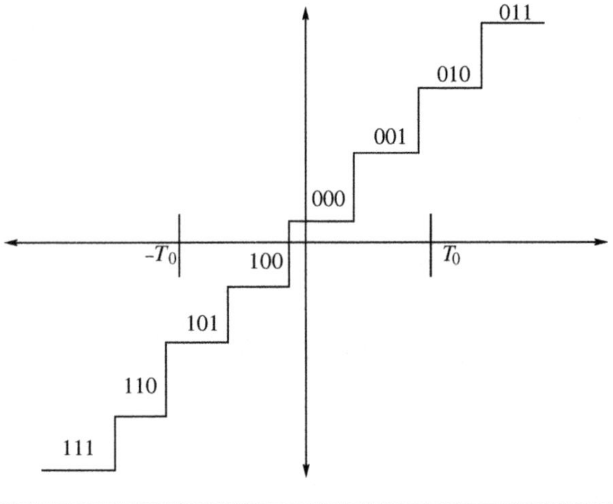

FIGURE 16.8 **A 3-bit midrise quantizer.**

Before we describe the EZW algorithm, we need to introduce some terminology. Given a threshold T, if a given coefficient has a magnitude greater than T, it is called a *significant* coefficient at level T. If the magnitude of the coefficient is less than T (it is insignificant) and all of its descendants have magnitudes less than T, then the coefficient is called a *zerotree root*. Finally, it might happen that the coefficient itself is less than T, but some of its descendants have a value greater than T. Such a coefficient is called an *isolated zero*.

The EZW algorithm is a multiple-pass algorithm with each pass consisting of two steps: *significance map encoding* or the *dominant pass*, and *refinement* or the *subordinate pass*. If c_{max} is the value of the largest coefficient, the initial value of the threshold T_0 is given by

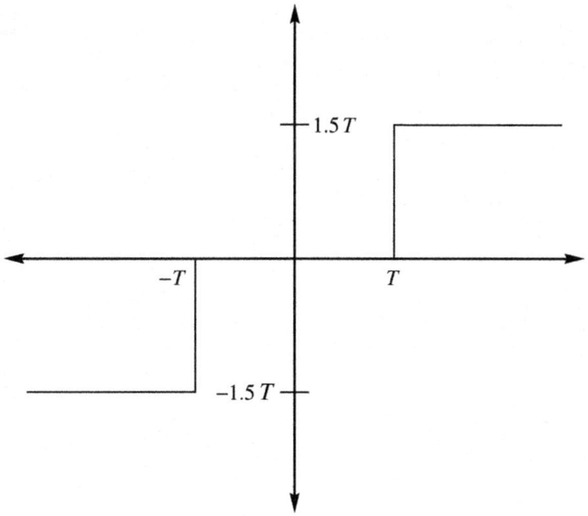

FIGURE 16.9 A three-level midtread quantizer.

$$T_0 = 2^{\lfloor \log_2 c_{\max} \rfloor} \tag{1}$$

This selection guarantees that the largest coefficient will lie in the interval $[T_0, 2T_0)$. In each pass, the threshold T_i is reduced to half the value it had in the previous pass:

$$T_i = \frac{1}{2} T_{i-1} \tag{2}$$

For a given value of T_i, we assign one of four possible labels to the coefficients: *significant positive (sp), significant negative (sn), zerotree root (zr)*, and *isolated zero (iz)*. If we use a fixed-length code, we need 2 bits to represent each of the labels. Note that when a coefficient is labeled a zerotree root, we do not need to label its descendants. This assignment is referred to as *significance map coding*.

We can view the significance map coding in part as quantization using a three-level midtread quantizer. This situation is shown in Figure 16.9. The coefficients labeled *significant* are simply those that fall in the outer levels of the quantizer and are assigned an initial reconstructed value of $1.5T_i$ or $-1.5T_i$, depending on whether the coefficient is positive or negative. Note that selecting T_i according to (1) and (2) guarantees the significant coefficients will lie in the interval $[T, 2T]$. Once a determination of significance has been made, the significant coefficients are included in a list for further refinement in the refinement or subordinate passes. In the refinement pass, we determine whether the coefficient lies in the upper or lower half of the interval $[T, 2T]$. In successive refinement passes, as the value of T is reduced, the interval containing the significant coefficient is narrowed still further and the reconstruction is updated accordingly. An easy way to perform the refinement is to take the difference between the coefficient value and its reconstruction and quantize it using a two-level quantizer with reconstruction values $\pm T/4$. This quantized value is then added on to the current reconstruction value as a correction term.

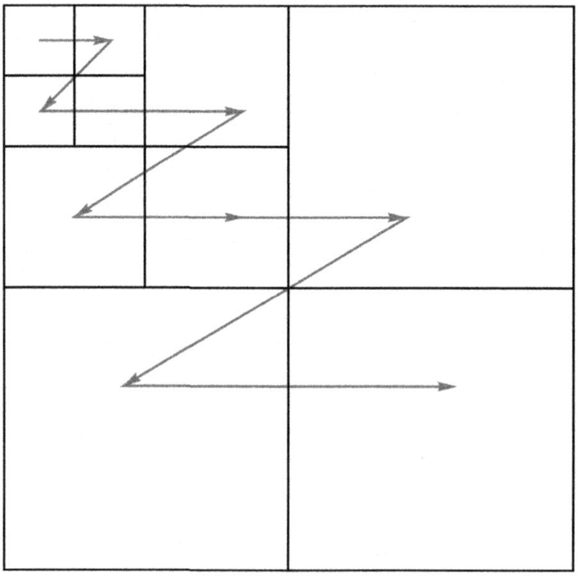

FIGURE 16.10 Scanning of wavelet coefficients for encoding using the EZW algorithm.

The wavelet coefficients not previously determined to be significant are scanned in the manner depicted in Figure 16.10, with each parent node in a tree scanned before its offspring. This makes sense because if the parent is determined to be a zerotree root, we would not need to encode the offspring.

Although this may sound confusing, in order to see how simple the encoding procedure actually is, let's use an example.

Example 16.3.1:

Let's use the following seven-level decomposition to demonstrate the various steps of EZW:

26	6	13	10
−7	7	6	4
4	−4	4	−3
2	−2	−2	0

To obtain the initial threshold value T_0, we find the maximum magnitude coefficient, which in this case is 26. Then

$$T_0 = 2^{\lfloor \log_2 26 \rfloor} = 16$$

Comparing the coefficients against 16, we find 26 is greater than 16 so we send *sp*. The next coefficient in the scan is 6, which is less than 16. Furthermore, its descendants (13, 10, 6, and 4) are all less than 16. Therefore, 6 is a zerotree root, and we encode this entire set with the label *zr*. The next coefficient in the scan is -7, which is also a zerotree root, as is 7, the final element in the scan. We do not need to encode the rest of the coefficients separately because they have already been encoded as part of the various zerotrees. The sequence of labels to be transmitted at this point is

$$sp\ zr\ zr\ zr$$

Since each label requires 2 bits (for fixed-length encoding), we have used up 8 bits from our bit budget. The only significant coefficient in this pass is the coefficient with a value of 26. We include this coefficient in our list to be refined in the subordinate pass. Calling the subordinate list L_S, we have

$$L_S = \{26\}$$

The reconstructed value of this coefficient is $1.5T_0 = 24$, and the reconstructed bands look like this:

24	0	0	0
0	0	0	0
0	0	0	0
0	0	0	0

The next step is the subordinate pass, in which we obtain a correction term for the reconstruction value of the significant coefficients. In this case, the list L_S contains only one element. The difference between this element and its reconstructed value is 26 – 24 = 2. Quantizing this with a two-level quantizer with reconstruction levels $\pm T_0/4$, we obtain a correction term of 4. Thus, the reconstruction becomes 24+4=28. Transmitting the correction term costs a single bit, therefore at the end of the first pass we have used up 9 bits. Using only these 9 bits, we obtain the following reconstruction:

28	0	0	0
0	0	0	0
0	0	0	0
0	0	0	0

We now reduce the value of the threshold by a factor of two and repeat the process. The value of T_1 is 8. We rescan the coefficients that have not yet been deemed significant. To emphasize the fact that we do not consider the coefficients that have been deemed significant in the previous pass, we replace them with $\star$:

⋆	6	13	10
− 7	7	6	4
4	− 4	4	− 3
2	− 2	− 2	0

The first coefficient we encounter has a value of 6. This is less than the threshold value of 8; however, the descendants of this coefficient include coefficients with values of 13 and 10. Therefore, this coefficient cannot be classified as a zerotree root. This is an example of what we defined as an isolated zero. The next two coefficients in the scan are -7 and 7. Both of these coefficients have magnitudes less than the threshold value of 8. Furthermore, all of their descendants also have magnitudes less than 8. Therefore, these two coefficients are coded as *zr*. The next two elements in the scan are 13 and 10, which are both coded as *sp*. The final two elements in the scan are 6 and 4. These are both less than the threshold, but they do not have any descendants. We code these coefficients as *iz*. Thus, this dominant pass is coded as

$$iz\ zr\ zr\ sp\ sp\ iz\ iz$$

which requires 14 bits, bringing the total number of bits used to 23. The significant coefficients are reconstructed with values $1.5T_1 = 12$. Thus, the reconstruction at this point is

28	0	12	12
0	0	0	0
0	0	0	0
0	0	0	0

We add the new significant coefficients to the subordinate list:

$$L_S = \{26, 13, 10\}$$

In the subordinate pass, we take the difference between the coefficients and their reconstructions and quantize these to obtain the correction or refinement values for these coefficients. The possible values for the correction terms are $\pm T_1/4 = \pm 2$:

$$\begin{aligned} 26 - 28 = -2 &\Rightarrow \text{Correction term} = -2 \\ 13 - 12 = 1 &\Rightarrow \text{Correction term} = 2 \\ 10 - 12 = -2 &\Rightarrow \text{Correction term} = -2 \end{aligned} \tag{3}$$

Each correction requires a single bit, bringing the total number of bits used to 26. With these corrections, the reconstruction at this stage is

26	0	14	10
0	0	0	0
0	0	0	0
0	0	0	0

If we go through one more pass, we reduce the threshold value to 4. The coefficients to be scanned are

⋆	6	⋆	⋆
− 7	7	6	4
4	− 4	4	− 3
2	− 2	− 2	0

The dominant pass results in the following coded sequence:

$$sp\ sn\ sp\ sp\ sp\ sp\ sn\ iz\ iz\ sp\ iz\ iz\ iz$$

This pass cost 26 bits, equal to the total number of bits used previous to this pass. The reconstruction upon decoding of the dominant pass is

26	6	14	10
−6	6	6	6
6	−6	6	0
0	0	0	0

The subordinate list is

$$L_S = \{26, 13, 10, 6 - 7, 7, 6, 4, 4, -4, 4\}$$

By now it should be reasonably clear how the algorithm works. We continue encoding until we have exhausted our bit budget or until some other criterion is satisfied. ♦

There are several observations we can make from this example. Notice that the encoding process is geared to provide the most bang for the bit at each step. At each step the bits are used to provide the maximum reduction in the reconstruction error. If at any time the encoding is interrupted, the reconstruction using this (interrupted) encoding is the best that the algorithm can provide using this many bits. The encoding improves as more bits are transmitted. This form of coding is called *embedded coding*. In order to enhance this aspect of the algorithm, we can also sort the subordinate list at the end of each pass using information available to both encoder and decoder. This increases the likelihood of larger coefficients being encoded first, thus providing for a greater reduction in the reconstruction error.

Finally, in the example we determined the number of bits used by assuming fixed-length encoding. In practice, arithmetic coding is used, providing a further reduction in rate.

16.4 Set Partitioning in Hierarchical Trees

The SPIHT (Set Partitioning in Hierarchical Trees) algorithm is a generalization of the EZW algorithm and was proposed by Amir Said and William Pearlman [216]. Recall that in EZW we transmit a lot of information for little cost when we declare an entire subtree to be insignificant and represent all the coefficients in it with a zerotree root label zr. The SPIHT algorithm uses a partitioning of the trees (which in SPIHT are called *spatial orientation trees*) in a manner that tends to keep insignificant coefficients together in larger subsets. The partitioning decisions are binary decisions that are transmitted to the decoder, providing a significance map encoding that is more efficient than EZW. In fact, the efficiency of the significance map encoding in SPIHT is such that arithmetic coding of the binary decisions provides very little gain. The thresholds used for checking significance are powers of two, so in essence the SPIHT algorithm sends the binary representation of the integer value of the wavelet coefficients. As in EZW, the significance map encoding, or set partitioning and ordering step, is followed by a refinement step in which the representations of the significant coefficients are refined.

Let's briefly describe the algorithm and then look at some examples of its operation. However, before we do that we need to get familiar with some notation. The data structure used by the SPIHT algorithm is similar to that used by the EZW algorithm—although not the same. The wavelet coefficients are again divided into trees originating from the lowest resolution band (Band I in our case). The coefficients are grouped into 2×2 arrays that, except for the coefficients in Band I, are offsprings of a coefficient of a lower resolution band. The coefficients in the lowest resolution band are also divided into 2×2 arrays. However, unlike the EZW case, all but one of them are root nodes. The coefficient in the top-left corner of the array does not have any offsprings. The data structure is shown pictorially in Figure 16.11 for a seven-band decomposition.

The trees are further partitioned into four types of sets, which are sets of coordinates of the coefficients:

- $\mathcal{O}(i, j)$ This is the set of coordinates of the offsprings of the wavelet coefficient at location (i, j). As each node can either have four offsprings or none, the size of $\mathcal{O}(i, j)$ is either zero or four. For example, in Figure 16.11 the set $\mathcal{O}(0, 1)$ consists of the coordinates of the coefficients b_1, b_2, b_3, and b_4.

- $\mathcal{D}(i, j)$ This is the set of all descendants of the coefficient at location (i, j). Descendants include the offsprings, the offsprings of the offsprings, and so on. For example, in Figure 16.11 the set $\mathcal{D}(0, 1)$ consists of the coordinates of the coefficients $b_1, \ldots, b_4, b_{11}, \ldots, b_{14}, \ldots, b_{44}$. Because the number of offsprings can either be zero

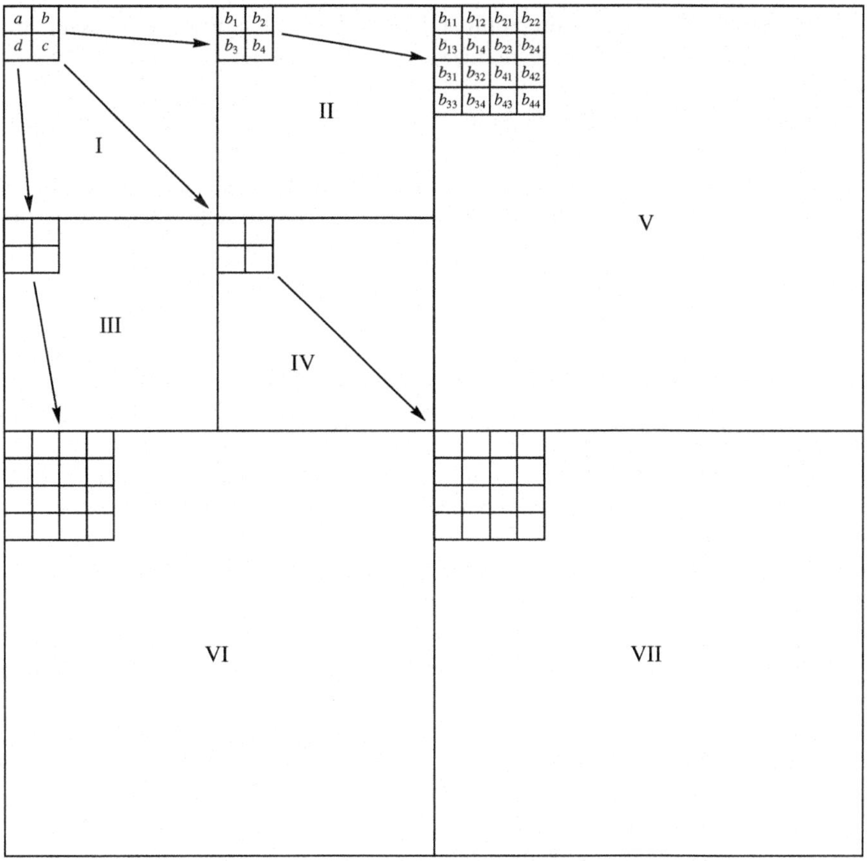

FIGURE 16.11 **Data structure used in the SPIHT algorithm.**

or four, the size of $\mathcal{D}(i, j)$ is either zero or a sum of powers of four.

- $\mathcal{H}$ This is the set of all root nodes—essentially Band I in the case of Figure 16.11.
- $\mathcal{L}(i, j)$ This is the set of coordinates of all the descendants of the coefficient at location (i, j) except for the immediate offsprings of the coefficient at location (i, j). In other words,

$$\mathcal{L}(i, j) = \mathcal{D}(i, j) - \mathcal{O}(i, j)$$

In Figure 16.11 the set $\mathcal{L}(0, 1)$ consists of the coordinates of the coefficients $b_{11}, \ldots, b_{14}, \ldots, b_{44}$.

A set $\mathcal{D}(i, j)$ or $\mathcal{L}(i, j)$ is said to be significant if any coefficient in the set has a magnitude greater than the threshold. Finally, thresholds used for checking significance are powers of two, so in essence the SPIHT algorithm sends the binary representation of the integer value of the wavelet coefficients. The bits are numbered with the least significant bit being the zeroth bit, the next bit being the first significant bit, and the kth bit being referred to as the $k-1$ most significant bit.

With these definitions under our belt, let us now describe the algorithm. The algorithm makes use of three lists: the *list of insignificant pixels* (LIP), the *list of significant pixels* (LSP), and the *list of insignificant sets* (LIS). The LSP and LIP will contain the coordinates of coefficients, while the LIS will contain the coordinates of the roots of sets of type $\mathcal{D}$ or $\mathcal{L}$. We start by determining the initial value of the threshold. We do this by calculating

$$n = \lfloor \log_2 c_{\max} \rfloor$$

where $c_{\max}$ is the maximum magnitude of the coefficients to be encoded. The LIP is initialized with set $\mathcal{H}$. Those elements of $\mathcal{H}$ that have descendants are also placed in the LIS as type $\mathcal{D}$ entries. The LSP is initially empty.

In each pass, we will first process the members of the LIP, then the members of the LIS. This is essentially the significance map encoding step. We then process the elements of the LSP in the refinement step.

We begin by examining each coordinate contained in the LIP. If the coefficient at that coordinate is significant (that is, it is greater than 2^n), we transmit a 1 followed by a bit representing the sign of the coefficient (we will assume 1 for positive, 0 for negative). We then move that coefficient to the LSP. If the coefficient at that coordinate is not significant, we transmit a 0.

After examining each coordinate in the LIP, we begin examining the sets in the LIS. If the set at the coordinate (i, j) is not significant, we transmit a 0. If the set is significant, we transmit a 1. What we do after that depends on whether the set is of type $\mathcal{D}$ or $\mathcal{L}$.

If the set is of type $\mathcal{D}$, we check each of the offsprings of the coefficient at that coordinate. In other words, we check the four coefficients whose coordinates are in $\mathcal{O}(i, j)$. For each coefficient that is significant, we transmit a 1, the sign of the coefficient, and then move the coefficient to the LSP. For the rest we transmit a 0 and add their coordinates to the LIP. Now that we have removed the coordinates of $\mathcal{O}(i, j)$ from the set, what is left is simply set $\mathcal{L}(i, j)$. If this set is not empty, we move it to the end of the LIS and mark it to be of type $\mathcal{L}$. Note that this new entry into the LIS has to be examined during *this* pass. If the set is empty, we remove the coordinate i, j from the list.

If the set is of type $\mathcal{L}$, we add each coordinate in $\mathcal{O}(i, j)$ to the end of the LIS as the root of a set of type $\mathcal{D}$. Again, note that these new entries in the LIS have to be examined during this pass. We then remove (i, j) from the LIS.

Once we have processed each of the sets in the LIS (including the newly formed ones), we proceed to the refinement step. In the refinement step we examine each coefficient that was in the LSP *prior to the current pass* and output the nth most significant bit of $|c_{i,j}|$. We ignore the coefficients that have been added to the list in this pass because, by declaring them significant at this particular level, we have already informed the decoder of the value of the nth most significant bit.

This completes one pass. Depending on the availability of more bits or external factors, if we decide to continue with the coding process, we decrement n by one and continue. Let's see the functioning of this algorithm in an example.

Example 16.4.1:

Let's use the same example we used for demonstrating the EZW algorithm:

26	6	13	10
−7	7	6	4
4	−4	4	−3
2	−2	−2	0

We will go through three passes at the encoder and generate the transmitted bitstream, then decode this bitstream.

First Pass The value for n in this case is 4. The three lists at the encoder are

$$\begin{aligned}
&\text{LIP}: \{(0,0) \rightarrow 26, (0,1) \rightarrow 6, (1,0) \rightarrow -7, (1,1) \rightarrow 7\}\\
&\text{LIS}: \{(0,1)\mathcal{D}, (1,0)\mathcal{D}, (1,1)\mathcal{D}\}\\
&\text{LSP}: \{\}
\end{aligned}$$

In the listing for the LIP, we have included the arrows to make it easier to follow the example. Beginning our algorithm, we examine the contents of the LIP. The coefficient at location (0, 0) is greater than 16. In other words, it is significant; therefore, we transmit a 1, then a 0 to indicate the coefficient is positive and move the coordinate to the LSP. The next three coefficients are all insignificant at this value of the threshold; therefore, we transmit a 0 for each coefficient and leave them in the LIP. The next step is to examine the contents of the LIS. Looking at the descendants of the coefficient at location (0, 1) (13, 10, 6, and 4), we see that none of them are significant at this value of the threshold so we transmit a 0. Looking at the descendants of c_{10} and c_{11}, we can see that none of these are significant at this value of the threshold. Therefore, we transmit a 0 for each set. As this is the first pass, there are no elements from the previous pass in the LSP; therefore, we do not do anything in the refinement pass. We have transmitted a total of 8 bits at the end of this pass (10000000), and the situation of the three lists is as follows:

$$\begin{aligned}
&\text{LIP}: \{(0,1) \rightarrow 6, (1,0) \rightarrow -7, (1,1) \rightarrow 7\}\\
&\text{LIS}: \{(0,1)\mathcal{D}, (1,0)\mathcal{D}, (1,1)\mathcal{D}\}\\
&\text{LSP}: \{(0,0) \rightarrow 26\}
\end{aligned}$$

Second Pass For the second pass we decrement n by 1 to 3, which corresponds to a threshold value of 8. Again, we begin our pass by examining the contents of the LIP. There are three elements in the LIP. Each is insignificant at this threshold so we transmit three 0s. The next step is to examine the contents of the LIS. The first element of the LIS is the set containing the descendants of the coefficient at location (0, 1). Of this set, both 13 and 10 are significant at this value of the threshold; in other words, the set $\mathcal{D}(0, 1)$ is significant. We signal this by sending a 1 and examine the offsprings of c_{01}. The first offspring has a value of 13, which is significant and positive, so we send a 1 followed by a 0. The same is true for the second offspring, which has a value of 10. So we send another 1 followed by a 0. We move the coordinates of these two to the LSP. The next two offsprings are both insignificant at this level; therefore, we move these to the LIP and transmit a 0 for each. As $\mathcal{L}(0, 1) = \{\}$, we remove $(0, 1)\mathcal{D}$ from the LIS. Looking at the other elements of the LIS, we can clearly see that both of these are insignificant at this level; therefore, we send a 0 for each. In the refinement pass we examine the contents of the LSP from the previous pass. There is only one element in there that is not from the current sorting pass, and it has a value of 26. The third MSB of 26 is 1; therefore, we transmit a 1 and complete this pass. In the second pass we have transmitted 13 bits: 0001101000001. The condition of the lists at the end of the second pass is as follows:

$$\begin{aligned}
&\text{LIP} : \{(0, 1) \rightarrow 6, (1, 0) \rightarrow -7, (1, 1) \rightarrow 7, (1, 2) \rightarrow 6, (1, 3) \rightarrow 4\} \\
&\text{LIS} : \{(1, 0)\mathcal{D}, (1, 1)\mathcal{D}\} \\
&\text{LSP} : \{(0, 0) \rightarrow 26, (0, 2) \rightarrow 13, (0, 3) \rightarrow 10\}
\end{aligned}$$

Third Pass The third pass proceeds with $n = 2$. As the threshold is now smaller, there are significantly more coefficients that are deemed significant, and we end up sending 26 bits. You can easily verify for yourself that the transmitted bitstream for the third pass is 10111010101101100110000010. The condition of the lists at the end of the third pass is as follows:

$$\begin{aligned}
\text{LIP} :\quad & \{(3, 0) \rightarrow 2, (3, 1) \rightarrow -2, (2, 3) \rightarrow -3, (3, 2) \rightarrow -2, (3, 3) \rightarrow 0\} \\
\text{LIS} :\quad & \{\} \\
\text{LSP} :\quad & \{(0, 0) \rightarrow 26, (0, 2) \rightarrow 13, (0, 3) \rightarrow 10, (0, 1) \rightarrow 6, (1, 0) \rightarrow -7, (1, 1) \rightarrow 7, \\
& (1, 2) \rightarrow 6, (1, 3) \rightarrow 4, (2, 0) \rightarrow 4, (2, 1) \rightarrow -4, (2, 2) \rightarrow 4\}
\end{aligned}$$

Now for decoding this sequence. At the decoder we also start out with the same lists as the encoder:

$$\begin{aligned}
&\text{LIP} : \{(0, 0), (0, 1), (1, 0), (1, 1)\} \\
&\text{LIS} : \{(0, 1)\mathcal{D}, (1, 0)\mathcal{D}, (1, 1)\mathcal{D}\} \\
&\text{LSP} : \{\}
\end{aligned}$$

We assume that the initial value of n is transmitted to the decoder. This allows us to set the threshold value at 16. Upon receiving the results of the first pass (10000000), we can see that the first element of the LIP is significant and positive and no other coefficient is significant at this level. Using the same reconstruction procedure as in EZW, we can reconstruct the coefficients at this stage as

24	0	0	0
0	0	0	0
0	0	0	0
0	0	0	0

and, following the same procedure as at the encoder, the lists can be updated as

$$\text{LIP} : \{(0, 1), (1, 0), (1, 1)\}$$
$$\text{LIS} : \{(0, 1)\mathcal{D}, (1, 0)\mathcal{D}, (1, 1)\mathcal{D}\}$$
$$\text{LSP} : \{(0, 0)\}$$

For the second pass we decrement n by one and examine the transmitted bitstream: 0001101000001. Since the first 3 bits are 0 and there are only three entries in the LIP, all of the entries in the LIP are still insignificant. The next 9 bits give us information about the sets in the LIS. The fourth bit of the received bitstream is 1. This means that the set with root at coordinate (0,1) is significant. Since this set is of type $\mathcal{D}$, the next bits relate to its offsprings. The 101000 sequence indicates that the first two offsprings are significant at this level and positive and the last two are insignificant. Therefore, we move the first two offsprings to the LSP and the last two to the LIP. We can also approximate these two significant coefficients in our reconstruction by $1.5 \times 2^3 = 12$. We also remove $(0, 1)\mathcal{D}$ from the LIS. The next two bits are both 0, indicating that the two remaining sets are still insignificant. The final bit corresponds to the refinement pass. It is a 1, so we update the reconstruction of the (0, 0) coefficient to $24 + 8/2 = 28$. The reconstruction at this stage is

28	0	12	12
0	0	0	0
0	0	0	0
0	0	0	0

and the lists are as follows:

$$\text{LIP} : \{(0, 1), (1, 0), (1, 1), (1, 2), (1, 3)\}$$
$$\text{LIS} : \{(1, 0)\mathcal{D}, (1, 1)\mathcal{D}\}$$
$$\text{LSP} : \{(0, 0), (0, 2), (0, 3)\}$$

For the third pass, we again decrement n, which is now 2, giving a threshold value of 4. Decoding the bitstream generated during the third pass (10111010101101100110000010), we

FIGURE 16.12 Reconstruction of Sinan image encoded using SPIHT at 0.5 bits per pixel.

update our reconstruction to

26	6	14	10
−6	6	6	6
6	−6	6	0
0	0	0	0

and our lists become

$$\text{LIP}: \{(3,0), (3,1)\}$$
$$\text{LIS}: \{\}$$
$$\text{LSP}: \{(0,0), (0,2), (0,3), (0,1), (1,0), (1,1), (1,2), (2,0), (2,1), (3,2)\}$$

At this stage, we do not have any sets left in the LIS, and we simply update the values of the coefficients. ♦

Finally, let's look at an example of an image coded using SPIHT. The image shown in Figure 16.12 is the reconstruction obtained from a compressed representation that used 0.5 bits per pixel. (The programs used to generate this image were obtained from the authors.) Comparing this with Figure 16.5, we can see a definite improvement in the quality.

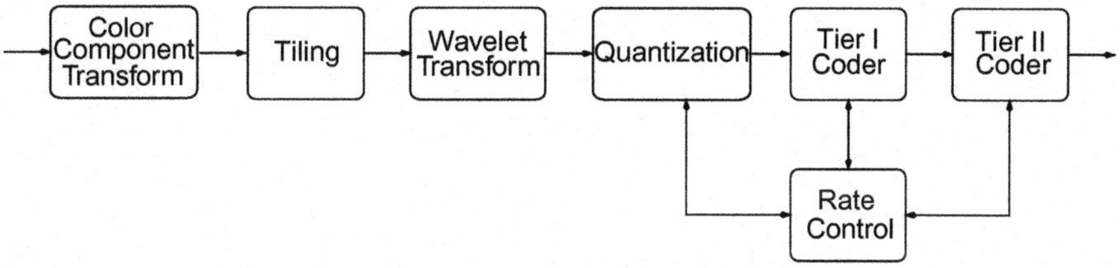

FIGURE 16.13 A block diagram for the JPEG2000 algorithm.

Wavelet decomposition has been finding its way into various standards. The earliest example was the FBI fingerprint image compression standard. The latest is the increasingly dominant image compression standard developed by the JPEG committee, commonly referred to as JPEG 2000.

16.5 JPEG 2000

The JPEG standard provides excellent performance at rates above 0.25 bits per pixel. However, at lower rates there is a sharp degradation in the quality of the reconstructed image. To correct this and other shortcomings, the JPEG committee developed another standard, commonly known as JPEG2000. The JPEG2000 standard is based on wavelet decomposition. The standard has multiple parts dealing with a variety of applications, from basic image compression to volumetric imaging and wireless applications. In this section we will look at Part 1 of the standard, which deals with the basic image compression algorithm.

A block diagram of the JPEG 2000 algorithm is shown in Figure 16.13.

The coding is based on a scheme originally proposed by Taubman [217] and Taubman and Zakhor [218] known as EBCOT. The acronym EBCOT stands for "Embedded Block Coding with Optimized Truncation," which nicely summarizes the technique. It is a block coding scheme that generates an embedded bitstream. The block coding is independently performed on nonoverlapping blocks within individual subbands. Within a subband all blocks that do not lie on the right or lower boundaries are required to have the same dimensions. A dimension cannot exceed 256.

Embedding and independent block coding seem inherently contradictory. The way EBCOT resolves this contradiction is to organize the bitstream in a succession of layers. Each layer corresponds to a certain distortion level. Within each layer, each block is coded with a variable number of bits (which could be zero). Rate control is obtained using a Lagrangian optimization, called Post Compression Rate Distortion (PCRD) optimization, that makes use of the partitioning or truncation points that are part of the EBCOT algorithm. The quality of the reproduction is proportional to the numbers of layers received. It should be noted that PCRD is not required by the standard.

In the following, we briefly describe each of these blocks in the JPEG 2000 block diagram. We will pay particular attention to the process of generating an embedded bitstream for a code-block. We then give a brief overview of the PCRD optimization used to obtain the quality layers. Finally we discuss the way all of this comes together to generate the JPEG 2000 bitstream. As with all standards, this is necessarily an overview with the details left to the standards documents themselves. The book by Taubman and Marcellin [220] also provides a detailed description of the JPEG 2000 algorithm.

16.5.1 Color Component Transform

The JPEG 2000 standard assumes the existence of multiple components in an image. These are likely color components though the standard does not require this. To improve the compression efficiency the standard provides the option of transforming the components. Prior to transformation, as in the JPEG standard, if the samples of the image are unsigned they are level-shifted by 2^{B-1} where B is the number of bits per sample used for each component of a pixel. There are two transformations available: a reversible component transform (RCT) that allows for lossless recovery of the original samples, and an irreversible component transform (ICT) that provides no such guarantee. Both transformations can be viewed as a means of going from red-green-blue components to a luminance and two chrominance components.

Referring to the three (level-shifted) components of the image by I_0, I_1, and I_2 the forward reversible transform is given by

$$
\begin{aligned}
Y_0(x, y) &= \left\lfloor \frac{I_0(x, y) + 2I_1(x, y) + I_2(x, y)}{4} \right\rfloor \\
Y_1(x, y) &= I_2(x, y) - I_1(x, y) \\
Y_2(x, y) &= I_0(x, y) - I_1(x, y)
\end{aligned}
$$

where Y_0 corresponds to the luminance component and Y_1 and Y_2 represent the chrominance components.

The inverse RCT can be derived as

$$
\begin{aligned}
I_1(x, y) &= Y_0(x, y) - \left\lfloor \frac{Y_1(x, y) + Y_2(x, y)}{4} \right\rfloor \\
I_0(x, y) &= Y_2(x, y) + I_1(x, y) \\
I_2(x, y) &= Y_1(x, y) + I_1(x, y)
\end{aligned}
$$

Where loss of information is acceptable the JPEG 2000 standard uses an irreversible transformation to go from what is probably, though not necessarily, the red, green, and blue components to the luminance and chrominance components using the ICT. The forward ICT is given by

$$
\begin{aligned}
Y_0(x, y) &= 0.299I_0(x, y) + 0.587I_1(x, y) + 0.144I_2(x, y) \\
Y_1(x, y) &= -0.16875I_0(x, y) - 0.33126I_1(x, y) + 0.5I_2(x, y) \\
Y_2(x, y) &= 0.5I_0(x, y) - 0.41869I_1(x, y) - 0.08131I_2(x, y)
\end{aligned}
$$

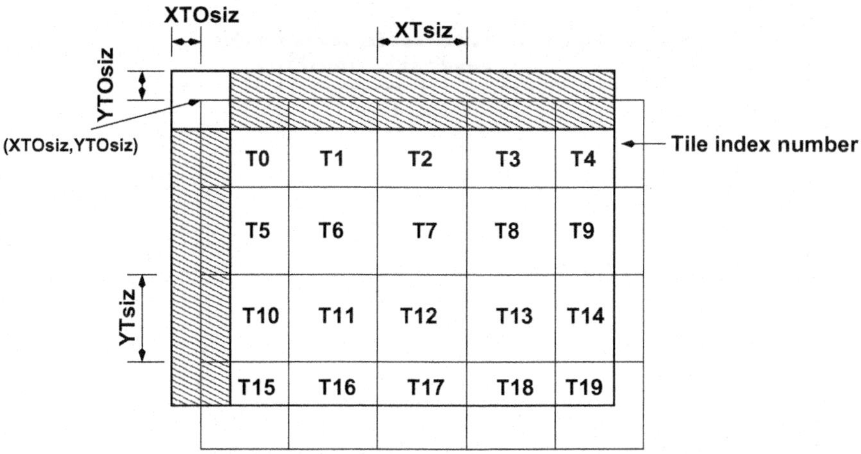

FIGURE 16.14 **Position of tiles with respect to a high-resolution reference grid [219].**

The inverse ICT is given by

$$I_0(x, y) = Y_0(x, y) + 1.402Y_2(x, y)$$
$$I_1(x, y) = Y_0(x, y) - 0.34413Y_1(x, y) - 0.71414Y_2(x, y)$$
$$I_2(x, y) = Y_0(x, y) + 1.772Y_1(x, y)$$

16.5.2 Tiling

One of the major strengths of the JPEG 2000 standard is its flexibility. The standard allows multiple resolutions, multiple levels of quality, and the ability for region-of-interest encoding and multiple access. The latter attributes are made available through options for spatial partitioning of the image. *Tiles* are rectangular partitions of the image, defined with respect to a high-resolution grid, which are coded independently allowing for random access as well as editing functions such as cropping. The tile components at each resolution can be further partitioned into *precincts* that allow for a much more complete random access capability. The relationships of the tile parameters with respect to the high-resolution grid can be seen in Figure 16.14. We will describe how the various parameters shown in Figure 16.14 are encoded in the bitstream in Section 16.5.7.

16.5.3 Wavelet Transform

There are two types of wavelet filters that are included in the standard. One type consists of filters that generate integer coefficients; this type is particularly useful when the wavelet decomposition is part of a lossless compression scheme as it makes the wavelet transform lossless. This decomposition is referred to as the *reversible* transform. The low-pass filter in the reversible transform contains five coefficients while the high-pass filter impulse response

TABLE 16.1 Filter coefficients for the reversible transform.

Low-Pass Filter	
k	h_k
0	$\frac{3}{4}$
± 1	$\frac{1}{4}$
± 2	$-\frac{1}{8}$
High-Pass Filter	
k	h_k
0	1
± 1	$-\frac{1}{2}$

TABLE 16.2 Filter coefficients for the 9/7 ireversible transform.

Low-Pass Filter	
k	h_k
0	0.602949018236
± 1	0.266864118443
± 2	−0.078223266529
± 3	−0.016864118443
± 4	0.026748757411
High-pass filter	
k	h_k
0	0.557543526229
± 1	−0.29635881557
± 2	−0.028771763114
± 3	0.045635881557

contains three nonzero coefficients. Because of this the filter is also referred to as the (5,3) filter. The coefficients are given in Table 16.1. Notice that the filters are symmetric, which means they have linear phase. This in turn means that we can use them without any coefficient expansion. This filter can be implemented using lifting. The lifting procedure consists of one prediction and one update step. Using the terminology of the previous chapter, these are given by

$$d_k = x_{2k+1} - \left\lfloor \frac{x_{2k} + x_{2k+2}}{2} \right\rfloor$$

and

$$s_k = x_{2k+2} + \left\lfloor \frac{d_{k-1} + d_k + 2}{4} \right\rfloor$$

The irreversible wavelet transform is the CDF(4,4) transform described in the previous chapter. We have already given the lifting implementation of this wavelet. The corresponding filters are given in Table 16.2. The filter again is symmetric and because the analysis filters contain nine and seven coefficients it is known as the 9/7 transform.

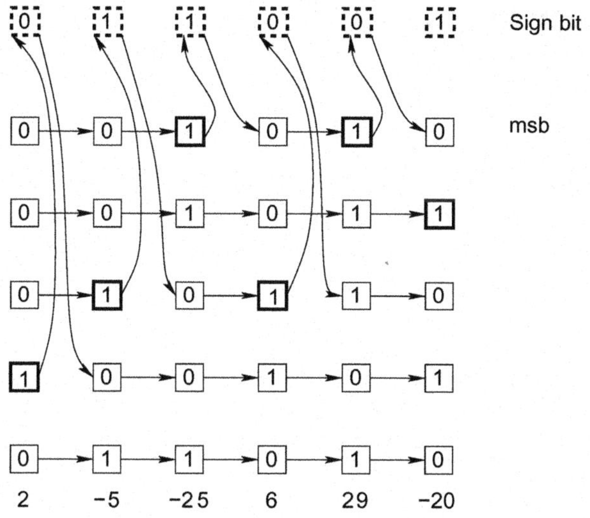

FIGURE 16.15 **Embedded coding of quantization values 2, −5, −25, 6, 29, and −20.**

16.5.4 Quantization

The coefficients are quantized using a midtread quantizer in which the step-sizes are uniform except for the middle step size, which is twice the size. The quantization index can be generated as

$$q_b(u, v) = sign\,(a_b(u, v))\,\lfloor\frac{|a_b(u, v)|}{\Delta_b}\rfloor\Delta_b$$

where $a_b(u, v)$ is the coefficient at location (u, v) in subband b, Δ_b is the step size, and $q_b(u, v)$ is the quantization index. The reconstructed value $\hat{a}_b(u, v)$ can be obtained from the quantization index by

$$\hat{a}_b(u, v) = (q_b(u, v) + \delta_b)$$

where δ_b can take on a value in the interval [0, 1) and is usually taken to be 0.5 [220]. The quantized values are encoded in an embedded manner as shown in Figure 16.15.

The assumption implicit in the example in Figure 16.15 is that the highest resolution or depth is five. The value of Δ_b is 1. The encoder scans the most significant bit of each quantized value, then the next most significant bit, and so on. When a quantized value first becomes significant, the encoder encodes a sign bit. Truncating the bitstream is effectively equivalent to multiplying the step size by $2^{discard}$ where *discard* is the number of bitplanes discarded. In this example, if we truncated the bitstream after the first two rows, we would obtain quantized values of 0, 0, −24, 0, 24, and −16. As we are discarding three bitplanes these would be the results one would expect with $\Delta_b = 8$.

16.5.5 Tier I Coding

Block Coding

The EBCOT algorithm is a bitplane coder that independently encodes the bitplanes of the coefficients in a code-block in multiple passes. Each code-block is first divided into subblocks. For each bitplane the subblocks that contain any coefficients that are significant at that level are deemed to be significant subblocks. A coefficient c_{ij} is said to be significant at level n if $|c_{ij}| \geqslant 2^n$. If we denote the significance of a subblock B_i at level n by $\sigma^n(B_i)$, then $\sigma^n(B_i) = 1$ if there is a coefficient c_{ij} in the block with a magnitude greater than or equal to 2^n, otherwise $\sigma^n(B_i) = 0$. These significance values are encoded using a quad-tree structure. The subblocks are identified with leaf nodes. Four neighboring subblocks are viewed as offsprings of an internal node that corresponds to the union of the four subblocks. These are, in turn, organized into sets of 2×2 *quads*, and so on. A node in this tree is said to be significant at level n if any of its descendants are significant at that level. The significant values of the nodes starting from the root node are encoded using an MQ coder with equiprobable contexts. The significance values that can be inferred from previously encoded significance values are not encoded. For example, if the significance value of the parent code is 0 then the significance values of all of the descendants are also going to be zero and, therefore, are not encoded. Similarly, if a node of the tree has been encoded as significant at a higher level then it is clearly significant at the lower level and, therefore, does not need to be encoded. Finally, if a quad is declared to be significant and three of its descendants have been coded as insignificant, then the fourth descendant is clearly significant and, therefore, the significance value for this descendant does not need to be encoded.

Each bitplane is encoded in three passes, namely, the *significance propagation pass*, the *magnitude refinement pass*, and the *cleanup pass*. In each pass the coefficients are encoded using either a *zero-coding* mode or a *run-length* mode. The first time a coefficient becomes significant, the encoder encodes its sign using a predictive coding approach. The results of the passes are encoded using an MQ coder (described in Chapter 4) with 18 different contexts. Nine of these contexts are used in the zero-coding mode, five are used for sign coding, three are used for magnitude refinement, and one is used for run-length coding. Each bit in a bitplane is encoded only once.

In the significance propagation pass, only the zero-coding mode is used. The nine zero-coding contexts are obtained by looking at the significance value of the two horizontal, two vertical, and four diagonal neighbors of the coefficient being encoded. The rule for context determination is shown in Table 16.3. Only coefficients with known significant neighbors are encoded in this pass. The idea is that neighbors of coefficients that were declared significant in previous bitplanes are most likely to become significant in subsequent bitplanes. Each coefficient with significant neighbors is encoded as either significant or insignificant in the appropriate context as determined by the rule in Table 16.3. If the coefficient is encoded as significant the significance status of the coefficient is followed by the sign of the bit. The sign is encoded as a prediction residual. The prediction for the sign is generated using the sign of the two horizontal and the two vertical neighbors. The sign of the two horizontal neighbors and the two vertical neighbors are each encoded into a single value s_x (where x is either h for

TABLE 16.3 Context determination for the zero-coding mode.

LL & LH			HL			HH		Context
σ_h	σ_v	σ_d	σ_h	σ_v	σ_d	σ_d	σ_{h+v}	
0	0	0	0	0	0	0	0	ZC0
0	0	1	0	0	1	0	1	ZC1
0	0	$\geqslant 2$	0	0	$\geqslant 2$	0	$\geqslant 2$	ZC2
0	1	x	1	0	x	1	0	ZC3
0	2	x	2	0	x	1	1	ZC4
1	0	0	0	1	0	1	$\geqslant 2$	ZC5
1	0	$\geqslant 1$	0	1	$\geqslant 1$	2	0	ZC6
1	$\geqslant 1$	x	$\geqslant 1$	1	x	2	$\geqslant 1$	ZC1
2	x	x	x	2	x	$\geqslant 3$	x	ZC1

TABLE 16.4 Prediction and context generation for the sign coding process.

s_h	s_v	Context	Prediction
1	1	SC4	1
1	0	SC3	1
1	−1	SC2	1
0	1	SC1	1
0	0	SC0	1
0	−1	SC1	−1
−1	1	SC2	−1
−1	0	SC3	−1
−1	−1	SC4	−1

horizontal or v for vertical) according to the following rule:

$$s_x = \begin{cases} 1 & \text{if both neighbors are positive or one is positive and one unknown} \\ -1 & \text{if both neighbors are negative or one is negative and one unknown} \\ 0 & \text{otherwise} \end{cases}$$

These values are then used according to the rule in Table 16.4 to generate both the context for the arithmetic coder and the prediction for the sign. If the prediction is accurate, the encoder encodes a 0 in the appropriate sign coding context, otherwise it encodes a 1. Note that in this pass coefficients that have previously been declared significant are not encoded.

In the magnitude refinement pass, the bits corresponding to coefficients that have been declared significant in previous passes are encoded using the contexts MR0, MR1, and MR2. The context MR0 is used when the magnitude refinement pass is applied to a coefficient for the first time, and there are no horizontal or vertical neighbors that are significant. If there are significant neighbors and the magnitude refinement pass is being applied to the coefficient for the first time, the context MR1 is used. Finally, if this is not the first time the magnitude refinement is being applied to a coefficient, the context MR2 is used.

The cleanup pass is a catch-all pass in which all bits not encoded in the significance propagation pass and the magnitude refinement pass are encoded. These bits are encoded in either the run-length mode or one of the nine zero-coding modes. The run-length mode is used if four consecutive samples are insignificant, the neighbors of all these coefficients are insignificant, the samples all reside in the same subblock, and the horizontal index of the first symbol is even. If all these conditions are satisfied, the encoder switches to the run-length mode and emits a 0 in the run-length context. Once in the run-length mode, if a significant coefficient is encountered, the encoder generates a 1 that is encoded in the run-length context. This is followed by the position of the coefficient encoded in the equiprobable context followed by the encoding of the sign as described previously. Once the encoder is out of the run-length mode, it encodes the bits in the zero-coding mode until the conditions again allow a transition into the run-length coding mode. At the beginning of the encoding of the highest bitplane in a subblock, none of the coefficients will have been declared significant. Therefore, even though for the other bitplanes the significance coding pass is the first pass, the very first bitplane is encoded using the cleanup pass.

We work through a toy example to see how these passes are used to encode a stripe in a code-block. (Another, more detailed example, by Xavier Delaunay, can be found at http://d.xav.free.fr/ebcot/.)

Example 16.5.2: JPEG 2000 Tier I Coding

The codeblock in JPEG 2000 is divided into stripes, where a stripe consists of four rows, except possibly the last stripe. We will work through the encoding of a stripe using the example shown in Figure 16.16. We will assume the rows above and below the stripe have coefficients that are not significant. The encoder first determines the highest significant bitplane in the stripe. In this particular example, the largest coefficient magnitude is 12, which means that the encoder will encode a total of four bitplanes. The bits from the most significant bitplane are shown in Figure 16.17.

Nominally, the first pass should be the significance propagation pass. However, the significance propagation pass relies on the existence of coefficients that have been declared significant in previous bitplanes. As this is the first pass through this particular stripe, none of the coefficients have been declared significant. Therefore, for the first bitplane, we use a cleanup pass. The first four bits in the bitplane are all zero. Furthermore, none of them are neighbors of any coefficients that have been declared significant. Therefore, we encode the first column as a 0 using the run-length context. The second column is also encoded with a zero in the run-length context. The second coefficient in the third row is significant. Therefore, we send a 1 in the run-length context to indicate that we are no longer in the run-length mode. We then encode the position of the coefficient (01) using the equiprobable context. As this is the first time this coefficient has been declared significant, the next step is to encode the sign of the coefficient. As none of the neighbors have a declared sign, the value of both s_h and s_v is 0. Looking up the corresponding row in Table 16.4, we find the predicted value of the sign to be positive and the context to be SC0. As the predicted sign is correct, a 0 is encoded in the context SC0.

We are no longer in the run-length mode; therefore, the next bit is encoded in the zero-coding mode. For the next coefficient, the only significant neighbor is the coefficient above it, so $\sigma_h = 0, \sigma_v = 1$ and $\sigma_d = 0$. In order to find the zero-coding context we need to know the

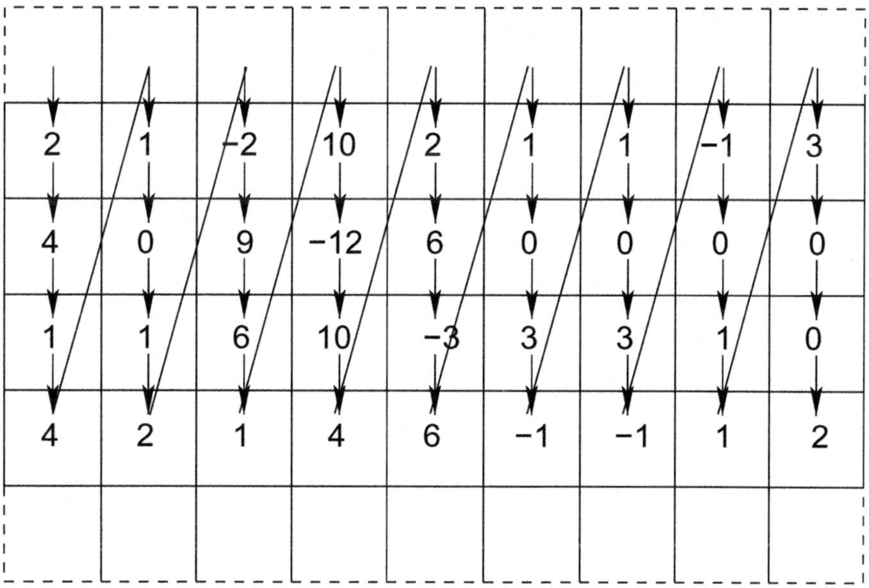

FIGURE 16.16 **An example stripe of coefficients from a codeblock.**

subband from which this stripe was obtained. Let's assume the subband was the LH subband. Looking up the corresponding row and column in Table 16.3, we see that the zero-coding context is ZC3. The coefficient being encoded is not significant at this bitplane so we encode a 0 in the ZC3 context. The next bit is encoded as a 0 in the ZC0 context as, for that bit, $\sigma_v = 0, \sigma_h = 0$ and $\sigma_d = 0$.

Continuing on to the next column, $\sigma_h = 0, \sigma_v = 0$, and $\sigma_d = 1$ and the zero-coding context is ZC1. The coefficient is significant so we encode a 1 in the ZC1 context. For this coefficient, the sign of the vertical and horizontal neighbors is unknown; therefore, $s_h = 0$ and $s_v = 0$ and the predicted sign is positive. This prediction is correct, so we encode a 0 in the context SC0. For the second coefficient in this row, $\sigma_h = 1, \sigma_v = 1$, and $\sigma_d = 0$. Thus, the zero-coding context from Table 16.3 is ZC1. The coefficient is significant, therefore, we encode a 1 in the ZC1 context. To encode the sign, we compute the values of s_h and s_v which are both 1. The prediction for these values is that the sign is positive. As this is not the case, we encode a 1 in the SC4 context. For the next coefficient, $\sigma_h = 0, \sigma_v = 1$, and $\sigma_d = 1$. From Table 16.3, we can see that this corresponds to the context ZC3. As this coefficient is significant we encode a 1 in the ZC3 context. The values of s_h and s_v are 0 and -1 respectively. From Table 16.4 this results in a prediction of a negative sign and a sign context of SC1. The prediction is incorrect so we encode a 1 in the SC1 context. To encode the last coefficient in the column, we note that $\sigma_h = 0, \sigma_v = 1$, and $\sigma_d = 0$ which, from Table 16.3, gives us a context of ZC3. The coefficient is insignificant; therefore, we encode a 0 in the ZC3 context.

The coefficients in the next column have neighbors which are significant, so we remain in the zero-coding mode. For the first coefficient in the next row, $\sigma_h = 1, \sigma_v = 0$, and $\sigma_d = 1$, which results in a zero-coding context of ZC6. As the coefficient is insignificant, we encode it

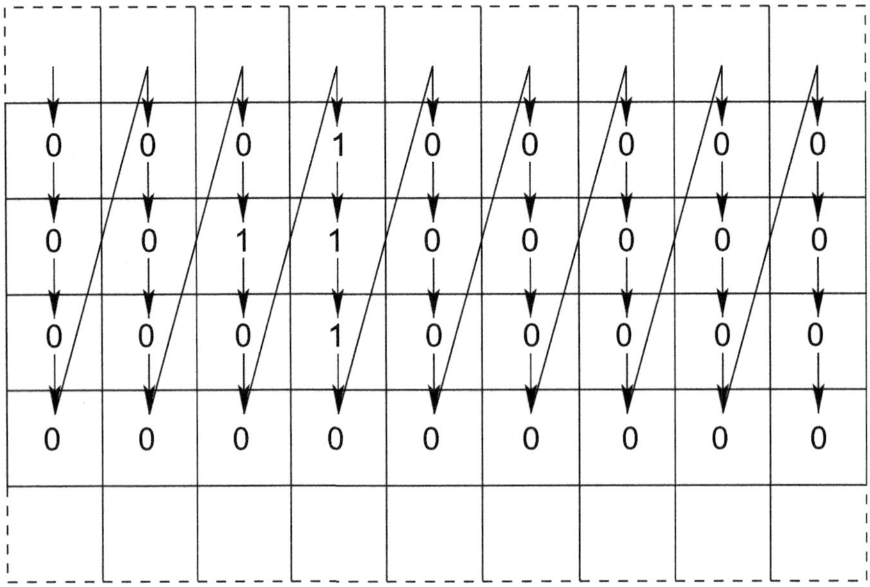

FIGURE 16.17 **The most significant bitplane of the example stripe of coefficients in Figure 16.16.**

as a 0 in this context. For the next coefficient, $\sigma_h = 1, \sigma_v = 0$, and $\sigma_d = 2$, which again results in a zero-coding context of ZC6. The coefficient is insignificant so we encode it as a zero in the ZC6 context. For the next coefficient $\sigma_h = 1, \sigma_v = 0$, and $\sigma_d = 1$. As this coefficient is also insignificant, we encode a 0 in the ZC6 context. The next column does not have any neighbors that are known to be significant; therefore, we switch to the run-length mode. As all coefficients in the column are insignificant, we encode a 0 in the run-length context. The same is true for the following three columns which brings us to the end of the stripe.

We now begin encoding the next bitplane. This is shown in Figure 16.18.

The coefficients that have been declared significant in the previous pass are marked with darker boxes. As we have some significant coefficients to work off of, we will go through each of the three passes. In the first pass, the significance propagation pass, we only examine those coefficients that have neighbors already declared to be significant. The first such coefficient is the first coefficient in the second column. This coefficient has a diagonal neighbor, which is significant. In order to encode this bit, we first determine the zero-coding context as all bits in the significance propagation pass are encoded in the zero-coding mode. For this bit $\sigma_h = 0, \sigma_v = 0$, and $\sigma_d = 1$. Using Table 16.3, we obtain the context as ZC1. As the bit is not significant, we encode a 0 in the ZC1 context. For the next bit, $\sigma_h = 1, \sigma_v = 0$, and $\sigma_d = 0$. This corresponds to a zero-coding context of ZC5. We encode a 0 in the ZC5 context. The next bit is also insignificant with $\sigma_h = 0, \sigma_v = 0$, and $\sigma_d = 1$; therefore, we encode a 0 in the ZC1 context. The last bit in the column does not have any known significant neighbors so we skip over it.

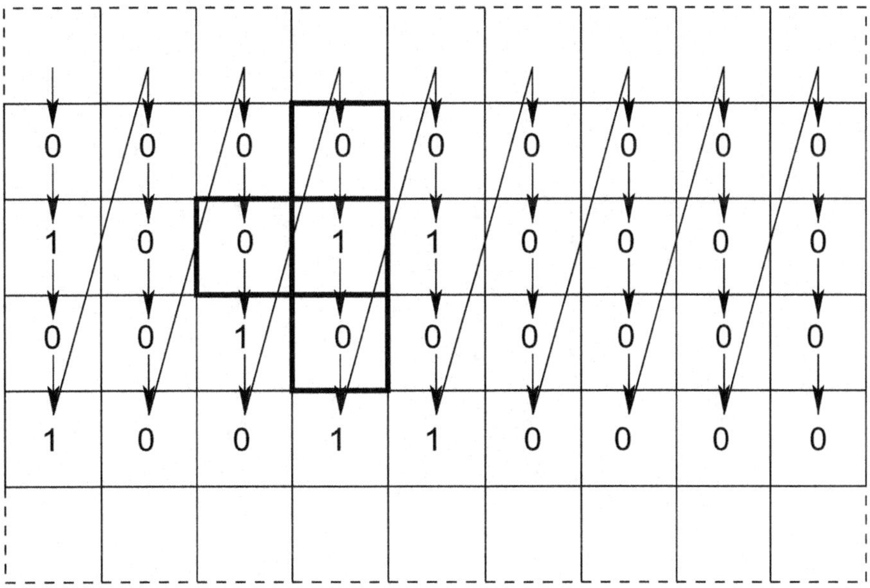

FIGURE 16.18 **The second most significant bitplane of the example stripe of coefficients in Figure 16.16.**

From Figure 16.18, we can see that the first bit in the third column has a known significant neighbor. For this bit, $\sigma_h = 1, \sigma_v = 1$, and $\sigma_d = 1$. The bit itself is 0 so we encode a 0 in the ZC1 context. The second coefficient in this column was declared significant while encoding the previous bitplane, so we defer its encoding to the magnitude refinement pass. The third coefficient in this column has known significant vertical and diagonal neighbors, $\sigma_h = 0, \sigma_v = 1$, and $\sigma_d = 1$. This corresponds to the ZC3 context. This coefficient is significant at this bitplane; therefore, we encode a 1 in the ZC3 context. We then compute s_h and s_v in order to determine the prediction for the sign and the sign context. As one of the vertical neighbors is positive and the other unknown, $s_v = 1$. Similarly, $s_h = 1$. From Table 16.4, the predicted sign is positive and the context is SC4. As the prediction is correct, we encode a 0 in the SC4 context. The last coefficient has a diagonal neighbor that is significant from the previous bitplane; therefore, it will be encoded in this pass. It also has a vertical neighbor that has been declared significant; therefore, $\sigma_h = 0, \sigma_v = 1$, and $\sigma_d = 1$. The coefficient is encoded in this bitplane with a 0 in the context ZC3.

The first three coefficients in the fourth column were declared significant in previous passes, so we skip those and encode the last bit in this column (in this stripe). For this bit, $\sigma_h = 0, \sigma_v = 1$, and $\sigma_d = 0$. The coefficient becomes significant at this bitplane, so we encode a 1 in the context ZC3. We compute the sign indicators as $s_h = 0$ and $s_v = 1$. This results in a predicted positive sign and a sign context SC1. As the prediction is correct, we encode this with a 0 in the SC1 context.

The coefficients in the fifth column all have neighbors that have been declared significant. The first coefficient is insignificant in this bitplane with $\sigma_h = 1, \sigma_v = 0$, and $\sigma_d = 1$ and is

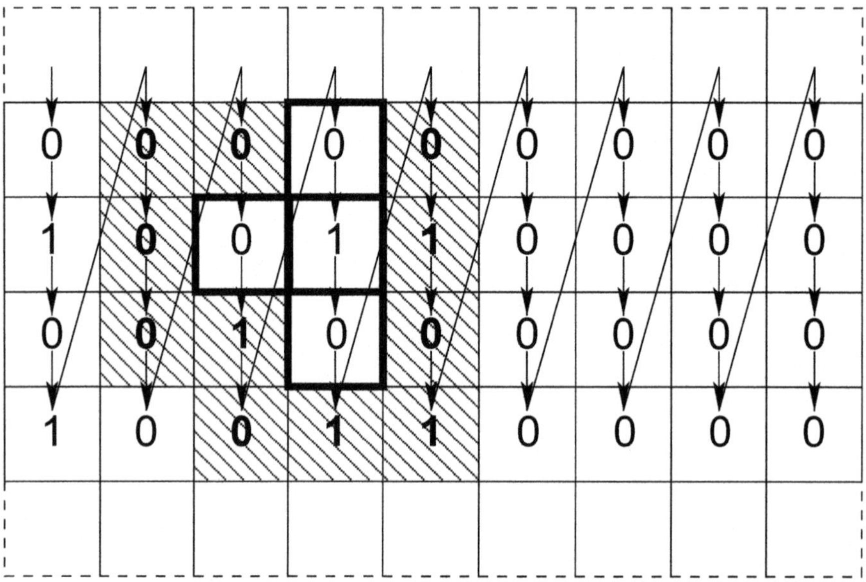

FIGURE 16.19 **The second most significant bitplane of the example stripe of coefficients in Figure 15.16 after the significance propagation pass. The encoded bits are shown in bold in shaded boxes.**

encoded as a 0 in the ZC6 context. The second coefficient becomes significant in this bitplane with $\sigma_h = 1, \sigma_v = 0$, and $\sigma_d = 2$. We encode a 1 in the ZC6 context and compute $s_h = -1$ and $s_v = 0$, which implies a sign context of SC3. The predicted sign is negative, which is incorrect, so we encode a 1 in the SC3 context. The next bit in this column is 0 with $\sigma_h = 1, \sigma_v = 1$, and $\sigma_d = 1$, so we encode a 0 in the ZC1 context. The final coefficient to be encoded in this pass is the fourth coefficient in this column. This coefficient becomes significant in this bitplane with $\sigma_h = 1, \sigma_v = 0$, and $\sigma_d = 1$. We encode a 1 in context ZC6. For this coefficient, $s_h = 1$ and $s_v = 0$. The predicted sign according to Table 16.4 is positive, so we encode a 0 in sign context SC3. There are no other coefficients in this bitplane that have significant neighbors so this concludes the significance propagation pass.

The bitplane at this point can be represented by Figure 16.19 where the bits that have been encoded are shown in bold in shaded boxes. The bits in the darker boxes belong to coefficients that were declared significant in the previous pass. These are the bits that are encoded in the magnitude refinement pass. For each of these coefficients, the magnitude refinement pass is being used for the first time; each of these coefficients have one or more significant neighbors, so we encode each bit using the MR1 context. We encode a 0 in the MR1 context, a 0 in the MR1 context, a 1 in the MR1 context, and a 1 in the MR1 context.

Finally, we use the cleanup pass to encode the remaining bits. The bits in the first column all have significant neighbors, so they are encoded using the zero-coding contexts. The same is true for the last bit in the second column and all of the bits in the sixth column. The bits in the seventh column are all 0; they do not have any significant neighbors, so we encode this column with a 0 in the run-length context. The eighth and ninth columns are also encoded using a 0 in the run-length context. This brings to an end the coding of this bitplane.

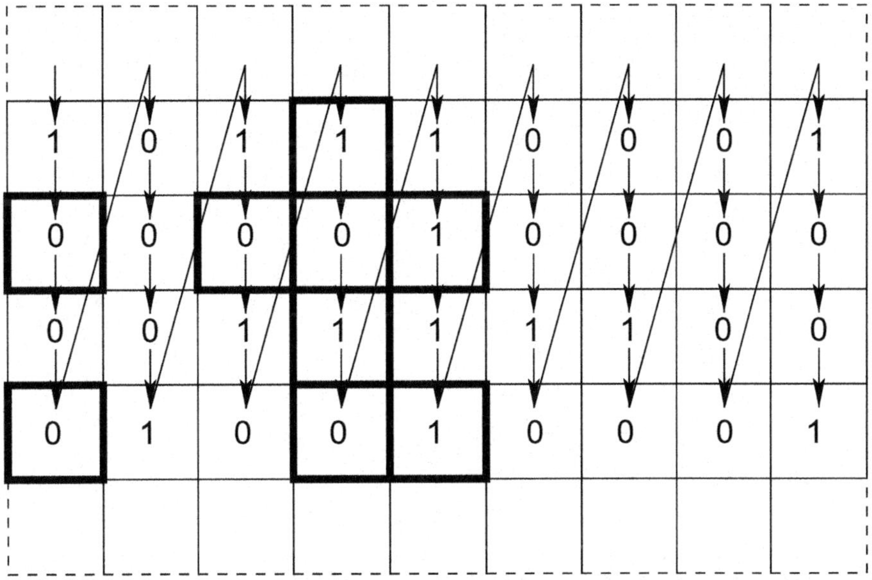

FIGURE 16.20 **The third most significant bitplane of the example stripe of coefficients in Figure 16.16. The coefficients declared significant are in bold boxes.**

As we begin encoding the next bitplane the significance status of the various coefficients is as shown in Figure 16.20. As we can see, the number of significant coefficients has increased. This means that this bitplane will be encoded in the zero-coding mode with no opportunity to use the more parsimonious run-length coding mode. This is in line with other algorithms that generate an embedded bitstream. Much of the coding resources are devoted to the last bits of precision. ♦

The example also illustrates another feature of JPEG 2000, namely the existence of a large number of truncation points. The end of each pass in the encoding presents the opportunity for truncating the bitstream at that point. This permits the natural construction of fractional bitplanes. This is done as part of the Tier II coding procedures.

16.5.6 Tier II Coding

Tier II coding in JPEG 2000 is focused on the rich scalability options characteristic of this standard. We defer much of the discussion of the organization of the data to Section 16.5.7. In this section we look at how the JPEG 2000 standard takes advantage of the truncation points described in the previous section. The PCRD algorithm makes use of this particular feature in building the quality levels that provide JPEG 2000 with its SNR scalability.

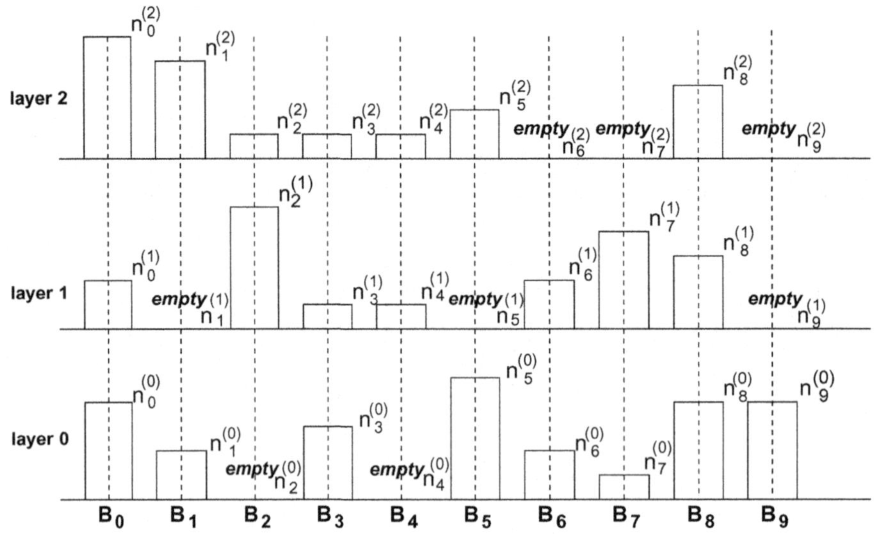

FIGURE 16.21 **An example of the collection of bits from subblocks B_k into quality levels.**

Rate Control

The embedded coding scheme in JPEG 2000 is similar in philosophy to the EZW and SPIHT algorithms; however, the data structures used are different. The EZW and SPIHT algorithms use trees of coefficients from the same spatial location across different bands. In the case of the EBCOT algorithm, each block resides entirely within a subband, and each block is coded independently of other blocks. This allows JPEG 2000 to organize bits from different embedded bitstreams from subblocks into what Taubman and Marcellin call a *pack-stream*. As the construction of the pack-stream occurs after the compression has been performed, the process has been called a post-compression process. In order to perform the post-compression process in an optimum manner, the EBCOT algorithm uses the PCRD-opt algorithm.

Let the truncation points for code-block i be denoted by $\{t_i^{(k)}\}$. A reasonable set of truncation points are the ends of the various coding passes described above. The EBCOT algorithm collects bits from different code-blocks into quality layers. The number of bits collected from various code-blocks are selected from the truncation points that lie on the convex hull of the operational rate-distortion curve. A conceptual view of the construction of quality layers is shown in Figure 16.21. The number of bits from subblock B_k in quality level Q_l is given by $n_k^{(l)}$ where the number of bits contributed by the kth code-block depends on the truncation points $\{t_k^{(j)}\}$:

$$n_k^{(l)} = t_k^{(j)} - \sum_{i=0}^{l-1} n_k^{(i)}$$

for some j such that

$$t_k^{(j)} \geqslant \sum_{i=0}^{l-1} n_k^{(i)}$$

As shown in Chapter 13, the operational rate distortion function can be obtained by minimizing the Lagrangian functional $D(\lambda) + \lambda R(\lambda)$. We assume the existence of an additive distortion metric such that for each quality level Q_l the corresponding distortion D^l is given by

$$D^l = \sum D_i^{n_i}$$

where i ranges over the code-blocks and n_i is based on the available truncation points. Given a rate constraint for each of the quality levels R_{max}^l we can set up the problem of finding the number of bits contributed by each code-block in a quality level as the problem of finding the set of n_i that minimize

$$\sum_i \left(D_i^{n_i,\lambda} + \lambda R_{i,\lambda} \right)$$

If for some λ we can find the truncation points that satisfy the rate constraint, then this set of truncation points will provide the minimum distortion for the given rate. From the discussion on operational rate distortion functions in Chapter 13 we know that

1. λ is the negative of the slope of the distortion rate function.
2. The slope is identical for each coefficient for an optimum bit allocation.

Therefore, for a given λ we can restrict our attention to individual code-blocks and for each code-block find the truncation point that minimizes the quantity $D_i^{n_i,\lambda} + \lambda R_{i,\lambda}$. Taubman and Marcellin [220] propose the following simple algorithm for finding an optimum set of truncation points for a given value of λ:

1. Initialize $n_i^{opt} = 0$.
2. For $t = 1, 2, \ldots T_i$
   - **(a)** Compute $\Delta R = R_i^{(t)} - R_i^{n_i^{opt}}$ and $\Delta D = -(D_i^{(t)} - D_i^{n_i^{opt}})$.
   - **(b)** If $\Delta D / \Delta R > \lambda$ set $n_i^{opt} = t$.

Note that at the end of the Tier I coding pass the set of valid truncation points are known for each of the code-blocks. This procedure is then a simple enumerative procedure for a given value of λ. The particular values of λ itself can be obtained using the bisection procedure described in Chapter 13.

16.5.7 JPEG 2000 bitstream

One of the major contributions of the JPEG 2000 standard is its flexibility [220]. We have seen that with EBCOT each subband of each component is partitioned into code-blocks. The bitplanes corresponding to the subband coefficients are encoded using a three-pass procedure

FIGURE 16.22 Organization of codestream [219].

that generates an embedded bitstream. Each packet contains the binary codestream for each code-block within a precinct, thus enabling a very high level of spatial random access. The context of the packet is provided in the packet header and the packet is indexed by layer, resolution, precinct, and component. The packets are combined into a *layer* that contains contributions from all code-blocks in a tile.

The bitstream is organized using two-byte markers with the first byte being 0xFF. There are six kinds of markers: delimiting markers, fixed information markers, functional markers, pointer markers, in bitstream markers, and informational markers. Some of the markers are followed by segments that contain the parameters corresponding to the marker.

Delimiting Markers The codestream is bracketed by the start of codestream (SOC) marker, which is the first marker in the main header and the end of codestream (EOC) marker, which terminates the codestream. The codestream can contain several tiles with each tile header bracketed by the start of tile-part (SOT) marker followed by the SOT marker segment, and the start of data (SOD) marker. The SOD marker separates the header from the data. This organization is shown in Figure 16.22.

Fixed Information Marker The fixed information marker provides required information about the image. There is only one such marker, is the image and tile size marker (SIZ). The segment following this marker contains information about the uncompressed image. The information contained in this segment and its organization is as follows:

Parameter	Size (bytes)	Description
SIZ	2	Marker (0xFF51)
Lsiz	2	Length of segment in bytes
Rsiz	2	
Xsiz	4	Width of reference grid
Ysiz	4	Height of reference grid
XOsiz	4	Horizontal offset from grid origin to left of image area
YOsiz	4	Vertical offset from grid origin to top of image area
XTsiz	4	Width of one reference tile with respect to grid
YTsiz	4	Height of one reference tile with respect to grid
XTOsiz	4	Horizontal offset from grid origin to left side of first tile
YTOsiz	4	Vertical offset from grid origin to top side of first tile
Csiz	2	Number of components in the image
$Ssiz^i$	1	Precision in bits and sign of the ith component
$XRSiz^i$	1	Horizontal separation of ith component with respect to grid
$YRSiz^i$	1	Vertical separation of ith component with respect to grid

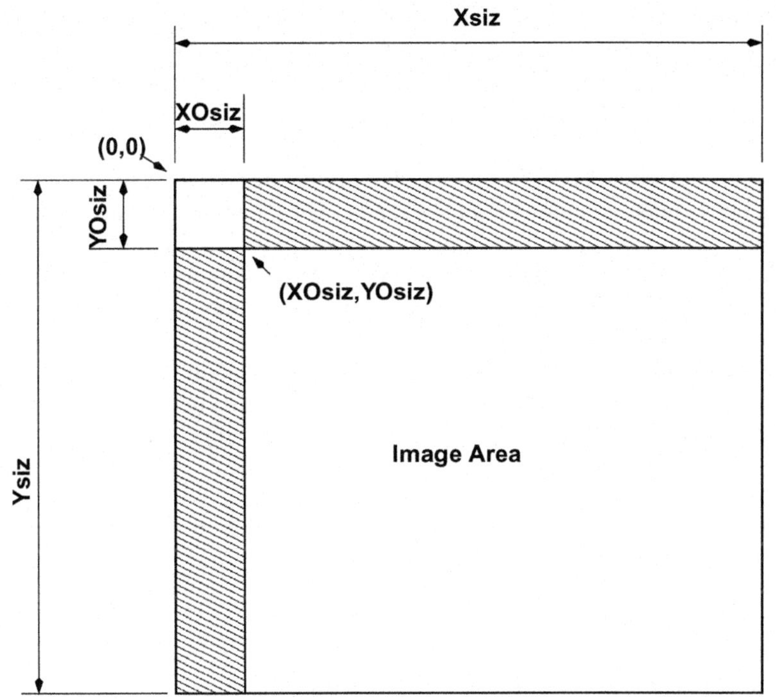

FIGURE 16.23 Some of the parameters in the SIZ marker with respect to a reference grid [219].

The parameters contained in the SIZ marker segment are shown with respect to a reference grid in Figure 16.23.

Functional Markers There are six functional marker segments. The same markers can be used in reference to the whole image or the tile depending on their location. If found in the main header they describe the functions used to code the entire image. If found in the tile header they describe the functions used to code the tile. The functional markers include the coding style (COD) marker, the coding style component (COC) marker, the region of interest (RGN) marker, the quantization default (QCD) marker, the quantization component (QCC) marker, and the progression order change default (POD) marker.

The coding style (COD) marker is followed by a segment that describes the type of entropy coding used, the number of decomposition levels (with zero implying no transform), the progression order, the number of layers, the code-block dimensions, the wavelet transform used, and the multiple component transform used. The information about the decomposition level, code-block dimensions, transform, and packet partition size contained in the COD marker can be overridden for a particular component by the coding style component (COC) marker.

The region of interest (RGN) marker allows for selectively targeting particular regions of interest to the user. Currently, the only available method is to scale up the value of coefficients of interest forcing the significant bits into higher bitplanes. This allows these coefficients to enter the bitstream earlier and allows for more refinement of the coefficients in the region of interest.

The quantization default (QCD) marker specifies the quantization style for all components and the quantizer stepsize values for all subbands. The QCD marker can be overridden for a particular component or set of components by the quantization component (QCC) marker.

The progression order change (POD) marker overrides the progression field of the COD marker segment. JPEG 2000 allows for five different orders of progression including by increasing resolution or increasing quality.

Pointer Markers There are three pointer markers that encode the lengths in the tile-parts. The TLM marker and the associated segment optionally occurs in the main header and has the following structure:

Parameter	Size (bytes)	Description
TLM	2	Marker (0xFF55)
Ltlm	2	Length of segment in bytes
Ztlm	1	Index relative to other TLM marker segments in the header
Stlm	1	Size of Ttlm and Ptlm parameters
$Ttlm^i$	0, 1, or 2	Tile number of the ith tile-part
$Ptlm^i$	2 or 4	Length in bytes from beginning of SOT marker to end of data for the ith tile-part

The packet length (PLM) marker is an optional marker that can appear in the main header. The PLM segment provides a list of packet lengths in the tile parts.

Parameter	Size (bytes)	Description
PLM	2	Marker (0xFF57)
Lplm	2	Length of segment in bytes
Zplm	1	Index relative to other PLM marker segments in the header
$Nplm^i$	1	Number of bytes of Iplm information for ith tile-part. One value for each tile-part.
$Iplm^{i,j}$	variable	Length of jth packet in the ith tile-part

The PLT marker also indicates the list of packet lengths but appears in the tile-part header rather than the main header. The PLT and PLM markers are both optional and can be used together or separately.

Parameter	Size (bytes)	Description
PLT	2	Marker (0xFF58)
Lplt	2	Length of segment in bytes
Zplt	1	Index relative to other PLT marker segments in the header
$Iplt^i$	variable	Length of ith packet

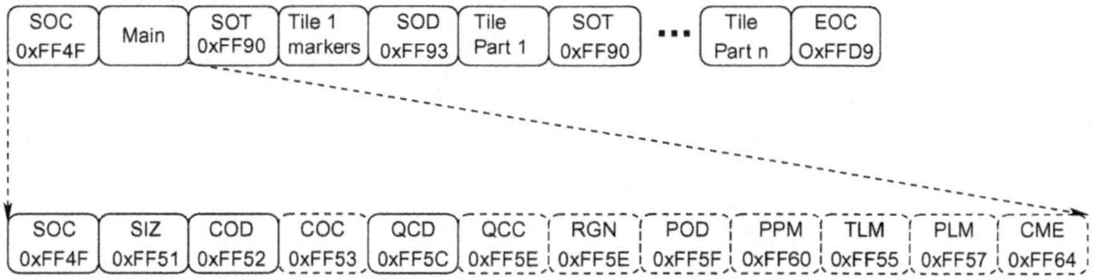

FIGURE 16.24 **Organization of markers in the main header. Optional markers are shown in dashed boxes [219].**

The packet headers can be organized in one of three ways. In the first approach, all the packet headers can be placed together using a PPM marker so that multiple reads are not required to decode the different headers. This organization can occur only in the main header and is as follows:

Parameter	Size (bytes)	Description
PPM	2	Marker (0xFF60)
Lppm	2	Length of segment in bytes
Zppm	1	Index relative to other PPM marker segments in the header
$Nppm^i$	4	Number of bytes of Ippm information for ith tile-part. One value for each tile-part.
$Ippm^{i,j}$	variable	Packet header for every packet in order in the tile-part. The component number, layer, and resolution are determined from method of progression or POD. One value for each packet in the tile-part.

The second approach is to place packed packet headers in the tile-part header using the marker PPT. The structure of the segment is the same as that of the PPM except only the headers of the packets in the tile-part are included.

Parameter	Size (bytes)	Description
PPT	2	Marker (0xFF61)
Lppt	2	Length of segment in bytes
Zppt	1	Index relative to other PPT marker segments in the header
Ippt^i	variable	Packet header for every packet in order in the tile-part. The component number, layer, and resolution are determined from method of progression or POD. One value for each packet in the tile-part.

The third approach is to distribute the packet headers in the bitstream.

The organization of various markers in the main header is shown in Figure 16.24. The optional markers are shown in dashed boxes while the required markers are shown in solid boxes.

In Bitstream Markers There are two marker segments that are found in the bitstream rather than in the headers. They are the start of packet (SOP) marker segment and the end of packet (EPH) marker. The SOP segment is organized as follows:

Parameter	Size (bytes)	Description
SOP	2	Marker (0xFF91)
Lsop	2	Length of segment in bytes
Nsop	2	Packet sequence number starting from 0. The numbering is modulo 65,536.

The EPH marker has a value of 0xFF92.

Informational Marker The informational marker, "Comment and Extension" (CME), can be used as many times as desired in the main or tile-part header. The marker is followed by the length of the segment in two bytes, two bytes of registration value and an undefined number of bytes of unstructured data.

For storing application specific metadata with the JPEG 2000 codestream the standard suggests the use of the JP2 file format. The JP2 file format is a binary container for the JPEG 2000 codestream and uses the concept of boxes to organize the information. A schematic representation of this organization is shown in Figure 16.25.

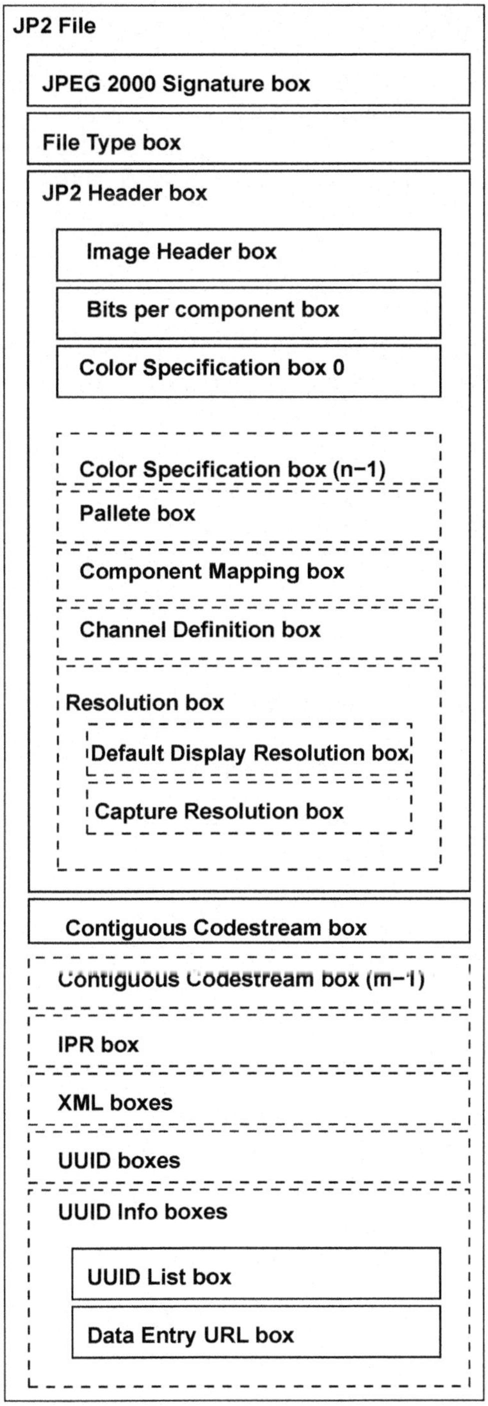

FIGURE 16.25 **Schematic representation of a JP2 file [219].**

16.6 Summary

In this chapter we have described several image compression techniques based on wavelet decomposition. In particular we have spent some time describing the JPEG 2000 algorithm. This algorithm is fast becoming the standard of choice in a number of applications. In particular the JPEG 2000 algorithm has been very effective in the compression of large high-resolution images.

Further Reading

1. The most comprehensive resource for the JPEG 2000 standard is *JPEG 2000: Image Compression Fundamentals, Standards and Practice*, by D. Taubman and M. Marcellin [220].

2. The SPIHT algorithm as well as the principles behind it are presented in a very accessible manner in a two-part monograph by W. Pearlman and A. Said: *Set Partition Coding: Part I of Set Partition Coding and Image Wavelet Coding Systems* and *Image Wavelet Coding Systems: Part II of Set Partition Coding and Image Wavelet Coding Systems.* [280, 281].

16.7 Projects and Problems

1. Consider the following seven-level decomposition:

21	6	15	12
−6	3	6	3
3	−3	0	−3
3	0	0	0

   (a) Find the bitstream generated by the EZW coder.
   (b) Decode the bitstream generated in the previous step. Verify that you get the original coefficient values.

2. Consider the coefficients from the seven-level decomposition in the previous problem:

   (a) Find the bitstream generated by the SPIHT coder.
   (b) Decode the bitstream generated in the previous step. Verify that you get the original coefficient values.

17

Audio Coding

17.1 Overview

Lossy compression schemes can be based on a source model, as in the case of speech compression, or a user or sink model, as is somewhat the case in image compression. In this chapter we look at audio compression approaches that are explicitly based on the model of the user. We will look at audio compression approaches in the context of audio compression standards. Principally, we will examine the different MPEG standards for audio compression. These include MPEG Layer I, Layer II, and Layer III (or MP3), and the Advanced Audio Coding Standard. As with other standards described in this book, the goal here is not to provide all the details required for implementation. Rather the goal is to provide the reader with enough familiarity so that they can then find it much easier to understand these standards.

17.2 Introduction

The various speech coding algorithms we studied earlier rely heavily on the speech production model to identify structures in the speech signal that can be used for compression. Audio compression systems have taken, in some sense, the opposite tack. Unlike speech signals, audio signals can be generated using a large number of different mechanisms. Lacking a unique model for audio production, the audio compression methods have focused on the unique model for audio perception, a psychoacoustic model for hearing. At the heart of the techniques described in this chapter is a psychoacoustic model of human perception.

By identifying what can and, more important, what cannot be heard, the schemes described in this chapter obtain much of their compression by discarding information that cannot be perceived. The motivation for the development of many of these perceptual coders was their potential application in broadcast multimedia. However, their major impact has been in the distribution of audio over the Internet.

We live in an environment rich in auditory stimuli. Even an environment described as quiet is filled with all kinds of natural and artificial sounds. The sounds are always present and come to us from all directions. Living in this stimulus-rich environment, it is essential that we have mechanisms for ignoring some of the stimuli and focusing on others. Over the course of our evolutionary history we have developed limitations on what we can hear. Some of these limitations are physiological, based on the machinery of hearing. Others are psychological, based on how our brain processes auditory stimuli. The insight of researchers in audio coding has been the understanding that these limitations can be useful in selecting information that needs to be encoded and information that can be discarded. The limitations of human perception are incorporated into the compression process through the use of psychoacoustic models. We briefly describe the auditory model used by the most popular audio compression approaches. Our description is necessarily superficial and we refer readers interested in more detail to [108,200].

The machinery of hearing is frequency dependent. The variation of what is perceived as equally loud at different frequencies was first measured by Fletcher and Munson at Bell Labs in the mid-1930s [107]. These measurements of perceptual equivalence were later refined by Robinson and Dadson. This dependence is usually displayed as a set of equal loudness curves, where the sound pressure level (SPL) is plotted as a function of frequency for tones perceived to be equally loud. Clearly, what two people think of as equally loud will be different. Therefore, these curves are actually averages and serve as a guide to human auditory perception. The particular curve that is of special interest to us is the threshold-of-hearing curve. This is the SPL curve that delineates the boundary of audible and inaudible sounds at different frequencies. In Figure 17.1 we show a plot of this audibility threshold in quiet. Sounds that lie below the threshold are not perceived by humans. Thus, we can see that a low amplitude sound at a frequency of 3 kHz may be perceptible while the same level of sound at 100 Hz would not be perceived.

17.2.1 Spectral Masking

Lossy compression schemes require the use of quantization at some stage. Quantization can be modeled as an additive noise process in which the output of the quantizer is the input plus the quantization noise. To hide quantization noise, we can make use of the fact that signals below a particular amplitude at a particular frequency are not audible. If we select the quantizer step size such that the quantization noise lies below the audibility threshold, the noise will not be perceived. Furthermore, the threshold of audibility is not absolutely fixed and typically rises when multiple sounds impinge on the human ear. This phenomenon gives rise to *spectral masking*. A tone at a certain frequency will raise the threshold in a *critical band* around that frequency. These critical bands have a constant Q, which is the ratio of frequency to bandwidth. Thus, at low frequencies the critical band can have a bandwidth as low as 100 Hz, while at

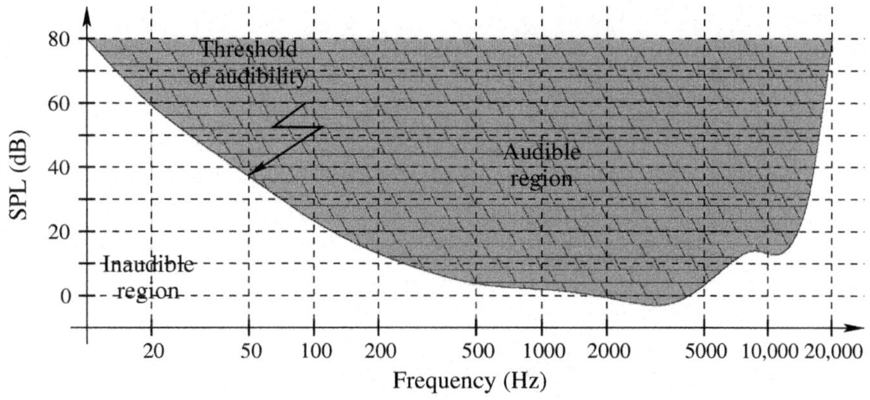

FIGURE 17.1 A typical plot of the audibility threshold.

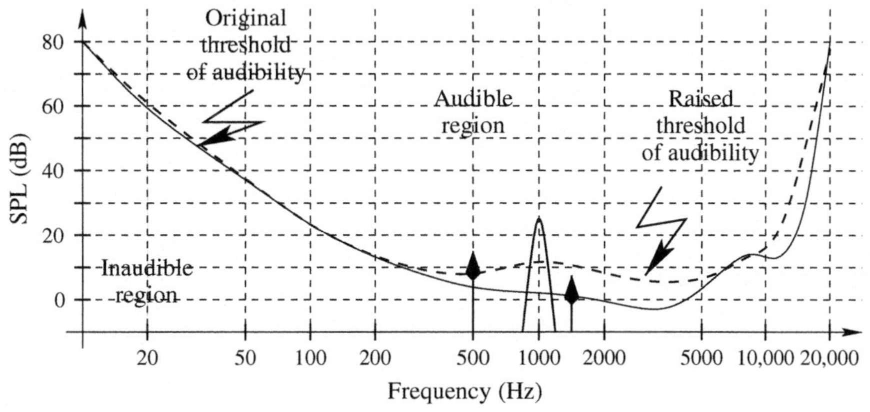

FIGURE 17.2 Change in the audibility threshold.

higher frequencies the bandwidth can be as large as 4 kHz. This increase of the threshold has major implications for compression. Consider the situation in Figure 17.2. Here a tone at 1 kHz has raised the threshold of audibility so that the adjacent tone above it in frequency is no longer audible. At the same time, while the tone at 500 Hz is audible, because of the increase in the threshold the tone can be quantized more crudely. This is because increase of the threshold will allow us to introduce more quantization noise at that frequency. The degree to which the threshold is increased depends on a variety of factors, including whether the signal is sinusoidal or atonal.

17.2.2 Temporal Masking

Along with spectral masking, the psychoacoustic coders also make use of the phenomenon of temporal masking. The temporal masking effect is the masking that occurs when a sound raises

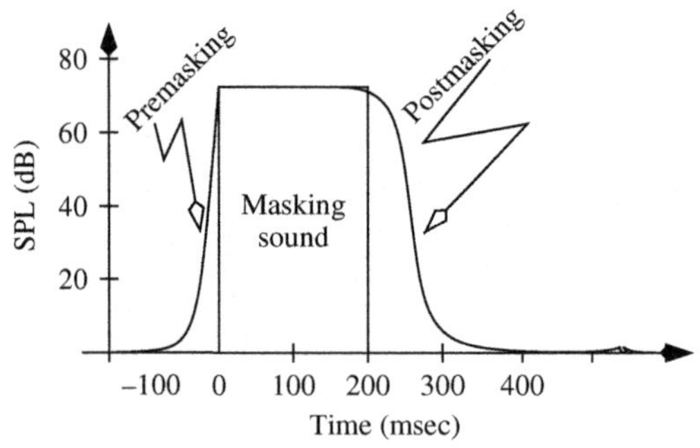

FIGURE 17.3 **Change in the audibility threshold in time.**

the audibility threshold for a brief interval preceding and following the sound. In Figure 17.3 we show the threshold of audibility close to a masking sound. Sounds that occur in an interval around the masking sound (both after and before the masking tone) can be masked. The fact that sounds that occur before the masking tone can get masked can be seen as evidence for the inferential nature of human perception. If the masked sound occurs prior to the masking tone, this is called premasking or backward masking, and if the sound being masked occurs after the masking tone, this effect is called postmasking or forward masking. Forward masking remains in effect for a much longer time interval than backward masking.

17.2.3 Psychoacoustic Model

These attributes of the ear are used by all algorithms that use a psychoacoustic model. There are two models used in the MPEG audio coding algorithms. Although they differ in some details, the general approach used in both cases is the same. The first step in the psychoacoustic model is to obtain a spectral profile of the signal being encoded. The audio input is windowed and transformed into the frequency domain using a filter bank or a frequency domain transform. The sound pressure level (SPL) is calculated for each spectral band. If the algorithm uses a subband approach, then the SPL for the band is computed from the SPL for each frequency coefficient X_k. Because tonal and nontonal components have different effects on the masking level, the next step is to determine the presence and location of these components. The presence of any tonal components is determined by first looking for local maxima where a local maximum is declared at location k if $|X_k|^2 > |X_{k-1}|^2$ and $|X_k|^2 \geqslant |X_{k+1}|^2$. A local maximum is determined to be a tonal component if

$$20 \log_{10} \frac{|X_k|}{|X_{k+j}|} \geqslant 7$$

where the values j depend on the frequency. The identified tonal maskers are removed from each critical band and the power of the remaining spectral lines in the band is summed to obtain the nontonal masking level. Once all the maskers are identified, those with SPL below the audibility threshold are removed. Furthermore, of those maskers that are very close to each other in frequency, the lower-amplitude masker is removed. The effects of the remaining maskers are obtained using a spreading function that models spectral masking. Finally, the masking due to the audibility level and the maskers is combined to give the final masking thresholds. These thresholds are then used in the coding process.

In the following sections we describe the various audio coding algorithms used in the MPEG standards. Although these algorithms provide audio that is perceptually noiseless, it is important to remember that even if we cannot perceive it, there is quantization noise distorting the original signal. This becomes especially important if the reconstructed audio signal goes through any postprocessing. Postprocessing may change some of the audio components, making the previously masked quantization noise audible. Therefore, if there is any kind of processing to be done, including mixing or equalization, the audio should be compressed only after the processing has taken place. This "hidden noise" problem also prevents multiple stages of encoding and decoding or tandem coding.

17.3 MPEG Audio Coding

We begin with the three separate, stand-alone audio compression strategies that are used in MPEG-1 and MPEG-2 and known as Layer I, Layer II, and Layer III. The Layer III audio compression algorithm is also referred to as mp3. Most standards have *normative* sections and *informative* sections. The *normative* actions are those that are required for compliance to the standard. Most current standards, including the MPEG standards, define the bitstream that should be presented to the decoder, leaving the design of the encoder to individual vendors. That is, the bitstream definition is normative, while most guidance about encoding is informative. Thus, two MPEG-compliant bitstreams that encode the same audio material at the same rate but on different encoders may sound very different. On the other hand, a given MPEG bitstream decoded on different decoders will result in essentially the same output.

A simplified block diagram representing the basic strategy used in all three layers is shown in Figure 17.4. The input, consisting of 16-bit PCM words, is first transformed to the frequency domain. The frequency coefficients are quantized, coded, and packed into an MPEG bitstream. Although the overall approach is the same for all layers, the details can vary significantly. Each layer is progressively more complicated than the previous layer and also provides higher compression. The three layers are backward compatible. That is, a decoder for Layer III should be able to decode Layer I- and Layer II-encoded audio. A decoder for Layer II should be able to decode Layer I-encoded audio. Notice the existence of a block labeled *Psychoacoustic model* in Figure 17.4

17.3.1 Layer I Coding

The Layer I coding scheme provides 4:1 compression. In Layer I coding the time frequency mapping is accomplished using a bank of 32 subband filters. The output of the subband filters is critically sampled. That is, the output of each filter is downsampled by 32. The samples are divided into groups of 12 samples each. Twelve samples from each of the 32 subband filters, or a total of 384 samples, make up one frame of the Layer I coder. Once the frequency components are obtained the algorithm examines each group of 12 samples to determine a

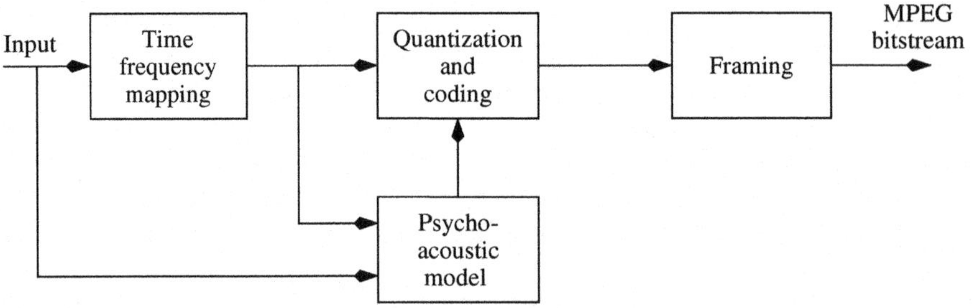

FIGURE 17.4 The MPEG audio coding algorithms.

scalefactor. The scalefactor is used to make sure that the coefficients make use of the entire range of the quantizer. The subband output is divided by the scalefactor before being linearly quantized. There are a total of 63 scalefactors specified in the MPEG standard. Specification of each scalefactor requires 6 bits.

To determine the number of bits to be used for quantization, the coder makes use of the psychoacoustic model. The inputs to the model include the fast Fourier transform (FFT) of the audio data as well as the signal itself. The model calculates the masking thresholds in each subband, which in turn determine the amount of quantization noise that can be tolerated and hence the quantization step size. As the quantizers all cover the same range, selection of the quantization step size is the same as selection of the number of bits to be used for quantizing the output of each subband. In Layer I the encoder has a choice of 14 different quantizers for each band (plus the option of assigning 0 bits). The quantizers are all midtread quantizers ranging from 3 levels to 65,535 levels. Each subband gets assigned a variable number of bits. However, the total number of bits available to represent all the subband samples is fixed. Therefore, the bit allocation can be an iterative process. The objective is to keep the noise-to-mask ratio more or less constant across the subbands.

The output of the quantization and bit allocation steps are combined into a frame as shown in Figure 17.5. Because MPEG audio is a streaming format, each frame carries a header, rather than having a single header for the entire audio sequence. The header is made up of 32 bits. The first 12 bits comprise a sync pattern consisting of all 1s. This is followed by a 1-bit version ID, a 2-bit layer indicator, and 1 bit to indicate CRC protection. The CRC protection bit is set to 0 if there is no CRC protection and is set to a 1 if there is CRC protection. If the layer and protection information is known, all 16 bits can be used for providing frame synchronization. The next 4 bits make up the bit rate index, which specifies the bit rate in kbits/sec. There are 14 specified bit rates to choose from. This is followed by 2 bits that indicate the sampling frequency. The sampling frequencies for MPEG-1 and MPEG-2 are different (one of the few differences between the audio coding standards for MPEG-1 and MPEG-2) and are shown in Table 17.1. These bits are followed by a single padding bit. If the bit is "1," the frame needs an additional bit to adjust the bit rate to the sampling frequency. The next two bits indicate the mode. The possible modes are "stereo," "joint stereo," "dual channel," and "single channel." The stereo mode consists of two channels that are encoded separately but intended to be played

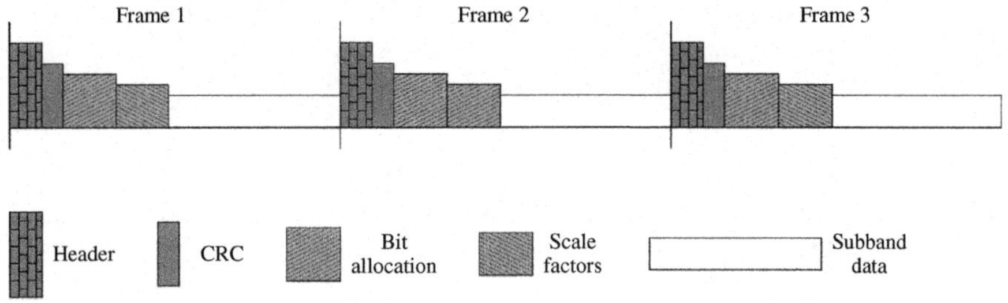

FIGURE 17.5 **Frame structure for Layer 1.**

TABLE 17.1 **Allowable sampling frequencies in MPEG-1 and MPEG-2.**

Index	MPEG-1	MPEG-2
00	44.1 kHz	22.05 kHz
01	48 kHz	24 kHz
10	32 kHz	16 kHz
11	Reserved	

together. The joint stereo mode consists of two channels that are encoded together. The left and right channels are combined to form a *mid* and a *side* signal as follows:

$$M = \frac{L+R}{2}$$
$$S = \frac{L-R}{2}$$

The dual channel mode consists of two channels that are encoded separately and are not intended to be played together, such as a translation channel. These are followed by two mode extension bits that are used in the joint stereo mode. The next bit is a copyright bit ("1" if the material is copyrighted, "0" if it is not). The next bit is set to "1" for original media and "0" for copy. The final two bits indicate the type of de-emphasis to be used.

If the CRC bit is set, the header is followed by a 16-bit CRC. This is followed by the bit allocations used by each subband and is in turn followed by the set of 6-bit scalefactors. The scalefactor data is followed by the 384 quantized samples.

17.3.2 Layer II Coding

The Layer II coder provides a higher compression rate by making some relatively minor modifications to the Layer I coding scheme. These modifications include how the samples are

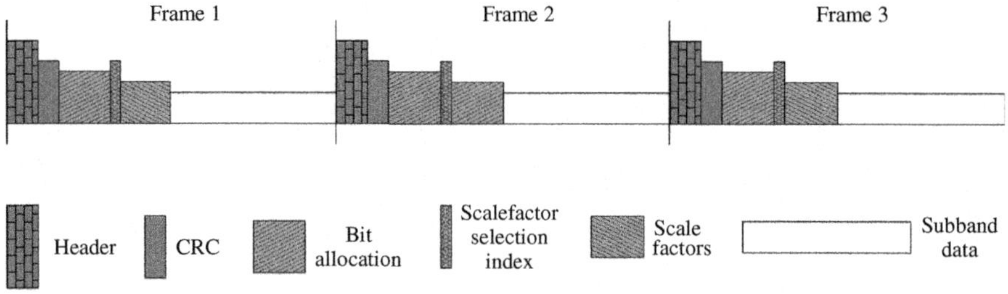

FIGURE 17.6 Frame structure for Layer 2.

grouped together, the representation of the scalefactors, and the quantization strategy. Where the Layer I coder puts 12 samples from each subband into a frame, the Layer II coder groups three sets of 12 samples from each subband into a frame. The total number of samples per frame increases from 384 samples to 1152 samples. This reduces the amount of overhead per sample. In Layer I coding a separate scalefactor is selected for each block of 12 samples. In Layer II coding the encoder tries to share a scalefactor among two or all three groups of samples from each subband filter. The only time separate scalefactors are used for each group of 12 samples is when not doing so would result in a significant increase in distortion. The particular choice used in a frame is signaled through the *scalefactor selection information* field in the bitstream.

The major difference between the Layer I and Layer II coding schemes is in the quantization step. In the Layer I coding scheme the output of each subband is quantized using one of 14 possibilities; the same 14 possibilities are available for each of the subbands. In Layer II coding the quantizers used for each of the subbands can be selected from a different set of quantizers depending on the sampling rate and the bit rates. For some sampling rate and bit rate combinations, many of the higher subbands are assigned 0 bits. That is, the information from those subbands is simply discarded. Where the quantizer selected has 3, 5, or 9 levels, the Layer II coding scheme uses one more enhancement. Notice that in the case of 3 levels we have to use 2 bits per sample, which would have allowed us to represent 4 levels. The situation is even worse in the case of 5 levels, where we are forced to use 3 bits, wasting three codewords, and in the case of 9 levels, where we have to use 4 bits, thus wasting 7 levels. To avoid this situation, the Layer II coder groups 3 samples into a *granule*. If each sample can take on 3 levels, a granule can take on 27 levels. This can be accommodated using 5 bits. If each sample had been encoded separately we would have needed 6 bits. Similarly, if each sample can take on 9 values, a granule can take on 729 values. We can represent 729 values using 10 bits. If each sample in the granule had been encoded separately, we would have needed 12 bits. Using all these savings, the compression ratio in Layer II coding can be increased from 4:1 to 8:1 or 6:1.

The frame structure for the Layer II coder can be seen in Figure 17.6. The only real difference between this frame structure and that of the Layer I coder is the scalefactor selection information field.

17.3.3 Layer III Coding—mp3

Layer III coding, which has become widely popular under the name mp3, is considerably more complex than the Layer I and Layer II coding schemes. One of the problems with the Layer I and II coding schemes was that with 32-band decomposition, the bandwidth of the subbands at lower frequencies is significantly larger than the critical bands. This makes it difficult to make an accurate judgement of the mask-to-signal ratio. If we get a high amplitude tone within a subband and if the subband is narrow enough, we can assume that it masked other tones in the band. However, if the bandwidth of the subband is significantly higher than the critical bandwidth at that frequency, it becomes more difficult to determine whether other tones in the subband will be masked.

A simple way to increase the spectral resolution would be to decompose the signal directly into a higher number of bands. However, one of the requirements on the Layer III algorithm is that it be backward compatible with Layer I and Layer II coders. To satisfy this backward compatibility requirement, the spectral decomposition in the Layer III algorithm is performed in two stages. First the 32-band subband decomposition used in Layer I and Layer II is employed. The output of each subband is then transformed using a modified discrete cosine transform (MDCT) with a 50% overlap. The Layer III algorithm specifies two sizes for the MDCT, 6 or 18. This means that the output of each subband can be decomposed into 18 frequency coefficients or 6 frequency coefficients.

The reason for having two sizes for the MDCT is that when we transform a sequence into the frequency domain, we lose time resolution even as we gain frequency resolution. The larger the block size the more we lose in terms of time resolution. The problem with this is that any quantization noise introduced into the frequency coefficients will get spread over the entire block size of the transform. Backward temporal masking occurs for only a short duration prior to the masking sound (approximately 20 msec). Therefore, quantization noise will appear as a *pre-echo*. Consider the signal shown in Figure 17.7. The sequence consists of 128 samples, the first 118 of which are 0, followed by a sharp increase in value. The 128-point DCT of this sequence is shown in Figure 17.8. Notice that many of these coefficients are quite large. If we were to send all these coefficients, we would have data expansion instead of data compression. If we keep only the 10 largest coefficients, the reconstructed signal is shown in Figure 17.9. Notice that not only are the nonzero signal values not well represented, there is also error in the samples prior to the change in value of the signal. If this were an audio signal and the large values had occurred at the beginning of the sequence, the forward masking effect would have reduced the perceptibility of the quantization error. In the situation shown in Figure 17.9, backward masking will mask some of the quantization error. However, backward masking occurs for only a short duration prior to the masking sound. Therefore, if the length of the block in question is longer than the masking interval, the distortion will be evident to the listener.

If we get a sharp sound that is very limited in time (such as the sound of castanets) we would like to keep the block size small enough that it can contain this sharp sound. Then, when we incur quantization noise it will not get spread out of the interval in which the actual sound occurred and will therefore get masked. The Layer III algorithm monitors the input and where necessary substitutes three short transforms for one long transform. What actually happens is that the subband output is multiplied by a window function of length 36 during

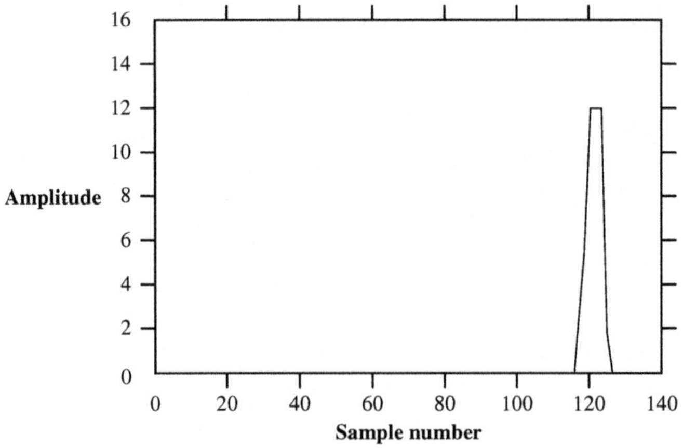

FIGURE 17.7 Source output sequence.

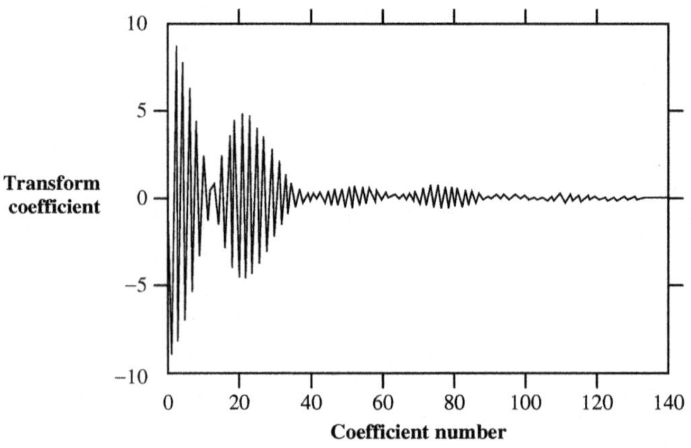

FIGURE 17.8 Transformed sequence.

the stationary periods (that is a block size of 18 plus 50% overlap from neighboring blocks). This window is called the *long window*. If a sharp attack is detected, the algorithm shifts to a sequence of three *short windows* of length 12 after a transition window of length 30. This initial transition window is called the *start* window. If the input returns to a more stationary mode, the short windows are followed by another transition window called the *stop* window of length 30 and then the standard sequence of long windows. The process of transitioning between windows is shown in Figure 17.10. A possible set of window transitions is shown in Figure 17.11. For the long windows we end up with 18 frequencies per subband, resulting in a total of 576 frequencies. For the short windows we get 6 coefficients per subband for a total of 192 frequencies. The standard allows for a mixed block mode in which the two lowest subbands use long windows while the remaining subbands use short windows. Notice that while the number of frequencies may change depending on whether we are using long or short

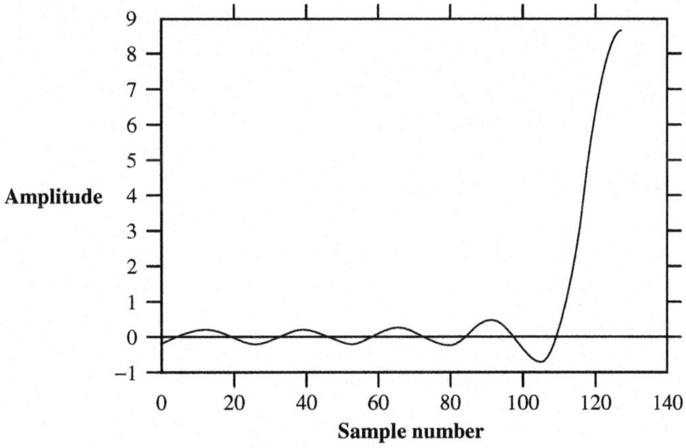

FIGURE 17.9 Reconstructed sequence from 10 DCT coefficients.

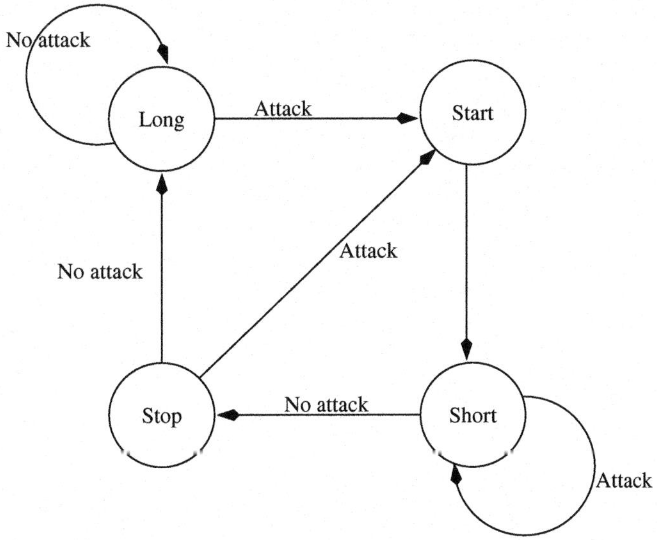

FIGURE 17.10 State diagram for the window switching process.

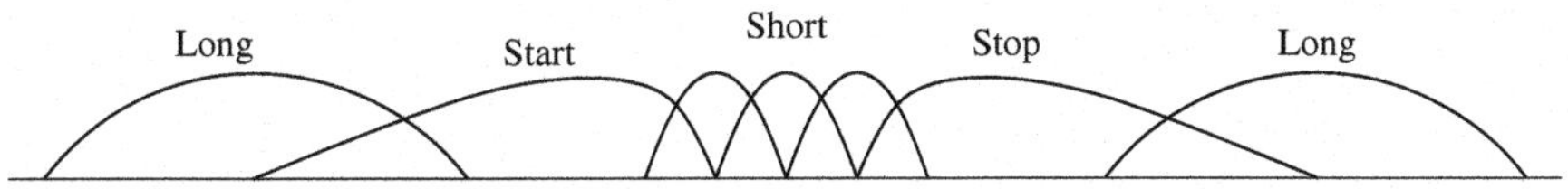

FIGURE 17.11 Sequence of windows.

windows, the number of samples in a frame stays at 1152. That is 36 samples, or 3 groups of 12, from each of the 32 subband filters.

The coding and quantization of the output of the MDCT is conducted in an iterative fashion using two nested loops. There is an outer loop called the *distortion control loop* whose purpose is to ensure that the introduced quantization noise lies below the audibility threshold. The scalefactors are used to control the level of quantization noise. In Layer III scalefactors are assigned to groups or "bands" of coefficients in which the bands are approximately the size of critical bands. There are 21 scalefactor bands for long blocks and 12 scalefactor bands for short blocks.

The inner loop is called the *rate control loop*. The goal of this loop is to make sure that a target bit rate is not exceeded. This is done by iterating between different quantizers and Huffman codes. The quantizers used in MP3 are companded nonuniform quantizers. The scaled MDCT coefficients are first quantized and organized into regions. Coefficients at the higher end of the frequency scale are likely to be quantized to zero. These consecutive zero outputs are treated as a single region and the run length is Huffman encoded. For frequency coefficients in the region below this region of zero coefficients, the encoder identifies the set of coefficients that are quantized to 0 or ±1. These coefficients are grouped into groups of four. This set of quadruplets is the second region of coefficients. Each quadruplet is encoded using a single Huffman codeword. The remaining coefficients are divided into two or three subregions. Each subregion is assigned a Huffman code based on its statistical characteristics. If the result of using this variable-length coding exceeds the bit budget, the quantizer is adjusted to increase the quantization stepsize. The process is repeated until the target rate is satisfied.

Once the target rate is satisfied, control passes back to the outer, distortion control loop. The psychoacoustic model is used to check whether the quantization noise in any band exceeds the allowed distortion. If it does, the scalefactor is adjusted to reduce the quantization noise. Once all scalefactors have been adjusted, control returns to the rate control loop. The iterations terminate either when the distortion and rate conditions are satisfied or the scalefactors cannot be adjusted any further.

There will be frames in which the number of bits used by the Huffman coder is less than the amount allocated. These bits are saved in a conceptual *bit reservoir*. In practice what this means is that the start of a block of data does not necessarily coincide with the header of the frame. Consider the three frames shown in Figure 17.12. In this example, the main data for the first frame (which includes scalefactor information and the Huffman coded data) does not occupy the entire frame. Therefore, the main data for the second frame starts before the second frame actually begins. The same is true for the remaining data. The main data can begin in the *previous frame*. However, the main data for a particular frame cannot spill over into the *following* frame.

All this complexity allows for a very efficient encoding of audio inputs. The typical MP3 audio file has a compression ratio of about 10:1. In spite of this high level of compression, most people cannot tell the difference between the original and the compressed representation. We say most because trained professionals can at times tell the difference between the original and compressed versions. People who can identify very minute differences between coded and original signals have played an important role in the development of audio coders. By identifying where distortion may be audible they have helped focus the efforts to improve the coding process. This development process has made MP3 the format of choice for compressed music.

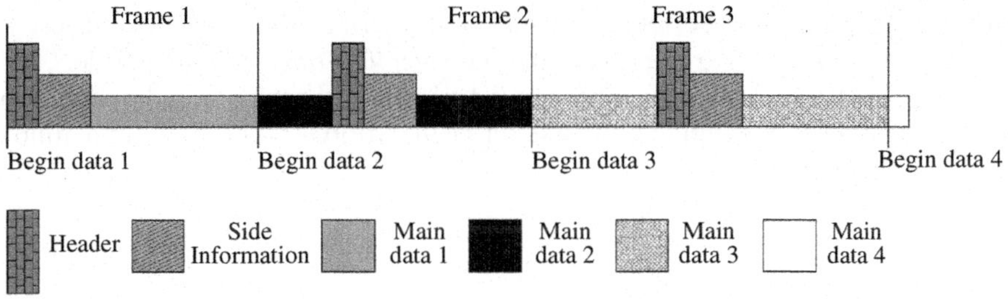

FIGURE 17.12 **Sequence of windows.**

17.4 MPEG Advanced Audio Coding

The MPEG Layer III algorithm has been highly successful. However, it has some built-in drawbacks because of the constraints under which it was designed. The principal constraint was the requirement that it be backward compatible. This requirement for backward compatibility forced the rather awkward decomposition structure involving a subband decomposition followed by an MDCT decomposition. The period immediately following the release of the MPEG specifications also saw major developments in hardware capability. The Advanced Audio Coding (AAC) standard was approved as a higher quality multichannel alternative to the backward compatible MPEG Layer III in 1997.

The AAC standard takes a modular approach based on a set of self-contained tools or modules. Some of these tools are taken from the earlier MPEG audio standard while others are new. As with previous standards, the AAC standard actually specifies the decoder. The decoder tools specified in the AAC standard are listed in Table 17.2. As shown in the table, some of these tools are required for all profiles while others are only required for some profiles.

TABLE 17.2 **AAC decoder tools. [221]**

Tool Name	
Bitstream Formatter	Required
Huffman Decoding	Required
Inverse Quantization	Required
Rescaling	Required
M/S	Optional
Interblock Prediction	Optional
Intensity	Optional
Dependently Switched Coupling	Optional
TNS	Optional
Block Switching / MDCT	Required
Gain Control	Optional
Independently Switched Coupling	Optional

By using some or all of these tools, the standard describes three profiles. These are the *main* profile, the *low complexity* profile, and the *sampling-rate-scalable* profile. The AAC approach used in MPEG-2 was later enhanced and modified to provide an audio coding option in MPEG-4. In the following section we first describe the MPEG-2 AAC algorithm, followed by the MPEG-4 AAC algorithm.

17.4.1 MPEG-2 AAC

A block diagram of an MPEG-2 AAC encoder is shown in Figure 17.13. Each block represents a tool. The psychoacoustic model used in the AAC encoder is the same as the model used in the MPEG Layer III encoder. As in the Layer III algorithm, the psychoacoustic model is used

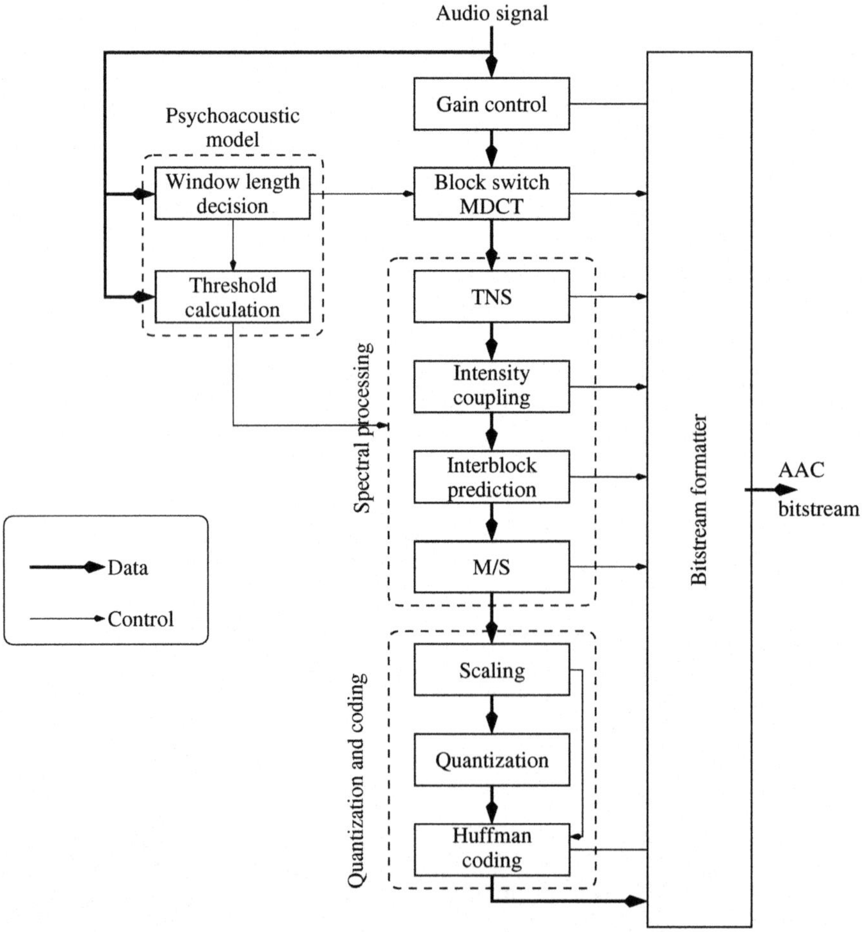

FIGURE 17.13 An MPEG-2 AAC encoder [221].

to trigger switching in the block length of the MDCT transform and to produce the threshold values used to determine scalefactors and quantization thresholds. The audio data is fed in parallel to both the acoustic model and to the modified discrete cosine transform.

Block Switching and MDCT

Because the AAC algorithm is not backward compatible it does away with the requirement of the 32-band filter bank. Instead, the frequency decomposition is accomplished by a modified discrete cosine transform (MDCT). The MDCT is described in Chapter 13. The AAC algorithm allows switching between a window length of 2048 samples and 256 samples. These window lengths include a 50% overlap with neighboring blocks. So 2048 time samples are used to generate 1024 spectral coefficients, and 256 time samples are used to generate 128 frequency coefficients. The kth spectral coefficient of block i, $X_{i,k}$, is given by

$$X_{i,k} = 2\sum_{n=0}^{N-1} z_{i,n} \cos\left(\frac{2\pi(n+n_o)}{N}\left(k+\frac{1}{2}\right)\right)$$

where $z_{i,n}$ is the nth time sample of the ith block, N is the window length, and

$$n_o = \frac{N/2+1}{2}$$

The longer block length allows the algorithm to take advantage of stationary portions of the input to get significant improvements in compression. The short block length allows the algorithm to handle sharp attacks without incurring substantial distortion and rate penalties. Short blocks occur in groups of eight in order to avoid framing issues. As in the case of MPEG Layer III, there are four kinds of windows: long, short, start, and stop. The decision about whether to use a group of short blocks is made by the psychoacoustic model. The coefficients are divided into scalefactor bands in which the number of coefficients in the bands reflects the critical bandwidth. Each scalefactor band is assigned a single scalefactor. The exact division of the coefficients into scalefactor bands for the different windows and different sampling rates is specified in the standard [221].

Spectral Processing

In MPEG Layer III coding the compression gain is mainly achieved through the unequal distribution of energy in the different frequency bands, the use of the psychoacoustic model, and Huffman coding. The unequal distribution of energy allows use of fewer bits for spectral bands with less energy. The psychoacoustic model is used to adjust the quantization step size in a way that masks the quantization noise. Huffman coding allows further reductions in the bit rate. All these approaches are also used in the AAC algorithm. In addition, the algorithm makes use of prediction to reduce the dynamic range of the coefficients and thus allow further reduction in the bit rate.

Recall that prediction is generally useful only in stationary conditions. By their very nature, transients are almost impossible to predict. Therefore, generally speaking, predictive coding would not be considered for signals containing significant amounts of transients. However, music signals have exactly this characteristic. Although they may contain long periods of stationary signals, they also generally contain a significant amount of transient signals. The AAC algorithm makes clever use of the time frequency duality to handle this situation. The standard contains two kinds of predictors: an intrablock predictor, referred to as Temporal Noise Shaping (TNS), and an interblock predictor. The interblock predictor is used during stationary periods. During these periods it is reasonable to assume that the coefficients at a certain frequency do not change their value significantly from block to block. Making use of this characteristic, the AAC standard implements a set of parallel DPCM systems. There is one predictor for each coefficient up to a maximum number of coefficients. The maximum is different for different sampling frequencies. Each predictor is a backward adaptive two-tap predictor. This predictor is really useful only in stationary periods. Therefore, the psychoacoustic model monitors the input and determines when the output of the predictor is to be used. The decision is made on a scalefactor band by scalefactor band basis. Because notification of the decision that the predictors are being used has to be sent to the decoder, this would increase the rate by one bit for each scalefactor band. Therefore, once the preliminary decision to use the predicted value has been made, further calculations are made to check if the savings will be sufficient to offset this increase in rate. If the savings are determined to be sufficient, a *predictor_data_present* bit is set to 1 and one bit for each scalefactor band (called the *prediction_used* bit) is set to 1 or 0 depending on whether prediction was deemed effective for that scalefactor band. If not, the *predictor_data_present* bit is set to 0 and the *prediction_used* bits are not sent. Even when a predictor is disabled, the adaptive algorithm is continued so that the predictor coefficients can track the changing coefficients. However, because this is a streaming audio format it is necessary from time to time to reset the coefficients. Resetting is done periodically in a staged manner and also when a short frame is used.

When the audio input contains transients, the AAC algorithm uses the intraband predictor. Recall that narrow pulses in time correspond to wide bandwidths. The narrower a signal in time, the broader its Fourier transform will be. This means that when transients occur in the audio signal, the resulting MDCT output will contain a large number of correlated coefficients. Thus, unpredictability in time translates to a high level of predictability in terms of the frequency components. The AAC uses neighboring coefficients to perform prediction. A target set of coefficients is selected in the block. The standard suggests a range of 1.5 kHz to the uppermost scalefactor band as specified for different profiles and sampling rates. A set of linear predictive coefficients is obtained using any of the standard approaches, such as the Levinson-Durbin algorithm described in Chapter 18. The maximum order of the filter ranges from 12 to 20 depending on the profile. The process of obtaining the filter coefficients also provides the expected prediction gain g_p. This expected prediction gain is compared against a threshold to determine if intrablock prediction is going to be used. The standard suggests a value of 1.4 for the threshold. The order of the filter is determined by the first partial correlation (PARCOR) coefficient with a magnitude smaller than a threshold (suggested to be 0.1). The PARCOR coefficients corresponding to the predictor are quantized and coded for transfer to the decoder. The reconstructed LPC coefficients are then used for prediction. In the time domain predictive coders, one effect of linear prediction is the spectral shaping of the quantization noise. The effect of prediction in the frequency domain is the *temporal* shaping of the quantization noise, hence the name Temporal Noise Shaping. The shaping of the noise means that the noise will be higher during time periods when the signal amplitude is high and lower when the signal amplitude is low. This is especially useful in audio signals because of the masking properties of human hearing.

Quantization and Coding

The quantization and coding strategy used in AAC is similar to what is used in MPEG Layer III. Scalefactors are used to control the quantization noise as part of an outer *distortion control loop*. The quantization step size is adjusted to accommodate a target bit rate in an inner *rate control loop*. The quantized coefficients are grouped into *sections*. The section boundaries have to coincide with scalefactor band boundaries. The quantized coefficients in each section are coded using the same Huffman codebook. The partitioning of the coefficients into sections is a dynamic process based on a greedy merge procedure. The procedure starts with the maximum number of sections. Sections are merged if the overall bit rate can be reduced by merging. Merging those sections will result in the maximum reduction in bit rate. This iterative procedure is continued until there is no further reduction in the bit rate.

Stereo Coding

The AAC scheme uses multiple approaches to stereo coding. Apart from independently coding the audio channels, the standard allows mid/side (M/S) coding and intensity stereo coding. Both stereo coding techniques can be used at the same time for different frequency ranges. Intensity coding makes use of the fact that at higher frequencies two channels can be represented by a single channel plus some directional information. The AAC standard suggests using this technique for scalefactor bands above 6 kHZ. The M/S approach is used to reduce noise imaging. As described previously in the joint stereo approach, the two channels (L and R) are combined to generate sum and difference channels.

Profiles

The main profile of MPEG-2 AAC uses all the tools except for the gain control tool of Figure 17.13. In the low complexity profile in addition to the gain control tool the interblock prediction tool is also dropped. In addition the maximum prediction order for intraband prediction (TNS) for long windows is 12 for the low complexity profile as opposed to 20 for the main profile.

The Scalable Sampling Rate profile does not use the coupling and interband prediction tools. However this profile does use the gain control tool. In the scalable-sampling profile the MDCT block is preceded by a bank of four equal width 96-tap filters. The filter coefficients are provided in the standard. The use of this filter bank allows for a reduction in rate and decoder complexity. By ignoring one or more of the filter bank outputs the output bandwidth can be reduced. This reduction in bandwidth and sample rate also leads to a reduction in the decoder complexity. The gain control allows for the attenuation and amplification of different bands in order to reduce perceptual distortion.

17.4.2 MPEG-4 AAC

MPEG-4 AAC adds a perceptual noise substitution (PNS) tool and substitutes a long term prediction (LTP) tool for the interband prediction tool in the spectral coding block. In the quantization and coding section MPEG-4 AAC adds the options of Transform-Domain Weighted Interleave Vector Quantization (TwinVQ) and Bit Sliced Arithmetic Coding (BSAC).

Perceptual Noise Substitution (PNS)

There are portions of music that sound like noise. Although this may sound like a harsh (or realistic) subjective evaluation, that is not what is meant here. What is meant by noise here is a portion of audio where the MDCT coefficients are stationary without containing tonal components [222]. This kind of noiselike signal is the hardest to compress. However, at the same time it is very difficult to distinguish one noiselike signal from another. MPEG-4 AAC makes use of this fact by not transmitting such noiselike scalefactor bands. Instead the decoder is alerted to this fact and the power of the noiselike coefficients in this band is sent. The decoder generates a noiselike sequence with the appropriate power and inserts it in place of the unsent coefficients.

Long Term Prediction

The interband prediction in MPEG-2 AAC is one of the more computationally expensive parts of the algorithm. MPEG-4 AAC replaces that with a cheaper long term prediction (LTP) module.

TwinVQ

The Transform-Domain Weighted Interleave Vector Quantization (TwinVQ) [282] option is suggested in the MPEG-4 AAC scheme for low bit rates. Developed at NTT in the early 1990s, the algorithm uses a two-stage process for flattening the MDCT coefficients. In the first stage, a linear predictive coding algorithm is used to obtain the LPC coefficients for the audio data corresponding to the MDCT coefficients. These coefficients are used to obtain the spectral envelope for the audio data. Dividing the MDCT coefficients with this spectral envelope results in some degree of "flattening" of the coefficients. The spectral envelope computed from the LPC coefficients reflects the gross features of the envelope of the MDCT coefficients. However, it does not reflect any of the fine structure. This fine structure is predicted from the previous frame and provides further flattening of the MDCT coefficients. The flattened coefficients are interleaved and grouped into subvectors and quantized. The flattening process reduces the dynamic range of the coefficients, allowing them to be quantized using a smaller VQ codebook than would otherwise have been possible. The flattening process is reversed in the decoder as the LPC coefficients are transmitted to the decoder.

Bit Sliced Arithmetic Coding (BSAC)

In addition to the Huffman coding approach used in the MPEG-2 AAC scheme, the MPEG-4 AAC scheme also provides the option of using binary arithmetic coding. The binary arithmetic coding is performed on the bitplanes of the magnitudes of the quantized MDCT coefficients. By bitplane we mean the corresponding bit of each coefficient. Consider the sequence of 4-bit coefficients x_n : 5, 11, 8, 10, 3, 1. The most significant bitplane would consist of the MSBs of these numbers, 011100. The next bitplane would be 100000. The next bitplane is 010110. The least significant bitplane is 110011.

The coefficients are divided into *coding bands* of 32 coefficients each. One probability table is used to encode each coding band. Because we are dealing with binary data, the probability table is simply the number of zeros. If a coding band contains only zeros, this is indicated to the decoder by selecting the probability table 0. The sign bits associated with the nonzero coefficients are sent after the arithmetic code when the coefficient has a 1 for the first time.

Arithmetic coding is also used on the scalefactor information. The maximum scalefactor is coded as an 8-bit integer. The differences between scalefactors are encoded using an arithmetic code. The first scalefactor is encoded using the difference between it and the maximum scalefactor.

17.5 Dolby AC-3 (Dolby Digital)

Unlike the MPEG algorithms described in the previous section, the Dolby AC-3 method became a de facto standard. It was developed in response to the standardization activities of the *Grand Alliance*, which was developing a standard for HDTV in the United States. However, even before it was accepted as the recommendation for HDTV audio, Dolby AC-3 had already made its debut in the movie industry. It was first released in a few theaters during the showing of *Star Trek IV* in 1991 and was formally released with the movie *Batman Returns* in 1992. It was accepted by the *Grand Alliance* in October of 1993 and became an Advanced Television Systems Committee (ATSC) standard in 1995. Dolby AC-3 had the multichannel capability required by the movie industry along with the ability to downmix the channels to accommodate the varying capabilities of different applications. The 5.1 channels include right, center, left, left rear, and right rear, and a narrowband low-frequency effects channel (the 0.1 channel). The scheme supports downmixing the 5.1 channels to 4, 3, 2, or 1 channel. It is now the standard used for DVDs as well as for Direct Broadcast Satellites (DBS) and other applications.

A block diagram of the Dolby AC-3 algorithm is shown in Figure 17.14. Much of the Dolby AC-3 scheme is similar to what we have already described for the MPEG algorithms. As in the MPEG schemes, the Dolby AC-3 algorithm uses the modified DCT (MDCT) with 50% overlap for frequency decomposition. As in the case of MPEG, there are two different sizes of windows used. For the stationary portions of the audio a window of size 512 is used to get 256 coefficients. A surge in the power of the high-frequency coefficients is used to indicate the presence of a transient and the 512 window is replaced by two windows of size 256. The one place where the Dolby AC-3 algorithm differs significantly from the algorithm described for MPEG is in the bit allocation.

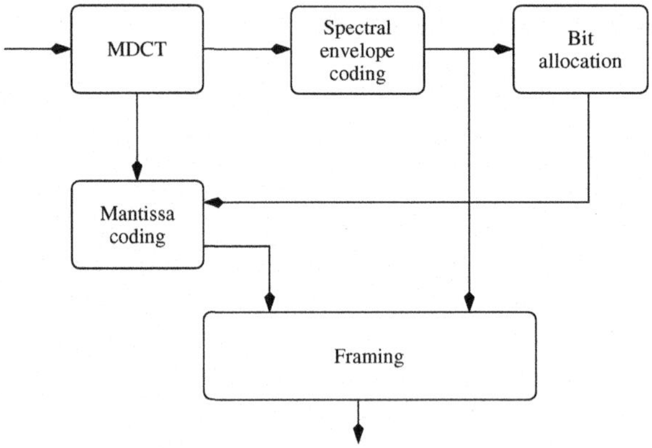

FIGURE 17.14 The Dolby AC-3 algorithm.

17.5.1 Bit Allocation

The Dolby AC-3 scheme has a very interesting method for bit allocation. Like the MPEG schemes, it uses a psychoacoustic model that incorporates the hearing thresholds and the presence of noise and tone maskers. However, the input to the model is different. In the MPEG schemes the audio sequence being encoded is provided to the bit allocation procedure and the bit allocation is sent to the decoder as side information. In the Dolby AC-3 scheme the signal itself is not provided to the bit allocation procedure. Instead a crude representation of the spectral envelope is provided to both the decoder and the bit allocation procedure. As the decoder then possesses the information used by the encoder to generate the bit allocation, the allocation itself is not included in the transmitted bitstream.

The representation of the spectral envelope is obtained by representing the MDCT coefficients in binary exponential notation. The binary exponential notation of a number 110.101 is 0.110101×2^3, where 110101 is called the mantissa and 3 is the exponent. Given a sequence of numbers, the exponents of the binary exponential representation provide an estimate of the relative magnitude of the numbers. The Dolby AC-3 algorithm uses the exponents of the binary exponential representation of the MDCT coefficients as the representation of the spectral envelope. This encoding is sent to the bit allocation algorithm, which uses this information in conjunction with a psychoacoustic model to generate the number of bits to be used to quantize the mantissa of the binary exponential representation of the MDCT coefficients. To reduce the amount of information that needs to be sent to the decoder, the spectral envelope coding is not performed for every audio block. Depending on how stationary the audio is, the algorithm uses one of three strategies [200].

The D15 Method

When the audio is relatively stationary, the spectral envelope is coded once for every six audio blocks. Because a frame in Dolby AC-3 consists of six blocks, during each block we get a new spectral envelope and hence a new bit allocation. The spectral envelope is coded differentially. The first exponent is sent as is. The difference between exponents is encoded using one of five values $\{0, \pm 1, \pm 2\}$. Three differences are encoded using a 7-bit word. Note that three differences can take on 125 different combinations. Therefore, using 7 bits, which can represent 128 different values, is highly efficient.

The D25 and D45 Methods

If the audio is not stationary, the spectral envelope is sent more often. To keep the bit rate down, the Dolby AC-3 algorithm uses one of two strategies. In the D25 strategy, which is used for moderate spectral activity, every other coefficient is encoded. In the D45 strategy, used during transients, every fourth coefficient is encoded. These strategies make use of the fact that during a transient the fine structure of the spectral envelope is not that important, allowing for a more crude representation.

17.6 Other Standards

We have described a number of audio compression approaches that make use of the limitations of human audio perception. These are by no means the only ones. Competitors to Dolby Digital include Digital Theater Systems (DTS) and Sony Dynamic Digital Sound (SDDS). Both of these proprietary schemes use psychoacoustic modeling. The Adaptive TRansform Acoustic Coding (ATRAC) algorithm [223] was developed for the MiniDisc by Sony in the early 1990s, followed by enhancements in ATRAC3 and ATRAC3plus. As with the other schemes described in this chapter, the ATRAC approach uses MDCT for frequency decomposition, though the audio signal is first decomposed into three bands using a two-stage decomposition. As in the case of the other schemes, the ATRAC algorithm recommends the use of the limitations of human audio perception in order to discard information that is not perceptible.

Another algorithm that also uses MDCT and a psychoacoustic model is the open source encoder Vorbis. The Vorbis algorithm also uses vector quantization and Huffman coding to reduce the bit rate.

17.7 Summary

The audio coding algorithms described in this chapter take, in some sense, the opposite tack from the speech coding algorithms described previously. Instead of focusing on the source of information, as is the case with a speech coding algorithm, the focus in an audio coding algorithm is on the sink, or user, of the information. By identifying the components of the source signal that are not perceptible, the algorithms reduce the amount of data that needs to be transmitted.

Further Reading

1. The book *Introduction to Digital Audio Coding and Standards*, by M. Bosi and R.E. Goldberg [200], provides a detailed accounting of the standards described here as well as a comprehensive look at the process of constructing a psychoacoustic model.
2. *The MPEG Handbook*, by J. Watkinson [222], is an accessible source of information about aspects of audio coding as well as the MPEG algorithms.
3. An excellent tutorial on the MPEG algorithms is the appropriately named *A Tutorial on MPEG/Audio Compression*, by D. Pan [224].
4. A thorough review of audio coding can be found in *Perceptual Coding of Digital Audio*, by T. Painter and A. Spanias [225].
5. The website http://www.tnt.uni-hannover.de/project/mpeg/audio/faq/ contains information about all the audio coding schemes described here as well as an overview of MPEG-7 audio.

18

Analysis/Synthesis and Analysis by Synthesis Schemes

18.1 Overview

Analysis/synthesis schemes rely on the availability of a parametric model of the source output generation. When such a model exists, the transmitter analyzes the source output and extracts the model parameters, which are transmitted to the receiver. The receiver uses the model along with the transmitted parameters to synthesize an approximation to the source output. The difference between this approach and the techniques we have looked at in previous chapters is that what is transmitted is not a direct representation of the samples of the source output; instead, the transmitter informs the receiver how to go about regenerating those outputs. For this approach to work, a good model for the source has to be available. Since good models for speech production exist, this approach has been widely used for the low-rate coding of speech. We describe several different analysis/synthesis techniques for speech compression. In recent years the fractal approach to image compression has been gaining in popularity. Because this approach is also one in which the receiver regenerates the source output using "instructions" from the transmitter, we describe it in this chapter.

18.2 Introduction

In previous chapters we presented a number of lossy compression schemes that provide an estimate of each source output value to the receiver. Historically, an earlier approach towards lossy compression is to model the source output and send the model parameters to the source instead of the estimates of the source output. The receiver tries to synthesize the source output based on the received model parameters.

Consider an image transmission system that works as follows. At the transmitter, we have a person who examines the image to be transmitted and comes up with a description of the image. At the receiver, we have another person who then proceeds to create that image. For example, suppose the image we wish to transmit is a picture of a field of sunflowers. Instead of trying to send the picture, we simply send the words "field of sunflowers." The person at the receiver paints a picture of a field of sunflowers on a piece of paper and gives it to the user. Thus, an image of an object is transmitted from the transmitter to the receiver in a highly compressed form. This approach towards compression should be familiar to listeners of sports broadcasts on radio. It requires that both transmitter and receiver work with the same model. In terms of sports broadcasting, this means that the viewer has a mental picture of the sports arena, and both the broadcaster and listener attach the same meaning to the same terminology.

This approach works for sports broadcasting because the source being modeled functions under very restrictive rules. In a basketball game, when the referee calls a dribbling foul, listeners generally don't picture a drooling chicken. If the source violates the rules, the reconstruction would suffer. If the basketball players suddenly decided to put on a ballet performance, the transmitter (sportscaster) would be hard pressed to represent the scene accurately to the receiver. Therefore, it seems that this approach to compression can only be used for artificial activities that function according to man-made rules. Of the sources that we are interested in, only text fits this description, and the rules that govern the generation of text are complex and differ widely from language to language.

Fortunately, while natural sources may not follow man-made rules, they are subject to the laws of physics, which can prove to be quite restrictive. This is particularly true of speech. No matter what language is being spoken, the speech is generated using machinery that is not very different from person to person. Moreover, this machinery has to obey certain physical laws that substantially limit the behavior of outputs. Therefore, speech can be analyzed in terms of a model, and the model parameters can be extracted and transmitted to the receiver. At the receiver the speech can be synthesized using the model. This analysis/synthesis approach was first employed by Homer Dudley at Bell Laboratories, who developed what is known as the channel vocoder (described in the next section). Actually, the synthesis portion had been attempted even earlier by Kempelen Farkas Lovag (1734–1804). He developed a "speaking machine" in which the vocal tract was modeled by a flexible tube whose shape could be modified by an operator. Sound was produced by forcing air through this tube using bellows [226].

Unlike speech, images are generated in a variety of different ways; therefore, the analysis/synthesis approach does not seem very useful for image or video compression. However, if we restrict the class of images to "talking heads" of the type we would encounter in a videoconferencing situation, we might be able to satisfy the conditions required for this approach. When we talk, our facial gestures are restricted by the way our faces are constructed and by the physics of motion. This realization has led to the new field of model-based video coding (see Chapter 19).

A totally different approach to image compression based on the properties of self-similarity is the *fractal coding* approach. While this approach does not explicitly depend on some physical

FIGURE 18.1 **A model for speech synthesis.**

limitations, it fits in with the techniques described in this chapter; that is, what is stored or transmitted is not the samples of the source output, but a method for synthesizing the output. We will study this approach in Section 18.6.

18.3 Speech Compression

A very simplified model of speech synthesis is shown in Figure 18.1. As we described in Chapter 7, speech is produced by forcing air first through an elastic opening, the vocal cords, and then through the laryngeal, oral, nasal, and pharynx passages, and finally through the mouth and the nasal cavity. Everything past the vocal cords is generally referred to as the vocal tract. The first action generates the sound, which is then modulated into speech as it traverses through the vocal tract.

In Figure 18.1, the excitation source corresponds to the sound generation, and the vocal tract filter models the vocal tract. As we mentioned in Chapter 7, there are several different sound inputs that can be generated by different conformations of the vocal cords and the associated cartilages.

Therefore, in order to generate a specific fragment of speech, we have to generate a sequence of sound inputs or excitation signals and the corresponding sequence of appropriate vocal tract approximations.

At the transmitter, the speech is divided into segments. Each segment is analyzed to determine an excitation signal and the parameters of the vocal tract filter. In some of the schemes, a model for the excitation signal is transmitted to the receiver. The excitation signal is then synthesized at the receiver and used to drive the vocal tract filter. In other schemes, the excitation signal itself is obtained using an analysis-by-synthesis approach. This signal is then used by the vocal tract filter to generate the speech signal.

Over the years many different analysis/synthesis speech compression schemes have been developed, and substantial research into the development of new approaches and the improvement of existing schemes continues. Given the large amount of information, we can only sample some of the more popular approaches in this chapter. See [283,284] for more detailed coverage and pointers to the vast literature on the subject.

The approaches we will describe in this chapter include *channel vocoders*, which are of special historical interest; the *linear predictive coder*, which is the U.S. Government standard at the rate of 2.4 kbps; *code-excited linear prediction* (CELP) based schemes; *sinusoidal coders*, which provide excellent performance at rates of 4.8 kbps and higher and are also a part of several national and international standards; and *mixed excitation linear prediction*, which is the new 2.4 kbps federal standard speech coder. In our description of these approaches, we will use the various national and international standards as examples.

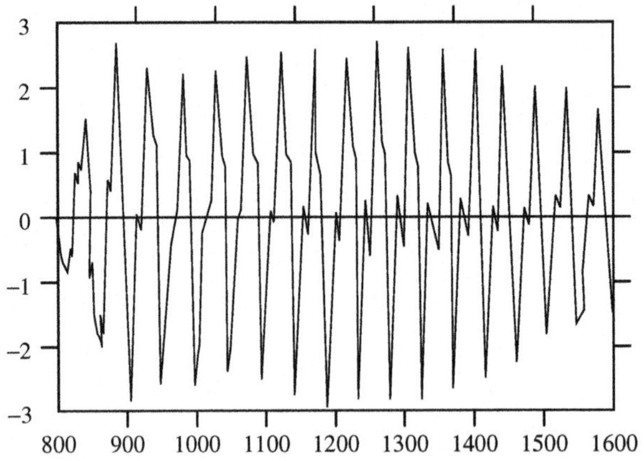

FIGURE 18.2 **The sound /e/ in *test*.**

18.3.1 The Channel Vocoder

In the channel vocoder [227], each segment of input speech is analyzed using a bank of band-pass filters called the *analysis filters*. The energy at the output of each filter is estimated at fixed intervals and transmitted to the receiver. In a digital implementation, the energy estimate may be the average squared value of the filter output. In analog implementations, this is the sampled output of an envelope detector. Generally, an estimate is generated 50 times every second. Along with the estimate of the filter output, a decision is made as to whether the speech in that segment is voiced, as in the case of the sounds */a/ /e/ /o/*, or unvoiced, as in the case for the sounds */s/ /f/*. Voiced sounds tend to have a pseudoperiodic structure, as seen in Figure 18.2, which is a plot of the /e/ part of a male voice saying the word *test*. The period of the fundamental harmonic is called the *pitch* period. The transmitter also forms an estimate of the pitch period, which is transmitted to the receiver.

Unvoiced sounds tend to have a noiselike structure, as seen in Figure 18.3, which is the /s/ sound in the word *test*.

At the receiver, the vocal tract filter is implemented by a bank of band-pass filters. The bank of filters at the receiver, known as the *synthesis filters*, is identical to the bank of analysis filters. Based on whether the speech segment is deemed to be voiced or unvoiced, either a pseudonoise source or a periodic pulse generator is used as the input to the synthesis filter bank. The period of the pulse input is determined by the pitch estimate obtained for the segment being synthesized at the transmitter. The input is scaled by the energy estimate at the output of the analysis filters. A block diagram of the synthesis portion of the channel vocoder is shown in Figure 18.4.

Since the introduction of the channel vocoder, a number of variations have been developed. The channel vocoder matches the frequency profile of the input speech. There is no attempt to reproduce the speech samples per se. However, not all frequency components of speech are equally important. In fact, as the vocal tract is a tube of nonuniform cross section, it resonates at a number of different frequencies. These frequencies are known as *formants* [119]. The formant values change with different sounds; however, we can identify ranges in which they

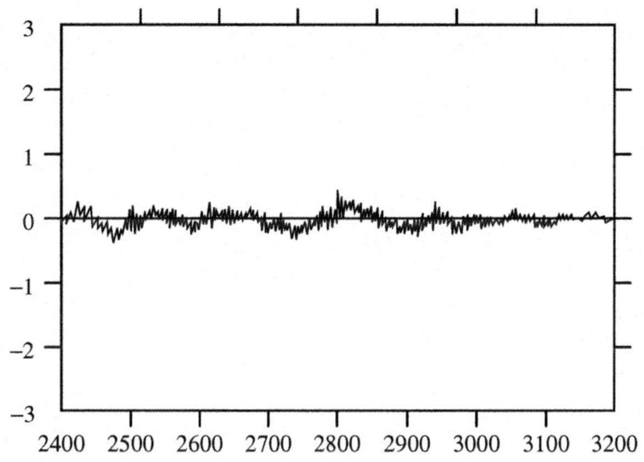

FIGURE 18.3 **The sound /s/ in *test*.**

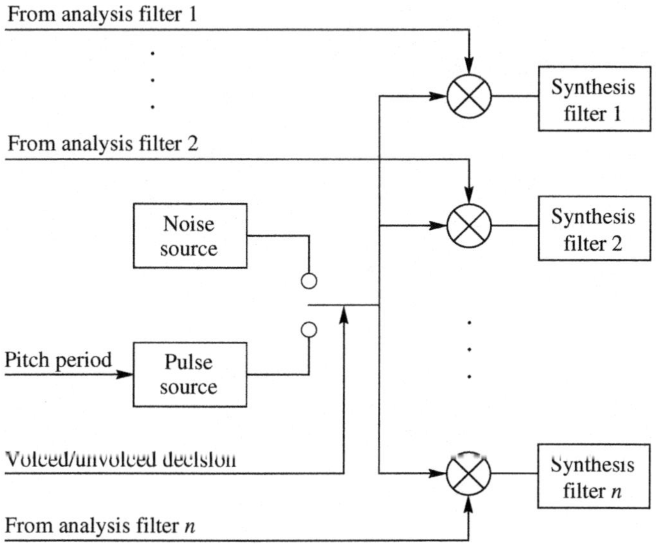

FIGURE 18.4 **The channel vocoder receiver.**

occur. For example, the first formant occurs in the range 200–800 Hz for a male speaker, and in the range 250–1000 Hz for a female speaker. The importance of these formants has led to the development of *formant vocoders*, which transmit an estimate of the formant values (usually four formants are considered sufficient) and an estimate of the bandwidth of each formant. At the receiver the excitation signal is passed through tunable filters that are tuned to the formant frequency and bandwidth.

An important development in the history of the vocoders was an understanding of the importance of the excitation signal. Schemes that require the synthesis of the excitation signal at the receiver spend a considerable amount of computational resources to obtain accurate voicing information and accurate pitch periods. This expense can be avoided through the use of voice excitation. In the voice-excited channel vocoder, the voice is first filtered using a narrowband low-pass filter. The output of the low-pass filter is sampled and transmitted to the receiver. At the receiver, this low-pass signal is passed through a nonlinearity to generate higher-order harmonics that, together with the low-pass signal, are used as the excitation signal. Voice excitation removes the problem of pitch extraction. It also removes the necessity for declaring every segment either voiced or unvoiced. As there are usually quite a few segments that are neither totally voiced or unvoiced, this can result in a substantial increase in quality. Unfortunately, the increase in quality is reflected in the high cost of transmitting the low-pass filtered speech signal.

The channel vocoder, although historically the first approach to analysis/synthesis—indeed the first approach to speech compression—is not as popular as some of the other schemes described here. However, all the different schemes can be viewed as descendants of the channel vocoder.

18.3.2 The Linear Predictive Coder (Government Standard LPC-10)

Of the many descendants of the channel vocoder, the most well-known is the linear predictive coder (LPC). Instead of the vocal tract being modeled by a bank of filters, in the linear predictive coder the vocal tract is modeled as a single linear filter whose output y_n is related to the input ϵ_n by

$$y_n = \sum_{i=1}^{M} b_i y_{n-i} + G\epsilon_n \tag{1}$$

where G is called the gain of the filter. As in the case of the channel vocoder, the input to the vocal tract filter is either the output of a random noise generator or a periodic pulse generator. A block diagram of the LPC receiver is shown in Figure 18.5.

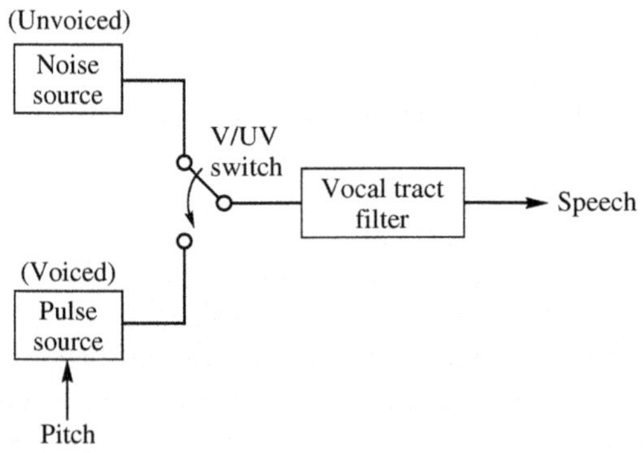

FIGURE 18.5 A model for speech synthesis.

At the transmitter, a segment of speech is analyzed. The parameters obtained include a decision as to whether the segment of speech is voiced or unvoiced, the pitch period if the segment is declared voiced, and the parameters of the vocal tract filter. In this section, we will take a somewhat detailed look at the various components that make up the linear predictive coder. As an example, we will use the specifications for the 2.4-kbit U.S. Government standard LPC-10.

The input speech is generally sampled at 8000 samples per second. In the LPC-10 standard, the speech is broken into 180 sample segments, corresponding to 22.5 milliseconds of speech per segment.

The Voiced/Unvoiced Decision

If we compare Figures 18.2 and 18.3, we can see there are two major differences. Notice that the samples of the voiced speech have larger amplitude; that is, there is more energy in the voiced speech. Also, the unvoiced speech contains higher frequencies. As both speech segments have average values close to zero, this means that the unvoiced speech waveform crosses the $x = 0$ line more often than the voiced speech sample. Therefore, we can get a fairly good idea about whether the speech is voiced or unvoiced based on the energy in the segment relative to background noise and the number of zero crossings within a specified window. In the LPC-10 algorithm, the speech segment is first low-pass filtered using a filter with a bandwidth of 1 kHz. The energy at the output relative to the background noise is used to obtain a tentative decision about whether the signal in the segment should be declared voiced or unvoiced. The estimate of the background noise is basically the energy in the unvoiced speech segments. This tentative decision is further refined by counting the number of zero crossings and checking the magnitude of the coefficients of the vocal tract filter. We will talk more about this latter point later in this section. Finally, it can be perceptually annoying to have a single voiced frame sandwiched between unvoiced frames. The voicing decision of the neighboring frames is considered in order to prevent this from happening.

Estimating the Pitch Period

Estimating the pitch period is one of the most computationally intensive steps of the analysis process. Over the years a number of different algorithms for pitch extraction have been developed. In Figure 18.2, it would appear that obtaining a good estimate of the pitch should be relatively easy. However, we should keep in mind that the segment shown in Figure 18.2 consists of 800 samples, which is considerably more than the samples available to the analysis algorithm. Furthermore, the segment shown here is noise-free and consists entirely of a voiced input. It can be a difficult undertaking for a machine to extract the pitch from a short noisy segment that may contain both voiced and unvoiced samples.

Several algorithms make use of the fact that the autocorrelation of a periodic function $R_{xx}(k)$ will have a maximum when k is equal to the pitch period. Coupled with the fact that the estimation of the autocorrelation function generally leads to a smoothing out of the noise, this makes the autocorrelation function a useful tool for obtaining the pitch period. Unfortunately, there are also some problems with the use of the autocorrelation. Voiced speech is not exactly periodic, which makes the maximum lower than we would expect from a periodic signal.

Generally, a maximum is detected by checking the autocorrelation value against a threshold; if the value is greater than the threshold, a maximum is declared to have occurred. When there is uncertainty about the magnitude of the maximum value, it is difficult to select a value for the threshold. Another problem occurs because of the interference due to other resonances in the vocal tract. There are a number of algorithms that resolve these problems in different ways (see [116,119] for details).

In this section, we will describe a closely related technique, employed in the LPC-10 algorithm, that uses the average magnitude difference function (AMDF). The AMDF is defined as

$$\text{AMDF}(P) = \frac{1}{N}\sum_{i=k_0+1}^{k_0+N} |y_i - y_{i-P}| \tag{2}$$

If a sequence $\{y_n\}$ is periodic with period P_0, samples that are P_0 apart in the $\{y_n\}$ sequence will have values close to each other, and therefore the AMDF will have a minimum at P_0. If we evaluate this function using the /e/ and /s/ sequences, we get the results shown in Figures 18.6 and 18.7. Notice that not only do we have a minimum when P equals the pitch period, but any spurious minimums we may obtain in the unvoiced segments are very shallow; that is, the difference between the minimum and average values is quite small. Therefore, the AMDF can serve a dual purpose: it can be used to identify the pitch period as well as the voicing condition.

The job of pitch extraction is simplified by the fact that the pitch period in humans tends to fall in a limited range. Thus, we do not have to evaluate the AMDF for all possible values of P. For example, the LPC-10 algorithm assumes that the pitch period is between 2.5 and 19.5 milliseconds. Assuming a sampling rate of 8000 samples a second, this means that P is between 20 and 160.

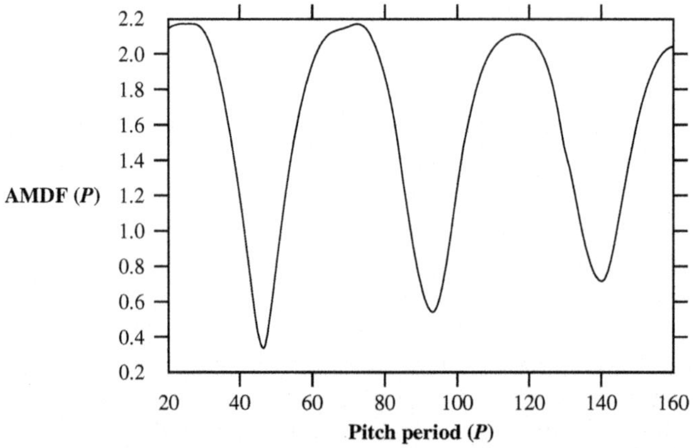

FIGURE 18.6 **AMDF function for the sound /e/ in *test*.**

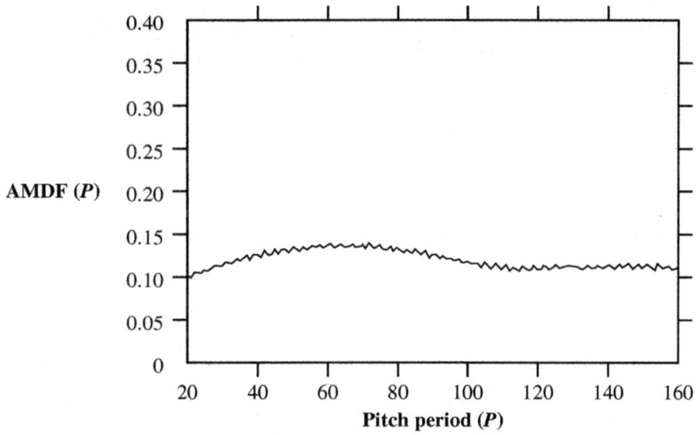

FIGURE 18.7 **AMDF function for the sound /s/ in *test*.**

Obtaining the Vocal Tract Filter

In linear predictive coding, the vocal tract is modeled by a linear filter with the input-output relationship shown in Equation (1). At the transmitter, during the analysis phase we obtain the filter coefficients that best match the segment being analyzed in a mean squared error sense. That is, if $\{y_n\}$ are the speech samples in that particular segment, then we want to choose $\{a_i\}$ to minimize the average value of e_n^2 where

$$e_n^2 = \left(y_n - \sum_{i=1}^{M} a_i y_{n-i} - G\epsilon_n \right)^2 \tag{3}$$

If we take the derivative of the expected value of e_n^2 with respect to the coefficients $\{a_j\}$, we get a set of M equations:

$$\frac{\partial}{\partial a_j} E\left[\left(y_n - \sum_{i=1}^{M} a_i y_{n-i} - G\epsilon_n \right)^2 \right] = 0 \tag{4}$$

$$\Rightarrow -2E\left[\left(y_n - \sum_{i=1}^{M} a_i y_{n-i} - G\epsilon_n \right) y_{n-j} \right] = 0 \tag{5}$$

$$\Rightarrow \sum_{i=1}^{M} a_i E\left[y_{n-i} y_{n-j} \right] = E\left[y_n y_{n-j} \right] \tag{6}$$

where in the last step we have made use of the fact that $E[\epsilon_n y_{n-j}]$ is zero for $j \neq 0$. In order to solve (6) for the filter coefficients, we need to be able to estimate $E[y_{n-i} y_{n-j}]$. There are two different approaches for estimating these values, called the *autocorrelation* approach and the *autocovariance* approach, each leading to a different algorithm. In the autocorrelation approach, we assume that the $\{y_n\}$ sequence is stationary and therefore

$$E\left[y_{n-i} y_{n-j} \right] = R_{yy}(|i - j|) \tag{7}$$

Furthermore, we assume that the $\{y_n\}$ sequence is zero outside the segment for which we are calculating the filter parameters. Therefore, the autocorrelation function is estimated as

$$R_{yy}(k) = \sum_{n=n_0+1+k}^{n_0+N} y_n y_{n-k} \tag{8}$$

and the M equations of the form of (6) can be written in matrix form as

$$\mathbf{R}A = P \tag{9}$$

where

$$\mathbf{R} = \begin{bmatrix} R_yy(0) & R_{yy}(1) & R_{yy}(2) & \cdots & R_{yy}(M-1) \\ R_{yy}(1) & R_{yy}(0) & R_{yy}(1) & \cdots & R_{yy}(M-2) \\ R_{yy}(2) & R_{yy}(1) & R_{yy}(0) & \cdots & R_{yy}(M-3) \\ \vdots & \vdots & \vdots & & \vdots \\ R_{yy}(M-1) & R_{yy}(M-2) & R_{yy}(M-3) & \cdots & R_{yy}(0) \end{bmatrix} \tag{10}$$

$$A = \begin{bmatrix} a_1 \\ a_2 \\ a_3 \\ \vdots \\ a_M \end{bmatrix} \tag{11}$$

and

$$P = \begin{bmatrix} R_{yy}(1) \\ R_{yy}(2) \\ R_{yy}(3) \\ \vdots \\ R_{yy}(M) \end{bmatrix} \tag{12}$$

This matrix equation can be solved directly to find the filter coefficients:

$$A = \mathbf{R}^{-1} P \tag{13}$$

However, the special form of the matrix $\mathbf{R}$ obviates the need for computing $\mathbf{R}^{-1}$. Note that not only is $\mathbf{R}$ symmetric, but also each diagonal of $\mathbf{R}$ consists of the same element. For example, the main diagonal contains only the element $R_{yy}(0)$, while the diagonals above and below the main diagonal contain only the element $R_{yy}(1)$. This special type of matrix is called a *Toeplitz matrix*, and there are a number of efficient algorithms available for the inversion of Toeplitz matrices [228]. Because $\mathbf{R}$ is Toeplitz, we can obtain a recursive solution to (9) that is computationally very efficient and that has an added attractive feature from the point of view of compression. This algorithm is known as the Levinson-Durbin algorithm [229,230]. We describe the algorithm without derivation. For details of the derivation, see [231,119].

In order to compute the filter coefficients of an Mth-order filter, the Levinson-Durbin algorithm requires the computation of all filters of order less than M. Furthermore, during the computation of the filter coefficients, the algorithm generates a set of constants k_i known as the *reflection* coefficients, or *partial correlation* (PARCOR) coefficients. In the algorithm description below, we denote the order of the filter using superscripts. Thus, the coefficients of the fifth-order filter would be denoted by $\{a_i^{(5)}\}$. The algorithm also requires the computation of the estimate of the average error $E[e_n^2]$. We will denote the average error using an mth-order filter by E_m. The algorithm proceeds as follows:

1. Set $E_0 = R_{yy}(0)$, $i = 0$.
2. Increment i by one.
3. Calculate $k_i = \left(\sum_{j=1}^{i-1} a_j^{(i-1)} R_{yy}(i-j+1) - R_{yy}(i)\right) / E_{i-1}$.
4. Set $a_i^{(i)} = k_i$.
5. Calculate $a_j^{(i)} = a_j^{(i-1)} + k_i a_{i-j}^{(i-1)}$ for $j = 1, 2, \ldots, i-1$.
6. Calculate $E_i = \left(1 - k_i^2\right) E_{i-1}$.
7. If $i < M$, go to step 2.

In order to get an effective reconstruction of the voiced segment, the order of the vocal tract filter needs to be sufficiently high. Generally, the order of the filter is 10 or more. Because the filter is an IIR filter, error in the coefficients can lead to instability, especially for the high orders necessary in linear predictive coding. As the filter coefficients are to be transmitted to the receiver, they need to be quantized. This means that quantization error is introduced into the value of the coefficients, and that can lead to instability.

This problem can be avoided by noticing that if we know the PARCOR coefficients, we can obtain the filter coefficients from them. Furthermore, PARCOR coefficients have the property that as long as the magnitudes of the coefficients are less than one, the filter obtained from them is guaranteed to be stable. Therefore, instead of quantizing the coefficients $\{a_i\}$ and transmitting them, the transmitter quantizes and transmits the coefficients $\{k_i\}$. As long as we make sure that all the reconstruction values for the quantizer have magnitudes less than one, it is possible to use relatively high-order filters in the analysis/synthesis schemes.

The assumption of stationarity that was used to obtain (7) is not really valid for speech signals. If we discard this assumption, the equations to obtain the filter coefficients change. The term $E[y_{n-i}y_{n-j}]$ is now a function of both i and j. Defining

$$c_{ij} = E[y_{n-i}y_{n-j}] \tag{14}$$

we get the equation

$$\mathbf{C}A = S \tag{15}$$

where

$$\mathbf{C} = \begin{bmatrix} c_{11} & c_{12} & c_{13} & \cdots & c_{1M} \\ c_{21} & c_{22} & c_{23} & \cdots & c_{2M} \\ \vdots & \vdots & \vdots & & \vdots \\ c_{M1} & c_{M2} & c_{M3} & \cdots & c_{MM} \end{bmatrix} \tag{16}$$

and

$$S = \begin{bmatrix} c_{10} \\ c_{20} \\ c_{30} \\ \vdots \\ c_{M0} \end{bmatrix} \tag{17}$$

The elements c_{ij} are estimated as

$$c_{ij} = \sum_{n=n_0+1}^{n_0+N} y_{n-i} y_{n-j} \tag{18}$$

Notice that we no longer assume that the values of y_n outside of the segment under consideration are zero. This means that in calculating the **C** matrix for a particular segment, we use samples from previous segments. This method of computing the filter coefficients is called the *covariance method.*

The **C** matrix is symmetric but no longer Toeplitz, so we can't use the Levinson-Durbin recursion to solve for the filter coefficients. The equations are generally solved using a technique called the *Cholesky decomposition.* We will not describe the solution technique here. (You can find it in most texts on numerical techniques; an especially good source is [182].) For an in-depth study of the relationship between the Cholesky decomposition and the reflection coefficients, see [232].

The LPC-10 algorithm uses the covariance method to obtain the reflection coefficients. It also uses the PARCOR coefficients to update the voicing decision. In general, for voiced signals the first two PARCOR coefficients have values close to one. Therefore, if both the first two PARCOR coefficients have very small values, the algorithm sets the voicing decision to unvoiced.

Transmitting the Parameters

Once the various parameters have been obtained, they need to be coded and transmitted to the receiver. There are a variety of ways this can be done. Let us look at how the LPC-10 algorithm handles this task.

The parameters that need to be transmitted include the voicing decision, the pitch period, and the vocal tract filter parameters. One bit suffices to transmit the voicing information. The pitch is quantized to 1 of 60 different values using a log-companded quantizer. The LPC-10 algorithm uses a 10th-order filter for voiced speech and a 4th-order filter for unvoiced speech. Thus, we have to send 11 values (10 reflection coefficients and the gain) for voiced speech and 5 for unvoiced speech.

The vocal tract filter is especially sensitive to errors in reflection coefficients that have magnitudes close to one. As the first few coefficients are most likely to have values close to one, the LPC-10 algorithm specifies the use of nonuniform quantization for k_1 and k_2. The nonuniform quantization is implemented by first generating the coefficients

$$g_i = \frac{1 + k_i}{1 - k_i} \tag{19}$$

which are then quantized using a 5-bit uniform quantizer. The coefficients k_3 and k_4 are both quantized using a 5-bit uniform quantizer. In the voiced segments, coefficients k_5 through k_8 are quantized using a 4-bit uniform quantizer, k_9 is quantized using a 3-bit uniform quantizer, and k_{10} is quantized using a 2-bit uniform quantizer. In the unvoiced segments, the 21 bits used to quantize k_5 through k_{10} in the voiced segments are used for error protection.

The gain G is obtained by finding the root mean squared (rms) value of the segment and quantized using 5-bit log-companded quantization. Including an additional bit for synchronization, we end up with a total of 54 bits per frame. Multiplying this by the total number of frames per second gives us the target rate of 2400 bits per second.

Synthesis

At the receiver, the voiced frames are generated by exciting the received vocal tract filter by a locally stored waveform. This waveform is 40 samples long. It is truncated or padded with zeros depending on the pitch period. If the frame is unvoiced, the vocal tract is excited by a pseudorandom number generator.

The LPC-10 coder provides intelligible reproduction at 2.4 kbits. The use of only two kinds of excitation signals gives an artificial quality to the voice. This approach also suffers when used in noisy environments. The encoder can be fooled into declaring segments of speech unvoiced because of background noise. When this happens, the speech information gets lost.

18.3.3 Code-Excited Linear Predicton (CELP)

As we mentioned earlier, one of the most important factors in generating natural-sounding speech is the excitation signal. As the human ear is especially sensitive to pitch errors, a great deal of effort has been devoted to the development of accurate pitch detection algorithms. However, no matter how accurate the pitch is in a system using the LPC vocal tract filter, the use of a periodic pulse excitation that consists of a single pulse per pitch period leads to a "buzzy twang" [233]. In 1982, Atal and Remde [234] introduced the idea of multipulse linear predictive coding (MP-LPC), in which several pulses were used during each segment. The spacing of these pulses is determined by evaluating a number of different patterns from a codebook of patterns.

A codebook of excitation patterns is constructed. Each entry in this codebook is an excitation sequence that consists of a few nonzero values separated by zeros. Given a segment from the speech sequence to be encoded, the encoder obtains the vocal tract filter using the LPC analysis described previously. The encoder then excites the vocal tract filter with the entries of the codebook. The difference between the original speech segment and the synthesized speech is fed to a perceptual weighting filter, which weights the error using a perceptual weighting criterion. The codebook entry that generates the minimum average weighted error is declared to be the best match. The index of the best-match entry is sent to the receiver along with the parameters for the vocal tract filter.

This approach was improved upon by Atal and Schroeder in 1984 with the introduction of the system that is commonly known as *code-excited linear prediction* (CELP). In CELP, instead of having a codebook of pulse patterns, we allow a variety of excitation signals. For each segment the encoder finds the excitation vector that generates synthesized speech that best matches the speech segment being encoded. This approach is closer in a strict sense to a waveform coding technique such as DPCM than to the analysis/synthesis schemes. However, as the ideas behind CELP are similar to those behind LPC, we included CELP in this chapter. The main components of the CELP coder include the LPC analysis, the excitation codebook, and the perceptual weighting filter. Each component of the CELP coder has been investigated in great detail by a large number of researchers. For a survey of some of the results, see [235]. In the rest of the section, we give two examples of very different kinds of CELP coders. The first algorithm is Federal Standard (FS) 1016 in the United States, a 4.8 kbps coder; the other is the CCITT (now ITU-T) G.728 standard, a low-delay 16 kbps coder.

Besides CELP, the MP-LPC algorithm had another descendant that has become a standard. In 1986, Kroon, Deprettere, and Sluyter [236] developed a modification of the MP-LPC algorithm. Instead of using excitation vectors in which the nonzero values are separated by an arbitrary number of zero values, they forced the nonzero values to occur at regularly spaced intervals. Furthermore, they allowed the nonzero values to take on a number of different values. They called this scheme *regular pulse excitation* (RPE) coding. A variation of RPE, called *regular pulse excitation with long-term prediction* (RPE-LTP) [237], was adopted as a standard for digital cellular telephony by the Group Speciale Mobile (GSM) subcommittee of the European Telecommunications Standards Institute at the rate of 13 kbps.

Federal Standard 1016

The vocal tract filter used by the CELP coder in FS 1016 is given by

$$y_n = \sum_{i=1}^{10} b_i y_{n-i} + \beta y_{n-P} + G\epsilon_n \tag{20}$$

where P is the pitch period and the term βy_{n-P} is the contribution due to the pitch periodicity. The input speech is sampled at 8000 samples per second and divided into 30-millisecond frames containing 240 samples. Each frame is divided into four subframes of length 7.5 milliseconds [238]. The coefficients $\{b_i\}$ for the 10th-order short-term filter are obtained using the autocorrelation method.

The pitch period P is calculated once every subframe. In order to reduce the computational load, the pitch value is assumed to lie between between 20 and 147 every odd subframe. In every even subframe, the pitch value is assumed to lie within 32 samples of the pitch value in the previous frame.

The FS 1016 algorithm uses two codebooks [239], a stochastic codebook and an adaptive codebook. An excitation sequence is generated for each subframe by adding one scaled element from the stochastic codebook and one scaled element from the adaptive codebook. The scale factors and indices are selected to minimize the perceptual error between the input and synthesized speech.

The stochastic codebook contains 512 entries. These entries are generated using a Gaussian random number generator, the output of which is quantized to -1, 0, or 1. If the input is less than -1.2, it is quantized to -1; if it is greater than 1.2, it is quantized to 1; and if it lies between -1.2 and 1.2, it is quantized to 0. The codebook entries are adjusted so that each entry differs from the preceding entry in only two places. This structure helps reduce the search complexity.

The adaptive codebook consists of the excitation vectors from the previous frame. Each time a new excitation vector is obtained, it is added to the codebook. In this manner, the codebook adapts to local statistics.

The FS 1016 coder has been shown to provide excellent reproductions in both quiet and noisy environments at rates of 4.8 kbps and above [239]. Because of the richness of the excitation signals, the reproduction does not suffer from the problem of sounding artificial. The lack of a voicing decision makes it more robust to background noise. The quality of the reproduction of this coder at 4.8 kbps has been shown to be equivalent to a delta modulator operating at 32 kbps [239]. The price for this quality is much higher complexity and a much longer coding delay. We will address this last point in the next section.

CCITT G.728 Speech Standard

By their nature, the schemes described in this chapter have some coding delay built into them. By "coding delay," we mean the time between when a speech sample is encoded to when it is decoded if the encoder and decoder were connected back-to-back (i.e., there were no transmission delays). In the schemes we have studied, a segment of speech is first stored in a buffer. We do not start extracting the various parameters until a complete segment of speech is available to us. Once the segment is completely available, it is processed. If the processing is real time, this means another segment's worth of delay. Finally, once the parameters have been obtained, coded, and transmitted, the receiver has to wait until at least a significant part of the information is available before it can start decoding the first sample. Therefore, if a segment contains 20 milliseconds' worth of data, the coding delay would be approximately somewhere between 40 to 60 milliseconds. This kind of delay may be acceptable for some applications; however, there are other applications where such long delays are not acceptable. For example, in some situations there are several intermediate tandem connections between the initial transmitter and the final receiver. In such situations, the total delay would be a multiple of the coding delay of a single connection. The size of the delay would depend on the number of tandem connections and could rapidly become quite large.

For such applications, CCITT approved recommendation G.728, a CELP coder with a coder delay of 2 milliseconds operating at 16 kbps. As the input speech is sampled at 8000 samples per second, this rate corresponds to an average rate of 2 bits per sample.

In order to lower the coding delay, the size of each segment has to be reduced significantly because the coding delay will be some multiple of the size of the segment. The G.728 recommendation uses a segment size of five samples. With five samples and a rate of 2 bits per sample, we only have 10 bits available to us. Using only 10 bits, it would be impossible to encode the parameters of the vocal tract filter as well as the excitation vector. Therefore, the algorithm obtains the vocal tract filter parameters in a backward adaptive manner; that is, the vocal tract filter coefficients used to synthesize the current segment are obtained by analyzing the previous decoded segments. The CCITT requirements for G.728 included the requirement that the algorithm operate under noisy channel conditions. It would be extremely difficult to extract the pitch period from speech corrupted by channel errors. Therefore, the G.728 algorithm does away with the pitch filter. Instead, the algorithm uses a 50th-order vocal tract filter. The order of the filter is large enough to model the pitch of most female speakers. Not being able to use pitch information for male speakers does not cause much degradation [240]. The vocal tract filter is updated every fourth frame, which is once every 20 samples or 2.5 milliseconds. The autocorrelation method is used to obtain the vocal tract parameters.

As the vocal tract filter is completely determined in a backward adaptive manner, we have all 10 bits available to encode the excitation sequence. Ten bits would be able to index 1024 excitation sequences. However, to examine 1024 excitation sequences every 0.625 milliseconds is a rather large computational load. In order to reduce this load, the G.728 algorithm uses a product codebook where each excitation sequence is represented by a normalized sequence and a gain term. The final excitation sequence is a product of the normalized excitation sequence and the gain. Of the 10 bits, 3 bits are used to encode the gain using a predictive encoding scheme, while the remaining 7 bits form the index to a codebook containing 127 sequences.

Block diagrams of the encoder and decoder for the CCITT G.728 coder are shown in Figure 18.8. The low-delay CCITT G.728 CELP coder operating at 16 kbps provides reconstructed speech quality superior to the 32 kbps CCITT G.726 ADPCM algorithm described in Chapter 10. Various efforts are under way to reduce the bit rate for this algorithm without compromising too much on quality and delay.

18.3.4 Sinusoidal Coders

A competing approach to CELP in the low-rate region is a relatively new form of coder called the sinusoidal coder [235]. Recall that the main problem with the LPC coder was the paucity of excitation signals. The CELP coder resolved this problem by using a codebook of excitation signals. The sinusoidal coders solve this problem by using an excitation signal that is the sum of sine waves of arbitrary amplitudes, frequencies, and phases. Thus, the excitation signal is of the form

$$e_n = \sum_{l=1}^{L} a_l \cos(n\omega_l + \psi_l) \tag{21}$$

where the number of sinusoids L required for each frame depends on the contents of the frame. If the input to a linear system is a sinusoid with frequency ω_l, the output will also be a sinusoid with frequency ω_l, albeit with different amplitude and phase. The vocal tract filter is a linear system. Therefore, if the excitation signal is of the form of (21), the synthesized speech $\{s_n\}$ will be of the form

$$s_n = \sum_{i=1}^{L} A_l \cos(n\omega_l + \phi_l). \tag{22}$$

Thus, each frame is characterized by a set of spectral amplitudes A_l, frequencies ω_l, and phase terms ϕ_l. The number of parameters required to represent the excitation sequence is the same as the number of parameters required to represent the synthesized speech. Therefore, rather than estimate and transmit the parameters of both the excitation signal and vocal tract filter and then synthesize the speech at the receiver by passing the excitation signal through the vocal tract filter, the sinusoidal coders directly estimate the parameters required to synthesize the speech at the receiver.

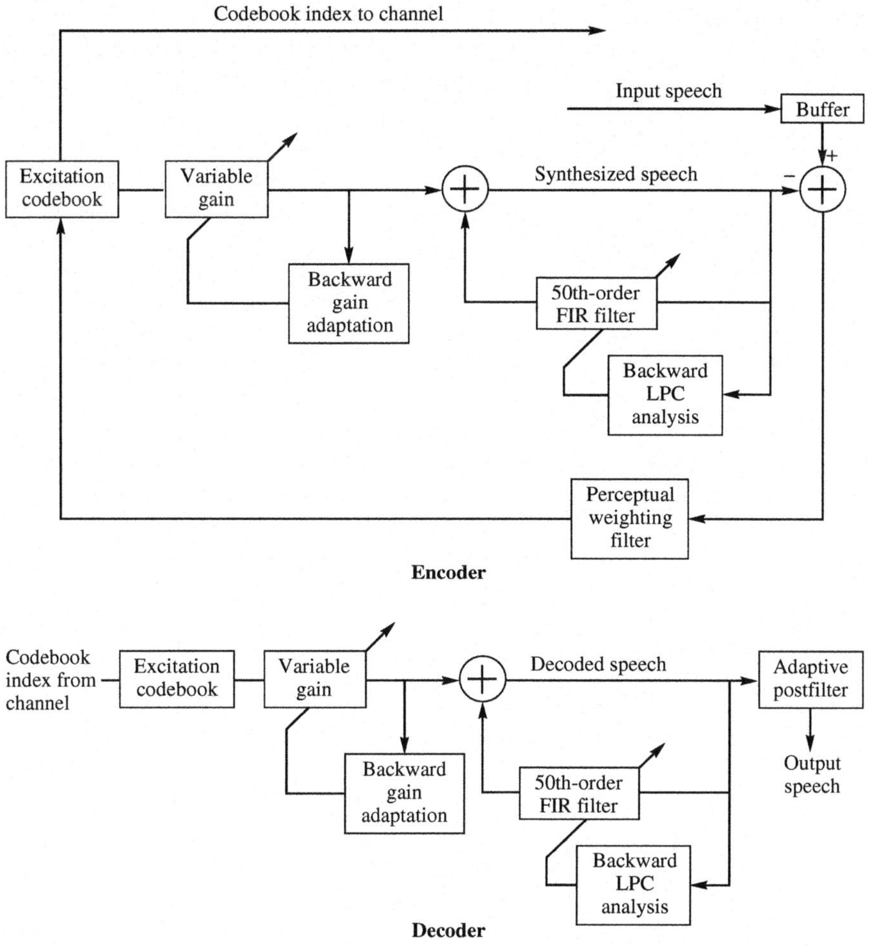

FIGURE 18.8 Encoder and decoder for the CCITT G.728 16 kbps speech coder.

Just like the coders discussed previously, the sinusoidal coders divide the input speech into frames and obtain the parameters of the speech separately for each frame. If we synthesized the speech segment in each frame independent of the other frames, we would get synthetic speech that is discontinuous at the frame boundaries. These discontinuities severely degrade the quality of the synthetic speech. Therefore, the sinusoidal coders use different interpolation algorithms to smooth the transition from one frame to another.

Transmitting all the separate frequencies ω_l would require significant transmission resources, so the sinusoidal coders obtain a fundamental frequency w_0 for which the approximation

$$\hat{y}_n = \sum_{k=1}^{K(\omega_0)} \hat{A}(k\omega_0)\cos(nk\omega_0 + \phi_k) \tag{23}$$

is close to the speech sequence y_n. Because this is a harmonic approximation, the approximate sequence $\{\hat{y}_n\}$ will be most different from the speech sequence $\{y_n\}$ when the segment of speech being encoded is unvoiced. Therefore, this difference can be used to decide whether the frame or some subset of it is unvoiced.

The two most popular sinusoidal coding techniques today are represented by the sinusoidal transform coder (STC) [241] and the multiband excitation coder (MBE) [242]. While the STC and MBE are similar in many respects, they differ in how they handle unvoiced speech. In the MBE coder, the frequency range is divided into bands, each consisting of several harmonics of the fundamental frequency ω_0. Each band is checked to see if it is unvoiced or voiced. The voiced bands are synthesized using a sum of sinusoids, while the unvoiced bands are obtained using a random number generator. The voiced and unvoiced bands are synthesized separately and then added together.

In the STC, the proportion of the frame that contains a voiced signal is measured using a "voicing probability" P_v. The voicing probability is a function of how well the harmonic model matches the speech segment. Where the harmonic model is close to the speech signal, the voicing probability is taken to be unity. The sine wave frequencies are then generated by

$$w_k = \begin{cases} kw_0 & \text{for } kw_0 \leqslant w_c P_v \\ k^* w_0 + (k - k^*) w_u & \text{for } kw_0 > w_c P_v \end{cases} \tag{24}$$

where w_c corresponds to the cutoff frequency (4 kHz), w_u is the unvoiced pitch corresponding to 100 Hz, and k^* is the largest value of k for which $k^* w_0 \leqslant w_c P_v$. The speech is then synthesized as

$$\hat{y}_n = \sum_{k=1}^{K} \hat{A}(w_k) \cos(n w_k + \phi_k) \tag{25}$$

Both the STC and the MBE coders have been shown to perform well at low rates. A version of the MBE coder known as the improved MBE (IMBE) coder was approved by the Association of Police Communications Officers (APCO) as the standard for law enforcement.

18.3.5 Mixed Excitation Linear Prediction (MELP)

The mixed excitation linear prediction (MELP) coder was selected to be the federal standard for speech coding at 2.4 kbps by the Defense Department Voice Processing Consortium (DDVPC). The MELP algorithm uses the same LPC filter to model the vocal tract. However, it uses a much more complex approach to the generation of the excitation signal.

A block diagram of the decoder for the MELP system is shown in Figure 18.9. As evident from the figure, the excitation signal for the synthesis filter is no longer simply noise or a periodic pulse but a multiband mixed excitation. The mixed excitation contains both a filtered signal from a noise generator as well as a contribution that depends directly on the input signal.

The first step in constructing the excitation signal is pitch extraction. The MELP algorithm obtains the pitch period using a multistep approach. In the first step an integer pitch value P_1 is obtained in the following manner:

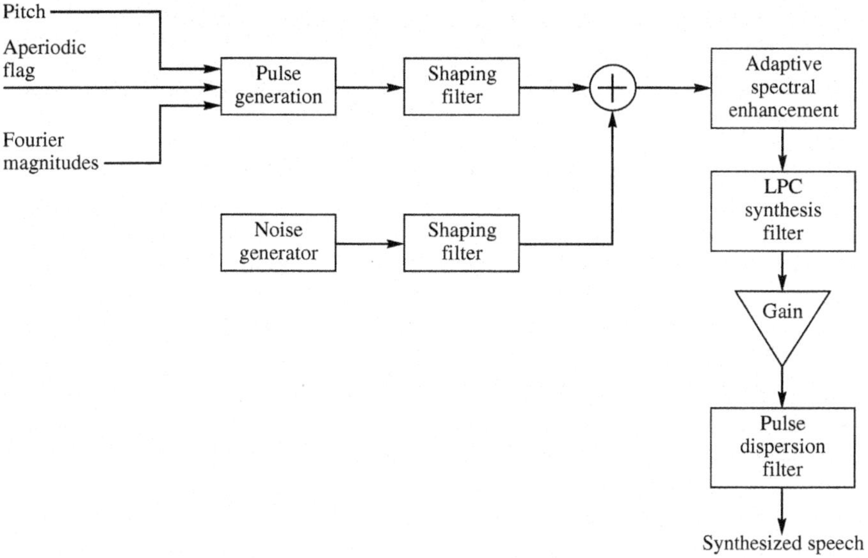

FIGURE 18.9 Block diagram of MELP decoder.

1. The input is first filtered using a low-pass filter with a cutoff of 1 kHz.
2. The normalized autocorrelation is then computed for lags between 40 and 160 samples.

The normalized autocorrelation $r(\tau)$ is defined as

$$r(\tau) = \frac{c_\tau(0,\tau)}{\sqrt{c_\tau(0,0)c_\tau(\tau,\tau)}}$$

where

$$c_\tau(m,n) = \sum_{-\lfloor \tau/2 \rfloor - 80}^{-\lfloor \tau/2 \rfloor + 79} y_{k+m} y_{k+n}$$

The first estimate of the pitch P_1 is obtained as the value of τ that maximizes the normalized autocorrelation function. This value is refined by looking at the signal filtered using a filter with a passband in the 0–500 Hz range. This stage uses two values of P_1, one from the current frame and one from the previous frame, as candidates. The normalized autocorrelation values are obtained for lags from five samples less to five samples more than the candidate P_1 values. The lags that provide the maximum normalized autocorrelation value for each candidate are used for *fractional pitch refinement*. The idea behind fractional pitch refinement is that if the maximum value of $r(\tau)$ is found for some $\tau = T$, then the maximum could be in the interval $(T-1, T]$ or $[T, T+1)$. The fractional offset is computed using

$$\Delta = \frac{c_T(0,T+1)c_T(T,T) - c_T(0,T)c_T(T,T+1)}{c_T(0,T+1)[c_T(T,T) - c_T(T,T+1)] + c_T(0,T)[c_T(T+1,T+1) - c_T(T,T+1)]} \quad (26)$$

The normalized autocorrelation at the fractional pitch values is given by

$$r(T+\Delta) = \frac{(1-\Delta)c_T(0,T) + \Delta c_T(0,T+1)}{\sqrt{c_T(0,0)[(1-\Delta)^2 c_T(T,T) + 2\Delta(1-\Delta)c_T(T,T+1) + \Delta^2 c_T(T+1,T+1)]}} \tag{27}$$

The fractional estimate that gives the higher autocorrelation is selected as the refined pitch value P_2.

The final refinements of the pitch value are obtained using the linear prediction residuals. The residual sequence is generated by filtering the input speech signal with the filter obtained using LPC analysis. For the purposes of pitch refinement the residual signal is filtered using a low-pass filter with a cutoff of 1 kHz. The normalized autocorrelation function is computed for this filtered residual signal for lags from five samples less to five samples more than the candidate P_2 value, and a candidate value of P_3 is obtained. If $r(P_3) \geqslant 0.6$, we check to make sure that P_3 is not a multiple of the actual pitch. If $r(P_3) < 0.6$, we do another fractional pitch refinement around P_3 using the input speech signal. If in the end $r(P_3) < 0.55$, we replace P_3 with a long-term average value of the pitch. The final pitch value is quantized on a logarithmic scale using a 99-level uniform quantizer.

The input is also subjected to a multiband voicing analysis using five filters with passbands 0–500, 500–1000, 1000–2000, 2000–3000, and 3000–4000 Hz. The goal of the analysis is to obtain the voicing strengths Vbp_i for each band used in the shaping filters. Noting that P_2 was obtained using the output of the lowest band filter, $r(P_2)$ is assigned as the lowest band voicing strength Vbp_1. For the other bands, Vbp_i is the larger of $r(P_2)$ for that band and the correlation of the envelope of the band-pass signal. If the value of Vbp_1 is small, this indicates a lack of low-frequency structure, which in turn indicates an unvoiced or transition input. Thus, if $Vbp_1 < 0.5$, the pulse component of the excitation signal is selected to be aperiodic, and this decision is communicated to the decoder by setting the aperiodic flag to 1. When $Vbp_1 > 0.6$, the values of the other voicing strengths are quantized to 1 if their value is greater than 0.6, and to 0 otherwise. In this way signal energy in the different bands is turned on or off depending on the voicing strength. There are several exceptions to this quantization rule. If Vbp_2, Vbp_3, and Vbp_4 all have magnitudes less than 0.6 and Vbp_5 has a value greater than 0.6, they are all (including Vbp_5) quantized to 0. Also, if the residual signal d_n means d sub n in math mode contains a few large values, indicating sudden transitions in the input signal, the voicing strengths are adjusted. In particular, the *peakiness* is defined as

$$\text{peakiness} = \frac{\sqrt{\frac{1}{160}\sum_{n=1}^{160} d_n^2}}{\frac{1}{160}\sum_{n=1}^{160} |d_n|} \tag{28}$$

If this value exceeds 1.34, Vbp_1 is forced to 1. If the peakiness value exceeds 1.6, Vbp_1, Vbp_2, and Vbp_3 are all set to 1.

In order to generate the pulse input, the algorithm measures the magnitude of the discrete Fourier transform coefficients corresponding to the first 10 harmonics of the pitch. The prediction residual is generated using the quantized predictor coefficients. The algorithm searches in a window of width $\lfloor 512/\hat{P}_3 \rfloor$ samples around the initial estimates of the pitch harmonics for the actual harmonics where $\hat{P}_3$ is the quantized value of P_3. The magnitudes of the harmonics are quantized using a vector quantizer with a codebook size of 256. The codebook is searched using a weighted Euclidean distance that emphasizes lower frequencies over higher frequencies.

At the decoder, using the magnitudes of the harmonics and information about the periodicity of the pulse train, the algorithm generates one excitation signal. Another signal is generated using a random number generator. Both are shaped by the multiband shaping filter before being combined. This mixture signal is then processed through an *adaptive spectral enhancement filter*, which is based on the LPC coefficients, to form the final excitation signal. Note that in order to preserve continuity from frame to frame, the parameters used for generating the excitation signal are adjusted based on their corresponding values in neighboring frames.

18.4 Wideband Speech Compression—ITU-T G.722.2

One of the earliest forms of (remote) speech communication was over the telephone. This experience set the expectations for quality rather low. When technology advanced, people still did not demand higher quality in their voice communications. However, the multimedia revolution is changing that. With ever-increasing quality in video and audio there is an increasing demand for higher quality in speech communication. Telephone-quality speech is limited to the band between 200 Hz and 3400 Hz. This range of frequency contains enough information to make speech intelligible and provide some degree of speaker identification. To improve the quality of speech, it is necessary to increase the bandwidth of speech. Wideband speech is bandlimited to 50–7000 Hz. The higher frequencies give more clarity to the voice signal while the lower frequencies contribute timbre and naturalness. The ITU-T G.722.2 standard, approved in January of 2002, provides a multirate coder for wideband speech coding.

Wideband speech is sampled at 16,000 samples per second. The signal is split into two bands, a lower band from 50–6400 Hz and a narrow upper band from 6400–7000 Hz. The coding resources are devoted to the lower band. The upper band is reconstructed at the receiver based on information from the lower band and using random excitation. The lower band is downsampled to 12.8 kHz.

The coding method is a code-excited linear prediction method that uses an algebraic codebook as the fixed codebook (the algebraic structure of the codebook is described in Section 18.5). The adaptive codebook contains low-pass interpolated past excitation vectors. The basic idea is the same as in CELP. A synthesis filter is derived from the input speech. An excitation vector consisting of a weighted sum of the fixed and adaptive codebooks is used to excite the synthesis filter. The perceptual closeness of the output of the filter to the input speech is used to select the combination of excitation vectors. The selection, along with the parameters of the synthesis filter, is communicated to the receiver, which then synthesizes the speech. A voice activity detector is used to reduce the rate during silence intervals. Let us examine the various components in slightly more detail.

The speech is processed in 20-ms frames. Each frame is composed of four 5-ms subframes. The LP analysis is conducted once per frame using an overlapping 30-ms window. Autocorrelation values are obtained for the windowed speech and the Levinson-Durbin algorithm is used to obtain the LP coefficients. These coefficients are transformed to Immittance Spectral

Pairs (ISP) (described below), which are quantized using a vector quantizer. The reason behind the transformation is that we will need to quantize whatever representation we have of the synthesis filters introducing quantization error. The elements of the ISP representation are uncorrelated if the underlying process is stationary, which means that error in one coefficient will not cause the entire spectrum to get distorted.

Given a set of sixteen LP coefficients $\{a_i\}$, define two polynomials

$$f_1'(z) = A(z) + z^{-16}A(z^{-1}) \tag{29}$$

$$f_2'(z) = A(z) - z^{-16}A(z^{-1}) \tag{30}$$

where

$$A(z) = \sum_i a_i z^{-i}$$

Clearly, if we know the polynomials their sum will give us $A(z)$. Instead of sending the polynomials, we can send the roots of these polynomials. These roots are known to all lie on the unit circle, and the roots of the two polynomials alternate. The polynomial $f_2'(z)$ has two roots at $z = 1$ and $z = -1$. These are removed and we get the two polynomials

$$f_1(z) = f_1'(z) \tag{31}$$

$$f_2(z) = \frac{f_2'(z)}{1 - z^{-2}} \tag{32}$$

These polynomials can now be factored as follows

$$f_1(z) = (1 + a_{16}) \prod_{i=0,2,\ldots,14} \left(1 - 2q_i z^{-i} + z^{-2}\right) \tag{33}$$

$$f_2(z) = (1 + a_{16}) \prod_{i=1,3,\ldots,13} \left(1 - 2q_i z^{-i} + z^{-2}\right) \tag{34}$$

where $q_i = \cos(\omega_i)$ and ω_i are the immittance spectral frequencies. The ISP coefficients are quantized using a combination of differential encoding and vector quantization. The vector of sixteen frequencies is split into subvectors and these vectors are quantized in two stages. The quantized ISPs are transformed to LP coefficients, which are then used in the fourth subframe for synthesis. The ISP coefficients used in the other three subframes are obtained by interpolating the coefficients in the neighboring subframes.

For each 5-ms subframe we need to generate an excitation vector. As in CELP, the excitation is a sum of vectors from two codebooks, a fixed codebook and an adaptive codebook. One of the problems with vector codebooks has always been the storage requirements. The codebook should be large enough to provide for a rich set of excitations. However, with a dimension of 64 samples (for 5 ms), the number of possible combinations can get enormous. The G.722.2 algorithm solves this problem by imposing an algebraic structure on the fixed codebook. The 64 positions are divided into four tracks. The first track consists of positions 0, 4, 8,…, 60. The second track consists of the positions 1, 5, 9,…, 61. The third track consists of positions 2, 6, 10,…, 62 and the final track consists of the remaining positions. We can place a single signed pulse in each track by using 4 bits to denote the position and a fifth bit for the sign.

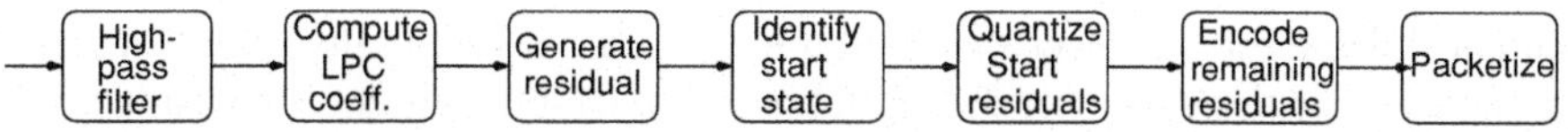

FIGURE 18.10 Steps in the encoding of speech for the Internet Low Bandwidth Coder.

This effectively gives us a 20-bit fixed codebook. This corresponds to a codebook size of 2^{20}. However, we do not need to store the codebook. By assigning more or fewer pulses per track we can dramatically change the "size" of the codebook and get different coding rates. The standard details a rapid search procedure to obtain the excitation vectors.

The voice activity detector allows the encoder to significantly reduce the rate during periods of speech pauses. During these periods the background noise is coded at a low rate by transmitting parameters describing the noise. This *comfort noise* is synthesized at the decoder.

18.5 Coding of Speech for Internet Applications

The Internet presents a very different set of problems and opportunities for speech compression. Data transmitted over the Internet has to be packetized. This can mean that the information used to encode speech contained in a packet has to come from within the packet. This in turn results in much higher bit rates than we have considered in previous sections. Consecutive packets can make their way along different routes from the encoder to the decoder. If the delay between packets is sufficiently long the delayed packet has to be considered lost and the decoder has to come up with strategies to ameliorate the effects of this loss. On the positive side the processing power available for the encoding and decoding can be significantly higher if the codec is implemented on computers rather than telephones (though the distinction may sometimes be academic).

In this section we will look at three different speech coding standards developed specifically for the Internet environment, the Internet Low Bitrate Codec (iLBC), the ITU-T G.729 standard, and SILK, the coder used by Skype.

18.5.1 iLBC

The iLBC was first proposed by Global IP Sound (later Global IP Solutions, which was bought by Google in 2011), and is used in a number of Voice over Internet Protocol (VoIP) applications including Google Talk, Skype, and Yahoo Messenger. The coder was standardized by the Internet Engineering Task Force in RFC 3951 [243] and RFC 3952 [244]. The iLBC allows speech coding of 8000 samples per second speech at two fixed rates, 15.2 kbits per second and 13.33 kbits per second, where the input is quantized using a 16-bit uniform quantizer. A block diagram of the encoding process is shown in Figure 18.10.

After an optional high-pass filter with a cutoff frequency of 90 Hz to remove low-frequency noise, such as a 60-cycle hum or a DC bias, the data are divided into blocks that are encoded independent of each other. This independence helps ameliorate the effects of packet loss. For the higher rate option the data are blocked into 20 millisecond blocks corresponding to

160 samples, while for the lower rate option the data are divided into 30 millisecond blocks containing 240 samples each. For the higher bit rate each block is encoded using 304 bits per block while for the lower bit rate each block is encoded using 400 bits. Each block is divided into subblocks of 40 samples each.

For the computation of the autocorrelation coefficients we use a portion of the samples from the previous block along with the samples from the current block. For the 15.2 kbit encoding we use 80 samples from the previous block, while for the 13.33 kbit encoding we use 60 samples from the previous block. The use of samples from the previous blocks helps in the continuity of the reconstruction. Because the computed LPC coefficients are transmitted to the decoder the use of the previous block in generating the LPC coefficients does not detract from the independent block coding feature of the iLBC algorithm. The samples from the current block along with the samples from the previous block are windowed before the autocorrelation coefficients are computed. For the high-rate case a single set of LPC coefficients are computed while in the low-rate case two sets of LPC coefficients are computed. For the high-rate case a single window of length 240 samples centered on the third subblock is used, while for the low-rate case two windows centered on the second and fifth subblocks are used.

To reduce problems with numerical precision the autocorrelation coefficients are smoothed using a window given by

$$w_{ac}[k] = \begin{cases} 1.0001 & k = 0 \\ e^{-\frac{1}{2}\left(\frac{120\pi k}{f_s}\right)^2} & k = 1, 2, \ldots, 10 \end{cases} \tag{35}$$

The multiplication of $w_{ac}[0]$ by 1.0001 corresponds to the addition of a white noise floor 40 dB below the signal power.

In each case the encoder computes LPC coefficients using the Durbin-Levinson algorithm that are then represented using the line spectral frequency (LSF) representation, described below. Let $A_k(z)$ be the tenth-order linear predictive filter

$$A_k(z) = 1 + \sum_{i=1}^{10} a_k(i) z^{-i}$$

Define the two functions $f_1'(z)$ and $f_2'(z)$ as

$$f_1'(z) = A_k(z) + z^{-11} A_k(z^{-1})$$
$$f_2'(z) = A_k(z) - z^{-11} A_k(z^{-1})$$

These polynomials have roots only on the unit circle. Clearly, the polynomial $f_1'(z)$ has a root at $z = -1$ while the polynomial $f_2'(z)$ has a root at $z = 1$. Removing these roots we obtain the polynomials

$$f_1(z) = \frac{f_1'(z)}{1 + z^{-1}}$$
$$f_2(z) = \frac{f_2'(z)}{1 - z^{-1}}$$

These polynomials have five complex conjugate roots each on the unit circle. The polynomials can be written as products of five second-order polynomials of the form $(1 - 2q_i z^{-1} + z^{-2})$ where $q_i = \cos(\omega_i)$ and ω_i are known as the line spectral frequencies (LSFs). Notice the similarity between this and the immittance spectral frequency described in Section 18.4. Each vector of ten LSF values is split into three vectors of dimensions 3, 3, and 4. The first three LSF coefficients are quantized using a three-dimensional vector quantizer with 64 output levels, the next three coefficients are quantized using a different three-dimensional vector quantizer with a codebook of size 128, while the last four coefficients are quantized using a four-dimensional vector quantizer with codebook size 128. Thus a total of 20 bits are used to encode each set of LSF coefficients; 6 bits for the first three coefficients, and 7 bits each for the next three and four coefficients.

For a stable reconstruction filter the LSF coefficients should be in increasing order. However, because we split the LSF vector prior to quantization it is possible that some of the coefficients might be out of place. RFC 3951 [243] specifies a procedure for correcting any discrepancies. In the low rate case we have two sets of quantized and unquantized LSF coefficients, one obtained by centering a window on the second subblock and one obtained by centering a window on the fifth subblock. These coefficients are used to generate specific LSF coefficients for each subblock. For subblocks 2 and 5 we use the generated LSF coefficients. The LSF coefficients used for blocks 3 and 4 are obtained using a linear interpolation of the computed LSF coefficients. The LSF coefficients for subblock 1 are computed using an interpolation between the LSF coefficients of subblock 2 and a predefined set of coefficients that are assumed to belong to the previous block. For subblock 6 we use the computed LSF coefficients for subblock 5. In each case we compute a set of coefficients from the unquantized LSF coefficients and another set from the quantized LSF coefficients. For the high-rate case LSF coefficients for the subblocks are obtained by linearly interpolating the LSF coefficients (quantized and unquantized) with the LSF coefficients of the previous block. After the LSF coefficients for each subblock have been generated, they are used to generate the LPC coefficients for each subblock. Thus, for each subblock k we get two analysis filters,

$$A_k(z) = 1 + \sum_i a_k(i) z^{-i}$$

from the unquantized coefficients, and

$$\widetilde{A}_k(z) = 1 + \sum_i \tilde{a}_k(i) z^{-i}$$

from the quantized coefficients. The analysis filters $\{\widetilde{A}_k(z)\}$ are used to generate the residual signal for each subblock, with each analysis filter used for the corresponding subblock. The transition between subblocks is managed by letting the filter memory carry over to the consecutive subblock.

The goal of the iLBC procedure is to encode the blocks independently. This means that the algorithm cannot use an adaptive codebook that relies on the past of the signal. As there is a significant probability of packet loss such an approach would lead to the development of different codebooks in the encoder and decoder. Therefore, the algorithm employs a scheme that uses the past decoded residuals from a block to generate the codebook. Because of this the order in which the various parts of the residual sequence are encoded becomes important. From a perceptual perspective it is important that the encoder be able to relatively accurately encode any voiced segments present in the block. In order to do this the algorithm uses the notion of a *start state*. The power in the residual signal for each block is computed (with optional weighting) and the two consecutive blocks with the highest residual signal power are identified as containing the start state segment. Which set of two subblocks contains the start state is encoded using 2 bits for the high-rate case and 3 bits for the low-rate case. Of the 80 samples contained in these two blocks 57 samples for the high-rate case and 58 samples for the low-rate case are selected as the start state segment by discarding either the first 23 or 22 samples from the first of these two subblocks or the last 23 or 22 samples from the second of the two subblocks based on which portion has lower residual power. This decision is encoded using one bit.

The start state segment is filtered using a set of all-pass filters $\{P_k(z)\}$ where

$$P_k(z) = \frac{\widetilde{A}_{rk}(z)}{\widetilde{A}_k(z)}$$

where

$$\widetilde{A}_{rk}(z) = z^{-11} + \sum_i \tilde{a}_k(i+1)z^{i-10}$$

and k corresponds to the subblocks containing the start state segment.

The maximum element of the all-pass sequence is identified and the base-10 logarithm of this value is quantized using a quantization table specified in RFC 3951. The quantized value q_{max} is used to generate a scaling factor α given by

$$\alpha = \frac{4.5}{10^{q_{max}}}$$

The all-pass filtered residual samples are multiplied by α to generate a normalized sequence, which is filtered using a perceptual weighting filter $W_k(z)$ given by

$$W_k(z) = \frac{1}{A_k(\frac{z}{0.4222})}$$

This filtered sequence is encoded using a DPCM system with a 3-bit quantizer and a predictor $\rho_k(z)$, where

$$\rho_k(z) = 1 - \frac{1}{W_k(z)}$$

In order to recover the residual sequence, first the effect of scaling is undone by dividing each sample by α. Then the effect of the all-pass filtering is undone by time-reversing the unscaled sequence, filtering it using the all-pass filter and circular convolution, and then undoing the time reversal.

The residual sequence from the start state segment is used to form a codebook that is used to encode the 23 or 22 samples discarded from the two subblocks when obtaining the start state. After this codebooks are generated for the encoding of each of the 40 samples in the remaining subblocks. First the subblocks following the start state are encoded, then the subblocks preceding the subblocks are encoded.

The codebook is generated using the residual sequences from the sequences already encoded. The residual sequence is stored in a codebook memory that is of size 85 samples for generating the codebook for encoding the 23 or 22 sample segment and is of size 147 samples for generating the codebook for encoding the complete subblocks. A base codebook is created by sliding a window the length of the target sequence over the perceptually weighted codebook memory resulting in 64 code vectors for the shorter sequence and 128 code vectors for the subblock length. The number of code vectors is doubled by filtering the code vectors in the base codebook using an FIR filter of length 8. The codebook is further augmented with 20 code vectors used to encode the subblocks using interpolations of the codebook memory.

Each target vector is encoded using a three-stage gain-shape vector quantization approach. Let x_i be the target vector to be quantized and let $\{c_n\}$ be the set of vectors in the codebook. In the first stage the normalized inner product

$$\sigma_n = \frac{(x_i^T c_n)^2}{|c_n|^2}$$

is computed and the vector c_m which gives the maximum value for σ_n over all elements of the codebook and also satisfies the constraints $|g_i| < 1.3$ and $x_i^T c_m > 0$ is obtained, where

$$g_i = \frac{(x_i^T c_m)}{|c_m|^2}$$

If none of the code vectors satisfy the two constraints the first code vector in the codebook is selected. The gain value g_i is quantized using a 5-bit quantizer specified by RFC 3951. The reconstruction vector c_m weighted by a scaled quantized value of the gain is then removed from the target vector to generate a new target vector for the second stage. The same process is followed to get the reconstruction vector for the third stage. The gain values in the second and third stages are quantized using 4 and 3 bits respectively. At the end of this process we have three quantized gain values and three indices into the codebook.

Notice that we have encoded the start state samples using a 3-bit DPCM while the remaining residuals are encoded using a three stage gain-shape vector quantizer. To smooth out discrepancies introduced due to the different treatment of samples within a block the encoder readjusts the gains. This completes the encoding process. The decoding process is a simple inverse of the encoding process. However, this presupposes the fact that the packet containing the encoder output will reach the decoder. In the Internet environment it is quite possible that the packet will not reach its destination in a timely manner. The iLBC algorithm contains a Packet Loss Concealment (PLC) strategy to deal with the inevitable loss of packets.

When a packet is lost the decoder needs to substitute a block of samples to replace the number of samples in the packet. This substitution is based on a pitch-synchronous repetition of the excitation signal used by the last reconstruction filter in the previous block. The autocorrelation values for the previous block are used to determine the level of voiced and unvoiced signal and the pitch periodicity. This information is used to generate an excitation signal using a periodic signal and a random signal. If several blocks in a row are lost the excitation signal is dampened to allow a graceful degradation in the decoded signal. After a block or several blocks have been lost the first correctly received block cannot be directly decoded. This is to prevent perceptible discontinuities in the decoded speech. A correlation analysis is performed to find the best phase match between the concealed packet and the received packet and the excitation signal of the received packet is modified using an overlap-add procedure.

18.5.2 G.729

The ITU-T G.729 standard was not developed specifically for the Internet. It provides relatively good quality at 8 kbits per second [245], however, its relative complexity has made it a less attractive option for standard telephone applications. In the Internet world this complexity is not as severe a drawback as the encoding and decoding is performed on relatively powerful computers, while the increase in quality afforded by the complexity is a definite plus. A block diagram of the encoder is shown in Figure 18.11. As in the previous techniques the encoder needs to determine the synthesis filter coefficients and the excitation vector, which are then used by the decoder to reconstruct the speech.

The various operations shown in the block diagram are performed either once per frame, or once per subframe. A frame corresponds to 10 milliseconds of speech, which at a sampling rate of 8000 samples per second corresponds to 80 samples. A subframe is a half frame made up of 40 samples corresponding to 5 milliseconds of speech.

The preprocessing step involves downscaling the input by two to reduce the possibility of overflow and high-pass filtering to remove low-frequency interference including DC bias and interference from the power source. The filter has two poles and two zeros with a cutoff of 140 Hz.

The computation of the synthesis filter is relatively straightforward; the identification of the excitation signal is somewhat more involved. The filtered signal is analyzed to obtain the short-term analysis and synthesis filters. Once per frame the signal is windowed using a 30 millisecond window given by

$$w_{lp}(k) = \begin{cases} 0.54 - 0.46\cos\left(\frac{2\pi k}{399}\right) & k = 0, 1, \ldots, 199 \\ \cos\left(\frac{2\pi(k-200)}{159}\right) & k = 200, 201, \ldots, 239 \end{cases}$$

The autocorrelation coefficients are obtained from the windowed signal and scaled to reduce arithmetic problems using the same scaling as the iLBC encoder (Equation (35)). The Levinson-Durbin algorithm is then used to obtain the linear predictive coefficients, which are then used to obtain the line spectral frequencies. A fourth order moving average predictor is used to predict the LSF coefficients in the current frame. The prediction residual is quantized using a two-stage vector quantizer. The first stage is a 10-dimensional VQ with a codebook of size 128. The quantized error is then further quantized using the second stage, which consists of two five-dimensional vector quantizers, one for the first five coefficients and one for the second five coefficients. Each five-dimensional vector quantizer has a codebook size of 32. Once the LSF coefficients are quantized they are adjusted in order to make sure that the order of the frequencies is preserved and also to prevent two frequencies from being too close to each other. The latter situation would lead to unwanted resonances in the synthesis filter. In order to preserve continuity between frames the computed and quantized LSF values are used for the second subframe. For the first subframe the standard requires using an interpolated set of LSF values. The interpolation is performed on the LSP values q_i. The LSP values of the

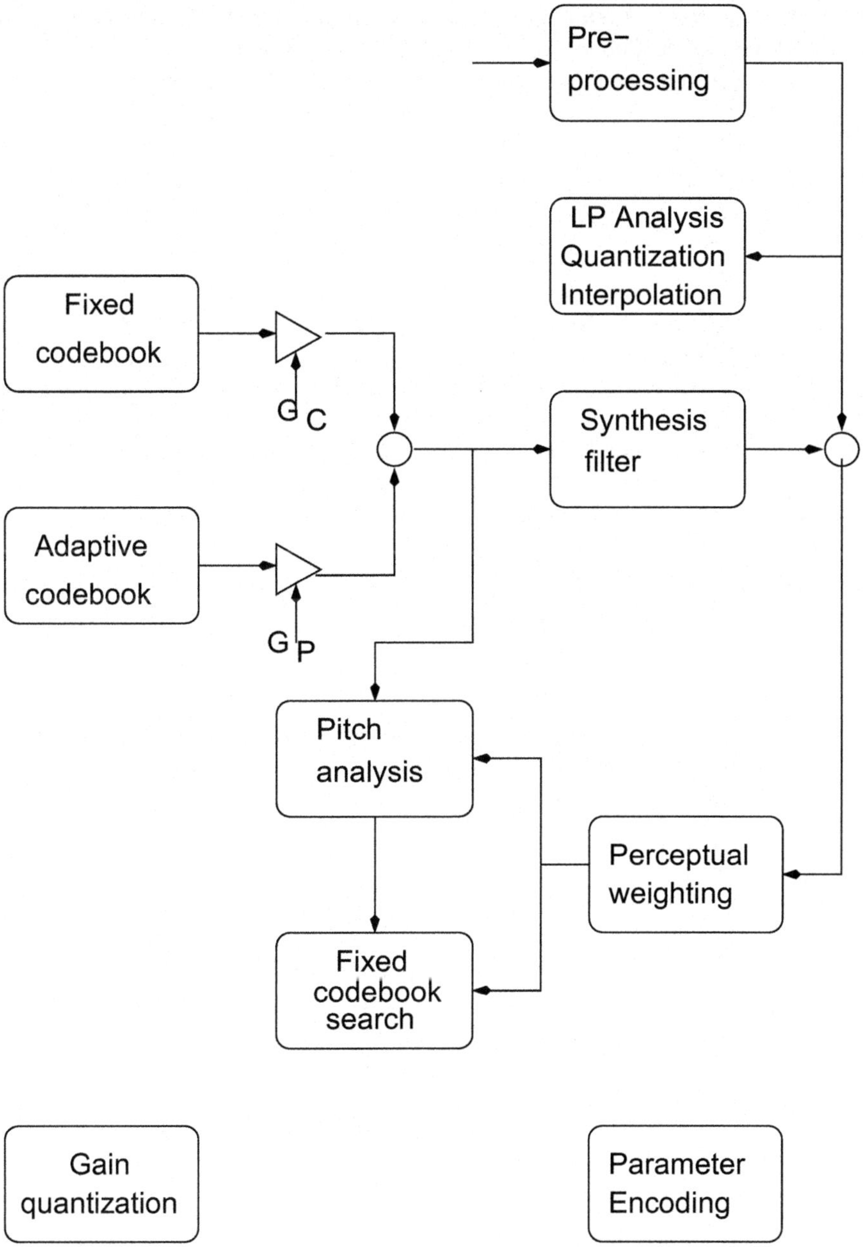

FIGURE 18.11 **Steps in the encoding of speech for the ITU-T G.729 encoder.**

first subframe are taken to be the average of the LSP values computed for this frame and the LSP values computed for the previous frame.

The identification of the excitation signal involves the selection of vectors from the adaptive codebook and the fixed codebook and the computation of the gains associated with each of them. In order to select the adaptive codebook entry we also need to identify the pitch and the target signal. The target signal $x(n)$ can be obtained in several ways. The approach suggested in ITU-T Rec G.729 is to filter the LP residual signal through the combination of the synthesis filter $\widehat{A}(z)$ and the weighting filter $W(z)$ given by

$$W(z) = \frac{A(z/\gamma_1)}{A(z/\gamma_2)}$$

where γ_1 and γ_2 are computed for each frame using the first two reflection coefficients. $\widehat{A}(z)$ is the filter constructed using the quantized linear predictive coefficients.

This weighting filter is also used to generate a weighted signal, which is used to estimate the pitch delay in the frame. The pitch is estimated in two steps; first a crude estimate is obtained and this is then refined. The crude estimate is obtained by finding the maxima of the autocorrelation function of the weighted signal in three ranges of the delay. A selection process that advantages smaller values of the delay over larger values is then used to select among the three estimates. The logic behind advantaging the smaller values is to avoid choosing pitch multiples. This crude estimate of the pitch is then refined using a closed loop analysis. The search maximizes

$$R(k) = \frac{\sum_{n=0}^{39} x(n) y_k(n)}{\sqrt{\sum_{n=0}^{39} y_k(n) y_k(n)}}$$

where $x(n)$ is the target signal and $y_k(n)$ is the past filtered excitation at delay k. The filter in this case is the weighted synthesis filter $W(z)/\widehat{A}(z)$. Once the integer delay k and the fractional delay t have been determined the adaptive codebook vector is obtained as

$$v(n) = \sum_{i=0}^{9} u(n-k+i) b_{30}(t+3i) + \sum_{i=0}^{9} u(n-k+1+i) b_{30}(3-t+3i)$$

where $u(n)$ is the past excitation signal and $b_{30}(n)$ is based on a Hamming windowed sinc function truncated at ± 29 and padded with zeros at ± 30. This filter has a 3 dB cutoff at 3600 Hz. The adaptive codebook delay G_P is computed as

$$G_P = \frac{\sum_{n=0}^{39} x(n) y(n)}{\sqrt{\sum_{n=0}^{39} y(n) y(n)}}$$

where $y(n)$ is the filtered adaptive codebook vector.

Once the adaptive codebook vector has been determined the vector multiplied by the gain is removed from the target vector to generate the target for the fixed codebook search. The fixed codebook is based on an algebraic structure in which each vector has four nonzero pulses with amplitude either $+1$ or -1. The positions available to these pulses are shown in Table 18.1. The codebook vectors $c(n)$ are constructed as

$$c(n) = s_0 \delta(n-m_0) + s_1 \delta(n-m_1) + s_2 \delta(n-m_2) + s_3 \delta(n-m_3) \quad n = 0, 1, \ldots, 39$$

TABLE 18.1 Pulses used to construct the fixed codebook.

Pulse	Sign	Positions
i_0	$s_0 = \pm 1$	m_0 : 0, 5, 10, 15, 20, 25, 30, 35
i_1	$s_1 = \pm 1$	m_1 : 1, 6, 11, 16, 21, 26, 31, 36
i_2	$s_2 = \pm 1$	m_2 : 2, 7, 12, 17, 22, 27, 32, 37
i_3	$s_3 = \pm 1$	m_3 : 3, 8, 13, 18, 23, 28, 33, 38
		4, 9, 14, 19, 24, 29, 34, 39

TABLE 18.2 Bit allocation per frame for the G.729 coder.

Description	Bits
Predictor of LSP quantizer	1
First stage vector quantizer for LSP	7
Second stage for lower five coefficients	5
Second stage for upper five coefficients	5
Pitch delay for first subframe	8
Parity bit for pitch delay	1
Fixed codebook first subframe	13
Signs of fixed codebook pulses for first subframe	4
Gain codebook first subframe	7
Pitch delay for second subframe	5
Fixed codebook second subframe	13
Signs of fixed codebook pulses for first subframe	4
Gain codebook second subframe	7

The selected code vector is filtered to enhance harmonic components in order to improve the quality of the reconstructed speech.

The number of bits used to encode the pitch information, index of the fixed code vector, and the respective gains are given in Table 18.2. These total to 80 bits per frame resulting in a coding rate of 8 kbits/sec. While this is substantially less than the rate of the G.722.2 coder, the quality is about the same [245] though with higher complexity.

18.5.3 SILK

The SILK coder developed by Skype is another example of how a speech coder designed for the Internet can leverage the greater amount of processing power available to the encoder and decoder when communicating between computers rather than between cellular phones. It uses variable length encoding and includes the ability to adaptively respond to changing network conditions. A block diagram of the various processes carried out in the SILK encoder is shown in Figure 18.12. We will use this figure to give a brief overview of some aspects of the SILK encoder.

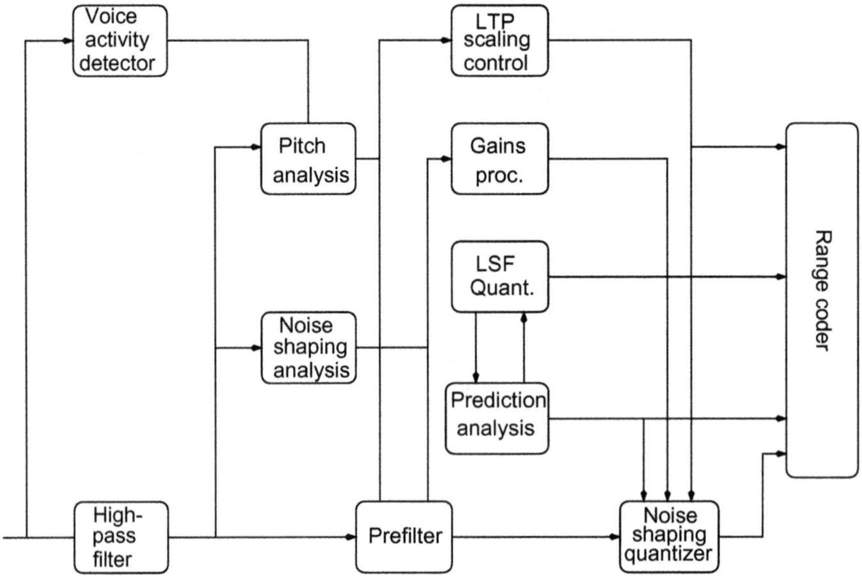

FIGURE 18.12 Steps in the encoding of speech for the SILK coder.

The SILK encoder can operate in one of four modes, a narrowband mode that accepts inputs sampled at 8 kHZ, a medium-band mode that supports inputs at 8 kHz and 12 kHz, a wideband mode that also supports inputs sampled at 16 kHz, and a super wideband mode that can support inputs sampled at 8kHz, 12 kHz, 16kHz, and 24 kHz. The encoder can accept inputs at all sampling rates and if necessary resample the input to a lower sampling rate. This ability to dynamically change the sampling rate of the input allows the encoder to adapt to changing channel conditions. The input is divided into 20 millisecond frames and the encoder can package one to five frames into a single packet. Including more frames in a packet lowers the overhead at the cost of increasing the latency. This is one of the many ways the SILK coder responds to network conditions.

The voice activity detector decomposes the input into four equal bandwidth subbands with additional differentiation in the lowest subband to estimate the level of speech activity. The coder has the option of drastically reducing transmission during silence intervals and increased background noise.

The signal is also filtered using a high-pass filter with a 70 Hz cutoff to get rid of any DC biases and 50- or 60-cycle hum. The filter signal is then provided to the pitch analysis block, which generates pitch lags every five milliseconds. The pitch analysis block performs an LPC analysis of the input and uses the coefficients in a whitening filter. The number of coefficients in the whitening filter can be 16, 12, or 8 depending on the complexity setting. Correlation of downsampled versions of the whitened sequence are used to estimate the speech type and the pitch.

Noise shaping analysis is used to generate the coefficients for the prefilter. The idea behind noise shaping is to shape the quantization noise spectrum in such a way that the highest amount

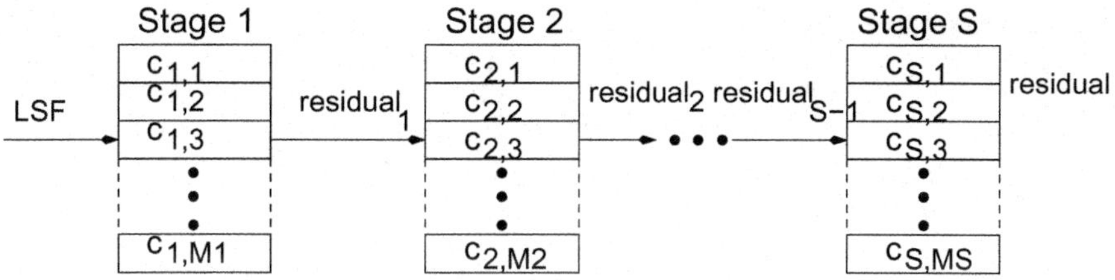

FIGURE 18.13 Multistage vector quantizer for the quantization of LSF coefficients in the SILK encoder.

of noise power occupies the same region as the highest amount of signal power. Thus even if the total noise power is not reduced the perceptual effect of the noise can be greatly reduced.

The LSF coefficients are obtained in a manner similar to the other methods discussed earlier. However, the quantization process is significantly more complex. The vector of LSF coefficients is quantized using a multistage quantizer as shown in Figure 18.13. Each stage of the quantizer operates on the residual of the previous stage. Thus the input of the second stage is the difference between the LSF coefficients and the code vectors of the first $M1$ level vector quantizer. A greedy approach to multistage vector quantization is to simply take the best representative at each stage. However, this reduces the efficacy of the quantization process. Ideally, we would like to keep the residual from each stage until the final stage at which time we could make a selection among all possible combinations. However, this would mean keeping track of an exponentially growing number of quantized representations. The total number of possible combinations in the SILK algorithm for sampling rates above 8 kHz is 2^{36}! The SILK algorithm allows this to be a dynamic process where at each stage the most representative code vectors, called the survivors, are allowed to proceed to the next stage. At each stage the weighted sum of the accumulated bit rate and accumulated distortion for each code vector is evaluated to determine the survivors for the next stage. The number of survivors is adjusted based on available resources.

The SILK encoder uses a variable-rate entropy coder to encode the various coding parameters. The range coder is an analog of the arithmetic coder that uses a nonbinary alphabet. This helps improve the rate distortion performance of the coder while significantly increasing the complexity of the coder.

18.6 Image Compression

Although there have been a number of attempts to mimic the linear predictive coding approach for image compression, they have not been overly successful. A major reason for this is that while speech can be modeled as the output of a linear filter, most images cannot. However, a totally different analysis/synthesis approach, conceived in the mid-1980s, has found some degree of success—fractal compression.

Fractal Compression

There are several different ways to approach the topic of fractal compression. Our approach is to use the idea of fixed-point transformation. A function $f(\cdot)$ is said to have a fixed point x_0 if $f(x_0) = x_0$. Suppose we restrict the function $f(\cdot)$ to be of the form $ax + b$. Then, except for when $a = 1$, this equation always has a fixed point:

$$\begin{aligned} ax_0 + b &= x_o \\ \Rightarrow \quad x_0 &= \frac{b}{1-a} \end{aligned} \tag{36}$$

This means that if we want to transmit the value of x_0, we can instead transmit the values of a and b and obtain x_0 at the receiver using (36). We do not have to solve this equation to obtain x_0. Instead, we can take a guess at what x_0 should be and then refine the guess using the recursion

$$x_0^{(n+1)} = ax_0^{(n)} + b \tag{37}$$

Example 18.6.1:

Suppose that instead of sending the value $x_0 = 2$, we send the values of a and b as 0.5 and 1.0. The receiver starts out with a guess for x_0 as $x_0^{(0)} = 1$. Then

$$\begin{aligned} x_0^{(1)} &= ax_0^{(0)} + b = 1.5 \\ x_0^{(2)} &= ax_0^{(1)} + b = 1.75 \\ x_0^{(3)} &= ax_0^{(2)} + b = 1.875 \\ x_0^{(4)} &= ax_0^{(3)} + b = 1.9375 \\ x_0^{(5)} &= ax_0^{(4)} + b = 1.96875 \\ x_0^{(6)} &= ax_0^{(5)} + b = 1.984375 \end{aligned} \tag{38}$$

and so on. As we can see, with each iteration we come closer and closer to the actual x_0 value of 2. This would be true no matter what our initial guess was, at least for this type of function. ♦

Thus, the value of x_0 is accurately identified by specifying the fixed-point equation. The receiver can retrieve the value either by the solution of (36) or via the recursion (37).

Let us generalize this idea. Suppose that for a given image $\mathcal{I}$ (treated as an array of integers), there exists a function $f(\cdot)$ such that $f(\mathcal{I}) = \mathcal{I}$. If it is cheaper in terms of bits to represent $f(\cdot)$ than it is to represent $\mathcal{I}$, we can treat $f(\cdot)$ as the compressed representation of $\mathcal{I}$.

This idea was first proposed by Michael Barnsley and Alan Sloan [246] based on the idea of self-similarity. Barnsley and Sloan noted that certain natural-looking objects can be obtained as the fixed point of a certain type of function. If an image can be obtained as a fixed point of

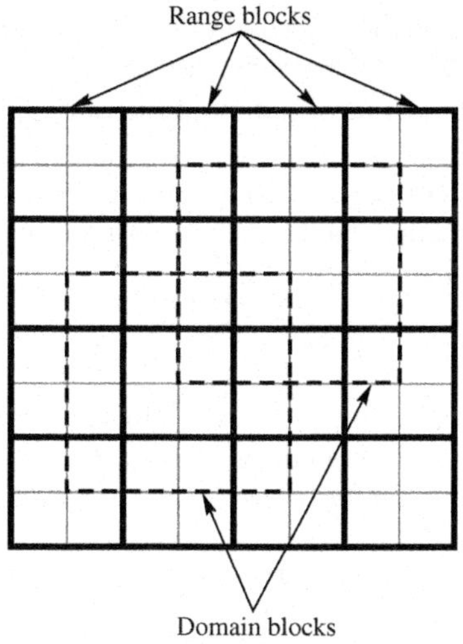

FIGURE 18.14 Range blocks and examples of domain blocks.

some function, can we then solve the *inverse* problem? That is, given an image, can we find the function for which the image is the fixed point? The first practical public answer to this came from Arnaud Jacquin in his Ph.D. dissertation [247] in 1989. The technique we describe in this section is from Jacquin's 1992 paper [248].

Instead of generating a single function directly for which the given image is a fixed point, we partition the image into blocks R_k, called *range* blocks, and obtain a transformation f_k for each block. The transformations f_k are not fixed-point transformations since they do not satisfy the equation

$$f_k(R_k) = R_k \tag{39}$$

Instead, they are a mapping from a block of pixels D_k from some other part of the image. While each individual mapping f_k is not a fixed-point mapping, we will see later that we can combine all these mappings to generate a fixed-point mapping. The image blocks D_k are called *domain* blocks, and they are chosen to be larger than the range blocks. In [248], the domain blocks are obtained by sliding a $K \times K$ window over the image in steps of $K/2$ or $K/4$ pixels. As long as the window remains within the boundaries of the image, each $K \times K$ block thus encountered is entered into the domain pool. The set of all domain blocks does not have to partition the image. In Figure 18.14 we show the range blocks and two possible domain blocks.

The transformations f_k are composed of a *geometric* transformation g_k and a *massic* transformation m_k. The geometric transformation consists of moving the domain block to the location of the range block and adjusting the size of the domain block to match the size of the range block. The massic transformation adjusts the intensity and orientation of the pixels in the domain block after it has been operated on by the geometric transform. Thus,

$$\widehat{R}_k = f_k(D_k) = m_k(g_k(D_k)) \tag{40}$$

We have used $\widehat{R}_k$ instead of R_k on the left-hand side of (40) because it is generally not possible to find an exact functional relationship between domain and range blocks. Therefore, we have to settle for some degree of loss of information. Generally, this loss is measured in terms of mean squared error.

The effect of all these functions together can be represented as the transformation $f(\cdot)$. Mathematically, this transformation can be viewed as a union of the transformations f_k:

$$f = \bigcup_k f_k \tag{41}$$

Notice that while each transformation f_k maps a block of different size and location to the location of R_k, looking at it from the point of view of the entire image, it is a mapping from the image to the image. As the union of R_k is the image itself, we could represent all the transformations as

$$\widehat{I} = f(\widehat{I}) \tag{42}$$

where we have used $\widehat{I}$ instead of I to account for the fact that the reconstructed image is an approximation to the original.

We can now pose the encoding problem as that of obtaining D_k, g_k, and m_k such that the difference $d(R_k, \widehat{R}_k)$ is minimized, where $d(R_k, \widehat{R}_k)$ can be the mean squared error between the blocks R_k and $\widehat{R}_k$.

Let us first look at how we would obtain g_k and m_k assuming that we already know which domain block D_k we are going to use. We will then return to the question of selecting D_k.

Knowing which domain block we are using for a given range block automatically specifies the amount of displacement required. If the range blocks R_k are of size $M \times M$, then the domain blocks are usually taken to be of size $2M \times 2M$. In order to adjust the size of D_k to be the same as that of R_k, we generally replace each 2×2 block of pixels with their average value. Once the range block has been selected, the geometric transformation is easily obtained.

Let's define $T_k = g_k(D_k)$, and t_{ij} as the ijth pixel in T_k, $i, j = 0, 1, \ldots, M-1$. The massic transformation m_k is then given by

$$m_k(t_{ij}) = i(\alpha_k t_{ij} + \Delta_k) \tag{43}$$

where $i(\cdot)$ denotes a shuffling or rearrangement of the pixels with the block. Possible rearrangements (or *isometries*) include the following:

1. Rotation by 90 degrees, $i(t_{ij}) = t_{j(M-1-i)}$
2. Rotation by 180 degrees, $i(t_{ij}) = t_{(M-1-i)(M-1-j)}$
3. Rotation by −90 degrees, $i(t_{ij}) = t_{(M-1-i)j}$
4. Reflection about midvertical axis, $i(t_{ij}) = t_{i(M-1-j)}$
5. Reflection about midhorizontal axis, $i(t_{ij}) = t_{(M-1-i)j}$
6. Reflection about diagonal, $i(t_{ij}) = t_{ji}$
7. Reflection about cross diagonal, $i(t_{ij}) = t_{(M-1-j)(M-1-i)}$
8. Identity mapping, $i(t_{ij}) = t_{ij}$

Therefore, for each massic transformation m_k, we need to find values of α_k, Δ_k, and an isometry. For a given range block R_k, in order to find the mapping that gives us the closest approximation $\widehat{R}_k$, we can try all possible combinations of transformations and domain blocks—a massive computation. In order to reduce the computations, we can restrict the number of domain blocks to search. However, in order to get the best possible approximation, we would like the pool of domain blocks to be as large as possible. Jacquin [248] resolves this situation in the following manner. First, he generates a relatively large pool of domain blocks by the method described earlier. The elements of the domain pool are then divided into *shade blocks, edge blocks*, and *midrange blocks*. The shade blocks are those in which the variance of pixel values within the block is small. The edge block, as the name implies, contains those blocks that have a sharp change of intensity values. The midrange blocks are those that fit into neither category—not too smooth but with no well-defined edges. The shade blocks are then removed from the domain pool. The reason is that, given the transformations we have described, a shade domain block can only generate a shade range block. If the range block is a shade block, it is much more cost effective simply to send the average value of the block rather than attempt any more complicated transformations.

The encoding procedure proceeds as follows. A range block is first classified into one of the three categories described above. If it is a shade block, we simply send the average value of the block. If it is a midrange block, the massic transformation is of the form $\alpha_k t_{ij} + \Delta_k$. The isometry is assumed to be the identity isometry. First α_k is selected from a small set of values—Jacquin [248] uses the values $(0.7, 0.8, 0.9, 1.0)$—such that $d(R_k, \alpha_k T_k)$ is minimized. Thus, we have to search over the possible values of α and the midrange domain blocks in the domain pool in order to find the (α_k, D_k) pair that will minimize $d(R_k, \alpha_k T_k)$. The value of Δ_k is then selected as the difference of the average values of R_k and $\alpha_k T_k$.

If the range block R_k is classified as an edge block, selection of the massic transformation is a somewhat more complicated process. The block is first divided into a bright and a dark region. The dynamic range of the block $r_d(R_k)$ is then computed as the difference of the average values of the light and dark regions. For a given domain block, this is then used to compute the value of α_k by

$$\alpha_k = \min_J \left\{ \frac{r_d(R_k)}{r_d(T_j)}, \alpha_{\max} \right\} \tag{44}$$

where $\alpha_{\max}$ is an upper bound on the scaling factor. The value of α_k obtained in this manner is then quantized to one of a small set of values. Once the value of α_k has been obtained, Δ_k is obtained as the difference of either the average values of the bright regions or the average values of the dark regions, depending on whether we have more pixels in the dark regions or the light regions. Finally, each of the isometries is tried out to find the one that gives the closest match between the transformed domain block and the range block.

Once the transformations have been obtained, they are communicated to the receiver in terms of the following parameters: the location of the selected domain block and a single bit denoting whether the block is a shade block or not. If it is a shade block, the average intensity value is transmitted; if it is not, the quantized scale factor and offset are transmitted along with the label of the isometry used.

The receiver starts out with some arbitrary initial image I_0. The transformations are then applied for each of the range blocks to obtain the first approximation. Then the transformations are applied to the first approximation to get the second approximation, and so on. Let us see an example of the decoding process.

FIGURE 18.15 Original Elif image.

Example 18.6.2:

The image Elif, shown in Figure 18.15, was encoded using the fractal approach. The original image was of size 256×256, and each pixel was coded using 8 bits. Therefore, the storage space required was 65,536 bytes. The compressed image consisted of the transformations described above. The transformations required a total of 4580 bytes, which translates to an average rate of 0.56 bits per pixel. The decoding process started with the transformations being applied to an all-zero image. The first six iterations of the decoding process are shown in Figure 18.16. The process converged in nine iterations. The final image is shown in Figure 18.17. Notice the difference in this reconstructed image and the low-rate reconstructed image obtained using the DCT. The blocking artifacts are for the most part gone. However, this does not mean that the reconstruction is free of distortions and artifacts. They are especially visible in the chin and neck region. ♦

In our discussion (and illustration) we have assumed that the size of the range blocks is constant. If so, how large should we pick the range block? If we pick the size of the range block to be large, we will have to send fewer transformations, thus improving the compression. However, if the size of the range block is large, it becomes more difficult to find a domain block that, after appropriate transformation, will be close to the range block, which in turn will increase the distortion in the reconstructed image. One compromise between picking a large or small value for the size of the range block is to start out with a large size and, if a good enough match is not found, to progressively reduce the size of the range block until we have either found a good match or reached a minimum size. We could also compute a weighted sum of the rate and distortion

$$J = D + \beta R$$

FIGURE 18.16 The first six iterations of the fractal decoding process.

FIGURE 18.17 **Final reconstructed Elif image.**

where D is a measure of the distortion, and R represents the number of bits required to represent the block. We could then either subdivide or not depending on the value of J.

We can also start out with range blocks that have the minimum size (also called the *atomic blocks*) and obtain larger blocks via merging smaller blocks.

There are a number of ways in which we can perform the subdivision. The most commonly known approach is *quadtree partitioning*, initially introduced by Samet [249]. In quadtree partitioning we start by dividing up the image into the maximum-size range blocks. If a particular block does not have a satisfactory reconstruction, we can divide it up into four blocks. These blocks in turn can also, if needed, be divided into four blocks. An example of quadtree partitioning can be seen in Figure 18.18. In this particular case there are three possible sizes for the range blocks. Generally, we would like to keep the minimum size of the range block small if fine detail in the image is of greater importance [250]. Since we have multiple sizes for the range blocks, we also need multiple sizes for the domain blocks.

Quadtree partitioning is not the only method of partitioning available. Another popular method of partitioning is the HV method. In this method we allow the use of rectangular regions. Instead of dividing a square region into four more square regions, rectangular regions are divided either vertically or horizontally in order to generate more homogeneous regions. In particular, if there are vertical or horizontal edges in a block, it is partitioned along these edges. One way to obtain the locations of partitions for a given $M \times N$ range block is to calculate the biased vertical and horizontal differences:

$$v_i = \frac{\min(i, N-i-1)}{N-1}\left(\sum_j \mathcal{I}_{i,j} - \sum_j \mathcal{I}_{i+1,j}\right)$$

$$h_j = \frac{\min(j, M-j-1)}{M-1}\left(\sum_j \mathcal{I}_{i,j} - \sum_j \mathcal{I}_{i,j+1}\right)$$

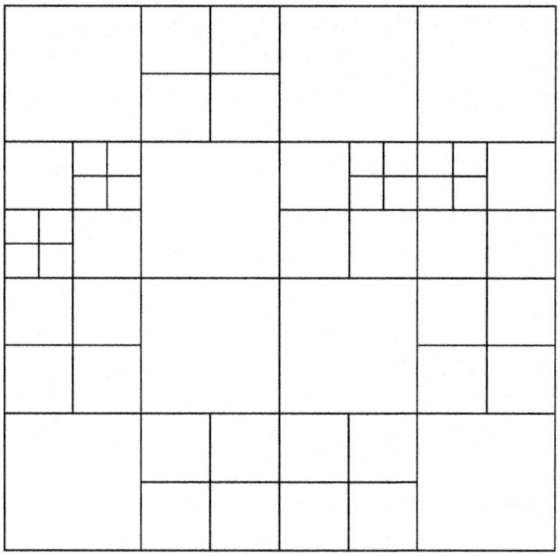

FIGURE 18.18 An example of quadtree partitioning.

The values of i and j for which $|v_i|$ and $|h_j|$ are the largest indicate the row and column for which there is maximum difference between two halves of the block. Depending on whether $|v_i|$ or $|h_j|$ is larger, we can divide the rectangle either vertically or horizontally.

Finally, partitioning does not have to be rectangular, or even regular. People have experimented with triangle partitions as well as irregular-shaped partitions [251].

The fractal approach is a novel way of looking at image compression. At present the quality of the reconstructions using the fractal approach is about the same as the quality of the reconstruction using the DCT approach employed in JPEG. However, the fractal technique is relatively new, and further research may bring significant improvements. The fractal approach has one significant advantage: decoding is simple and fast. This makes it especially useful in applications where compression is performed once and decompression is performed many times.

18.7 Summary

We have looked at two very different ways of using the analysis/synthesis approach. In speech coding the approach works because of the availability of a mathematical model for the speech generation process. We have seen how this model can be used in a number of different ways, depending on the constraints of the problem. Where the primary objective is to achieve intelligible communication at the lowest rate possible, the LPC algorithm provides a very nice solution. If we also want the quality of the speech to be high, CELP and the different sinusoidal techniques provide higher quality at the cost of more complexity and processing delay. If delay also needs to be kept below a threshold, one particular solution is the low-delay CELP algorithm in the G.728 recommendation. With the arrival of the internet the constraints on speech coding have changed. We have looked at several algorithms that have been designed to operate in this changed environment. For images, fractal coding provides a very different way to look at the problem. Instead of using the physical structure of the system to generate the source output, it uses a more abstract view to obtain an analysis/synthesis technique.

Further Reading

1. For information about various aspects of speech processing, *Voice and Speech Processing,* by T. Parsons [119], is a very readable source.
2. The classic tutorial on linear prediction is "Linear Prediction," by J. Makhoul [252], which appeared in the April 1975 issue of the *Proceedings of the IEEE.*
3. For a thorough review of recent activity in speech compression, see "Speech Coding Methods, Standards, and Applications," by J.D. Gibson [245], which appeared in *IEEE Circuits and Systems Magazine.*
4. An excellent source for information about speech coders is *Digital Speech: Coding for Low Bit Rate Communication Systems*, by A. Kondoz [285].
5. An excellent description of the G.728 algorithm can be found in "A Low Delay CELP Coder for the CCITT 16 kb/s Speech Coding Standard," by J.-H. Chen, R.V. Cox, Y.-C. Lin, N. Jayant, and M.J. Melchner [240], in the June 1992 issue of the *IEEE Journal on Selected Areas in Communications.*
6. A good introduction to fractal image compression is *Fractal Image Compression: Theory and Application*, Y. Fisher (ed.) [250].
7. The October 1993 issue of the *Proceedings of the IEEE* contains a special section on fractals with a tutorial on fractal image compression by A. Jacquin.

18.8 Projects and Problems

1. Write a program for the detection of voiced and unvoiced segments using the AMDF function. Test your algorithm on the `test.snd` sound file.
2. The `testf.raw` file is a female voice saying the word *test.* Isolate 10 voiced and unvoiced segments from the `testm.raw` file and the `testf.snd` file. (Try to pick the same segments in the two files.) Compute the number of zero crossings in each segment and compare your results for the two files.
3.
   (a) Select a voiced segment from the `testf.raw` file. Find the fourth-, sixth-, and tenth-order LPC filters for this segment using the Levinson-Durbin algorithm.
   (b) Pick the corresponding segment from the `testf.snd` file. Find the fourth-, sixth-, and tenth-order LPC filters for this segment using the Levinson-Durbin algorithm.
   (c) Compare the results of (a) and (b).
4. Select a voiced segment from the `test.raw` file. Find the fourth-, sixth-, and tenth-order LPC filters for this segment using the Levinson-Durbin algorithm. For each of the filters, find the multipulse sequence that results in the closest approximation to the voiced signal.

19

Video Compression

19.1 Overview

Video compression can be viewed as image compression with a temporal component since video consists of a time sequence of images. From this point of view, the only "new" technique introduced in this chapter is a strategy to take advantage of this temporal correlation. However, there are different situations in which video compression becomes necessary, each requiring a solution specific to its peculiar conditions. In this chapter we briefly look at video compression algorithms and standards developed for different video communications applications.

19.2 Introduction

Of all the different sources of data, perhaps the one that produces the largest amount of data is video. Consider a video sequence generated using the CCIR 601 format (Section 19.4). Each image frame is made up of more than a quarter million pixels. At the rate of 30 frames per second and 16 bits per pixel, this corresponds to a data rate of about 21 Mbytes or 168 Mbits per second. This is certainly a change from the data rates of 2.4, 4.8, and 16 kbits per second that are the targets for speech coding systems discussed in Chapter 18.

Video compression can be viewed as the compression of a sequence of images; in other words, image compression with a temporal component. This is essentially the approach we will take in this chapter. However, there are limitations to this approach. We do not perceive motion video in the same manner as we perceive still images. Motion video may mask coding artifacts that would be visible in still images. On the other hand, artifacts that may not be visible in reconstructed still images can be very annoying in reconstructed motion video sequences. For example, consider a compression scheme that introduces a modest random amount of change in the average intensity of the pixels in the image. Unless a reconstructed still image was being compared side by side with the original image, this artifact may go totally unnoticed. However, in a motion video sequence, especially one with low activity, random intensity changes can be quite annoying. As another example, poor reproduction of edges can be a serious problem in the compression of still images. However, if there is some temporal activity in the video sequence, errors in the reconstruction of edges may go unnoticed.

Although a more holistic approach might lead to better compression schemes, it is more convenient to view video as a sequence of correlated images. Most of the video compression algorithms make use of the temporal correlation to remove redundancy. The previous reconstructed frame is used to generate a prediction for the current frame. The difference between the prediction and the current frame, the prediction error or residual, is encoded and transmitted to the receiver. The previous reconstructed frame is also available at the receiver. Therefore, if the receiver knows the manner in which the prediction was performed, it can use this information to generate the prediction values and add them to the prediction error to generate the reconstruction. The prediction operation in video coding has to take into account motion of the objects in the frame, which is known as motion compensation (described in the next section).

We will also describe a number of different video compression algorithms. For the most part, we restrict ourselves to discussions of techniques that have found their way into international standards. Because there are a significant number of products that use proprietary video compression algorithms, it is difficult to find or include descriptions of them.

We can classify the algorithms based on the application area. While attempts have been made to develop standards that are "generic," the application requirements can play a large part in determining the features to be used and the values of parameters. When the compression algorithm is being designed for two-way communication, it is necessary for the coding delay to be minimal. Furthermore, compression and decompression should have about the same level of complexity. The complexity can be unbalanced in a broadcast application, where there is one transmitter and many receivers, and the communication is essentially one-way. In this case, the encoder can be much more complex than the receiver. There is also more tolerance for encoding delays. In applications where the video is to be decoded on workstations and personal computers, the decoding complexity has to be extremely low in order for the decoder to decode a sufficient number of images to give the illusion of motion. However, as the encoding is generally not done in real time, the encoder can be quite complex. When the video is to be transmitted over packet networks, the effects of packet loss have to be taken into account when designing the compression algorithm. Thus, each application will present its own unique requirements and demand a solution that fits those requirements.

We will assume that you are familiar with the particular image compression technique being used. For example, when discussing transform-based video compression techniques, we assume that you have reviewed Chapter 13 and are familiar with the descriptions of transforms and the JPEG algorithm contained in that chapter.

19.3 Motion Compensation

In most video sequences there is little change in the contents of the image from one frame to the next. Even in sequences that depict a great deal of activity, there are significant portions of the image that do not change from one frame to the next. Most video compression schemes take advantage of this redundancy by using the previous frame to generate a prediction for the current frame. We have used prediction previously when we studied differential encoding schemes. If we try to apply those techniques blindly to video compression by predicting the value of each pixel by the value of the pixel at the same location in the previous frame, we will run into trouble because we would not be taking into account the fact that objects tend to move between frames. Thus, the object in one frame that was providing the pixel at a certain location (i_0, j_0) with its intensity value might be providing the same intensity value in the next frame to a pixel at location (i_1, j_1). If we don't take this into account, we can actually increase the amount of information that needs to be transmitted.

Example 19.3.1:

Consider the two frames of a motion video sequence shown in Figure 19.1. The only differences between the two frames are that the devious looking individual has moved slightly downward and to the right of the frame, while the triangular object has moved to the left. The differences between the two frames are so slight, you would think that if the first frame was available to both the transmitter and receiver, not much information would need to be transmitted to the receiver in order to reconstruct the second frame. However, if we simply take the difference between the two frames, as shown in Figure 19.2, the displacement of the objects in the frame results in an image that contains more detail than the original image. In other words, instead of the differencing operation reducing the information, there is actually more information that needs to be transmitted. ♦

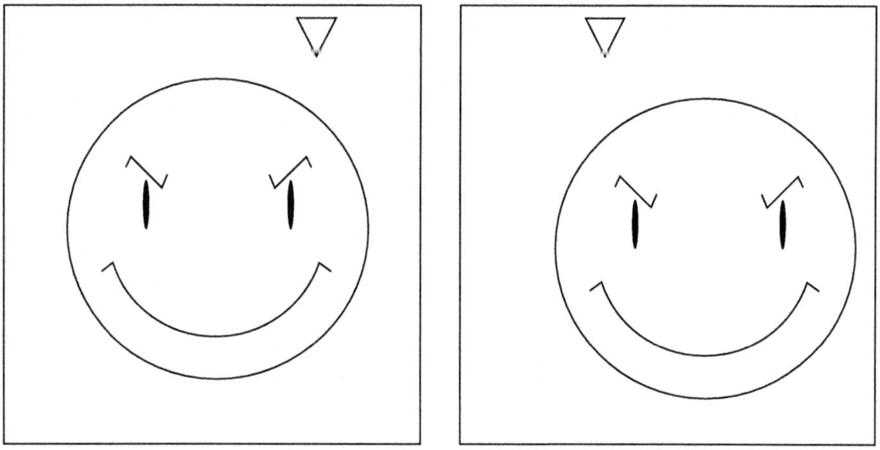

FIGURE 19.1 Two frames of a video sequence.

FIGURE 19.2 Difference between the two frames.

In order to use a previous frame to predict the pixel values in the frame being encoded, we have to take the motion of objects in the image into account. Although a number of approaches have been investigated, the method that has worked best in practice is a simple approach called *block-based motion compensation*. In this approach, the frame being encoded is divided into blocks of size $M \times M$. For each block, we search the previous reconstructed frame for the block of size $M \times M$ that most closely matches the block being encoded. We can measure the closeness of a match, or distance, between two blocks by the sum of absolute differences between corresponding pixels in the two blocks. We would obtain the same results if we used the sum of squared differences between the corresponding pixels as a measure of distance. Generally, if the distance from the block being encoded to the closest block in the previous reconstructed frame is greater than some prespecified threshold, the block is declared uncompensable and is encoded without the benefit of prediction. This decision is also transmitted to the receiver. If the distance is below the threshold, then a *motion vector* is transmitted to the receiver. The motion vector is the relative location of the block to be used for prediction obtained by subtracting the coordinates of the upper-left corner pixel of the block being encoded from the coordinates of the upper-left corner pixel of the block being used for prediction.

Suppose the block being encoded is an 8×8 block between pixel locations (24, 40) and (31, 47); that is, the upper-left corner pixel of the 8×8 block is at location (24, 40). If the block that best matches it in the previous frame is located between pixels at location (21, 43) and (28, 50), then the motion vector would be $(-3, 3)$. The motion vector was obtained by subtracting the location of the upper-left corner of the block being encoded from the location of the upper-left corner of the best matching block. Note that the blocks are numbered starting from the top-left corner. Therefore, a positive x component means that the best matching block in the previous frame is to the right of the location of the block being encoded. Similarly, a positive y component means that the best matching block is at a location below that of the location of the block being encoded.

Example 19.3.2:

Let us again try to predict the second frame of Example 10.3.1 using motion compensation. We divide the image into blocks and then predict the second frame from the first in the manner described above. Figure 19.3 shows the blocks in the previous frame that were used to predict some of the blocks in the current frame.

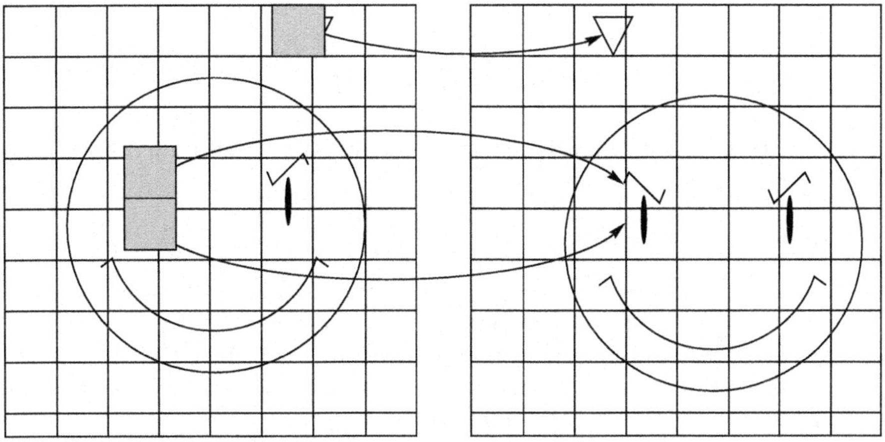

FIGURE 19.3 Motion-compensated prediction.

Notice that in this case all that needs to be transmitted to the receiver are the motion vectors. The current frame is completely predicted by the previous frame. ♦

We have been describing motion compensation where the displacement between the block being encoded and the best matching block is an integer number of pixels in the horizontal and vertical directions. There are algorithms in which the displacement is measured in half pixels. In order to do this, pixels of the coded frame being searched are interpolated to obtain twice as many pixels as in the original frame. This "doubled" image is then searched for the best matching block.

The doubled image is obtained as follows: Consider Table 19.1. In this image A, B, C, and D are the pixels of the original frame. The pixels h_1, h_2, v_1, and v_2 are obtained by interpolating between the two neighboring pixels:

TABLE 19.1 "Doubled" image.

A	h_1	B
v_1	c	v_2
C	h_2	D

$$
\begin{aligned}
h_1 &= \left\lfloor \frac{A+B}{2} + 0.5 \right\rfloor \\
h_2 &= \left\lfloor \frac{C+D}{2} + 0.5 \right\rfloor \\
v_1 &= \left\lfloor \frac{A+C}{2} + 0.5 \right\rfloor \\
v_2 &= \left\lfloor \frac{B+D}{2} + 0.5 \right\rfloor
\end{aligned} \tag{1}
$$

The pixel c is obtained as the average of the four neighboring pixels from the coded original:

$$
c = \left\lfloor \frac{A+B+C+D}{4} + 0.5 \right\rfloor
$$

We have described motion compensation in very general terms in this section. The various schemes in this chapter use specific motion compensation schemes that differ from each other. The differences generally involve the region of search for the matching block and the search procedure. We will look at the details with the study of the compression schemes. But before we begin our study of compression schemes, we briefly discuss how video signals are represented in the next section.

19.4 Video Signal Representation

The development of different representations of video signals has depended a great deal on past history. We will also take a historical view, starting with black-and-white television and proceeding up to digital video formats. The history of the development of analog video signal formats is different for the United States than for Europe. Although we will show the development using the formats used in the United States, the basic ideas are the same for all formats.

A black-and-white analog television picture is generated by exciting the phosphor on the television screen using an electron beam whose intensity is modulated to generate the image we see. The path that the modulated electron beam traces is shown in Figure 19.4. The line created by the horizontal traversal of the electron beam is called a line of the image. In order to trace a second line, the electron beam has to be deflected back to the left of the screen. During this period, the gun is turned off in order to prevent the retrace from becoming visible. The image generated by the traversal of the electron gun has to be updated rapidly enough for persistence of vision to make the image appear stable. However, higher rates of information transfer require higher bandwidths, which translate to higher costs.

In order to keep the cost of bandwidth low it was decided to send 525 lines 30 times a second. These 525 lines are said to constitute a *frame*. However, a thirtieth of a second between frames is long enough for the image to appear to flicker. To avoid the flicker, it was decided to divide the image into two interlaced fields. A field is sent once every sixtieth of a second.

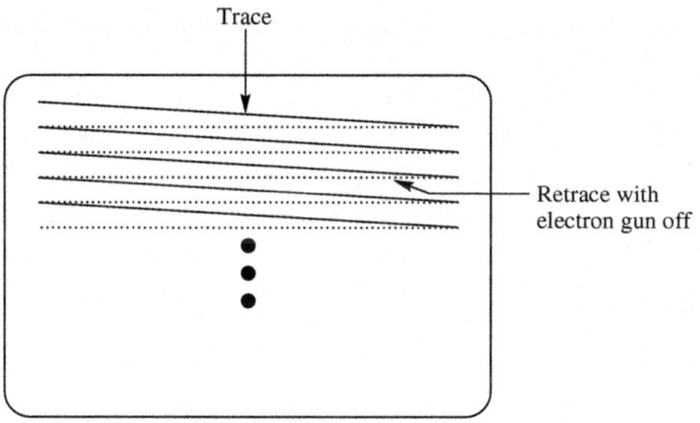

FIGURE 19.4 **The path traversed by the electron beam in a television.**

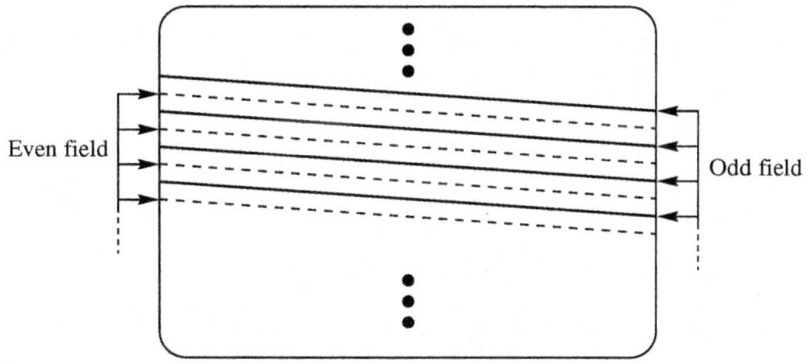

FIGURE 19.5 **A frame and its constituent fields.**

First, one field consisting of 262.5 lines is traced by the electron beam. Then, the second field consisting of the remaining 262.5 lines is traced *between* the lines of the first field. The situation is shown schematically in Figure 19.5. The first field is shown with solid lines while the second field is shown with dashed lines. The first field begins on a full line and ends on a half line while the second field begins on a half line and ends on a full line. Not all 525 lines are displayed on the screen. Some are lost due to the time required for the electron gun to position the beam from the bottom to the top of the screen. We actually see about 486 lines per frame.

In an analog color television, instead of a single electron gun, we have three electron guns that act in unison. These guns excite red, green, and blue phosphor dots embedded in the screen. The beam from each gun strikes only one kind of phosphor, and the gun is named according to the color of the phosphor it excites. Thus, the red gun strikes only the red phosphor, the blue gun strikes only the blue phosphor, and the green gun strikes only the green phosphor. (Each gun is prevented from hitting a different type of phosphor by an aperture mask.)

In order to control the three guns we need three signals: a red signal, a blue signal, and a green signal. If we transmitted each of these separately, we would need three times the bandwidth. With the advent of color television, there was also the problem of backward compatibility. Most people had black-and-white television sets, and television stations did not want to broadcast using a format that most of the viewing audience could not see on their existing sets. Both issues were resolved with the creation of a composite color signal. In the United States, the specifications for the composite signal were created by the National Television Systems Committee, and this signal is often called an NTSC signal. The corresponding signals in Europe are PAL (Phase Alternating Lines), developed in Germany, and SECAM (Séquential Couleur avec Mémoire), developed in France. There is some (hopefully) good-natured rivalry between proponents of the different systems. Some problems with color reproduction in the NTSC signal have led to the name *Never Twice the Same Color*, while the idiosyncracies of the SECAM system have led to the name *Système Essentiellement Contre les Américains* (system essentially against the Americans).

The composite color signal consists of a *luminance* component, corresponding to the black-and-white television signal, and two *chrominance* components. The luminance component is denoted by Y:

$$Y = 0.299R + 0.587G + 0.114B \tag{2}$$

where R is the red component, G is the green component, and B is the blue component. The weighting of the three components was obtained through extensive testing with human observers. The two chrominance signals are obtained as

$$C_b = B - Y \tag{3}$$

$$C_r = R - Y \tag{4}$$

These three signals can be used by the color television set to generate the red, blue, and green signals needed to control the electron guns. The luminance signal can be used directly by the black-and-white televisions.

Because the eye is much less sensitive to changes of the chrominance in an image, the chrominance signal does not need to have higher frequency components. Thus, lower bandwidth of the chrominance signals along with a clever use of modulation techniques permits all three signals to be encoded without need of any bandwidth expansion. (A simple and readable explanation of television systems can be found in [253].)

The early efforts toward digitization of the video signal were devoted to sampling the composite signal, and in the United States the Society of Motion Picture and Television Engineers developed a standard that required sampling the NTSC signal at a little more than 14 million times a second. In Europe, the efforts at standardization of video were centered around the characteristics of the PAL signal. Because of the differences between NTSC and PAL, this would have resulted in different "standards." In the late 1970s, this approach was dropped in favor of sampling the components and the development of a worldwide standard. This standard was developed under the auspices of the International Consultative Committee on Radio (CCIR) and was called CCIR recommendation 601-2. CCIR is now known as ITU-R, and the recommendation is officially known as ITU-R recommendation BT.601-2. However, the standard is generally referred to as recommendation 601 or CCIR 601.

The standard proposes a family of sampling rates based on the sampling frequency of 3.725 MHz (3.725 million samples per second). Multiples of this sampling frequency permit samples on each line to line up vertically, thus generating the rectangular array of pixels necessary for digital processing. Each component can be sampled at an integer multiple of 3.725 MHz, up to a maximum of four times this frequency. The sampling rate is represented

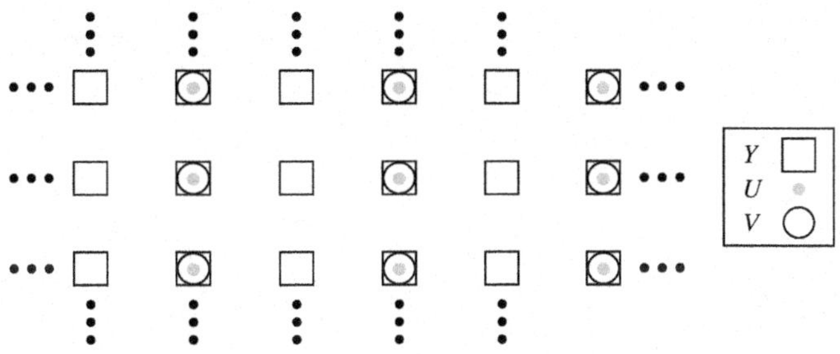

FIGURE 19.6 **Recommendation 601 4:2:2 sampling format.**

as a triple of integers, with the first integer corresponding to the sampling of the luminance component and the remaining two corresponding to the chrominance components. Thus, 4:4:4 sampling means that all components were sampled at 13.5 MHz. The most popular sampling format is the 4:2:2 format, in which the luminance signal is sampled at 13.5 MHz, while the lower-bandwidth chrominance signals are sampled at 6.75 MHz. If we ignore the samples of the portion of the signal that do not correspond to active video, the sampling rate translates to 720 samples per line for the luminance signal and 360 samples per line for the chrominance signal. The sampling format is shown in Figure 19.6. The luminance component of the digital video signal is also denoted by Y, while the chrominance components are denoted by U and V. The sampled analog values are converted to digital values as follows. The sampled values of YC_bC_r are normalized so that the sampled Y values, Y_s, take on values between 0 and 1, and the sampled chrominance values, C_{rs} and C_{bs}, take on values between $\frac{-1}{2}$ and $\frac{1}{2}$. These normalized values are converted to 8-bit numbers according to the transformations

$$Y = 219Y_s + 16 \tag{5}$$

$$U = 224C_{bs} + 128 \tag{6}$$

$$V = 224C_{rs} + 128 \tag{7}$$

Thus, the Y component takes on values between 16 and 235, and the U and V components take on values between 16 and 240.

An example of the Y component of a CCIR 601 frame is shown in Figure 19.7. In the top image we show the fields separately, while in the bottom image the fields have been interlaced. Notice that in the interlaced image the smaller figure looks blurred. This is because the individual moved in the sixtieth of a second between the two fields. (This is also proof—if any was needed—that a three-year-old cannot remain still, even for a sixtieth of a second!)

The YUV data can also be arranged in other formats. In the Common Interchange Format (CIF), which is used for videoconferencing, the luminance of the image is represented by an array of 288 × 352 pixels, and the two chrominance signals are represented by two arrays consisting of 144 × 176 pixels. In the QCIF (Quarter CIF) format, we have half the number of pixels in both the rows and columns.

The MPEG-1 algorithm, which was developed for encoding video at rates up to 1.5 Mbits per second, uses a different subsampling of the CCIR 601 format to obtain the MPEG-SIF format, where SIF stands for Source Input Format. Starting from a 4:2:2, 480-line CCIR 601

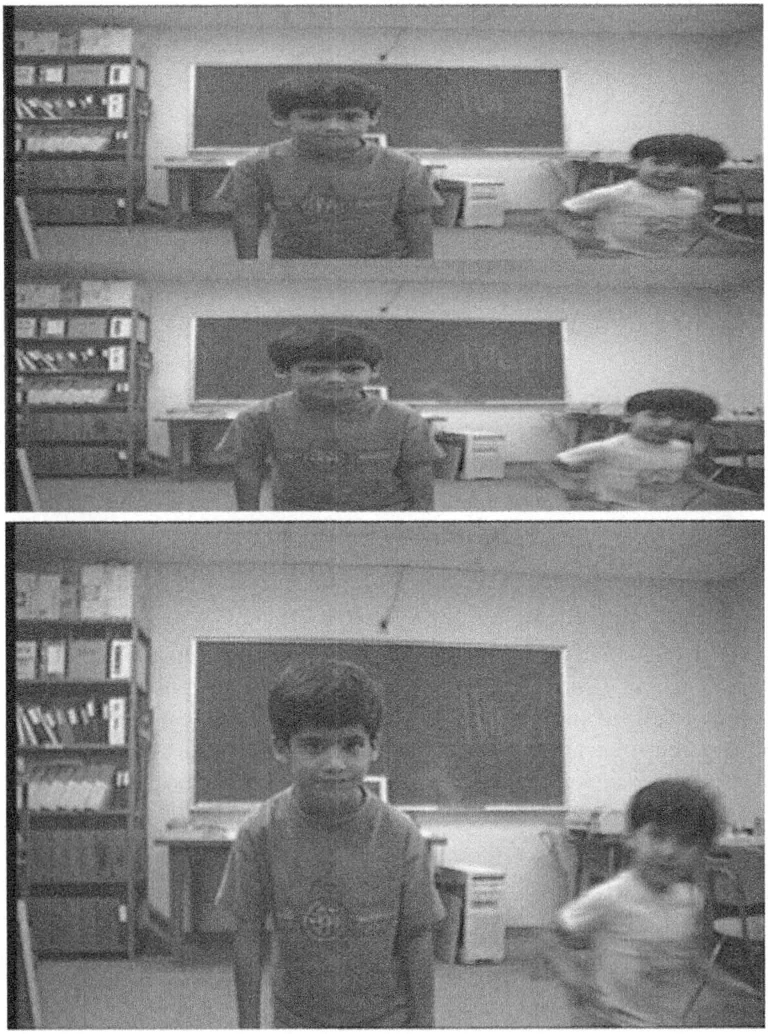

FIGURE 19.7 **Top: Fields of a CCIR 601 frame. Bottom: An interlaced CCIR 601 frame.**

format, the vertical resolution is first reduced by taking only the odd field for both the luminance and the chrominance components. The horizontal resolution is then reduced by filtering (to prevent aliasing) and then subsampling by a factor of two in the horizontal direction. This results in 360×240 samples of Y and 180×240 samples each of U and V. The vertical resolution of the chrominance samples is further reduced by filtering and subsampling in the vertical direction by a factor of two to obtain 180×120 samples for each of the chrominance signals. The process is shown in Figure 19.8, and the resulting format is shown in Figure 19.9.

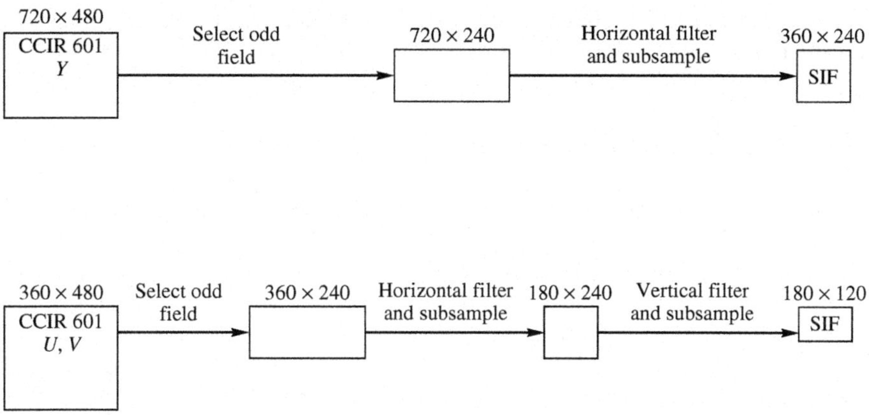

FIGURE 19.8 Generation of an SIF frame.

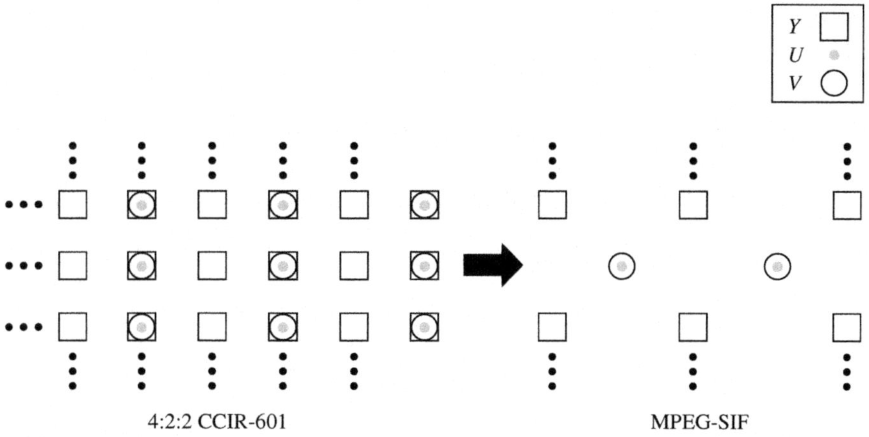

FIGURE 19.9 CCIR 601 to MPEG-SIF.

In the following we describe several of the video coding standards in existence today. Our order of description follows the historical development of the standards. As each standard has built upon features of previous standards this seems like a logical plan of attack. As in the case of image compression, most of the standards for video compression are based on the discrete cosine transform (DCT). The standard for teleconferencing applications, ITU-T recommendation H.261, is no exception. Most systems currently in use for videoconferencing use proprietary compression algorithms. However, in order for the equipment from different manufacturers to communicate with each other, these systems also offer the option of using H.261. We will describe the compression algorithm used in the H.261 standard in the next section. We will follow that with a description of the MPEG algorithms used in Video CDs, DVDs, and HDTV, and a discussion of the latest joint offering from ITU and MPEG.

We will also describe a new approach towards compression of video for videophone applications called three-dimensional model-based coding. This approach is far from maturity, and our description will be rather cursory. The reason for including it here is the great promise it holds for the future.

19.5 ITU-T Recommendation H.261

The earliest DCT-based video coding standard is the ITU-T H.261 standard. This algorithm assumes one of two formats, CIF and QCIF. A block diagram of the H.261 video coder is shown in Figure 19.10. The basic idea is simple. An input image is divided into blocks of 8×8 pixels. For a given 8×8 block, we subtract the prediction generated using the previous frame. (If there is no previous frame or the previous frame is very different from the current frame, the prediction might be zero.) The difference between the block being encoded and the prediction is transformed using a DCT. The transform coefficients are quantized and the quantization label encoded using a variable-length code. In the following discussion, we will take a more detailed look at the various components of the compression algorithm.

19.5.1 Motion Compensation

Motion compensation requires a large amount of computation. Consider finding a matching block for an 8×8 block. Each comparison requires taking 64 differences and then computing the sum of the absolute value of the differences. If we assume that the closest block in the previous frame is located within 20 pixels in either the horizontal or vertical direction of the block to be encoded, we need to perform 1681 comparisons. There are several ways we can reduce the total number of computations.

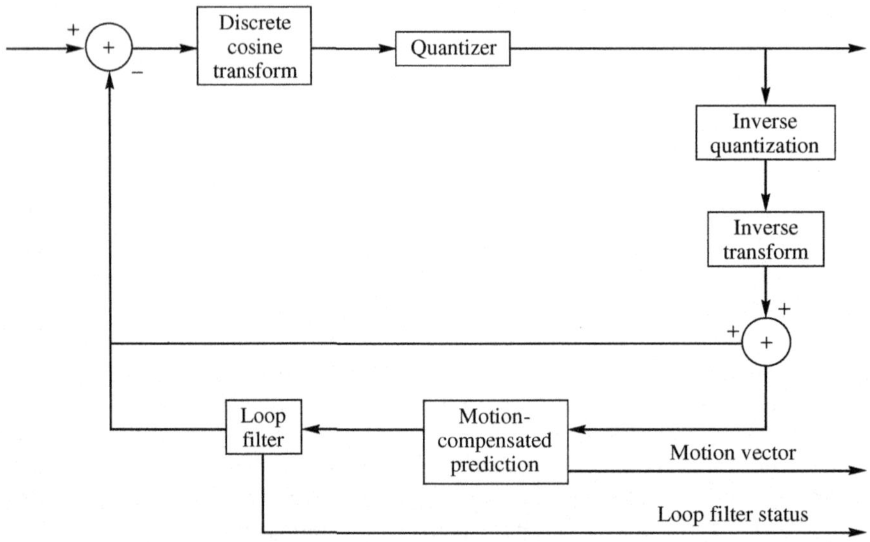

FIGURE 19.10 Block diagram of the ITU-T H.261 encoder.

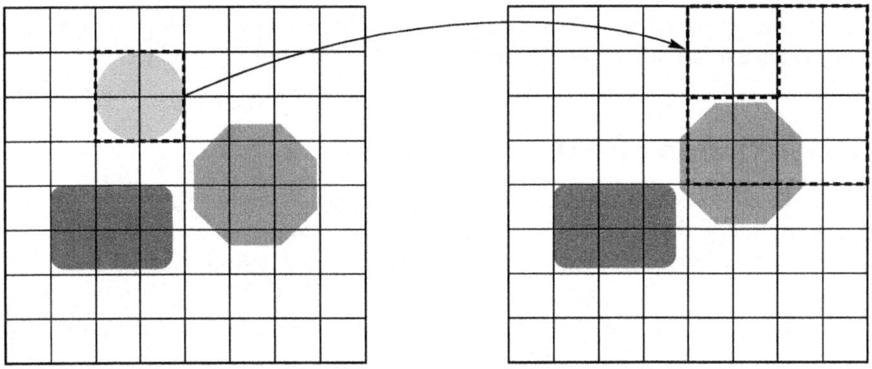

FIGURE 19.11 Effect of block size on motion compensation.

One way is to increase the size of the block. Increasing the size of the block means more computations per comparison. However, it also means that we will have fewer blocks per frame, so the number of times we have to perform the motion compensation will decrease. However, different objects in a frame may be moving in different directions. The drawback to increasing the size of the block is that the probability that a block will contain objects moving in different directions increases with size. Consider the two images in Figure 19.11. If we use blocks that are made up of 2×2 squares, we can find a block that exactly matches the 2×2 block that contains the circle. However, if we increase the size of the block to 4×4 squares, the block that contains the circle also contains the upper part of the octagon. We cannot find a similar 4×4 block in the previous frame. Thus, there is a trade-off involved. Larger blocks reduce the amount of computation; however, they can also result in poor prediction, which in turn can lead to poor compression performance.

Another way we can reduce the number of computations is by reducing the search space. If we reduce the size of the region in which we search for a match, the number of computations will be reduced. However, reducing the search region also increases the probability of missing a match. Again, we have a trade-off between computation and the amount of compression.

The H.261 standard has balanced the trade-offs in the following manner. The 8×8 blocks of luminance and chrominance pixels are organized into *macroblocks*, which consist of four luminance blocks, and one each of the two types of chrominance blocks. The motion-compensated prediction (or motion compensation) operation is performed on the macroblock level. For each macroblock, we search the previous reconstructed frame for the macroblock that most closely matches the macroblock being encoded. In order to further reduce the amount of computations, only the luminance blocks are considered in this matching operation. The motion vector for the prediction of the chrominance blocks is obtained by halving the component values of the motion vector for the luminance macroblock. Therefore, if the motion vector for the luminance blocks is $(-3, 10)$, then the motion vector for the chrominance blocks would be $(-1, 5)$.

The search area is restricted to ± 15 pixels of the macroblock being encoded in the horizontal and vertical directions. That is, if the upper-left corner pixel of the block being encoded is

(x_c, y_c), and the upper-left corner of the best matching macroblock is (x_p, y_p), then (x_c, y_c) and (x_p, y_p) have to satisfy the constraints $|x_c - x_p| < 15$ and $|y_c - y_p| < 15$.

19.5.2 The Loop Filter

Sometimes sharp edges in the block used for prediction can result in the generation of sharp changes in the prediction error. This in turn can cause high values for the high-frequency coefficients in the transforms, which can increase the transmission rate. To avoid this, prior to taking the difference, the prediction block can be smoothed by using a two-dimensional spatial filter. The filter is separable; it can be implemented as a one-dimensional filter that first operates on the rows, then on the columns. The filter coefficients are $\frac{1}{4}, \frac{1}{2}, \frac{1}{4}$, except at block boundaries where one of the filter taps would fall outside the block. To prevent this from happening, the block boundaries remain unchanged by the filtering operation.

Example 19.5.3:

Let's filter the 4×4 block of pixel values shown in Table 19.2 using the filter specified for the H.261 algorithm. From the pixel values we can see that this is a gray square with a white L in it. (Recall that small pixel values correspond to darker pixels and large pixel values correspond to lighter pixels, with 0 corresponding to black and 255 corresponding to white.)

TABLE 19.2 Original block of pixels.

110	218	116	112
108	210	110	114
110	218	210	112
112	108	110	116

Let's filter the first row. We leave the first pixel value the same. The second value becomes

$$\frac{1}{4} \times 110 + \frac{1}{2} \times 218 + \frac{1}{4} \times 116 = 165$$

where we have assumed integer division. The third filtered value becomes

$$\frac{1}{4} \times 218 + \frac{1}{2} \times 116 + \frac{1}{4} \times 112 = 140$$

The final element in the first row of the filtered block remains unchanged. Continuing in this fashion with all four rows, we get the 4×4 block shown in Table 19.3.

Now repeat the filtering operation along the columns. The final 4×4 block is shown in Table 19.4. Notice how much more homogeneous this last block is compared to the original block. This means that it will most likely not introduce any sharp variations in the difference block, and the high-frequency coefficients in the transform will be closer to zero, leading to compression. ♦

TABLE 19.3 After filtering the rows.

110	165	140	112
108	159	135	114
110	188	187	112
112	109	111	116

This filter is either switched on or off for each macroblock. The conditions for turning the filter on or off are not specified by the recommendations.

19.5.3 The Transform

The transform operation is performed with a DCT on an 8×8 block of pixels or pixel differences. If the motion compensation operation does not provide a close match, then the transform operation is performed on an 8×8 block of pixels. If the transform operation is performed on a block level, either a block or the difference between the block and its predicted value is quantized and transmitted to the receiver. The receiver performs the inverse operations to reconstruct the image. The receiver operation is also simulated at the transmitter, where the reconstructed images are obtained and stored in a frame store. The encoder is said to be in *intra* mode if it operates directly on the input image without the use of motion compensation. Otherwise, it is said to be in *inter* mode.

19.5.4 Quantization and Coding

Depending on how good or poor the prediction is, we can get a wide variation in the characteristics of the coefficients that are to be quantized. In the case of an intra block, the DC coefficients will take on much larger values than the other coefficients. Where there is little motion from frame to frame, the difference between the block being encoded and the prediction will be small, leading to small values for the coefficients.

In order to deal with this wide variation, we need a quantization strategy that can be rapidly adapted to the current situation. The H.261 algorithm does this by switching between 32 different quantizers, possibly from one macroblock to the next. One quantizer is reserved for the intra DC coefficient, while the remaining 31 quantizers are used for the other coefficients.

TABLE 19.4 Final block.

110	165	140	112
108	167	148	113
110	161	154	113
112	109	111	116

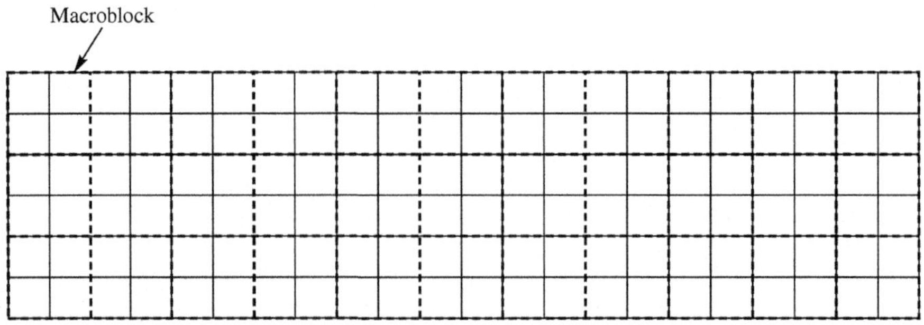

FIGURE 19.12 A GOB consisting of 33 macroblocks.

The intra DC quantizer is a uniform midrise quantizer with a step size of 8. The other quantizers are midtread quantizers with a step size of an even value between 2 and 62. Given a particular block of coefficients, if we use a quantizer with a smaller step size, we are likely to get a larger number of nonzero coefficients. Because of the manner in which the labels are encoded, the number of bits that will need to be transmitted will increase. Therefore, the availability of transmission resources will have a major impact on the quantizer selection. We will discuss this aspect further when we talk about the transmission buffer. Once a quantizer is selected, the receiver has to be informed about the selection. In H.261, this is done in one of two ways. Each macroblock is preceded by a header. The quantizer being used can be identified as part of this header. When the amount of activity or motion in the sequence is relatively constant, it is reasonable to expect that the same quantizer will be used for a large number of macroblocks. In this case, it would be wasteful to identify the quantizer being used with each macroblock. The macroblocks are organized into *groups of blocks* (GOBs), each of which consists of three rows of 11 macroblocks. This hierarchical arrangement is shown in Figure 19.12. Only the luminance blocks are shown. The header preceding each GOB contains a 5-bit field for identifying the quantizer. Once a quantizer has been identified in the GOB header, the receiver assumes that quantizer is being used, unless this choice is overridden using the macroblock header.

The quantization labels are encoded in a manner similar to, but not exactly the same as, JPEG. The labels are scanned in a zigzag fashion like JPEG. The nonzero labels are coded along with the number, or run, of coefficients quantized to zero. The 20 most commonly occurring combinations of (run, label) are coded with a single variable-length codeword. All other combinations of (run, label) are coded with a 20-bit word, made up of a 6-bit escape sequence, a 6-bit code denoting the run, and an 8-bit code for the label.

In order to avoid transmitting blocks that have no nonzero quantized coefficient, the header preceding each macroblock can contain a variable-length code called the *coded block pattern* (CBP) that indicates which of the six blocks contain nonzero labels. The CBP can take on one of 64 different pattern numbers, which is then encoded by a variable-length code. The pattern number is given by

$$\text{CBP} = 32P_1 + 16P_2 + 8P_3 + 4P_4 + 2P_5 + P_6$$

where P_1 through P_6 correspond to the six different blocks in the macroblock, and is one if the corresponding block has a nonzero quantized coefficient and zero otherwise.

19.5.5 Rate Control

The binary codewords generated by the transform coder form the input to a transmission buffer. The function of the transmission buffer is to keep the output rate of the encoder fixed. If the buffer starts filling up faster than the transmission rate, it sends a message back to the transform coder to reduce the output from the quantization. If the buffer is in danger of becoming emptied because the transform coder is providing bits at a rate lower than the transmission rate, the transmission buffer can request a higher rate from the transform coder. This operation is called *rate control*.

The change in rate can be accomplished in two different ways. First, the quantizer being used will affect the rate. If a quantizer with a large step size is used, a larger number of coefficients will be quantized to zero. Also, there is a higher probability that those not quantized to zero will be one of the values that have a shorter variable-length codeword. Therefore, if a higher rate is required, the transform coder selects a quantizer with a smaller step size, and if a lower rate is required, the transform coder selects a quantizer with a larger step size. The quantizer step size is set at the beginning of each GOB, but can be changed at the beginning of any macroblock. If the rate cannot be lowered enough and there is a danger of buffer overflow, the more drastic option of dropping frames from transmission is used.

The ITU-T H.261 algorithm was primarily designed for videophone and videoconferencing applications. Therefore, the algorithm had to operate with minimal coding delay (less than 150 milliseconds). Furthermore, for videophone applications, the algorithm had to operate at very low bit rates. In fact, the title for the recommendation is "Video Codec for Audiovisual Services at $p \times 64$ kbit/s," where p takes on values from 1 to 30. A p value of 2 corresponds to a total transmission rate of 128 kbps, which is the same as two voice-band telephone channels. These are very low rates for video, and the ITU-T H.261 recommendations perform relatively well at these rates.

19.6 Model-Based Coding

In speech coding, a major decrease in rate is realized when we go from coding waveforms to an analysis/synthesis approach. An attempt at doing the same for video coding is described in the next section. A technique that has not yet reached maturity but shows great promise for use in videophone applications is an analysis/synthesis technique. The analysis/synthesis approach requires that the transmitter and receiver agree on a model for the information to be transmitted. The transmitter then analyzes the information to be transmitted and extracts the model parameters, which are transmitted to the receiver. The receiver uses these parameters to synthesize the source information. While this approach has been successfully used for speech compression for a long time (see Chapter 18), the same has not been true for images. In a delightful book, *Signals, Systems, and Noise—The Nature and Process of Communications*, published in 1961, J.R. Pierce [11] described his "dream" of an analysis/synthesis scheme for what we would now call a videoconferencing system:

> Imagine that we had at the receiver a sort of rubbery model of the human face. Or we might have a description of such a model stored in the memory of a huge electronic computer.... Then, as the person before the transmitter talked, the transmitter would have to follow the movements of his eyes, lips, and jaws, and other muscular movements and transmit these so that the model at the receiver could do likewise.

Pierce's dream is a reasonably accurate description of a three-dimensional model-based approach to the compression of facial image sequences. In this approach, a generic wireframe model, such as the one shown in Figure 19.13, is constructed using triangles. When encoding the movements of a specific human face, the model is adjusted to the face by matching features and the outer contour of the face. The image textures are then mapped onto this wireframe model to synthesize the face. Once this model is available to both transmitter and receiver, only changes in the face are transmitted to the receiver. These changes can be classified as *global motion* or *local motion* [254]. Global motion involves movement of the head, while local motion involves changes in the features—in other words, changes in facial expressions. The global motion can be modeled in terms of movements of rigid bodies. The facial expressions can be represented in terms of relative movements of the vertices of the triangles in the wireframe model. In practice, separating a movement into global and local components can be difficult because most points on the face will be affected by both the changing position of the head and the movement due to changes in facial expression. Different approaches have been proposed to separate these effects [255, 254, 256].

The global movements can be described in terms of rotations and translations. The local motions, or facial expressions, can be described as a sum of *action units* (AU), which are a set of 44 descriptions of basic facial expressions [257]. For example, AU1 corresponds to the raising of the inner brow and AU2 corresponds to the raising of the outer brow; therefore, AU1 + AU2 would mean raising the brow.

Although the synthesis portion of this algorithm is relatively straightforward, the analysis portion is far from simple. Detecting changes in features, which tend to be rather subtle, is a very difficult task. There is a substantial amount of research in this area, and if this problem is resolved, this approach promises rates comparable to the rates of the analysis/synthesis voice coding schemes. A good starting point for exploring this fascinating area is [258].

19.7 Asymmetric Applications

There are a number of applications in which it is cost effective to shift more of the computational burden to the encoder. For example, in multimedia applications where a video sequence is stored on a CD-ROM, the decompression will be performed many times and has to be performed in real time. However, the compression is performed only once, and there is no need for it to be in real time. Thus, the encoding algorithms can be significantly more complex. A similar situation arises in broadcast applications, where for each transmitter there might be thousands of receivers. In the following sections we will look at the standards developed for such asymmetric applications.

FIGURE 19.13 Generic wireframe model.

These standards have been developed by a joint committee of the International Standards Organization (ISO) and the International Electrotechnical Society (IEC), which is best known as MPEG (Moving Picture Experts Group). MPEG was initially set up in 1988 to develop a set of standard algorithms, at different rates, for applications that required storage of video and audio on digital storage media. Originally, the committee had three work items, nicknamed MPEG-1, MPEG-2, and MPEG-3, targeted at rates of 1.5, 10, and 40 Mbits per second, respectively. Later, it became clear that the algorithms developed for MPEG-2 would accommodate the MPEG-3 rates, and the third work item was dropped [259]. The MPEG-1 work item resulted in a set of standards, ISO/IEC IS 11172, "Information Technology—Coding of Moving Pictures and Associated Audio for Digital Storage Media Up to about 1.5 Mbit/s" [260]. During the development of the standard, the committee felt that the restriction to digital storage media was not necessary, and the set of standards developed under the second work item, ISO/IEC 13818 or MPEG-2, was issued under the title "Information Technology—Generic Coding of Moving Pictures and Associated Audio Information" [261]. In July 1993 the MPEG committee began working on MPEG-4, the third and most ambitious of its standards. The goal of MPEG-4 was to provide an object-oriented framework for the encoding of multimedia. It took two years for the committee to arrive at a satisfactory definition of the scope of MPEG-4, and the call for proposals was finally issued in 1996. The standard ISO/IEC 14496 was finalized in 1998 and approved as an international standard in 1999. We examined the audio standard in Chapter 18. In the following sections we look at the video standards.

19.8 The MPEG-1 Video Standard

The basic structure of the compression algorithm proposed by MPEG is very similar to that of ITU-T H.261. Blocks (8×8 in size) of either an original frame or the difference between a frame and the motion-compensated prediction are transformed using the DCT. The blocks are organized in macroblocks, which are defined in the same manner as in the H.261 algorithm, and the motion compensation is performed at the macroblock level. The transform coefficients are quantized and transmitted to the receiver. A buffer is used to smooth delivery of bits from the encoder and also for rate control.

The basic structure of the MPEG-1 compression scheme may be viewed as very similar to that of the ITU-T H.261 video compression scheme; however, there are significant differences in the details of this structure. The H.261 standard has videophone and videoconferencing as the main application areas; the MPEG standard at least initially had applications that require digital storage and retrieval as a major focus. This does not mean that use of either algorithm is precluded in applications outside its focus, but simply that the features of the algorithm may be better understood if we keep in mind the target application areas. In videoconferencing a call is set up, conducted, and then terminated. This set of events always occurs together and in sequence. When accessing video from a storage medium, we do not always want to access the video sequence starting from the first frame. We want the ability to view the video sequence starting at, or close to, some arbitrary point in the sequence. A similar situation exists in broadcast situations. Viewers do not necessarily tune into a program at the beginning. They may do so at any random point in time. In H.261 each frame, after the first frame, may contain blocks that are coded using predictions from the previous frame. Therefore, to decode a particular frame in the sequence, it is possible that we may have to decode the sequence starting at the first frame. One of the major contributions of MPEG-1 was the provision of a random access capability. This capability is provided rather simply by requiring that there be frames periodically that are coded without any reference to past frames. These frames are referred to as **I** frames.

In order to avoid a long delay between the time a viewer switches on the TV to the time a reasonable picture appears on the screen, or between the frame that a user is looking for and the frame at which decoding starts, the **I** frames should occur quite frequently. However, because the **I** frames do not use temporal correlation, the compression rate is quite low compared to the frames that make use of the temporal correlations for prediction. Thus, the number of frames between two consecutive **I** frames is a trade-off between compression efficiency and convenience.

In order to improve compression efficiency, the MPEG-1 algorithm contains two other kinds of frames, the *predictive coded* (**P**) frames and the *bidirectionally predictive coded* (**B**)

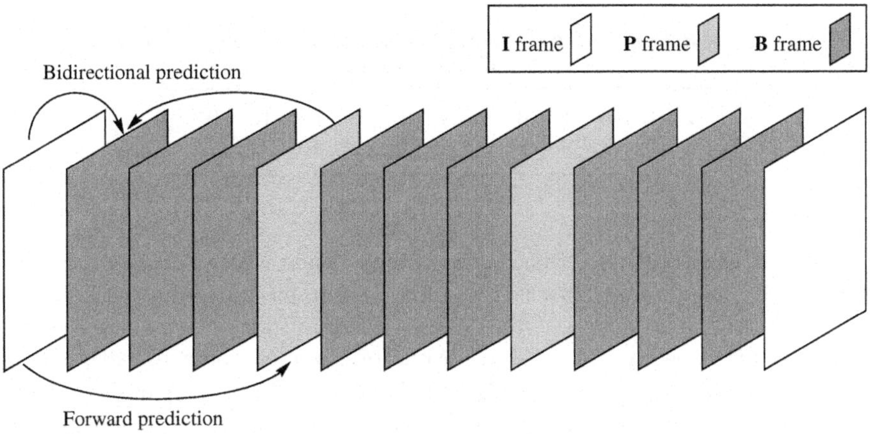

FIGURE 19.14 A possible arrangement for a group of pictures.

frames. The **P** frames are coded using motion-compensated prediction from the last **I** or **P** frame, whichever happens to be closest. Generally, the compression efficiency of **P** frames is substantially higher than **I** frames. The **I** and **P** frames are sometimes called *anchor* frames, for reasons that will become obvious.

To compensate for the reduction in the amount of compression due to the frequent use of **I** frames, the MPEG standard introduced **B** frames. The **B** frames achieve a high level of compression by using motion-compensated prediction from the most recent anchor frame and the closest future anchor frame. By using both past and future frames for prediction, generally we can get better compression than if we only used prediction based on the past. For example, consider a video sequence in which there is a sudden change between one frame and the next. This is a common occurrence in TV advertisements. In this situation, prediction based on the past frames may be useless. However, predictions based on future frames would have a high probability of being accurate. Note that a **B** frame can only be generated after the future anchor frame has been generated. Furthermore, the **B** frame is not used for predicting any other frame. This means that **B** frames can tolerate more error because this error will not be propagated by the prediction process.

The different frames are organized together in a *group of pictures* (GOP). A GOP is the smallest random access unit in the video sequence. The GOP structure is set up as a trade-off between the high compression efficiency of motion-compensated coding and the fast picture acquisition capability of periodic intra-only processing. As might be expected, a GOP has to contain at least one **I** frame. Furthermore, the first **I** frame in a GOP is either the first frame of the GOP, or is preceded by **B** frames that use motion-compensated prediction only from this **I** frame. A possible GOP is shown in Figure 19.14.

Because of the reliance of the **B** frame on future anchor frames, there are two different sequence orders. The *display order* is the sequence in which the video sequence is displayed to the user. A typical display order is shown in Table 19.5. Let us see how this sequence was generated. The first frame is an **I** frame, which is compressed without reference to any previous frame. The next frame to be compressed is the fourth frame. This frame is compressed using

TABLE 19.5 A typical sequence of frames in display order.

I	B	B	P	B	B	P	B	B	P	B	B	I
1	2	3	4	5	6	7	8	9	10	11	12	13

motion-compensated prediction from the first frame. Then we compress frame 2, which is compressed using motion-compensated prediction from frame 1 and frame 4. The third frame is also compressed using motion-compensated prediction from the first and fourth frames. The next frame to be compressed is frame 7, which uses motion-compensated prediction from frame 4. This is followed by frames 5 and 6, which are compressed using motion-compensated predictions from frames 4 and 7. Thus, there is a processing order that is quite different from the display order. The MPEG document calls this the *bitstream order*. The bitstream order for the sequence shown in Table 19.5 is given in Table 19.6. In terms of the bitstream order, the first frame in a GOP is always the **I** frame.

As we can see, unlike the ITU-T H.261 algorithm, the frame being predicted and the frame upon which the prediction is based are not necessarily adjacent. In fact, the number of frames between the frame being encoded and the frame upon which the prediction is based is variable. When searching for the best matching block in a neighboring frame, the region of search depends on assumptions about the amount of motion. More motion will lead to larger search areas than a small amount of motion. When the frame being predicted is always adjacent to the frame upon which the prediction is based, we can fix the search area based on our assumption about the amount of motion. When the number of frames between the frame being encoded and the prediction frame is variable, we make the search area a function of the distance between the two frames. While the MPEG standard does not specify the method used for motion compensation, it does recommend using a search area that grows with the distance between the frame being coded and the frame being used for prediction.

Once motion compensation has been performed, the block of prediction errors is transformed using the DCT and quantized, and the quantization labels are encoded. This procedure is the same as that recommended in the JPEG standard and is described in Chapter 13. The quantization tables used for the different frames are different and can be changed during the encoding process.

Rate control in the MPEG standard can be performed at the sequence level or at the level of individual frames. At the sequence level, any reduction in bit rate first occurs with the **B** frames because they are not essential for the encoding of other frames. At the level of the individual frames, rate control takes place in two steps. First, as in the case of the H.261

TABLE 19.6 A typical sequence of frames in bitstream order.

I	P	B	B	P	B	B	P	B	B	I	B	B
1	4	2	3	7	5	6	10	8	9	13	11	12

algorithm, the quantizer step sizes are increased. If this is not sufficient, then the higher-order frequency coefficients are dropped until the need for rate reduction is past.

The format for MPEG is very flexible. However, the MPEG committee has provided some suggested values for the various parameters. For MPEG-1 these suggested values are called the *constrained parameter bitstream* (CPB). The horizontal picture size is constrained to be less than or equal to 768 pixels, and the vertical size is constrained to be less than or equal to 576 pixels. More importantly, the pixel rate is constrained to be less than 396 macroblocks per frame if the frame rate is 25 frames per second or less, and 330 macroblocks per frame if the frame rate is 30 frames per second or less. The definition of a macroblock is the same as in the ITU-T H.261 recommendations. Therefore, this corresponds to a frame size of 352×288 pixels at the 25-frames-per-second rate, or a frame size of 352×240 pixels at the 30-frames-per-second rate. Keeping the frame at this size allows the algorithm to achieve bit rates of between 1 and 1.5 Mbits per second. When referring to MPEG-1 parameters, most people are actually referring to the CPB.

The MPEG-1 algorithm provides reconstructed images of VHS quality for moderate- to low-motion video sequences, and worse than VHS quality for high-motion sequences at rates of around 1.2 Mbits per second. As the algorithm was targeted to applications such as CD-ROM, there is no consideration of interlaced video. In order to expand the applicability of the basic MPEG algorithm to interlaced video, the MPEG committee provided some additional recommendations, the MPEG-2 recommendations.

19.9 The MPEG-2 Video Standard—H.262

While MPEG-1 was specifically proposed for digital storage media, the idea behind MPEG-2 was to provide a generic, application-independent standard. To this end, MPEG-2 takes a "tool kit" approach, providing a number of subsets, each containing different options from the set of all possible options contained in the standard. For a particular application, the user can select from a set of *profiles* and *levels*. The profiles define the algorithms to be used, while the levels define the constraints on the parameters. There are five profiles: *simple, main, snr-scalable* (where *snr* stands for signal-to-noise ratio), *spatially scalable*, and *high*. There is an ordering of the profiles; each higher profile is capable of decoding video encoded using all profiles up to and including that profile. For example, a decoder designed for profile *snr-scalable* could decode video that was encoded using profiles *simple, main*, and *snr-scalable*. The *simple* profile eschews the use of **B** frames. Recall that the **B** frames require the most computation to generate (forward and backward prediction), require memory to store the coded frames needed for prediction, and increase the coding delay because of the need to wait for "future" frames for both generation and reconstruction. Therefore, removal of the **B** frames makes the requirements simpler. The *main* profile is very much the algorithm we have discussed in the previous section. The *snr-scalable, spatially scalable*, and *high* profiles may use more than one bitstream to encode the video. The base bitstream is a lower-rate encoding of the video sequence. This bitstream could be decoded by itself to provide a reconstruction of the video sequence. The other bitstream is used to enhance the quality of the reconstruction. This layered approach is useful when transmitting video over a network, where some connections may only permit a lower rate. The base bitstream can be provided to these connections while providing the base and enhancement layers for a higher-quality reproduction over the links that can accommodate the higher bit rate. To understand the concept of layers, consider the following example.

Example 19.9.1:

Suppose after the transform we obtain a set of coefficients θ_{ij}, the first eight of which are

$$29.75 \quad 6.1 \quad -6.03 \quad 1.93 \quad -2.01 \quad 1.23 \quad -0.95 \quad 2.11$$

Let us suppose we quantize this set of coefficients using a step size of 4. For simplicity we will use the same step size for all coefficients. Recall that the quantizer label is given by

$$l_{ij} = \left\lfloor \frac{\theta_{ij}}{Q^t_{ij}} + 0.5 \right\rfloor \tag{8}$$

and the reconstructed value is given by

$$\hat{\theta}_{ij} = l_{ij} \times Q^t_{ij} \tag{9}$$

Using these equations and the fact that $Q^t_{ij} = 4$, the reconstructed values of the coefficients are

$$28 \quad 8 \quad -8 \quad 0 \quad -4 \quad 0 \quad -0 \quad 4$$

The error in the reconstruction is

$$1.75 \quad -1.9 \quad 1.97 \quad 1.93 \quad 1.99 \quad 1.23 \quad -0.95 \quad -1.89$$

Now suppose we have some additional bandwidth made available to us. We can quantize the difference and send that to enhance the reconstruction. Suppose we used a step size of 2 to quantize the difference. The reconstructed values for this enhancement sequence would be

$$2 \quad -2 \quad 2 \quad 2 \quad 2 \quad 2 \quad 0 \quad -2$$

Adding this to the previous base-level reconstruction, we get an enhanced reconstruction of

$$30 \quad 6 \quad -6 \quad 2 \quad -2 \quad 2 \quad 0 \quad 2$$

which results in an error of

$$-0.25 \quad 0.1 \quad -0.03 \quad -0.07 \quad -0.01 \quad -0.77 \quad -0.95 \quad 0.11$$

The layered approach allows us to increase the accuracy of the reconstruction when bandwidth is available, while at the same time permitting a lower-quality reconstruction when there is not sufficient bandwidth for the enhancement. In other words, the quality is *scalable*. In this particular case, the error between the original and reconstruction decreases because of the enhancement. Because the signal-to-noise ratio is a measure of error, this can be called *snr-scalable*. If the enhancement layer contains a coded bitstream corresponding to frames that occur between frames of the base layer, the system can be called *temporally scalable*. If the enhancement allows an upsampling of the base layer, the system is *spatially scalable*. ♦

TABLE 19.7 Allowable profile-level combinations in MPEG-2.

	Simple Profile	Main Profile	SNR-Scalable Profile	Spatially Scalable Profile	High Profile
High Level		Allowed			Allowed
High 1440		Allowed		Allowed	Allowed
Main Level	Allowed	Allowed	Allowed		Allowed
Low Level		Allowed	Allowed		

The levels are *low, main, high 1440*, and *high*. The *low* level corresponds to a frame size of 352 × 240, the *main* level corresponds to a frame size of 720 × 480, the *high 1440* level corresponds to a frame size of 1440 × 1152, and the *high* level corresponds to a frame size of 1920 × 1080. All levels are defined for a frame rate of 30 frames per second. There are many possible combinations of profiles and levels, not all of which are allowed in the MPEG-2 standard. Table 19.7 shows the allowable combinations [259]. A particular profile-level combination is denoted by *XX@YY* where *XX* is the two-letter abbreviation for the profile and *YY* is the two-letter abbreviation for the level. There are a large number of issues, such as bounds on parameters and decodability between different profile-level combinations, that we have not addressed here because they do not pertain to our main focus, compression (see the international standard [261] for these details).

Because MPEG-2 has been designed to handle interlaced video, there are field-based alternatives to the **I**, **P**, and **B** frames. The **P** and **B** frames can be replaced by two **P** fields or two **B** fields. The **I** frame can be replaced by two **I** fields or an **I** field and a **P** field where the **P** field is obtained by predicting the bottom field by the top field. Because an 8 × 8 field block actually covers twice the spatial distance in the vertical direction as an 8 frame block, the zigzag scanning is adjusted to adapt to this imbalance. The scanning pattern for an 8 × 8 field block is shown in Figure 19.15.

The most important addition from the point of view of compression in MPEG-2 is the addition of several new motion-compensated prediction modes: the field prediction and the dual prime prediction modes. MPEG-1 did not allow interlaced video. Therefore, there was no need for motion-compensation algorithms based on fields. In the **P** frames, field predictions are obtained using one of the two most recently decoded fields. When the first field in a frame is being encoded, the prediction is based on the two fields from the previous frame. However, when the second field is being encoded, the prediction is based on the second field from the previous frame and the first field from the current frame. Information about which field is to be used for prediction is transmitted to the receiver. The field predictions are performed in a manner analogous to the motion-compensated prediction described earlier.

In addition to the regular frame and field prediction, MPEG-2 also contains two additional modes of prediction. One is 16 × 8 motion compensation. In this mode, two predictions are generated for each macroblock, one for the top half and one for the bottom half. The other is called dual prime motion compensation. In this technique, two predictions are formed for each field from the two recent fields. These predictions are averaged to obtain the final prediction.

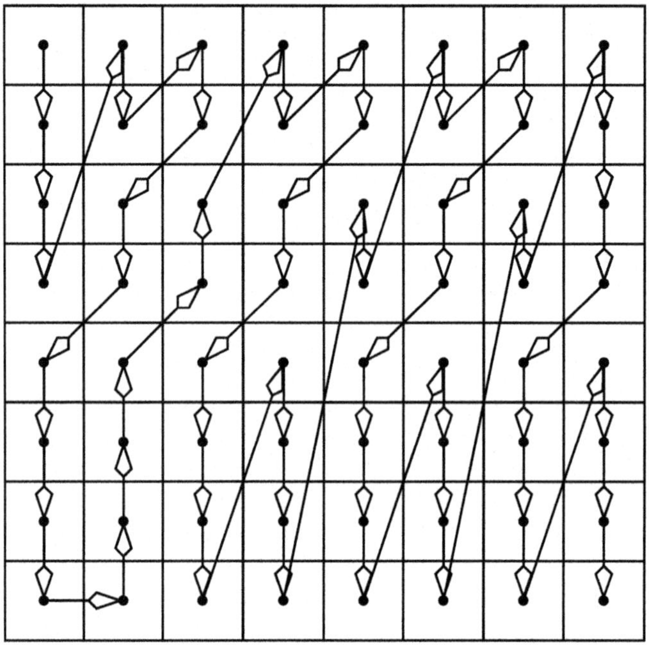

FIGURE 19.15 Scanning pattern for the DCT coefficients of a field block.

When the Federal Communications Commission (FCC) requested proposals for the HDTV standard, they received four proposals for digital HDTV from four consortia. After the evaluation phase, the FCC declined to pick a winner among the four, and instead suggested that all these consortia join forces and produce a single proposal. The resulting partnership had the exalted title of the "Grand Alliance." The Grand Alliance system used the *main* profile of the MPEG-2 standard implemented at the *high* level.

19.10 ITU-T Recommendation H.263

The H.263 standard was developed to update the H.261 video conferencing standard with the experience acquired in the development of the MPEG and H.262 algorithms. The initial algorithm provided incremental improvement over H.261. After the development of the core algorithm, several optional updates were proposed that significantly improved compression performance. The standard with these optional components is sometimes referred to as H.263+ (or H.263 + +).

In the following sections we first describe the core algorithm and then describe some of the options. The standard focuses on noninterlaced video. The different picture formats addressed

TABLE 19.8 The standardized H.263 formats.

Picture format	Number of luminance pixels (columns)	Number of luminance lines (rows)	Number of chrominance pixels (columns)	Number of chrominance lines (rows)
sub-QCIF	128	96	64	48
QCIF	176	144	88	72
CIF	352	288	176	144
4CIF	704	576	352	288
16CIF	1408	1152	704	576

by the standard are shown in Table 19.8. The picture is divided into *groups of blocks* (GOBs) or slices. A group of blocks is a strip of pixels across the picture with a height that is a multiple of 16 lines. The number of multiples depends on the size of the picture, and the bottommost GOB may have less than 16 lines. Each GOB is divided into macroblocks, which are defined as in the H.261 recommendation.

A block diagram of the baseline video coder is shown in Figure 19.16. It is very similar to Figure 19.10, the block diagram for the H.261 encoder. The only major difference is the ability to work with both predicted or **P** frames and intra or **I** frames. As in the case of H.261, the motion-compensated prediction is performed on a macroblock basis. The vertical and horizontal components of the motion vector are restricted to the range $[-16, 15.5]$. The transform used for representing the prediction errors in the case of the **P** frame and the pixels in the case of the **I** frames is the discrete cosine transform. The transform coefficients are quantized using uniform midtread quantizers. The DC coefficient of the intra block is quantized using a uniform quantizer with a step size of 8. There are 31 quantizers available for the quantization of all other coefficients with step sizes ranging from 2 to 62. Apart from the DC coefficient of the intra block, all coefficients in a macroblock are quantized using the same quantizer.

The motion vectors are differentially encoded. The prediction is the median of the motion vectors in the neighboring blocks. The H.263 recommendation allows half pixel motion compensation as opposed to only integer pixel compensation (which is all that is allowed in H.261). Notice that the sign of the component is encoded in the last bit of the variable length code, a "0" for positive values and a "1" for negative values. Two values that differ only in their sign differ only in the least significant bit.

The code for the quantized transform coefficients is indexed by three indicators. The first indicates whether the coefficient being encoded is the last nonzero coefficient in the zigzag scan. The second indicator is the number of zero coefficients preceding the coefficient being encoded, and the last indicates the absolute value of the quantized coefficient level. The sign bit is appended as the last bit of the variable-length code.

Here we describe some of the optional modes of the H.263 recommendation. The first four options were part of the initial H.263 specification. The remaining options were added later and the resulting standard is sometimes referred to as the H.263+ standard.

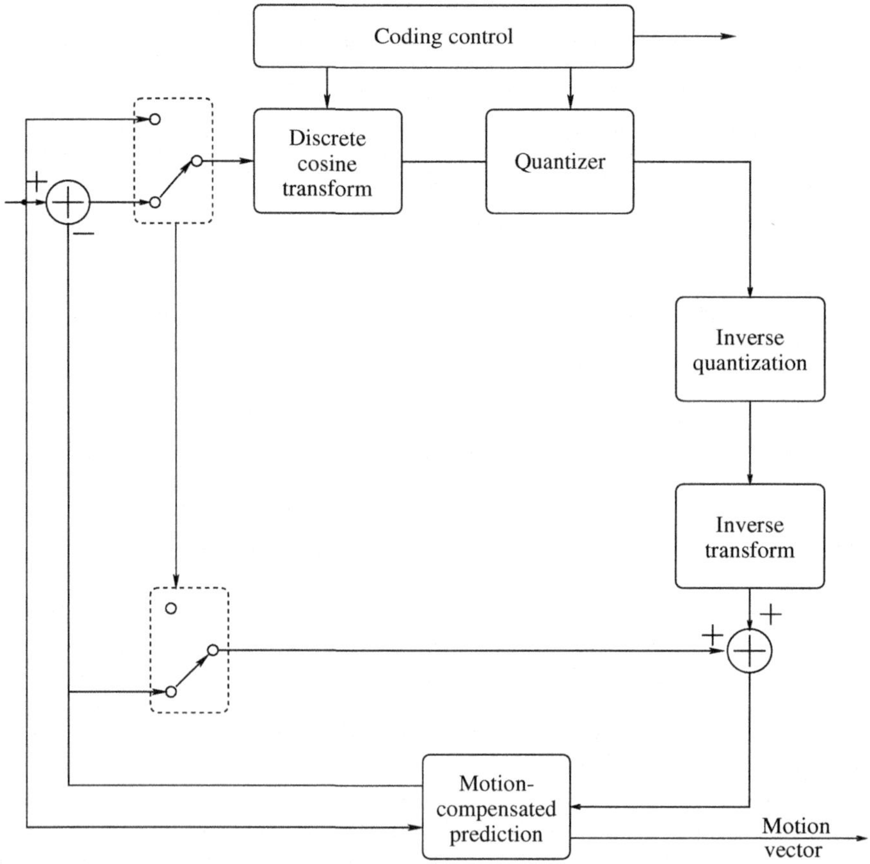

FIGURE 19.16 **A block diagram of the H.263 video compression algorithm.**

19.10.1 Unrestricted Motion Vector Mode

In this mode the motion vector range is extended to $[-31.5, 31.5]$, which is particularly useful in improving the compression performance of the algorithm for larger picture sizes. The mode also allows motion vectors to point outside the picture. This is done by repeating the edge pixels to create the picture beyond its boundary.

19.10.2 Syntax-Based Arithmetic Coding Mode

In this mode the variable-length codes are replaced with an arithmetic coder. The word length for the upper and lower limits is 16. The option specifies several different Cum_Count tables that can be used for arithmetic coding. There are separate Cum_Count tables for encoding motion vectors, intra DC components, and intra and inter coefficients.

19.10.3 Advanced Prediction Mode

In the baseline mode a single motion vector is sent for each macroblock. Recall that a macroblock consists of four 8×8 luminance blocks and two chrominance blocks. In the advanced prediction mode the encoder can send four motion vectors, one for each luminance block. The chrominance motion vectors are obtained by adding the four luminance motion vectors and dividing by 8. The resulting values are adjusted to the nearest half pixel position. This mode also allows for *Overlapped Block Motion Compensation (OBMC)*. In this mode the motion vector is obtained by taking a weighted sum of the motion vector of the current block and two of the four vertical and horizontal neighboring blocks.

19.10.4 PB-Frames and Improved PB-Frames Mode

The PB frame consists of a **P** picture and a **B** picture in the same frame. The blocks for the **P** frame and the **B** frame are interleaved so that a macroblock consists of six blocks of a **P** picture followed by six blocks of a **B** picture. The motion vector for the **B** picture is derived from the motion vector for the **P** picture by taking into account the time difference between the **P** picture and the **B** picture. If the motion cannot be properly derived, a delta correction is included. The improved PB-frame mode updates the PB-frame mode to include forward, backward, and bidirectional prediction.

19.10.5 Advanced Intra Coding Mode

The coefficients for the **I** frames are obtained directly by transforming the pixels of the picture. As a result, there can be significant correlation between some of the coefficients of neighboring blocks. For example, the DC coefficient represents the average value of a block. It is very likely that the average value will not change significantly between blocks. The same may be true, albeit to a lesser degree, for the low-frequency horizontal and vertical coefficients. The advanced intra coding mode allows the use of this correlation by using coefficients from neighboring blocks for predicting the coefficients of the block being encoded. The prediction errors are then quantized and coded.

When this mode is used, the quantization approach and variable-length codes have to be adjusted to adapt to the different statistical properties of the prediction errors. Furthermore, it might also become necessary to change the scan order. The recommendation provides alternate scanning patterns as well as alternate variable length codes and quantization strategies.

19.10.6 Deblocking Filter Mode

This mode is used to remove blocking effects from the 8×8 block edges. This smoothing of block boundaries allows for better prediction. This mode also permits the use of four motion vectors per macroblock and motion vectors that point beyond the edge of the picture.

19.10.7 Reference Picture Selection Mode

This mode is used to prevent error propagation by allowing the algorithm to use a picture other than the previous picture to perform prediction. The mode permits the use of a back-channel that the decoder uses to inform the encoder about the correct decoding of parts of the picture. If a part of the picture is not correctly decoded, it is not used for prediction. Instead, an alternate frame is selected as the reference frame. The information about which frame was selected as the reference frame is transmitted to the decoder. The number of possible reference frames is limited by the amount of frame memory available.

19.10.8 Temporal, SNR, and Spatial Scalability Mode

This is very similar to the scalability structures defined earlier for the MPEG-2 algorithm. Temporal scalability is achieved by using separate **B** frames, as opposed to the PB frames. SNR scalability is achieved using the kind of layered coding described earlier. Spatial scalability is achieved using upsampling.

19.10.9 Reference Picture Resampling

Reference picture resampling allows a reference picture to be "warped" in order to permit the generation of better prediction. It can be used to adaptively alter the resolution of pictures during encoding.

19.10.10 Reduced-Resolution Update Mode

This mode is used for encoding highly active scenes. The macroblock in this mode is assumed to cover an area twice the height and width of the regular macroblock. The motion vector is assumed to correspond to this larger area. Using this motion vector a predicted macroblock is created. The transform coefficients are decoded and then upsampled to create the expanded texture block. The predicted and texture blocks are then added to obtain the reconstruction.

19.10.11 Alternative Inter VLC Mode

The variable-length codes for inter and intra frames are designed with different assumptions. In the case of the inter frames it is assumed that the values of the coefficients will be small and there can be large numbers of zero values between nonzero coefficients. This is a result of prediction that, if successfully employed, reduces the magnitude of the differences, and hence the coefficients, and also leads to large numbers of zero-valued coefficients. Therefore, coefficients indexed with large runs and small coefficient values are assigned shorter codes. In the case of the intra frames, the opposite is generally true. There is no prediction, therefore there is a much smaller probability of runs of zero-valued coefficients. Also, large-valued coefficients are quite possible. Therefore, coefficients indexed by small run values and larger coefficient values are assigned shorter codes. During periods of increased temporal activity, prediction is generally not as good and therefore the assumptions under which the variable-length codes for the inter frames were created are violated. In these situations it is likely that the variable-length codes designed for the intra frames are a better match. The alternative inter VLC mode allows for the use of the intra codes in these sitations, improving compression performance. Note that the codewords used in intra and inter frame coding are the same. What is different is the interpretation. To detect the proper interpretation, the decoder first decodes the block assuming an inter frame codebook. If the decoding results in more than 64 coefficients it switches its interpretation.

19.10.12 Modified Quantization Mode

In this mode, along with changes in the signalling of changes in quantization parameters, the quantization process is improved in several ways. In the baseline mode, both the luminance and chrominance components in a block are quantized using the same quantizer. In the modified quantization mode, the quantizer used for the luminance coefficients is different from the quantizer used for the chrominance component. This allows the quantizers to be more closely matched to the statistics of the input. The modified quantization mode also allows for the quantization of a wider range of coefficient values, preventing significant overload. If the coefficient exceeds the range of the baseline quantizer, the encoder sends an escape symbol followed by an 11-bit representation of the coefficient. This relaxation of the structured representation of the quantizer outputs makes it more likely that bit errors will be accepted as valid quantizer outputs. To reduce the chances of this happening, the mode prohibits "unreasonable" coefficient values.

19.10.13 Enhanced Reference Picture Selection Mode

Motion-compensated prediction is accomplished by searching the previous picture for a block similar to the block being encoded. The enhanced reference picture selection mode allows the encoder to search more than one picture to find the best match and then use the best-suited picture to perform motion-compensated prediction. Reference picture selection can be accomplished on a macroblock level. The selection of the pictures to be used for motion compensation can be performed in one of two ways. A sliding window of M pictures can be used and the last M decoded, with reconstructed pictures stored in a multipicture buffer. A more complex adaptive memory (not specified by the standard) can also be used in place of the simple sliding window. This mode significantly enhances the prediction, resulting in a reduction in the rate for equivalent quality. However, it also increases the computational and memory requirements on the encoder. This memory burden can be mitigated to some extent by assigning an unused label to pictures or portions of pictures. These pictures, or portions of pictures, then do not need to be stored in the buffer. This unused label can also be used as part of the adaptive memory control to manage the pictures that are stored in the buffer.

19.11 ITU-T Recommendation H.264, MPEG-4 Part 10, Advanced Video Coding

As described in the previous section, the H.263 recommendation started out as an incremental improvement over H.261 and ended up with a slew of optional features, which in fact make the improvement over H.261 more than incremental. In H.264 we have a standard that started out with a goal of significant improvement over the MPEG-1/2 standards and achieved those goals. The standard, while initiated by ITU-T's Video Coding Experts Group (VCEG), ended up being a collaboration of the VCEG and ISO/IEC's MPEG committees, which joined to form the Joint Video Team (JVT) in December 2001 [262]. The collaboration of various groups in the development of this standard has also resulted in the richness of names. It is variously known as ITU-T H.264, MPEG-4 Part 10, MPEG-4 Advanced Video Coding (AVC), as well as the name under which it started its life, H.26L. We will just refer to it as H.264.

The basic block diagram looks very similar to the previous schemes. There are intra and inter pictures. The inter pictures are obtained by subtracting a motion-compensated prediction from the original picture. The residuals are transformed into the frequency domain. The transform coefficients are scanned, quantized, and encoded using variable-length codes. A local decoder reconstructs the picture for use in future predictions. The intra picture is coded without reference to past pictures.

While the basic block diagram is very similar to the previous standards the details are quite different. We will look at these details in the following sections. We begin by looking at the basic structural elements, then look at the decorrelation of the inter frames. The decorrelation process includes motion-compensated prediction and transformation of the prediction error. We then look at the decorrelation of the intra frames. This includes intra prediction modes and transforms used in this mode. We finally look at the different binary coding options.

The macroblock structure is the same as used in the other standards. Each macroblock consists of four 8×8 luminance blocks and two chrominance blocks. An integer number of sequential macroblocks can be put together to form a slice. In the previous standards the smallest subdivision of the macroblock was into its 8×8 component blocks. The H.264 standard allows 8×8 macroblock partitions to be further divided into sub-macroblocks of size 8×4, 4×8, and 4×4. These smaller blocks can be used for motion-compensated prediction, allowing for tracking of much finer details than is possible with the other standards. Along with the 8×8 partition, the macroblock can also be partitioned into two 8×16 or 16×8 blocks. In field mode the H.264 standard groups 16×8 blocks from each field to form a 16×16 macroblock.

19.11.1 Motion-Compensated Prediction

The H.264 standard uses its macroblock partitions to develop a tree-structured motion compensation algorithm. One of the problems with motion-compensated prediction has always been the selection of the size and shape of the block used for prediction. Different parts of a video scene will move at different rates in different directions or stay put. A smaller-size block allows tracking of diverse movement in the video frame, leading to better prediction and hence lower bit rates. However, more motion vectors need to be encoded and transmitted, using up valuable bit resources. In fact, in some video sequences the bits used to encode the motion vectors may make up most of the bits used. If we use small blocks, the number of motion vectors goes up, as does the bit rate. Because of the variety of sizes and shapes available to it, the H.264 algorithm provides a high level of accuracy and efficiency in its prediction. It uses small block sizes in regions of activity and larger block sizes in stationary regions. The availability of rectangular shapes allows the algorithm to focus more precisely on regions of activity.

The motion compensation is accomplished using quarter-pixel accuracy. To do this the reference picture is "expanded" by interpolating twice between neighboring pixels. This results in a much smoother residual. The prediction process is also enhanced by the use of filters on the four block edges. The standard allows for searching of up to 32 pictures to find the best matching block. The selection of the reference picture is done on the macroblock partion level, so all sub-macroblock partitions use the same reference picture.

As in H.263, the motion vectors are differentially encoded. The basic scheme is the same. The median values of the three neighboring motion vectors are used to predict the current motion vector. This basic strategy is modified if the block used for motion compensation is a 16×16, 16×8, or 8×16 block.

For **B** pictures, as in the case of the previous standards, two motion vectors are allowed for each macroblock or sub-macroblock partition. The prediction for each pixel is the weighted average of the two prediction pixels.

Finally, a $\mathbf{P}_{skip}$ type macroblock is defined for which 16×16 motion compensation is used and the prediction error is not transmitted. This type of macroblock is useful for regions of little change as well as for slow pans.

19.11.2 The Transform

Unlike the previous video coding schemes, the transform used is not an 8×8 DCT. For most blocks the transform used is a 4×4 integer DCT-like matrix. The transform matrix is given by

$$H = \begin{bmatrix} 1 & 1 & 1 & 1 \\ 2 & 1 & 1 & 2 \\ 1 & -1 & -1 & 1 \\ 1 & -2 & 2 & -1 \end{bmatrix}$$

The inverse transform matrix is given by

$$H^I = \begin{bmatrix} 1 & 1 & 1 & \frac{1}{2} \\ 1 & \frac{1}{2} & -1 & -1 \\ 1 & -\frac{1}{2} & -1 & 1 \\ 1 & -1 & 1 & -\frac{1}{2} \end{bmatrix}$$

The transform operations can be implemented using addition and shifts. Multiplication by 2 is a single-bit left shift and division by 2 is a single-bit right shift. However, there is a price for the simplicity. Notice that the norm of the rows is not the same and the product of the forward and inverse transforms does not result in the identity matrix. This discrepancy is compensated for by the use of scale factors during quantization. There are several advantages to using a smaller integer transform. The integer nature makes the implementation simple and also avoids error accumulation in the transform process. The smaller size allows better representation of small stationary regions of the image. The smaller blocks are less likely to contain a wide range of values. Where there are sharp transitions in the blocks, any ringing effect is contained within a small number of pixels.

19.11.3 Intra Prediction

In the previous standards the **I** pictures were transform coded without any decorrelation. This meant that the number of bits required for the **I** frames was substantially higher than for the other pictures. When asked why he robbed banks, the notorious robber Willie Sutton is supposed to have said simply, "because that's where the money is." Because most of the bits in video coding are expended in encoding the **I** frame, it made a lot of sense for the JVT to look at improving the compression of the **I** frame in order to substantially reduce the bitrate.

The H.264 standard contains a number of spatial prediction modes. For 4×4 blocks there are nine prediction modes. Eight of these are summarized in Figure 19.17. The 16 pixels in the block $a - p$ are predicted using the 13 pixels on the boundary (and extending from it).[1] The arrows corresponding to the mode numbers show the direction of prediction. For example, mode 0 corresponds to the downward pointing arrow. In this case pixel A is used to predict pixels a, e, i, m, pixel B is used to predict pixels b, f, j, n, pixel C is used to predict pixels c, g, k, o, and pixel D is used to predict pixels d, h, l, p. In mode 3, also called the diagonal down/left mode pixel, B is used to predict pixel a, pixel C is used to predict pixels b, e, pixel D is used to predict pixels c, f, i, pixel E is used to predict pixels d, g, j, m, pixel F is used to predict pixels h, k, n, pixel G is used to predict pixels l, o, and pixel H is used to predict pixel p. If pixels E, F, G, and H are not available, pixel D is repeated four times. Notice that no direction is available for mode 2. This is called the DC mode, in which the average of the left and top boundary pixels is used as a prediction for all 16 pixels in the 4×4 block. In most cases the prediction modes used for all the 4×4 blocks in a macroblock are heavily correlated. The standard uses this correlation to efficiently encode the mode information.

In smooth regions of the picture it is more convenient to use prediction on a macroblock basis. In the case of a full macroblock there are four prediction modes. Three of them correspond to modes 0, 1, and 2 (vertical, horizontal, and DC). The fourth prediction mode is called the planar prediction mode. In this mode a three-parameter plane is fitted to the pixel values of the macroblock.

19.11.4 Quantization

The H.264 standard uses a uniform scalar quantizer for quantizing the coefficients $\theta_{i,j}$. There are 52 scalar quantizers indexed by Q_{step}. The step size doubles for every sixth Q_{step}. The quantization incorporates scaling necessitated by the approximations used to make the transform simple. If $\alpha_{i,j}(Q_{step})$ are the weights for the (i, j)th coefficient then the quantization

[1] The jump from pixel L to Q is a historical artifact. In an earlier version of the standard, pixels below L were also used in some prediction modes.

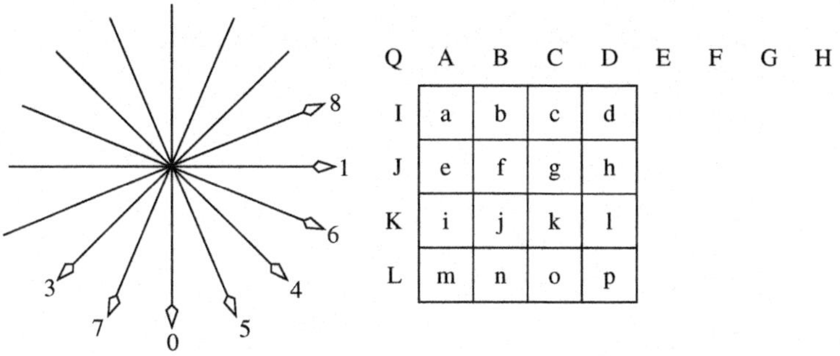

FIGURE 19.17 Prediction modes for 4 × 4 intra prediction.

label is given by

$$l_{i,j} = sign(\theta_{i,j})\left\lfloor \frac{|\theta_{i,j}|\alpha_{i,j}(Q_{step})}{Q_{step}} \right\rfloor$$

In order to broaden the quantization interval around the origin we add a small value in the numerator:

$$l_{i,j} = sign(\theta_{i,j})\left\lfloor \frac{|\theta_{i,j}|\alpha_{i,j}(Q_{step}) + f(Q_{step})}{Q_{step}} \right\rfloor$$

In actual implementation we do away with divisions and the quantization is implemented as [262]

$$l_{i,j} = sign(\theta_{i,j})[|\theta_{i,j}|M(Q_M, r) + f2^{17+Q_E}] >> 17 + Q_E$$

where

$$Q_M = Q_{step}(mod\ 6)$$
$$Q_E = \left\lfloor \frac{Q_{step}}{6} \right\rfloor$$
$$r = \begin{cases} 0 & i, j \text{ even} \\ 1 & i, j \text{ odd} \\ 2 & \text{otherwise} \end{cases}$$

$>>$ denotes a binary right-shift, and M is given in Table 19.9.

The inverse quantization is given by

$$\hat{\theta}_{i,j} = l_{i,j}S(Q_M, r) \ll Q_E$$

where $<<$ denotes a left-shift, and S is given in Table 19.10

Prior to quantization, the transforms of the 16 × 16 luminance residuals and the 8 × 8 chrominance residuals of the macroblock-based intra prediction are processed to further remove redundancy. Recall that macroblock-based prediction is used in smooth regions of the **I** picture. Therefore, it is very likely that the DC coefficients of the 4 × 4 transforms are heavily correlated. To remove this redundancy, a discrete Walsh-Hadamard transform is used on the

TABLE 19.9 *$M(Q_M, r)$ values in H.264.*

Q_M	$r = 0$	$r = 1$	$r = 2$
0	13107	5243	8066
1	11916	4660	7490
2	10082	4194	6554
3	9362	3647	5825
4	8192	3355	5243
5	7282	2893	4559

TABLE 19.10 *$S(Q_M, r)$ values in H.264.*

Q_M	$r = 0$	$r = 1$	$r = 2$
0	10	16	13
1	11	18	14
2	13	20	16
3	14	23	18
4	16	25	20
5	18	29	23

DC coefficients in the macroblock. In the case of the luminance block, this is a 4×4 transform for the 16 DC coefficients. The smaller chrominance block contains four DC coefficients, so we use a 2×2 discrete Walsh-Hadamard transform.

19.11.5 Coding

The H.264 standard contains two options for binary coding. The first uses exponential Golomb codes to encode the parameters and a context-adaptive variable-length code (CAVLC) to encode the quantizer labels [262]. The second binarizes all the values and then uses a context-adaptive binary arithmetic code (CABAC) [44].

An exponential Golomb code for a positive number x can be obtained as the unary code for $M = \lfloor \log_2(x+1) \rfloor$ concatenated with the M-bit natural binary code for $x + 1$. The unary code for a number x is given as x zeros followed by a 1. The exponential Golomb code for zero is 1.

The quantizer labels are first scanned in a zigzag fashion. In many cases the last nonzero labels in the zigzag scan have a magnitude of 1. The number N of nonzero labels and the number T of trailing 1s are used as an index into a codebook that is selected based on the values of N and T for the neighboring blocks. The maximum allowed value of T is 3. If the number of trailing labels with a magnitude of 1 is greater than 3, the remaining are encoded in the same manner as the other nonzero labels. The nonzero labels are then coded in reverse order. That is, the quantizer labels corresponding to the higher-frequency coefficients are encoded first. First the signs of the trailing 1s are encoded with 0s signifying positive values and 1s signifying negative values. Then the remaining quantizer labels are encoded in reverse scan order. After this, the total number of 0s in the scan between the beginning of the scan and the last nonzero label is encoded. This will be a number between 0 and $16 - N$. Then the run of zeros before each label starting with the last nonzero label is encoded until we run out of zeros or coefficients. The number of bits used to code each zero run will depend on the number of zeros remaining to be assigned.

In the second technique, which provides higher compression, all values are first converted to binary strings. This binarization is performed, depending on the data type, using unary codes, truncated unary codes, exponential Golomb codes, and fixed-length codes, plus five specific binary trees for encoding macroblock and sub-macroblock types. The binary string is encoded in one of two ways. Redundant strings are encoded using a context-adaptive binary arithmetic code. Binary strings that are random, such as the suffixes of the exponential Golomb codes, bypass the arithmetic coder. The arithmetic coder has 399 contexts available to it, with 325 of these contexts used for encoding the quantizer labels. These numbers include contexts for both frame and field slices. In a pure frame or field slice only 277 of the 399 context models are used. These context models are simply Cum_Count tables for use with a binary arithmetic coder. The H.264 standard recommends a multiplication-free implementation of binary arithmetic coding.

The H.264 standard is substantially more flexible than previous standards, with a much broader range of applications. In terms of performance, it claims a 50% reduction in bit rate over previous standards for equivalent perceptual quality [262].

19.12 MPEG-4 Part 2

The MPEG-4 standard provides a more abstract approach to the coding of multimedia. The standard views a multimedia "scene" as a collection of objects. These objects can be visual, such as a still background or a talking head, or aural, such as speech, music, background noise, and so on. Each of these objects can be coded independently using different techniques to generate separate elementary bitstreams. These bitstreams are multiplexed along with a scene description. A language called the Binary Format for Scenes (BIFS) based on the Virtual Reality Modeling Language (VRML) has been developed by MPEG for scene descriptions. The decoder can use the scene description and additional input from the user to combine or compose the objects to reconstruct the original scene or create a variation on it. The protocol for managing the elementary streams and their multiplexed version, called the Delivery Multimedia Integration Framework (DMIF), is an important part of MPEG-4. However, as our focus in this book is on compression, we will not discuss the protocol (for details, see the standard [221]).

A block diagram for the basic video coding algorithm is shown in Figure 19.18. Although shape coding occupies a very small portion of the diagram, it is a major part of the algorithm. The different objects that make up the scene are coded and sent to the multiplexer. The information about the presence of these objects is also provided to the motion-compensated predictor, which can use object-based motion-compensation algorithms to improve the compression efficiency. What is left after the prediction can be transmitted using a DCT-based coder. The

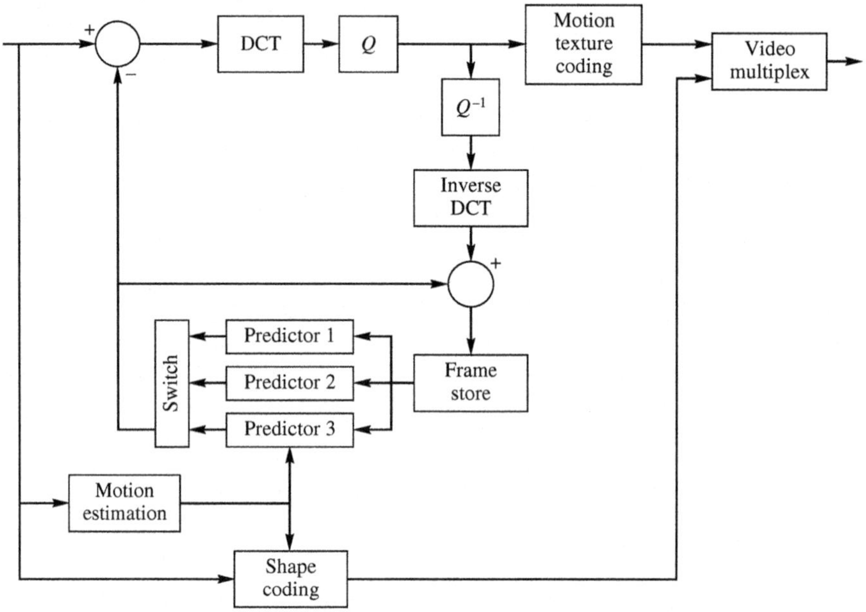

FIGURE 19.18 A block diagram for video coding.

video coding algorithm can also use a background "sprite"—generally a large panoramic still image that forms the background for the video sequence. The sprite is transmitted once, and the moving foreground video objects are placed in front of different portions of the sprite based on the information about the background provided by the encoder.

The MPEG-4 standard also envisions the use of model-based coding, where a triangular mesh representing the moving object is transmitted followed by texture information for covering the mesh. Information about movement of the mesh nodes can then be transmitted to animate the video object. The texture coding technique suggested by the standard is the embedded zerotree wavelet (EZW) algorithm. In particular, the standard envisions the use of a facial animation object to render an animated face. The shape, texture, and expressions of the face are controlled using facial definition parameters (FDPs) and facial action parameters (FAPs). BIFS provides features to support custom models and specialized interpretation of FAPs.

The MPEG-2 standard allows for SNR and spatial scalability. The MPEG-4 standard also allows for object scalability, in which certain objects may not be sent in order to reduce the bandwidth requirement.

19.13 Packet Video

The increasing popularity of communication over networks has led to increased interest in the development of compression schemes for use over networks. In this section we look at some of the issues involved in developing video compression schemes for use over networks.

19.13.1 ATM Networks

With the explosion of information, we have also seen the development of new ways of transmitting the information. One of the most efficient ways of transferring information among a large number of users is the use of asynchronous transfer mode (ATM) technology. In the past, communication has usually taken place over dedicated channels; that is, in order to communicate between two points, a channel was dedicated only to transferring information between those two points. Even if there was no information transfer going on during a particular period, the channel could not be used by anyone else. Because of the inefficiency of this approach, there is an increasing movement away from it. In an ATM network, the users divide their information into packets, which are transmitted over channels that can be used by more than one user.

We can draw an analogy between the movement of packets over a communication network and the movement of automobiles over a road network. If we break up a message into packets, then the movement of the message over the network is like the movement of a number of cars on a highway system going from one point to the other. Although two cars may not occupy the same position at the same time, they can occupy the same road at the same time. Thus, more than one group of cars can use the road at any given time. Furthermore, not all the cars in the group have to take the same route. Depending on the amount of traffic on the various roads that run between the origin of the traffic and the destination, different cars can take different routes. This is a more efficient utilization of the road than if the entire road was blocked off until the first group of cars completed its traversal of the road.

Using this analogy, we can see that the availability of transmission capacity, that is, the number of bits per second that we can transmit, is affected by factors that are outside our control. If at a given time there is very little traffic on the network, the available capacity will be high. On the other hand, if there is congestion on the network, the available capacity will be low. Furthermore, the ability to take alternate routes through the network also means that some of the packets may encounter congestion, leading to a variable amount of delay through the network. In order to prevent congestion from impeding the flow of vital traffic, networks will prioritize the traffic, with higher-priority traffic being permitted to move ahead of lower-priority traffic. Users can negotiate with the network for a fixed amount of guaranteed traffic. Of course, such guarantees tend to be expensive, so it is important that the user have some idea about how much high-priority traffic they will be transmitting over the network.

19.13.2 Compression Issues in ATM Networks

In video coding, this situation provides both opportunities and challenges. In the video compression algorithms discussed previously, there is a buffer that smooths the output of the compression algorithm. Thus, if we encounter a high-activity region of the video and generate more than the average number of bits per second, in order to prevent the buffer from overflowing, this period has to be followed by a period in which we generate fewer bits per second than the average. Sometimes this may happen naturally, with periods of low activity following periods of high activity. However, it is quite likely that this would not happen, in which case we have to reduce the quality by increasing the step size or dropping coefficients, or maybe even entire frames.

The ATM network, if it is not congested, will accommodate the variable rate generated by the compression algorithm. But if the network is congested, the compression algorithm will have to operate at a reduced rate. If the network is well designed, the latter situation will not happen too often, and the video coder can function in a manner that provides uniform quality. However, when the network is congested, it may remain so for a relatively long period. Therefore, the compression scheme should have the ability to operate for significant periods of time at a reduced rate. Furthermore, congestion might cause such long delays that some packets arrive after they can be of any use; that is, the frame they were supposed to be a part of might have already been reconstructed.

In order to deal with these problems, it is useful if the video compression algorithm provides information in a layered fashion, with a low-rate high-priority layer that can be used to reconstruct the video, even though the reconstruction may be poor, and low-priority enhancement layers that enhance the quality of the reconstruction. This is similar to the idea of progressive transmission, in which we first send a crude but low-rate representation of the image, followed by higher-rate enhancements. It is also useful if the bit rate required for the high-priority layer does not vary too much.

19.13.3 Compression Algorithms for Packet Video

Almost any compression algorithm can be modified to perform in the ATM environment, but some approaches seem more suited to this environment. We briefly present two approaches (see the original papers for more details).

One compression scheme that functions in an inherently layered manner is subband coding. In subband coding, the lower-frequency bands can be used to provide the basic reconstruction, with the higher-frequency bands providing the enhancement. As an example, consider the compression scheme proposed for packet video by Karlsson and Vetterli [263]. In their scheme, the video is divided into 11 bands. First, the video signal is divided into two temporal bands. Each band is then split into four spatial bands. The low-low band of the temporal low-frequency band is then split into four spatial bands. A graphical representation of this splitting is shown in Figure 19.19. The subband denoted 1 in the figure contains the basic information about the video sequence. Therefore, it is transmitted with the highest priority. If the data in all the other subbands are lost, it will still be possible to reconstruct the video using only the information in this subband. We can also prioritize the output of the other bands, and if the network starts getting congested and we are required to reduce our rate, we can do so by not transmitting the information in the lower-priority subbands. Subband 1 also generates the least variable data rate. This is very helpful when negotiating with the network for the amount of priority traffic.

Given the similarity of the ideas behind progressive transmission and subband coding, it should be possible to use progressive transmission algorithms as a starting point in the design of layered compression schemes for packet video. Chen, Sayood, and Nelson [264] use a DCT-based progressive transmission scheme [265] to develop a compression algorithm for packet video. In their scheme, they first encode the difference between the current frame and the prediction for the current frame using a 16×16 DCT. They only transmit the DC coefficient and the three lowest-order AC coefficients to the receiver. The coded coefficients make up the highest-priority layer.

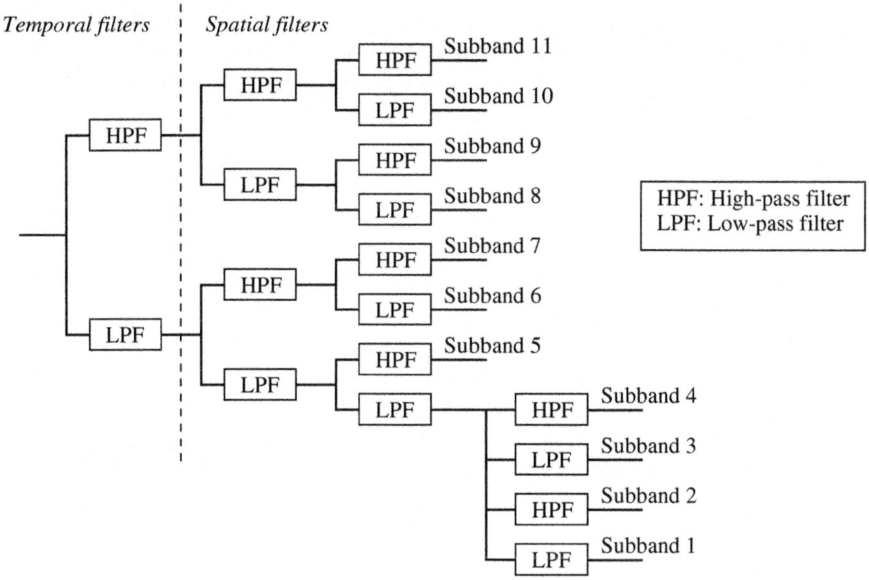

FIGURE 19.19 Analysis filter bank.

The reconstructed frame is then subtracted from the original. The sum of squared errors is calculated for each 16×16 block. Blocks with squared error greater than a prescribed threshold are subdivided into four 8×8 blocks, and the coding process is repeated using an 8×8 DCT. The coded coefficients make up the next layer. Because only blocks that fail to meet the threshold test are subdivided, information about which blocks are subdivided is transmitted to the receiver as side information.

The process is repeated with 4×4 blocks, which make up the third layer, and 2×2 blocks, which make up the fourth layer. Although this algorithm is a variable-rate coding scheme, the rate for the first layer is constant. Therefore, the user can negotiate with the network for a fixed amount of high-priority traffic. In order to remove the effect of delayed packets from the prediction, only the reconstruction from the higher-priority layers is used for prediction.

This idea can be used with many different progressive transmission algorithms to make them suitable for use over ATM networks.

19.14 Summary

In this chapter we described a number of different video compression algorithms. The only new information in terms of compression algorithms was the description of motion-compensated prediction. While the compression algorithms themselves have already been studied in previous chapters, we looked at how these algorithms are used under different requirements. The three scenarios that we looked at are teleconferencing, asymmetric applications such as broadcast video, and video over packet networks. Each application has slightly different requirements, leading to different ways of using the compression algorithms. We have by no means attempted to cover the entire area of video compression. However, by now you should have sufficient background to explore the subject further using the following list as a starting point.

Further Reading

1. An excellent source for information about the technical issues involved with digital video is the book *The Art of Digital Video*, by J. Watkinson [266].
2. The MPEG-1 standards document [260], "Information Technology—Coding of Moving Pictures and Associated Audio for Digital Storage Media Up to about 1.5 Mbit/s," has an excellent description of the video compression algorithm.
3. Detailed information about the MPEG-1 and MPEG-2 video standards can also be found in *MPEG Video Compression Standard*, by J.L. Mitchell, W.B. Pennebaker, C.E. Fogg, and D.J. LeGall [267].
4. To find more on model-based coding, see "Model Based Image Coding: Advanced Video Coding Techniques for Very Low Bit-Rate Applications," by K. Aizawa and T.S. Huang [258], in the February 1995 issue of the *Proceedings of the IEEE.*
5. A good place to begin exploring the various areas of research in packet video is the June 1989 issue of the *IEEE Journal on Selected Areas of Communication.*
6. The MPEG 1/2 and MPEG 4 standards are covered in an accesible manner in *The MPEG Handbook*, by J. Watkinson [266]. Focal Press 2001.
7. A good source for information about H.264 and MPEG-4 is *H.264 and MPEG-4 Video Compression*, by I.E.G. Richardson. Wiley, 2003 [286].

19.15 Projects and Problems

1. **(a)** Take the DCT of the Sinan image and plot the average squared value of each coefficient.
   (b) Circularly shift each line of the image by eight pixels. That is, $new\_image[i, j] = old\_image[i, j + 8 \pmod{256}]$. Take the DCT of the difference and plot the average squared value of each coefficient.
   (c) Compare the results in parts (a) and (b) above. Comment on the differences.

A

Probability and Random Processes

In this appendix we will look at some of the concepts relating to probability and random processes that are important in the study of systems. Our coverage will be highly selective and somewhat superficial, but enough to use probability and random processes as a tool in understanding data compression systems.

A.1 Probability

There are several different ways of defining and thinking about probability. Each approach has some merit; perhaps the best approach is the one that provides the most insight into the problem being studied.

A.1.1 Frequency of Occurrence

The most common way that most people think about probability is in terms of outcomes, or sets of outcomes, of an experiment. Let us suppose we conduct an experiment E that has N possible outcomes. We conduct the experiment n_T times. If the outcome ω_i occurs n_i times, we say that the frequency of occurrence of the outcome ω_i is $\frac{n_i}{n_T}$. We can then define the probability of occurrence of the outcome ω_i as

$$P(\omega_i) = \lim_{n_T \to \infty} \frac{n_i}{n_T}$$

In practice we do not have the ability to repeat an experiment an infinite number of times, so we often use the frequency of occurrence as an approximation to the probability. To make this more concrete consider a specific experiment. Suppose we turn on a television 1,000,000 times. Of these times, 800,000 times we turn the television on during a commercial and 200,000 times we turn it on and don't get a commercial. We could say the frequency of occurrence, or the estimate of the probability, of turning on a television set in the middle of a commercial is 0.8. Our experiment E here is turning on a television set, and the outcomes are *commercial* and *no commercial*. We could have been more careful with noting what was on when we turned on the television set and noticed whether the program was a news program (2000 times), a newslike program (20,000 times), a comedy program (40,000 times), an adventure program (18,000 times), a variety show (20,000 times), a talk show (90,000 times), or a movie (10,000 times), and whether the commercial was for products or services. In this case the outcomes would be *product commercial, service commercial, comedy, adventure, news, pseudonews, variety, talk show*, and *movie*. We could then define an *event* as a set of outcomes. The event *commercial* would consist of the outcomes *product commercial*, and *service commercial*; the event *no commercial* would consist of the outcomes *comedy, adventure, news, pseudonews, variety, talk show*, and *movie*. We could also define other events such as *programs that may contain news*. This set would contain the outcomes *news, pseudonews*, and *talk shows*, and the frequency of occurrence of this set is 0.112.

Formally, when we define an experiment E, we also define a *sample space* S associated with the experiment that consists of the *outcomes* $\{\omega_i\}$. We can then combine these outcomes into sets that are called *events*, and assign probabilities to these events. The largest subset of S (event) is S itself, and the probability of the event S is simply the probability that the experiment will have an outcome. The way we have defined things, this probability is one; that is, $P(S) = 1$.

A.1.2 A Measure of Belief

Sometimes the idea that the probability of an event is obtained through the repetitions of an experiment runs into trouble. What, for example, is the probability of your getting from Logan Airport to a specific address in Boston in a specified period of time? The answer depends on a lot of different factors, including your knowledge of the area, the time of day, the condition of your transport, and so on. You cannot conduct an experiment and get your answer because the moment you conduct the experiment, the conditions have changed, and the answer will now be different. We deal with this situation by defining a priori and a posteriori probabilities. The a priori probability is what you think or believe the probability to be before certain information is received or certain events take place; the a posteriori probability is the probability after you have received further information. Probability is no longer as rigidly defined as in the frequency of occurrence approach but is a somewhat more fluid quantity, the value of which changes with changing experience. For this approach to be useful we have to have a way of describing how the probability evolves with changing information. This is done through the use of *Bayes' rule*, named after the person who first described it. If $P(A)$ is the a priori probability of the event A and $P(A|B)$ is the a posteriori probability of the event A given that the event B has occurred, then

$$P(A|B) = \frac{P(A, B)}{P(B)} \tag{A.1}$$

where $P(A, B)$ is the probability of the event A *and* the event B occurring. Similarly,

$$P(B|A) = \frac{P(A, B)}{P(A)} \tag{A.2}$$

Combining (1) and (2) we get

$$P(A|B) = \frac{P(B|A)P(A)}{P(B)} \tag{A.3}$$

If the events A and B do not provide any information about each other, it would be reasonable to assume that

$$P(A|B) = P(A)$$

and therefore from (1),

$$P(A, B) = P(A)P(B) \tag{A.4}$$

Whenever (4) is satisfied, the events A and B are said to be *statistically independent*, or simply *independent*.

Example A.1.1:

A very common channel model used in digital communication is the *binary symmetric channel*. In this model the input is a random experiment with outcomes 0 and 1. The output of the channel is another random event with two outcomes 0 and 1. Obviously, the two outcomes are connected in some way. To see how, let us first define some events:

A: Input is 0
B: Input is 1
C: Output is 0
D: Output is 1

Let's suppose the input is equally likely to be a 1 or a 0. So $P(A) = P(B) = 0.5$. If the channel is perfect, that is, you get out of the channel what you put in, then we have

$$P(C|A) = P(D|B) = 1$$

and

$$P(C|B) = P(D|A) = 0$$

With most real channels this system is seldom encountered, and generally there is a small probability ϵ that the transmitted bit will be received in error. In this case, our probabilities are

$$P(C|A) = P(D|B) = 1 - \epsilon$$
$$P(C|B) = P(D|A) = \epsilon$$

How do we interpret $P(C)$ and $P(D)$? These are simply the probability that at any given time the output is a 0 or a 1. How would we go about computing these probabilities given the available information? Using (A.1) we can obtain $P(A, C)$ and $P(B, C)$ from $P(C|A)$,

$P(C|B)$, $P(A)$, and $P(B)$. These are the probabilities that the input is 0 and the output is 1, and the input is 1 and the output is 1. The event C—that is, the output is 1—will occur only when one of the two *joint* events occurs, therefore,

$$P(C) = P(A, C) + P(B, C)$$

Similarly,

$$P(D) = P(A, D) + P(B, D)$$

Numerically, this comes out to be

$$P(C) = P(D) = 0.5$$

♦

A.1.3 The Axiomatic Approach

Finally, there is an approach that simply defines probability as a measure, without much regard for physical interpretation. We are very familiar with measures in our daily lives. We talk about getting a 9-foot cable or a pound of cheese. Just as length and width measure the extent of certain physical quantities, probability measures the extent of an abstract quantity, a set. The thing that probability measures is the "size" of the event set. The probability measure follows similar rules to those followed by other measures. Just as the length of a physical object is always greater than or equal to zero, the probability of an event is always greater than or equal to zero. If we measure the length of two objects that have no overlap, then the combined length of the two objects is simply the sum of the lengths of the individual objects. In a similar manner the probability of the union of two events that do not have any outcomes in common is simply the sum of the probability of the individual events. So as to keep this definition of probability in line with the other definitions, we normalize this quantity by assigning the largest set, which is the sample space S, the size of 1. Thus, the probability of an event always lies between 0 and 1. Formally, we can write these rules down as the three *axioms* of probability.

Given a sample space S:

- *Axiom 1:* If A is an event in S, then $P(A) \geqslant 0$.
- *Axiom 2:* The probability of the sample space is 1; that is, $P(S) = 1$.
- *Axiom 3:* If A and B are two events in S and $A \cap B = \phi$, then $P(A \cup B) = P(A) + P(B)$.

Given these three axioms we can come up with all the other rules we need. For example, suppose A^c is the complement of A. What is the probability of A^c? We can get the answer by using Axiom 2 and Axiom 3. We know that

$$A^c \cup A = S$$

and Axiom 2 tells us that $P(S) = 1$, therefore,

$$P(A^c \cup A) = 1 \tag{A.5}$$

We also know that $A^c \cap A = \phi$, therefore, from Axiom 3

$$P(A^c \cup A) = P(A^c) + P(A) \tag{A.6}$$

Combining Equations (A.5) and (A.6), we get

$$P(A^c) = 1 - P(A) \tag{A.7}$$

Similarly, we can use the three axioms to obtain the probability of $A \cup B$ when $A \cap B \neq \phi$ as

$$P(A \cup B) = P(A) + P(B) - P(A \cap B) \tag{A.8}$$

In all of the above we have been using two events A and B. We can easily extend these rules to more events.

Example A.1.2:

Find $P(A \cup B \cup C)$ when $A \cap B = A \cap C = \phi$, and $B \cup C \neq \phi$.

Let

$$D = B \cup C$$

Then

$$A \cap C = \phi, \quad A \cap B = \phi \quad \Rightarrow \quad A \cap D = \phi$$

Therefore, from Axiom 3,

$$P(A \cup D) = P(A) + P(D)$$

and using (A.8)

$$P(D) = P(B) + P(C) - P(B \cap C)$$

Combining everything, we get

$$P(A \cup B \cup C) = P(A) + P(B) + P(C) - P(B \cap C)$$ ♦

The axiomatic approach is especially useful when an experiment does not have discrete outcomes. For example, if we are looking at the voltage on a telephone line, the probability of any specific value of the voltage is zero because there are an uncountably infinite number of different values that the voltage can take, and we can assign nonzero values to only a countably infinite number. Using the axiomatic approach, we can view the sample space as the range of voltages, and events as subsets of this range.

We have given three different interpretations of probability, and in the process described some rules by which probabilities can be manipulated. The rules described here (such as Bayes' rule, the three axioms, and the other rules we came up with) work the same way regardless of which interpretation you hold dear. The purpose of providing you with three different interpretations is to provide you with a variety of perspectives with which to view a given situation. For example, if someone says that the probability of a head when you flip a coin is 0.5, you might interpret that number in terms of repeated experiments (*if I flipped the coin 1000 times, I would expect to get 500 heads*). However, if someone tells you that the probability of your getting killed while crossing a particular street is 0.1, you might wish to interpret this information in a more subjective manner. The idea is to use the interpretation that gives you the most insight into a particular problem, while remembering that your interpretation will not change the mathematics of the situation.

Now that we have expended a lot of verbiage to say what probability is, let's spend a few lines saying what it is not. Probability does not imply certainty. When we say that the probability of an event is one, this does not mean that event *will* happen. On the other hand, when we say that the probability of an event is zero, that does not mean that event *won't* happen. Remember, mathematics only models reality, it is *not* reality.

A.2 Random Variables

When we are trying to mathematically describe an experiment and its outcomes, it is much more convenient if the outcomes are numbers. A simple way to do this is to define a mapping or function that assigns a number to each outcome. This mapping or function is called a *random variable*. To put that more formally: Let S be a sample space with outcomes $\{\omega_i\}$. Then the random variable X is a mapping

$$X : S \rightarrow \mathcal{R} \tag{A.9}$$

where $\mathcal{R}$ denotes the real number line. Another way of saying the same thing is

$$X(\omega) = x \qquad \omega \in S, x \in \mathcal{R} \tag{A.10}$$

The random variable is generally represented by an uppercase letter, and this is the convention we will follow. The value that the random variable takes on is called the *realization* of the random variable and is represented by a lowercase letter.

Example A.2.1:

Let's take our television example and rewrite it in terms of a random variable X:

$$\begin{aligned}
X(productcommercial) &= 0 \\
X(servicecommercial) &= 1 \\
X(news) &= 2 \\
X(pseudonews) &= 3 \\
X(talkshow) &= 4 \\
X(variety) &= 5 \\
X(comedy) &= 6 \\
X(adventure) &= 7 \\
X(movie) &= 8
\end{aligned}$$

Now, instead of talking about the probability of certain programs, we can talk about the probability of the random variable X taking on certain values or ranges of values. For example, $P(X(\omega) \leqslant 1)$ is the probability of seeing a commercial when the television is turned on (generally, we drop the argument and simply write this as $P(X \leqslant 1)$). Similarly, $P(programsthatmaycontainnews)$ can be written as $P(1 < X \leqslant 4)$, which is substantially less cumbersome. ♦

A.3 Distribution Functions

Defining the random variable in the way that we did allows us to define a special probability $P(X \leqslant x)$. This probability is called the *cumulative distribution function (cdf)* and is denoted by $F_X(x)$, where the random variable is the subscript and the realization is the argument. One of the primary uses of probability is the modeling of physical processes, and we will find the cumulative distribution function very useful when we try to describe or model different random processes. We will see more on this later.

For now, let us look at some of the properties of the *cdf*:

Property 1: $0 \leqslant F_X(x) \leqslant 1$. This follows from the definition of the *cdf*.

Property 2: The *cdf* is a monotonically nondecreasing function. That is,

$$x_1 \geqslant x_2 \quad \Rightarrow \quad F_X(x_1) \geqslant F_X(x_2)$$

To show this simply write the *cdf* as the sum of two probabilities:

$$\begin{aligned} F_X(x_1) &= P(X \leqslant x_1) = P(X \leqslant x_2) + P(x_2 < X \leqslant x_1) \\ &= F_X(x_2) + P(x_1 < X \leqslant x_2) \geqslant F_X(x_2) \end{aligned}$$

Property 3:

$$\lim_{n\to\infty} F_X(x) = 1$$

Property 4:

$$\lim_{n\to-\infty} F_X(x) = 0$$

Property 5: If we define

$$F_X(x^-) = P(X < x)$$

then

$$P(X = x) = F_X(x) - F_X(x^-)$$

Example A.3.1:

Assuming that the frequency of occurrence was an accurate estimate of the probabilities, let us obtain the *cdf* for our television example:

$$F_X(x) = \begin{cases} 0 & x < 0 \\ 0.4 & 0 \leqslant x < 1 \\ 0.8 & 1 \leqslant x < 2 \\ 0.802 & 2 \leqslant x < 3 \\ 0.822 & 3 \leqslant x < 4 \\ 0.912 & 4 \leqslant x < 5 \\ 0.932 & 5 \leqslant x < 6 \\ 0.972 & 6 \leqslant x < 7 \\ 0.99 & 7 \leqslant x < 8 \\ 1.00 & 8 \leqslant x \end{cases}$$

♦

Notice a few things about this *cdf*. First, the *cdf* consists of step functions. This is characteristic of discrete random variables. Second, the function is continuous from the right. This is due to the way the *cdf* is defined.

The *cdf* is somewhat different when the random variable is a continuous random variable. For example, if we sample a speech signal and then take the differences of the samples, the resulting random process would have a *cdf* that looks something like this:

$$F_X(x) = \begin{cases} \frac{1}{2}e^{2x} & x \leqslant 0 \\ 1 - \frac{1}{2}e^{-2x} & x > 0 \end{cases}$$

The thing to notice in this case is that because $F_X(x)$ is continuous

$$P(X = x) = F_X(x) - F_X(x^-) = 0$$

We can also have processes that have distributions that are continuous over some ranges and discrete over others.

Along with the cumulative distribution function, another function that also comes in very handy is the *probability density function (pdf)*. The *pdf* corresponding to the *cdf* $F_X(x)$ is written as $f_X(x)$. For continuous *cdfs*, the *pdf* is simply the derivative of the *cdf*. For discrete random variables, taking the derivative of the *cdf* introduces delta functions, which have problems of their own. So in the discrete case, we obtain the *pdf* through differencing. It is somewhat awkward to have different procedures for obtaining the same function for different types of random variables. It is possible to define a rigorous unified procedure for getting the *pdf* from the *cdf* for all kinds of random variables. However, in order to do so, we need some familiarity with measure theory, which is beyond the scope of this appendix. Let us look at some examples of *pdfs*.

Example A.3.2:

For our television scenario,

$$f_X(x) = \begin{cases} 0.4 & \text{if} X = 0 \\ 0.4 & \text{if} X = 1 \\ 0.002 & \text{if} X = 2 \\ 0.02 & \text{if} X = 3 \\ 0.09 & \text{if} X = 4 \\ 0.02 & \text{if} X = 5 \\ 0.04 & \text{if} X = 6 \\ 0.018 & \text{if} X = 7 \\ 0.01 & \text{if} X = 8 \\ 0 & \text{otherwise} \end{cases}$$

♦

Example A.3.3:

For our speech example, the *pdf* is given by

$$f_X(x) = \frac{1}{2}e^{-2|x|}$$

♦

A.4 Expectation

When dealing with random processes, we often deal with average quantities, like the signal power and noise power in communication systems, and the mean time between failures in various design problems. To obtain these average quantities, we use something called an *expectation operator*. Formally, the expectation operator $E[\,]$ is defined as follows: The *expected value* of a random variable X is given by

$$E[X] = \sum_i x_i P(X = x_i) \tag{A.11}$$

when X is a discrete random variable with realizations $\{x_i\}$ and by

$$E[X] = \int_{-\infty}^{\infty} x f_X(x) dx \tag{A.12}$$

where $f_X(x)$ is the *pdf* of X.

The expected value is very much like the average value and, if the frequency of occurrence is an accurate estimate of the probability, is identical to the average value. Consider the following example:

Example A.4.1:

Suppose in a class of 10 students the grades on the first test were

$$10, 9, 8, 8, 7, 7, 7, 6, 6, 2$$

The average value is $\frac{70}{10}$, or 7. Now let's use the frequency of occurrence approach to estimate the probabilities of the various grades. (Notice in this case the random variable is an identity mapping, i.e., $X(\omega) = \omega$.) The probability estimate of the various values the random variable can take on is

$$P(10) = P(9) = P(2) = 0.1, \quad P(8) = P(6) = 0.2, \quad P(7) = 0.3,$$
$$P(6) = P(5) = P(4) = P(3) = P(1) = P(0) = 0$$

The expected value is therefore given by

$$E[X] = (0)(0) + (0)(1) + (0.1)(2) + (0)(3) + (0)(4) + (0)(5) + (0.2)(6)$$
$$+(0.3)(7) + (0.2)(8) + (0.1)(9) + (0.1)(10) = 7 \qquad \blacklozenge$$

It seems that the expected value and the average value *are* exactly the same! But we have made a rather major assumption about the accuracy of our probability estimate. In general the relative frequency is not exactly the same as the probability, and the average expected values are different. To emphasize this difference and similarity, the expected value is sometimes referred to as the *statistical average*, while our everyday average value is referred to as the *sample average*.

We said at the beginning of this section that we are often interested in things such as signal power. The average signal power is often defined as the average of the signal squared. If we say that the random variable is the signal value, then this means that we have to find the expected value of the square of the random variable. There are two ways of doing this. We could define a new random variable $Y = X^2$, then find $f_Y(y)$ and use (A.12) to find $E[Y]$. An easier approach is to use the *fundamental theorem of expectation*, which is

$$E[g(X)] = \sum_i g(x_i)P(X = x_i) \tag{A.13}$$

for the discrete case, and

$$E[g(X)] = \int_{-\infty}^{\infty} g(x)f_X(x)dx \tag{A.14}$$

for the continuous case.

The expected value, because of the way it is defined, is a linear operator. That is,

$$E[\alpha X + \beta Y] = \alpha E[X] + \beta E[Y], \qquad \alpha \text{ and } \beta \text{ are constants}$$

You are invited to verify this for yourself.

There are several functions $g()$ whose expectations are used so often that they have been given special names.

A.4.1 Mean

The simplest and most obvious function is the identity mapping $g(X) = X$. The expected value $E(X)$ is referred to as the *mean* and is symbolically referred to as μ_X. If we take a random variable X and add a constant value to it, the mean of the new random process is simply the old mean plus the constant. Let

$$Y = X + a$$

where a is a constant value. Then

$$\mu_Y = E[Y] = E[X + a] = E[X] + E[a] = \mu_X + a$$

A.4.2 Second Moment

If the random variable X is an electrical signal, the total power in this signal is given by $E[X^2]$, which is why we are often interested in it. This value is called the *second moment* of the random variable.

A.4.3 Variance

If X is a random variable with mean μ_X, then the quantity $E[(X - \mu_X)^2]$ is called the *variance* and is denoted by σ_X^2. The square root of this value is called the *standard deviation* and is denoted by σ. The variance and the standard deviation can be viewed as a measure of the "spread" of the random variable. We can show that

$$\sigma_X^2 = E[X^2] - \mu_X^2$$

If $E[X^2]$ is the total power in a signal, then the variance is also referred to as the total AC power.

A.5 Types of Distribution

There are several specific distributions that are very useful when describing or modeling various processes.

A.5.1 Uniform Distribution

This is the distribution of ignorance. If we want to model data about which we know nothing except its range, this is the distribution of choice. This is not to say that there are not times when the uniform distribution is a good match for the data. The *pdf* of the uniform distribution is given by

$$f_X(x) = \begin{cases} \frac{1}{b-a} & \text{for } a \leqslant X \leqslant b \\ 0 & \text{otherwise} \end{cases} \tag{A.15}$$

The mean of the uniform distribution can be obtained as

$$\mu_X = \int_a^b x \frac{1}{b-a} dx = \frac{b+a}{2}$$

Similarly, the variance of the uniform distribution can be obtained as

$$\sigma_X^2 = \frac{(b-a)^2}{12}$$

Details are left as an exercise.

A.5.2 Gaussian Distribution

This is the distribution of choice in terms of mathematical tractability. Because of its form, it is especially useful with the squared error distortion measure. The probability density function for a random variable with a Gaussian distribution is

$$f_X(x) = \frac{1}{\sqrt{2\pi\sigma^2}} \exp -\frac{(x-\mu)^2}{2\sigma^2} \tag{A.16}$$

where the mean of the distribution is μ and the variance is σ^2.

A.5.3 Laplacian Distribution

Many sources that we will deal with will have probability density functions that are quite peaked at zero. For example, speech consists mainly of silence; therefore, samples of speech will be zero or close to zero with high probability. Image pixels themselves do not have any attraction to small values. However, there is a high degree of correlation among pixels. Therefore, a large number of the pixel-to-pixel differences will have values close to zero. In these situations, a Gaussian distribution is not a very close match to the data. A closer match is the Laplacian distribution, which has a *pdf* that is peaked at zero. The density function for a zero mean random variable with Laplacian distribution and variance σ^2 is

$$f_X(x) = \frac{1}{\sqrt{2\sigma^2}} \exp \frac{-\sqrt{2}\,|x|}{\sigma} \tag{A.17}$$

A.5.4 Gamma Distribution

A distribution with a *pdf* that is even more peaked, though considerably less tractable than the Laplacian distribution, is the gamma distribution. The density function for a gamma-distributed random variable with zero mean and variance σ^2 is given by

$$f_X(x) = \frac{\sqrt[4]{3}}{\sqrt{8\pi\sigma\,|x|}} \exp \frac{-\sqrt{3}\,|x|}{2\sigma} \tag{A.18}$$

A.6 Stochastic Process

We are often interested in experiments whose outcomes are a function of time. For example, we might be interested in designing a system that encodes speech. The outcomes are particular patterns of speech that will be encountered by the speech coder. We can mathematically describe this situation by extending our definition of a random variable. Instead of the random variable mapping an outcome of an experiment to a number, we map it to a function of time. Let S be a sample space with outcomes $\{\omega_i\}$. Then the random or stochastic process X is a mapping

$$X : S \rightarrow \mathcal{F} \tag{A.19}$$

where $\mathcal{F}$ denotes the set of functions on the real number line. In other words,

$$X(\omega) = x(t) \qquad \omega \in S,\ x \in \mathcal{F},\ -\infty < t < \infty \tag{A.20}$$

The functions $x(t)$ are called the *realizations* of the random process, and the collection of functions $\{x_\omega(t)\}$ indexed by the outcomes ω is called the *ensemble* of the stochastic process. We can define the mean and variance of the ensemble as

$$\mu(t) = E[X(t)] \tag{A.21}$$

$$\sigma^2(t) = E[(X(t) - \mu(t))^2] \tag{A.22}$$

If we sample the ensemble at some time t_0, we get a set of numbers $\{x_\omega(t_0)\}$ indexed by the outcomes ω, which by definition is a random variable. By sampling the ensemble at different times t_i, we get different random variables $\{x_\omega(t_i)\}$. For simplicity we often drop the ω and t and simply refer to these random variables as $\{x_i\}$.

We will have a distribution function associated with each of these random variables. We can also define a joint distribution function for two or more of these random variables. Given a set of random variables $\{x_1, x_2, \ldots, x_N\}$, the *joint* cumulative distribution function is defined as

$$F_{X_1X_2\cdots X_N}(x_1, x_2, \ldots, x_N) = P(X_1 < x_1, X_2 < x_2, \ldots, X_N < x_N) \tag{A.23}$$

Unless it is clear from the context what we are talking about, we will refer to the *cdf* of the individual random variables X_i as the *marginal cdf* of X_i.

We can also define the joint probability density function for these random variables $f_{X_1X_2\cdots X_N}(x_1, x_2, \ldots, x_N)$ in the same manner as we defined the *pdf* in the case of the single random variable. We can classify the relationships between these random variables in a number of different ways. In the following we define some relationships between two random variables. The concepts are easily extended to more than two random variables.

Two random variables X_1 and X_2 are said to be *independent* if their joint distribution function can be written as the product of the marginal distribution functions of each random variable; that is,

$$F_{X_1X_2}(x_1, x_2) = F_{X_1}(x_1)F_{X_2}(x_2) \tag{A.24}$$

This also implies that

$$f_{X_1X_2}(x_1, x_2) = f_{X_1}(x_1)f_{X_2}(x_2) \tag{A.25}$$

If all the random variables $X_1, X_2, \ldots$ are independent and they have the same distribution, they are said to be *independent, identically distributed* (*iid*).

Two random variables X_1 and X_2 are said to be *orthogonal* if

$$E[X_1 X_2] = 0 \tag{A.26}$$

Two random variables X_1 and X_2 are said to be *uncorrelated* if

$$E[(X_1 - \mu_1)(X_2 - \mu_2)] = 0 \tag{A.27}$$

where $\mu_1 = E[X_1]$ and $\mu_2 = E[X_2]$.

The *autocorrelation function* of a random process is defined as

$$R_{xx}(t_i, t_2) = E[X_1 X_2] \tag{A.28}$$

For a given value of N, suppose we sample the stochastic process at N times $\{t_i\}$ to get the N random variables $\{X_i\}$ with *cdf* $F_{X_1 X_2 \ldots X_N}(x_1, x_2, \ldots, x_N)$, and another N times $\{t_i + T\}$ to get the random variables $\{X'_i\}$ with *cdf* $F_{X'_1 X'_2 \ldots X'_N}(x'_1, x'_2, \ldots, x'_N)$. If

$$F_{X_1 X_2 \ldots X_N}(x_1, x_2, \ldots, x_N) = F_{X'_1 X'_2 \ldots X'_N}(x'_1, x'_2, \ldots, x'_N) \tag{A.29}$$

for all N and T, the process is said to be *stationary*.

The assumption of stationarity is a rather important assumption because it is a statement that the statistical characteristics of the process under investigation do not change with time. Thus, if we design a system for an input based on the statistical characteristics of the input today, the system will still be useful tomorrow because the input will not change its characteristics. The assumption of stationarity is also a very strong assumption, and we can usually make do quite well with a weaker condition, *wide sense* or *weak sense* stationarity.

A stochastic process is said to be wide sense or weak sense stationary if it satisfies the following conditions:

1. The mean is constant; that is, $\mu(t) = \mu$ for all t.
2. The variance is finite.
3. The autocorrelation function $R_{xx}(t_1, t_2)$ is a function only of the difference between t_1 and t_2, and not of the individual values of t_1 and t_2; that is,

$$R_{xx}(t_1, t_2) = R_{xx}(t_1 - t_2) = R_{xx}(t_2 - t_1). \tag{A.30}$$

Further Reading

1. The classic books on probability are the two-volume set *An Introduction to Probability Theory and Its Applications,* by W. Feller [287,288].
2. A commonly used text for an introductory course on probability and random processes is *Probability, Random Variables, and Stochastic Processes*, by A. Papoulis [289].

A.7 Projects and Problems

1. If $A \cap B \neq \phi$, show that

$$P(A \cup B) = P(A) + P(B) - P(A \cap B)$$

2. Show that expectation is a linear operator in both the discrete and the continuous case.
3. If a is a constant, show that $E[a] = a$.
4. Show that for a random variable X,

$$\sigma_X^2 = E[X^2] - \mu_X^2$$

5. Show that the variance of the uniform distribution is given by

$$\sigma_X^2 = \frac{(b-a)^2}{12}$$

B

A Brief Review of Matrix Concepts

In this appendix we will look at some of the basic concepts of matrix algebra. Our intent is simply to familiarize you with some basic matrix operations that we will need in our study of compression. Matrices are very useful for representing linear systems of equations, and matrix theory is a powerful tool for the study of linear operators. In our study of compression techniques we will use matrices both in the solution of systems of equations and in our study of linear transforms.

B.1 A Matrix

A collection of real or complex elements arranged in M rows and N columns is called a matrix of order $M \times N$:

$$\mathbf{A} = \begin{bmatrix} a_{00} & a_{01} & \cdots & a_{0N-1} \\ a_{10} & a_{11} & \cdots & a_{1N-1} \\ \vdots & \vdots & & \vdots \\ a_{(M-1)0} & a_{(M-1)1} & \cdots & a_{M-1N-1} \end{bmatrix} \tag{B.1}$$

where the first subscript denotes the row that an element belongs to and the second subscript denotes the column. For example, the element a_{02} belongs in row 0 and column 2, and the element a_{32} belongs in row 3 and column 2. The generic ijth element of a matrix $\mathbf{A}$ is sometimes represented as $[\mathbf{A}]_{ij}$. If the number of rows is equal to the number of columns $(N = M)$, then the matrix is called a *square matrix*. A special square matrix that we will be using is the *identity matrix* $\mathbf{I}$, in which the elements on the diagonal of the matrix are 1 and all other elements are 0:

$$[\mathbf{I}]_{ij} = \begin{cases} 1 & i = j \\ 0 & i \neq j \end{cases} \tag{B.2}$$

If a matrix consists of a single column ($N = 1$), it is called a *column matrix* or *vector* of dimension M. If it consists of a single row ($M = 1$), it is called a *row matrix* or *vector* of dimension N.

The *transpose* $\mathbf{A}^T$ of a matrix $\mathbf{A}$ is the $N \times M$ matrix obtained by writing the rows of the matrix as columns and the columns as rows:

$$\mathbf{A}^T = \begin{bmatrix} a_{00} & a_{10} & \cdots & a_{(M-1)0} \\ a_{01} & a_{11} & \cdots & a_{(M-1)1} \\ \vdots & \vdots & & \vdots \\ a_{0(N-1)} & a_{1(N-1)} & \cdots & a_{M-1N-1} \end{bmatrix} \tag{B.3}$$

The transpose of a column matrix is a row matrix and vice versa.

Two matrices $\mathbf{A}$ and $\mathbf{B}$ are said to be equal if they are of the same order and their corresponding elements are equal; that is,

$$\mathbf{A} = \mathbf{B} \quad \Leftrightarrow \quad a_{ij} = b_{ij}, i = 0, 1, \ldots M - 1; j = 0, 1, \ldots N - 1 \tag{B.4}$$

B.2 Matrix Operations

You can add, subtract, and multiply matrices, but since matrices come in all shapes and sizes, there are some restrictions as to what operations you can perform with what kind of matrices. In order to add or subtract two matrices, their dimensions have to be identical—same number of rows and same number of columns. In order to multiply two matrices, the order in which they are multiplied is important. In general $\mathbf{A} \times \mathbf{B}$ is not equal to $\mathbf{B} \times \mathbf{A}$. Multiplication is only defined for the case where the number of columns of the first matrix is equal to the number of rows of the second matrix. The reasons for these restrictions will become apparent when we look at how the operations are defined.

When we add two matrices, the resultant matrix consists of elements that are the sum of the corresponding entries in the matrices being added. Let us add two matrices $\mathbf{A}$ and $\mathbf{B}$ where

$$\mathbf{A} = \begin{bmatrix} a_{00} & a_{01} & a_{02} \\ a_{10} & a_{11} & a_{12} \end{bmatrix}$$

and

$$\mathbf{B} = \begin{bmatrix} b_{00} & b_{01} & b_{02} \\ b_{10} & b_{11} & b_{12} \end{bmatrix}$$

The sum of the two matrices, $\mathbf{C}$, is given by

$$\mathbf{C} = \begin{bmatrix} c_{00} & c_{12} & c_{13} \\ c_{21} & c_{22} & c_{23} \end{bmatrix} = \begin{bmatrix} a_{00} + b_{00} & a_{01} + b_{01} & a_{02} + b_{02} \\ a_{10} + b_{10} & a_{11} + b_{11} & a_{12} + b_{12} \end{bmatrix} \tag{B.5}$$

Notice that each element of the resulting matrix $\mathbf{C}$ is the sum of corresponding elements of the matrices $\mathbf{A}$ and $\mathbf{B}$. In order for the two matrices to have corresponding elements, the dimension of the two matrices has to be the same. Therefore, addition is only defined for matrices with identical dimensions (i.e., same number of rows and same number of columns). Subtraction is defined in a similar manner. The elements of the difference matrix are made up of term-by-term subtraction of the matrices being subtracted.

We could have generalized matrix addition and matrix subtraction from our knowledge of addition and subtraction of numbers. Multiplication of matrices is another kettle of fish entirely. It is easiest to describe matrix multiplication with an example. Suppose we have two different matrices **A** and **B** where

$$\mathbf{A} = \begin{bmatrix} a_{00} & a_{01} & a_{02} \\ a_{10} & a_{11} & a_{12} \end{bmatrix}$$

and

$$\mathbf{B} = \begin{bmatrix} b_{00} & b_{01} \\ b_{10} & b_{11} \\ b_{20} & b_{21} \end{bmatrix} \tag{B.6}$$

The product is given by

$$\mathbf{C} = \mathbf{AB} = \begin{bmatrix} c_{00} & c_{01} \\ c_{10} & c_{11} \end{bmatrix} = \begin{bmatrix} a_{00}b_{00} + a_{01}b_{10} + a_{02}b_{20} & a_{00}b_{01} + a_{01}b_{11} + a_{02}b_{21} \\ a_{10}b_{00} + a_{11}b_{10} + a_{12}b_{20} & a_{10}b_{01} + a_{11}b_{11} + a_{12}b_{21} \end{bmatrix}$$

You can see that the i, j element of the product is obtained by adding term by term the product of elements in the ith row of the first matrix with those of the jth column of the second matrix. Thus, the element c_{10} in the matrix **C** is obtained by summing the term-by-term products of row 1 of the first matrix **A** with column 0 of the matrix **B**. We can also see that the resulting matrix will have as many rows as the matrix to the left and as many columns as the matrix to the right.

What happens if we reverse the order of the multiplication? By the rules above we will end up with a matrix with three rows and three columns.

$$\begin{bmatrix} b_{00}a_{00} + b_{01}a_{10} & b_{00}a_{01} + +b_{01}a_{11} & b_{00}a_{02} + b_{01}a_{12} \\ b_{10}a_{00} + b_{11}a_{10} & b_{10}a_{01} + +b_{11}a_{11} & b_{10}a_{02} + b_{11}a_{12} \\ b_{20}a_{00} + b_{21}a_{10} & b_{20}a_{01} + +b_{21}a_{11} & b_{20}a_{02} + b_{21}a_{12} \end{bmatrix}$$

The elements of the two product matrices are different, as are the dimensions.

As we can see, multiplication between matrices follows some rather different rules than multiplication between real numbers. The sizes have to match up—the number of columns of the first matrix has to be equal to the number of rows of the second matrix, and the order of multiplication is important. Because of the latter fact we often talk about premultiplying or postmultiplying. Premultiplying **B** by **A** results in the product **AB**, while postmultiplying **B** by **A** results in the product **BA**.

We have three of the four elementary operations. What about the fourth elementary operation, division? The easiest way to present division in matrices is to look at the formal definition of division when we are talking about real numbers. In the real number system, for every number a different from zero, there exists an inverse, denoted by $1/a$ or a^{-1}, such that the product of a with its inverse is one. When we talk about a number b divided by a number a, this is the same as the *multiplication* of b with the inverse of a. Therefore, we could define division by a matrix as the multiplication with the inverse of the matrix. **A/B** would be given by $\mathbf{AB}^{-1}$. Once we have the definition of an inverse of a matrix, the rules of multiplication apply.

So how do we define the inverse of a matrix? Following the definition for real numbers, in order to define the inverse of a matrix we need to have the matrix counterpart of 1. In matrices this counterpart is called the *identity matrix*. The identity matrix is a square matrix with diagonal elements being 1 and off-diagonal elements being 0. For example, a 3×3 identity matrix is given by

$$\mathbf{I} = \begin{bmatrix} 1 & 0 & 0 \\ 0 & 1 & 0 \\ 0 & 0 & 1 \end{bmatrix} \tag{B.7}$$

The identity matrix behaves like the number in the matrix world. If we multiply any matrix with the identity matrix (of appropriate dimension), we get the original matrix back. Given a square matrix $\mathbf{A}$, we define its inverse, $\mathbf{A}^{-1}$, as the matrix that when premultiplied or postmultiplied by $\mathbf{A}$ results in the identity matrix. For example, consider the matrix

$$\mathbf{A} = \begin{bmatrix} 3 & 4 \\ 1 & 2 \end{bmatrix} \tag{B.8}$$

The inverse matrix is given by

$$\mathbf{A}^{-1} = \begin{bmatrix} 1 & -2 \\ -0.5 & 1.5 \end{bmatrix} \tag{B.9}$$

To check that this is indeed the inverse matrix, let us multiply them:

$$\begin{bmatrix} 3 & 4 \\ 1 & 2 \end{bmatrix} \begin{bmatrix} 1 & -2 \\ -0.5 & 1.5 \end{bmatrix} = \begin{bmatrix} 1 & 0 \\ 0 & 1 \end{bmatrix} \tag{B.10}$$

and

$$\begin{bmatrix} 1 & -2 \\ -0.5 & 1.5 \end{bmatrix} \begin{bmatrix} 3 & 4 \\ 1 & 2 \end{bmatrix} = \begin{bmatrix} 1 & 0 \\ 0 & 1 \end{bmatrix} \tag{B.11}$$

If $\mathbf{A}$ is a vector of dimension M, we can define two specific kinds of products. If $\mathbf{A}$ is a column matrix, then the *inner product* or *dot product* is defined as

$$\mathbf{A}^T\mathbf{A} = \sum_{i=0}^{M-1} a_{i0}^2 \tag{B.12}$$

and the *outer product* or *cross product* is defined as

$$\mathbf{AA}^T = \begin{bmatrix} a_{00}a_{00} & a_{00}a_{10} & \cdots & a_{00}a_{(M-1)0} \\ a_{10}a_{00} & a_{10}a_{10} & \cdots & a_{10}a_{(M-1)0} \\ \vdots & \vdots & & \vdots \\ a_{(M-1)0}a_{00} & a_{\{(M-1)1\}}a_{10} & \cdots & a_{(M-1)0}a_{(M-1)0} \end{bmatrix} \tag{B.13}$$

Notice that the inner product results in a scalar, while the outer product results in a matrix.

In order to find the inverse of a matrix, we need the concepts of determinant and cofactor. Associated with each square matrix is a scalar value called the *determinant* of the matrix. The determinant of a matrix $\mathbf{A}$ is denoted as $|\mathbf{A}|$. To see how to obtain the determinant of an $N \times N$ matrix, we start with a 2×2 matrix. The determinant of a 2×2 matrix is given as

$$|\mathbf{A}| = \begin{vmatrix} a_{00} & a_{01} \\ a_{10} & a_{11} \end{vmatrix} = a_{00}a_{11} - a_{01}a_{10} \tag{B.14}$$

Finding the determinant of a 2×2 matrix is easy. To explain how to get the determinants of larger matrices, we need to define some terms.

The *minor* of an element a_{ij} of an $N \times N$ matrix is defined to be the determinant of the $N - 1 \times N - 1$ matrix obtained by deleting the row and column containing a_{ij}. For example, if $\mathbf{A}$ is a 4×4 matrix

$$\mathbf{A} = \begin{bmatrix} a_{00} & a_{01} & a_{02} & a_{03} \\ a_{10} & a_{11} & a_{12} & a_{13} \\ a_{20} & a_{21} & a_{22} & a_{23} \\ a_{30} & a_{31} & a_{32} & a_{33} \end{bmatrix} \tag{B.15}$$

then the minor of the element a_{12}, denoted by M_{12}, is the determinant

$$M_{12} = \begin{vmatrix} a_{00} & a_{01} & a_{03} \\ a_{20} & a_{21} & a_{23} \\ a_{30} & a_{31} & a_{33} \end{vmatrix} \tag{B.16}$$

The cofactor of a_{ij}, denoted by $\mathbf{A}_{ij}$, is given by

$$\mathbf{A}_{ij} = (-1)^{i+j} M_{ij} \tag{B.17}$$

Armed with these definitions we can write an expression for the determinant of an $N \times N$ matrix as

$$|\mathbf{A}| = \sum_{i=0}^{N-1} a_{ij}\mathbf{A}_{ij} \tag{B.18}$$

or

$$|\mathbf{A}| = \sum_{j=0}^{N-1} a_{ij}\mathbf{A}_{ij} \tag{B.19}$$

where the a_{ij} is taken from a single row or a single column. If the matrix has a particular row or column that has a large number of zeros in it, we would need fewer computations if we picked that particular row or column.

Equations (B.18) and (B.19) express the determinant of an $N \times N$ matrix in terms of determinants of $N - 1 \times N - 1$ matrices. We can express each of the $N - 1 \times N - 1$ determinants in terms of $N - 2 \times N - 2$ determinants, continuing in this fashion until we have everything expressed in terms of 2×2 determinants, which can be evaluated using (B.14).

Now that we know how to compute a determinant, we need one more definition before we can define the inverse of a matrix. The *adjoint* of a matrix $\mathbf{A}$, denoted by ($\mathbf{A}$), is a matrix whose ijth element is the cofactor $\mathbf{A}_{ji}$. The inverse of a matrix $\mathbf{A}$, denoted by $\mathbf{A}^{-1}$, is given by

$$\mathbf{A}^{-1} = \frac{1}{|\mathbf{A}|}(\mathbf{A}) \tag{B.20}$$

Notice that for the inverse to exist the determinant has to be nonzero. If the determinant for a matrix is zero, the matrix is said to be singular. The method we have described here works well with small matrices; however, it is highly inefficient if N becomes greater than 4. There are a number of efficient methods for inverting matrices; see the books in the Further Reading section for details.

Corresponding to a square matrix $\mathbf{A}$ of size $N \times N$ are N scalar values called the *eigenvalues* of $\mathbf{A}$. The eigenvalues are the N solutions of the equation $|\lambda\mathbf{I} - \mathbf{A}| = 0$. This equation is called the *characteristic equation*.

Example B.2.1: Let us find the eigenvalues of the matrix

$$\begin{bmatrix} 4 & 5 \\ 2 & 1 \end{bmatrix}$$

$$|\lambda\mathbf{I} - \mathbf{A}| = 0$$

$$\left| \begin{bmatrix} \lambda & 0 \\ 0 & \lambda \end{bmatrix} - \begin{bmatrix} 4 & 5 \\ 2 & 1 \end{bmatrix} \right| = 0$$

$$(\lambda - 4)(\lambda - 1) - 10 = 0$$

$$\lambda_1 = -1 \qquad \lambda_2 = 6 \tag{B.21}$$

♦

The eigenvectors V_k of an $N \times N$ matrix are the N vectors of dimension N that satisfy the equation

$$\mathbf{A}V_k = \lambda_k V_k \tag{B.22}$$

Further Reading

1. The subject of matrices is covered at an introductory level in a number of textbooks. A good one is *Advanced Engineering Mathematics*, by E. Kreyszig [290].
2. Numerical methods for manipulating matrices (and a good deal more) are presented in *Numerical Recipes in C*, by W.H. Press, S.A. Teukolsky, W.T. Vetterling, and B.P. Flannery [182].

C

The Root Lattices

Define $\mathbf{e}_i^L$ to be a vector in L dimensions whose ith component is 1 and all other components are 0. Some of the root systems that are used in lattice vector quantization are given as follows:

$$
\begin{array}{lll}
D_L & \pm\mathbf{e}_i^L \pm \mathbf{e}_j^L, & i \neq j, i, j = 1, 2, \ldots, L \\
A_L & \pm(\mathbf{e}_i^{L+1} - \mathbf{e}_j^{L+1}), & i \neq j, i, j = 1, 2, \ldots, L \\
E_L & \pm\mathbf{e}_i^L \pm \mathbf{e}_j^L, & i \neq j, i, j = 1, 2, \ldots, L-1, \\
 & \frac{1}{2}(\pm\mathbf{e}_1 \pm \mathbf{e}_2 \cdots \pm \mathbf{e}_{L-1} \pm \sqrt{2 - \frac{(L-1)}{4}}\mathbf{e}_L) & L = 6, 7, 8
\end{array}
$$

Let us look at each of these definitions a bit closer and see how they can be used to generate lattices.

D_L Let us start with the D_L lattice. For $L = 2$, the four roots of the D_2 algebra are $\mathbf{e}_1^2 + \mathbf{e}_2^2$, $\mathbf{e}_1^2 - \mathbf{e}_2^2$, $-\mathbf{e}_1^2 + \mathbf{e}_2^2$, and $-\mathbf{e}_1^2 - \mathbf{e}_2^2$, or (1, 1), (1, −1), (−1, 1), and (−1, −1). We can pick any two independent vectors from among these four to form the basis set for the D_2 lattice. Suppose we picked (1, 1) and (1, −1). Then any integral combination of these vectors is a lattice point. The resulting lattice is shown in Figure 10.24 in Chapter 10. Notice that the sums of the coordinates are all even numbers. This makes finding the closest lattice point to an input a relatively simple exercise.

A_L The roots of the A_L lattices are described using $L+1$-dimensional vectors. However, if we select any L independent vectors from this set, we will find that the points that are generated all lie in an L-dimensional slice of the $L + 1$-dimensional space. This can be seen from Figure C.1

We can obtain an L-dimensional basis set from this using a simple algorithm described in [151]. In two dimensions, this results in the generation of the vectors (1,0) and $(-\frac{1}{2}, \frac{\sqrt{3}}{2})$. The resulting lattice is shown in Figure 10.25 in Chapter 10. To find the closest point to the A_L lattice, we use the fact that in the embedding of the lattice in $L + 1$ dimensions, the sum

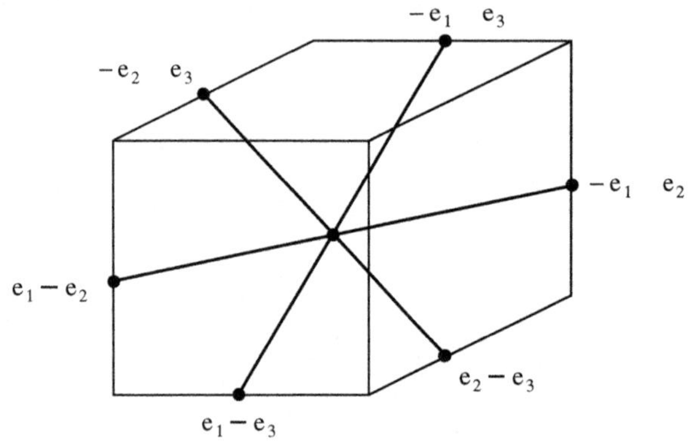

FIGURE C.1 The A_2 roots embedded in three dimensions.

of the coordinates is always zero. The exact procedure can be found in [153,152]. E_L As we can see from the definition, the E_L lattices go up to a maximum dimension of 8. Each of these lattices can be written as unions of the A_L and D_L lattices and their translated version. For example, the E_8 lattice is the union of the D_8 lattice and the D_8 lattice translated by the vector $(\frac{1}{2}, \frac{1}{2}, \frac{1}{2}, \frac{1}{2}, \frac{1}{2}, \frac{1}{2}, \frac{1}{2}, \frac{1}{2})$. Therefore, to find the closest E_8 point to an input $\mathbf{x}$, we find the closest point of D_8 to $\mathbf{x}$, and the closest point of D_8 to $\mathbf{x} - (\frac{1}{2}, \frac{1}{2}, \frac{1}{2}, \frac{1}{2}, \frac{1}{2}, \frac{1}{2}, \frac{1}{2}, \frac{1}{2})$, and pick the one that is closest to $\mathbf{x}$.

There are several advantages to using lattices as vector quantizers. There is no need to store the codebook, and finding the closest lattice point to a given input is a rather simple operation. However, the quantizer codebook is only a subset of the lattice. How do we know when we have wandered out of this subset, and what do we do about it? Furthermore, how do we generate a binary codeword for each of the lattice points that lie within the boundary? The first problem is easy to solve. Earlier we discussed the selection of a boundary to reduce the effect of the overload error. We can check the location of the lattice point to see if it is within this boundary. If not, we are outside the subset. The other questions are more difficult to resolve. Conway and Sloane [154] have developed a technique that functions by first defining the boundary as one of the quantization regions (expanded many times) of the root lattices. The technique is not very complicated, but it takes some time to set up, so we will not describe it here (see [154] for details).

We have given a sketchy description of lattice quantizers. For a more detailed tutorial review, see [152].

Bibliography

[1] T. C. Bell, J. G. Cleary, and I. H. Witten. *Text Compression*. Advanced Reference Series. Prentice Hall, Englewood Cliffs, New Jersey, 1990.

[2] B.L. van der Waerden. *A History of Algebra*. Springer-Verlag, 1985.

[3] C.E. Shannon. A mathematical theory of communication. *Bell System Technical Journal*, 27:379–423, 623–656, 1948.

[4] C.E. Shannon. Prediction and entropy of printed English. *Bell System Technical Journal*, 30:50–64, January 1951.

[5] R.W. Hamming. *Coding and Information Theory*. Prentice-Hall, 2nd edition, 1986.

[6] W.B. Pennebaker and J.L. Mitchell. *JPEG Still Image Data Compression Standard*. Van Nostrand Reinhold, 1993.

[7] R.G. Gallagher. *Information Theory and Reliable Communication*. Wiley, 1968.

[8] A.A. Sardinas and G.W. Patterson. A necessary and sufficient condition for the unique decomposition of coded messages. In *IRE Convention Records*, pages 104–108. IRE, 1953.

[9] R.J. McEliece. *The Theory of Information and Coding*, volume 3 of *Encyclopedia of Mathematics and its Application*. Addison-Wesley, 1977.

[10] J. Rissanen. Modeling by the shortest data description. Automatica, 14:465–471, 1978.

[11] J.R. Pierce. *Symbols, Signals, and Noise - The Nature and Process of Communications*. Harper, 1961.

[12] M.Li and P.Vitanyi. *An Introduction to Kolmogorov Complexity and its Applications*. Springer, 1997

[13] S. Tate. Complexity Measures. In K. Sayood, editor, *Lossless Compression Handbook*, pages 35–54. Academic Press, 2003.

[14] P. Grunwald. Minimum Description Length Tutorial. In P. Grunwald, I.J. Myung, and M.A. Pitt, editors, Advances in Minimum Description Length, pages 23–80. MIT Press, 2005.

[15] P. Grunwald, I.J. Myung, and M.A. Pitt. *Advances in Minimum Description Length.* MIT Press, 2005.

[16] D.A. Huffman. A method for the construction of minimum redundancy codes. *Proc. IRE,* 40:1098–1101, 1951.

[17] P. Deutsch. RFC 1951 - DEFLATE Compressed Data Format Specification Version 1.3, 1996. <http://www.faqs.org/rfcs/rfc1951.htm>.

[18] Lawrence L. Larmore and Daniel S. Hirschberg. A fast algorithm for optimal length-limited Huffman codes. *J. ACM,* 37:464–473, July 1990.

[19] A. Turpin and A. Moffat. Practical length-limited coding for large alphabets. *The Computer Journal,* 38(5):339–347, 1995.

[20] R.M. Fano. *Transmission of Information.* MIT Press, Cambridge, MA, 1961.

[21] R.G. Gallager. Variations on a theme by Huffman. *IEEE Transactions on Information Theory,* IT-24(6):668–674, November 1978.

[22] J.-L. Gailly and M. Adler. <http://www.gzip.org/algorithm.txt>.

[23] N. Faller. An adaptive system for data compression. In *Record of the 7th Asilomar Conference on Circuits, Systems, and Computers,* pages 593–597. IEEE, 1973.

[24] D.E. Knuth. Dynamic Huffman coding. *Journal of Algorithms,* 6:163–180, 1985.

[25] J.S. Vitter. Design and analysis of dynamic Huffman codes. *Journal of ACM,* 34(4):825–845, October 1987.

[26] P. Elias. Universal codeword sets and representations of the integers. *IEEE Transactions on Information Theory,* 21(2):194–203, 1975.

[27] S.W. Golomb. Run-length encodings. *IEEE Transactions on Information Theory,* IT-12:399–401, July 1966.

[28] R.F. Rice. Some Practical Universal Noiseless Coding Techniques. Technical Report, JPL Publication 79-22, JPL, March 1979.

[29] R.F. Rice, P.S. Yeh, and W. Miller. Algorithms for a very high speed universal noiseless coding module. Technical Report 91-1, Jet Propulsion Laboratory, California Institute of Technology, Pasadena, California, February 1991.

[30] P.S. Yeh, R.F. Rice, and W.Miller. On the optimality of code options for a universal noiseless coder. Technical Report 91-2, Jet Propulsion Laboratory, California Institute of Technology, Pasadena, California, February 1991.

[31] B.P. Tunstall. *Synthesis of Noiseless Compression Codes.* PhD thesis, Georgia Institute of Technology, September 1967.

[32] T. Robinson. SHORTEN: Simple Lossless and Near-Lossless Waveform Compression, 1994. Cambridge Univ. Eng. Dept., Cambridge, UK. Technical Report 156.

[33] T. Liebchen and Y.A. Reznik. MPEG-4 ALS: An emerging standard for lossless audio coding. In *Proceedings of the Data Compression Conference, DCC '04.* IEEE, 2004.

[34] M. Hans and R.W. Schafer. AudioPak—An integer arithmetic lossless audio code. In *Proceedings of the Data Compression Conference, DCC '98.* IEEE, 1998.

[35] S. Pigeon. Huffman Coding. In K. Sayood, editor, *Lossless Compression Handbook,* pages 79–100. Academic Press, 2003.

[36] D.A. Lelewer and D.S. Hirchberg. Data Compression. *ACM Computing Surveys,* September 1987.

[37] J.A. Storer. *Data Compression—Methods and Theory.* Computer Science Press, 1988.

[38] T.M. Cover and J.A. Thomas. *Elements of Information Theory.* Wiley Series in Telecommunications. John Wiley and Sons Inc., 1991.

[39] N. Abramson. *Information Theory and Coding*. McGraw-Hill, 1963.

[40] F. Jelinek. *Probabilistic Information Theory*. McGraw-Hill, 1968.

[41] R. Pasco. *Source Coding Algorithms for Fast Data Compression*. PhD thesis, Stanford University, 1976.

[42] J.J. Rissanen. Generalized Kraft inequality and arithmetic coding. *IBM Journal of Research and Development*, 20:198–203, May 1976.

[43] J.J. Rissanen and G.G. Langdon. Arithmetic coding. *IBM Journal of Research and Development*, 23(2):149–162, March 1979.

[44] D. Marpe, H. Schwarz, and T. Wiegand. Context based adaptive binary arithmetic coding in the H.264/AVC video coding standard. *IEEE Transaction on Circuits and Systems for Video Technology*, 13:620–636, 2003.

[45] J. Rissanen and K.M. Mohiuddin. A multiplication-free multialphabet arithmetic code. *IEEE Transactions on Communications*, 37:93–98, February 1989.

[46] I.H. Witten, R. Neal, and J.G. Cleary. Arithmetic coding for data compression. *Communications of the Association for Computing Machinery*, 30:520–540, June 1987.

[47] A. Said. Arithmetic Coding. In K. Sayood, editor, *Lossless Compression Handbook*, pages 101–152. Academic Press, 2003.

[48] G.G. Langdon Jr. An introduction to arithmetic coding. *IBM Journal of Research and Development*, 28:135–149, March 1984.

[49] J.J. Rissanen and G.G. Langdon. Universal modeling and coding. *IEEE Transactions on Information Theory*, IT-27(1):12–22, 1981.

[50] G.G. Langdon and J.J. Rissanen. Compression of black-white images with arithmetic coding. *IEEE Transactions on Communications*, 29(6):858–867, 1981.

[51] T. Bell, I.H. Witten, and J.G. Cleary. Modeling for text compression. ACM Computing Surveys, 21:557–591, December 1989.

[52] J. Ziv and A. Lempel. A universal algorithm for data compression. *IEEE Transactions on Information Theory*, IT-23(3):337–343, May 977.

[53] J. Ziv and A. Lempel. Compression of individual sequences via variable-rate coding. *IEEE Transactions on Information Theory*, IT-24(5):530–536, September 1978.

[54] J.A. Storer and T.G. Syzmanski. Data compression via textual substitution. *Journal of the ACM*, 29:928–951, 1982.

[55] T.C. Bell. Better OPM/L text compression. *IEEE Transactions on Communications*, COM-34:1176–1182, December 1986.

[56] T.A. Welch. A technique for high-performance data compression. *IEEE Computer*, pages 8–19, June 1984.

[57] A. Lempel and J. Ziv. On the complexity of finite sequences. *IEEE Transactions on Information Theory*, IT-22:75–81, 1976.

[58] D. Benedetto, E. Cagliotti, and V. Loreto. Language trees and zipping. *Physical Review Letters*, 88, January 2002.

[59] Ming Li, Jonathan H. Badger, Xin Chen, Sam Kwong, Paul Kearney, and Haoyong Zhang. An information-based sequence distance and its application to whole mitochondrial genome phylogeny. *Bioinformatics*, 17(2):149–154, 2001.

[60] H.H. Otu and K. Sayood. A new sequence distance measure for phylogenetic tree construction. *Bioinformatics*, 19(16):2122, 2003.

[61] Xuan Xiao, Shi-Huang Shao, Zheng-De Huang, and Kuo-Chen Chou. Using pseudo amino acid composition to predict protein structural classes: Approached with complexity measure factor. *Journal of Computational Chemistry*, 27(4):478–482, 2006.

[62] D.J. Russell, H.H. Otu, and K. Sayood. Grammar-based distance in progressive multiple sequence alignment. *BMC Bioinformatics*, 9, 2008.

[63] A. Albayrak, H. Otu, and U. Sezerman. Clustering of protein families into functional subtypes using relative complexity measure with reduced amino acid alphabets. *BMC Bioinformatics*, 11(1):428, 2010.

[64] M. Aboy, R. Hornero, D. Abasolo, and D. Alvarez. Interpretation of the Lempel-Ziv complexity measure in the context of biomedical signal analysis. *IEEE Transactions on Biomedical Engineering*, 53(11):2282–2288, Nov. 2006.

[65] N. Radhakrishnan and B.N. Gangadhar. Estimating regularity in epileptic seizure time-series data. *IEEE Engineering in Medicine and Biology Magazine*, 17:89–94, 1998.

[66] X.-S. Zhang, R.J. Roy, and E.W. Jensen. EEG complexity as a measure of depth of anesthesia for patients. *IEEE Transactions on Biomedical Engineering*, 48(12):1424–1433, Dec. 2001.

[67] Daniel Abasolo, Roberto Hornero, Carlos Gomez, Maria Garcia, and Miguel Lopez. Analysis of EEG background activity in Alzheimer's disease patients with Lempel-Ziv complexity and central tendency measure. *Medical Engineering Physics*, 28(4):315–322, 2006.

[68] H. Zhang, Y. Zhu, and Z. Wang. Complexity measure and complexity rate information based detection of ventricular tachycardia and fibrillation. *Medical and Biological Engineering and Computing*, 38:553–557, 2000.

[69] M. Nelson and J.-L. Gailly. *The Data Compression Book*. M&T Books, California, 1996.

[70] G. Held and T.R. Marshall. *Data Compression*. Wiley, third edition, 1991.

[71] G. Roelofs. PNG Lossless Compression. In K. Sayood, editor, *Lossless Compression Handbook*, pages 371–390. Academic Press, 2003.

[72] S.C. Sahinalp and N.M. Rajpoot. Dictionary-Based Data Compression: An Algorithmic Perspective. In K. Sayood, editor, *Lossless Compression Handbook*, pages 153–168. Academic Press, 2003.

[73] N. Chomsky. *The Minimalist Program*. MIT Press, 1995.

[74] J.G. Cleary and I.H. Witten. Data compression using adaptive coding and partial string matching. *IEEE Transactions on Communications*, 32(4):396–402, 1984.

[75] A. Moffat. Implementing the PPM data compression scheme. *IEEE Transactions on Communications*, Vol COM-38:1917–1921, November 1990.

[76] J.G. Cleary and W.J. Teahan. Unbounded length contexts for PPM. *The Computer Journal*, Vol 40:x30–74, February 1997.

[77] M. Burrows and D.J. Wheeler. A Block Sorting Data Compression Algorithm. Technical Report SRC 124, Digital Systems Research Center, 1994.

[78] P. Fenwick. Burrows-Wheeler Compression. In K. Sayood, editor, *Lossless Compression Handbook*, pages 169–194. Academic Press, 2003.

[79] S.L. Salzberg, A.J. Salzberg, A.R. Kerlavage, and J.F. Tomb. Skewed oligomers and origins of replication. *Gene*, 217:57–67, 1998.

[80] G.V. Cormack and R.N.S. Horspool. Data compression using dynamic Markov modelling. *The Computer Journal*, Vol 30:541–337, June 1987.

[81] P. Fenwick. Symbol-Ranking and ACB Compression. In K. Sayood, editor, *Lossless Compression Handbook*, pages 195–204. Academic Press, 2003.

[82] G.K. Wallace. The JPEG still picture compression standard. *Communications of the ACM*, 34:31–44, April 1991.

[83] X. Wu, N.D. Memon, and K. Sayood. A Context Based Adaptive Lossless/Nearly-Lossless Coding Scheme for Continuous Tone Images. ISO Working Document ISO/IEC SC29/WG1/N256, 1995.

[84] X. Wu and N.D. Memon. CALIC—A Context Based Adaptive Lossless Image Coding Scheme. *IEEE Transactions on Communications*, May 1996.

[85] K. Sayood and S. Na. Recursively indexed quantization of memoryless sources. *IEEE Transactions on Information Theory*, IT-38:1602–1609, November 1992.

[86] S. Na and K. Sayood. Recursive Indexing Preserves the Entropy of a Memoryless Geometric Source, 1996.

[87] M. Weinberger, G. Seroussi, and G. Sapiro. The LOCO-I Lossless Compression Algorithm: Principles and Standardization into JPEG-LS. Technical Report HPL-98-193, Hewlett-Packard Laboratory, November 1998.

[88] S.A. Martucci. Reversible compression of HDTV images using median adaptive prediction and arithmetic coding. In *IEEE International Symposium on Circuits and Systems,* pages 1310–1313. IEEE Press, 1990.

[89] S.D. Babacan and K. Sayood. Predictive image compression using conditional averages. In *Proceedings of the Data Compression Conference, DCC '04.* IEEE, 2004.

[90] Q. Wang, G.M. Garrity, J.M. Tiedje, and J.R. Cole. Naive Bayesian classifier for rapid assignment of rRNA sequences into the new bacterial taxonomy. *Applied and Environmental Microbiology,* 73:5261–5267, 2007.

[91] M. Rabbani and P.W Jones. *Digital Image Compression Techniques,* volume TT7 of *Tutorial Texts Series.* SPIE Optical Engineering Press, 1991.

[92] I.H. Witten, A. Moffat, and T.C. Bell. *Managing Gigabytes: Compressing and Indexing Documents and Images.* Van Nostrand Reinhold, New York, 1994.

[93] S.L. Tanimoto. Image transmission with gross information first. *Computer Graphics and Image Processing,* 9:72–76, January 1979.

[94] K.R. Sloan Jr. and S.L. Tanimoto. Progressive refinement of raster images. *IEEE Transactions on Computers,* C-28:871–874, November 1979.

[95] P.J. Burt and E.H. Adelson. The Laplacian pyramid as a compact image code. *IEEE Transactions on Communications,* COM-31:532–540, April 1983.

[96] K. Knowlton. Progressive transmission of grey-scale and binary pictures by simple, efficient, and lossless encoding schemes. *Proceedings of the IEEE,* 68:885–896, July 1980.

[97] H. Dreizen. Content-driven progressive transmission of grey-scale images. *IEEE Transactions on Communications,* COM-35:289–296, March 1987.

[98] J. Capon. A probabilistic model for run-length coding of pictures. *IRE Transactions on Information Theory,* pages 157–163, 1959.

[99] Y. Yasuda. Overview of digital facsimile coding techniques in Japan. *IEEE Proceedings,* 68:830–845, July 1980.

[100] R. Hunter and A.H. Robinson. International digital facsimile coding standards. *IEEE Proceedings,* 68:854–867, July 1980.

[101] ISO/IEC 11544:1993 Information technology-coded representation of picture and audio information-progressive bi-level image compression. 1994.

[102] G.G. Langdon Jr. and J.J. Rissanen. A simple general binary source code. *IEEE Transactions on Information Theory,* IT-28:800–803, September 1982.

[103] R.B. Arps and T.K. Truong. Comparison of international standards for lossless still image compression. *Proceedings of the IEEE,* 82:889–899, June 1994.

[104] W.K. Pratt. *Digital Image Processing*. Wiley-Interscience, 1978.

[105] F.W. Campbell. The human eye as an optical filter. *Proceedings of the IEEE*, 56:1009–1014, June 1968.

[106] J.L. Mannos and D.J. Sakrison. The effect of a visual fidelity criterion on the encoding of images. *IEEE Transactions on Information Theory*, IT-20:525–536, July 1974.

[107] H. Fletcher and W.A. Munson. Loudness, its measurement, definition, and calculation. *Journal of the Acoustical Society of America*, 5:82–108, 1933.

[108] B.C.J. Moore. *An Introduction to the Psychology of Hearing*. Academic Press, third edition, 1989.

[109] S.S. Stevens and H. Davis. *Hearing—Its Psychology and Physiology*. American Inst. of Physics, 1938.

[110] M. Mansuripur. *Introduction to Information Theory*. Prentice-Hall, 1987.

[111] C.E. Shannon. Coding Theorems for a Discrete Source with a Fidelity Criterion. In *IRE International Convention Records*, volume 7, pages 142–163. IRE, 1959.

[112] T. Berger. *Rate Distortion Theory: A Mathematical Basis for Data Compression*. Prentice-Hall, Englewood Cliffs, NJ, 1971.

[113] S. Arimoto. An algorithm for computing the capacity of arbitrary discrete memoryless channels. *IEEE Transactions on Information Theory*, IT-18:14–20, January 1972.

[114] R.E. Blahut. Computation of channel capacity and rate distortion functions. *IEEE Transaction on Information Theory*, IT-18:460–473, July 1972.

[115] A.M. Law and W.D. Kelton. *Simulation Modeling and Analysis*. McGraw Hill, 1982.

[116] L.R. Rabiner and R.W. Schafer. *Digital Processing of Speech Signals*. Signal Processing. Prentice-Hall, 1978.

[117] R.B. Ash. *Information Theory*. Dover, 1990. (Originally published by Interscience Publishers in 1965.)

[118] R.M. Gray. *Entropy and Information Theory*. Springer-Verlag, 1990.

[119] Thomas Parsons. *Voice and Speech Processing*. McGraw Hill, 1987.

[120] E.F. Abaya and G.L. Wise. On the existence of optimal quantizers. *IEEE Transactions on Information Theory*, IT-28:937–940, November 1982.

[121] N. Jayant and L. Rabiner. The application of dither to the quantization of speech signals. *Bell System Technical Journal*, 51:1293–1304, June 1972.

[122] J. Max. Quantizing for minimum distortion. *IRE Transactions on Information Theory*, IT-6:7–12, January 1960.

[123] W.C. Adams Jr. and C.E. Geisler. Quantizing characteristics for signals having Laplacian amplitude probability density function. *IEEE Transactions on Communications*, COM-26:1295–1297, August 1978.

[124] N.S. Jayant. Adaptive qunatization with one word memory. *Bell System Technical Journal*, pages 1119–1144, September 1973.

[125] D. Mitra. Mathematical analysis of an adaptive quantizer. *Bell Systems Technical Journal*, pages 867–898, May-June 1974.

[126] A. Gersho and D.J. Goodman. A training mode adaptive quantizer. *IEEE Transactions on Information Theory*, IT-20:746–749, November 1974.

[127] A. Gersho. Quantization. *IEEE Communications Magazine*, September 1977.

[128] J. Lukaszewicz and H. Steinhaus. On measuring by comparison. *Zastos. Mat.*, pages 225–231, 1955. (in Polish).

[129] S.P. Lloyd. Least squares quantization in PCM. *IEEE Transactions on Information Theory*, IT-28:127–135, March 1982.

[130] W.R. Bennett. Spectra of quantized signals. *Bell System Technical Journal*, 27:446–472, July 1948.

[131] T. Berger, F. Jelinek, and J. Wolf. Permutation codes for sources. *IEEE Transactions on Information Theory*, IT-18:166–169, January 1972.

[132] N. Farvardin and J.W. Modestino. Optimum quantizer performance for a class of non-gaussian memoryless sources. *IEEE Transactions on Information Theory*, pages 485–497, May 1984.

[133] H. Gish and J.N. Pierce. Asymptotically efficient quantization. *IEEE Transactions on Information Theory*, IT-14:676–683, September 1968.

[134] N.S. Jayant and P. Noll. *Digital Coding of Waveforms*. Prentice-Hall, 1984.

[135] W. Mauersberger. Experimental results on the performance of mismatched quantizers. *IEEE Transactions on Information Theory*, pages 381–386, July 1979.

[136] A. Gersho and R.M. Gray. *Vector Quantization and Signal Compression.* Kluwer Academic Publishers, 1991.

[137] Y. Linde, A. Buzo, and R.M. Gray. An algorithm for vector quantization design. *IEEE Transactions on Communications,* COM-28:84–95, Jan. 1980.

[138] E.E. Hilbert. Cluster Compression Algorithm—A Joint Clustering Data Compression Concept. Technical Report JPL Publication 77-43, NASA, 1977.

[139] W.H. Equitz. A New vector quantization clustering algorithm. *IEEE Transactions on Acoustics, Speech, and Signal Processing,* 37:1568–1575, October 1989.

[140] P.A. Chou, T. Lookabaugh, and R.M. Gray. Optimal pruning with applications to tree-structured source coding and modeling. *IEEE Transactions on Information Theory,* 35:31–42, January 1989.

[141] L. Breiman, J.H. Freidman, R.A. Olshen, and C.J. Stone. *Classification and Regression Trees.* Wadsworth, California, 1984.

[142] E.A. Riskin. Pruned tree structured vector quantization in image coding. In *Proceedings International Conference on Acoustics Speech and Signal Processing,* pages 1735–1737. IEEE, 1989.

[143] D.J. Sakrison. A geometric treatment of the source encoding of a Gaussian random variable. *IEEE Transactions on Information Theory,* IT-14:481–486, May 1968.

[144] T.R. Fischer. A pyramid vector quantizer. *IEEE Transactions on Information Theory,* IT-32:568–583, July 1986.

[145] M.J. Sabin and R.M. Gray. Product code vector quantizers for waveform and voice coding. *IEEE Transactions on Acoustics, Speech, and Signal Processing,* ASSP-32:474–488, June 1984.

[146] W.A. Pearlman. Polar quantization of a complex Gaussian random variable. *IEEE Transactions on Communications,* COM-27:892–899, June 1979.

[147] S.G. Wilson. Magnitude phase quantization of independent Gaussian variates. *IEEE Transactions on Communications,* COM-28:1924–1929, November 1980.

[148] P.F. Swaszek and J.B. Thomas. Multidimensional spherical coordinates quantization. *IEEE Transactions on Information Theory,* IT-29:570–575, July 1983.

[149] D.J. Newman. The hexagon theorem. *IEEE Transactions on Information Theory,* IT-28:137–139, March 1982.

[150] J.H. Conway and N.J.A. Sloane. Voronoi regions of lattices, second moments of polytopes, and quantization. *IEEE Transactions on Information Theory,* IT-28:211–226, March 1982.

[151] K. Sayood, J.D. Gibson, and M.C. Rost. An algorithm for uniform vector quantizer design. *IEEE Transactions on Information Theory,* IT-30:805–814, November 1984.

[152] J.D. Gibson and K. Sayood. Lattice Quantization. In P.W. Hawkes, editor, *Advances in Electronics and Electron Physics,* pages 259–328. Academic Press, 1990.

[153] J.H. Conway and N.J.A. Sloane. Fast quantizing and decoding algorithms for lattice quantizers and codes. *IEEE Transactions on Information Theory,* IT-28:227–232, March 1982.

[154] J.H. Conway and N.J.A. Sloane. A fast encoding method for lattice codes and quantizers. *IEEE Transactions on Information Theory,* IT-29:820–824, November 1983.

[155] H. Abut, editor. *Vector Quantization.* IEEE Press, 1990.

[156] A. Buzo, A.H. Gray, R.M. Gray, and J.D. Markel. Speech coding based upon vector quantization. *IEEE Transactions on Acoustics, Speech, and Signal Processing,* ASSP-28:562–574, October 1980.

[157] B. Ramamurthi and A. Gersho. Classified vector quantization of images. *IEEE Transactions on Communications,* COM-34:1105–1115, November 1986.

[158] V. Ramamoorthy and K. Sayood. A hybrid LBG/lattice vector quantizer for high quality image coding. In E. Arikan, editor, *Proc. 1990 Bilkent International Conference on New Trends in Communication, Control and Signal Processing.* Elsevier, 1990.

[159] B.H. Juang and A.H. Gray. Multiple stage vector quantization for speech coding. In *Proceedings IEEE International Conference on Acoustics, Speech, and Signal Processing,* pages 597–600. IEEE, April 1982.

[160] C.F. Barnes and R.L. Frost. Residual vector quantizers with jointly optimized code books. In *Advances in Electronics and Electron Physics,* pages 1–59. Elsevier, 1992.

[161] C.F. Barnes and R.L. Frost. Vector quantizers with direct sum codebooks. *IEEE Transactions on Information Theory,* 39:565–580, March 1993.

[162] A. Gersho and V. Cuperman. A pattern matching technique for speech coding. *IEEE Communications Magazine,* pages 15–21, December 1983.

[163] A.G. Al-Araj and K. Sayood. Vector quantization of nonstationary sources. In *Proceedings International Conference on Telecommunications - 1994,* pages 92–95. IEEE, 1994.

[164] A.G. Al-Araj. *Recursively Indexed Vector Quantization.* PhD thesis, University of Nebraska—Lincoln, 1994.

[165] S. Panchanathan and M. Goldberg. Adaptive algorithm for image coding using vector quantization. *Signal Processing: Image Communication,* 4:81–92, 1991.

[166] A.J. Viterbi and J.K. Omura. *Principles of Digital Communications and Coding.* McGraw Hill, 1979.

[167] J. Makhoul, S. Roucos, and H. Gish. Vector quantization in speech coding. *Proceedings of the IEEE,* 73:1551–1588, 1985.

[168] N.M. Nasrabadi and R.A. King. Image coding using vector quantization. *IEEE Transactions on Communications,* August 1988.

[169] C.C. Cutler. Differential Quantization for Television Signals. *U.S. Patent 2 605 361,* July 29, 1952.

[170] N.L. Gerr and S. Cambanis. Analysis of adaptive differential PCM of a stationary Gauss-Markov input. *IEEE Transactions on Information Theory,* IT-33:350–359, May 1987.

[171] H. Stark and J.W. Woods. *Probability, Random Processes, and Estimation Theory for Engineers.* Prentice-Hall, second edition, 1994.

[172] J.D. Gibson. Adaptive prediction in speech differential encoding systems. *Proceedings of the IEEE,* pages 488–525, April 1980.

[173] P.A. Maragos, R.W. Schafer, and R.M. Mersereau. Two dimensional linear prediction and its application to adaptive predictive coding of images. *IEEE Transactions on Acoustics, Speech, and Signal Processing,* ASSP-32:1213–1229, December 1984.

[174] J.D. Gibson, S.K. Jones, and J.L. Melsa. Sequentially adaptive prediction and coding of speech signals. *IEEE Transactions on Communications,* COM-22:1789–1797, November 1974.

[175] B. Widrow, J.M. McCool, M.G. Larimore, and C.R. Johnson Jr. Stationary and nonstationary learning characteristics of the LMS adaptive filter. *Proceedings of the IEEE,* pages 1151–1162, August 1976.

[176] N.S. Jayant. Adaptive delta modulation with one-bit memory. *Bell System Technical Journal,* pages 321–342, March 1970.

[177] R. Steele. *Delta Modulation Systems.* Halstead Press, 1975.

[178] R.L. Auger, M.W. Glancy, M.M. Goutmann, and A.L. Kirsch. The Space Shuttle Ground Terminal Delta Modulation System. *IEEE Transactions on Communications,* COM-26:1660–1670, November 1978. Part I of two parts.

[179] M.J. Shalkhauser and W.A. Whyte Jr. Digital CODEC for Real Time Signal Processing at 1.8 bpp. In *Global Telecommunication Conference,* 1989.

[180] D.G. Luenberger. *Optimization by Vector Space Methods.* Series in Decision and Control. John Wiley & Sons Inc., 1969.

[181] B.B. Hubbard. *The World According to Wavelets.* Series in Decision and Control. A K Peters, 1996.

[182] W.H. Press, S.A. Teukolsky, W.T. Vettering, and B.P. Flannery. *Numerical Recipes in C.* Cambridge University Press, 2nd edition, 1992.

[183] H. Hotelling. Analysis of a complex of statistical variables into principal components. *Journal of Educational Psychology,* 24, 1933.

[184] H. Karhunen.Über Lineare Methoden in der Wahrscheinlich-Keitsrechunung. *Annales Academiae Fennicae, Series A,* 1947.

[185] M. Loéve. Fonctions Aléatoires de Seconde Ordre. In P. Lévy, editor, *Processus Stochastiques et Mouvement Brownien.* Hermann, 1948.

[186] H.P. Kramer and M.V. Mathews. A linear encoding for transmitting a set of correlated signals. *IRE Transactions on Information Theory,* IT-2:41–46, September 1956.

[187] J.-Y. Huang and P.M. Schultheiss. Block quantization of correlated Gaussian random variables. *IEEE Transactions on Communication Systems,* CS-11:289–296, September 1963.

[188] K. Sayood and J.D. Gibson. Explicit additive noise models for uniform and nonuniform MMSE quantization. *Signal Processing,* 7:407–414, 1984.

[189] N. Ahmed and K.R. Rao. *Orthogonal Transforms for Digital Signal Processing.* Springer-Verlag, 1975.

[190] J.A. Saghri, A.J. Tescher, and J.T. Reagan. Terrain adaptive transform coding of multispectral data. In *Proceedings International Conference on Geosciences and Remote Sensing (IGARSS '94),* pages 313–316. IEEE, 1994.

[191] P.M. Farrelle and A.K. Jain. Recursive block coding—a new approach to transform coding. *IEEE Transactions on Communications,* COM-34:161–179, February 1986.

[192] M. Bosi and G. Davidson. High quality, low rate audio transform coding for transmission and multimedia application. In *Preprint 3365, Audio Engineering Society.* AES, October 1992.

[193] F.J. MacWilliams and N.J.A. Sloane. *The Theory of Error Correcting Codes.* North-Holland, 1977.

[194] M.M. Denn. *Optimization by Variational Methods.* McGraw-Hill, 1969.

[195] P.A. Wintz. Transform picture coding. *Proceedings of the IEEE,* 60:809–820, July 1972.

[196] W.-H. Chen and W.K. Pratt. Scene adaptive coder. *IEEE Transactions on Communications,* COM-32:225–232, March 1984.

[197] Y. Shoham and A. Gersho. Efficient bit allocation for an arbitrary set of quantizers. *IEEE Transactions on Acoustics, Speech, and Signal Processing,* ASSP-36:1445–1453, September 1988.

[198] H.S. Malvar. *Signal Processing with Lapped Transforms.* Artech House, Norwood, MA, 1992.

[199] J.P. Princen and A.P. Bradley. Analysis/synthesis filter design based on time domain aliasing cancellation. *IEEE Transactions on Acoustics Speech and Signal Processing,* ASSP-34:1153–1161, October 1986.

[200] M. Bosi and R.E. Goldberg. *Introduction to Digital Audio Coding and Standards.* Kluwer Academic Press, 2003.

[201] D.F. Elliot and K.R. Rao. *Fast Transforms—Algorithms, Analysis, Applications.* Academic Press, 1982.

[202] A.K. Jain. *Fundamentals of Digital Image Processing.* Prentice Hall, 1989.

[203] A. Crosier, D. Esteban, and C. Galand. Perfect channel splitting by use of interpolation/decimation techniques. In *Proc. International Conference on Information Science and Systems*, Patras, Greece, 1976. IEEE.

[204] J.D. Johnston. A filter family designed for use in quadrature mirror filter banks. In *Proceedings ICASSP*, pages 291–294. IEEE, April 1980.

[205] M.J.T. Smith and T.P. Barnwell III. A procedure for designing exact reconstruction filter banks for tree structured subband coders. In *Proceedings IEEE International Conference on Acoustics Speech and Signal Processing*. IEEE, 1984.

[206] P.P. Vaidyanathan. *Multirate Systems and Filter Banks*. Prentice Hall, 1993.

[207] H. Caglar. *A Generalized Parametric PR-QMF/Wavelet Transform Design Approach for Multiresolution Signal Decomposition*. PhD thesis, New Jersey Institute of Technology, May 1992.

[208] A.K. Jain and R.E. Crochiere. Quadrature mirror filter design in the time domain. *IEEE Transactions on Acoustics, Speech, and Signal Processing*, 32:353–361, April 1984.

[209] F. Mintzer. Filters for distortion-free two-band multirate filter banks. *IEEE Transactions on Acoustics, Speech, and Signal Processing*, ASSP-33:626–630, June 1985.

[210] J.W. Woods and T. Naveen. A filter based bit allocation scheme for subband compression of HDTV. *IEEE Transactions on Image Processing*, IP-1:436–440, July 1992.

[211] M. Vetterli. Multirate Filterbanks for Subband Coding. In J.W. Woods, editor, *Subband Image Coding*, pages 43–100. Kluwer Academic Publishers, 1991.

[212] C.S. Burrus, R.A. Gopinath, and H. Guo. *Introduction to Wavelets and Wavelet Transforms.* Prentice Hall, 1998.

[213] W. Sweldens. Lifting scheme: a new philosophy in biorthogonal wavelet construction. In *Wavelet Applications in Signal and Image Processing,* pages 68–79.

[214] I. Daubechies and W. Sweldens. Factoring wavelet transforms into lifting steps. *Journal of Fourier Analysis and Applications,* 4(3), 1996.

[215] J.M. Shapiro. Embedded image coding using zerotrees of wavelet coefficients. *IEEE Transactions on Signal Processing*, SP-41:3445–3462, December 1993.

[216] A. Said and W.A. Pearlman. A new fast and efficient coder based on set partitioning in hierarchical trees. *IEEE Transactions on Circuits and Systems for Video Technologies*, pages 243–250, June 1996.

[217] D. Taubman. *Directionality and Scalability in Image and Video Compression*. PhD thesis, University of California at Berkeley, May 1994.

[218] D. Taubman and A. Zakhor. Multirate 3-D subband coding with motion compensation. *IEEE Transactions on Image Processing*, IP-3:572–588, September 1994.

[219] ISO/IEC JTC1/SC29 WG1. JPEG 2000 Image coding system: Core coding system, March 2004.

[220] D. Taubman and M. Marcellin. *JPEG2000: Image Compression Fundamentals, Standards and Practice*. Kluwer Academic Press, 2001.

[221] ISO/IEC IS 14496. Coding of Moving Pictures and Audio.

[222] J. Watkinson. *The MPEG Handbook*. Focal Press, 2001.

[223] K. Tsutsui, H. Suzuki, O. Shimoyoshi, M. Sonohara, K. Agagiri, and R.M. Heddle. ATRAC: Adaptive Transform Acoustic Coding for MiniDisc. In *Conference Records Audio Engineering Society Convention*. AES, October 1992.

[224] D. Pan. A tutorial on MPEG/audio compression. *IEEE Multimedia*, 2:60–74, 1995.

[225] T. Painter and A. Spanias. Perceptual coding of digital audio. *Proceedings of the IEEE*, 88:451–513, 2000.

[226] H. Dudley and T.H. Tarnoczy. Speaking machine of Wolfgang Von Kempelen. *Journal of the Acoustical Society of America*, 22:151–166, March 1950.

[227] H. Dudley. Remaking speech. *Journal of the Acoustical Society of America*, 11:169–177, 1939.

[228] D.C. Farden. Solution of a Toeplitz set of linear equations. *IEEE Transactions on Antennas and Propogation*, 1977.

[229] N. Levinson. The Weiner RMS error criterion in filter design and prediction. *Journal of Mathematical Physics*, 25:261–278, 1947.

[230] J. Durbin. The fitting of time series models. *Review of the Institute Inter. Statist.*, 28:233–243, 1960.

[231] P.E. Papamichalis. *Practical Approaches to Speech Coding*. Prentice-Hall, 1987.

[232] J.D. Gibson. On reflection coefficients and the Cholesky decomposition. *IEEE Transactions on Acoustics, Speech, and Signal Processing*, ASSP-25:93–96, February 1977.

[233] M.R. Schroeder. Linear predictive coding of speech: review and current directions. *IEEE Communications Magazine*, 23:54–61, August 1985.

[234] B.S. Atal and J.R. Remde. A new model of LPC excitation for producing natural sounding speech at low bit rates. In *Proceedings IEEE International Conference on Acoustics, Speech, and Signal Processing*, pages 614–617. IEEE, 1982.

[235] A. Gersho. Advances in speech and audio compression. *Proceedings of the IEEE*, 82:900–918, 1994.

[236] P. Kroon, E.F. Deprettere, and R.J. Sluyter. Regular-pulse excitation—a novel approach to effective and efficient multipulse coding of speech. *IEEE Transactions on Acoustics, Speech, and Signal Processing*, ASSP-34:1054–1063, October 1986.

[237] K. Hellwig, P. Vary, D. Massaloux, and J.P. Petit. Speech codec for European Mobile Radio System. In *Conference Record, IEEE Global Telecommunication Conference*, pages 1065–1069. IEEE, 1989.

[238] J.P. Campbell, V.C. Welch, and T.E. Tremain. An expandable error protected 4800 bps CELP coder (US Federal Standard 4800 bps voice coder). In *Proceedings International Conference on Acoustics, Speech and Signal Processing*, pages 735–738. IEEE, 1989.

[239] J.P.Campbell Jr., T.E. Tremain, and V.C. Welch. The DOD 4.8 KBPS Standard (Proposed Federal Standard 1016). In B.S. Atal, V. Cuperman, and A. Gersho, editors, *Advances in Speech Coding*, pages 121–133. Kluwer, 1991.

[240] J.-H. Chen, R.V. Cox, Y.-C. Lin, N. Jayant, and M. Melchner. A low-delay CELP coder 16 kb/s speech coding standard. *IEEE Journal on Selected Areas in Communications*, 10:830–849, 1992.

[241] R.J. McAulay and T.F. Quatieri. Low-Rate Speech Coding Based on the Sinusoidal Model. In S. Furui and M.M. Sondhi, editors, *Advances in Speech Signal Processing*, chapter 6, pages 165–208. Marcel-Dekker, 1992.

[242] D.W. Griffin and J.S. Lim. Multi-band excitation vocoder. *IEEE Transactions on Acoustics, Speech and Signal Processing*, 36:1223–1235, August 1988.

[243] S. Anderson, A. Duric, H. Astrom, R. Hagen, W. Kleijn, J. Linden, M. Murthi, F. Galshiodt, J. Spittka, and J. Skoglund. RFE 3951–Internet Low Bit Rate Coded (iLBC), December 2004.

[244] A. Duric and S. Anderson. RFC 3952–Real Time Transport Protocol (RTP) Payload Format for Internet Low Bit Rate Codec (iLBC) Speech, 2004.

[245] J.D. Gibson. Speech coding methods, standards, and applications. *IEEE Circuits and Systems Magazine*, pages 30–49, 2005.

[246] D. Barry and JA Hartigan. Asynchronous distance between homologous dna sequences. *Biometrics*, pages 261–276, 1987.

[247] A.E. Jacquin. *A Fractal Theory of Iterated Markov Operators with Applications to Digital Image Coding*. PhD thesis, Georgia Institute of Technology, August 1989.

[248] A.E. Jacquin. Image coding based on a fractal theory of iterated contractive image transformations. *IEEE Transactions on Image Processing*, 1:18–30, January 1992.

[249] H. Samet. *The Design and Analysis of Spatial Data Structures*. Addison-Wesley, Reading, MA, 1990.

[250] Y.Fisher ed. *Fractal Image Compression*. Springer-Verlag, 1995.

[251] D. Saupe, M. Ruhl, R. Hamzaoui, L. Grandi, and D. Marini. Optimal hierarchical partitions for fractal image compression. In *Proc. IEEE International Conference on Image Processing*. IEEE, 1998.

[252] J. Makhoul. Linear prediction: a tutorial review. *Proceedings of the IEEE*, 63:561–580, April 1975.

[253] T. Adamson. *Electronic Communications*. Delmar, 1988.

[254] C.S. Choi, K. Aizawa, H. Harashima, and T. Takebe. Analysis and synthesis of facial image sequences in model-based image coding. *IEEE Transactions on Circuits and Systems for Video Technology*, 4:257–275, June 1994.

[255] H. Li and R. Forchheimer. Two-view facial movement estimation. *IEEE Transactions on Circuits and Systems for Video Technology*, 4:276–287, June 1994.

[256] G. Bozdaği, A.M. Tekalp, and L. Onural. 3-D motion estimation and wireframe adaptation including photometric effects for model-based coding of facial image sequences. *IEEE Transactions on Circuits and Systems for Video Technology*, 4:246–256, June 1994.

[257] P. Ekman and W.V. Friesen. *Facial Action Coding System*. Consulting Psychologists Press, 1977.

[258] K. Aizawa and T.S. Huang. Model-based image coding: advanced video coding techniques for very low bit-rate applications. *Proceedings of the IEEE*, 83:259–271, February 1995.

[259] L. Chiariglione. The development of an integrated audiovisual coding standard: MPEG. *Proceedings of the IEEE*, 83:151–157, February 1995.

[260] ISO/IEC IS 11172. Information technology—coding of moving pictures and associated audio for digital storage media up to about 1.5 Mbits/s.

[261] ISO/IEC IS 13818. Information technology—generic coding of moving pictures and associated audio information.

[262] T. Wiegand, G.J. Sullivan, G. Bjontegaard, and A. Luthra. Overview of the H.264/AVC video coding standard. *IEEE Transaction on Circuits and Systems for Video Technology*, 13:560–576, 2003.

[263] G. Karlsson and M. Vetterli. Packet video and its integration into the network architecture. *IEEE Journal on Selected Areas in Communications*, 7:739–751, June 1989.

[264] Y-C. Chen, K. Sayood, and D.J. Nelson. A robust coding scheme for packet video. *IEEE Transactions on Communications*, 40:1491–1501, September 1992.

[265] M.C. Rost and K. Sayood. A progressive data compression scheme based on adaptive transform coding. In *Proceedings 31st Midwest Symposium on Circuits and Systems*, pages 912–915. Elsevier, 1988.

[266] J. Watkinson. *The Art of Digital Video*. Focal Press, 1990.

[267] J.L. Mitchell, W.B. Pennebaker, C.E. Fogg, and D.J. LeGall. *MPEG Video Compression Standard*. Chapman and Hall, 1997.

[268] K.P. Subbalakshmi. Lossless Image Compression. In K. Sayood, editor, *Lossless Compression Handbook*, pages 207–226. Academic Press, 2003.

[269] M.W. Hoffman. JPEG-LS Lossless and Near Lossless Image Compression. In K. Sayood, editor, *Lossless Compression Handbook*, Academic Press, 2003.

[270] J.A. Bucklew and N.C. Gallagher Jr. A Note on Optimal Quantization. *IEEE Transactions on Information Theory*, IT-25:365–366, May 1979.

[271] J.A. Bucklew and N.C. Gallagher Jr. Some Properties of Uniform Step Size Quantizers. *IEEE Transactions on Information Theory*, IT-26:610–613, September 1980.

[272] K. Sayood and J.D. Gibson. Explicit Additive Noise Models for Uniform and Nonuniform MMSE Quantization. *Signal Processing*, 7:407–414, 1984.

[273] R.M. Gray. Vector Quantization. *IEEE Acoustics, Speech, and Signal Processing Magazine*, 1:4–29, April 1984.

[274] P. Swaszek. Vector Quantization. In I.F. Blake and V. Poor, editor, *Communications and Networks: A survey of Recent Advances*, pages 362–389. Springer-Verlag, 1986.

[275] B.P. Lathi. *Signal Processing and Linear Systems*. Berkeley Cambridge Press, 1998.

[276] S.K. Mitra and J.F. Kaiser. *Handbook for Digital Signal Processing*. Wiley-Interscience, 1993.

[277] K. Brandenburg and G. Stoll. ISO-MPEG-1 Audio: A Generic Standard for Coding High-Quality Digital Audio. *Journal of the Audio Engineering Society,* 42:780–792, October 1994.

[278] A. Ortega and K. Ramachandran. Rate Distortion Methods for Image and Video Compression. *Signal Processing Magazine,* 15:23–50, November 1998.

[279] I. Daubechies. *Ten lectures on wavelets*. SIAM, 1992.

[280] William A. Pearlman and Amir Said. *Set Partition Coding: Part I of Set Partition Coding and Image Wavelet Coding Systems.* Foundations and trends in signal processing. Now, 2008.

[281] William A. Pearlman and Amir Said. *Image Wavelet Coding Systems: Part II of Set Partition Coding and Image Wavelet Coding Systems.* Foundations and trends in signal processing. Now, 2008.

[282] N. Iwakami, T. Moriya, and S. Miki. High Quality Audio-Coding at Less than 64 kbit/s by Using Transform Domain weighted Interleave Vector Quantization TwinVQ. *Proceedings ICASSP '95, IEEE*, 5:3095–3098, 1985.

[283] B.S. Atal, V. Cuperman, and A. Gersho. *Speech and Audio Coding for Wireless and Network Applications*. Kluwer Academic Publishers, 1993.

[284] S. Furui and M.M. Sondhi. *Advances in Speech Signal Processing*. Marcel Dekker Inc., 1991.

[285] A.M. Kondoz. *Digital Speech: Coding for Low Bit Rate Communication Systems*. Wiley, 2004.

[286] I.E. Richardson. *H.264 and MPEG 4 Video Compression*. Wiley, 2003.

[287] W. Feller. *An Introduction to Probability Theory and Its Applications*, Vol. 1. Wiley, 3rd Edition, 1968.

[288] W. Feller. *An Introduction to Probability Theory and Its Applications*, Vol. 2. Wiley, 3rd Edition, 1971.

[289] A. Papoulis. *Probability, Random Variables, and Stochastic Process*. McGraw-Hill, 1984.

[290] E. Kreyszig. *Advanced Engineering Mathematics*. Wiley, 10th edition, 2011.

Index

A

AAC. *See* Advanced Audio Coding
ACB. *See* Associative coder of Buyanovsky
Action units (AU), 650
Adaptive arithmetic coding, 119, 130
- cumulative count table, 119
- static arithmetic code, 120

Adaptive codebook, 605, 612–613
Adaptive dictionary, 139
- LZ77 approach, 139
  - example, 140
  - implementation of, 150–151
  - theme variations, 143
- LZ78 approach, 143
  - example, 144
  - LZW algorithm, 145, 147
  - theme variations, 145

Adaptive differential PCM (ADPCM), 484
Adaptive Huffman coding, 67–68
- characteristics, 67–68
- decoding procedure, 73–74
- encoding procedure, 71–72
  - example, 72
  - flowchart, 72
- update procedure, 66, 70
  - adaptive Huffman coding algorithm, 69
  - adaptive Huffman tree, 71
  - example, 70
  - external node, 69–70
  - NYT node, 70

Adaptive model, 19
Adaptive prediction
- in DPCM, 358
  - backward adaptive prediction (DPCM-APB), 360–361
  - forward adaptive prediction (DPCM-APF), 359

Adaptive quantization, 268
- approaches, 268–269
- backward adaptive quantization, 271
  - Jayant quantizer, 273, 275
- in DPCM, 358
  - example, 358
- forward adaptive quantization, 269

Adaptive spectral enhancement filter, 611
Adaptive vector quantization, 335–336
- distortion, 336–337
- large codebook, 336
- recursively indexed vector quantizer, 336

A/D converter. *See* Analog-to-digital converter
Adjoint, of matrix, 679–680
ADPCM. *See* Adaptive differential PCM
ADSL. *See* Asymmetric Digital Subscriber Lines
Advanced Audio Coding (AAC), 581. *See also* MPRG AAC
- decoder tools, 581
- standard profiles, 581–582

Advanced intra coding mode, 661
Advanced prediction mode, 661
Advanced Television Systems Committee (ATSC), 587
Advanced video coding (AVC), 663
Admissibility condition, 503
AEP. *See* Asymptotic equipartition property
Algorithmic information theory, 37
- Kolmogorov complexity, 37–38

Aliasing, 394
All pole filter, 243
Alphabet, 16–17, 29
Alternative inter VLC mode, 662–663
AMDF. *See* Average magnitude difference function
Analog-to-digital converter (A/D converter), 252
Analog color television, 639
Analysis/synthesis scheme, 592
Analysis and synthesis filters, 462
Anchor frames, 652–653
APCO. *See* Association of Police Communications Officers
AR(N) model. *See* Nth order autoregressive model
Arithmetic coding, 43–44, 205
- adaptive arithmetic coding. *See* Adaptive arithmetic coding
- advantages, 78
- application, 77
- binary arithmetic coding. *See* Binary arithmetic coding
- binary code generating. *See* Binary code generation
- coding sequence, 93
  - tag generation, 94–100
  - tag deciphering, 101
- example, 44–48
- and Huffman coding, comparison, 127–128
- low-resolution layer, 205–207
- neighborhoods symbols, 207
- parameters for, 51
- QM coder, 207
- using contexts, 207–208

Arithmetic coding application

advantage of, 78
using adaptive coding, 52, 63, 77
ARJ, 143
ARMA model. *See* Autoregressive moving average model
ASCII code, 29, 137
Association of Police Communications Officers (APCO), 608
Associative coder of Buyanovsky (ACB), 178
Asymmetric applications, 650
generic wireframe model, 651
MPEG, 651–652
Asymmetric Digital Subscriber Lines (ADSL), 2
Asymptotic equipartition property (AEP), 326–327
Asynchronous transfer mode (ATM), 671
ATM. *See* Asynchronous transfer mode
ATM networks, 671
compression issues in, 671–672
transmission capacity availability, 671
ATSC. *See* Advanced Television Systems Committee
AU. *See* Action units
Audio coding
application, 485
audibility threshold, 570–571
Dolby AC-3, 587
algorithm, 587–588
bit allocation, 588–589
MPEG advanced audio coding
AAC, 581
MPEG-2 AAC, 582–583
MPEG-4 AAC, 586
MPEG audio coding
algorithms, 574
informative sections, 573
normative sections, 573
psychoacoustic model
nontonal components, 572–573
postprocessing, 573
tonal components, 572–573
spectral masking, 570–571
audibility threshold changes, 571
critical band, 570–571
temporal masking, 571–572
audibility threshold changes in time, 572
Audio compression, 85
application, 85–86
audio material, 85
FLAC, 86
MDCT, 439
source output sequence, 441
Audiophile, 220–221
Auditory perception
ear, 224
masking, 224
role, 224
Autocovariance approach, 599–600
Autoregressive moving average model (ARMA model), 243
AVC. *See* Advanced video coding
Average information derivation, 20
example, 20
formula, 23–24
monotonically increasing function, 23
occurrence identification, 21–22
properties, 20
rational probabilities, 23–24
Average magnitude difference function (AMDF), 598
for sound /e/in test, 598
for sound /s/in test, 599
Average mutual information, 228–229
Axiomatic approach, 678
example, 679

B

B frames. *See* Bidirectionally predictive coded frames
Backward adaptive quantization, 271
example, 273
Jayant quantizer. *See* Jayant quantizer
Band-pass filters, 224, 389–390, 452
high-pass filter, 389–390
low-pass filter, 389–390
Basic algorithm, 165–166, 348, 351
encoder and decoder generation, 349
example, 166–171, 350
lossless compression schemes, 348
quantization and reconstruction process, 349–350
Basis vectors, for subset, 374–375, 377–378
Bayes' rule, 679
Bennett integral, 286–287, 290–291
Bidirectionally predictive coded frames (B frames), 652–653
BIFS. *See* Binary Format for Scenes
Binary arithmetic coding, 120–124
example, 120
M coder, 124, 126–127
LPS probability, 127
MQ coder, 125–126
QM coder, 124–125
JBIG algorithm, 125
LPS probability, 125
scaling and rescaling process, 125
Binary code generation, 102–103
algorithm implementation, 106
interval possibilities, 106
tag generation with scaling, 107, 109
decoder implementation, 117
algorithm, 117–118
encoder implementation, 111–112
active interval, 112–113
example, 114–115
Total Count, 112
Total_Count, 114
using pseudocode, 113–114
for four-letter alphabet, 103
integer implementation, 111
decoder implementation, 117
encoder implementation, 111–112
uniqueness and efficiency, 103
Binary entropy function, 237
Binary Format for Scenes (BIFS), 669

Binary source function, 120
  rate distortion function for, 236
Biorthogonal wavelets, 516–517
  input sequence periodic extension
    periodic, 521
    symmetric, 521–523
Bit allocation, 482, 588
  map, 426–427
  operational rate distortion, 428, 483–484
  per frame, for G.729 coder, 621
  procedure, 461–462
  rate and distortion, 482
  scheme, 462
  two rate distortion functions, 482–483
Bit Sliced Arithmetic Coding (BSAC), 586–587
Bitstream markers, 566
Black-and-white analog television picture, 638
Block-based motion compensation, 636
Boundary gain, 325–326, 328
Braille coding, 2–3
BSAC. *See* Bit Sliced Arithmetic Coding
Burrows-Wheeler transform algorithm (BWT algorithm), 174
  decoding process, 177
  example, 175–176
  lexicographic ordering, 176
  move to front (mtf) coding, 177–178
  permutations, 175
  using mtf approach, 174–175
BWT algorithm. *See* Burrows-Wheeler transform algorithm

C

CABAC. *See* Context adaptive binary arithmetic coder; Context-adaptive binary arithmetic code
CALIC scheme. *See* Context Adaptive Lossless Image Compression scheme
Canadian Space Agency (CSA), 2
Canonical Huffman code, 50
  binary tree of depth four, 50–51
  codewords, 50–51
  conditions, 51
  generating code, 50–51
  procedure, 51–52
  tree structure, 52
  two Huffman trees, 49
cap. *See* Conditional average prediction
CAT scans. *See* Computerized axial tomography scans
CAVLC. *See* Context-adaptive variable-length code
CBP. *See* Coded block pattern
CCIR. *See* International Consultative Committee on Radio
CCIR recommendation 601–602, 640
CCIR 601, 640
CCITT. *See* Comité Consultatif International Télephoniqué et Télégraphique
CCITT G.728 Speech Standard, 605–606
  CCITT approved recommendation G.728, 605
  encoder and decoder, 606–607
CCITT group 3 and 4, recommendations, 200
  group 4 encoding algorithm, 203
  horizontal mode, 202
  MH, 200
  one-dimensional coding scheme, 200
  pass mode, 202
  relative element address designate, 201
  for T.4 and T.6, 200, 202–203
  two-dimensional coding definitions, 201–202
  two-dimensional scheme, 201
  vertical mode, 202
CCSDS. *See* Consultative Committee on Space Data Standards
cdf. *See* Cumulative distribution function
CELP. *See* Code-excited linear prediction
CFDM. *See* Constant factor adaptive delta modulation
Chaitin, G., 37
Channel, 219
Channel vocoder, 592–594
  excitation signal, 595–596
  formants, 594
  receiver, 595
  sound /e/in test, 594
  sound /s/in test, 594–595
  synthesis filters, 594
Cholesky decomposition, 602
Chrominance components, 548, 640
Classified vector quantization, 333
  three-stage vector quantizer, 334
  variation, 333
Cluster compression algorithm, 306
CME. *See* Comment and Extension
COC marker. *See* Coding style component marker
Code-excited linear prediction (CELP), 593, 603–604. *See also* Mixed excitation linear prediction (MELP)
  CCITT G.728 Speech Standard, 605–606
    CCITT approved recommendation G.728, 605
    encoder and decoder, 606–607
  coder, 603–604
  Federal Standard 1016, 604–605
  MP-LPC algorithm, 604
  RPE coding, 604
Code vectors, 296–297
Codebook, 296, 304
Coded block pattern (CBP), 648–649
Codewords, 29
  for eight-level quantizer, 256
Coding, 29, 668–669
  for four-letter alphabet, 30
  Kraft-McMillan inequality, 34–35, 37
    full binary tree of depth four, 36
    theorems, 34–35
  prefix codes, 33
    binary trees for different codes, 33
    internal nodes, 33–34
    root node, 33–34
  uniquely decodable codes, 30
    examples, 32–33
    test for unique decodability, 32
Coding of transform coefficients
  quantization and, 424
    operational rate-distortion bit allocation, 428

zonal sampling, 426–427
Coding sequence, 93
cdf usage, 93–94
examples, 95–100
tag deciphering, 101
tag generation, 94–95
using mapping concept, 94
Coding style component marker (COC marker), 564
Coding style marker (COD marker), 564
COD marker. *See* Coding style marker
Color component transform
forward reversible transform, 548
ICT, 548
forward ICT, 548
inverse ICT, 549
RCT, 548
Column matrix, 682
Comfort noise, 613
Comité Consultatif International Télephoniqué et Télégraphique (CCITT), 198–199
facsimile transmission, apparatus classification, 199
Comment and extension (CME), 566
Communication system, 219
Companded quantization, 282–283
compressor characteristics, 282
compressor function, 285
compressor mapping, 283
expander mapping, 283
nonuniform companded quantizer, 284
log-companded quantization, 282
Composite source model, 29
Compression algorithms, for packet video, 672
analysis filter bank, 673
progressive transmission algorithms, 672–673
reconstructed frame, 673
splitting, 672
Compression packages, 143
Compression ratio, 5–6
Compression techniques, 3
compression and reconstruction, 4
lossless compression techniques, 4
compression requirements, 5
pData, 5
radiological image, 4
reconstructed data, 5
text compression, 4
lossy compression techniques, 5
applications, 5
reconstruction, 5
modeling and coding, 6
code with codewords, 9
data value sequence, 7–8
examples, 7–9
reconstruction requirements, 6
performance measures, 5
compression ratio, 5–6
distortion, 6
fidelity and quality, 6
rate, 6
using lossy compression, 6
Compressor function, 282–283, 285
Computerized axial tomography scans (CAT scans), 193–194
Conditional average prediction (cap), 192–193
pixels, 192–193
ppm approach, 192
prediction error entropies, 193
Conditional entropy, 225
example, 225
Cones, 223
Constant factor adaptive delta modulation (CFDM), 363–364
granular condition, 363–364
speech encoding, 364
Constrained parameter bitstream (CPB), 655
Consultative Committee on Space Data Standards (CCSDS) recommendations, 77
fundamental sequence, 78
mapping, 77
preprocessor functions, 77
second extension option, 78
split sample option, 78
zero block option, 78
Context-adaptive binary arithmetic code (CABAC), 668–669
Context-adaptive variable-length code (CAVLC), 668
Context-based compression, 163
ACB, 178–179
BWT algorithm, 174
example, 175–176
move-to-front coding, 177–178
DMC, 179
cloning process, 181
cloning state, 179–180
encoding process, 180–181
implement algorithm issues, 180
three-state model obtained by cloning. 180
two-state model for, 179
experiments, 163–164
ppm algorithm, 165
basic algorithm, 165–166
context length, 172
escape symbol, 170
exclusion principle, 173
Context adaptive binary arithmetic coder (CABAC), 120
Context Adaptive Lossless Image Compression scheme (CALIC scheme), 186
alphabet representation, 189–190
CALIC algorithm, 190
grayscale images, 186
labeling neighbors, 186–187
pseudocode, 187
recursive indexing, 189
using pixel, 186–188
using prediction, 188–189
Continuously variable slope delta modulation (CVSD modulation), 364
adaptation logic, 365
autocorrelation function, 365
syllabically companded, 364
Continuous wavelet transform (CWT), 503–504
Contours of constant probability, 299, 325

for Gaussian distribution, 327–328
Convolution theorem, 385
for Fourier transform
of impulse response, 387–388
of simple function, 392–393
Covariance method, 601–602
CPB. *See* Constrained parameter bitstream
Critical band, 224
in spectral masking, 570–571
Critically (maximally) decimated filter bank, 477
Cross product, 683
CSA. *See* Canadian Space Agency
Csiz, 562
Cum–Count array, 112, 114, 120
Cumulative distribution function (cdf), 93–94, 681
Cutoff frequency, 452
CVSD modulation. *See* Continuously variable slope delta modulation
CWT. *See* Continuous wavelet transform

D

D/A converter. *See* Digital-to-analog converter
Data compression, 1–2
algorithms, 1–2
compression techniques, 3
lossless compression schemes, 4
lossy compression schemes, 4
measures of performance, 5
modeling and coding, 7–9
Morse code, 2–3
and reconstruction, 4
space agencies, 2
statistical structure, 3
without using compression, 2
voice coder, 3
DCT. *See* Discrete cosine transform
DDVPC. *See* Defense Department Voice Processing Consortium
Deblocking filter mode, 661
Decimation, 459–460, 462
Decision boundaries, 255
Decoder algorithm, 117–118
Decoder implementation, 117
Decoding procedure, 73–74, 493
example, 73
flowchart, 74
generic procedure. *See* Generic decoding procedures
Defense Department Voice Processing Consortium (DDVPC), 608
Deflate algorithm, 50–51
LZ77 algorithm, 150–151
Delimiting markers, 562
Delivery Multimedia Integration Framework (DMIF), 669
Delta modulation (DM), 361–362. *See also* Differential pulse code modulation system (DPCM system)
adaptive, 363
CFDM, 363–364
granular condition, 363–364
speech encoding, 364
CVSD modulation, 364
adaptation logic, 365
autocorrelation function, 365
syllabically companded, 364
linear, 362
signal sampling, 361–362
source output sampling and coding, 363
Determinant, of matrix, 678–679
Deterministic prediction, 204–205
DFS. *See* Discrete Fourier series
DFT. *See* Discrete Fourier transform
Dictionary compression scheme, 9
Dictionary order. *See* Lexicographic ordering
Dictionary techniques, 135–136
adaptive dictionary, 139
LZ77 approach, 139–140, 143
LZ78 Approach, 143–145, 147
applications, 150–151
GIF, 151
PNG, 152
UNIX compress, 151
V.42 bis, 153
Lempel-Ziv complexity, 156–158
DNA sequence, 158
usage, 158
static dictionary, 136–137
digram coding, 137
Difference distortion measures, 221
average of absolute difference, 222
Differential encoding, 351. *See also* Delta modulator (DM)
adaptive DPCM, 357
adaptive predictions in, 358
adaptive quantization in, 358
basic algorithm, 348, 351
encoder and decoder generation, 349
lossless compression schemes, 348
quantization and reconstruction process, 349–350
delta modulation. *See* Delta modulation (DM)
DPCM, prediction in, 352
example, 354
fine quantization assumption, 352
image coding. *See* Image coding
Sinan image, 346–347
pixel-to-pixel differences, 347
sinusoid, 346
quantization process, 350
quantizer designing, 345–346
and reconstructions, 350–351
sample-to-sample differences, 346
speech coding. *See* Speech coding
techniques, 346
Differential entropy, 229
discrete random variable, 229–230
of Gaussian pdf, 231
of random variable, 230
self-information, 229
Differential pulse code modulation system (DPCM system), 351. *See also* Differential encoding
adaptive prediction, 358
adaptive quantization, 357–358

fine quantization assumption, 352
forward adaptation, 357
performance, 356
prediction, 352–353
predictors, 355
reconstructed sequence, 356–357, 359
residual sequence, 355
sequence variance, 352
SNR and SPER, 355–356
speech sequence, 354, 356–357
using uniform quantizer, 355–356
Digital-to-analog converter (D/A converter), 253–254
Digram coding, 137
encoder, 137
example, 137
pairs, 138–139
sample dictionary, 137
Dilation, 501
Dirac delta function, 387–388,
Discrete convolution, 404, 405
Discrete cosine transform (DCT), 420, 422, 529, 642–643
basis matrices for, 421
Discrete delta function, 402
impulse response of, 404
Discrete Fourier series (DFS), impulse response of, 395, 404
Discrete Fourier transform (DFT), 304, 420. *See also* Fourier transform; Z-transform
basis matrices, 421
DCT of sequence, 422
DFS representation, DCT of sequence, 395, 422
FFT, 395
Fourier series coefficients, FFT, 395
of sequence, 422
Discrete sine transform (DST), 423
Discrete time Fourier transform, 396
Discrete time Markov chain, 26
Discrete time wavelet transform (DTWT), 504
Discrete Walsh-Hadamard transform (DWHT), 423, 667–668
Distortion, 6
Distortion criteria, 220
absolute difference, 222
applications, 222
auditory perception. *See* Auditory perception
design feedback, 221
difference distortion measures, 221
human visual system. *See* Human visual system (HVS)
mean squared error, 221–222
PSNR, 222
Distribution
gamma, 686
Gaussian, 686
Laplacian, 686
uniform, 685
Distribution functions, 681
examples, 682–683
DM. *See* Delta modulator
DMC. *See* Dynamic Markov compression
DMIF. *See* Delivery Multimedia Integration Framework
Dolby Digital. *See* Dolby AC-3
Dot product, 375, 683
Downsampling, 459–460, 462, 463
low-pass filter output, 466
DPCM-APB. *See* DPCM with Backward Adaptive Prediction
DPCM-APF. *See* DPCM with Forward Adaptive Prediction
DPCM system. *See* Differential pulse code modulation system
DPCM with Backward Adaptive Prediction (DPCM-APB), 360. *See also* Differential encoding
LMS algorithm, 361
optimum predictor coefficients, 360
residual value, 360
squared prediction error, 361
DPCM with Forward Adaptive Prediction (DPCM-APF), 359. *See also* Differential encoding
autocorrelation function, 359
parcor coefficients, 359
residual squared versus predictor coefficient, 360
DST. *See* Discrete sine transform
DTWT. *See* Discrete time wavelet transform
Dual prime motion compensation, 657–658
DWHT. *See* Discrete Walsh-Hadamard transform
Dynamic Markov compression (DMC), 179
cloning process, 181
cloning state, 179–180
encoding process, 180–181
implement algorithm issues, 180
three-state model obtained by cloning. 180
two-state model for, 179

E

Ear, 224
Eardrum. *See* Tympanic membrane
EBCOT. *See* Embedded Block Coding with Optimized Truncation
Eigenvalues, 677, 680
18-tap Coiflet low-pass filter coefficients, 519
Embedded Block Coding with Optimized Truncation (EBCOT), 547, 552
rate control, 560–561
Embedded coding, 539–540
in JPEG2000, 560
of quantization values, 551
Embedded zerotree wavelet coder (EZW coder), 532. *See also* Set Partitioning in Hierarchical Trees (SPIHT)
algorithm, 670
data structure in, 532, 534
embedded coding, 539–540
isolated zero, 534
multiple-pass algorithm, 534–535
seven-level decomposition, 536–539
significance map coding, 534–535
ten-band decomposition, 532–533
3-bit quantizer, 533–534
three-level midtread quantizer, 535–536
wavelet coefficient scanning, 536

Empty cell problem, 315
Encoder implementation, 111– 112
active interval, 112–113
encoding algorithm using pseudocode, 113–114
example, 114–115
Total Count, 112
Total_Count, 114
Encoding procedure, 71–72
example, 72
flowchart, 72
End of codestream marker (EOC marker), 562
End of packet marker (EPH marker), 566
Enhanced reference picture selection mode, 661
Entropy-coded quantization, 247
entropy-constrained quantization, 289
nonlinear equations, 289
high-rate optimum quantization, 289–290
example, 291
inverse mapping, 291–292
optimum quantizer, 290–291
recursive indexing, 291
using Bennett integral, 290
using boundary conditions, 290
Lloyd-Max quantizer outputs, 288
output entropies of, 288–289
Entropy-constrained quantization, 249
nonlinear equations, 289
Entropy, 15–16
conditional entropy, 225
example, 225
differential entropy, 229
example, 230
EOC marker. *See* End of codestream marker
EPH marker. *See* End of packet marker
ESA. *See* European Space Agency
Escape symbol, 170–171
method C, 172
methods A and B, 170–172
European Space Agency (ESA), 2
Expander function, 282–283, 287
Expectation, 683–684
Expectation operator, 683
Extended Huffman codes, 58–59
examples, 59, 61
extended alphabet, 59–61
Huffman code, 61
External nodes, 33–34
Eye, 223
EZW coder. *See* Embedded zerotree wavelet coder

F

Facial action parameter (FAP), 670
Facial definition parameter (FDP), 670
Facsimile coding algorithms
MMR, 208
modified Huffman, 208
MR, 208
Facsimile encoding, 198
CCITT group 3 and 4, 200
group 4 encoding algorithm, 203
horizontal mode, 202
MH, 200
one-dimensional coding scheme, 200
pass mode, 202
recommendations T.4 and T.6, 200, 202–203
relative element address designate, 201
two-dimensional coding definitions, 201–202
two-dimensional scheme, 201
vertical mode, 202
group 1 apparatus, 199
group 2 apparatus, 199
group 3 apparatus, 199
group 4 apparatus, 199
JBIG, 203
arithmetic coding, 205
MH and MR, and MMR and JBIG comparison, 208
redundancy removal, 204, 205
resolution reduction, 203–204
JBIG2-T.88, 209
generic decoding procedures, 209–210
halftone region decoding, 211
symbol region decoding, 211
run-length coding, 199–200
Capon model for, 200
Fano, Robert, 43
FAP. *See* Facial action parameter
Fast Fourier transform (FFT), 395, 574
FDP. *See* Facial definition parameter
Federal Standard (FS) 1016, 603–605
FFT. *See* Fast Fourier transform
Filter, 248, 388–389, 452
aliasing, 453
band-pass, 389–390
characteristics, 453
eight-band filter bank, 457
eight-tap Johnston low-pass filter coefficients,457
eight-tap Smith-Barnwell low-pass filter coefficients, 458
FIR filters, 454
high-pass, 389–390
IIR filters, 454
low-pass, 388–389
magnitude transfer function, 452
magnitudes, 389–390
QMF, 456
ripple, 452
16-tap Johnston low-pass filter coefficients, 458
16-tap Smith-Barnwell low-pass filter coefficients, 459
32-tap Johnston low-pass filter coefficients, 458
Filter banks design
downsampling, 462–464
decomposition into two bands, 464
spectrum of downsampled low-pass filter output, 466
spectrum of outputs of ideal filters, 464
spectrum of source output, 464
input sequence decomposition, 463
input sequence reconstruction, 463
upsampling, 462, 465

analysis filters, 465–467
antialiasing filters, 465–467
imaging, 465
interpolation filters, 465–467
synthesis filters, 465–467
upsampled signal spectrum, 467
Fine quantization assumption, 352
Finite context model, 27
Finite impulse response (FIR), 454
FIR. *See* Finite impulse response
First-order Markov model, 27–28
Fixed-length code, 29
Fixed codebook, 612–613
Fixed information marker, 562
FLAC. *See* Free Lossless Audio Codec
Fletcher-Munson curves, 224
Forward adaptive quantization, 269
examples, 269
compressed 3-bit speech sequences, 270
Sena image, 271
16-bit speech sequence, 270
variance estimation procedure, 269
Forward transform, 414
Four-tap Daubechies filter, 530–532
4-tap Daubechies low-pass filter coefficients, 517
4 loading, 265–266
Fourier series, 380
coefficient, 382
expansion, 390
exponential form, 381
inner product definition, 381–382
representation, 382
trigonometric, 380
Fourier transform, 362. *See also* Discrete Fourier transform (DFT); Z-transform
convolution theorem, 385
expansion, 382–383
function, 391
inverse, 384
modulation property, 384
Parseval's theorem, 384
periodic extension, 382–383
time function, 383
Fovea, 223
Fractal coding approach, 592–593
Fractal compression, 624
decoding process, 627
domain blocks, 625
domain pool elements, 626–627
encoding procedure, 626–627
final reconstructed Elif image, 630
massic transformation, 625–626
original Elif image, 628
range blocks, 625
self-similarity, 624–625
six iterations, 629
union of transformations, 626
Free Lossless Audio Codec (FLAC), 86
Frequency of occurrence, 675, 676
FS. *See* Federal Standard
Fundamental sequence, unary code, 78
Functional markers, 563–564

G

G.722, 484
ADPCM system, 484
QMF coefficient values, 484–485
quantizer adaptation, 485
G.726 recommendation, 366
predictor
adaptive algorithms, 368–369
backward adaptive, 368
quantizer, 367
adaptation algorithm, 367
backward adaptive, 367
fixed, 367
locked scale factor, 367–368
unlocked scale factor, 367–368
Gabor transform, 498–499
Gain-shape vector quantization, 332
Gain-shape vector quantizers, 327
Gamma distribution, 241–242, 686
Gaussian distribution, 241–242, 686
Gaussian source function, 238
for Gaussian random variable, 240
Generalized BFOS algorithm, 324
Generalized Lloyd algorithm (GLA), 306
Generic compression scheme, 220
Generic decoding procedures, 209–210
generic refinement decoding procedure, 210
generic region decoding, 209–210
using adaptive arithmetic coding, 210
using contexts, 210
Generic regions, 209
GIF. *See* Graphics Interchange Format
GLA. *See* Generalized Lloyd algorithm
GOB. *See* Groups of block
Golomb codes, 75–76
example, 76
Golomb-Rice codes, 75
GOP. *See* Group of pictures
Government Standard LPC-10. *See* Linear predictive coder (LPC)
Grade 1 Braille, 2–3
Grade 2 Braille, 2–3
Grand Alliance HDTV Proposal, 658
Granular error, 263–264
Granular noise. *See* Granular error
Granular regions, 362
Graphics Interchange Format (GIF), 151
comparison of, 152
PNG, 152
Group of pictures (GOP), 653
Groups of block (GOB), 647–648, 658–659
Group Speciale Mobile (GSM), 604
GSM. *See* Group Speciale Mobile
gzip, 62, 143

H

H.262. *See* MPEG-2 video standard
H.264, 663. *See also* ITU-T recommendation H.263; MPEG-1 video standard; MPEG-2 video standard; Video compression
- block diagram, 663
- coding, 668–669
- discrete Walsh-Hadamard transform, 667–668
- intra prediction, 666
- inverse transform matrix, 665–666
- macroblock structure, 663
- motion-compensated prediction, 663–665
- prediction modes, 667
- quantization, 666–668
- transform matrix, 665–666

Halftone region, 209
- decoding, 211

Hartleys, 14
Hewlett-Packard (HP), 190
HH image, 529–530
Hierarchical INTerpolation (HINT), 193
High-pass filters, 389–390
High-rate optimum quantization, 289–290
- example, 291
- inverse mapping, 291–292
- optimum quantizer, 290–291
- recursive indexing, 291
- using Bennett integral, 290
- using boundary conditions, 290

HINT. *See* Hierarchical INTerpolation
HL image, 529–530
Homogeneity, 385
Horizontal mode, 202
Hotelling transform, 418–419. *See also* Karhunen-Loéve transform
HP. *See* Hewlett-Packard
Huffman, David, 43
Huffman and arithmetic coding comparison, 127
- arithmetic code for two-symbol sequences, 128
- average length, 127–128
- example, 128

Huffman code design, 44
- alternative way, 47
- average length for code, 46–47
- codewords, 44, 45
  - assignment, 46
- Huffman code for original five-letter alphabet, 46
- Huffman encoding procedure, 47
- initial five-letter alphabet, 44
- reduced four-letter alphabet, 45
- reduced three-letter alphabet, 45
- reduced two-letter alphabet, 46

Huffman code implementation, 61–62
- decoding process states, 62–64
- nine-letter alphabet, 62
- state machine description, 63

Huffman code length, 56, 58
- example, 56
- using Jensen's inequality, 57
- using Kraft-McMillan inequality, 56–57

Huffman code optimality, 55
- necessary conditions, 55–56

Huffman codes, 43
Huffman coding
- adaptive Huffman coding, 67
  - decoding procedure, 73–74
  - encoding procedure, 71–72
  - update procedure, 66, 70
- algorithm, 43
  - optimum prefix codes observations, 43–44
  - requirement, 44
  - second observation holds, 44
- alternative way, 47
- average length for code, 46–47
- canonical Huffman codes, 50
  - binary tree, 50–51
  - codewords, 50–51
  - conditions, 51
  - generating code, 50–51
  - procedure, 51–52
  - tree structure, 52
  - two Huffman trees, 49
- codewords, 44–46
- extended Huffman codes, 58–59
  - examples, 59, 61
  - extended alphabet, 59–61
  - Huffman code, 61
- for five-letter alphabet, 44, 46
- Golomb codes, 75–76
- Huffman code length. *See* Huffman code length
- Huffman code optimality. *See* Huffman code optimality
- Huffman encoding procedure, 47
- implementation. *See* Huffman code implementation
- length-limited Huffman codes, 54–55
  - example, 54
- minimum variance Huffman codes, 47
  - binary Huffman tree building, 48
  - binary tree of depth four, 50
  - buffer, purpose of, 48–49
  - Huffman encoding procedure, 48
  - minimum variance Huffman code, 49
  - reduced four-letter alphabet, 48
  - reduced three-letter alphabet, 49
  - reduced two-letter alphabet, 49
  - two Huffman trees, 48–49
  - variable-length code, 48–49
- nonbinary Huffman codes, 65–66
  - code tree, 67
  - example, 65
  - reduced five-letter alphabet, 66
  - reduced three-letter alphabet, 67
  - sorted six-letter alphabet, 66
  - ternary code for six-letter alphabet, 67
- reduced four-letter alphabet, 45
- reduced three-letter alphabet, 45
- reduced two-letter alphabet, 46
- Rice codes, 76–77
  - CCSDS recommendation, 77

Tunstall codes, 79–80
Huffman coding application, 81. *See also* Huffman coding
audio compression, 85
application, 85–86
audio material, 85
FLAC, 86
lossless image compression, 81–82
adaptive Huffman codes, 83
Huffman codes on pixel difference values, 83
Huffman codes on pixel values, 82–83
test images, 82–83
text compression, 83–84
using Huffman codes, 84
Huffman encoding procedure, 47
minimum variance, 48
Huffman table marker, 439
Human visual system (HVS), 222–223
description, 224
disheartening prospects, 223
eye, 223
eye intensity, 223–224
monochromatic vision model, 224
weber ratio, 223
HVS. *See* Human visual system

I

I frames, 652
ICT. *See* Irreversible component transform
Identity matrix, 675, 678
IEC. *See* International Electrotechnical Commission
Ignorance model, 25
iid. *See* Independent, identically distributed
IIR. *See* Infinite impulse response
iLBC coder, 613
all-pass filters, 616
autocorrelation coefficient computation, 614
codebook generation, 617
goal, 615–616
LSF representation, 614–615
packet loss effects, 613–614
perceptual weighting filter, 616
PLC, 617–618
residual sequence, 616
using DPCM system, 616
Image and tile size marker (SIZ marker), 562
Image coding, 369. *See also* Speech coding
differential encoding scheme, 369
predicted value, 369–370
reconstructed images, 370
signal-to-noise ratio, 369–370
Image compression, 623. *See also* Speech compression; Video compression
atomic blocks, 630
comparison of, 152, 155
fractal compression, 624
decoding process, 627
domain blocks, 625
domain pool elements, 626–627
encoding procedure, 626–627
final reconstructed Elif image, 630
massic transformation, 625–626
original Elif image, 628
range blocks, 625
self-similarity, 624–625
six iterations, 629
union of transformations, 626
GIF, 151
Huffman codes for, 155
PNG, 152–153
quadtree partitioning, 630–631
representations codes, 154
using Huffman code, 153
using pixel values, 153
Image compression application, 486
alternate four subimages, 488
filtered and decimated output, 486–487
four subimages, 486–487
image decomposition
eight-tap Johnston filter decomposition, 488–489
eight-tap Smith-Barnwell filter decomposition, 489–490
16-tap Johnston filter decomposition, 488–489
JPEG, 432
coding, 434
JPEG File Interchange Format, 437–438
quantization, 433–434
transform, 432
sample image, 486–487
subband coding, 490
DPCM and scalar quantization, 492
eight-tap John-Ston filter, 490, 491
eight-tap Smith-Barnwell filter, 491
Image transmission system, 592
IMBE. *See* Improved MBE
Immittance Spectral Pairs (ISP), 611–612
Improved MBE (IMBE), 608
Improved PB-frames mode, 661
Impulse function, 387–388
Impulse response
Dirac delta function, 387–388
impulse function, 387–388
sifting property, 388
time function, 388
Independent, identically distributed (iid), 687
Inflate algorithm, 62
Infinite impulse response (IIR), 454
Informational marker, 566
Information theory, 13–15, 225
average information derivation, 20
classic paper, 21
example, 20
formula, 23–24
identifying occurrence, 21–22
monotonically increasing function, 23
properties, 20
rational probabilities, 23–24
average mutual information, 228–229
example, 229
mutual information, 228

conditional entropy, 225–226
differential entropy, 229
discrete random variable, 229–230
of Gaussian pdf, 231
of random variable, 230
self-information, 229
lengths, sequences of, 16–17
lossy compression approach, 225
mathematical definition, 14
using logarithm, 15
wealth of nations, 18
Inner product, 379, 683
Instantaneous code, 31
Integer implementation, 111
decoder implementation, 117
encoder implementation, 111–112
Integer pixel compensation, 659
Internal nodes, 33–34
International Consultative Committee on Radio (CCIR), 640
International Electrotechnical Commission (IEC), 130, 651–652
International Standards Organization (ISO), 130, 184, 651–652
International Telecommunication Union (ITU), 130, 184
Intra prediction, 666
Inverse Fourier transform, 384
Inverse transform matrix, 665–666
Irreversible component transform (ICT), 548
ISO. *See* International Standards Organization
Isolated zero, 534
ISP. *See* Immittance Spectral Pairs
ITU-R recommendation BT.601–602, 640
ITU-T G.722.2. *See* Wideband speech compression
ITU-T G.729 standard, 618
autocorrelation coefficients, 618–619
bit allocation per frame, 621
codebook vectors, 620–621
encoding of speech for, 619
excitation signal identification, 620
pulses, 621
weighting filter, 620
ITU-T H.264. *See* H. 264
ITU-T recommendation H. 261, 644. *See also* Video compression
ITU-T H.261 encoder, 644
motion compensation, 644
block size effect on, 645
CBP, 648–649
GOB macroblocks, 647–648
ITU-TH. 261 algorithm, 649
loop filter, 646
macroblock pixel encoding, 645–646
quantization and coding, 647
rate control, 649
trade-off balancing, 645
transform, 647
ITU-T recommendation H.263, 658. *See also* Video compression
advanced intra coding mode, 661
advanced prediction mode, 661
alternative inter VLC mode, 662–663
deblocking filter mode, 661
enhanced reference picture selection mode, 663
GOB, 658–659
I frames, 659
improved PB-frames mode, 661
last nonzero coefficient, 659
median, of motion vectors, 659
modified quantization mode, 663
P frames, 659
PB-frames mode, 661
reduced-resolution update mode, 662
reference picture
resampling, 662
selection mode, 662
SNR scalability mode, 662
spatial scalability mode, 662
standardized H.263 formats, 659
syntax-based arithmetic coding mode, 660
temporal scalability mode, 662
unrestricted motion vector mode, 660
video compression algorithm, 660
ITU. *See* International Telecommunication Union

J

Japan Aerospace Exploration Agency (JAXA), 2
JAXA. *See* Japan Aerospace Exploration Agency
Jayant quantizer, 271–273, 276
multiplier, 272
multiplier functions for 2-bit quantizer, 275
operation of, 273
output levels for, 272–273
performance of, 276
JBIG2-T.88, 209
decoder requirements, 209
encoder types, 209
generic decoding procedures, 209–210
refinement decoding procedure, 210
region decoding, 209–210
using adaptive arithmetic coding, 210
using contexts, 210
halftone region decoding, 211
symbol region decoding, 211
JBIG algorithm. *See* Joint Bi-level Image Experts Group algorithm
JFIF. *See* JPEG File Interchange Format
jnd. *See* Just noticeable difference
Joint Bi-level Image Experts Group algorithm (JBIG algorithm), 125, 203
arithmetic coding, 205
low-resolution layer, 205–207
neighborhoods symbols, 207
QM coder, 207
using contexts, 207–208
image coding schemes comparison, 208–209
JBIG comparison, 208–209

MH comparison, 208–209
MMR comparison, 208–209
MR comparison, 208–209
redundancy removal, 204
deterministic prediction, 205
typical prediction, 204
resolution reduction, 203–204
expression, 204
JBIG specification, 204
using pixels, 204
Joint cumulative distribution function, 687
Joint Photographic Experts Group (JPEG), 184, 428
coding, 434
JPEG File Interchange Format, 437–438
quantization, 433–434
transform, 432
Journal of Educational Psychology, 413
JP2 file format, 566–567
JPEG-LS, 190–191
coding procedures, 190
comparison, 191–192
prediction algorithm, 190
prediction error, 191
SIGN variable, 191
JPEG. *See* Joint Photographic Experts Group
JPEG 2000 bitstream, 561–562. *See also* JPEG 2000 standard
bitstream markers, 566
codestream organization, 562
delimiting markers, 562
fixed information marker, 562
functional markers, 563–564
informational marker, 566
JP2 file format, 566–567
organization of markers, 565
PLT and PLM markers, 564–565
pointer markers, 564
PPM marker, 565–566
SIZ marker parameter, 563
JPEG 2000 standard, 547. *See also* Wavelet-based image compression
algorithm, 547
color component transform
forward ICT, 548
forward reversible transform, 548
ICT, 548
inverse ICT, 549
RCT, 548
EBCOT, 547
PCRD optimization, 547–548
quantization, 551
discarding bitplanes, 551
embedded coding, 551
tiling, 549
wavelet transform
wavelet filter types, 550
reversible, 549–550
irreversible, 550
9/7 transform, 549–550
JPEG File Interchange Format (JFIF), 437–438
syntax of, 438
JPEG predictor, 185
JPEG standard, 184
comparison of, 185–186
compressed file size, 185
predictor, 185
predictive schemes, 184–185
Just noticeable difference (jnd), 223

K

Karhunen-Loéve transform, 418
example, 419
kbits per second (kbps), 194–195
kbps. *See* kbits per second
KLT. *See* Karhunen-Loéve transform
Kolmogorov, A.N., 37
Kolmogorov complexity, 37
Kraft-McMillan inequality, 34–35, 37
full binary tree of depth four, 36
theorems, 34–35
Kraft-McMillan sum, 53–54

L

Laplacian distribution, 241–242, 686
Lapped orthogonal transform (LOT), 448
LaTeX, character pairs in, 138
Lattice vector quantizer, 328–329
A2 lattice, 330
average squared error, 328
D2 lattice, 330–331
dimensions, 329
example, 331
hexagonal lattice, 329
possible quantization regions, 328
problems, 331
square and circular quantization regions, 328–329
using lattices, 329
Lattices, 329
LBG algorithm. *See* Linde-Buzo-Gray algorithm
Least mean squared algorithm (LMS algorithm), 361
Leaves. *See* External nodes
Lempel-Ziv complexity (LZ complexity), 156–158
DNA sequence, 158
usage, 158
Length-limited Huffman codes, 52–54
example, 54
Huffman code, 54
length-limited Huffman code, 55
Less Probable Symbol (LPS), 120–124
Letters, 15–16, 29
Levinson-Durbin algorithm, 600–601
Lexicographic ordering, 97, 176
LH image, 529–530
LHarc, 143
Lifting, 523
high-frequency difference sequence generation, 524

implementation coefficients, 526–527
odd- and even-indexed component decomposition, 524–525
Linde-Buzo-Gray algorithm (LBG algorithm), 297, 306, 309
cluster compression algorithm, 306
codebook
design of, 304
progression of, 311
compression measures, 317–318
empty cell problem, 315
examples, 303, 311, 316
final codebook, 310
final state, 310
for image compression, 315–316
initial codebook effects, 313
initial output points, 314
initialization, 309
k-means algorithm, 304, 306
Lloyd algorithm functions, 304–305
Sinan image, 316, 319
splitting approach, 309–312
two-level vector quantizer using splitting approach, 312
using PNN algorithm, 314
vector quantizer, 306
after one iteration, 308
alternate initial set of, 309
final state of, 309
initial set of output points, 307
initial state of, 308
training set for, 307
Linear predictive coder (LPC), 596312597. *See also* Speech compression
AMDF, 598, 599
parameter transmission, 602–603
pitch period estimation, 597–598
speech synthesis model, 596
synthesis, 603
vocal tract filter, 599
autocovariance approach, 599–600
Cholesky decomposition, 602
covariance method, 601–602
filter coefficient change, 601–602
Levinson-Durbin algorithm, 600–601
PARCOR coefficients, 600–601
Toeplitz matrix, 600
voiced/unvoiced decision, 597
Linear system models, 243
AR process, 244
AR source, 244
autocorrelation function of, 245
sample function, 245–247
digital signal-processing terminology, 243
equations, 243
filters, 388–389
band-pass, 389–390
high-pass, 389–390
low-pass, 388–389
magnitudes, 389–390
homogeneity, 385
impulse response
Dirac delta function, 387–388
impulse function, 387–388
sifting property, 388
time function, 388
sample-to-sample correlation, 243–244
scaling, 386
superposition, 386
time invariance, 386
transfer function, 386–387
using ARMA model, 243
Line spectral frequency (LSF), 614–615
LIP. *See* List of insignificant pixels
LIS. *See* List of insignificant sets
List of insignificant pixels (LIP), 542
List of insignificant sets (LIS), 542
List of significant pixels (LSP), 542
LL image, 529–530
LMS algorithm. *See* Least mean squared algorithm
LNT Pn, 205
Locked scale factor, 367–368
LOCO. *See* Low complexity profile
Long division, 403–404
Long term prediction (LTP), 586
Lookahead buffer, 139–140
Lossless compression, 4
algorithmic information theory, 37
Kolmogorov complexity, 37–38
coding, 29
Kraft-McMillan inequality, 34
prefix codes, 33
uniquely decodable codes, 30
composite source model, 29
compression requirements, 5
information theory, 13–14
average information derivation, 20
example, 15
Markov models, 26
mathematical preliminaries for, 13
MDL, 38–39
pData, 5
physical models, 25
probability models, 25
radiological image, 4
reconstructed data, 5
text compression, 4
Lossless image compression, 81–82
adaptive Huffman codes, 83
CALIC scheme, 186
algorithm, 190
alphabet representation, 189–190
grayscale images, 186
labeling neighbors of, 186–187
pseudocode, 187
recursive indexing, 189
using pixel, 186–188
using prediction, 188–189
conditional average prediction, 192–193
pixels, 192–193
ppm approach, 192
prediction error entropies, 193

facsimile encoding, 198–199
CCITT Group 3 and 4, 208
JBIG, 199, 203–205, 209–211
JBIG2-T. 88, 209
run-length coding, 200
Huffman codes
on pixel difference values, 83
on pixel values, 82–83
JPEG-LS. *See* JPEG-LS
JPEG standard, 184
comparison of, 185–186
compressed file size, 185
JPEG predictor, 185
predictive schemes, 184–185
MRC-T.44. *See* MRC-T.44
multiresolution approaches, 193
progressive image transmission, 193–194
test images, 82–83
Lossy coding, mathematical preliminaries for
distortion criteria, 220–221
auditory perception, 224
human visual system, 223
information theory, 225
average mutual information, 228–229
conditional entropy, 225–226
differential entropy, 229–231
example, 225
models, 240
linear system models, 243–244
physical models, 248
probability models, 240
rate distortion theory, 232
binary source, 236
examples, 233, 235
Gaussian source, 238
Lossy compression techniques, 4–6
applications, 5
reconstruction, 5
LOT. *See* Lapped orthogonal transform
Low-pass filters, 388–389
Low complexity (LOCO) profile, 190
LPC. *See* Linear predictive coder
LPS. *See* Less Probable Symbol
LSF. *See* Line spectral frequency
Lsiz, 562
LSP. *See* List of significant pixels
LTP. *See* Long term prediction
Luminance component, 640
LZ algorithms, 145–146
LZ complexity. *See* Lempel-Ziv complexity
LZ77 approach, 139–140
Achilles heel of, 143
decoding triple, 142
encoding process, 141
encoding using approach, 140
lookahead buffer, 139–140
LZSS, 143
offset, 139–140
possibilities during coding process, 140
search buffer, 139–140
theme, variations on, 143
LZ78 approach, 144
Achilles' heel of LZ77, 143
decoding, 147
completion of fifth entry, 150
constructing dictionary while decoding, 149
constructing fifth entry, 149–150
constructing LZW dictionary decoding, 148
final dictionary for, 149
initial dictionary, 148
dictionary, development of, 145
encoding, 145
constructing, LZW dictionary, 146
encoding, LZW dictionary for, 147
initial LZW dictionary, 146
initial dictionary, 144
LZ78 theme-LZW algorithm, variations on, 145–146
using LZ77 approach, 143
LZW algorithm, 145–146
decoding, 147
constructing dictionary while decoding, 149
constructing fifth entry, 149–150
constructing LZW dictionary decoding, 148
fifth entry, completion of, 150
final dictionary for, 149
initial dictionary, 148
encoding, 145
constructing, LZW dictionary, 146
encoding, LZW dictionary for, 147
initial LZW dictionary, 146
LZW applications, 150–151
GIF, 151
comparison of, 152
PNG, 152
PNG, 153
comparison of, 155
Huffman codes for, 152–153
representations codes, 154
using Huffman code, 153
using pixel values, 155
UNIX compress, 151
V.42 bis, 153, 155–156
CCITT recommends, 156
control codewords in, 155
encoder STEPUP, 156
using compression algorithm, 153, 155

M

M coder, 124, 126–127
LPS probability, 127
M-band QMF filter banks
equivalent structures, 475–477
input sequence decomposition, 474–475
spectral characteristics, 475–476
Magnetic resonance images (MRI), 193–194
Magnitude transfer function, 452
Make-up codes, 200
Markov models
discrete time Markov chain, 26

example, 27
finite state process, 26
linear filter, 26
in text compression, 27–28
context model, 28
ppm algorithm, 28
using first-order Markov model, 27–28
using second-order model, 28
two-state Markov model for binary images, 27
uses, 26
Masking, 224
Matrix, 675
adjoint, 679–680
characteristic equation, 680
column, 682
determinant, 678–679
eigenvalues, 677, 680
elementary operations, 678
identity, 675, 678
inner product, 683
operations, 675–677
outer product, 683
row, 682
square, 675
transpose, 682–683
MBE. *See* Multiband excitation coder
MDCT. *See* Modified discrete cosine transform
MDCT frames, 580
bit reservoir, 580
distortion control loop, 580
following frame, 580
previous frame, 580
rate control loop, 580
MDCT window function, 577–579
long window, 577–579
sequence of windows, 577–579, 588
short windows, 577–579
start window, 577–579
stop window, 577–579
window switching process, 577–579
MDL principle. *See* Minimum description length principle
Mean-removed vector quantization, 332–333
Sinan image using codebook, 332–333
Mean, 685
Mean squared error (mse), 221–222
Mean squared quantization error (msqe), 254–255
Measure of belief, 676
MELP. *See* Mixed excitation linear prediction
Method of principal components, 413
MH. *See* Modified Huffman
MH comparison, 208–209
Midrise quantizer, 257–258
Midtread quantizer, 257–258
Minimum description length principle (MDL principle), 38–39
Minimum variance Huffman codes, 47
binary Huffman tree building, 46
binary tree of depth four, 50
buffer, purpose of, 48–49
Huffman encoding procedure, 48
minimum variance Huffman code, 49
reduced four-letter alphabet, 48
reduced three-letter alphabet, 49
reduced two-letter alphabet, 49
two Huffman trees, 48–49
variable-length code, 48–49
Mirror condition, 471
Mismatch effects, 266
Demonstration, 267
msqe function of, 268
step size, 268
types, 266–267
variance mismatch on, 267
Mixed excitation linear prediction (MELP), 608. *See also* Code-excited linear prediction (CELP)
adaptive spectral enhancement filter, 611
decoder, 608–609
fractional offset, 609–610
normalized autocorrelation, 609–610
peakiness, 610
pitch period, 608
prediction residual, 610–611
Mixed excitation linear prediction, 593
Mixed Raster Content (MRC), 211
MMR. *See* Modified modified READ
Model-based coding, 649
AU, 650
global motion and local motion, 650
three-dimensional, 650
Modified discrete cosine transform (MDCT), 439, 577, 587
frames, 580
reconstructed sequence from 10 DCT coefficients, 579
source output sequence, 578
transformed sequence, 578
window function, 577–579
Modified Huffman (MH), 200
Modified modified READ (MMR), 203
Modified quantization mode, 663
Modified READ (MR), 201
Modulation property, 384
More Probable Symbol (MPS), 120–124
Morse code, 2–3, 29
Most significant bit (MSB), 113
Mother function, 501–502
Mother wavelet, 500–501
Motion-compensated prediction, 657, 663–665
Motion compensation, 634–635, 644
block-based, 636
block size effect on, 645
CBP, 648–649
frames, difference between, 636
doubled image, 637–638
GOB macroblocks, 647–648
ITU-TH. 261 algorithm, 649
loop filter, 646
macroblock pixel encoding, 645–646
motion-compensated prediction, 637
motion vector, 636
quantization and coding, 647
rate control, 649

trade-off balancing, 645
transform, 647
video sequence frames, 635
Motion vector, 636
Motion video, 633–634
Move-to-front coding (mtf coding), 174, 177–178
Moving Picture Experts Group (MPEG), 485, 651–652
MP-LPC. *See* Multipulse linear predictive coding
MPEG-1 algorithm, 641
MPEG-1 video standard, 652. *See also* MPEG-2 video standard; Video compression
anchor frames, 652–653
B frames, 652–653
bitstream order, 653–654
CPB, 655
display order, 653–654
GOP, 653
I frames, 652
P frames, 652–653
rate control, 654–655
typical sequence of frames, 654
VHS quality images, 655
MPEG-2 AAC. *See also* MPEG-4 AAC
block switching and MDCT, 583
encoder, 582
profiles, 585
quantization and coding, 585
spectral processing, 583
prediction_data_present bit, 583–584
prediction_used bit, 583–584
predictor, 583–584
temporal noise shaping, 584–585
TNS, 583–584
stereo coding, 585
MPEG-2 video standard. *See also* MPEG-1 video standard; Video compression
base bitstream, 655–656
dual prime motion compensation, 657–658
Grand Alliance HDTV Proposal, 658
layered approach, 656
motion-compensated prediction modes, 657
profile-level combinations, 657
profiles, 655–656
scanning pattern for DCT coefficients, 658
16×8 motion compensation, 657–658
MPEG-4 AAC, 586
BSAC, 587
LTP, 586
PNS, 586
TwinVQ, 586
MPEG-4 advanced video coding. *See* H.264
MPEG-4 Part 10. *See* H.264
MPEG-4 Part 2, 669. *See also* H.264; MPEG-1 video standard; MPEG-2 video standard; Video compression
EZW algorithm, 670
motion-compensation algorithms, 669–670
video coding algorithm, 669–670
MPEG. *See* Moving Picture Experts Group
MPEG audio coding
layer I coding
frame structure for, 574–575
MPEG-1 and MPEG-2, frequencies in, 573
scale factor, 573–574
layer II coding
frame structure, 576
layer I and II coding scheme difference, 576
layer III coding
MDCT, 577
window function, 577–579
MPRG AAC, 581–582
decoder tools, 581
MPS. *See* More Probable Symbol
MR. *See* Modified READ
MRA. *See* Multiresolution analysis
MRC. *See* Mixed Raster Content
MRC-T.44, 211–212
background layer, 212
data types, 213
foreground layer, 213
mask layer, 212
stripes, 212–213
T.44 recommendation, 211–212
MRI. *See* Magnetic resonance images
MSB. *See* Most significant bit
mse. *See* Mean squared error
msqe. *See* Mean squared quantization error
mtf coding. *See* Move-to-front coding
Multiband excitation coder (MBE), 608
Multiple-pass algorithm, 534–535
Multipulse linear predictive coding (MP-LPC), 603
Multiresolution analysis (MRA), 507–508
Multiresolution approaches, 193
HINT, 193, 194
progressive image transmission, 193–194
Multistage vector quantization, 334
different vector quantizer, 334–335
quantization rule, 335
RIVQ, 335
three-stage vector quantizer, 334
using LBG vector quantizers, 334–335
Multistage vector quantizer, 623
Mutual information, 228

N

NASA. *See* National Aeronautics and Space Administration
National Aeronautics and Space Administration (NASA), 2
NFC. *See* Noise feedback coding
Nine 9/7 transform, 550
Noise feedback coding (NFC), 365–366
Noise shaping analysis, 622–623
Nonbinary Huffman codes, 65–66
code tree, 67
example, 65
reduced five-letter alphabet, 66
reduced three-letter alphabet, 67
sorted six-letter alphabet, 66

ternary code for six-letter alphabet, 67
Nonstationary signal, 498
Nonuniform quantization, 277–278
companded quantization, 282–283
nonuniform midrise quantizer, 277
symmetric, 279
nonuniform quantizer, 278
pdf-optimized quantization, 278
decision boundary, 278–280
equation, 279–280
mismatch effects, 281
using Leibniz integral rule, 278
quantizer boundary, 280
reconstruction levels, 280
Nonuniform quantizer, 277–278
Nonuniform sources, 261–262
example, 262
overload and granular regions for, 265
quantization noise for, 264
uniform midrise quantizer, 263
quantization error for, 265
Non-Tunstall code, 80
North pixel, 186
Not yet transmitted (NYT), 66
*N*th-order autoregressive model (AR(N) model), 243
Nyquist rule. *See* Nyquist theorem
Nyquist theorem, 453
NYT. *See* Not yet transmitted

O

OBMC. *See* Overlapped Block Motion Compensation
Offline adaptive approach, 268–269
Offset, 139–140
Ondelettes, 501–502. *See also* Wavelets
One-dimensional coding scheme, 200
One-layer stripe (1LS), 212–213
Online adaptive approach, 268–269
1LS. *See* One-layer stripe
Operational distortion-rate function, 428
Operational rate-distortion function, 428
Orthogonal basis set, 379–380
Orthonormal basis set, 379–380
Orthonormal transform, 415
Outer product, 683
Overdecimated filter bank, 477
Overlapped Block Motion Compensation (OBMC), 661
Overload error, 263–264
Overload noise. *See* Overload error
Overload probability, 263–264

P

Packet length marker (PLM), 564–565
Packet Loss Concealment (PLC), 617
Packet video, 670. *See also* Video compression
ATM networks, 671
compression issues in, 671–672
transmission capacity availability, 671
compression algorithms, 672
analysis filter bank, 673
progressive transmission algorithms, 672–673
reconstructed frame, 673
splitting, 672
Pairwise nearest neighbor algorithm (PNN algorithm), 313–314
PARCOR coefficients. *See* Partial correlation coefficients
Parcor coefficients, 359
Parkinson's First Law, 2
Parseval's theorem, 384
Partial correlation coefficients (PARCOR coefficients), 600–601
Partial fraction expansion, 399–400
Pass mode, 202
PB-frames mode, 661
PCRD. *See* Post Compression Rate Distortion
Pdf-optimized quantization, 278
decision boundary, 278–280
equation, 279–280
mismatch effects, 281–282
properties, 280, 281
using Leibniz integral rule, 278
pdf. *See* Probability density function
Peak-signal-to-noise-ratio (PSNR), 222
Perceptual noise substitution (PNS), 586
Perfect reconstruction (PR), 467
applications, 469–470
conditions, 467
output of the low-pass filter, 468
power symmetric FIR filters, 472–474
two-channel PR quadrature mirror filters
characteristics, 472
mirror condition, 471
two-channel subband decimation and interpolation, 468
Periodic extension, 382–383, 390, 391
P frames. *See* Predictive coded frames
Physical models, 25, 248
data generation, 25
speech production, 248
filter, 248
vocal cords, 248
Pitch period, 365, 597–598
estimation, 580–581
PKZip, 143
PLC. *See* Packet Loss Concealment
PLM. *See* Packet length marker
PNG, 143. *See also* Portable network graphics
PNN algorithm. *See* Pairwise nearest neighbor algorithm
PNS. *See* Perceptual noise substitution
POD marker. *See* Progression order change default marker
Pointer markers, 564
Polar vector quantizers, 327–328
Polyphase decomposition, 477
two-band subband coder
analysis portion, 477, 479
synthesis portion, 479–481
Portable network graphics (PNG), 152
comparison of, 155
Huffman codes for, 155
representations codes, 154

using Huffman code, 153
using pixel values, 153
Post Compression Rate Distortion (PCRD), 547
Power symmetric FIR filters, 472–473
ppma algorithm, 170
ppm algorithm. *See* Prediction with partial match algorithm
PPM marker, 565–566
PR. *See* Perfect reconstruction
Prediction approach, 190
Prediction modes, 667
Prediction with partial match algorithm (ppm algorithm), 28, 165
basic algorithm, 165–171
context length, 172
using ppm algorithm, 172–173
escape symbol, 170–171
method C, 172
methods A and B, 170–172
exclusion principle
basic principle, 174
encode process, 173, 174
unit interval into subintervals, 173
Predictive coded frames (P frames), 652–653
Predictive coding schemes, 8
Predictor
adaptive algorithms, 368–369
backward adaptive, 368
Prefix codes, 33
binary trees for different codes, 33
internal nodes, 33–34
root node, 33–34
Probability
axiomatic approach, 678–679
Bayes' rule, 679
binary symmetric channel, 677
distribution functions, 681–683
expectation, 683–684
frequency of occurrence, 675–676
mean, 685
measure of belief, 676
random variables, 680
realization, 680
second moment, 685
statistically independent, 677
variance, 685
Probability density function (pdf), 251, 682
Probability models, 240–243
assumption, 25
candidate distributions, 242
estimate of distribution, 242
gamma distribution, 241–242
Gaussian distribution, 241, 242
ignorance model, 25
Laplacian distribution, 241, 242
probability model, 25
uniform distribution, 241, 242
Product code vector quantizers. *See* Gain-shape vector quantizers
Progression order change default marker (POD marker), 564
Progressive compatible sequential mode, 203
Progressive image transmission, 193–195
example, 194
comparison between, 197
Sena image, 195
pyramid structure for, 198
Progressive transmission algorithms, 672–673
Pruned TSVQ, 324
Pruning, 324
Psychoacoustic model, 572
nontonal components, 572–573
postprocessing, 573
tonal components, 572–573
PSNR. *See* Peak-signal-to-noise-ratio
Pyramid schemes, 196
for progressive transmission, 198
Pyramid vector quantization, 326–327
gain-shape vector quantizers, 327
SNR value, 327

Q

QCC marker. *See* Quantization component marker
QCD marker. *See* Quantization default marker
QCIF (Quarter CIF), 641
QM coder, 124, 125
JBIG algorithm, 125
LPS probability, 125
scaling and rescaling process, 125
QMF. *See* Quadrature mirror filters
Quadrature mirror filters (QMF), 456, 469–470
Quadtree partitioning, 630, 631
Quantization, 251, 252
error, 255
table marker, 439
transform coefficients, 424, 426
Quantization component marker (QCC marker), 564
Quantization default marker (QCD marker), 564
Quantization error, 255
for uniform midrise quantizer, 265
Quantization noise. *See* Quantization error
Quantization problem, 252, 253
additive noise model of quantizer, 256
codeword assignment, 256
D/A converter, 253–254
digitizing sine wave, 253
encoder mapping for, 252–253
3-bit D/A converter, 253–254
3-bit encoder, 252
quantizer, 252
input-output map, 254
Quantization table, 433, 434
Quantizer, 367
adaptation algorithm, 367
backward adaptive, 367
fixed, 367
locked scale factor, 367, 368

unlocked scale factor, 367, 368
Quantizer distortion. *See* Quantization error
Quarter CIF. *See* QCIF

R

Random variables, 680
Rate, 6
Rate dimension product, 318
Rate distortion theory, 217, 218, 232
binary entropy function, 237
binary source function, 236
compression scheme for, 234
example, 233, 235
Gaussian source function, 238, 240
height and weight measurements, 233
RCT. *See* Reversible component transform
READ. *See* Relative Element Address Designate
readau.c, 356
Reconstruction levels, 255
Recursive bit allocation algorithm, 426
Recursive indexing, 189, 291
Recursively indexed vector quantizer (RIVQ), 335
Reduced-resolution update mode, 662
Redundancy removal, 204
deterministic prediction, 205
typical prediction, 204–205
usage, 205
using pixels, 205
Reference picture resampling, 662
Reference picture selection mode, 662
Region of interest marker (RGN marker), 564
Regular pulse excitation with long-term prediction (RPE-LTP), 604
Relative Element Address Designate (READ), 201
Residual model, 6–7
Residual sequence, 184
Residual vector quantizers, 334
Resolution reduction, 203–204
expression, 204
JBIG specification, 204
using pixels, 204
Reversible component transform (RCT), 548
RGN marker. *See* Region of interest marker
Rice, Robert F., 76–77
Rice codes, 76–78
CCSDS recommendation, 77
fundamental sequence, 78
mapping, 77
preprocessor functions, 77
second extension option, 78
split sample option, 78
zero block option, 78
Risannen, Jorma, 38
RIVQ. *See* Recursively indexed vector quantizer
Rods, 223
Root lattices, 691, 692, 698
Row matrix, 682
RPE-LTP. *See* Regular pulse excitation with long-term prediction
RPE. *See* Scheme regular pulse excitation
Rsiz, 562
Run-length coding, 199–200
Capon model for, 200
Run-length mode, 552

S

Sample average, 684
Sampling theorem, 390
frequency domain view
Fourier series expansion, 390
Fourier transform function, 391
function reconstruction, 391–392
periodic extension, 390, 391
time domain view
aliased reconstruction, 393
aliasing, 394
Fourier transform, 393
sampled function, 392–393
sampling effect, 393
signal samples, 394
Scalar multiplication, 376
Scalar quantization
adaptive quantization, 268
backward adaptive quantization, 271, 273, 275
forward adaptive quantization, 269
entropy-coded quantization, 287–288
entropy coding of Lloyd-Max quantizer outputs, 288
entropy-constrained quantization, 289
high-rate optimum quantization, 289–291
nonuniform quantization, 277–278
companded quantization, 282–283
pdf-optimized quantization, 278, 281
quantization problem, 252, 253
uniform quantizer, 257–258
mismatch effects, 266
nonuniform sources, 261–262
uniformly distributed source, 258–260
Scalar quantization, and advantages, 298
individual height and weight, 298
eight-level, representations of, 301
height-weight vector quantizer, 300
scalar quantizers, 299
two-dimensional vector quantizer, 300
input-output map for, 302
modified two-dimensional vector quantizer, 303
Scalar quantizers, 252
Scaling, 386
Scaling function, 504
approximations of function, 506–507
Haar scaling function, 507
sample function, 506
triangle scaling function, 508
Scheme regular pulse excitation (RPE), 604
Search buffer, 139–140
Second extension option, 78
Second moment, 685
Self-information, 13–14
Separable transform, 415

Set Partitioning in Hierarchical Trees (SPIHT), 540. *See also* Embedded zerotree wavelet coder (EZW coder)
coordinate sets of coefficients, 540, 541
data structure, 540, 541
LSP and LIP versus LIS, 542, 543
Sinan image reconstruction, 546
Seven-level decomposition, 536–539
Shannon, Claude Elwood, 13–14
Shannon lower bound, 239
Shifting theorem, 405, 406
Short-term Fourier transform (STFT), 498–499
basis functions, 499
nonstationary signal, 498
problems, 499
three wavelet basis functions, 499, 500
Side information, 268–269
Signal-to-noise ratio (SNR), 221–222, 355–356
Signal-to-prediction error ratio (SPER), 355–356
Significance map coding, 534–535
SILK coder, 621
encoding of speech, 622
LSF coefficients, 623
multistage vector quantizer, 623
noise shaping analysis, 622–623
operating modes, 622
signal filtering, 622
variable-rate entropy coder, 623
Single-letter context. *See* First-order Markov model
Sinusoid, 346
encoding system, 351
quantization process, 350
quantizer designing, 345, 346
and reconstructions, 350, 351
sample-to-sample differences, 346
Sinusoidal coder, 606. *See also* Speech compression
coding techniques, 608
frequency transmission, 607–608
MBE, 608
STC, 608
Sinusoidal transform coder (STC), 608
6-tap Coiflet low-pass filter coefficients, 518
16 × 8 motion compensation, 657–658
SIZ marker. *See* Image and tile size marker
Skewed, 163
SLNT Pn, 205
Slope overload regions, 362
SNR. *See* Signal-to-noise ratio
SNR scalability mode, 662
SOC marker. *See* Start of codestream marker
SOD marker. *See* Start of data marker
Solomonoff, R., 37
SOP marker. *See* Start of packet marker
SOT marker. *See* Start of tile-part marker
Sound pressure level (SPL), 572–573
Source coder, 219, 220
Spatial scalability mode, 662
Spatial orientation trees, 540
Spectral masking, 571
audibility threshold changes, 571
critical band, 570–571
Spectral processing, 583
prediction_data_present bit, 583–584
prediction_used bit, 583–584
predictor, 583–584
temporal noise shaping (TNS), 583–585
Speech coding. *See also* Image coding
DPCM structure with pitch predictor, 365, 366
G.726 recommendation, 366
predictor, 368–369
quantizer, 367, 368
NFC, 365–366
pitch period, 365
residual sequence, 365, 366
Speech coding for internet applications, 613. *See also* Speech compression
iLBC coder, 613
all-pass filters, 616
autocorrelation coefficients computation, 614
codebook generation, 617
goal, 615–616
LSF representation, 614–615
packet loss effects, 613–614
perceptual weighting filter, 616
PLC, 617–618
residual sequence, 616
using DPCM system, 616
ITU-TG.729 standard, 618
autocorrelation coefficients, 618–619
bit allocation per frame, 621
codebook vectors, 620–621
encoding of speech for, 619
excitation signal identification, 620
pulses, 621
weighting filter, 620
SILK coder, 621
encoding of speech, 622
LSF coefficients, 623
multistage vector quantizer, 623
noise shaping analysis, 622–623
operating modes, 622
signal filtering, 622
variable-rate entropy coder, 623
Speech compression
channel vocoder, 592–594
excitation signal, 595–596
formants, 594
receiver, 595
sound /e/in test, 594
sound /s/in test, 594, 595
synthesis filters, 594
speech synthesis, model for, 593
vocal tract, 593
Speech encoding, 364
Speech production, 248
filter, 248
vocal cords, 248
Speech synthesis model
LPC receiver, 596
speech compression, 593
SPER. *See* Signal-to-prediction error ratio

Spherical vector quantizers, 327–328
SPIHT. *See* Set Partitioning in Hierarchical Trees
SPL. *See* Sound pressure level
Split sample option, 78
Splitting technique, 309–311, 672
Square matrix, 675
Ssiz, 562
Start of codestream marker (SOC marker), 562
Start of data marker (SOD marker), 562
Start of packet marker (SOP marker), 566
Start of tile-part marker (SOT marker), 562
Static dictionary, 136–137
  digram coding, 137
    digram encoder, 137
    example, 137
    frequently occurring pairs, 138, 139
    sample dictionary, 137
Static model, 19
Statistical average, 684
STC. *See* Sinusoidal transform coder
Stereo coding, 585
STFT. *See* Short-term Fourier transform
Stochastic process, 687
  autocorrelation function, 688
  iid, 687
  joint cdf, 687
  realizations, 687
  stationarity, 688
Structured vector quantizers
  lattice vector quantizers, 328
    *A*2 lattice, 330
    average squared error, 328
    *D*2 lattice, 330, 331
    dimensions, 329
    example, 331
    hexagonal lattice, 329
    possible quantization regions, 328
    problems, 331
    square and circular quantization regions, 328–329
    using lattices, 329
  polar and spherical vector quantizers, 327–328
  pyramid vector quantization, 326–327
    gain-shape vector quantizers, 327
    SNR value, 327
  tree-structured vector quantizer, 324
  two-dimensional uniform quantizer, 325
    contours of constant probability, 326
  using examples, 324–325
Subband coding
  algorithm, 458, 462
  analysis, 459
    block diagram, 460
    decimation or downsampling, 459–460
    magnitude transfer functions, 460
    nonoverlapping overlapping filter banks, 461
    overlapping filter banks, 461
  bit allocation, 482
  coding, 461
  difference sequences, 450
  filter banks design, 462–464
  filters, 452
  original set of samples, 449
  quantization, 461
  synthesis, 461
    analysis and synthesis filters, 462
    bit allocation scheme, 462
    encoding scheme, 462
  using filters, 456
Subband decomposition
  approaches, 529
  first-level decomposition, 530
  four-tap Daubechies filter, 530–532
  of *N*×*M* image, 529–530
  subband structures, 531
Subspace, 377
Superposition, 386
Syllabically companded, 364
Symbol region, 209
  decoding, 211
Syntax-based arithmetic coding mode, 660

T

Tabular method, 399
Tag deciphering, 101
Tag generating, 94–100
TCM. *See* Trellis-coded modulation
TCQ. *See* Trellis-coded quantization
Temporal masking, 571
Temporal noise shaping (TNS), 583–585
Temporal scalability mode, 662
Terminating codes, 200
test.snd, autocorrelation function for, 365
testm.raw, 356
Text compression, 4, 83–84
  using Huffman codes, 84
Three-dimensional model-based coding, 643
Three-layer stripes (3LS), 212–213
3LS. *See* Three-layer stripes
Threshold coding, 427
Tier I coding, 554. *See also* JPEG 2000 standard
  block coding, 552
  cleanup pass, 552, 554, 558
  context determination, 553
  coefficients, example stripe of, 555
  magnitude refinement pass, 552, 553
  prediction and context generation, 553
  significance propagation pass, 552–553
  significant bitplanes, 556–559
Tier II coding, 559. *See also* JPEG 2000 standard
  collection of bits, 560
  operational rate distortion function, 561
  rate control, 560–561
Tiling, 549
Time invariance, 386
TNS. *See* Temporal noise shaping
Toeplitz matrix, 600
Total Count, 65
Total_Count, 114, 120
Training set, 304

Transfer function, 386, 387
Transform, 414
  forward, 414
  matrix, 665–666
  orthonormal, 415
  separable, 415
Transform coding
  original sequence, 410
  reconstructed sequence, 412
  source output sequence, 410
  transformed sequence, 411, 412
  transform process
    geometric view, 413
    steps, 413
Transform-Domain Weighted Interleave Vector Quantization (TwinVQ), 586
Transforms of interest, 418
  DCT, 420
  DST, 423
  DWHT, 423, 424
  KLT, 418–419
Transpose, 682–683
Tree-structured vector quantizers (TSVQ), 320, 323
  decision tree for quantization, 323
  design of, 323–324
  method breakdown using quadrant approach, 321, 322
  output points, division of, 322
  pruned, 324
  symmetrical vector quantizer, 320, 321
Trellis-coded modulation (TCM), 337
Trellis-coded quantization (TCQ), 337
  selection process. 338
    state diagram, 339
    trellis diagram for, 339, 340
  TCM, 337
  trellis diagram, 338
  2-bit trellis-coded quantizer, 337–338
  using vector quantizer, 337
  Viterbi algorithm works, 338
Trellis diagram, 338
Trignometric Fourier series representation, 380
TSVQ. *See* Tree-structured vector quantizers
Tunstall codes, 79
  alphabet and probabilities, 80
  codebook, 81
  examples, 79, 80
  3-bit Tunstall code, 81
  2-bit non-Tunstall code, 80
  2-bit Tunstall code, 79
12-tap Coiflet low-pass filter coefficients, 519
12-tap Daubechies low-pass filter coefficients, 517
20-tap Daubechies low-pass filter coefficients, 518
TwinVQ. *See* Transform-DomainWeighted Interleave Vector Quantization
2-bit trellis-coded quantizer, 337–338
Two-dimensional vector quantizer. 300
  input-output map for, 302
  modified two-dimensional vector quantizer, 303
  two representations of, 301
Two-layer stripes (2LS), 212–213
2LS. *See* Two-layer stripes
Tympanic membrane, 224
Typical prediction, 204–205
  usage, 205
  using pixels, 205
Typical prediction, 209–210

U

Unary code, 75
Underdecimated filter bank, 477
Uniform distribution, 241, 242, 685
Uniformly distributed source, 258–259
  image compression, 260, 261
  quantization error for, 259–260
Uniform quantizer, 257–258
  midtread quantizer, 258
  mismatch effects, 266
    demonstration, 267
    msqe function of, 268
    step size, 268
    types, 266–267
    variance mismatch on, 267
  nonuniform sources, 261–262
    example, 262
    overload and granular regions for, 265
    quantization noise for, 264
    uniform midrise quantizer, 263, 265
  uniformly distributed source, 258–259
    image compression, 260, 261
    quantization error for, 259–260
Uniquely decodable codes, 30
  average length, 30
  code 1, 30
  code 2, 30
  code 3, 4, 30–31
  code 5, 31, 32
  code 6, 32, 33
  instantaneous code, 31
  small and large codes, 31
  unique decodability, test for, 32
    prefix and dangling suffix, 32
    procedures, 32
    using codewords, 32–33
UNIX compress, 151
Unlocked scale factor, 367, 368
Unrestricted motion vector mode, 660
Update procedure, 66
  adaptive Huffman coding algorithm, 69
  adaptive Huffman tree, 71
  example, 70
  external node, 69–70
  NYT node, 70
Upsampling, 462, 465
  analysis filters, 465–467
  antialiasing filters, 465–467
  imaging, 465
  interpolation filters, 465–467
  synthesis filters, 465–467
  upsampled signal spectrum, 467

V

V. 42 bis, 153, 155, 156
  CCITT recommends, 156
  control codewords in, 155
  encoder STEPUP, 156
  using compression algorithm, 153, 155
Variance, 685
Variations on theme, 332. *See also* Vector quantization
  adaptive vector quantization, 335–336
    distortion, 336–337
    indexed vector quantizer, 336
    large codebook, 336
  gain-shape vector quantization, 332
  mean-removed vector quantization, 332–333
    Sinan image using codebook, 332, 333
  multistage vector quantization, 334
    different vector quantizer, 334–335
    quantization rule, 335
    RIVQ, 335
    three-stage vector quantizer, 334
    using LBG vector quantizers, 334–335
  vector quantization, 333
    three-stage vector quantizer, 334
    variation, 333
Vector, 375
Vector addition, 376
Vector quantization, 296–297
  advantages and scalar quantization, 298, 300
  LBG algorithm, 297, 304
    empty cell problem, 315
    example, 306
    image compression, uses for, 315–316
    initializing, 303, 309, 311, 314
  procedure, 296
  structured vector quantizers, 324
    example, 325
    Lattice vector quantizers, 328, 331
    polar and spherical vector quantizers, 327–328
    pyramid vector quantization, 326–327
  tree-structured vector quantizers, 320
    design of, 323
    example, 320
    pruned, 324
  trellis-coded quantization, 337, 338
Vector quantizer, 252, 306
  alternate initial set of, 309
  final state of, 309
  initial set of, output points, 307
  initial state of, 308
  after one iteration, 308
  training set for, 307
Vector spaces, 375, 376
  basis, 377, 378
  basis vectors, 374–375
  dimension, 378
  dot product, 375
  examples, 376–378
  inner product, 379
  orthogonal basis set, 379–380
  orthonormal basis set, 379–380
  scalar multiplication, 376
  subspace, 377
  vector addition, 376
  vector in two-dimensional space, 374
Vertical mode, 202
Video compression, 633–634. *See also* Image compression
  algorithms, 634
  asymmetric applications, 650
    generic wireframe model, 651
    MPEG, 651–252
  model-based coding, 649
    AU, 650
    global motion and local motion, 650
    three-dimensional, 650
  motion compensation, 634, 635
    block-based, 636
    difference between frames, 636
    doubled image, 637–638
    motion vector, 636
    motion-compensated prediction, 637
    video sequence frames, 635
  motion video, 633–634
  video signal representation, 638
    analog color television, 639
    black-and-white analog television picture, 638
    CCIR 601 frame fields, 643
    CCIR 601 to MPEG-SIF, 643
    CCIR recommendations, 640–641
    chrominance components, 640
    composite color signals, 639–640
    frame and fields, 638–639
    line of image, 638, 639
    luminance component, 640
    MPEG-1 algorithm, 641
    recommendation sampling format, 641
    SIF frame generation, 643
    three-dimensional model-based coding, 642
    video signal digitization, 640
Video signal representation, 638
  analog color television, 639
  black-and-white analog television picture, 638
  CCIR 601 frame fields, 641, 642
  CCIR 601 to MPEG-SIF, 643
  CCIR recommendations, 640–641
  chrominance components, 640
  composite color signals, 639–640
  frame and fields, 638–639
  line of image, 638, 639
  luminance component, 640
  MPEG-1 algorithm, 641
  recommendation sampling format, 641
  SIF frame generation, 643
  three-dimensional model-based coding, 643
  video signal digitization, 640
Virtual Reality Modeling Language (VRML), 669
Viterbi algorithm, 337–338
Vocal cords, 248
Vocal tract, 248, 593

Vocal tract filter, 599
autocovariance approach, 599–600
Cholesky decomposition, 602
covariance method, 601–602
filter coefficients change, 601–602
Levinson-Durbin algorithm, 600–601
PARCOR coefficients, 600–601
Toeplitz matrix, 600
Vocoder. *See* Voice coder
Voice coder, 3
Voiced/unvoiced decision, 597
Voice over Internet Protocol (VoIP), 613
Voicing probability, 608
VoIP. *See* Voice over Internet Protocol
VRML. *See* Virtual Reality Modeling Language

W

Wavelet-based image compression, 529
EZW coder, 532
data structure in, 532, 534
embedded coding, 539–540
isolated zero, 534
multiple-pass algorithm, 534–535
seven-level decomposition, 536–539
significance map coding, 534–535
ten-band decomposition, 532, 533
3-bit quantizer, 533, 534
three-level midtread quantizer, 535–536
wavelet coefficient scanning, 536
SPIHT algorithm, 540
coordinate sets of coefficients, 540, 541
data structure, 540, 541
LSP and LIP versus LIS, 542, 543
Sinan image reconstruction, 546
subband decomposition
approaches, 529
first-level decomposition, 530
four-tap Daubechies filter, 530–532
of $N \times M$ image, 529–530
subband structures, 531
Wavelet coefficient scanning, 536
Wavelet implementation
scaling and wavelet coefficients, 513–515
three-level wavelet decomposition, 512, 513
using filters, 510–511
wavelets families
18-tap Coiflet low-pass filter coefficients, 519
4-tap Daubechies low-pass filter coefficients, 517
6-tap Coiflet low-pass filter coefficients, 518
20-tap Daubechies low-pass filter coefficients, 518
12-tap Daubechies low-pass filter coefficients, 517
12-tap Coiflet low-pass filter coefficients, 519
Wavelets, 500–501
admissibility condition, 503
biorthogonal wavelets, 516–517
CWT, 503–504
DTWT, 504
families, 516
function, 501, 502
mother wavelet, 500–501
ondelettes or wavelets, 501–502
scaled and translated function, 501, 502
Wavelet transform. *See also* Discrete Fourier transform (DFT); Fourier transform; Z-transform
irreversible, 550
filter coefficients for, 550
9/7 transform, 550
reversible, 549–550
filter coefficients for, 550
wavelet filter types, 549–550
Wealth of Nation (Smith), 16–18
Weber fraction. *See* Weber ratio
Weber ratio, 223
Wideband speech compression, 611. *See also* Speech compression
adaptive codebook, 612–613
coding method, 611
comfort noise, 613
encoding of speech, 613
fixed codebook, 612–613
LP coefficients, 612
speech processing, 611–612

X

Xsiz, 562, 563
XOsiz, 562, 563
XTsiz, 549, 562
XTOsiz, 549, 562
XRSiz*i*, 562

Y

Ysiz, 562, 563
YOsiz, 562, 563
YTsiz, 549, 562
YTOsiz, 549, 562
YRSiz*i*, 562

Z

Z-transform, 396. *See also* Discrete Fourier transform (DFT); Fourier transform
discrete convolution, 404–405
examples, 397, 398
inverse, 398, 402
long division, 403–404
pairs, 399
partial fraction expansion, 399–400
properties, 404
region of convergence, 396–397
shifting theorem, 405, 406
tabular method, 399
transfer function, 406
Zero block option, 78
Zero frequency problem, 28
Zero-coding mode, 552
Zigzag scanning pattern, 427
Zip, 143
zlib, 62
Zonal sampling, 426–427

Printed by Printforce, the Netherlands